Color Atlas of BASIC HISTOPATHOLOGY

first edition

Color Atlas of BASIC HISTOPATHOLOGY

first edition

Clara Milikowski, MD
Assistant Professor
Department of Pathology
University of Miami School of Medicine
Miami VA Medical Center
Miami, Florida

Irwin Berman, PhD
Professor
Department of Cell Biology & Anatomy
University of Miami School of Medicine
Miami, Florida

Appleton & Lange
Stamford, Connecticut

Notice: The authors and the publisher of this volume have taken care to make certain that the doses of drugs and schedules of treatment are correct and compatible with the standards generally accepted at the time of publication. Nevertheless, as new information becomes available, changes in treatment and in the use of drugs become necessary. The reader is advised to carefully consult the instruction and information material included in the package insert of each drug or therapeutic agent before administration. This advice is especially important when using, administering, or recommending new or infrequently used drugs. The authors and publisher disclaim all responsibility for any liability, loss, injury, or damage incurred as a consequence, directly or indirectly, of the use and application of any of the contents of this volume.

97 98 99 00 01 / 10 9 8 7 6 5 4 3 2 1

Prentice Hall International (UK) Limited, *London*
Prentice Hall of Australia Pty. Limited, *Sydney*
Prentice Hall Canada, Inc., *Toronto*
Prentice Hall Hispanoamericana, S.A., *Mexico*
Prentice Hall of India Private Limited, *New Delhi*
Prentice Hall of Japan, Inc., *Tokyo*
Simon & Schuster Asia Pte. Ltd., *Singapore*
Editora Prentice Hall do Brasil Ltda., *Rio de Janeiro*
Prentice Hall, *Upper Saddle River, New Jersey*

ISBN 0-8385-1382-4
ISSN 1088-8675

Acquisitions Editor: John Dolan
Development Editor: Cara Lyn Coffey
Production: Princeton Editorial Associates
Designer: Libby Schmitz

PRINTED IN HONG KONG

I dedicate this book to the memory of my father, Solomon Milikowski, who taught me that hard work is a means to accomplishment, and to Robert L. Hellman, MD, my mentor and friend. They both encouraged me throughout this endeavor. Also, I thank my husband, David, and children, Adam and Jason Burstyn, for their support and patience.

CM

CONTENTS

Preface ix
1. General Concepts 2
2. Inflammation and Repair 15
3. Fluid and Hemodynamic Derangements 24
4. Neoplasia 32
5. Cardiovascular System 60
6. Hematopoietic System 94
7. Lymphoid System 116
8. Immune System 144
9. Respiratory System 172
10. Oral Cavity and Salivary Glands 208
11. Gastrointestinal Tract 229
12. Liver, Pancreas, and Biliary Tract 276
13. Urinary System 315
14. Female Reproductive System 358
15. Breast 405
16. Male Reproductive System 426
17. Endocrine System 452
18. Skin and Subcutaneous Tissue 470
19. Musculoskeletal System 518
20. Nervous System 548
21. Sensory Organs 594

Index 605

PREFACE

The purpose of this atlas is to provide students with a source of photomicrographs of basic and systemic pathologic processes. The atlas is geared primarily to second-year medical students, although it should prove valuable to postgraduate pathology residents and others in the health-related professions.

The atlas contains a compilation of color photomicrographs taken primarily at the light microscopic level from routine hematoxalin- and eosin-stained material. A unique feature of this atlas is the presentation of additional photomicrographs of slides treated with special stains.

Because pathology encompasses such a broad array of disease processes, it is impossible to depict every pathologic entity and variant thereof. We have therefore chosen cases that we consider the "basic" or "core" pathologic entities which students should be familiar with. If entities not depicted here are considered important for students to see, we would be pleased to include them in the next edition.

ACKNOWLEDGMENTS

We express our appreciation to the following individuals from the Department of Pathology for their contributions to the atlas: Luis L. Alvarez, MD; Jocelyn H. Bruce, MD; Gerald E. Byrne, Jr, MD; Francisco Civantos, MD; Robert L. Hellman, MD; Carolyn Mies, MD; Azorides R. Morales, MD; Mehrdad Nadji, MD; Jorge L. Perez, MD; K. Rajendar Reddy, MD (Department of Medicine, Division of Hepatology); Maria M. Rodriguez, MD; Phillip Ruiz, MD, PhD; Raphael Valenzuela, MD, PhD; Ana L. Viciana, MD; and Clarence C. Whitcomb, MD.

We also express our appreciation to John J. Dolan, Senior Medical Editor; Cara Lyn Coffey, Development Editor; and Karen W. Davis, Production Editor, for their interest, enthusiasm, and excellent guidance in the preparation of this atlas. We wish to thank Appleton & Lange for their grant, which helped defray the cost of photographic materials used in the preparation of the atlas.

■ TISSUE CHANGES

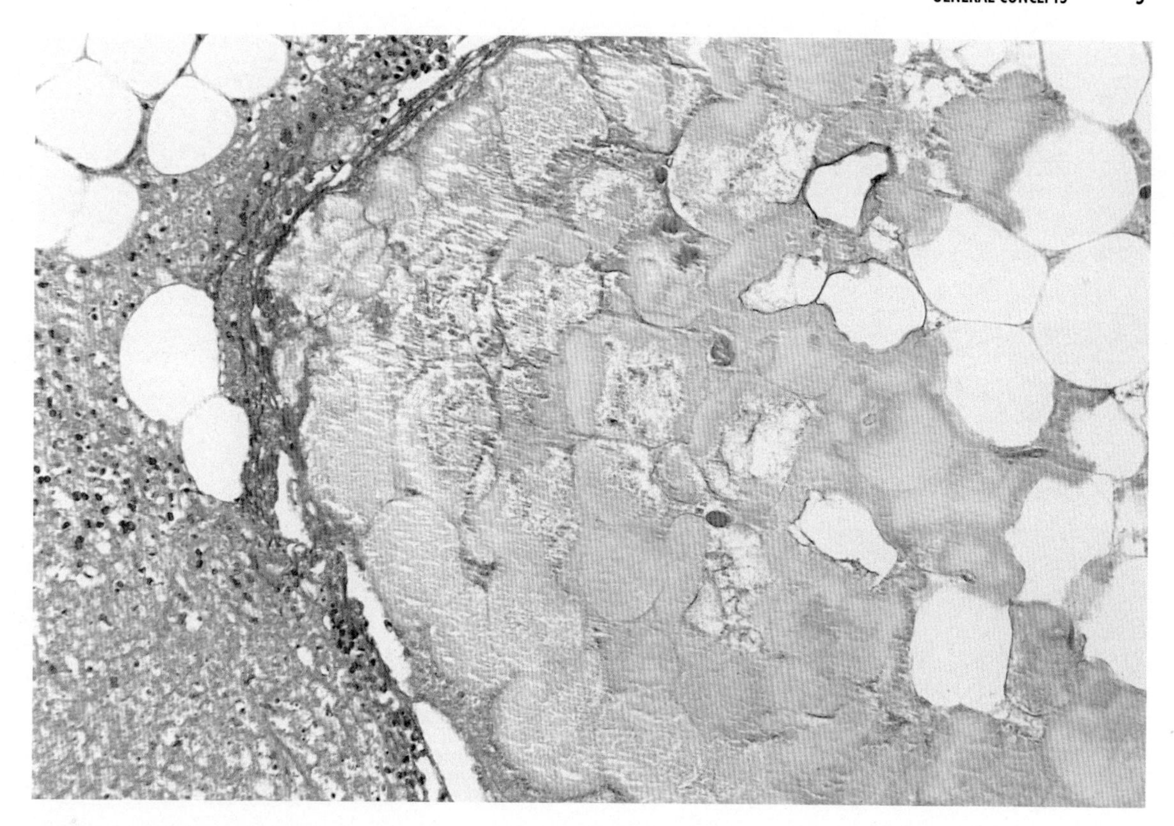

FIGURE 1–5. ENZYMATIC FAT NECROSIS: In enzymatic fat necrosis, lipases act on adipose tissue and convert the cells to ghostlike outlines.

FIGURE 1–6. CASEOUS NECROSIS: In caseous necrosis, the affected area is seen grossly as a cheesy white area of necrosis. Microscopically, caseous necrosis is characterized by a granuloma with central necrosis.

■ EXTRACELLULAR ACCUMULATION

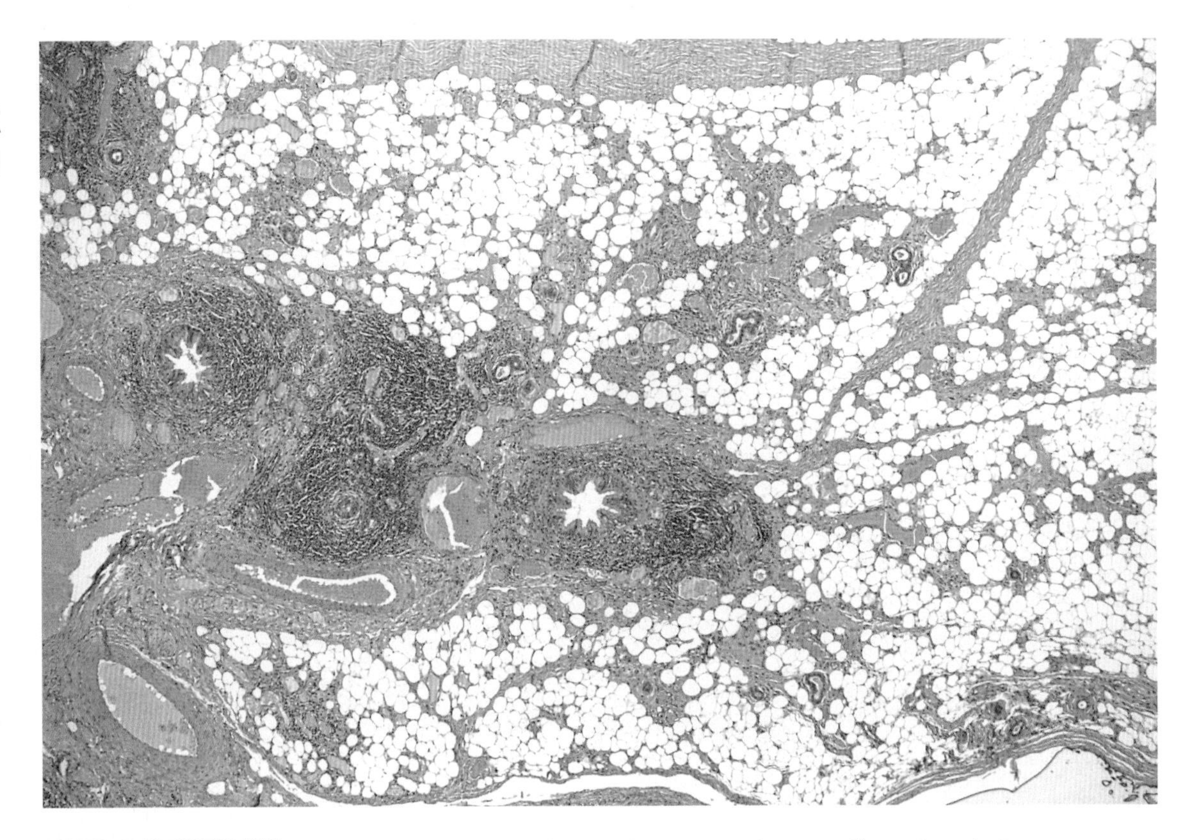

FIGURE 1–7. FATTY INFILTRATION: Fatty infiltration is characterized by the presence of mature adipose tissue in the stroma, between parenchymal cells.

■ INTRACELLULAR ACCUMULATIONS

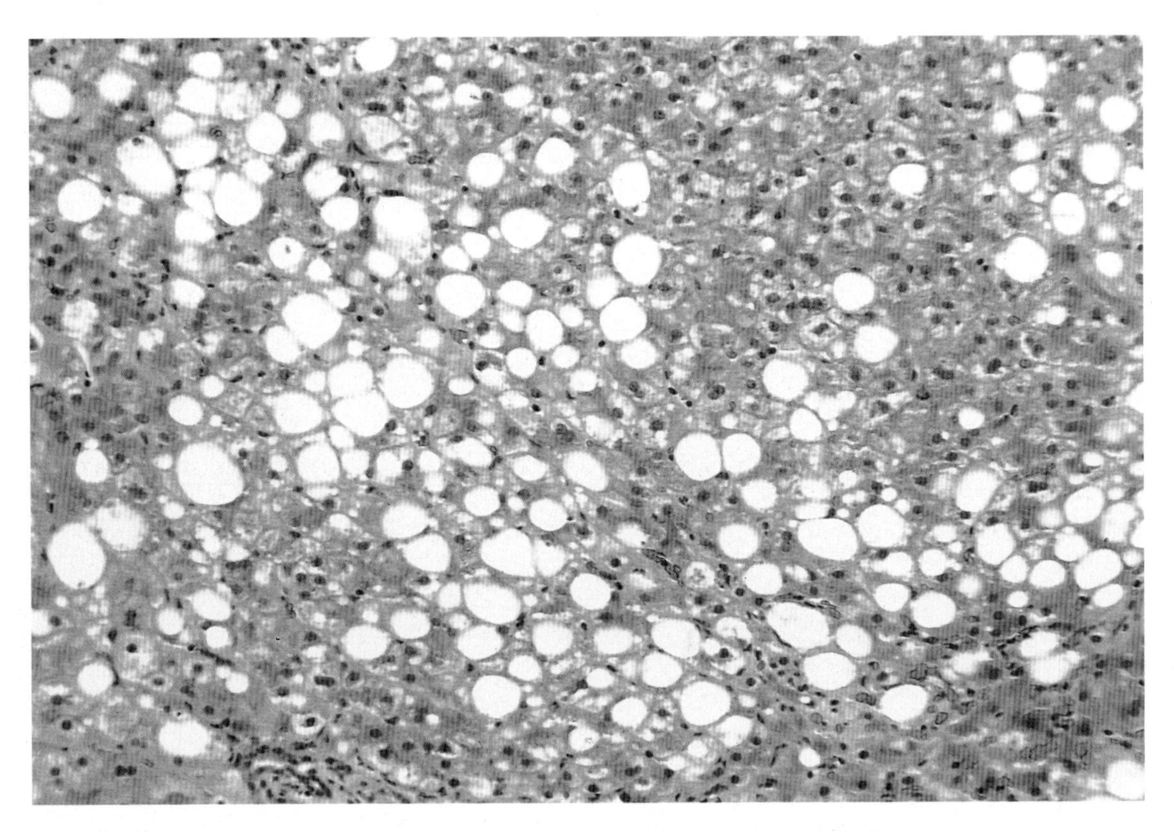

FIGURE 1–8. FAT: Fatty change is characterized by the accumulation of fat droplets within parenchymal cells.

■ INTRACELLULAR ACCUMULATIONS

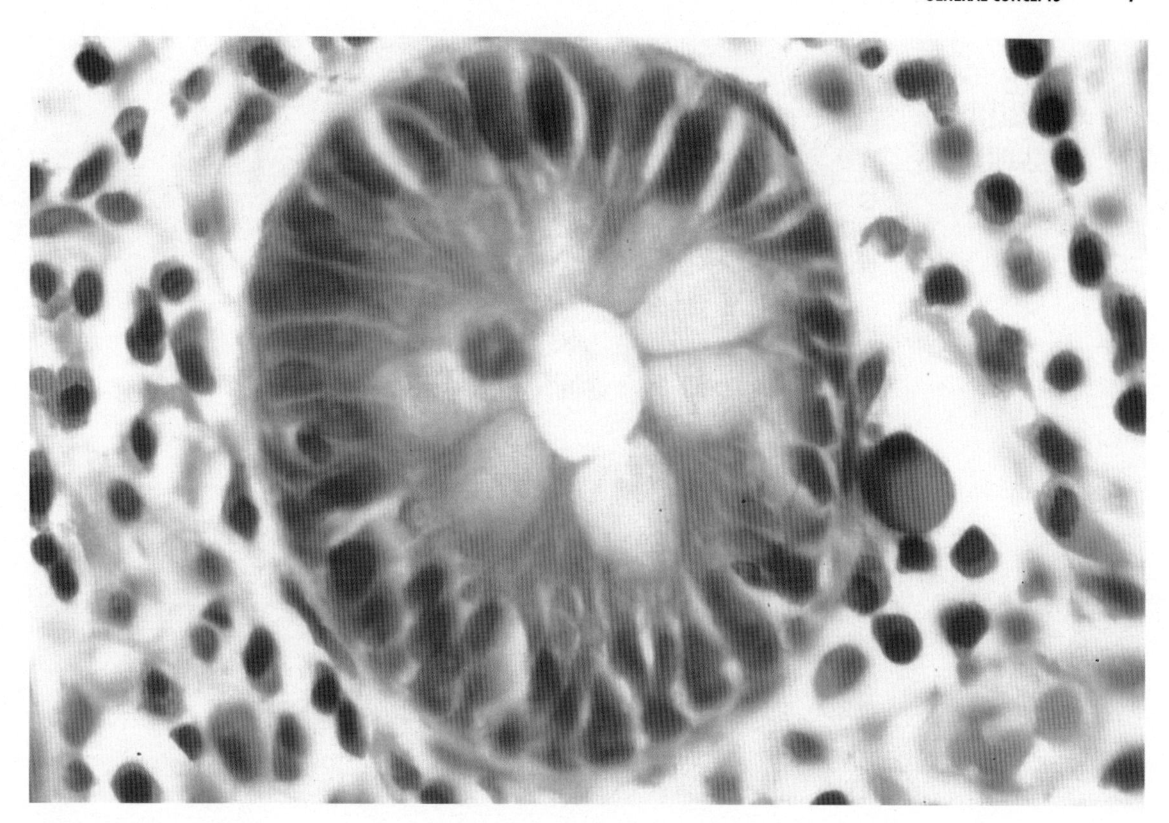

FIGURE 1–9. PROTEIN: Russell bodies in plasma cells are eosinophilic cytoplasmic droplets composed of excessive accumulations of immunoglobulins.

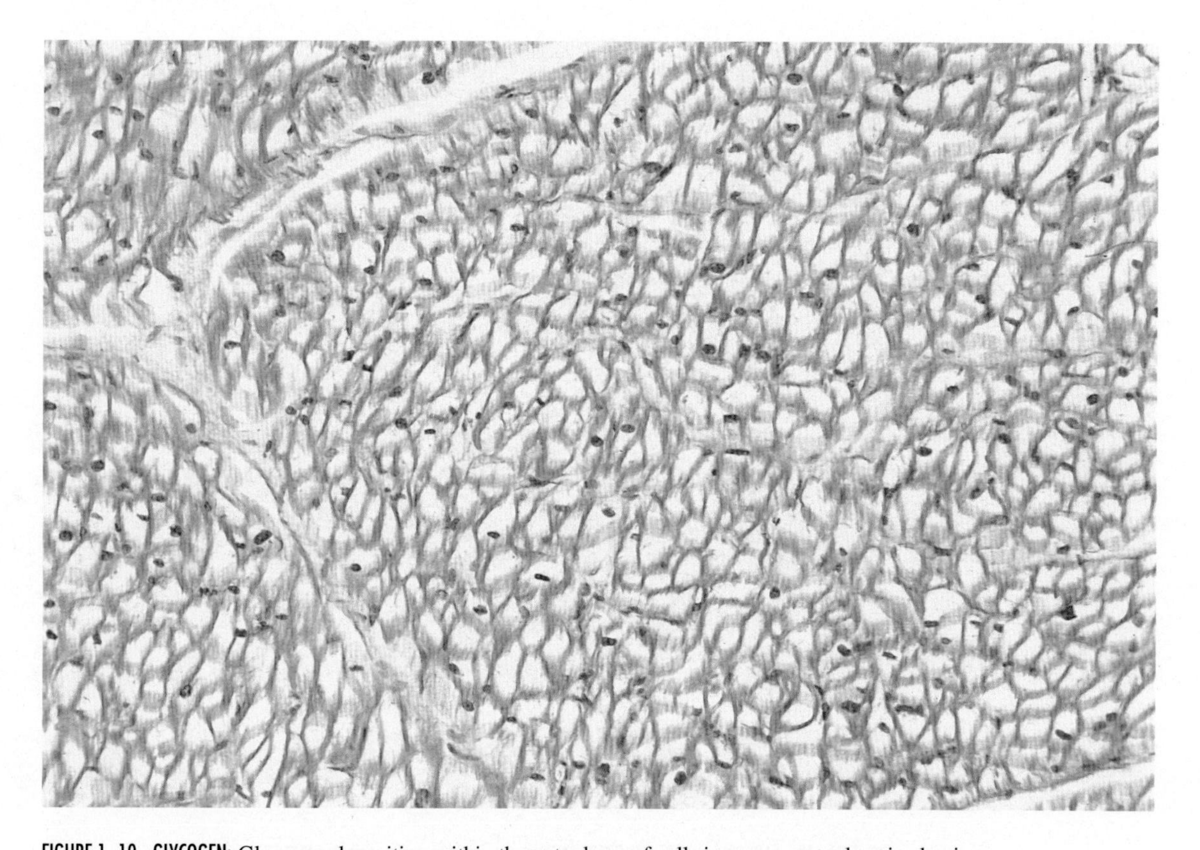

FIGURE 1–10. GLYCOGEN: Glycogen deposition within the cytoplasm of cells is seen as cytoplasmic clearing.

■ INTRACELLULAR ACCUMULATIONS

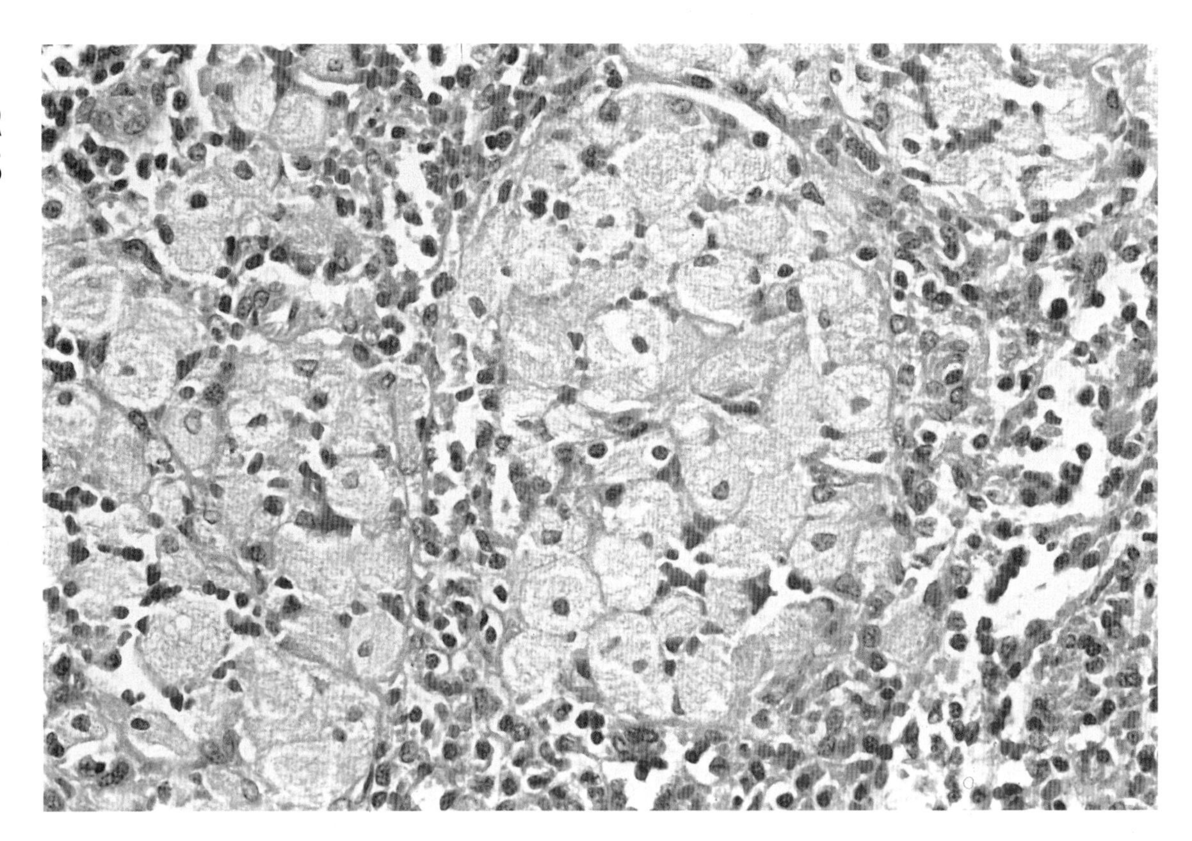

FIGURE 1–11. LIPID: Lipid accumulation as seen in **Gaucher's cells** has a characteristic "wrinkled tissue paper" appearance.

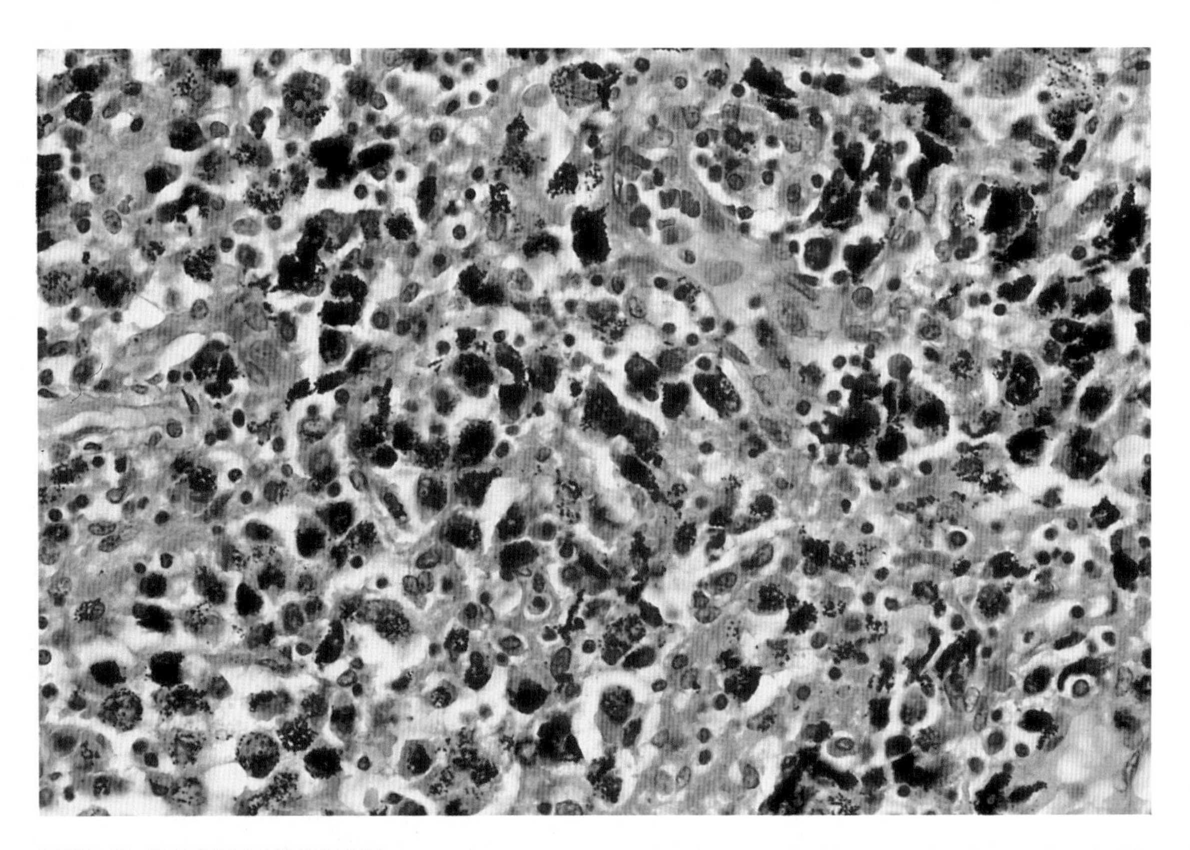

FIGURE 1–12. CARBON (ANTHRACOTIC PIGMENT): Carbon dust that has been inhaled is seen as a black granular pigment deposited in macrophages of the lungs and their draining lymph nodes.

■ INTRACELLULAR ACCUMULATIONS

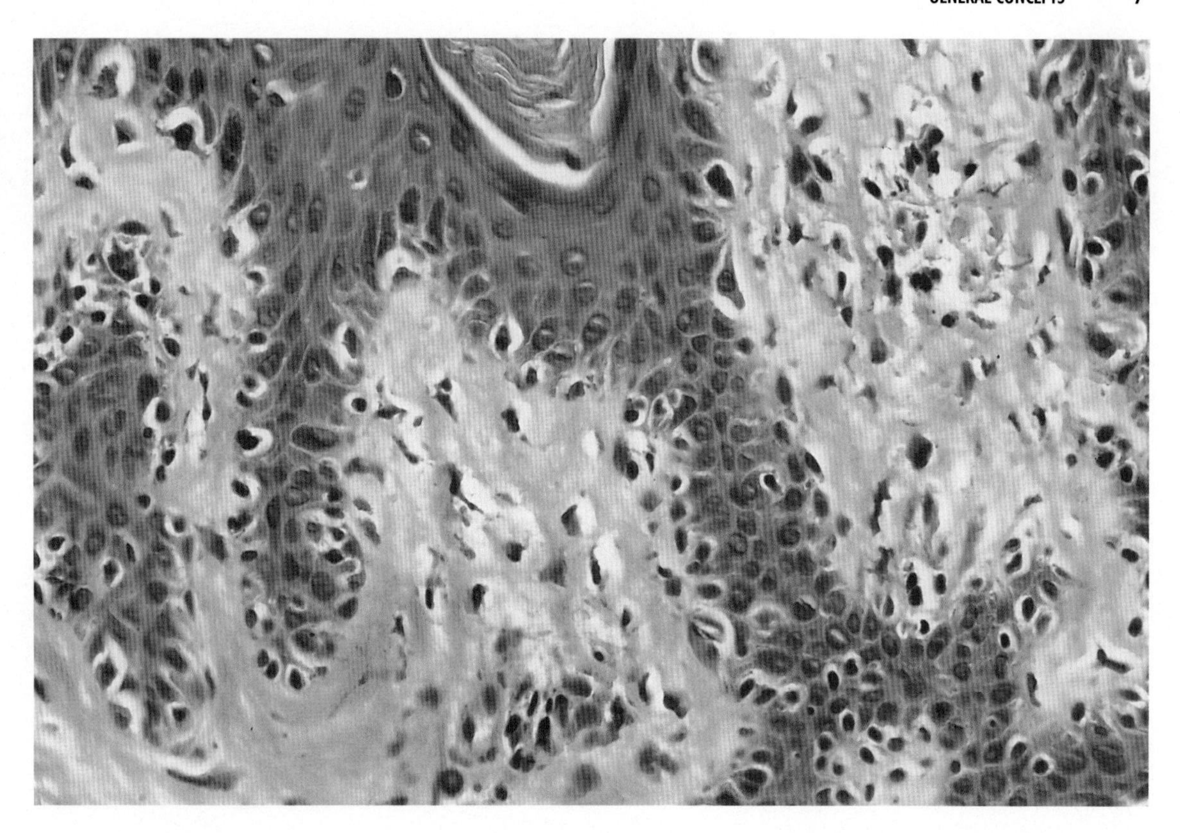

FIGURE 1–13. MELANIN: Melanin is seen as a brown-black granular pigment in the melanocytes of the epidermis and melanophages of the dermis.

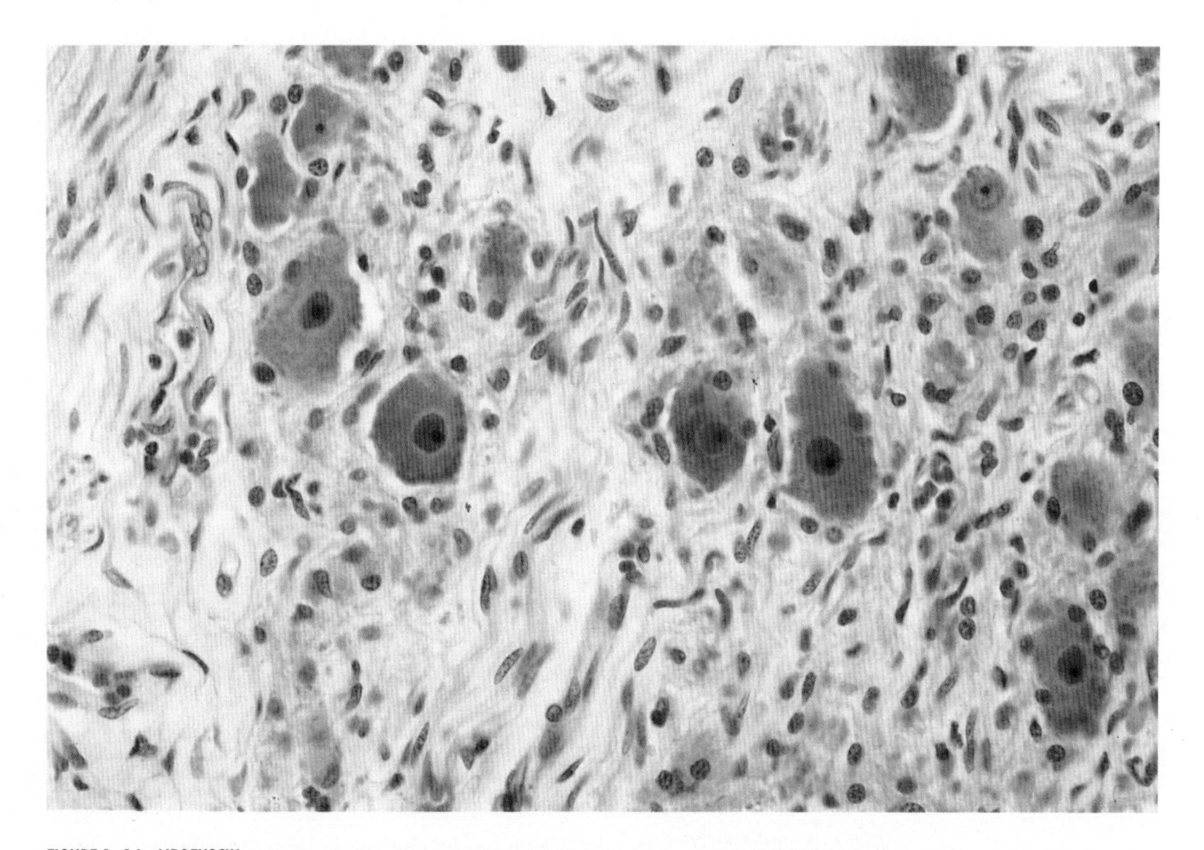

FIGURE 1–14. LIPOFUSCIN: Lipofuscin, the "wear and tear" pigment, is seen as a granular golden-brown intracytoplasmic accumulation.

■ INTRACELLULAR ACCUMULATIONS

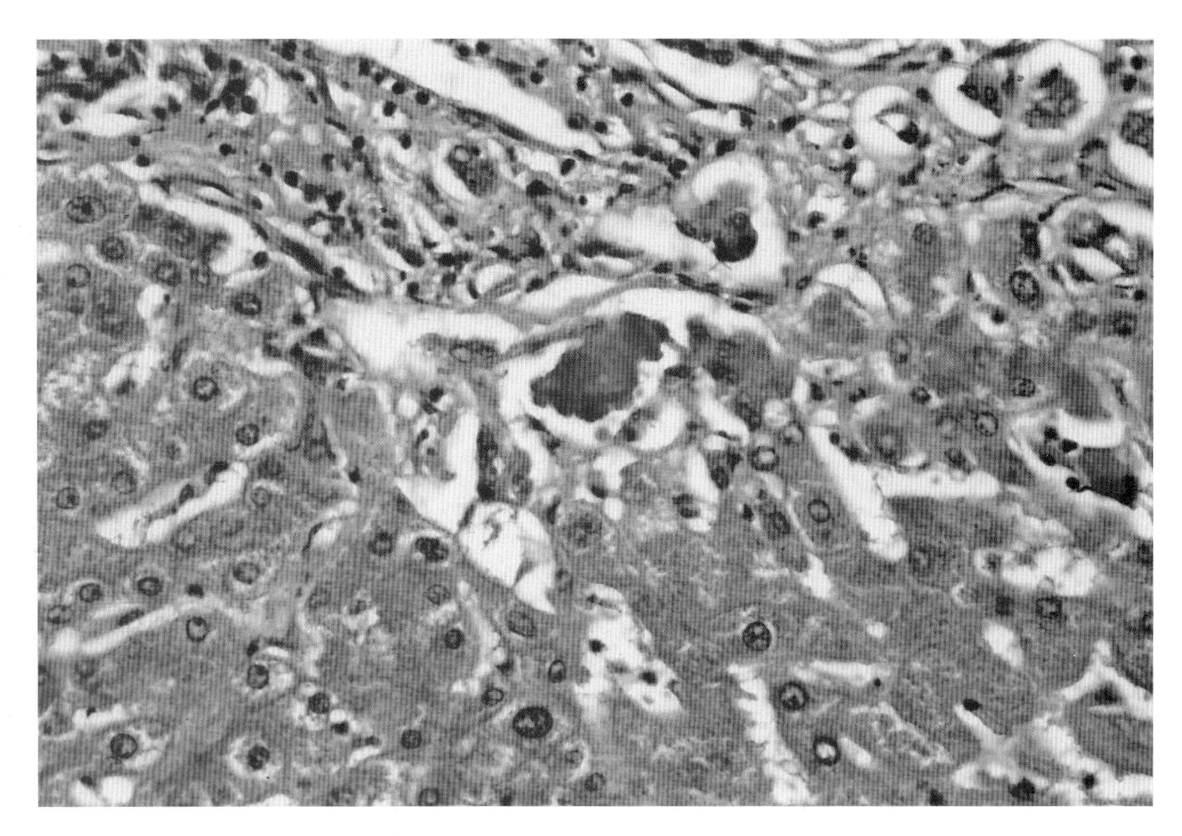

FIGURE 1–15. BILIRUBIN: Bilirubin is a non-iron–containing derivative of hemoglobin. It is present in hepatocytes, Kupffer cells, and bile sinusoids as a green-brown, amorphous, and globular material.

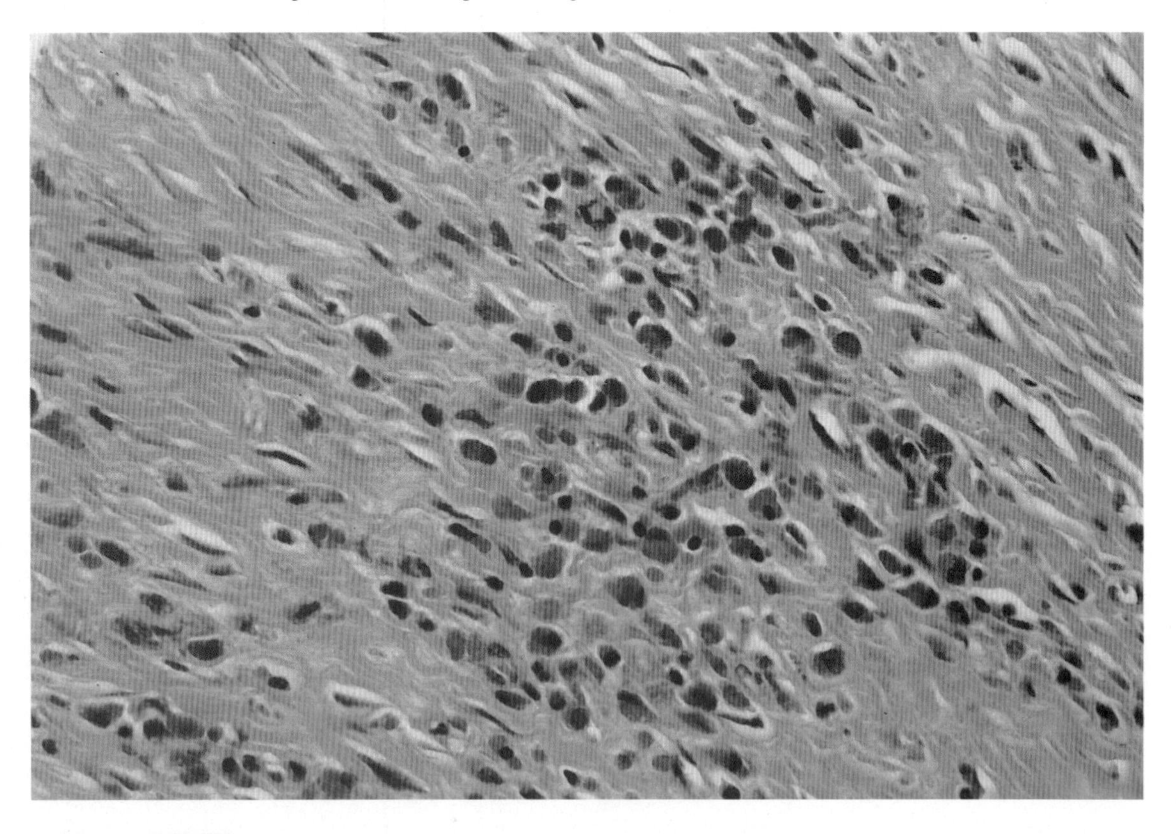

FIGURE 1–16. HEMOSIDERIN: Hemosiderin is a hemoglobin-derived pigment containing iron. In areas of hemorrhage, hemosiderin, a golden-yellow to brown granular pigment, is phagocytized by macrophages.

■ INTRACELLULAR ACCUMULATIONS

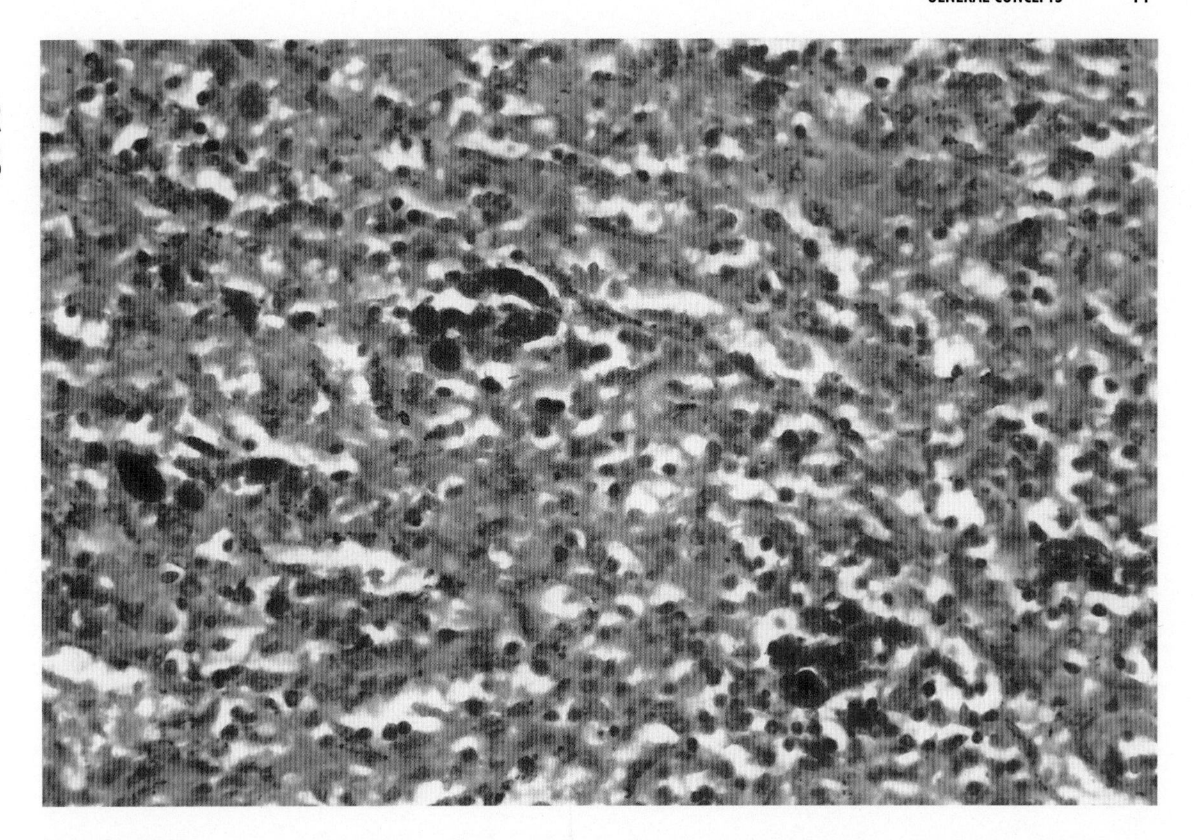

Figure 1–17 A

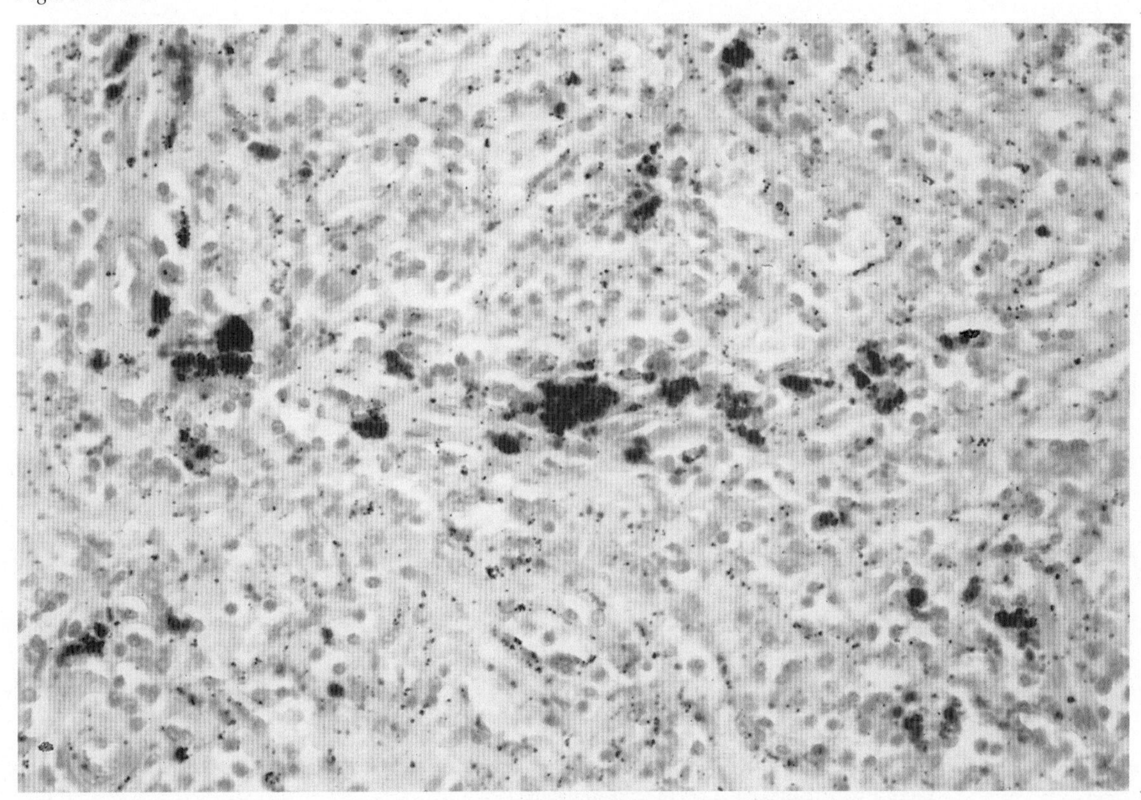

Figure 1–17 B

FIGURE 1–17. HEMOSIDEROSIS: A: Hemosiderosis is a systemic accumulation of brown-black granular iron particles in parenchymal cells. **B:** Prussian blue staining is used to confirm that the pigment present in hemosiderosis is iron.

■ CELL ADAPTATION

FIGURE 1–18. ATROPHY: Atrophy is a decrease in cell size. Interspersed between the normal myofibers of skeletal muscle are atrophic myofibers.

FIGURE 1–19. HYPERTROPHY: Hypertrophy is an increase in cell size. The cell and nucleus are increased in size in this example of myocardial hypertrophy.

FIGURE 1–20. HYPERPLASIA: Hyperplasia is an increase in cell number. In intraductal hyperplasia of the breast, there are increased numbers of cells in the duct.

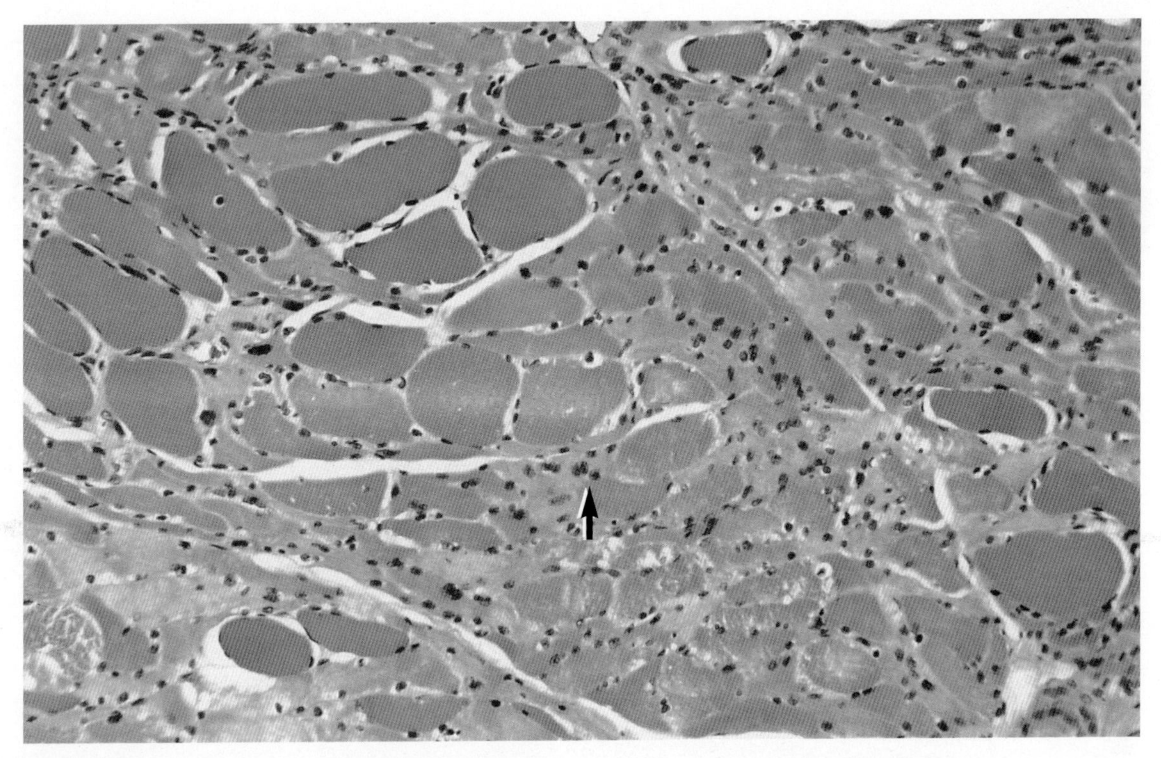

Figure 1–18

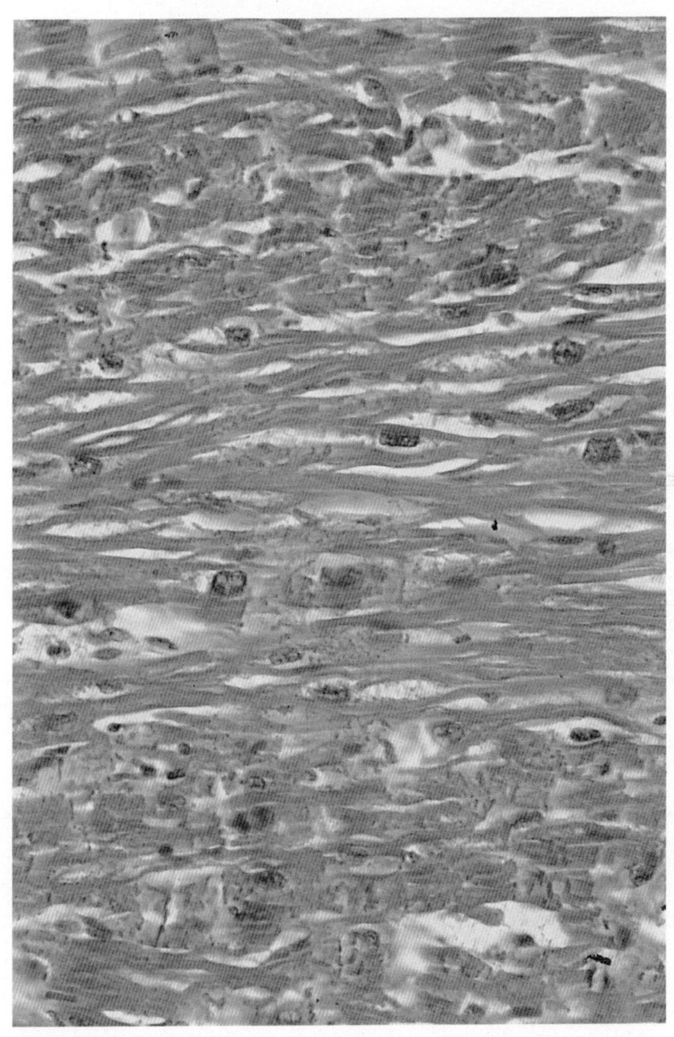

Figure 1–19

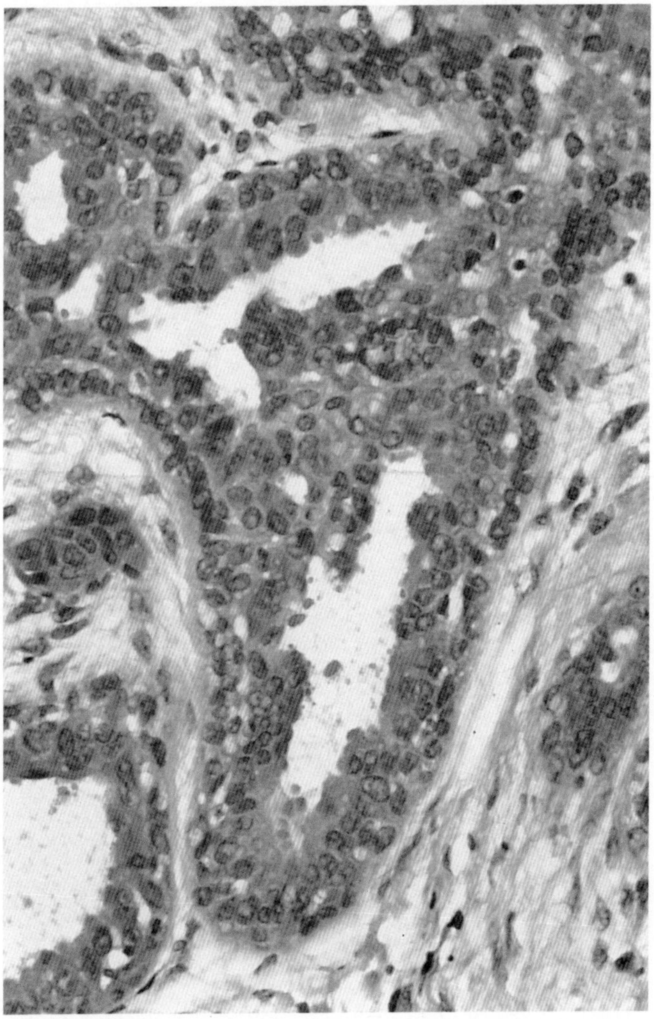

Figure 1–20

■ CELL ADAPTATION

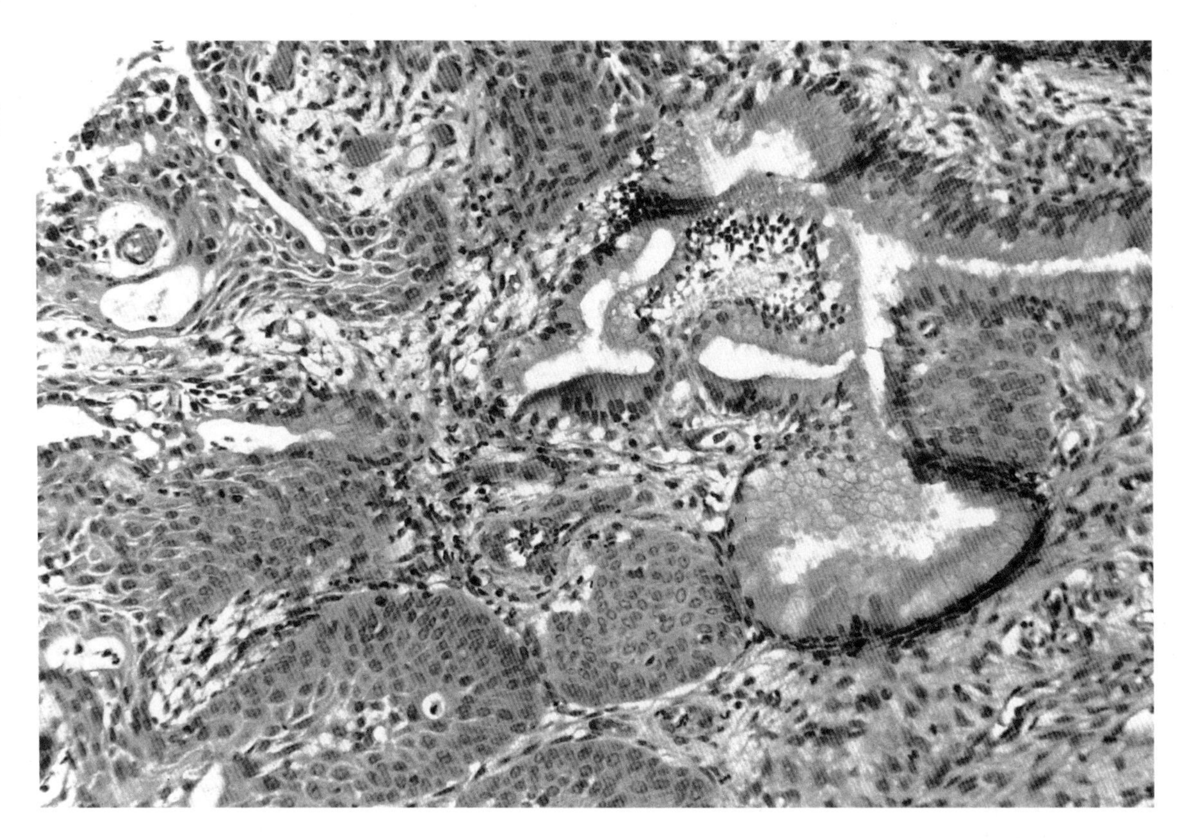

FIGURE 1–21. METAPLASIA: Metaplasia is a change of one adult cell type for another. In this photomicrograph, the mucus-secreting cells of the endocervix are replaced by squamous cells.

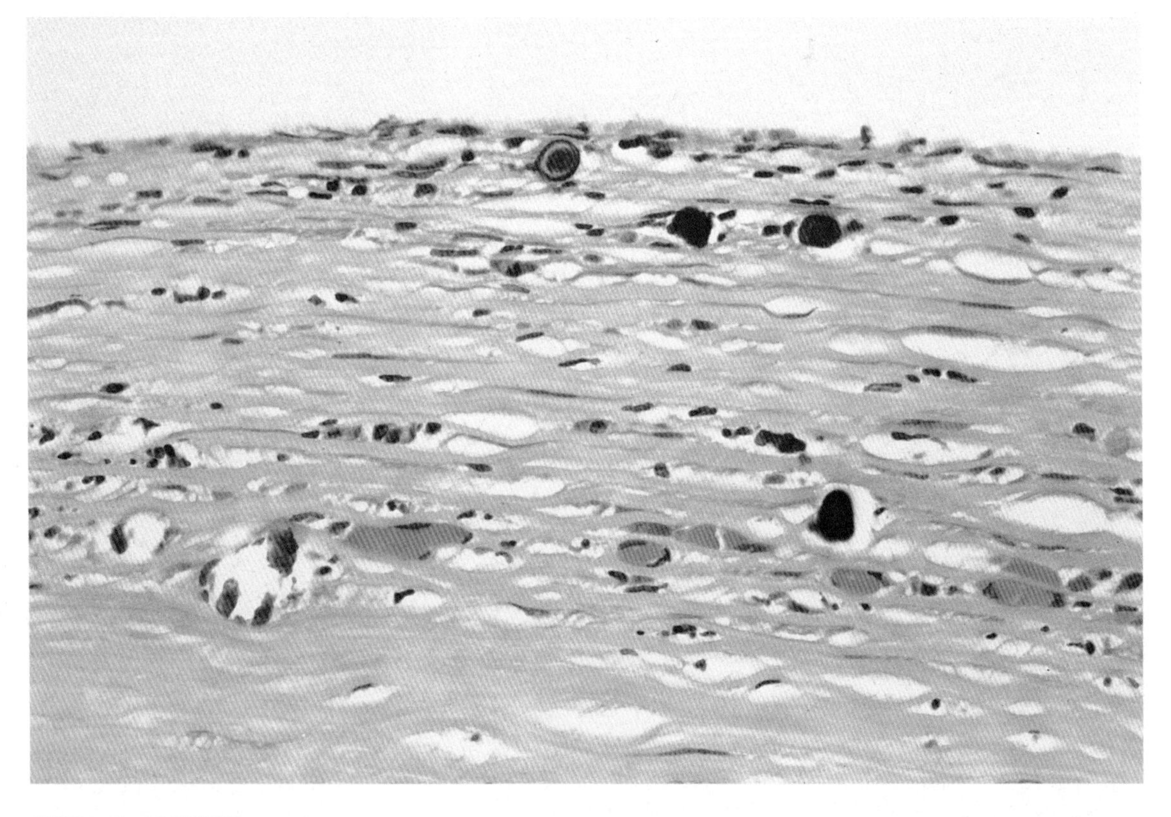

FIGURE 1–22. CALCIFICATION: In dystrophic calcification, the serum levels of calcium are normal; however, there is calcium deposition in abnormal tissues.

2. INFLAMMATION AND REPAIR

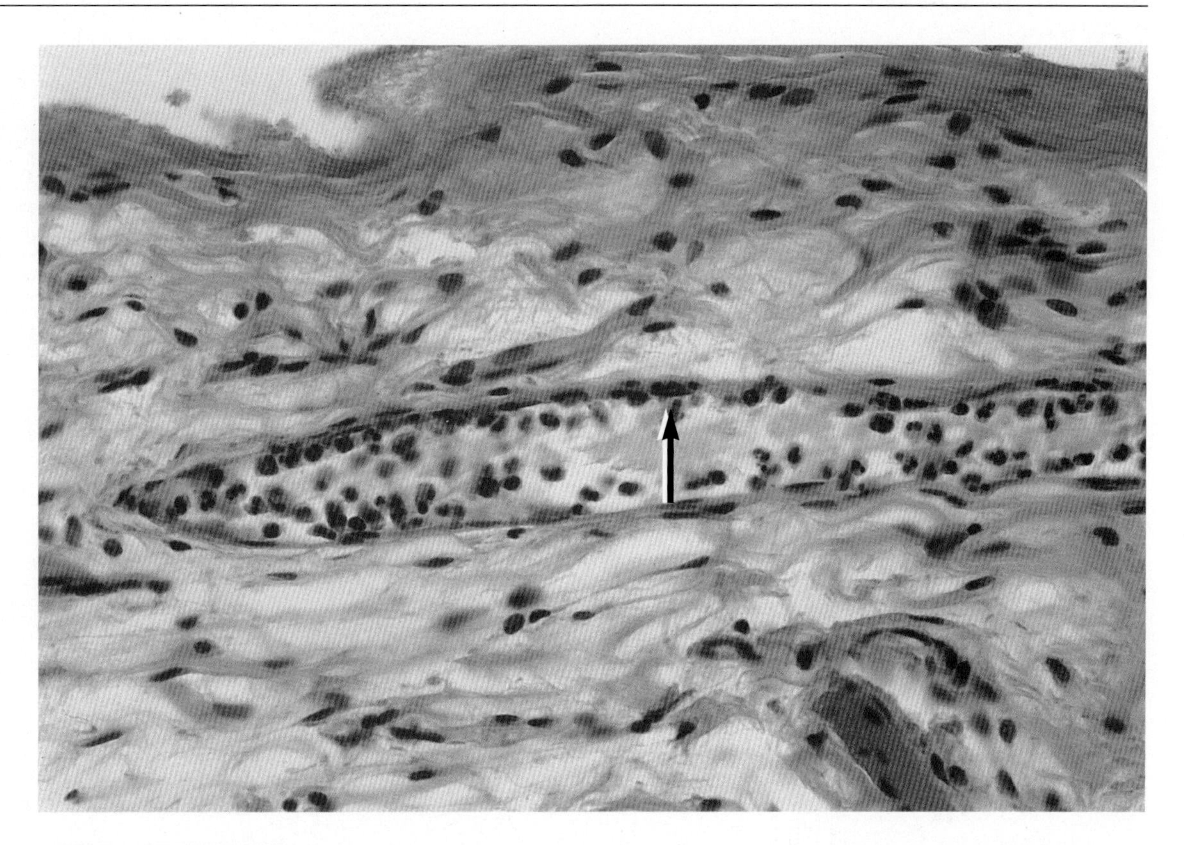

FIGURE 2–1. ACUTE INFLAMMATION: This photomicrograph of a blood vessel shows the early changes that take place in the inflammatory process. Neutrophils are seen **marginating** along the blood vessel wall. Others are seen **emigrating** (*arrow*) through the vessel wall into the interstitial spaces.

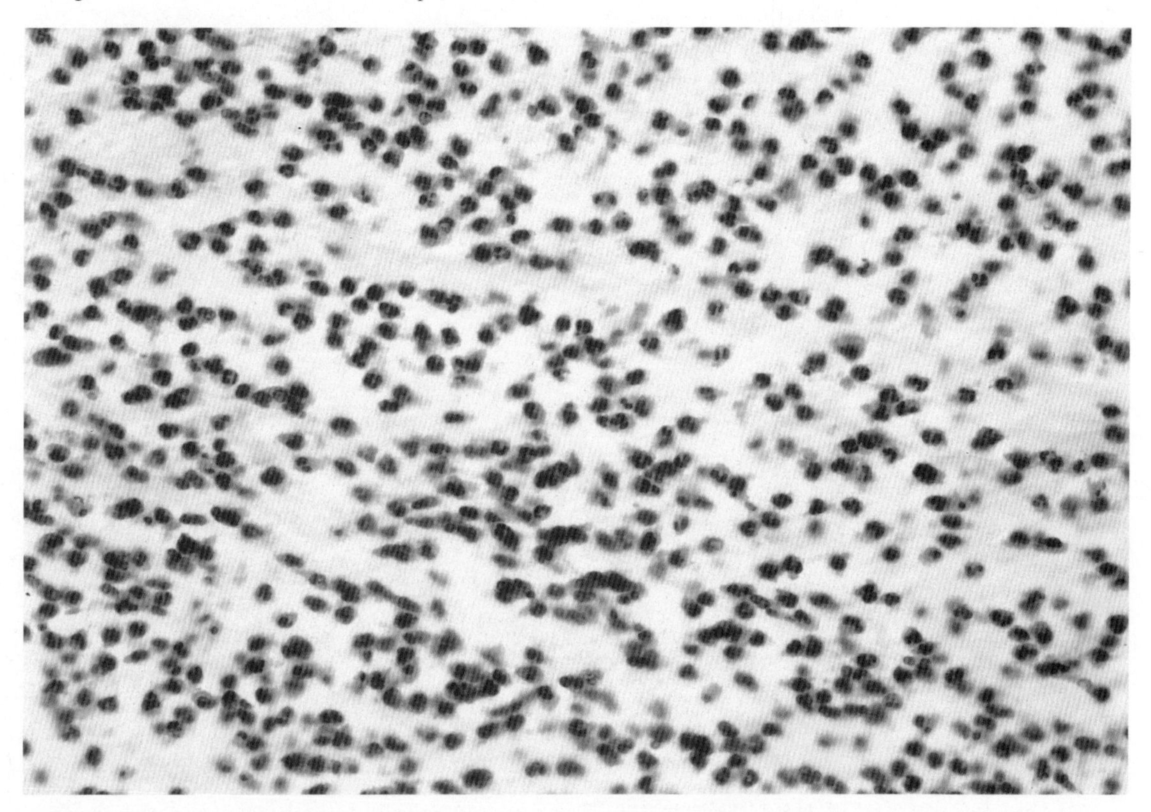

FIGURE 2–2. PURULENT (SUPPURATIVE) INFLAMMATION: Purulent inflammation is characterized by an accumulation of neutrophils. This is also known as pus.

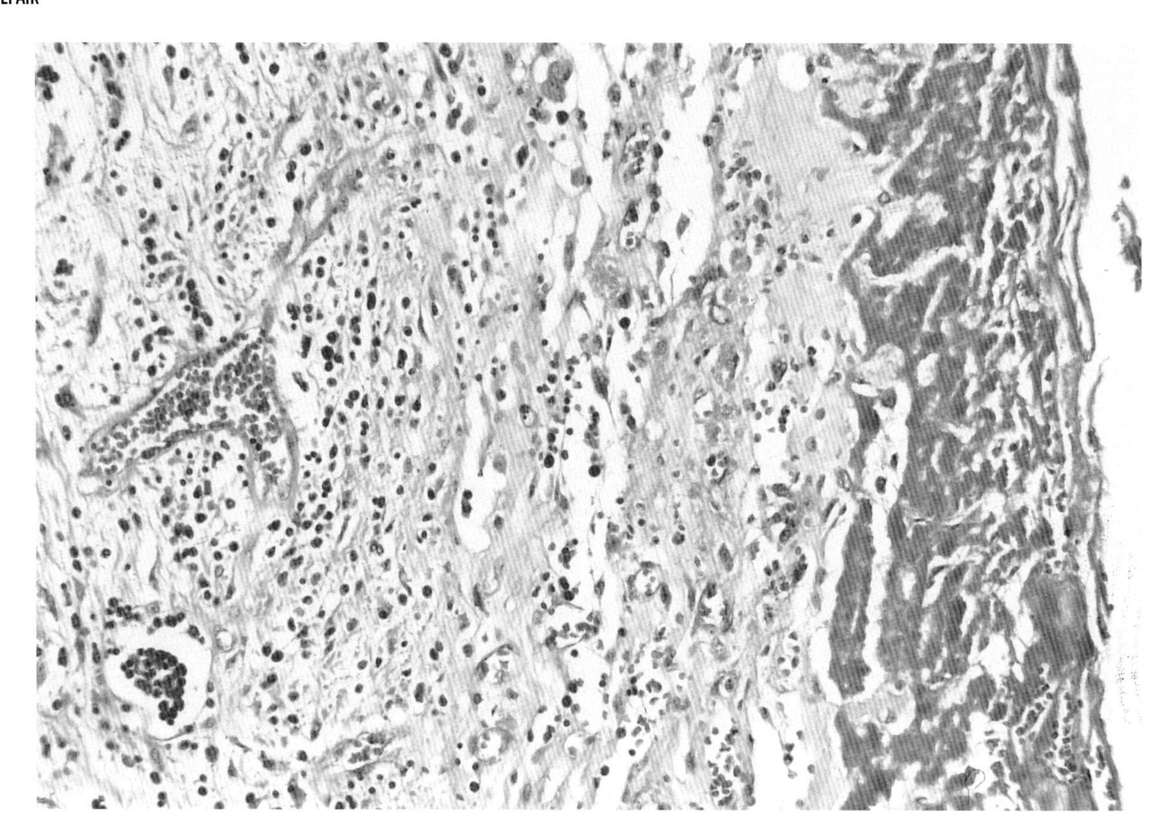

FIGURE 2–3. FIBRINOUS INFLAMMATION: Fibrin is seen as a tangled threadlike amorphous eosinophilic deposit characteristically overlying inflamed mesothelium.

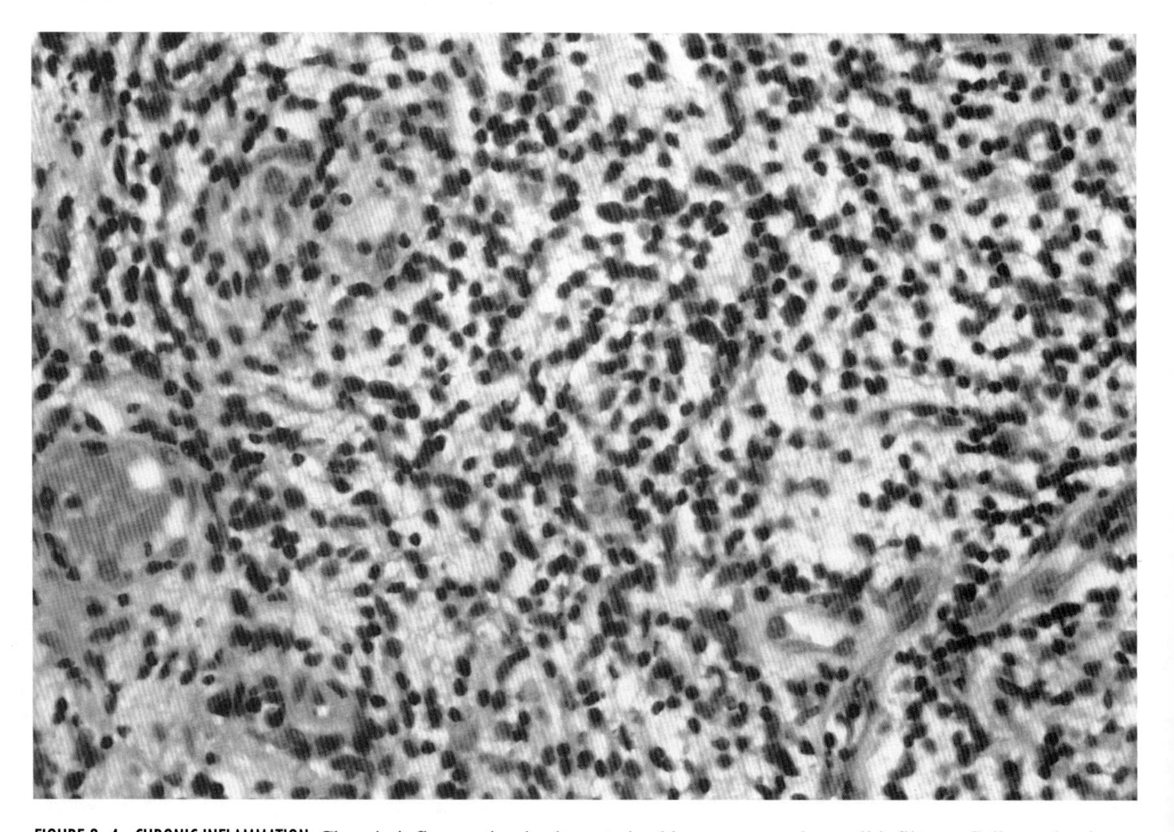

FIGURE 2–4. CHRONIC INFLAMMATION: Chronic inflammation is characterized by a mononuclear cell infiltrate. Cells predominating are lymphocytes and plasma cells, with variable numbers of macrophages.

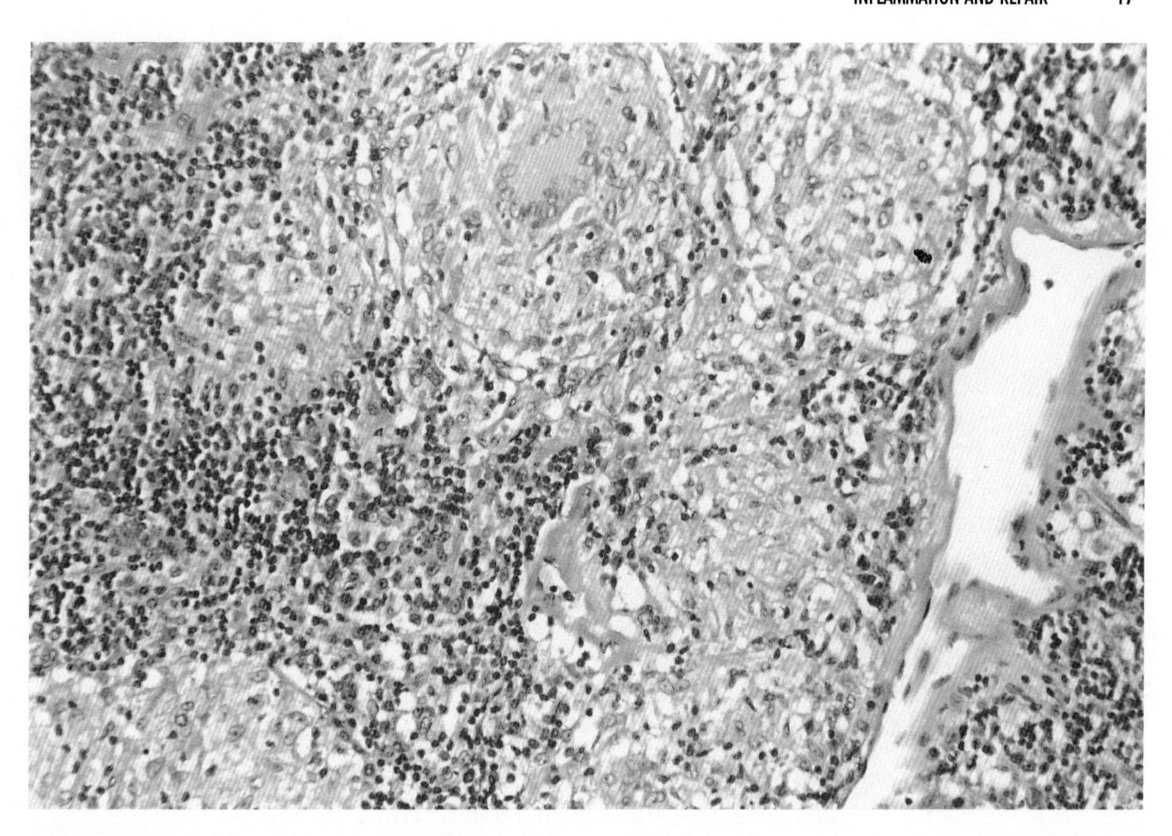

Figure 2–5 **A**

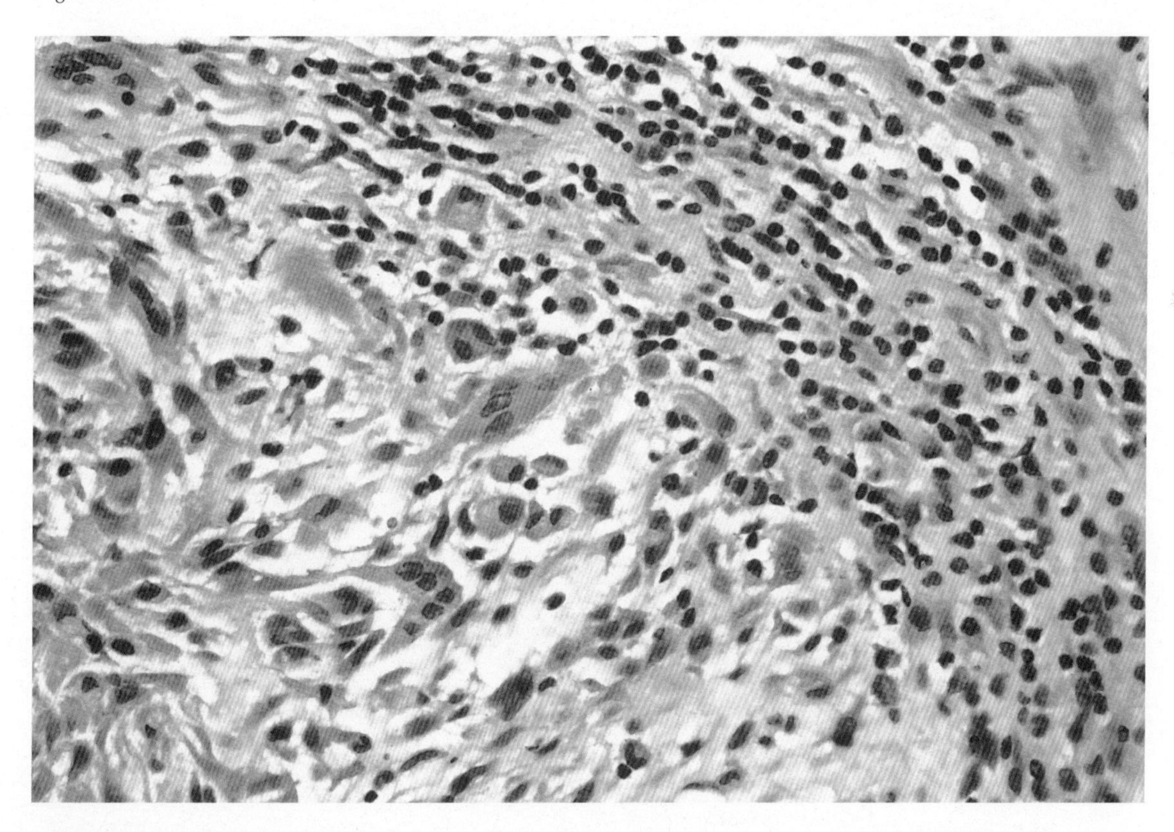

Figure 2–5 **B**

FIGURE 2–5. GRANULOMA: A: This is a low-power photomicrograph of granulomatous inflammation. **B:** A granuloma is characterized by a collection of epithelioid macrophages with abundant pale eosinophilic cytoplasm and indistinct cell borders surrounded by a rim of lymphocytes.

■ CHRONIC GRANULOMATOUS INFLAMMATION

FIGURE 2–6. FOREIGN BODY GIANT CELL: In this example of a granulomatous reaction is a foreign body giant cell. The giant cell contains suture material.

FIGURE 2–7. LANGHANS' GIANT CELL: A: In this photomicrograph of an immune granuloma, there is a Langhans' giant cell. **B:** Mycobacteria, which are acid-fast organisms, are present within the giant cell and stain red with a Kinyoun stain.

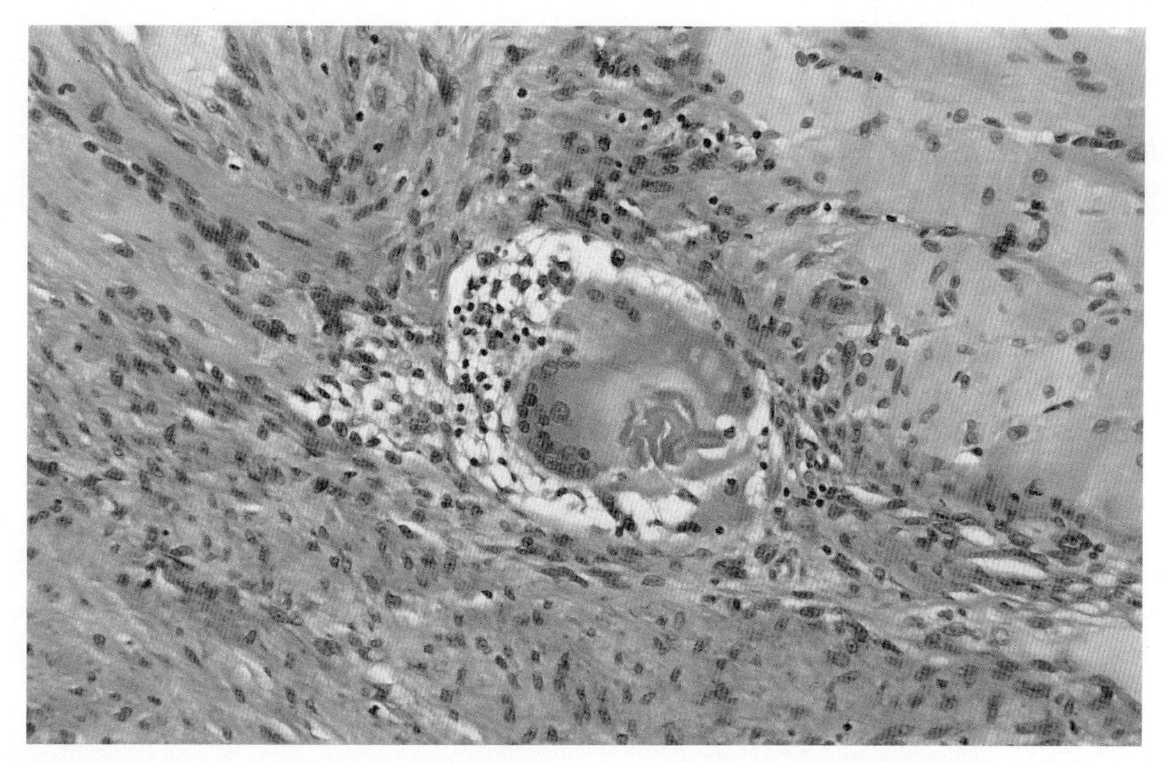

Figure 2–6

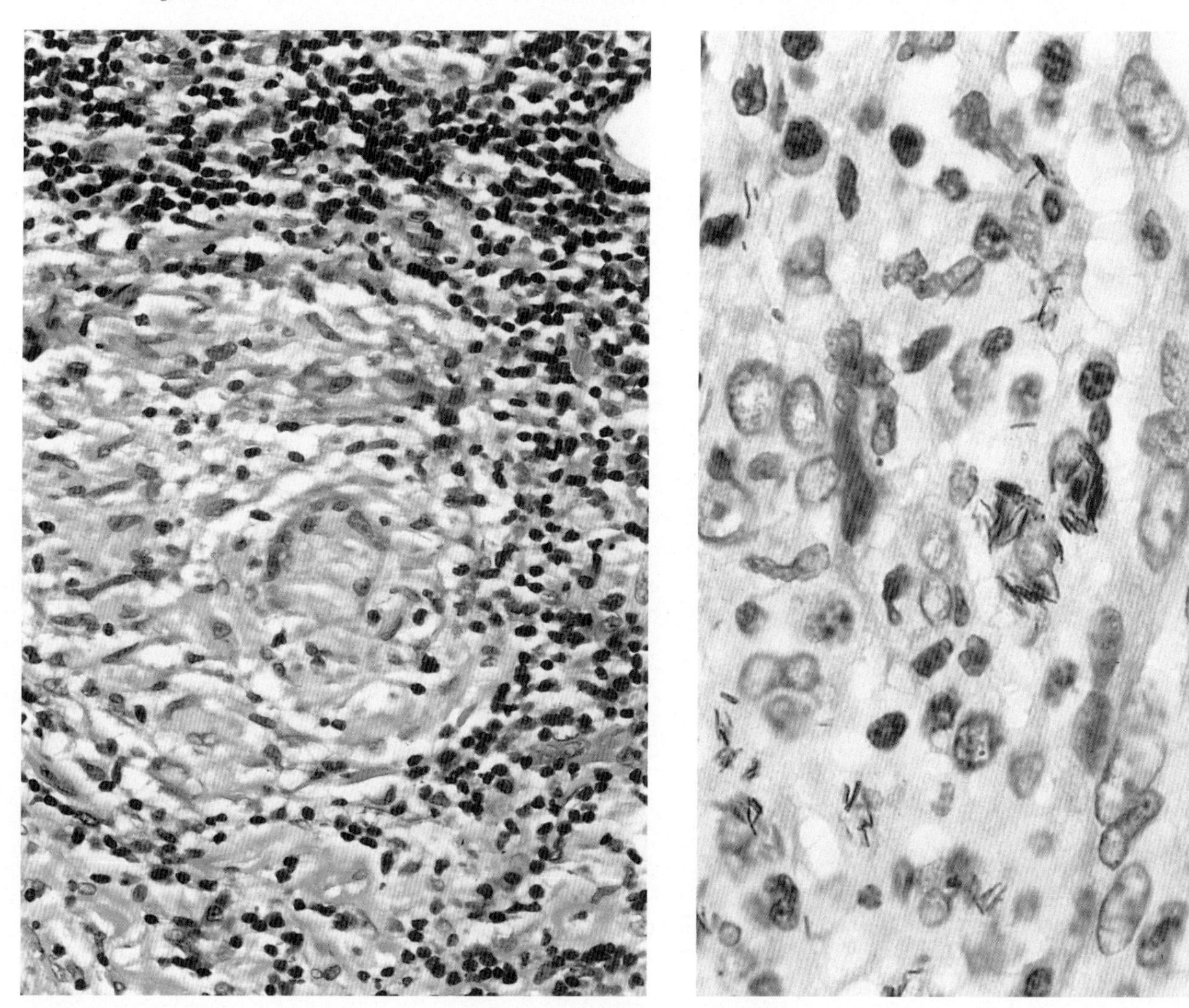

Figure 2–7 A

Figure 2–7 B

■ CHRONIC GRANULOMATOUS INFLAMMATION

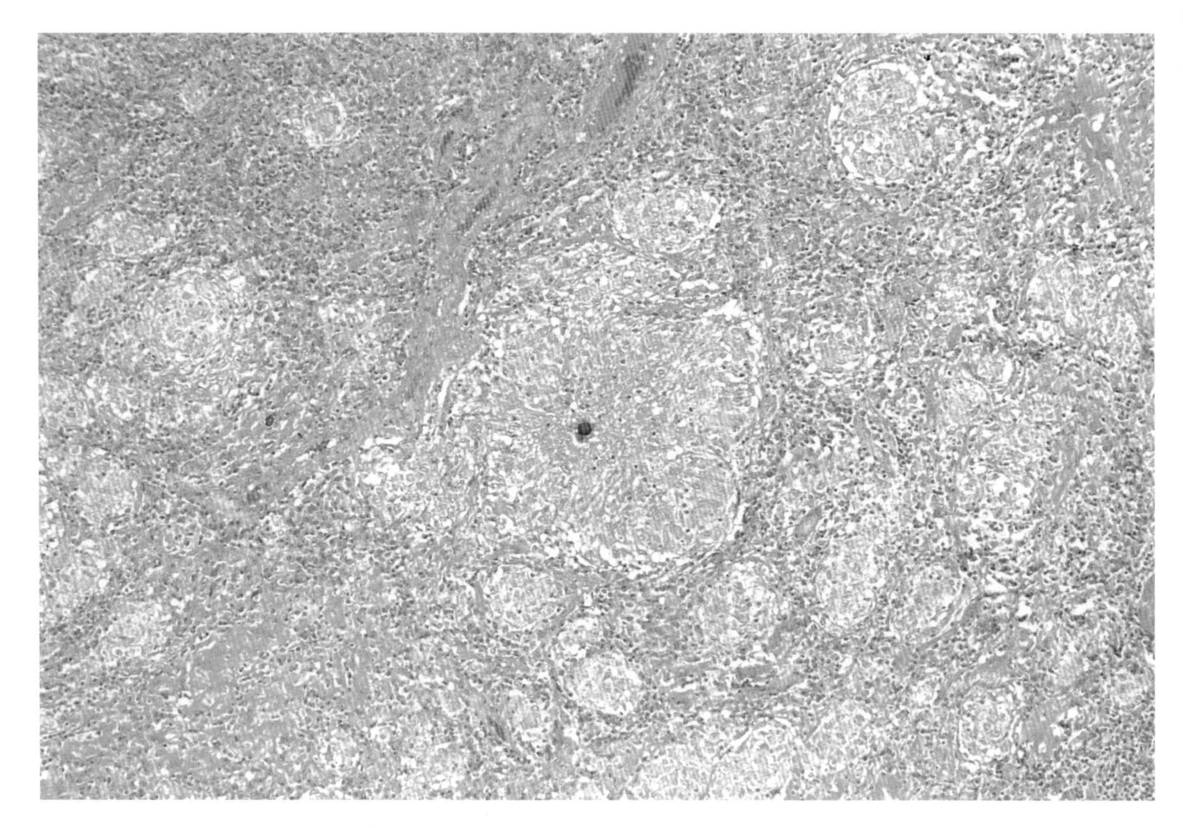

Figure 2–8 A

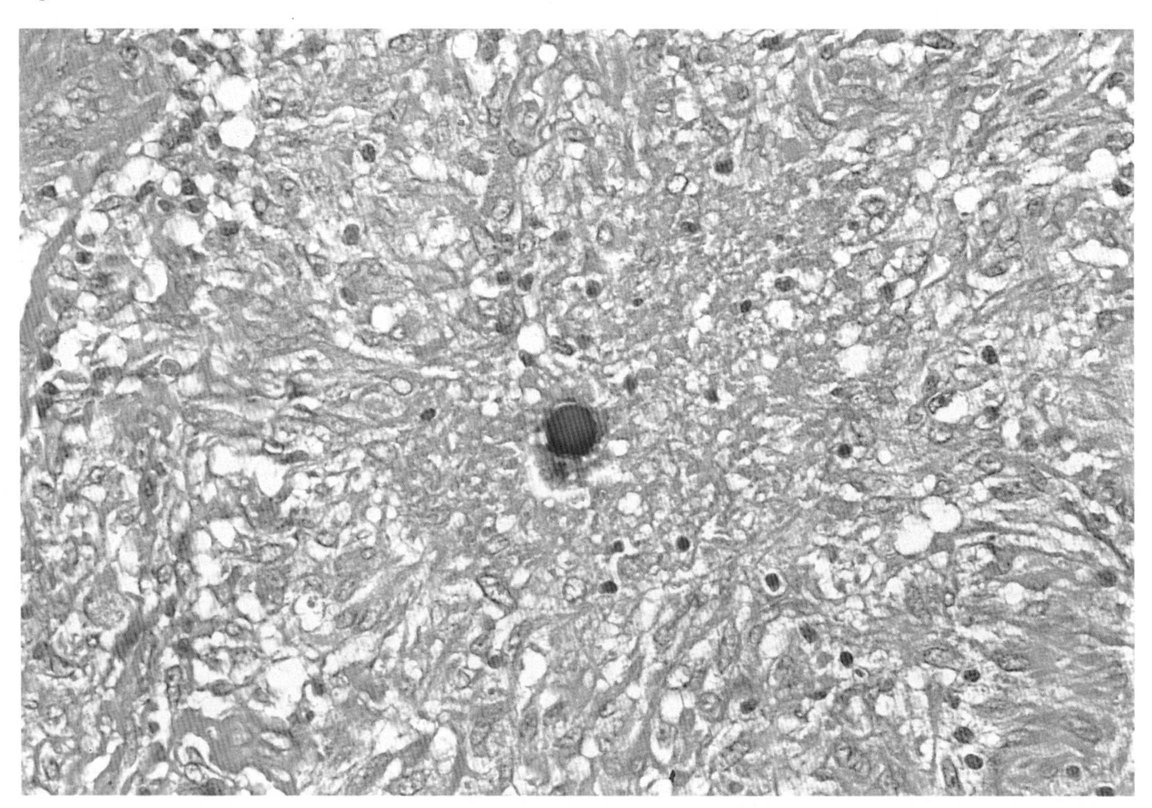

Figure 2–8 B

FIGURE 2–8. NONCASEATING GRANULOMA: A: Sarcoidosis is the classic example of noncaseating granulomatous inflammation. B: A **Schaumann's body,** seen as a concentric calcification, may be present.

■ CHRONIC GRANULOMATOUS INFLAMMATION

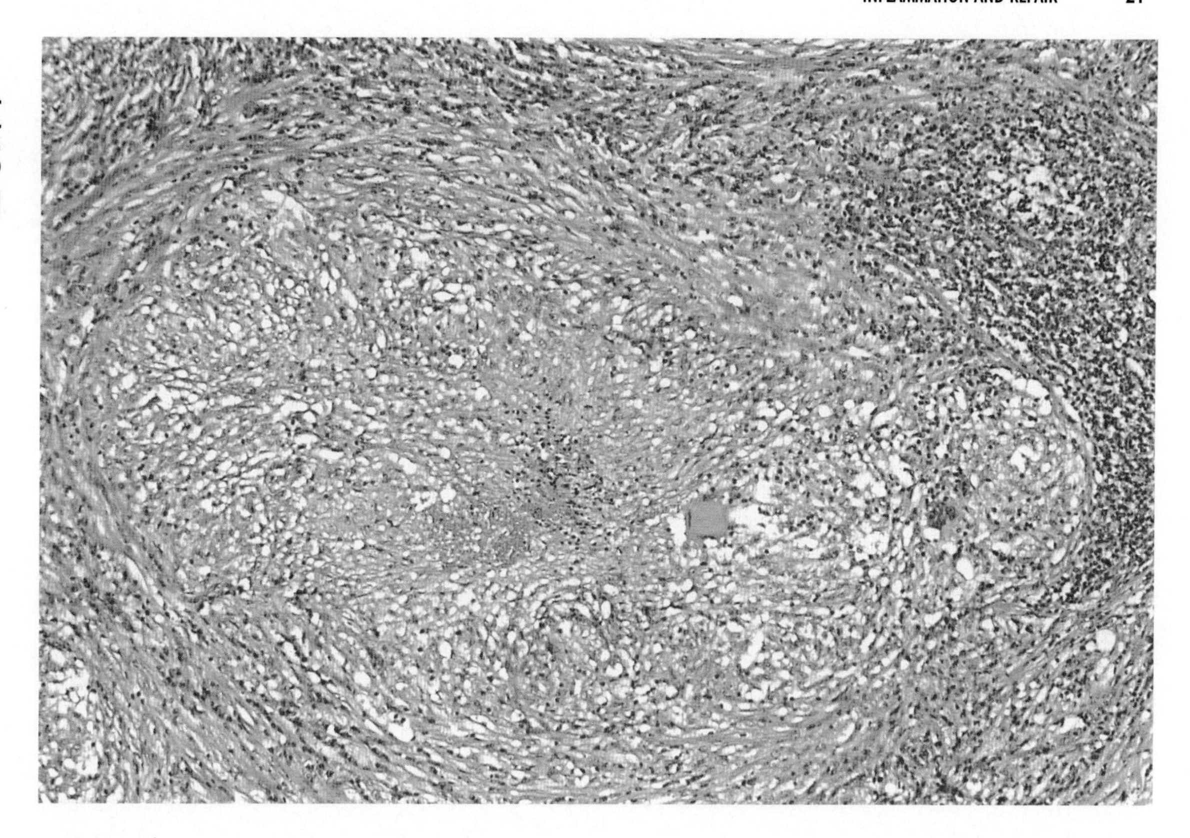

Figure 2–9 A

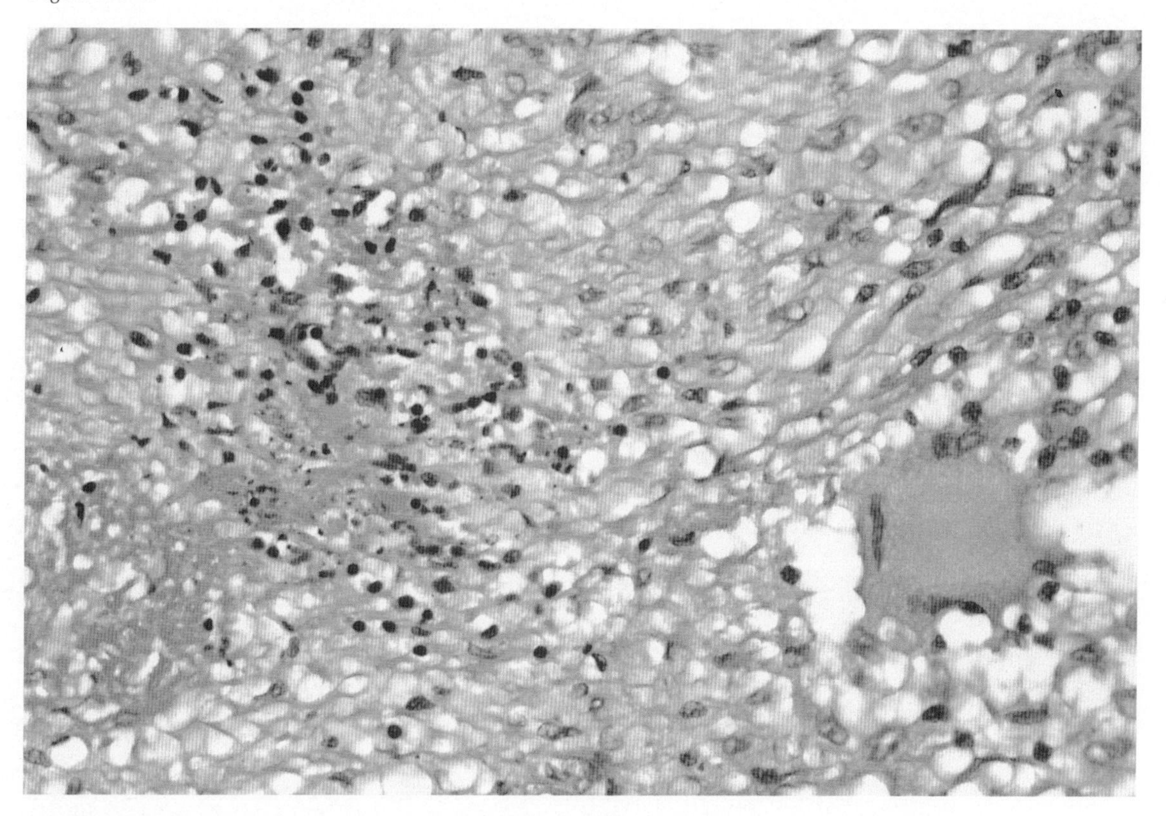

Figure 2–9 B

FIGURE 2–9. CASEATING GRANULOMA: A: Tuberculosis is the classic example of caseating granulomatous inflammation. Centrally, the granuloma contains amorphous granular debris. **B:** In this photomicrograph, a Langhan's-type giant cell is present adjacent to the necrotic center.

■ REPAIR

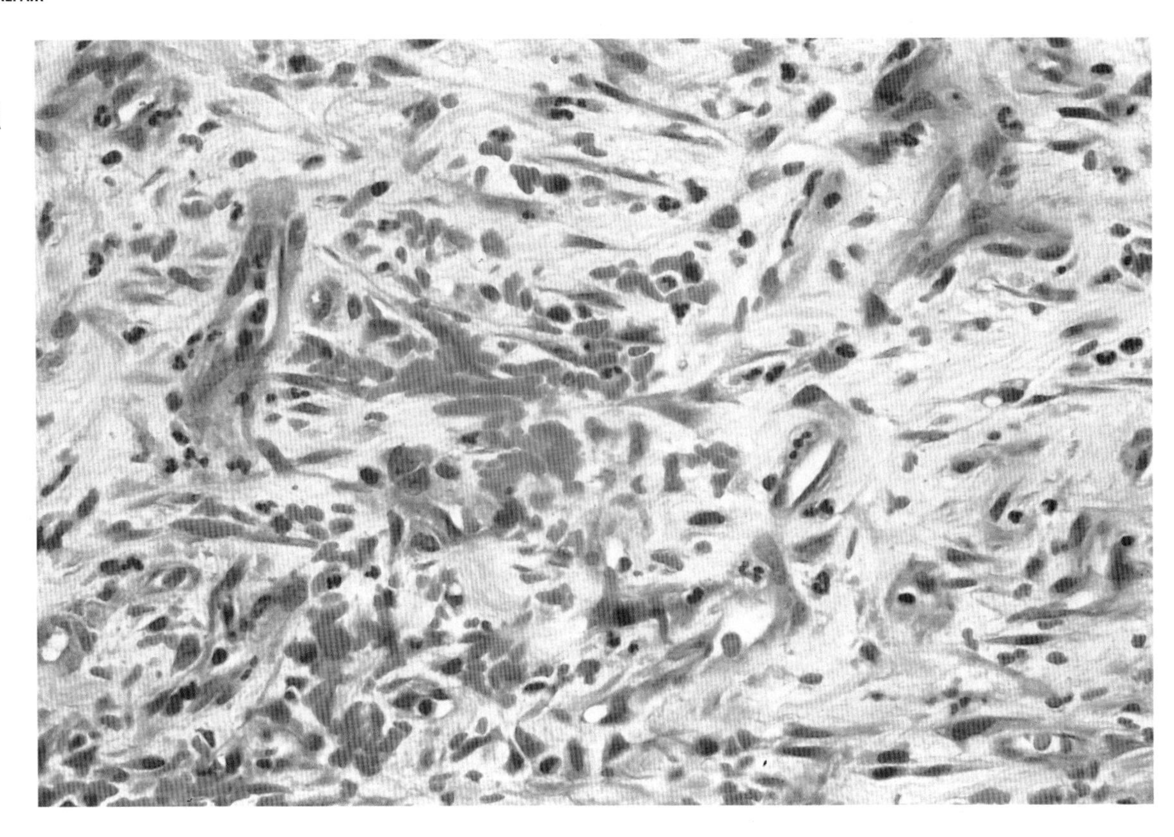

Figure 2–10 **A**

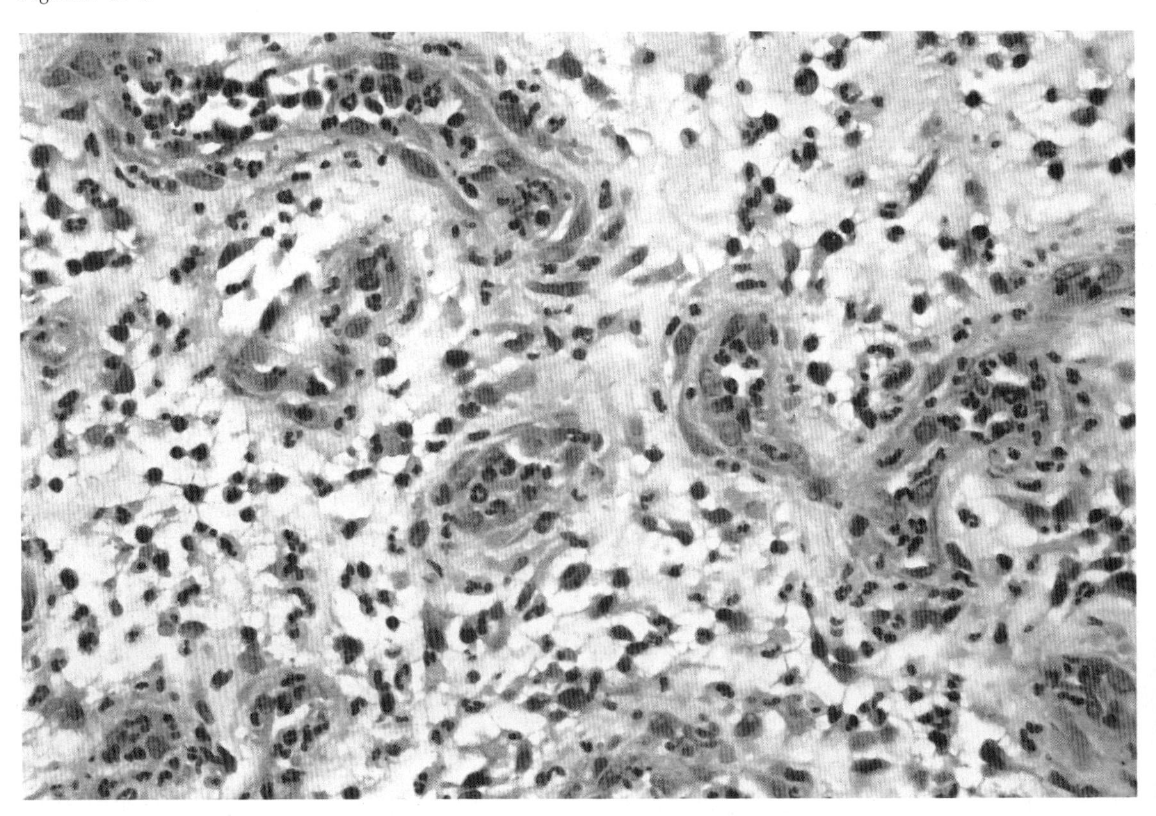

Figure 2–10 **B**

FIGURE 2–10. GRANULATION TISSUE: Granulation tissue is the hallmark of repair. **A:** Young granulation tissue is composed of a proliferation of blood vessels and fibroblasts in an edematous stroma. Red blood cell extravasation is prominent because of leaky blood vessels. **B:** Older granulation tissue is composed of well-developed blood vessels with an inflammatory infiltrate in the interstitial tissue.

■ REPAIR

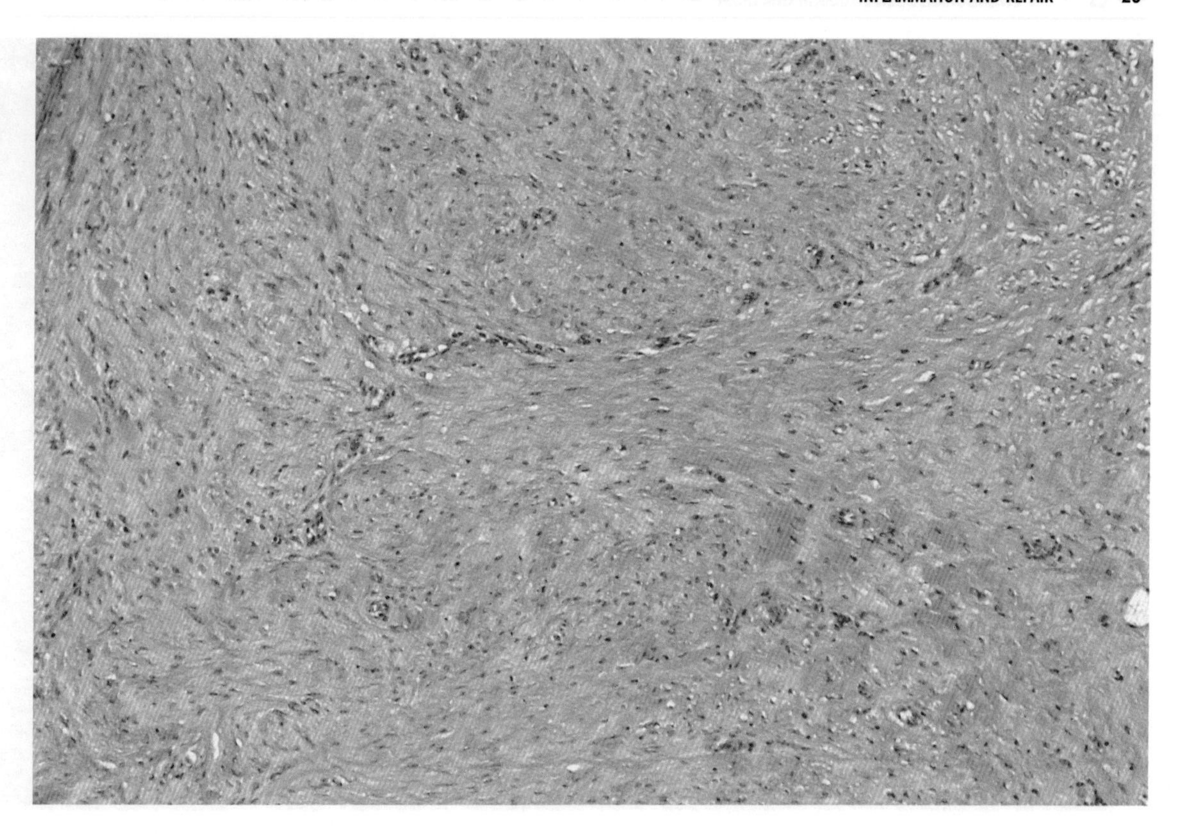

FIGURE 2–11. SCAR: A scar is composed of parallel bundles of collagen with few fibroblasts and few blood vessels.

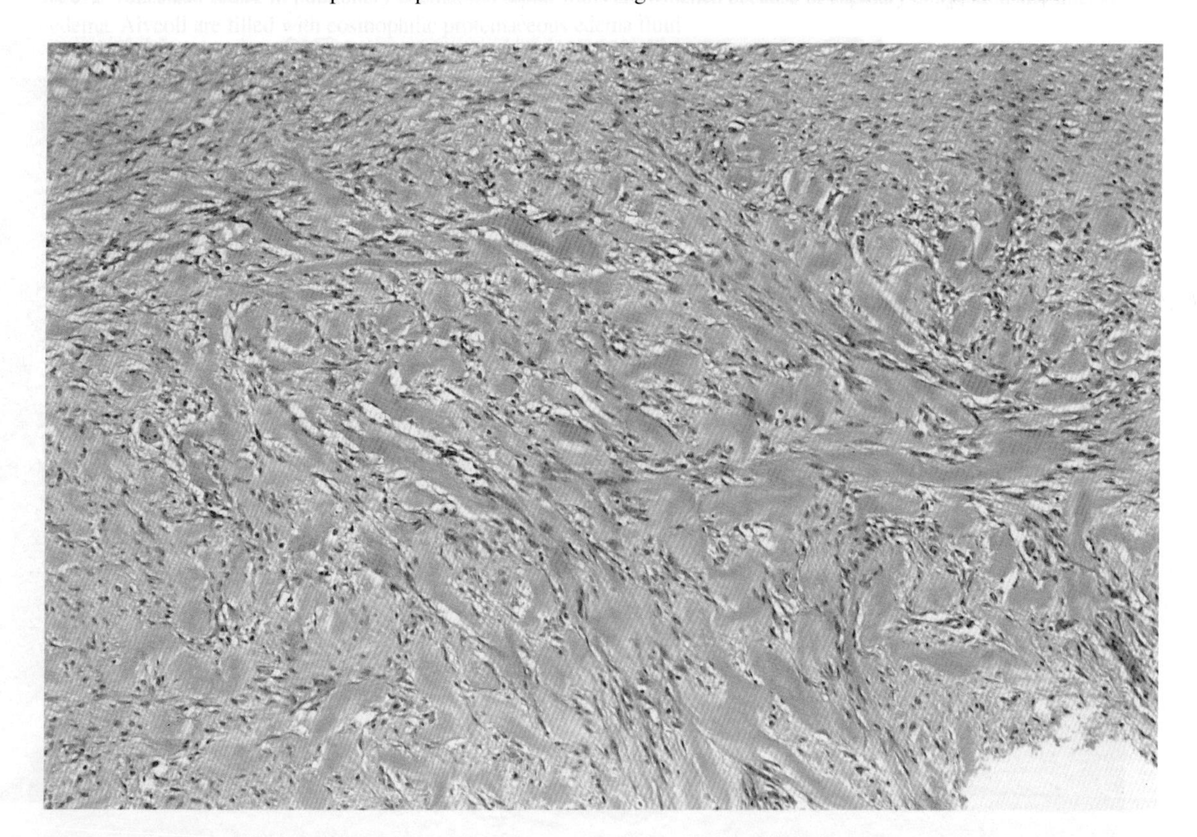

FIGURE 2–12. KELOID: A keloid is composed of thick collagen bundles in a whorled arrangement.

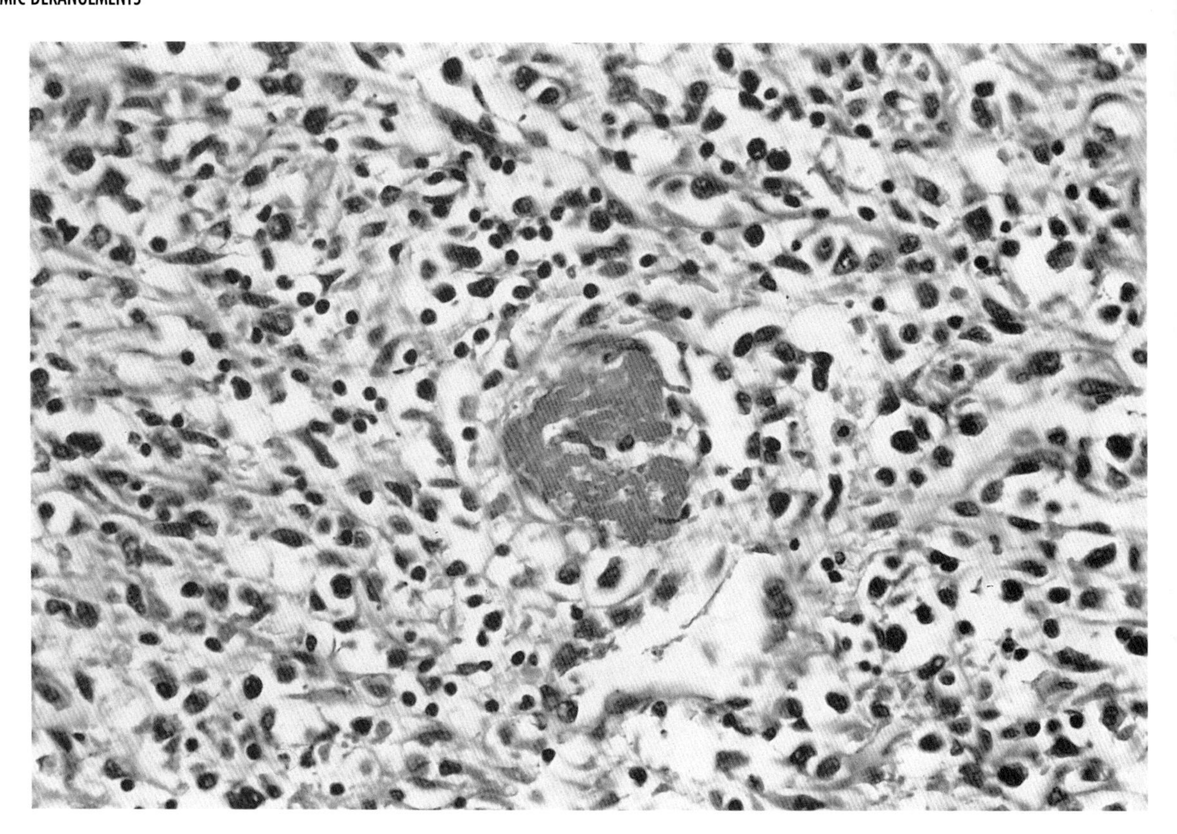

FIGURE 3–8. DISSEMINATED INTRAVASCULAR COAGULATION (DIC): In DIC, a fibrin microthrombus is present in a small blood vessel.

An **embolus** is an intravascular tissue detached from its point of origin that is carried to a distant site in the blood stream.

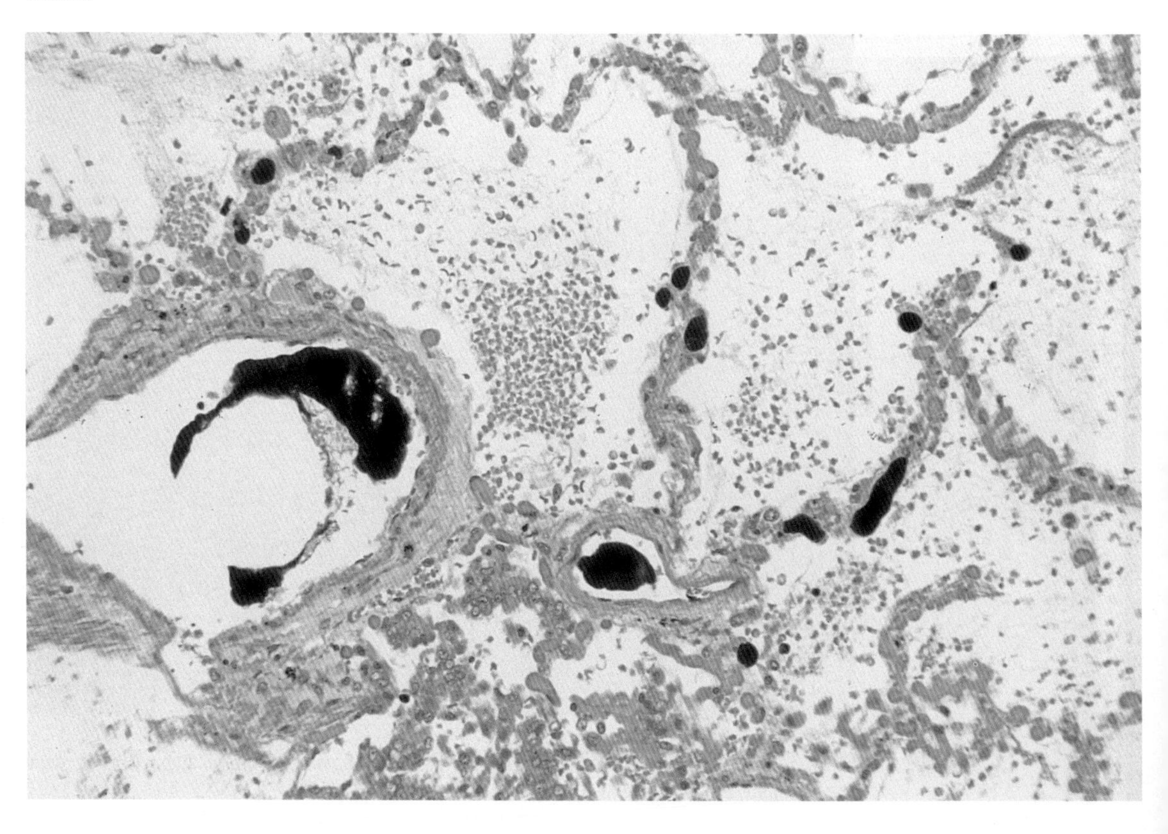

FIGURE 3–9. FAT EMBOLUS: In this photomicrograph of a lung fixed in osmium and stained with toluidine blue, note the black intravascular fat emboli. In addition, pulmonary edema is seen.

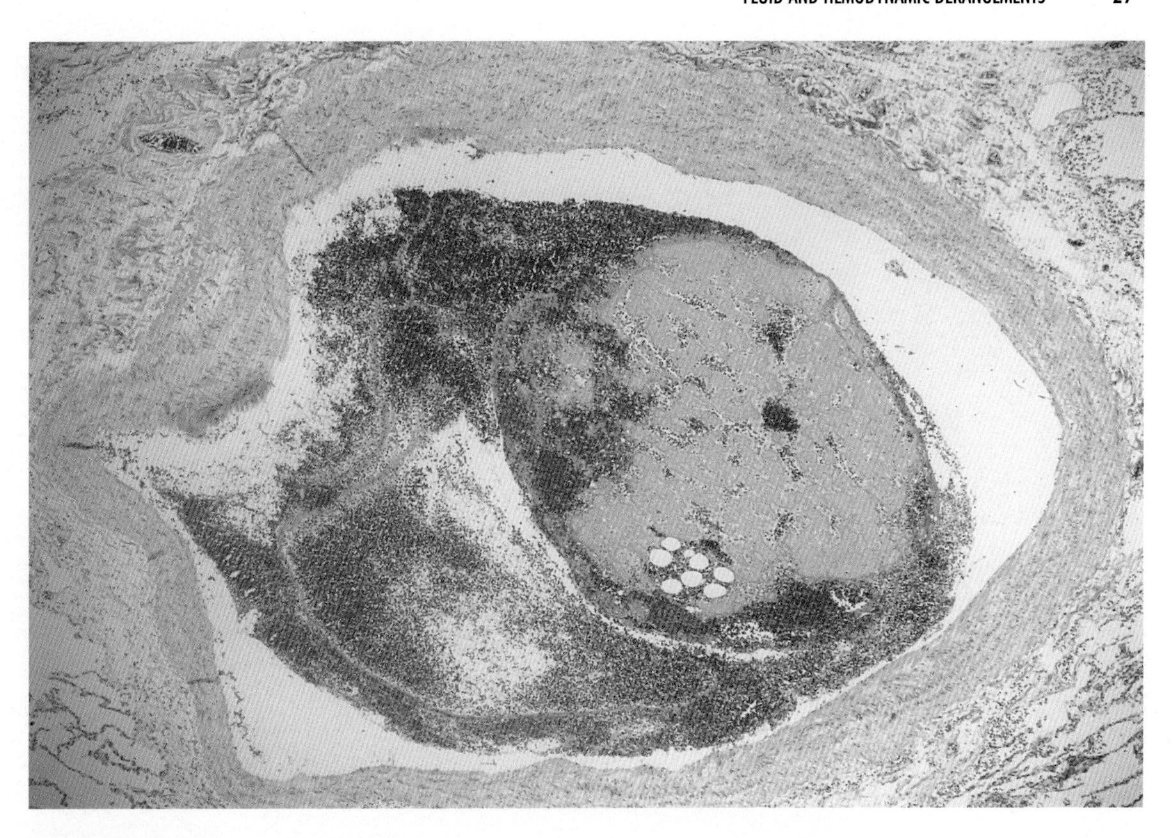

Figure 3–10 A

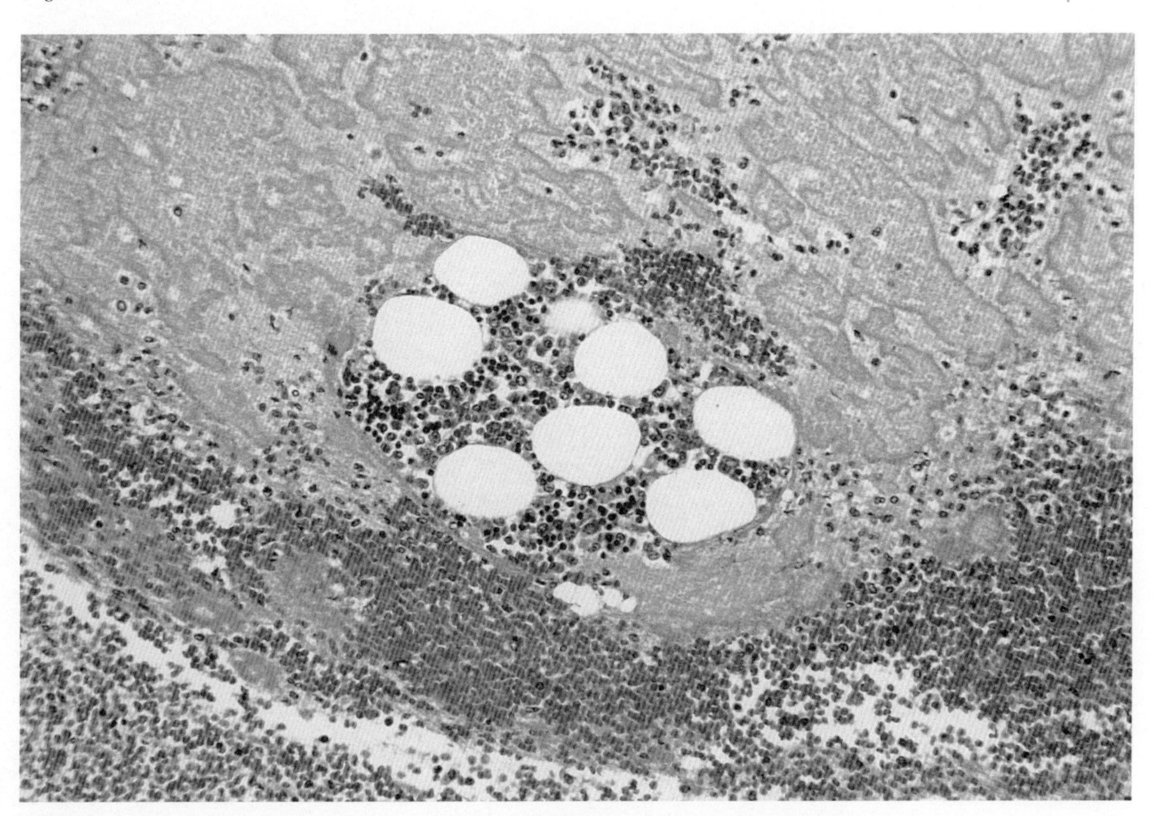

Figure 3–10 B

FIGURE 3–10. BONE MARROW EMBOLUS: A: In this photomicrograph of a blood vessel, a bone marrow embolus is present, secondary to resuscitation efforts. **B:** At higher power magnification, the bone marrow elements are more apparent.

An **infarct** is an area of ischemic necrosis secondary to decreased blood flow from the arterial supply or decreased venous drainage.

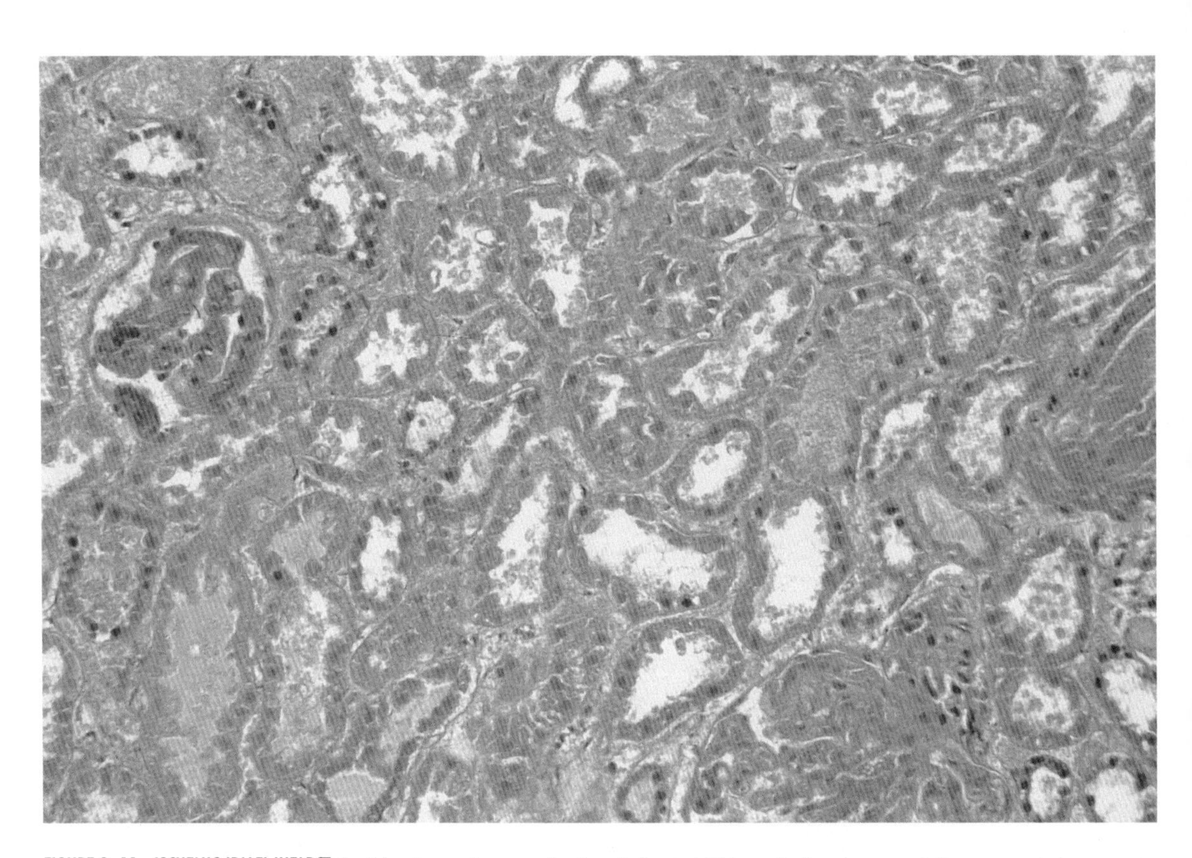

FIGURE 3–11. ISCHEMIC (PALE) INFARCT: In this photomicrograph of an infarcted kidney, ischemic coagulative necrosis is present.

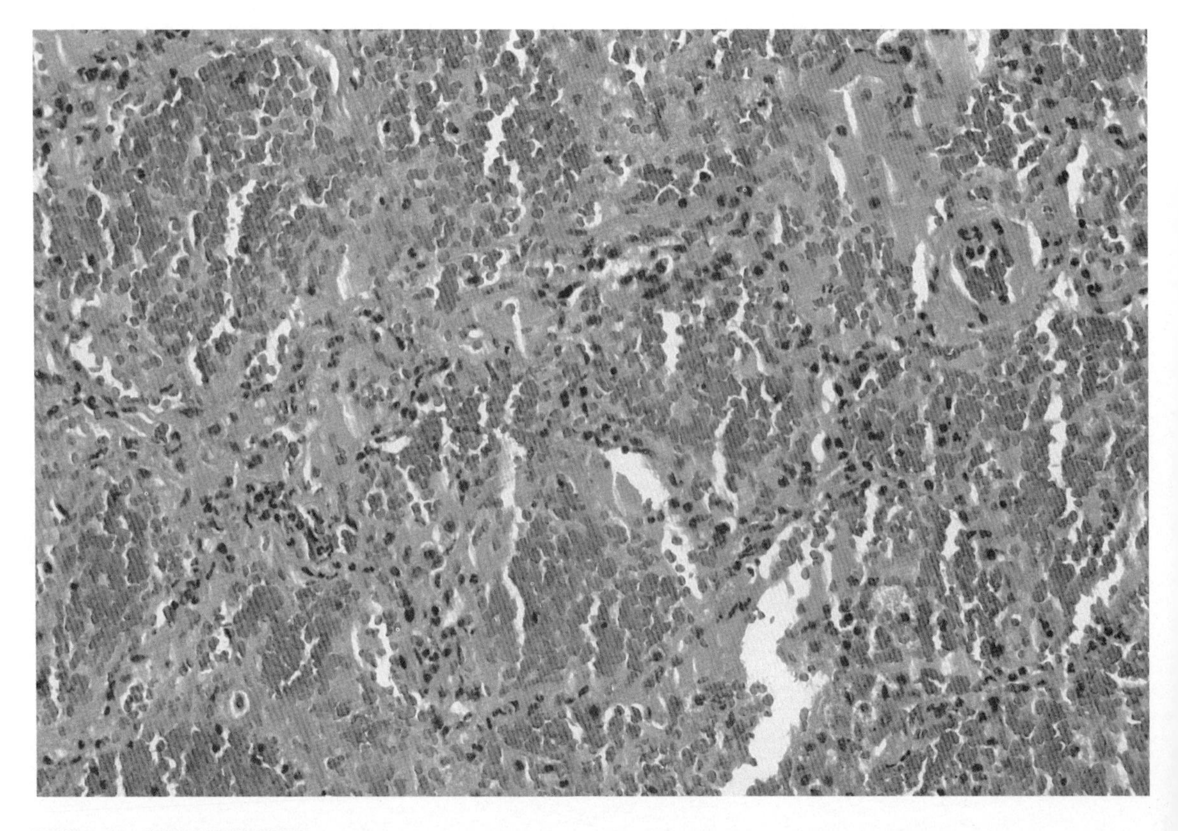

FIGURE 3–12. HEMORRHAGIC INFARCT: In this photomicrograph of an infarcted lung, blood is present in the alveolar spaces.

■ SHOCK

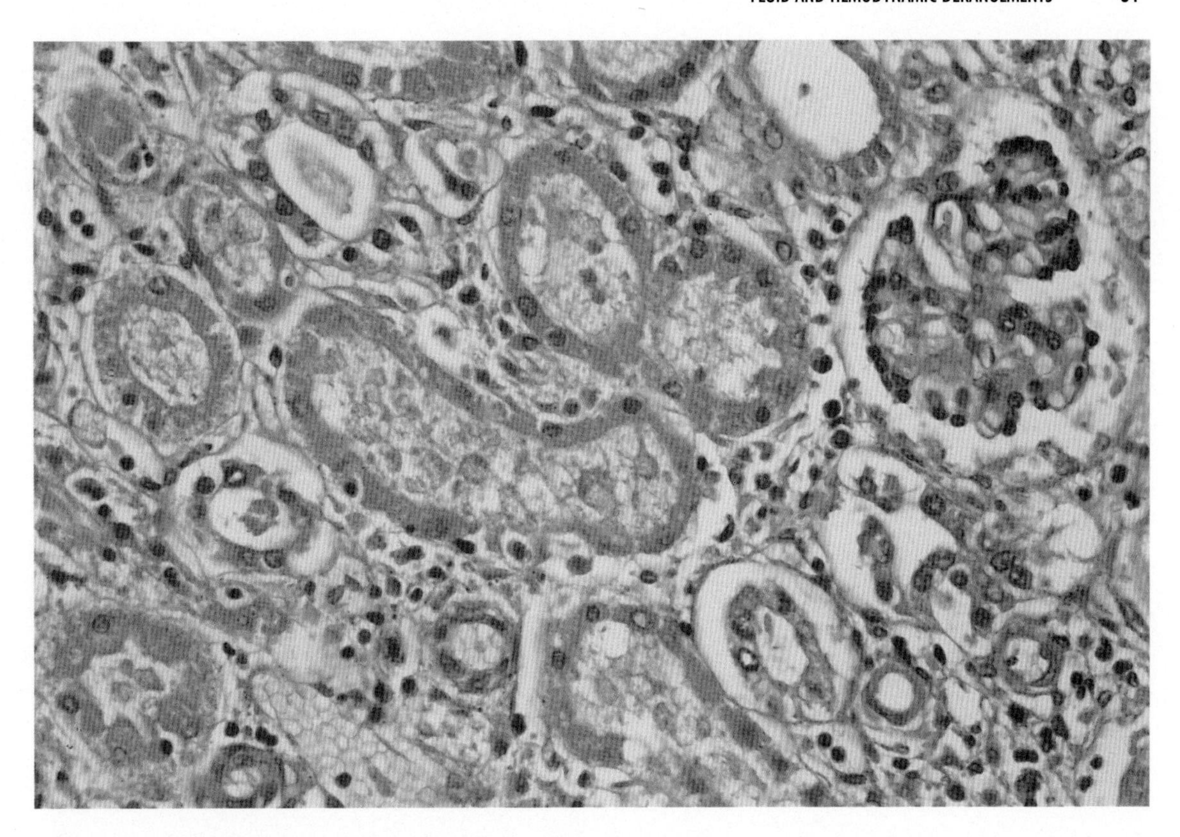

FIGURE 3–13. SHOCK KIDNEY (ACUTE TUBULAR NECROSIS [ATN]): In ATN, cells of the proximal convoluted tubules and ascending limbs display focal coagulative necrosis. Glomeruli and distal convoluted tubules are not affected.

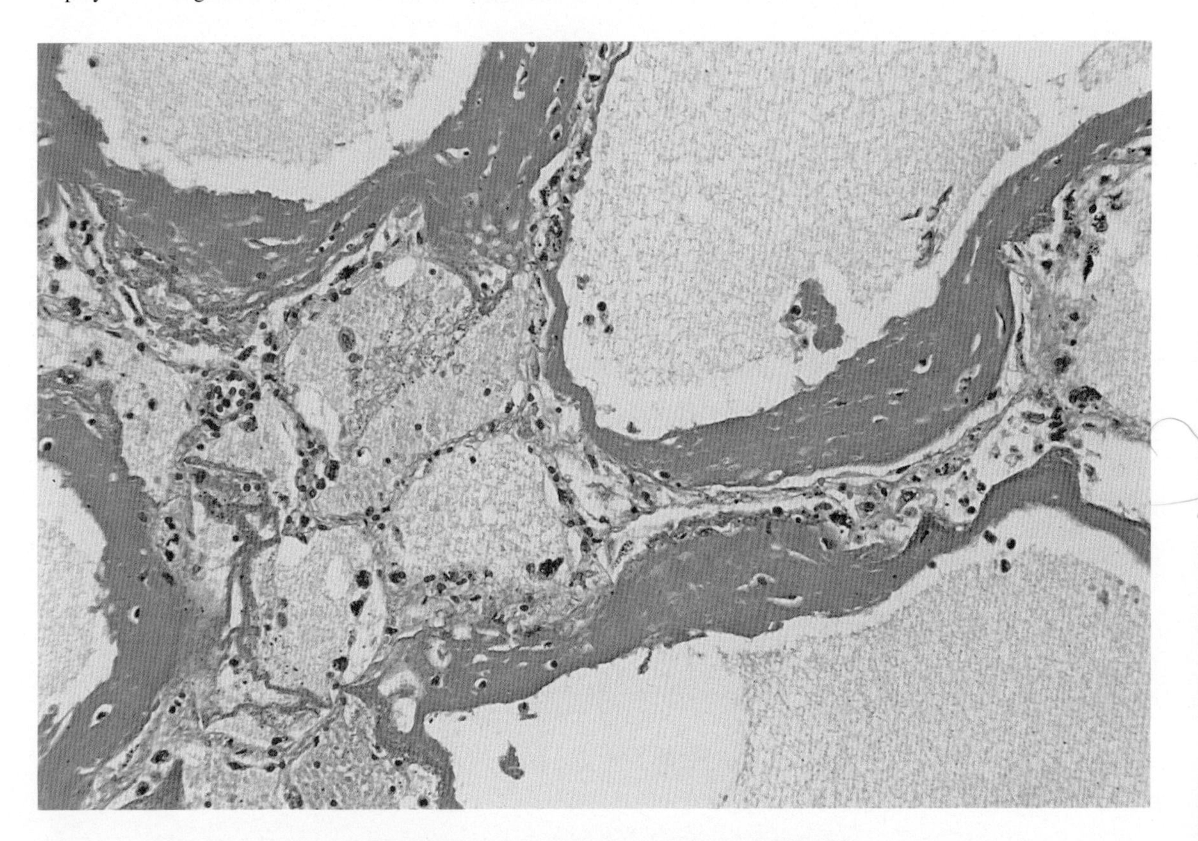

FIGURE 3–14. SHOCK LUNG (DIFFUSE ALVEOLAR DAMAGE [DAD], ADULT RESPIRATORY DISTRESS SYNDROME [ARDS]): The most characteristic feature of DAD is the presence of hyaline membranes in distended alveolar ducts.

NEOPLASIA

■ DEFINITIONS

Neoplasia means new growth. Such growth can be benign or malignant.

FIGURE 4–1. CHORISTOMA: A choristoma is a rest of normal tissue in an aberrant location. This photomicrograph is an example of thyroid tissue in a sublingual location.

FIGURE 4–2. HAMARTOMA: A hamartoma is a group of disorganized mature cells or tissues inherent in an organ. **A:** This is a photomicrograph of a bile duct hamartoma in the liver. **B:** This is a photomicrograph of a chondrolipoma in the lung.

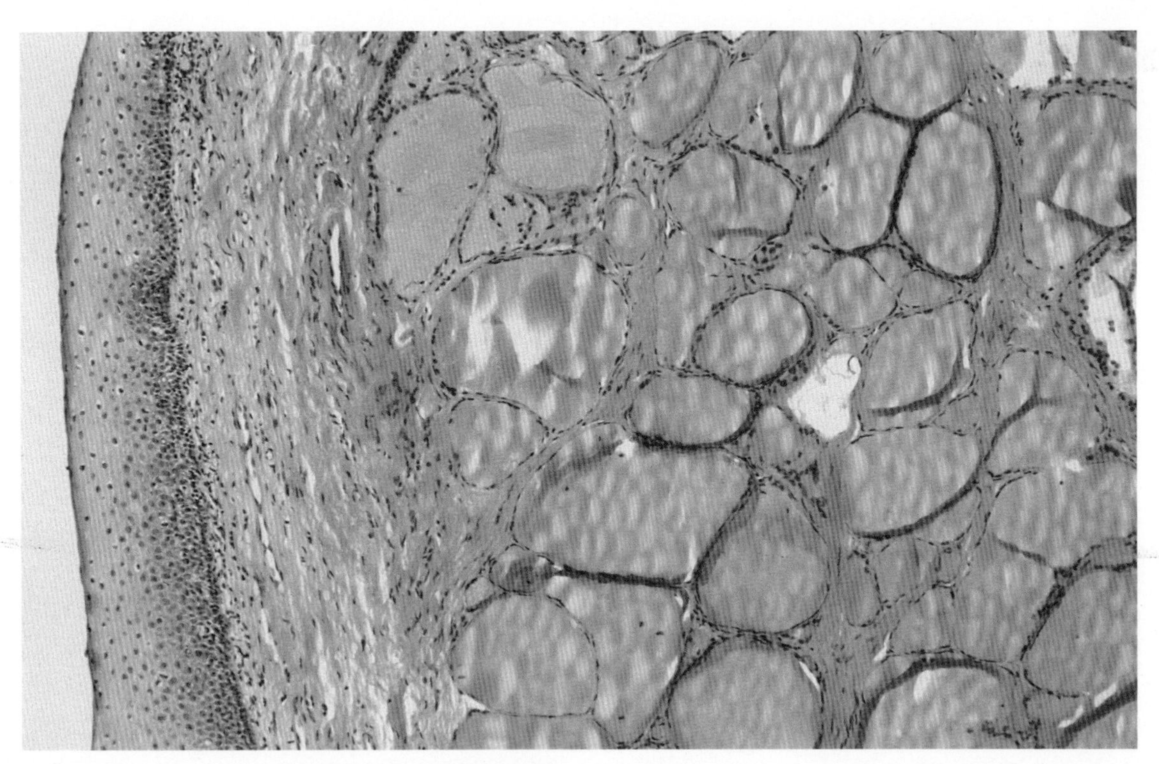

Figure 4–1

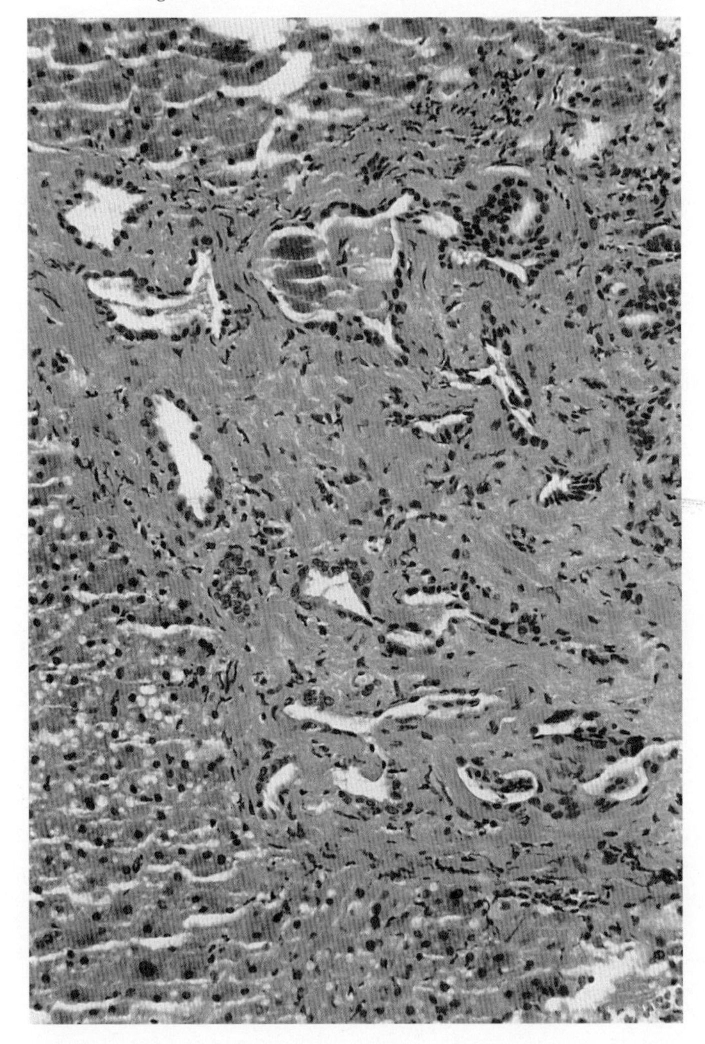

Figure 4–2 A

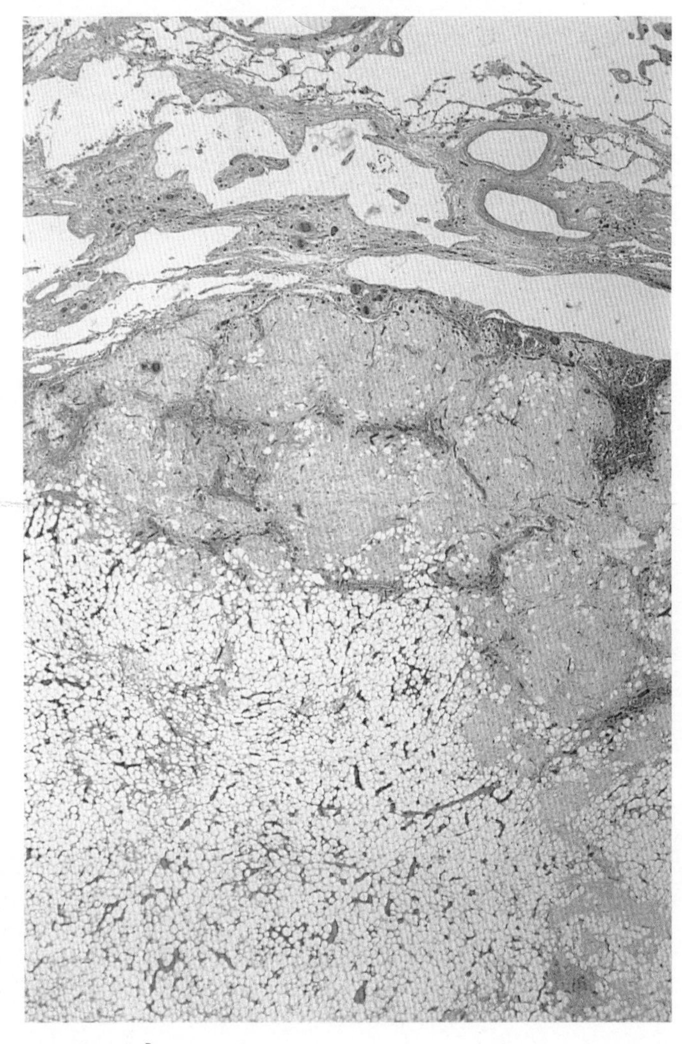

Figure 4–2 B

■ DEFINITIONS

FIGURE 4–3. PAPILLOMA: A papilloma is a benign epithelial neoplasm forming fingerlike projections. This is a photomicrograph of a squamous papilloma.

FIGURE 4–4. POLYP: A polyp is any projection above a mucosal surface. It may be benign or malignant. This is a photomicrograph of a benign colonic polyp in familial polyposis coli.

FIGURE 4–5. ADENOMA: Adenoma is a benign epithelial neoplasm derived from glands or forming a glandular pattern. One clue to the benign nature of this monomorphic adenoma from the parotid gland is the delimitation of the tumor by a capsule.

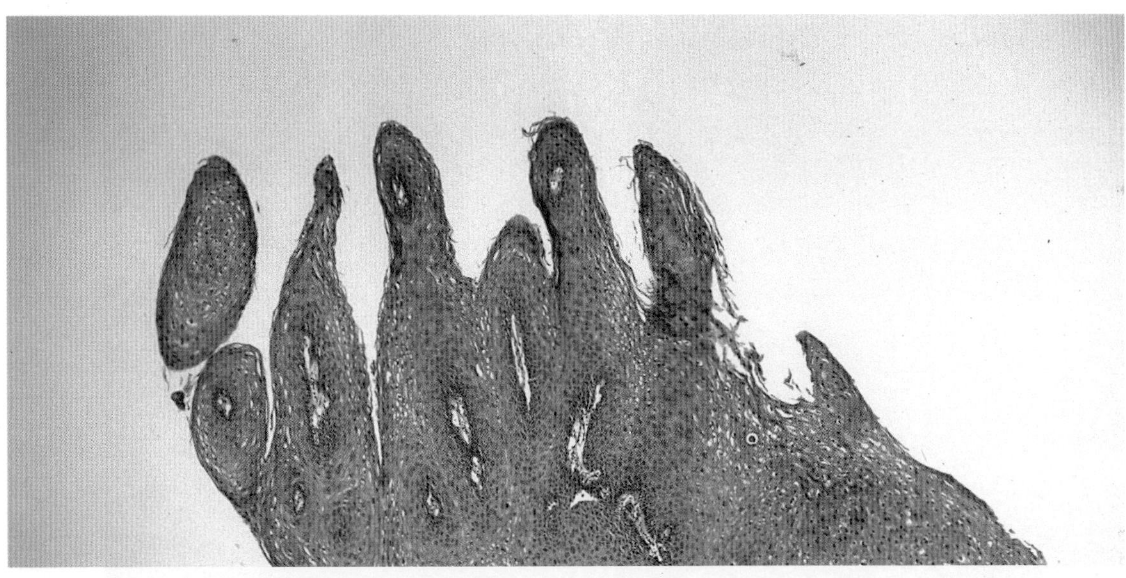

Figure 4–3

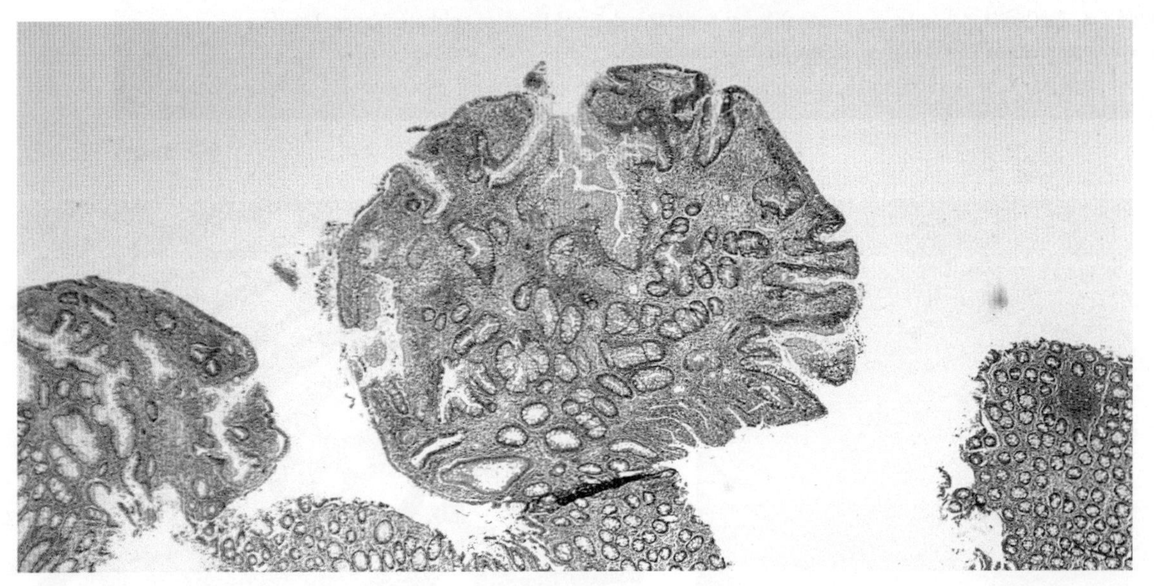

Figure 4–4

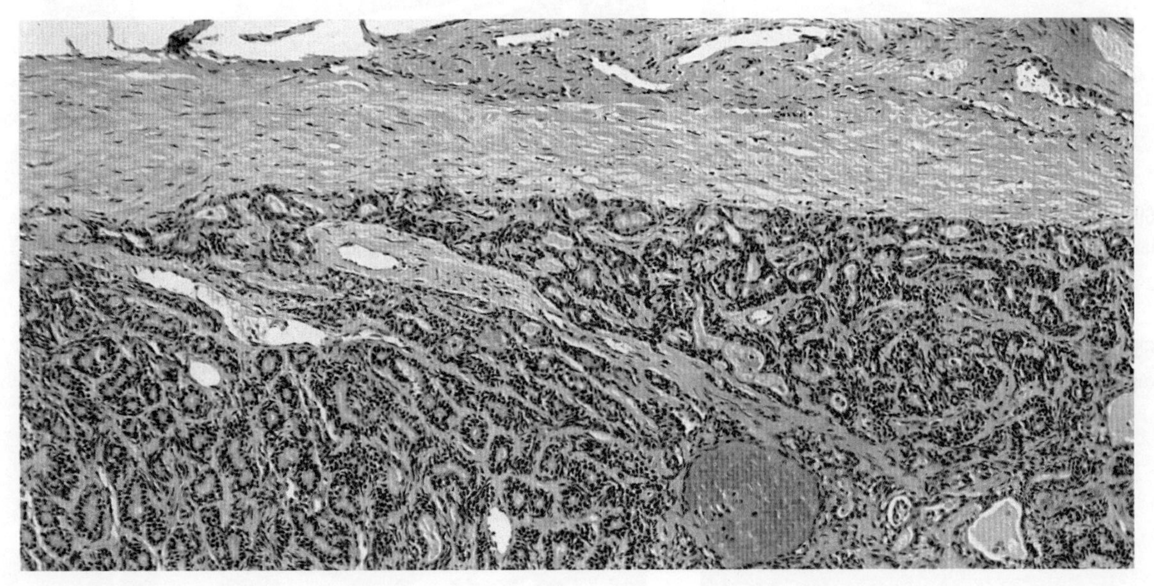

Figure 4–5

■ CHARACTERISTICS OF MALIGNANCY

FIGURE 4–9. LOSS OF POLARITY: When cells lose polarity, they do not retain their normal relationship to one another and they grow in a disorganized fashion. Immature cells are found in the superficial epithelial layers, and mitotic figures are found at levels above the basal cell layer.

FIGURE 4–10. HYPERCHROMASIA: In hyperchromasia, the nucleus of a neoplastic cell is darker than that in a normal cell because of an increase in the amount of deoxyribonucleic acid (DNA). Contrast the neoplastic cells rimming the papilla with the normal cells in the center of the papilla.
INCREASE IN THE NUCLEAR-TO-CYTOPLASMIC (N:C) RATIO: The normal N:C ratio of approximately 1:4 becomes approximately 1:1 in neoplastic cells. Contrast the neoplastic cells rimming the papilla with the normal cells in the center of the papilla.

FIGURE 4–11. PLEOMORPHISM: Pleomorphism is variation in size and shape of both a cell and its nucleus. This is a photomicrograph of a malignant neoplasm in which the cells are anaplastic. In addition, the malignant neoplasm contains **tumor giant cells.**

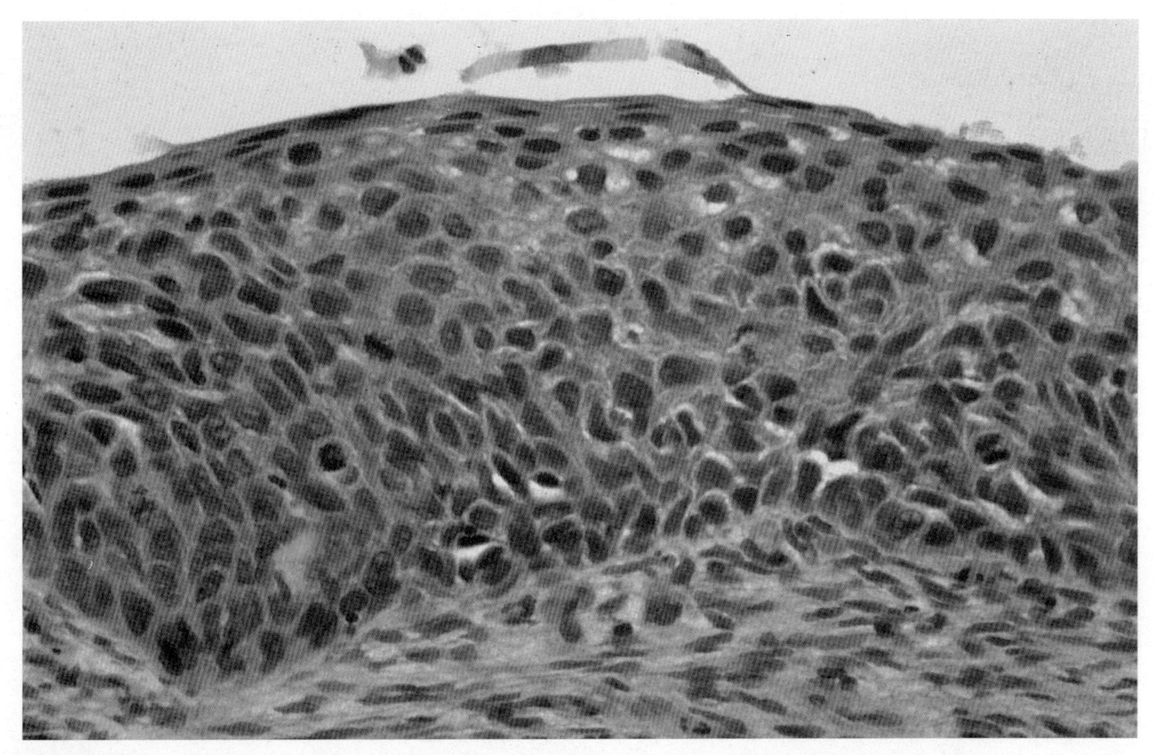

Figure 4–9

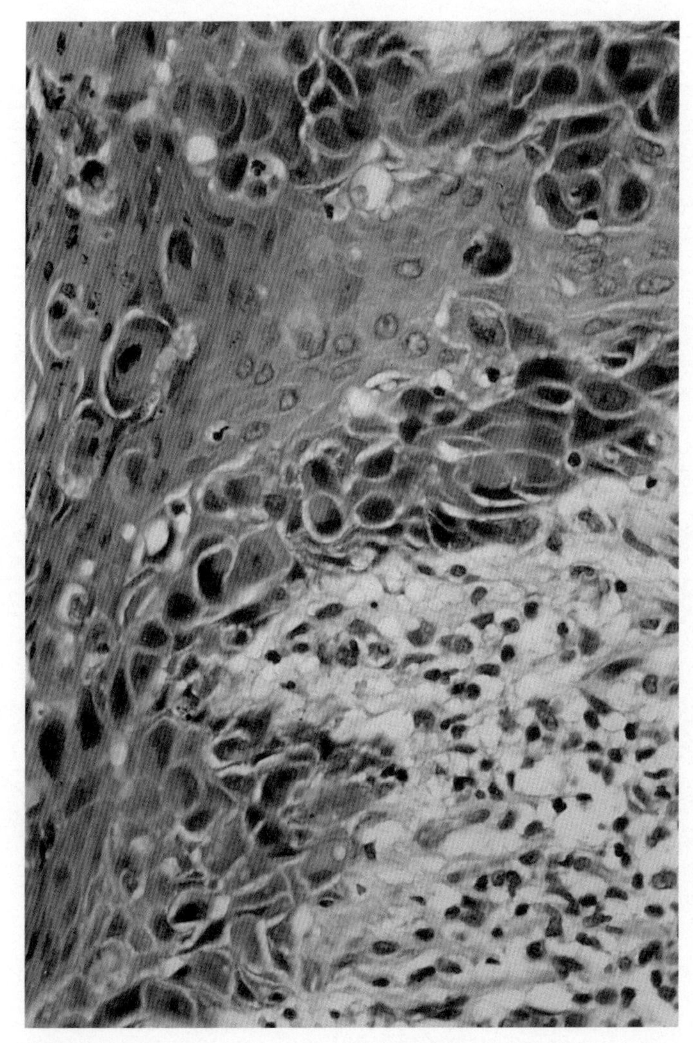

Figure 4–10

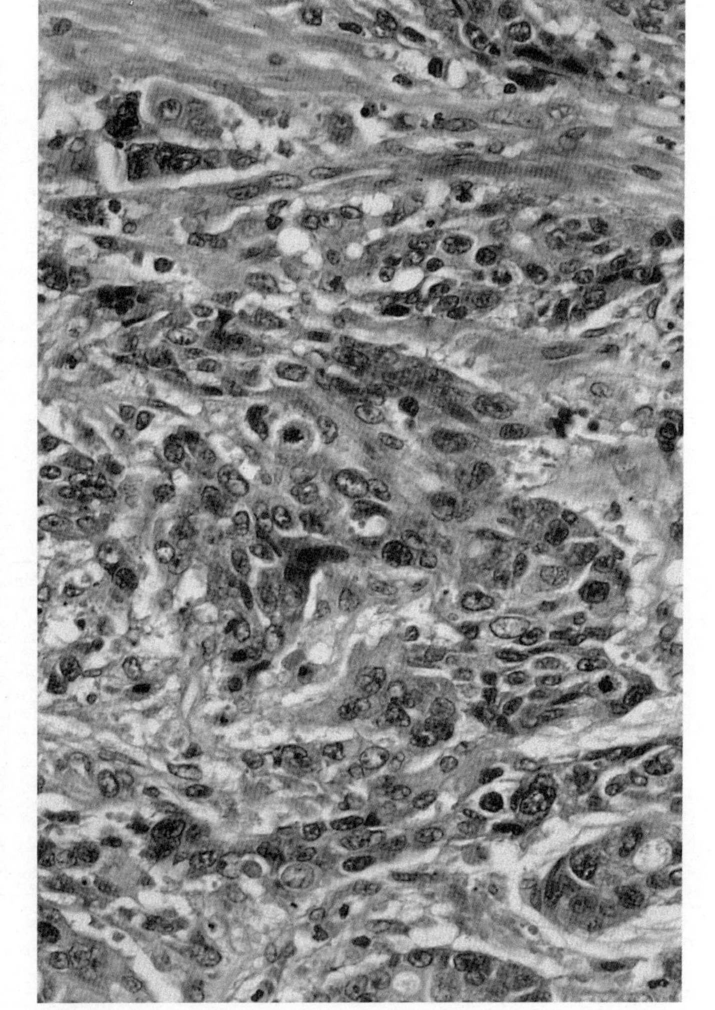

Figure 4–11

■ CHARACTERISTICS OF MALIGNANCY

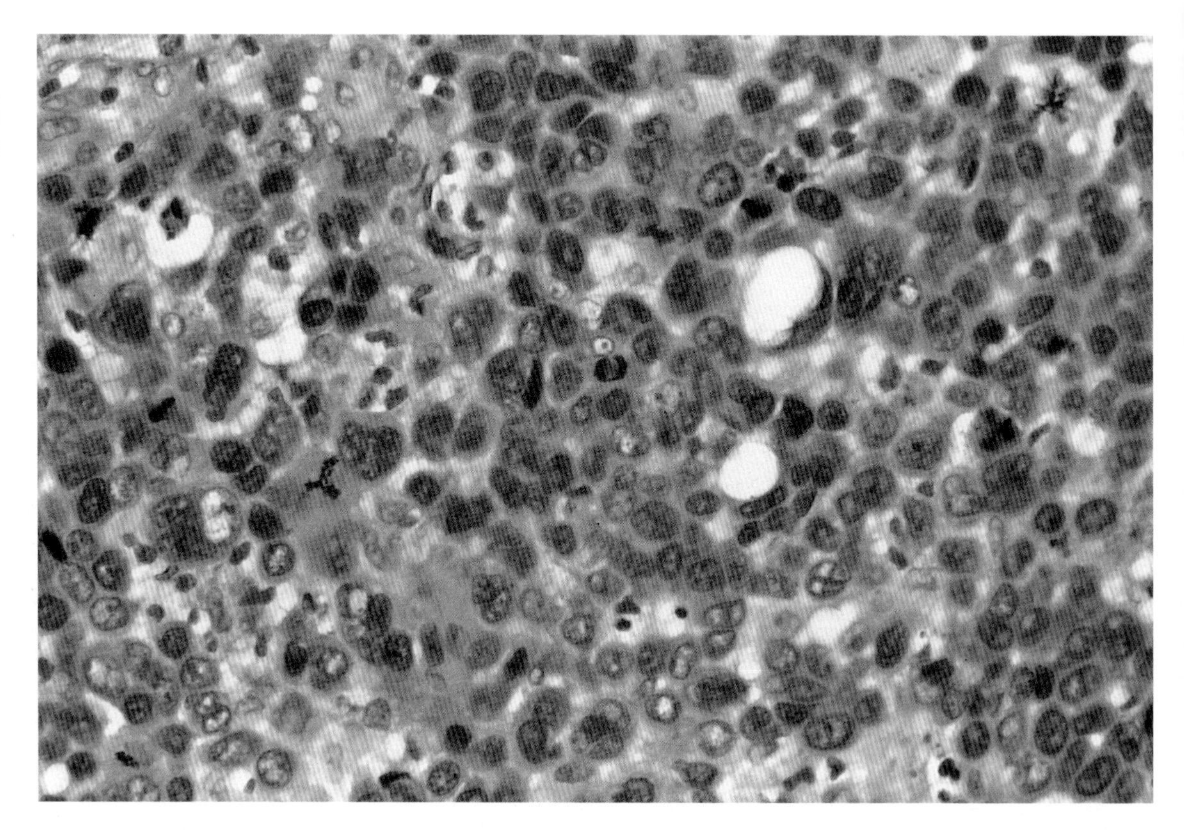

FIGURE 4–12. MITOSES: In malignancy, mitotic figures are increased in number and may be atypical and multipolar.

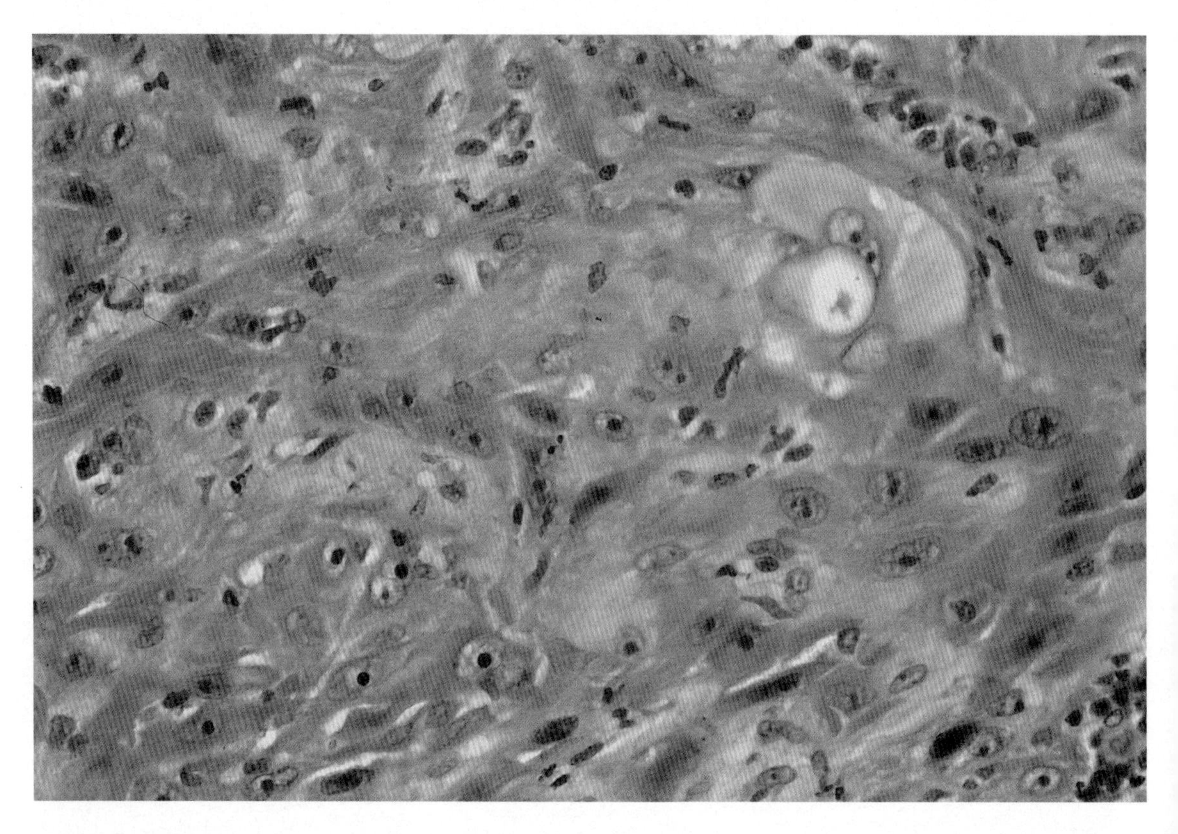

FIGURE 4–13. NUCLEOLI: Nucleoli are prominent in neoplastic cells.

■ CHARACTERISTICS OF MALIGNANCY

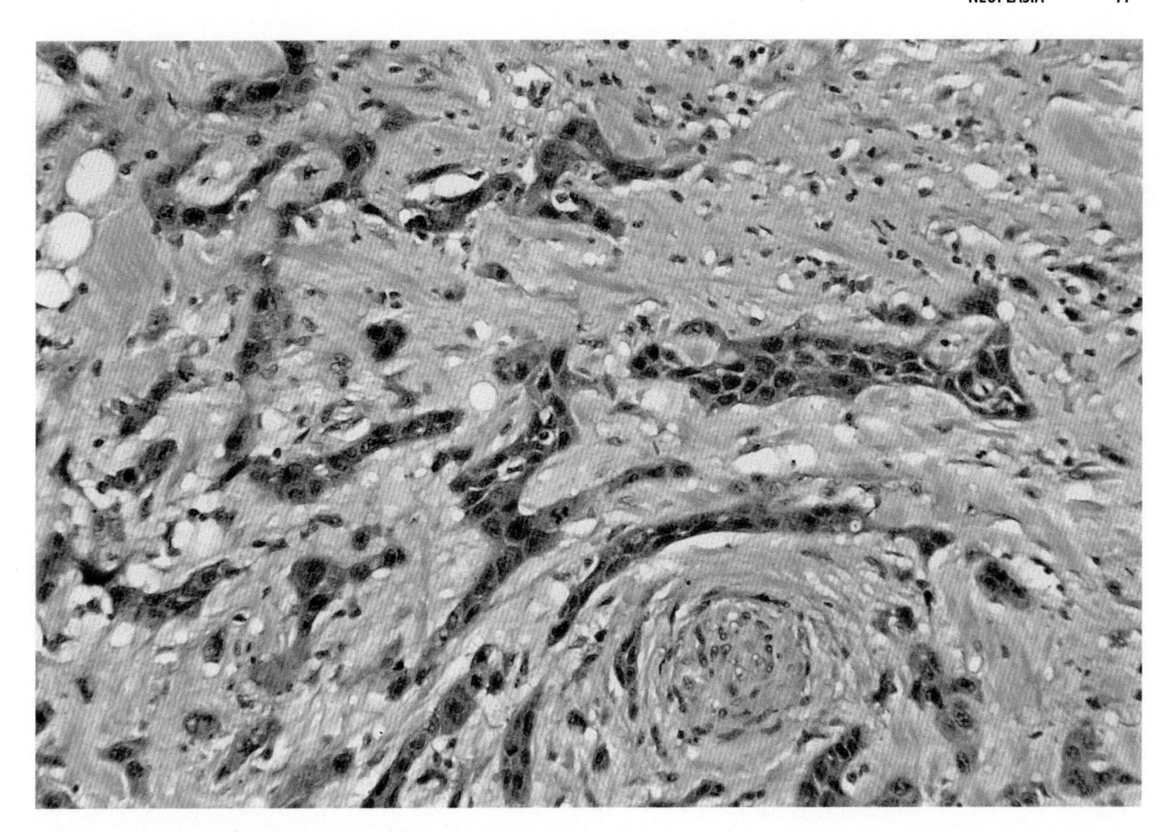

FIGURE 4–14. INFILTRATIVE BORDERS: Unlike benign neoplasms, which are encapsulated (Figure 4–5), malignant tumors infiltrate the surrounding stroma.

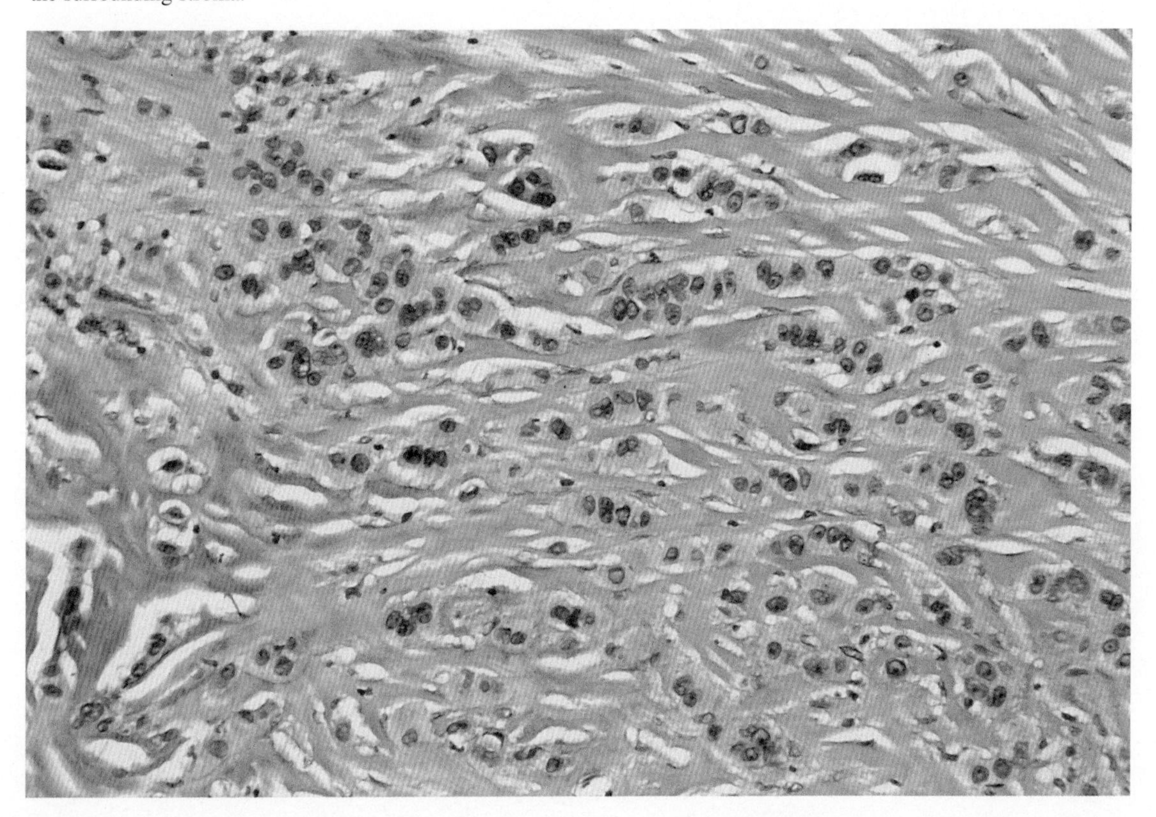

FIGURE 4–15. DESMOPLASIA: Desmoplasia is characterized by a dense collagenous reaction of the stroma in response to infiltrating neoplastic cells.

■ CHARACTERISTICS OF MALIGNANCY

METASTASIS

A **metastasis** is a tumor implant at a location distant from the site of the main tumor mass.

FIGURE 4–16. METASTASIS: This is a photomicrograph of an adenocarcinoma metastatic to a lymph node.

FIGURE 4–17. LYMPHATIC INVOLVEMENT: A tumor embolus from a breast carcinoma is present in a lymphatic channel.

FIGURE 4–18. PERINEURAL INVASION: In this photomicrograph, glands from a prostate adenocarcinoma are present in a perineural location.

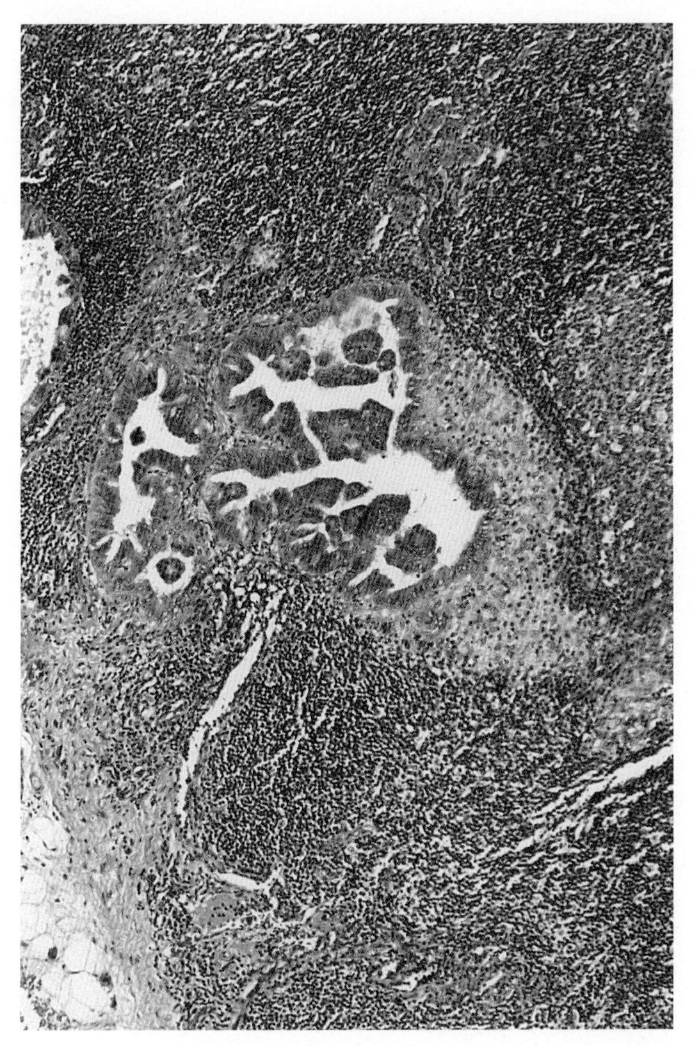

Figure 4–16

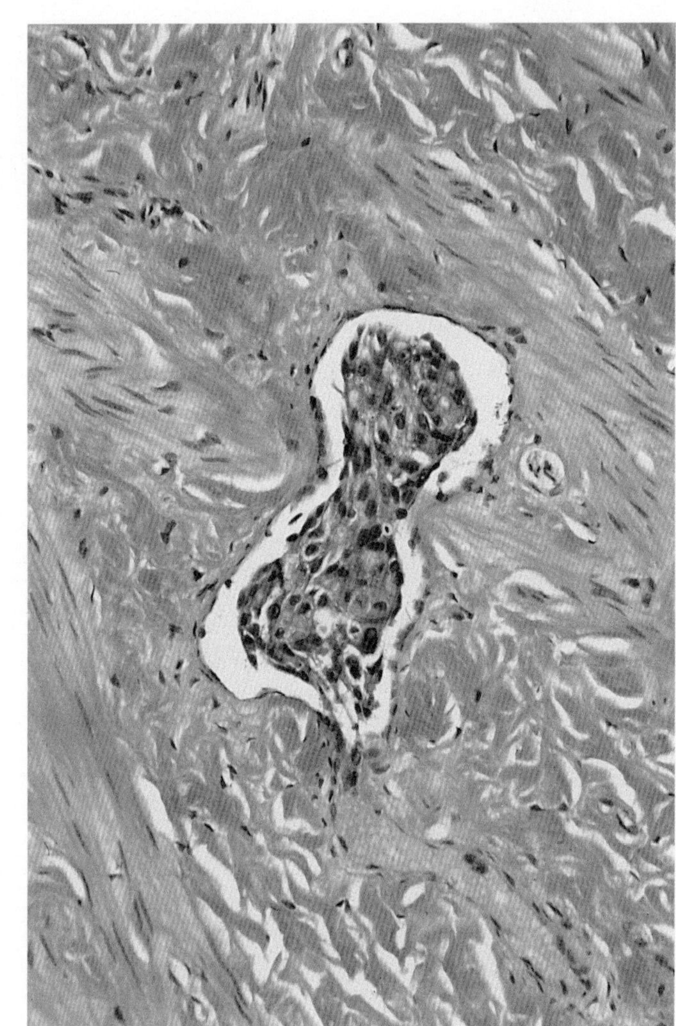

Figure 4–17

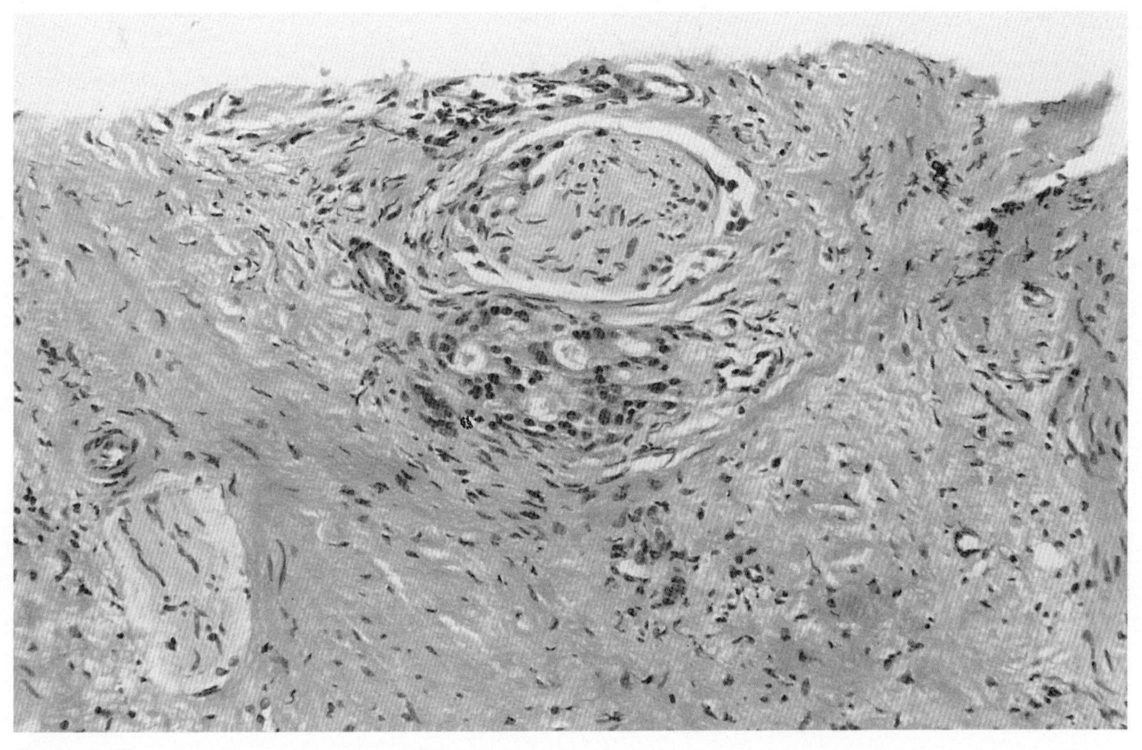

Figure 4–18

■ CHARACTERISTICS OF MALIGNANCY

FIGURE 4–19. BLOOD VESSEL INVOLVEMENT: A tumor embolus from a breast carcinoma is present in a blood vessel.

FIGURE 4–20: A: This is a photomicrograph of a squamous cell carcinoma breaking through the wall of the jugular vein. **B:** Confirming this is an elastic von Gieson stain showing disruption of the elastica by tumor cells.

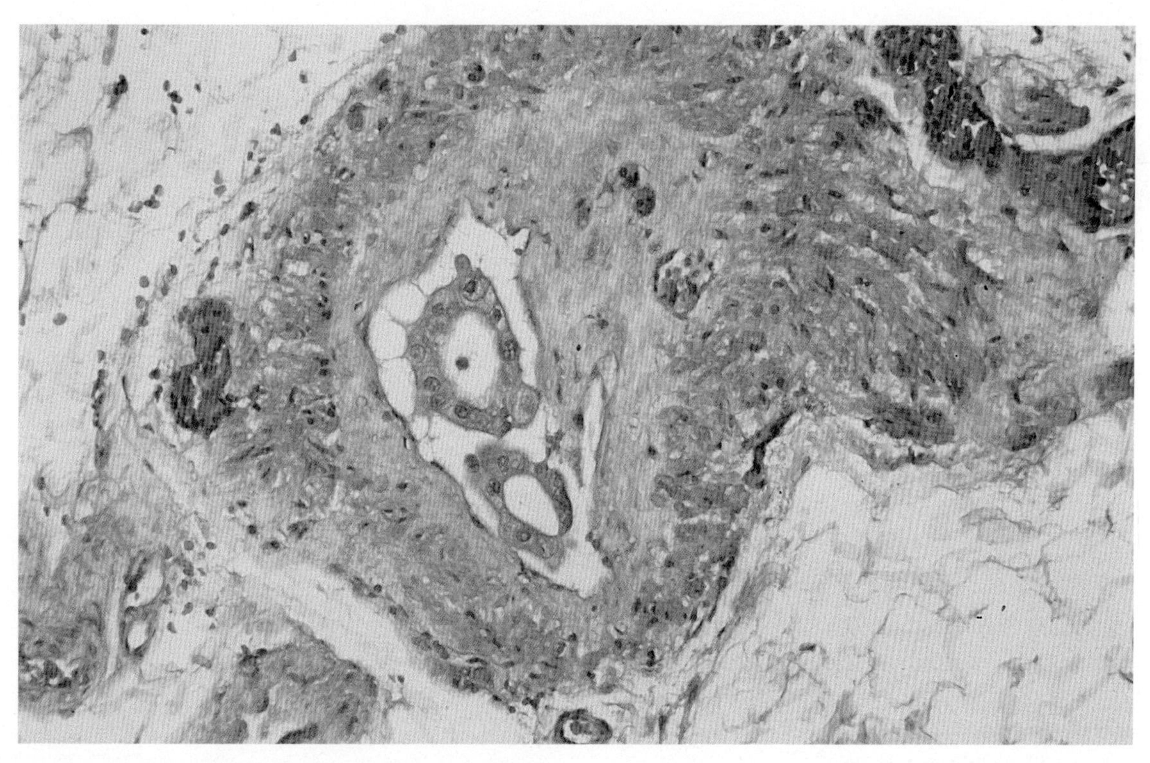

Figure 4–19

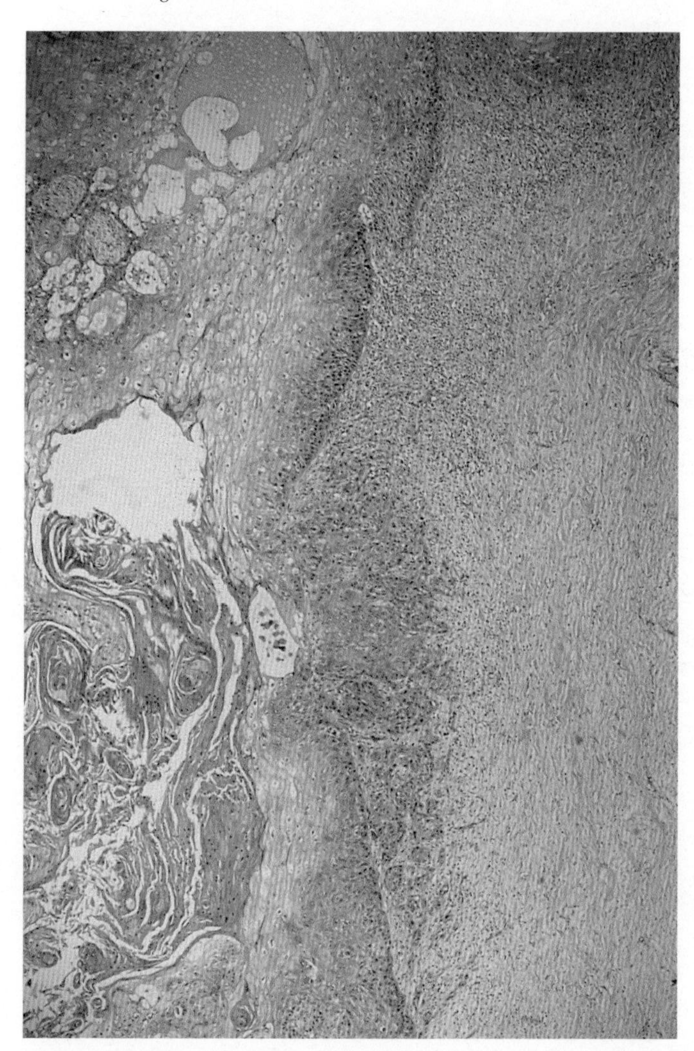

Figure 4–20 A

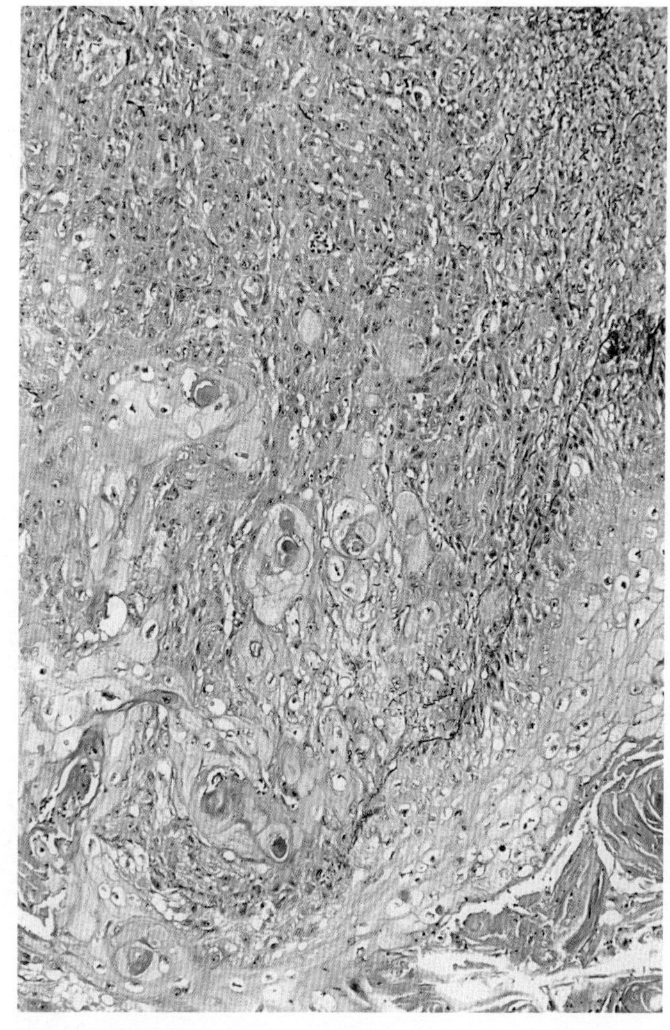

Figure 4–20 B

■ CHARACTERISTICS OF MALIGNANCY

Differentiation refers to the extent to which a neoplasm resembles its normal counterpart. Well- differentiated and moderately differentiated tumors resemble the cell of origin. Poorly differentiated tumors do not resemble their normal counterparts and may require special stains to aid in diagnosis.

FIGURE 4–21. MODERATELY DIFFERENTIATED SQUAMOUS CELL CARCINOMA: This is a photomicrograph of a moderately differentiated squamous cell carcinoma. **Keratinization** is one characteristic which indicates that this tumor has a squamous cell origin. Another characteristic identifying the squamous cell origin of this tumor is intercellular bridges (*arrow*).

FIGURE 4–22. POORLY DIFFERENTIATED CARCINOMA: A: This is a photomicrograph of a poorly differentiated carcinoma. There are no distinguishing histologic features that identify its cell of origin. B: Immunoperoxidase staining with cytokeratin confirms the epithelial nature of this tumor.

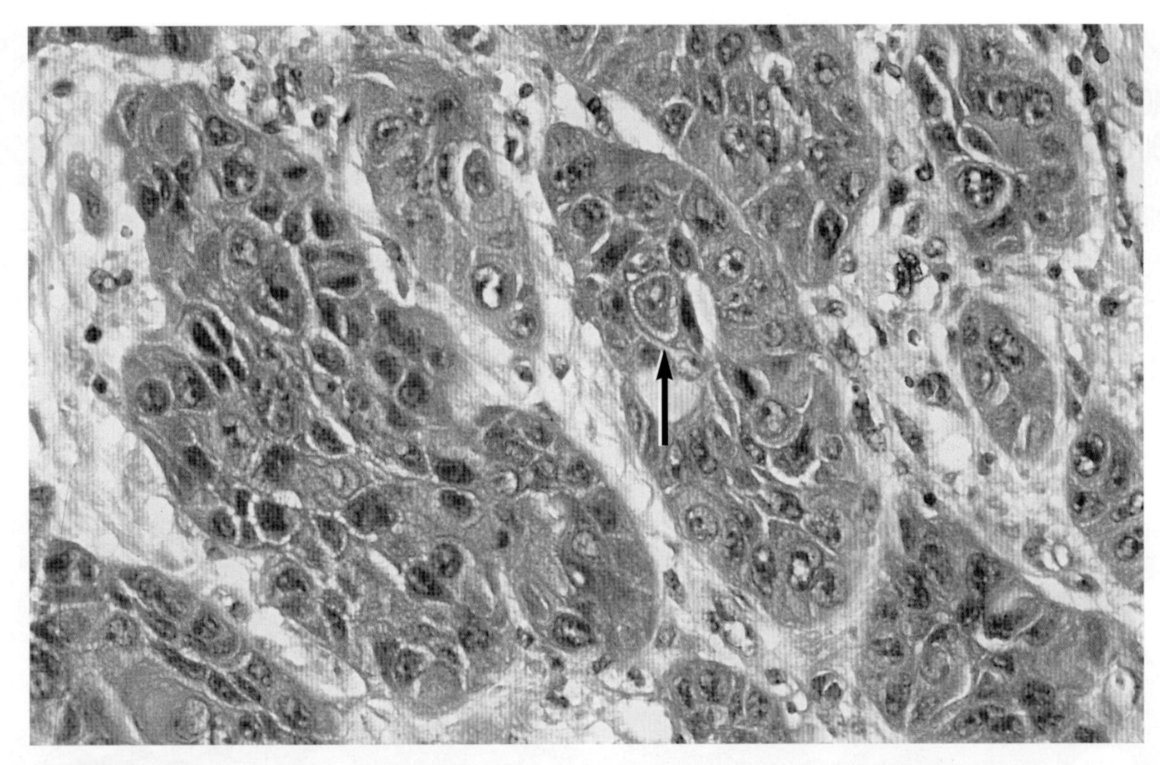

Figure 4–21

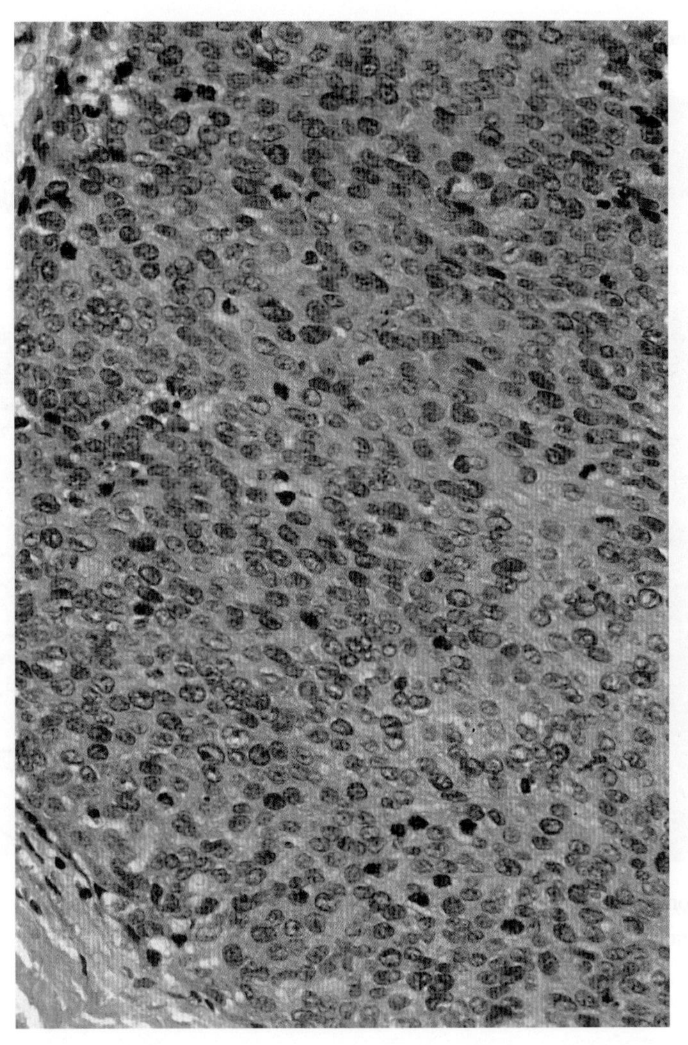

Figure 4–22 A

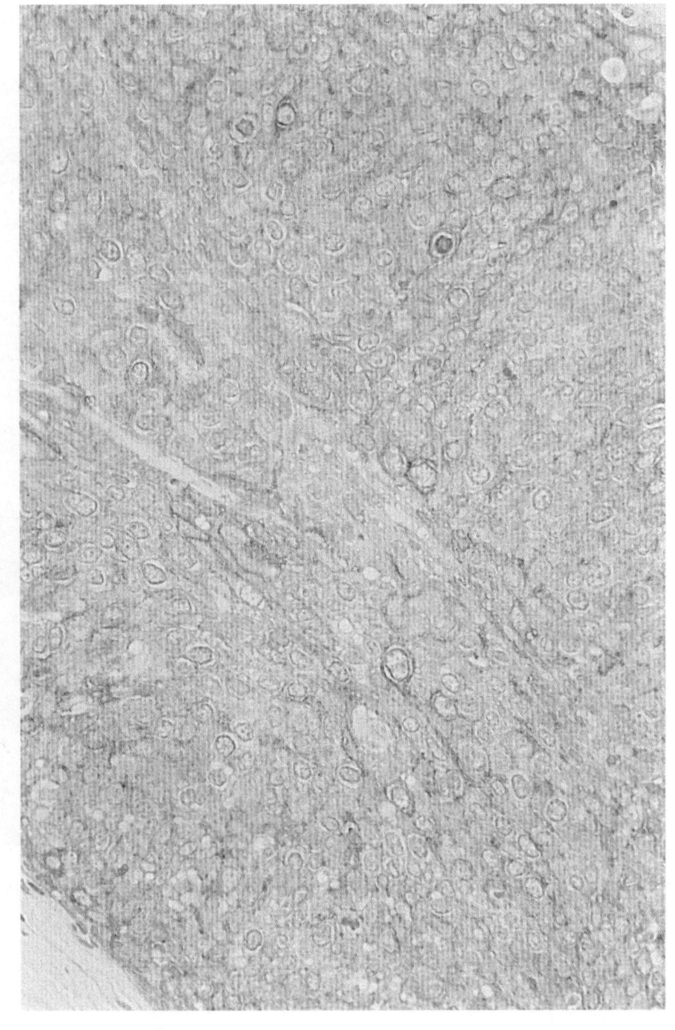

Figure 4–22 B

■ TUMORS DERIVED FROM FIBROUS TISSUE

BENIGN TUMORS

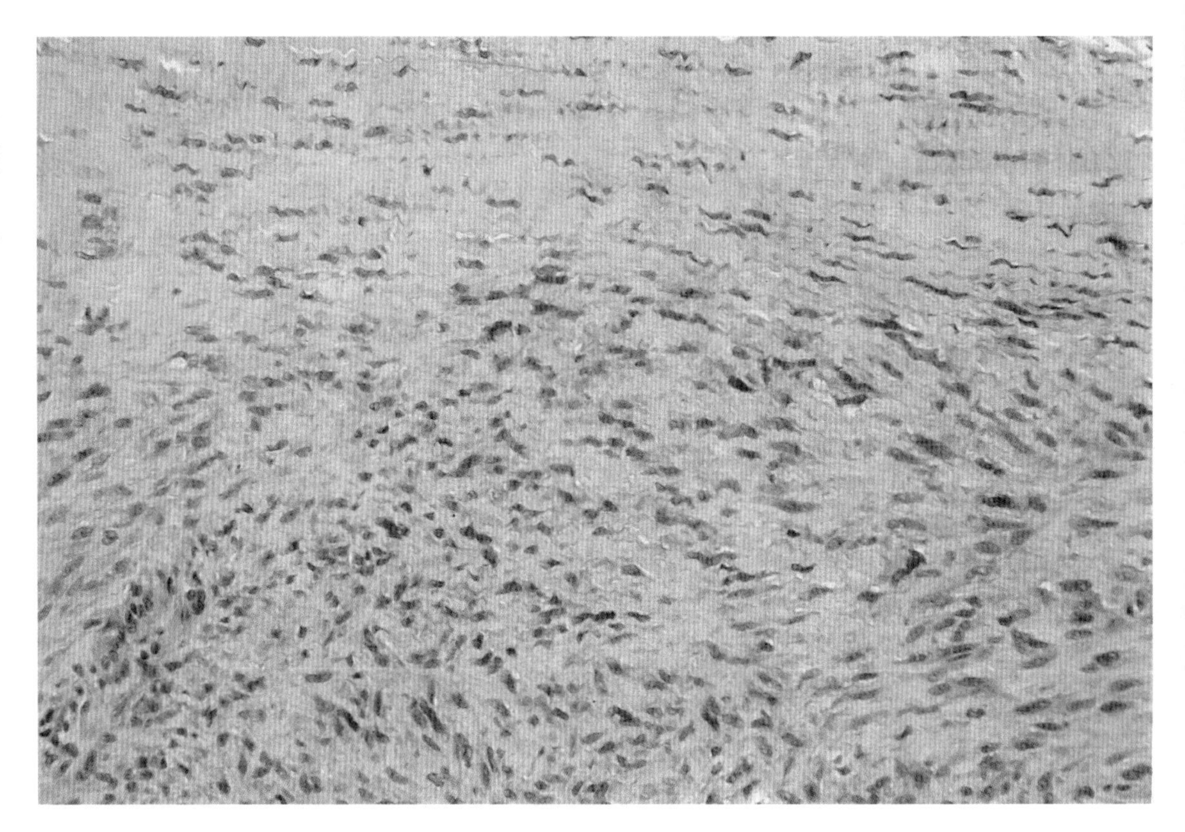

Figure 4–27 A

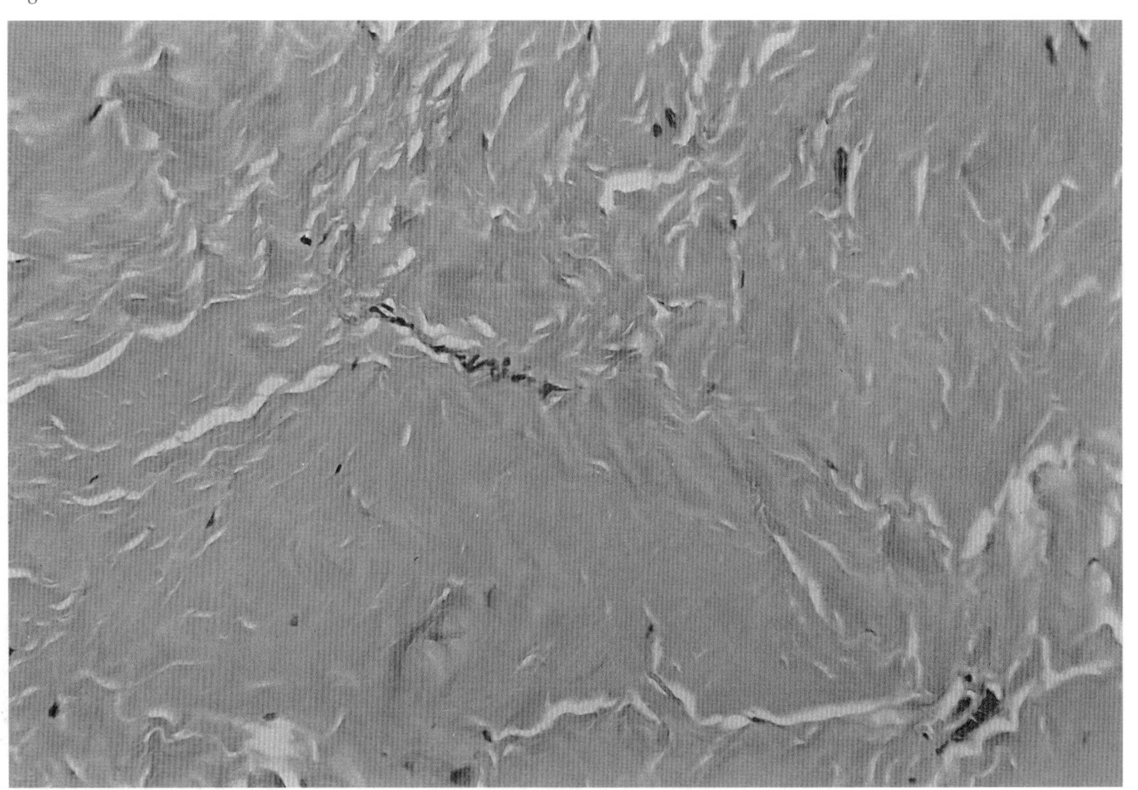

Figure 4–27 B

FIGURE 4–27. FIBROMATOSIS: Fibromatoses arise from the fascia of aponeurosis and are not encapsulated. A: In this example of a young Dupuytren's contracture, or palmar fibromatosis, the lesion is cellular, fibroblasts are plump, and collagen is minimal. B: This is an example of an old lesion of **Peyronie's disease,** or penile fibromatosis. The fibroblasts are sparse, and the connective tissue is thick and hyalinized.

■ TUMORS DERIVED FROM FIBROUS TISSUE

MALIGNANT TUMORS

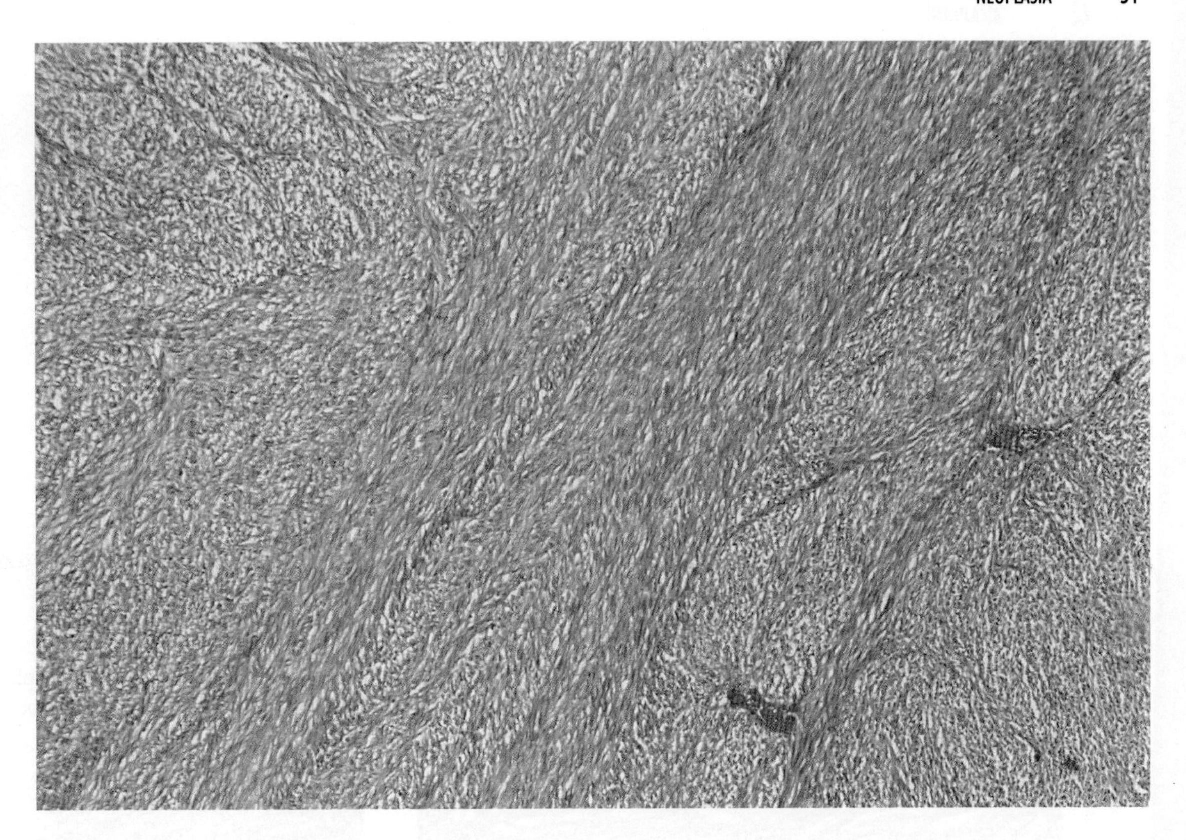

Figure 4–28 A

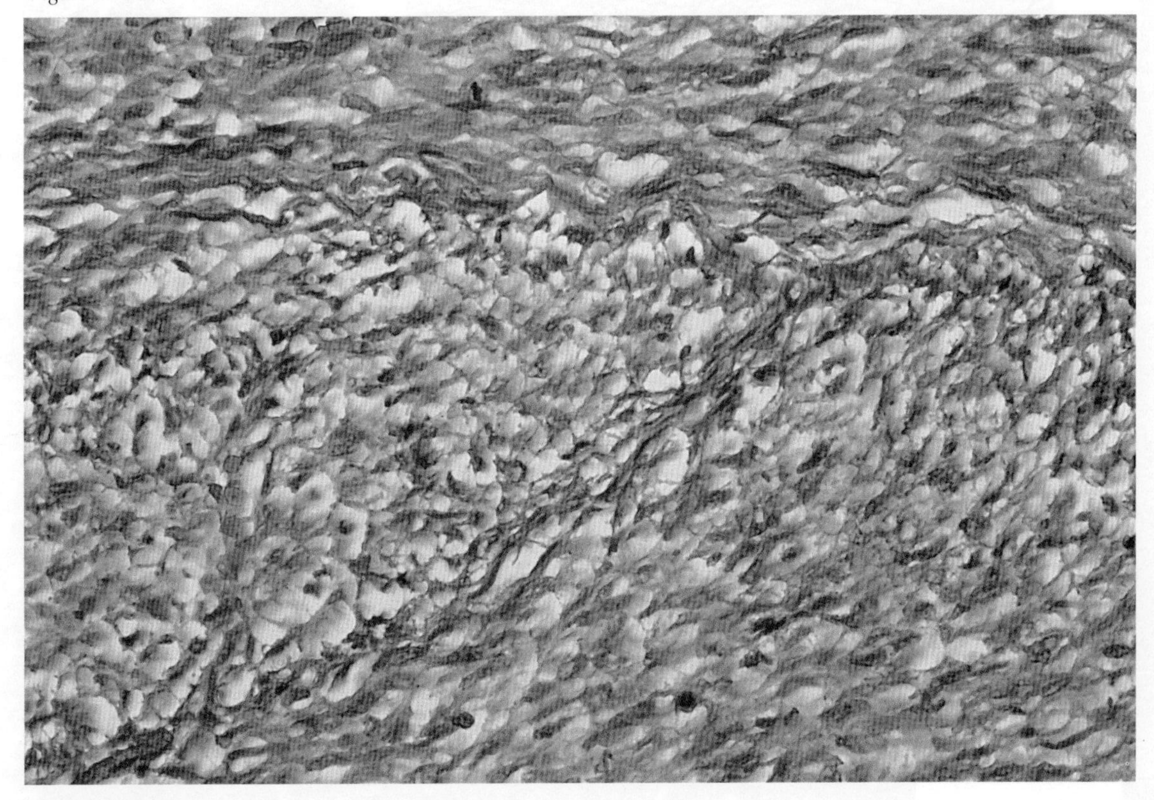

Figure 4–28 B

FIGURE 4–28. FIBROSARCOMA: A: A fibrosarcoma is characterized by interlacing bundles of spindle cells arranged in a herringbone pattern. **B:** The cells are uniform and mitotic figures are present.

■ TUMORS DERIVED FROM SKELETAL MUSCLE

BENIGN TUMORS

FIGURE 4–31. RHABDOMYOMA: A rhabdomyoma is composed of large glycogen-filled cells. A characteristic feature of a rhabdomyoma is **spider cells,** which are large cells with a centrally located cytoplasmic condensation and nucleus. Myofibrils radiate from the cell's center to its periphery.

MALIGNANT TUMORS

FIGURE 4–32. RHABDOMYOSARCOMA: A: This is a photomicrograph of an embryonal rhabdomyosarcoma. The cells are pleomorphic with hyperchromatic nuclei. Multinucleated tumor cells are present. B: Immunoperoxidase staining for desmin is positive in the neoplastic cells.

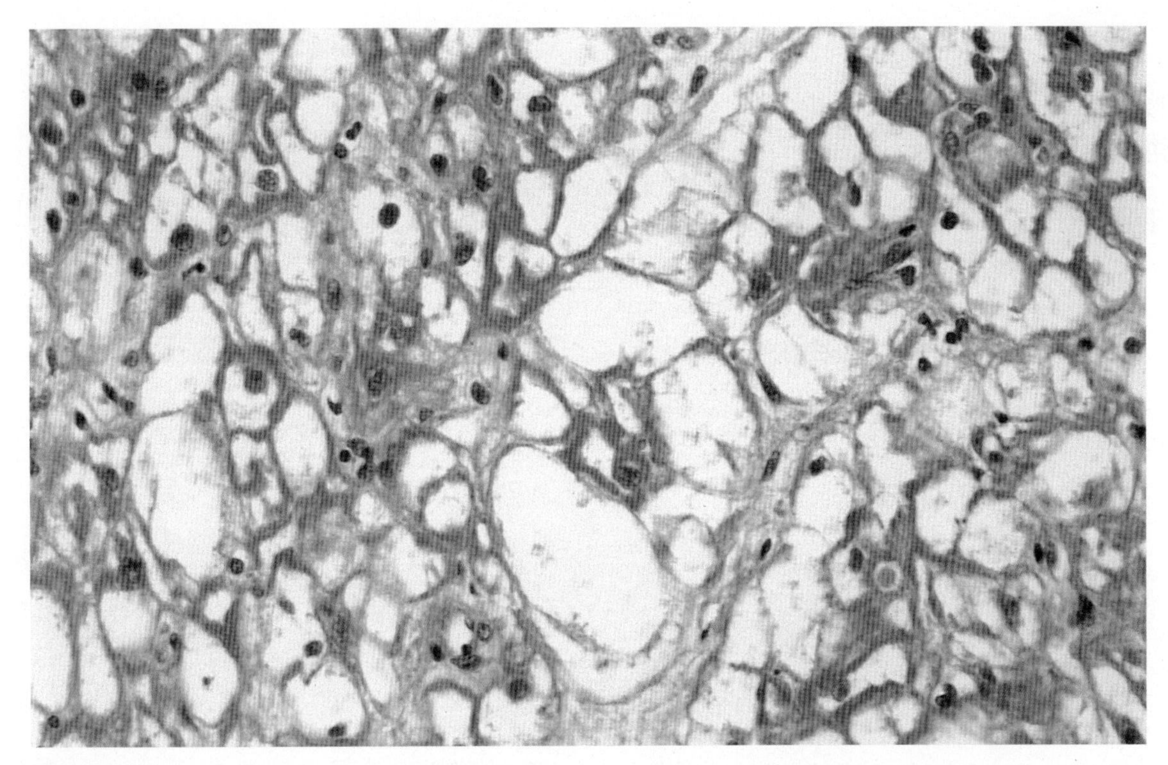

Figure 4–31

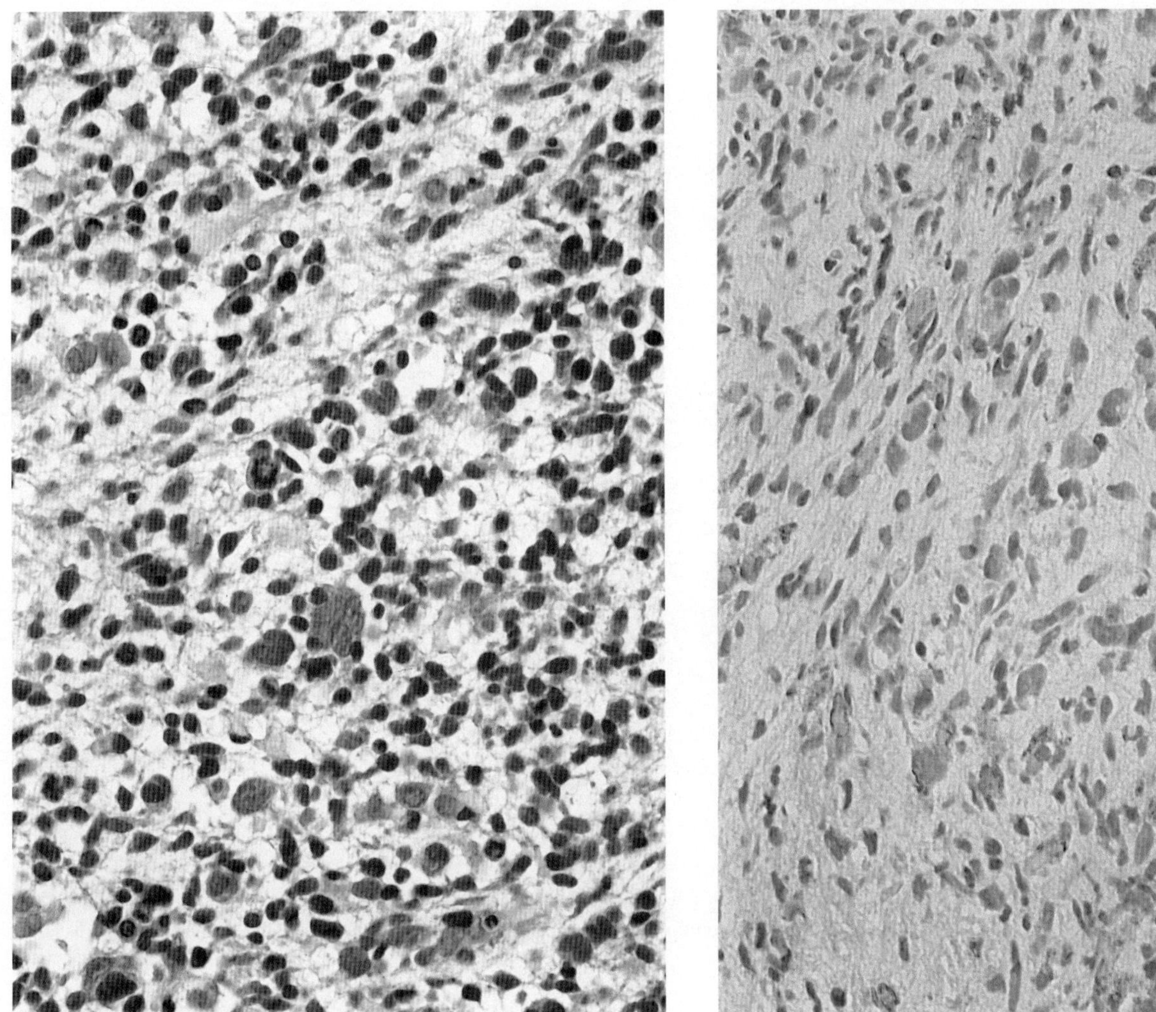

Figure 4–32 **A**

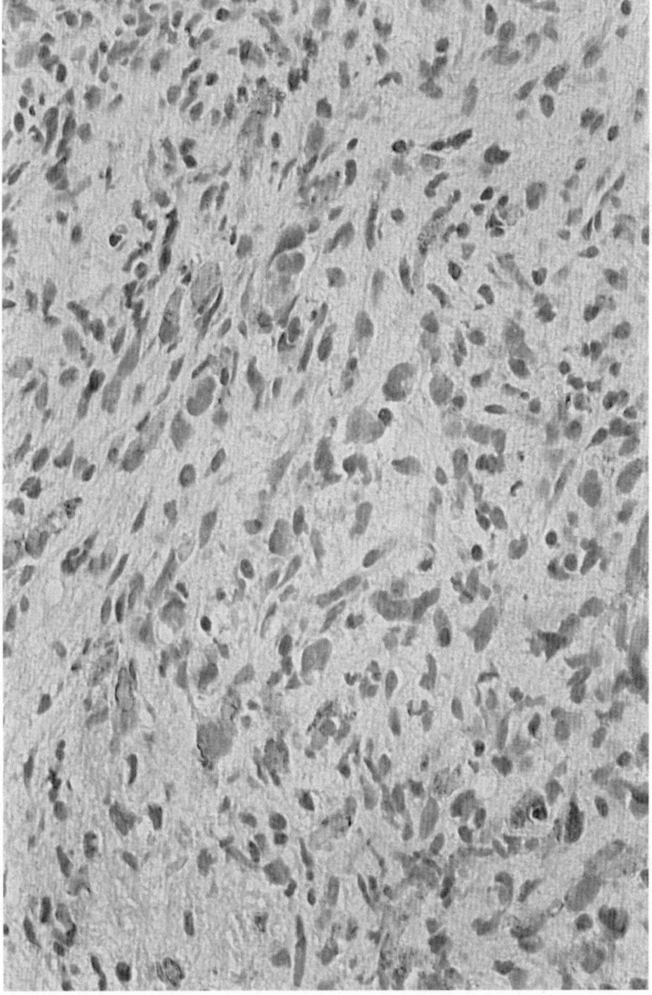

Figure 4–32 **B**

■ TUMORS DERIVED FROM SMOOTH MUSCLE

BENIGN TUMORS

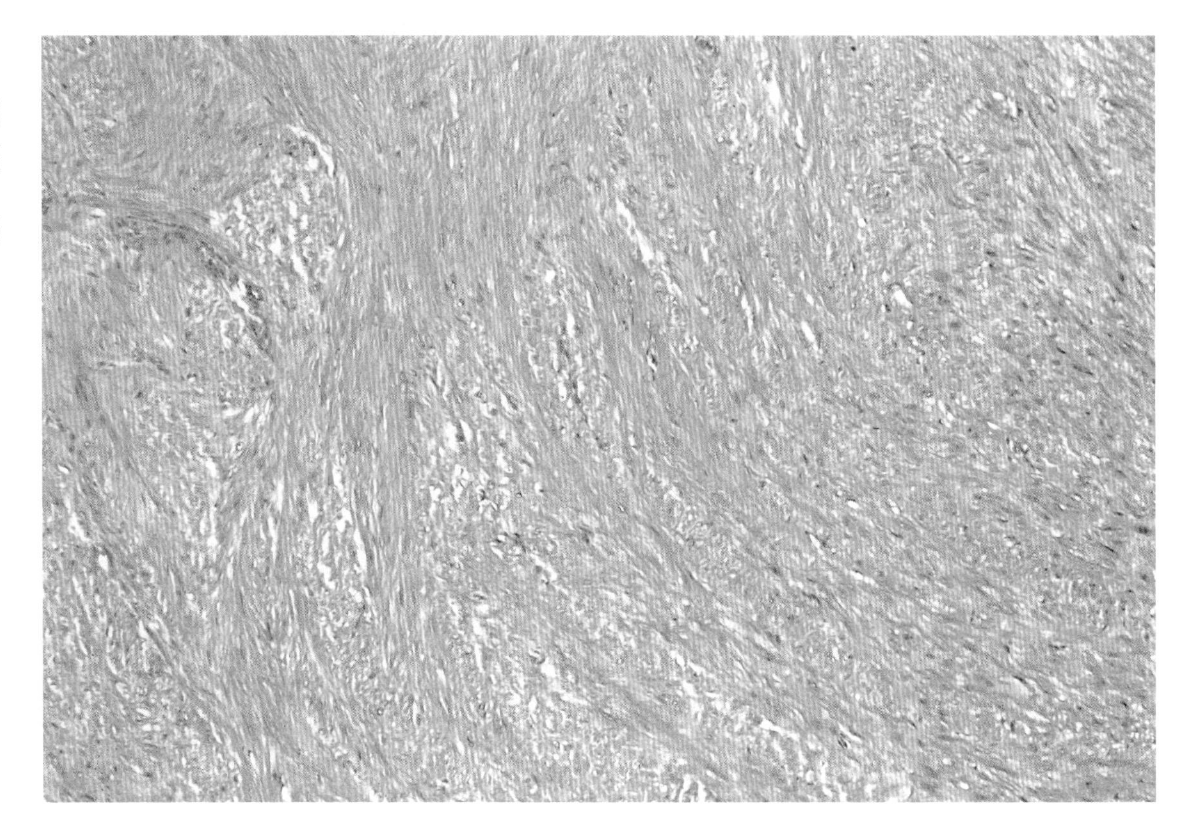

Figure 4–33 A

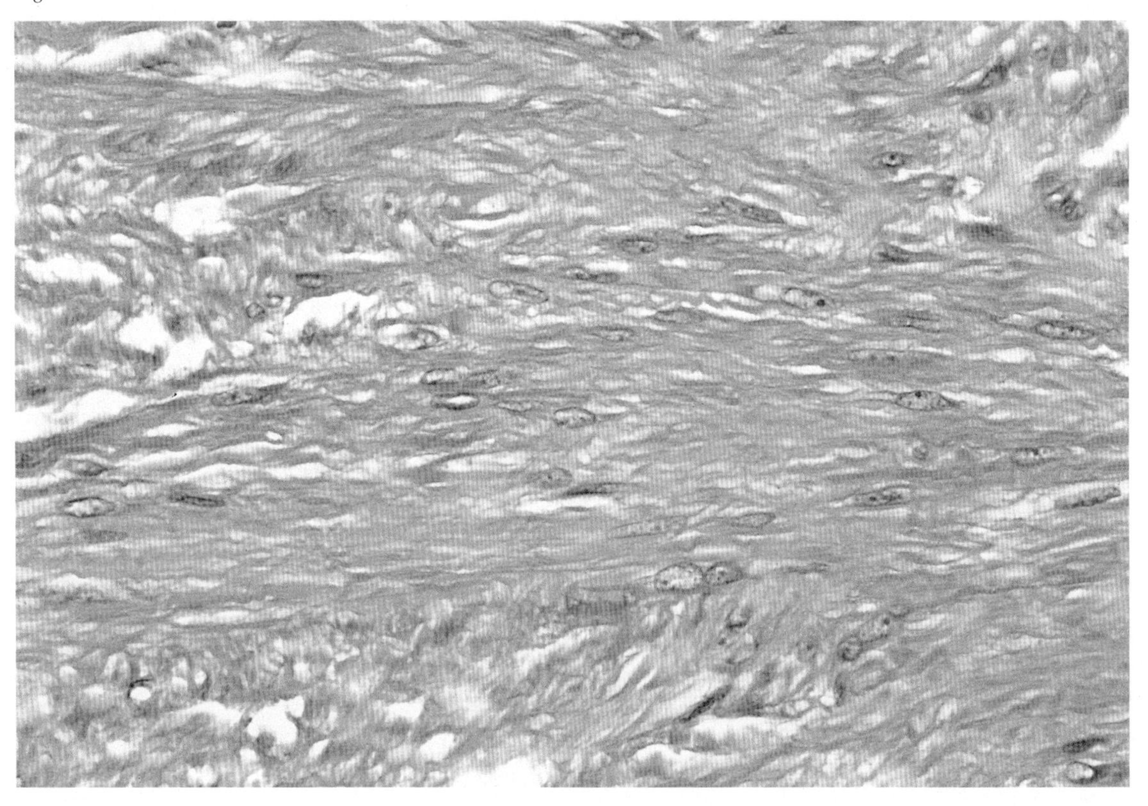

Figure 4–33 B

FIGURE 4–33. LEIOMYOMA: A: Leiomyomas are composed of interlacing bundles of eosinophilic smooth muscle cells. **B:** The cells are elongated with blunt-ended nuclei.

■ TUMORS DERIVED FROM SMOOTH MUSCLE

MALIGNANT TUMORS

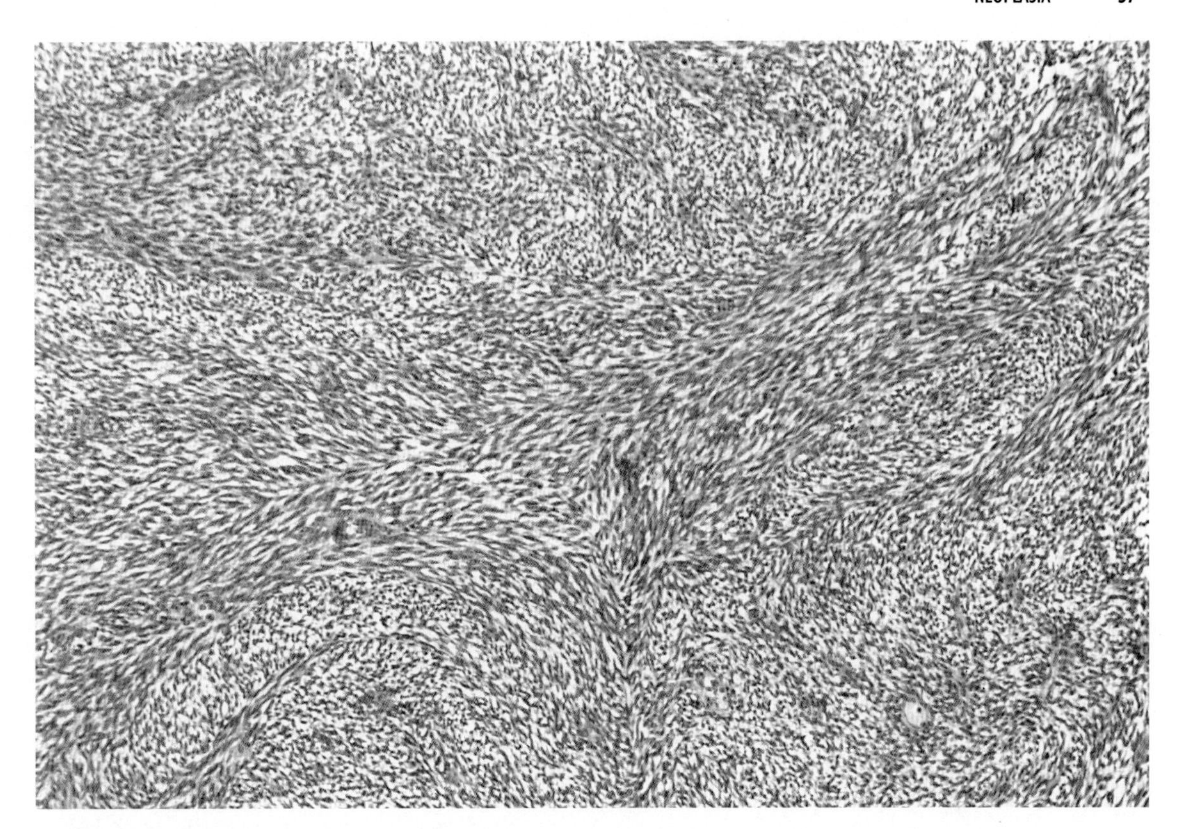

Figure 4–34 A

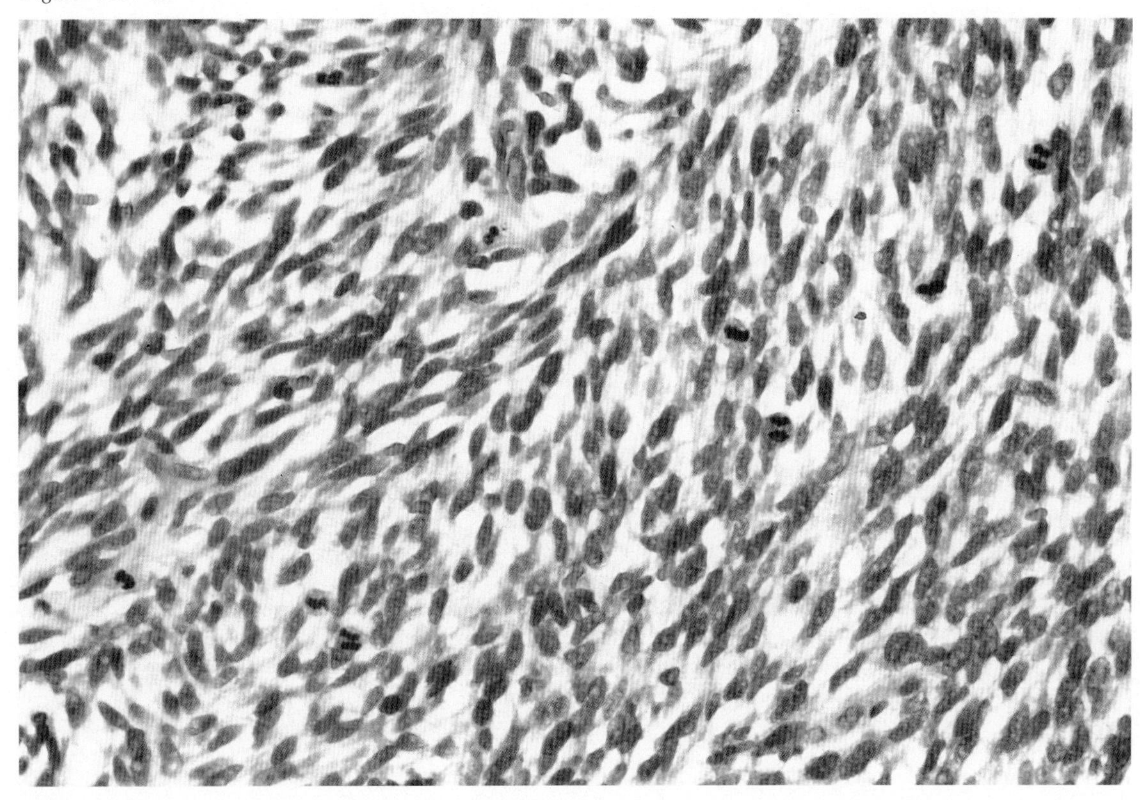

Figure 4–34 B

FIGURE 4–34. LEIOMYOSARCOMA: A: Leiomyosarcomas are cellular tumors composed of interlacing bundles of spindle cells. **B**: The cells are elongated and pleomorphic, with hyperchromatic nuclei. Mitotic figures are abundant.

■ TUMORS DERIVED FROM NEURAL TISSUE

BENIGN TUMORS

FIGURE 4–35. NEUROFIBROMA: A neurofibroma is a subcutaneous tumor composed of round cells, wavy cells, or both in a fine fibrillar collagen background.

MALIGNANT TUMORS

FIGURE 4–36. NEUROFIBROSARCOMA (MALIGNANT SCHWANNOMA): A: A neurofibrosarcoma resembles a fibrosarcoma (Figure 4–28 **A** and **B**) except that the cells have irregular contours. Nuclei are wavy and when cut en face appear oval. **B:** Immunoperoxidase staining with S-100 protein confirms the neurogenic origin of the tumor and enhances the visualization of the wavy nuclei.

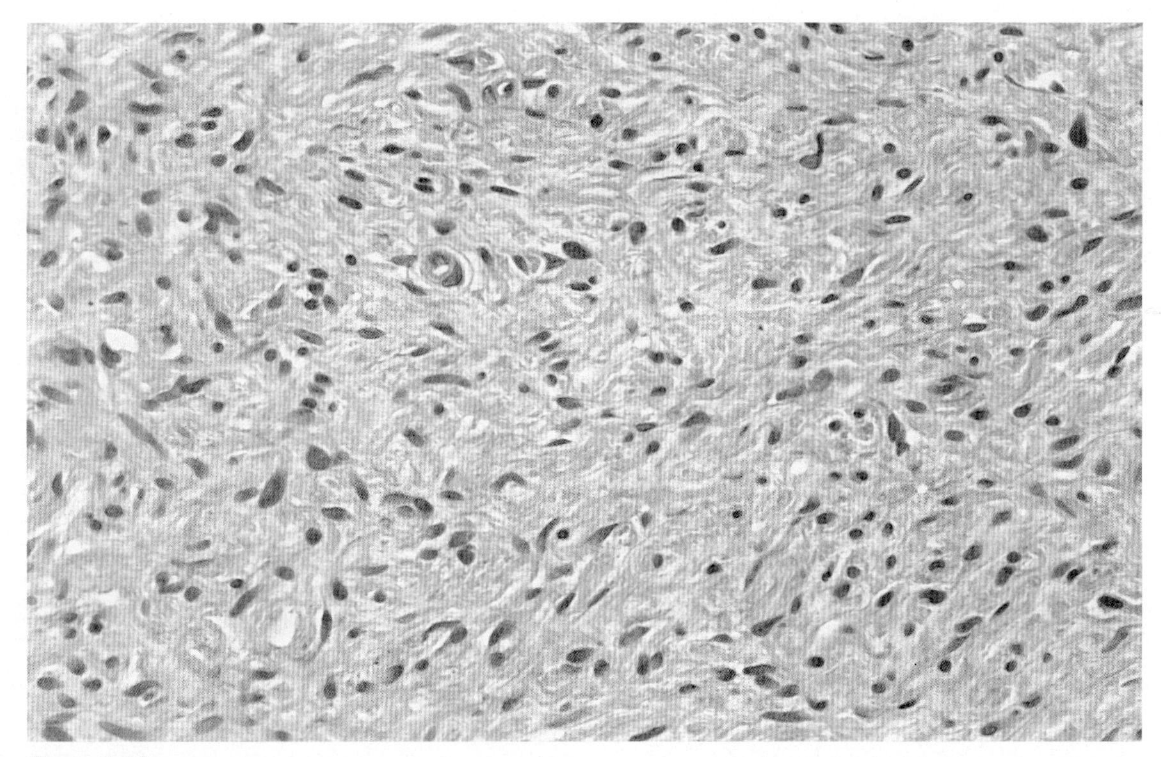

Figure 4–35

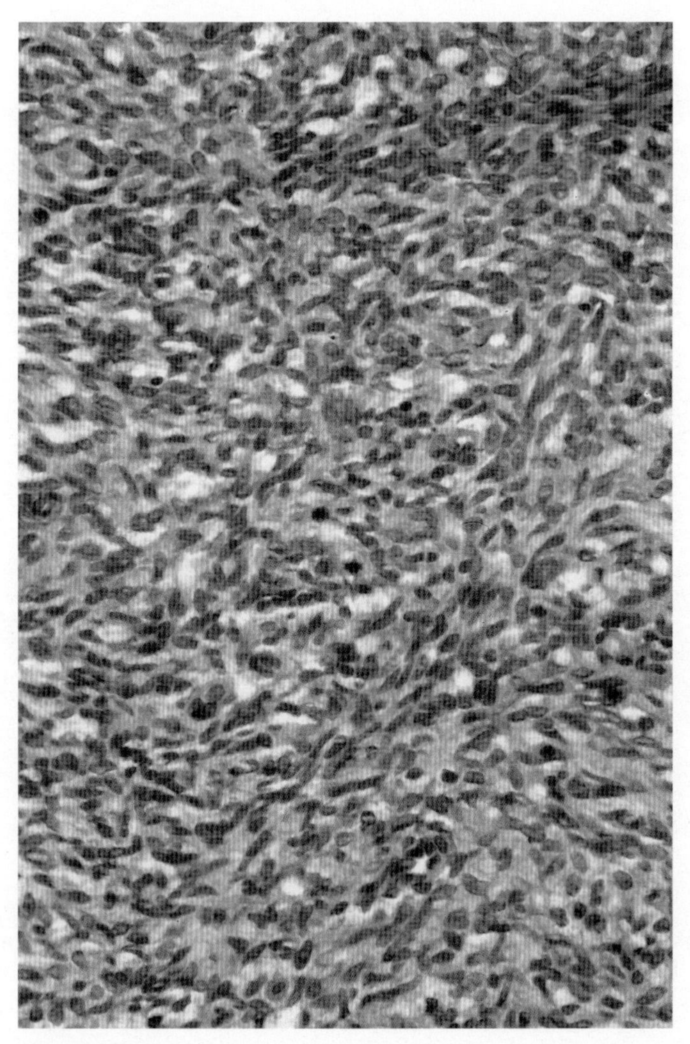

Figure 4–36 A

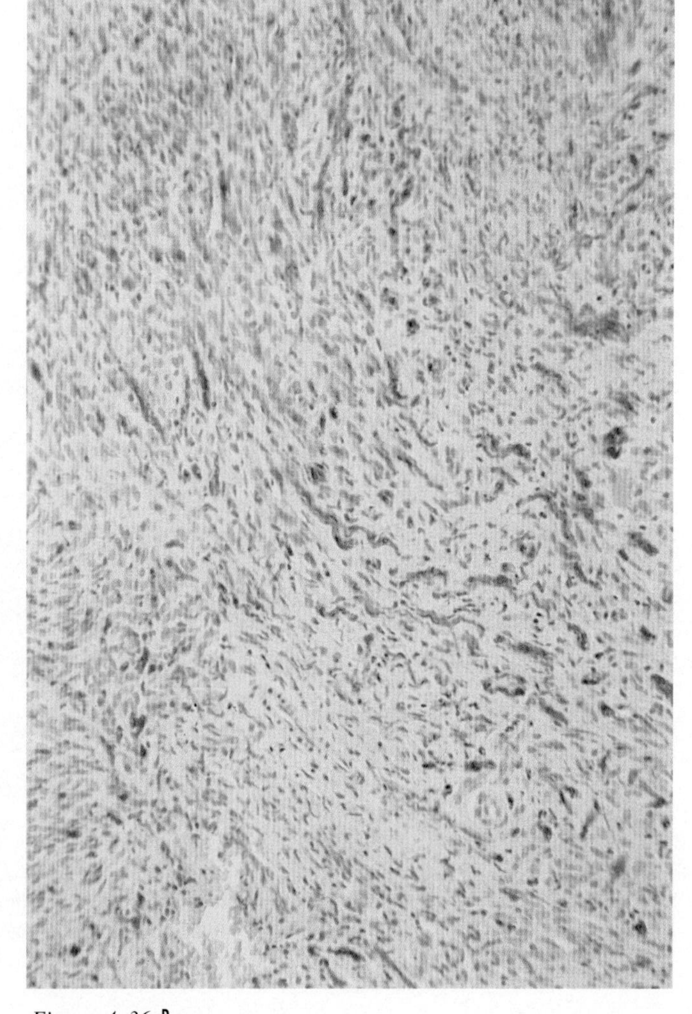

Figure 4–36 B

■ HEART

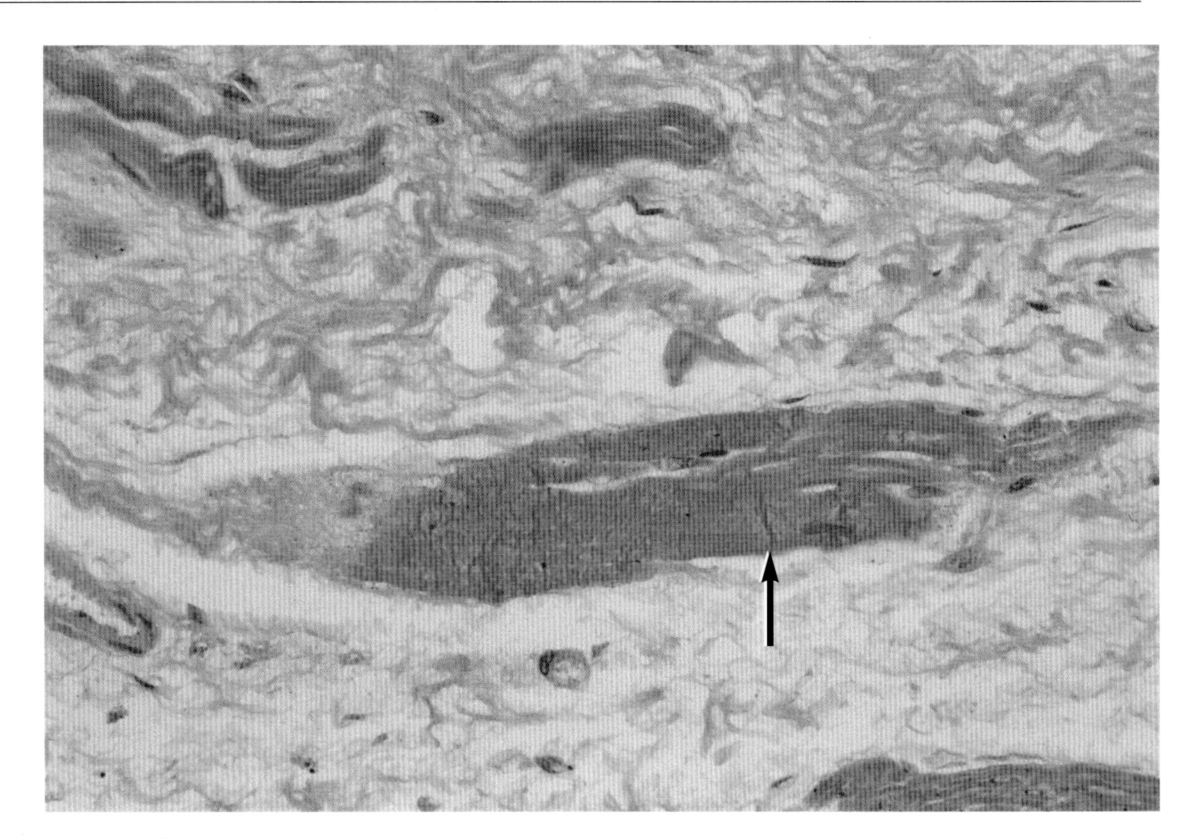

Figure 5–1 A

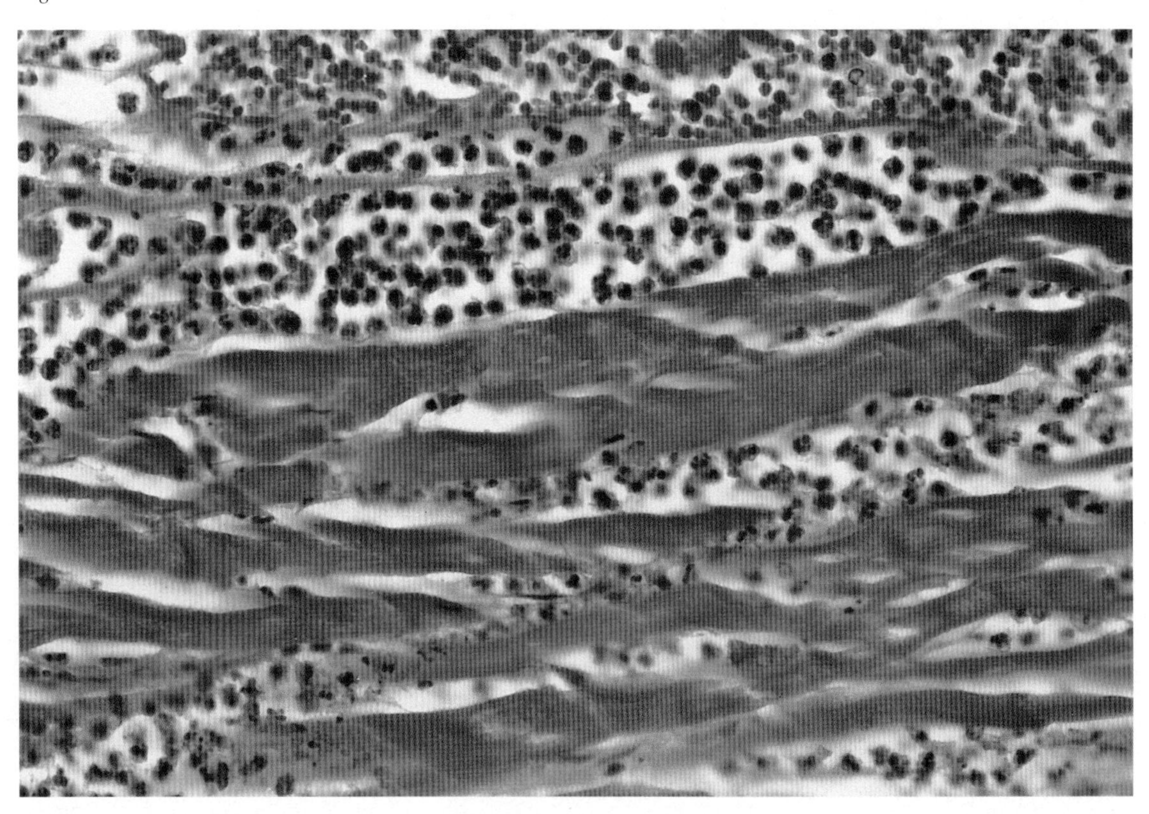

Figure 5–1 B

FIGURE 5–1. MYOCARDIAL INFARCTION: A: At 1–2 hours, one of the earliest changes seen by light microscopy of an **acute myocardial infarction** is **contraction band necrosis**. B: Beginning at approximately 12 hours and continuing to 72 hours, an acute myocardial infarction is characterized by coagulative necrosis of myocytes and an infiltration by neutrophils.

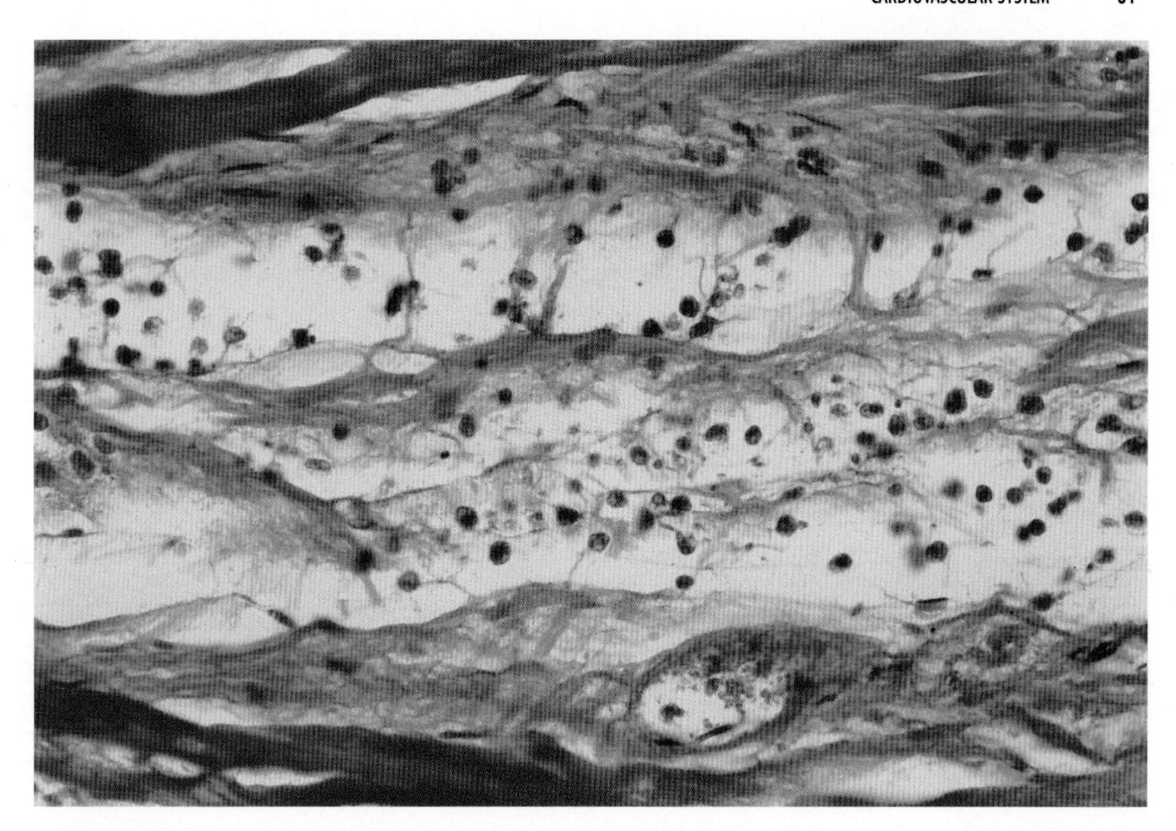

Figure 5–1 C

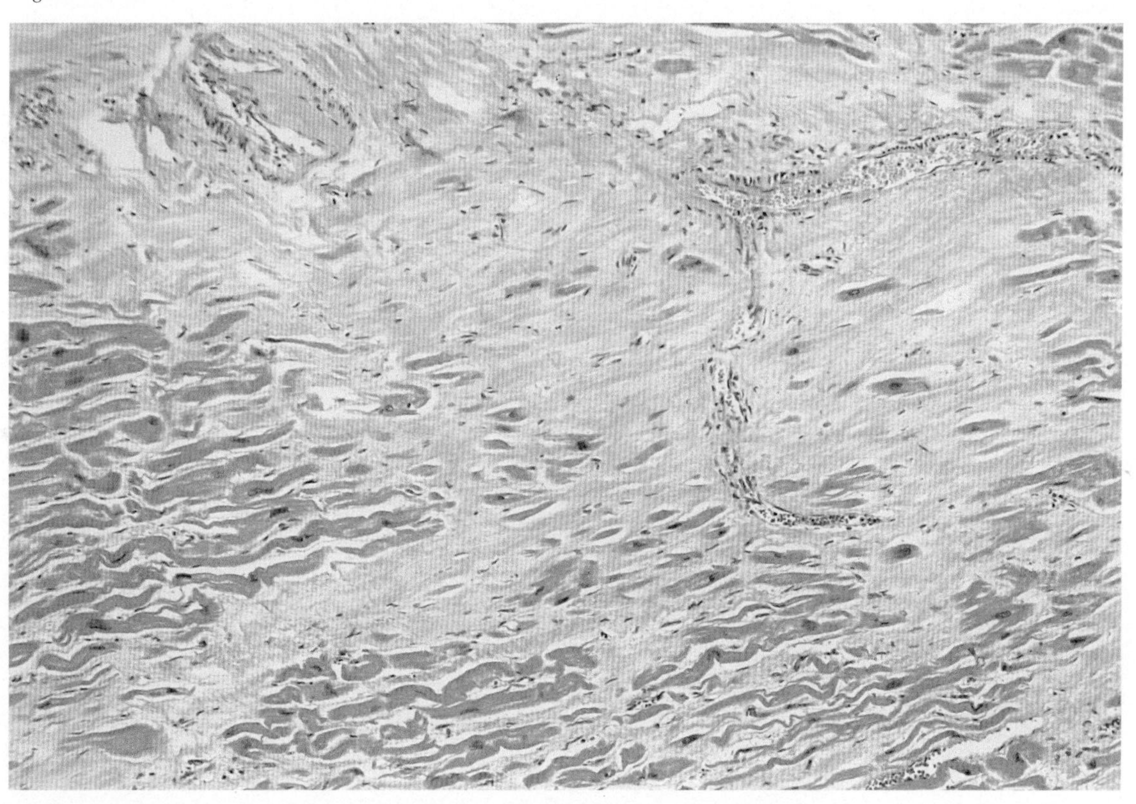

Figure 5–1 D

FIGURE 5–1. MYOCARDIAL INFARCTION: C: By 3–7 days, the neutrophils are replaced by macrophages and the necrotic myocytes are resorbed. At about 10 days, the healing process begins and fibroblasts replace the necrotic areas. D: By several weeks, a **healed myocardial infarction** is characterized by fibrosis and scarring.

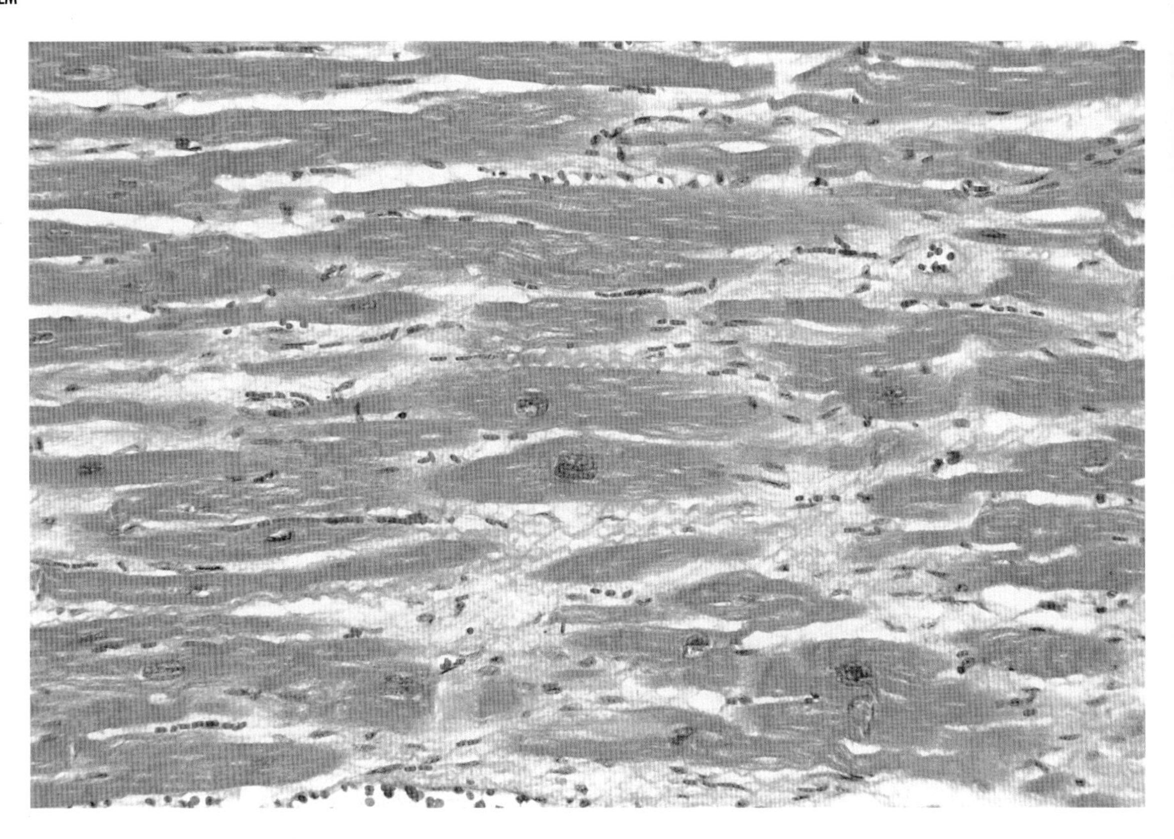

FIGURE 5–2. MYOCARDIAL HYPERTROPHY: Hypertrophy of myocytes is most commonly seen in patients with hypertension. Hypertrophy of myocytes is characterized by enlargement of individual myocytes as well as by enlarged "boxcar" nuclei. Compare the hypertrophic myocytes to the surrounding normal myocytes.

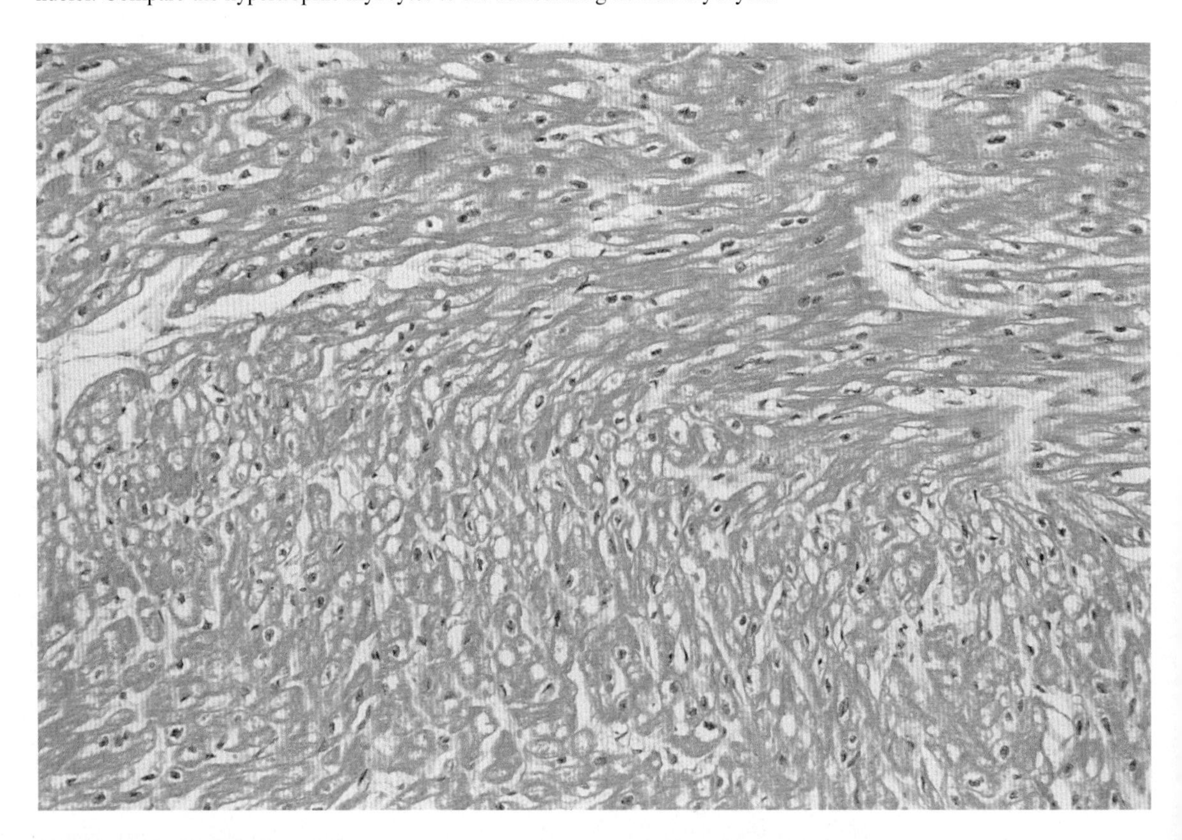

FIGURE 5–3. CARDIOMYOPATHY: In cardiomyopathy, muscle fibers are present in a haphazard orientation.

INFLAMMATORY LESIONS

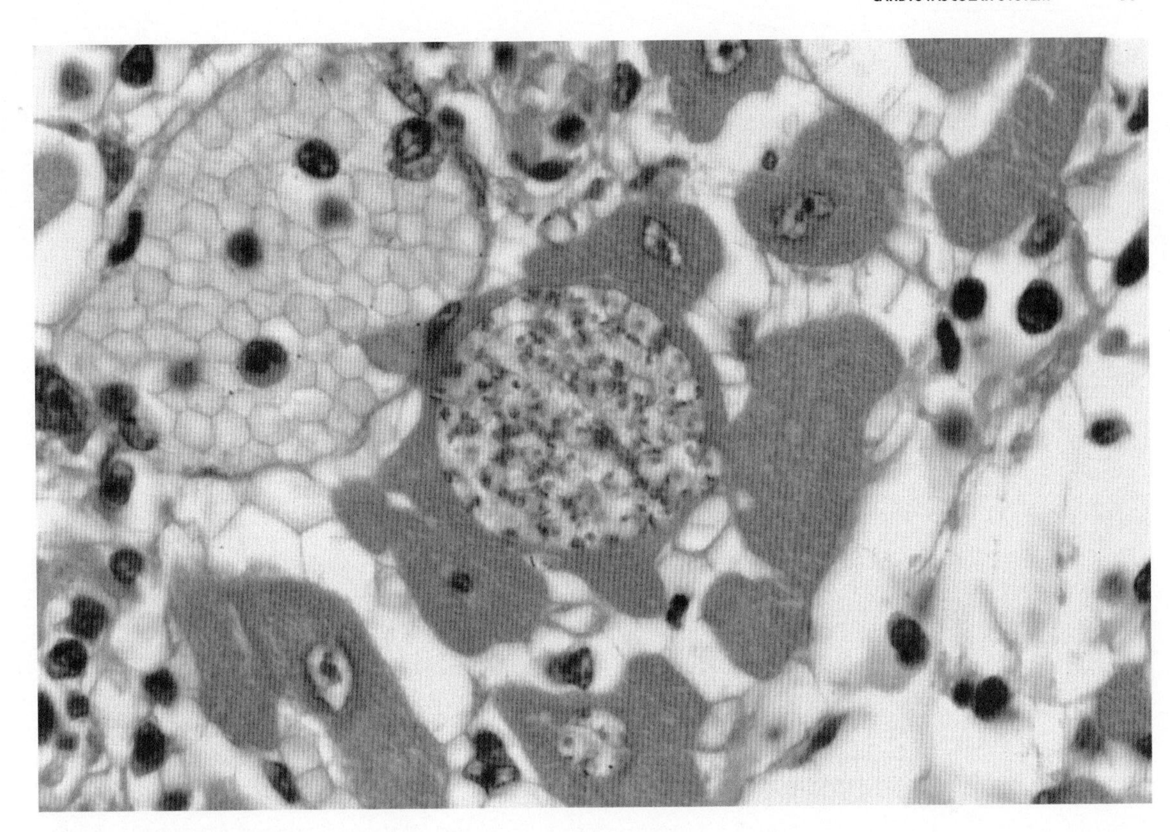

FIGURE 5–4. CHAGAS' DISEASE: Chagas' disease is caused by the parasite *Trypanosoma cruzi.* In the acute phase, amastigotes are present in myocardial fibers.

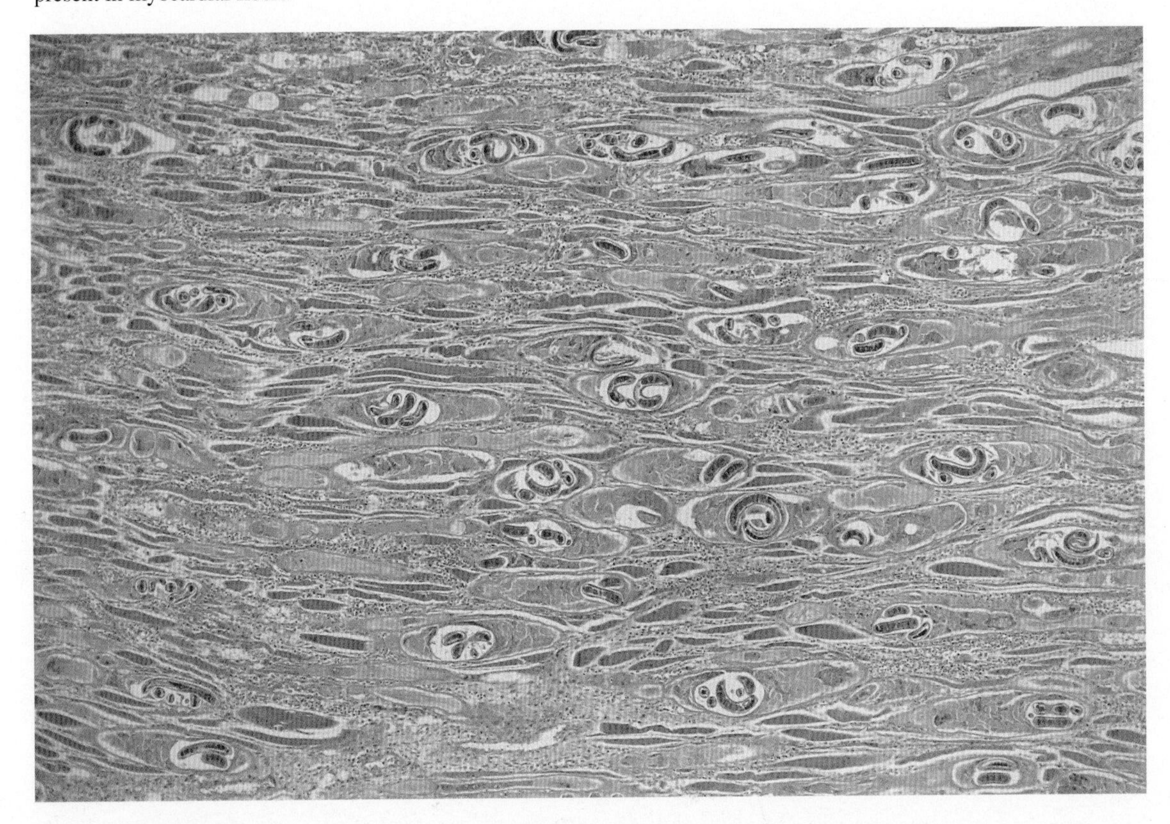

FIGURE 5–5. TRICHINOSIS: Trichinosis, caused by the pig tapeworm *Taenia solium,* affects striated muscle. In this example of trichinosis infestation of the myocardium, the worms are not encysted.

INFLAMMATORY LESIONS

FIGURE 5–6. RHEUMATIC HEART DISEASE: A: In acute rheumatic heart disease, there is necrosis and an associated inflammatory reaction in the myocardial interstitium. **B:** The **Aschoff body** is pathognomonic for rheumatic heart disease and is composed of a necrotic focus infiltrated by mononuclear and multinucleated giant cells, Anitschkow cells. **C:** Healed rheumatic valvulitis is characterized by blood vessel proliferation and an associated inflammatory reaction.

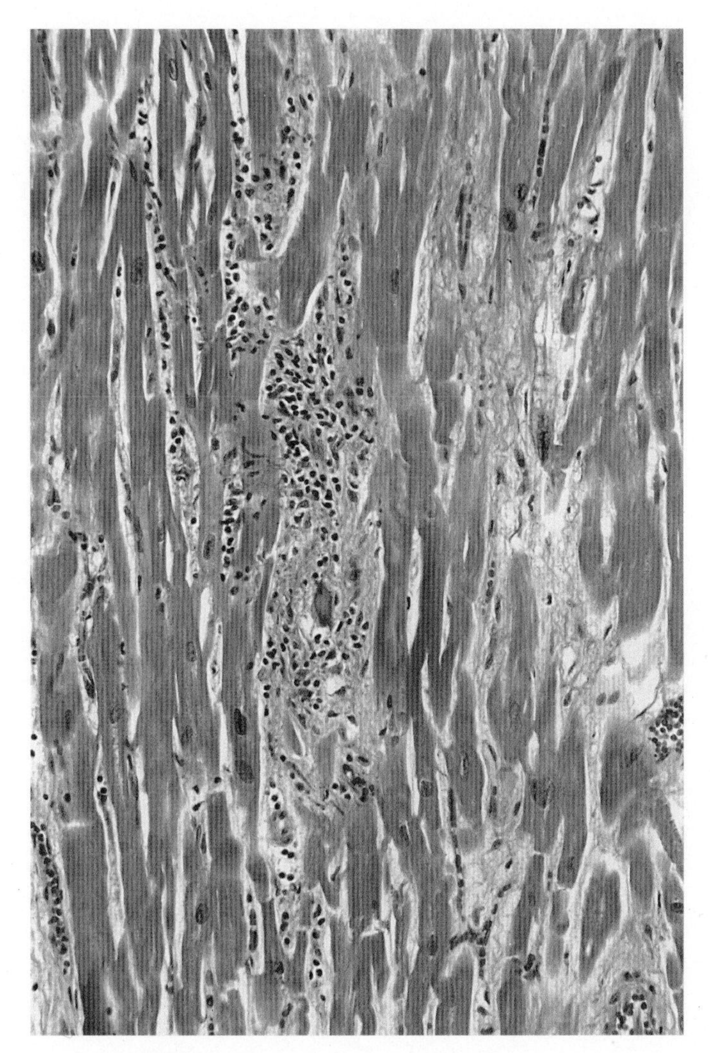

Figure 5–6 **A**

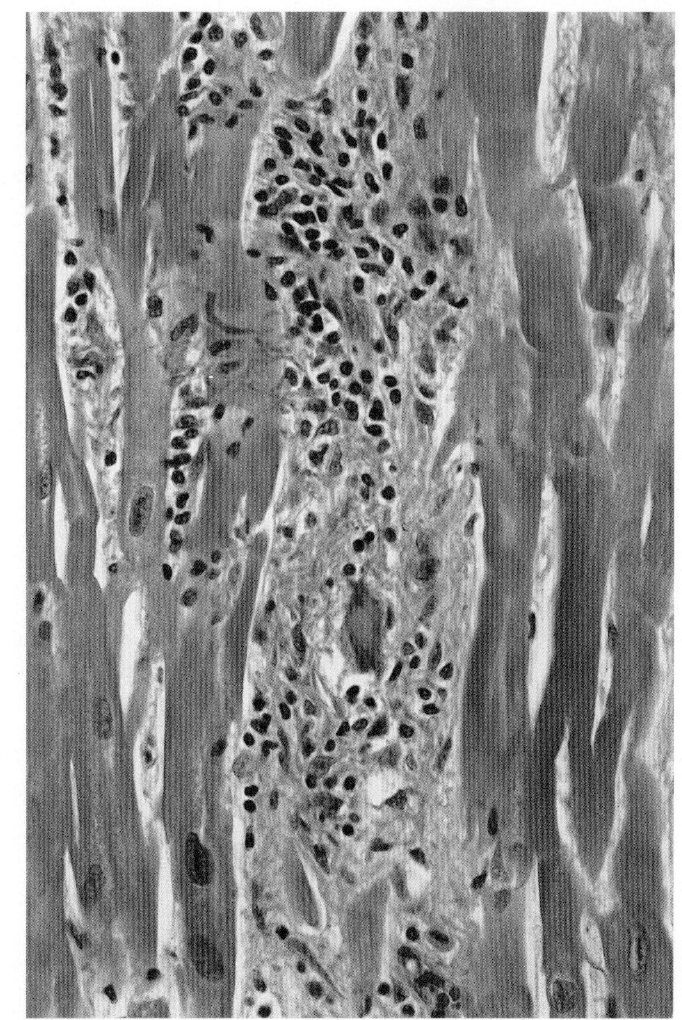

Figure 5–6 **B**

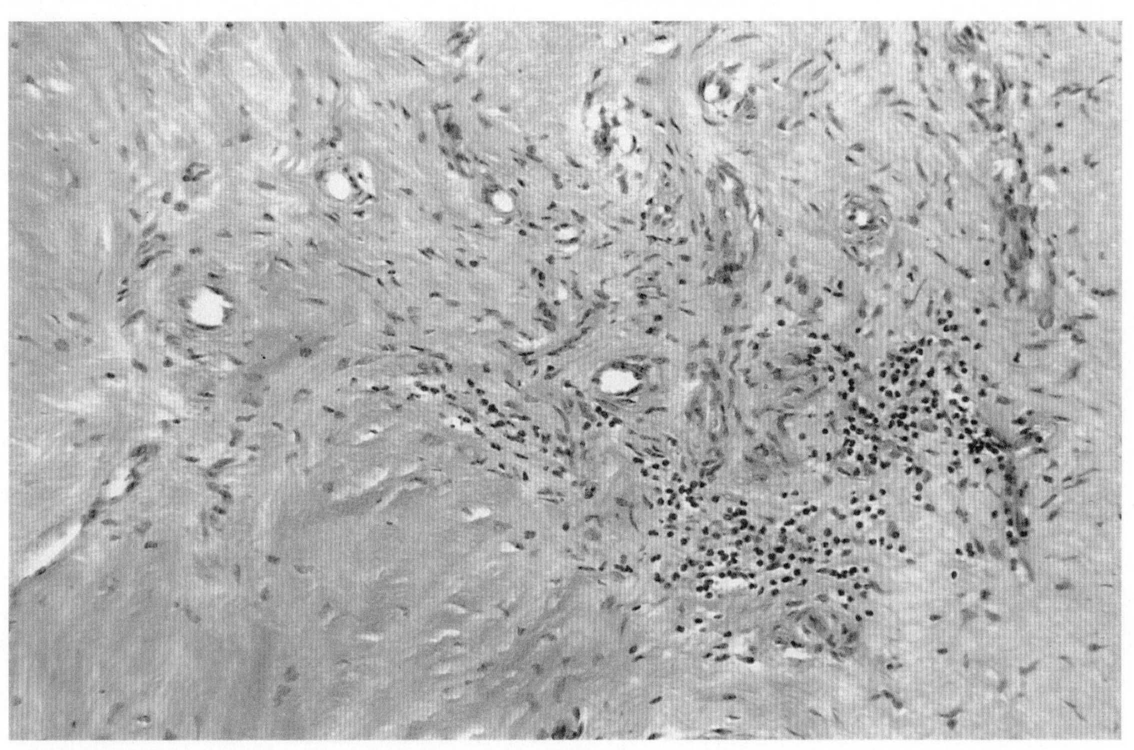

Figure 5–6 **C**

INFLAMMATORY LESIONS

FIGURE 5–7. INFECTIVE ENDOCARDITIS: A: Infective endocarditis is characterized by **vegetations** on the valve leaflets. The vegetations are composed of fibrin, blood cells, platelets, and microorganisms. The most common organisms causing infective endocarditis are bacteria (**B**, Gram stain) and fungi (**C**).

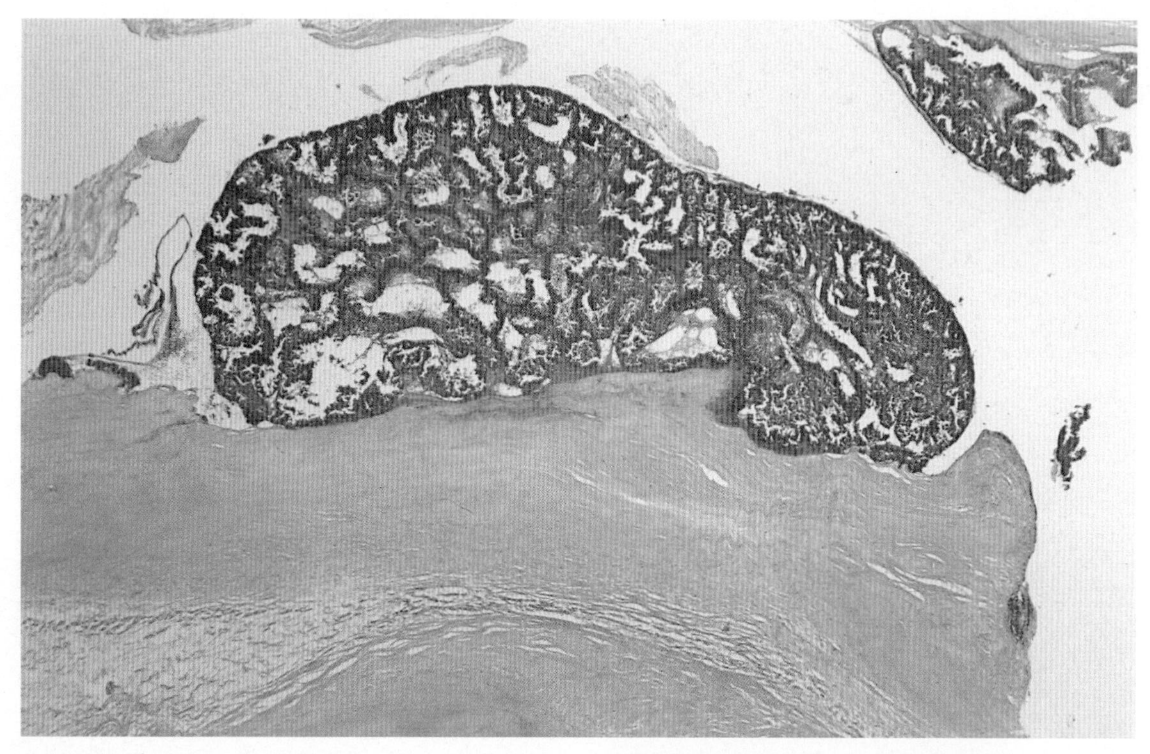

Figure 5–7 A

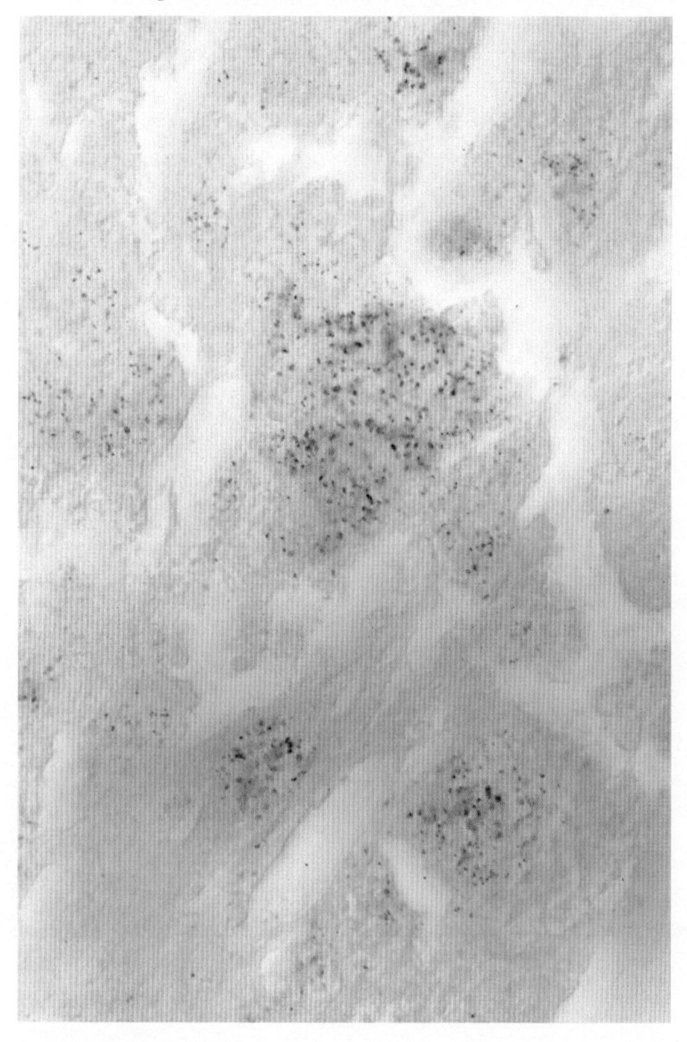

Figure 5–7 B

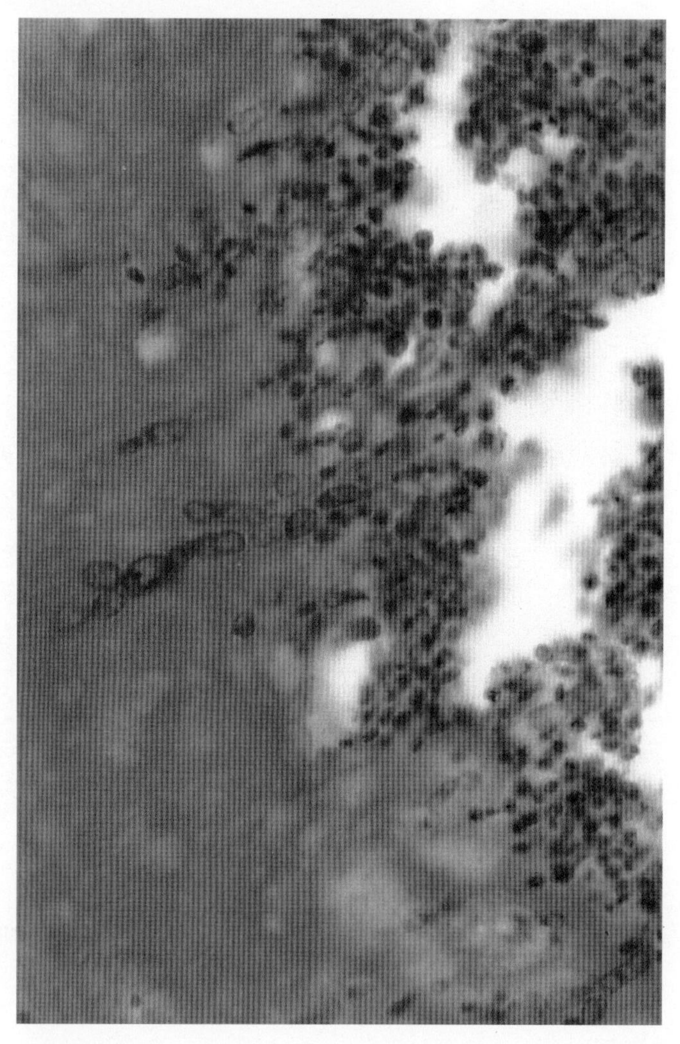

Figure 5–7 C

INFLAMMATORY LESIONS

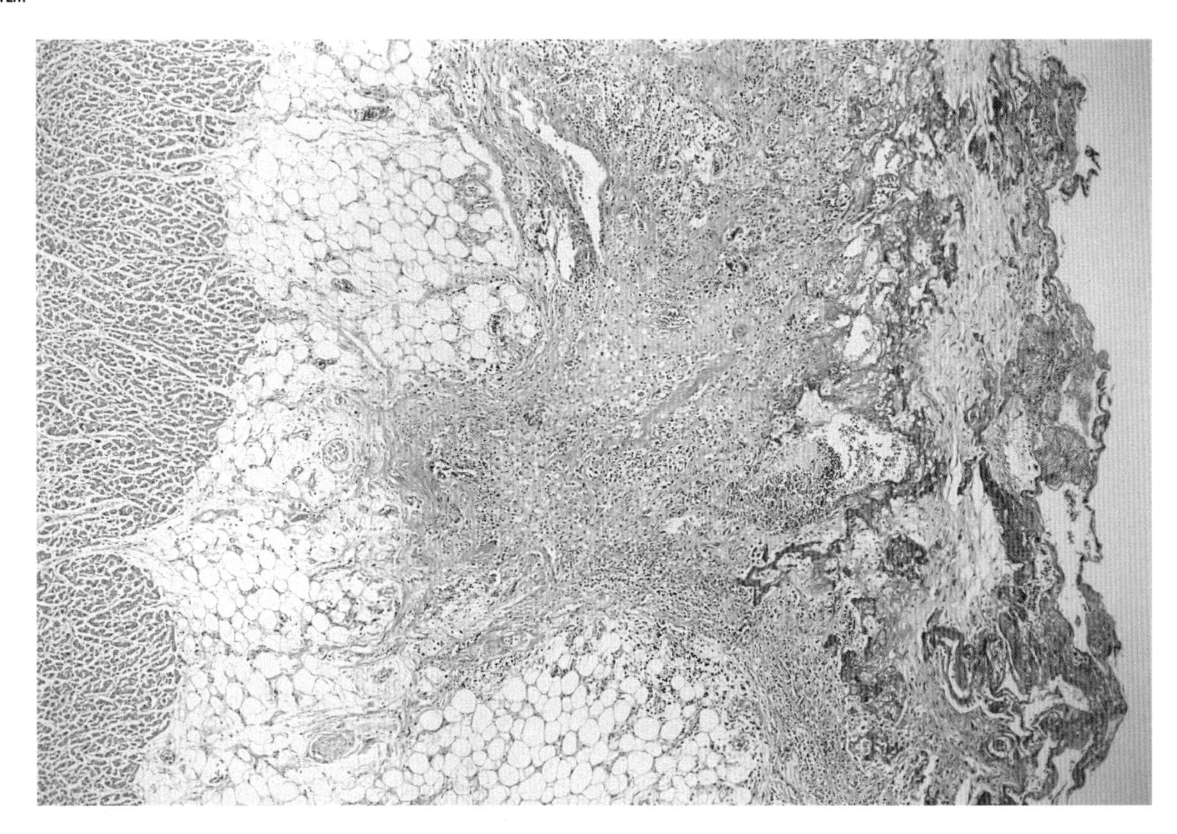

FIGURE 5–8. FIBRINOUS PERICARDITIS: Fibrinous pericarditis is characterized by fibrin deposition on the pericardial surface. Microscopically, fibrin is seen as pink acellular amorphous material.

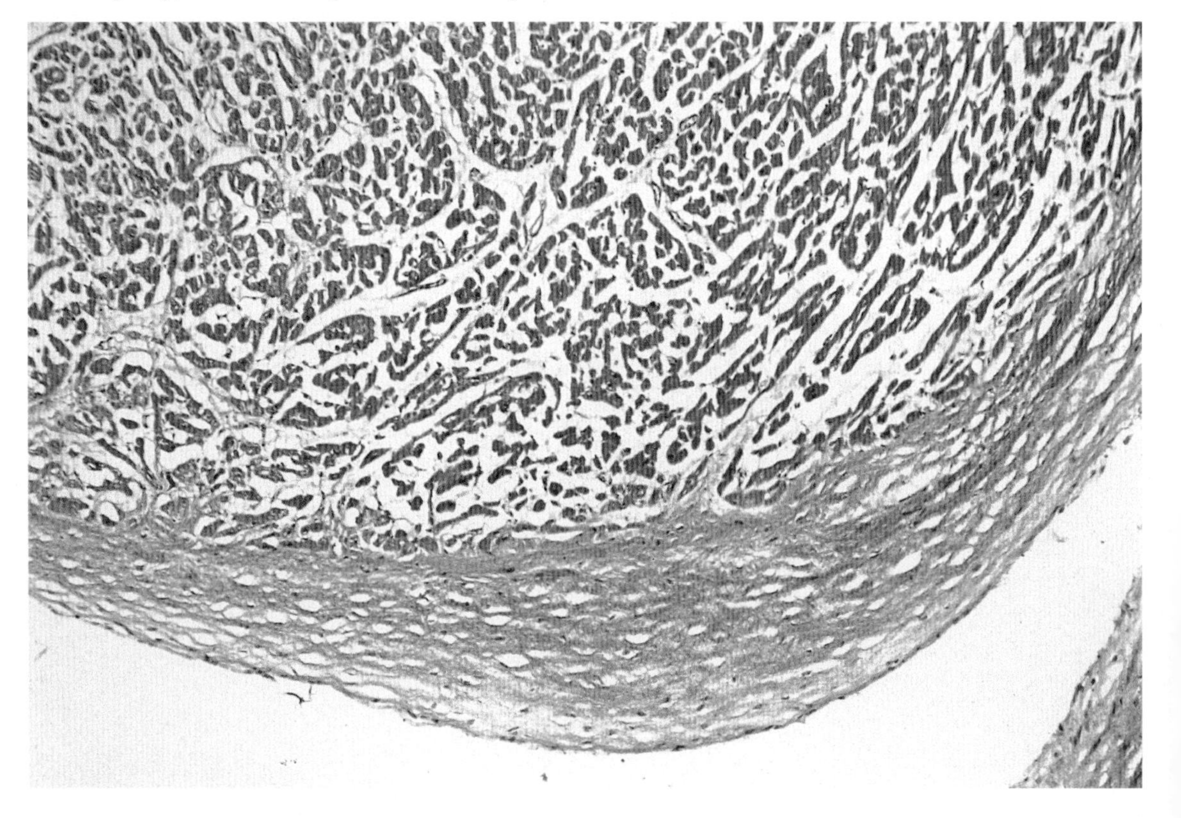

FIGURE 5–9. SUBENDOCARDIAL FIBROELASTOSIS: In subendocardial fibroelastosis, there is a focal or diffuse cartilagelike fibroelastic thickening of the mural endocardium.

DEGENERATIVE PROCESSES

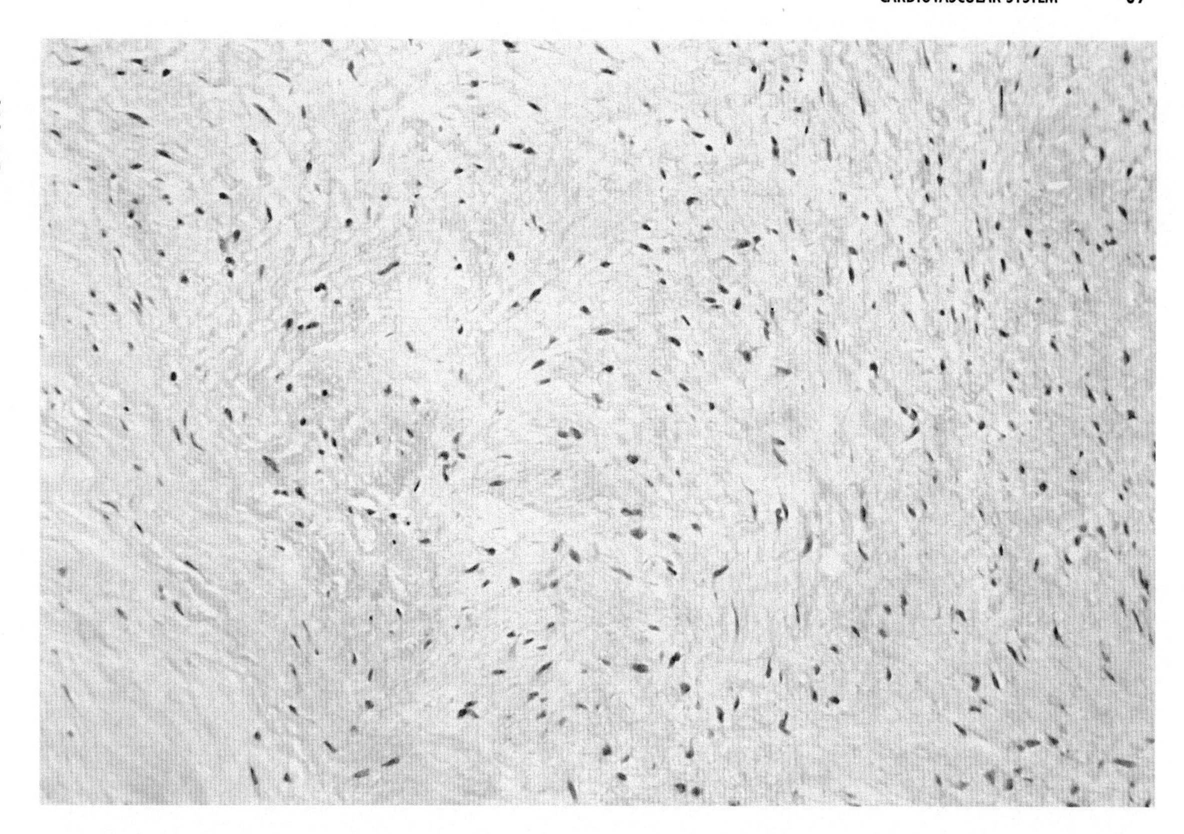

Figure 5–10 **A**

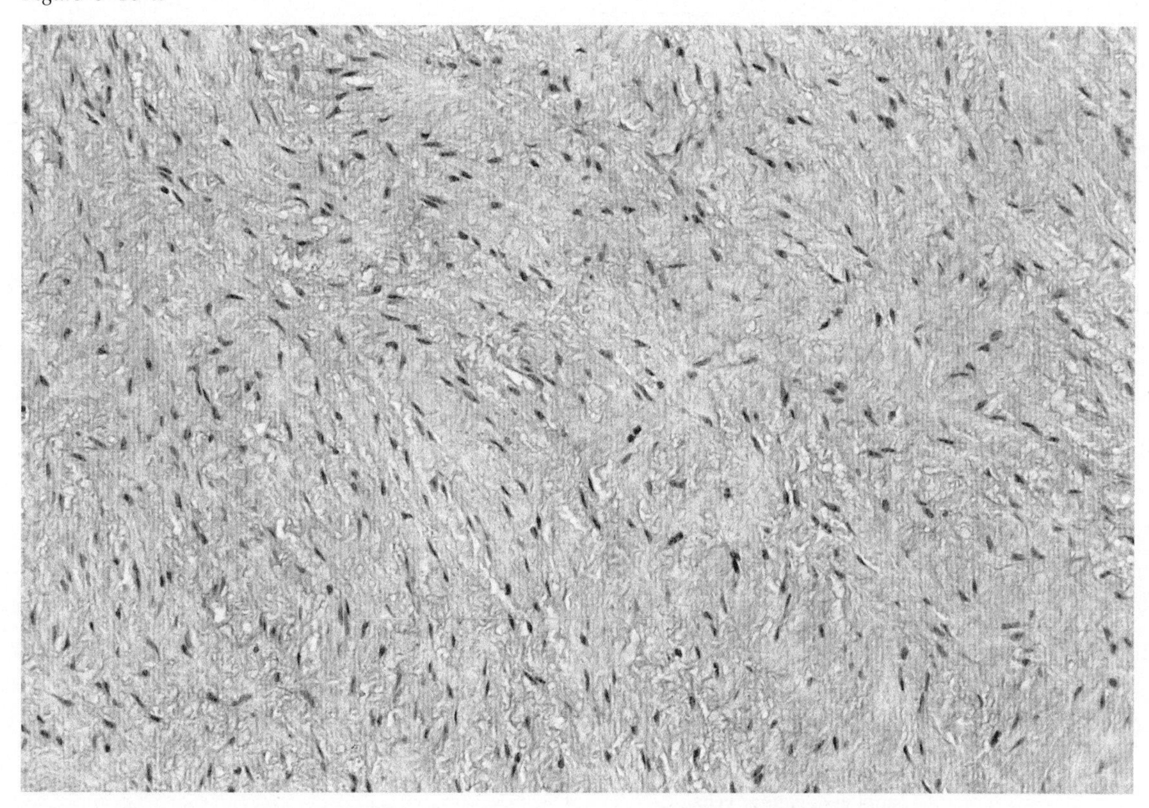

Figure 5–10 **B**

FIGURE 5–10. MYXOID DEGENERATION OF THE MITRAL VALVE (FLOPPY MITRAL VALVE, MITRAL VALVE PROLAPSE): A: Accumulation of acid mucopolysaccharides in the heart valve is a characteristic feature of myxoid degeneration. **B:** The mucopolysaccharides are best seen with alcian blue staining.

TUMORS

FIGURE 5–11. CARDIAC MYXOMA: A cardiac myxoma is the most common primary tumor affecting the heart. The myxoma is composed of stellate myxoma cells, endothelial cells, and macrophages in an acid mucopolysaccharide ground substance. The mucopolysaccharides are produced by the indigenous proliferating myxoma cells.

FIGURE 5–12. RHABDOMYOMA: Rhabdomyomas are characterized by a mixed population of large round or polygonal glycogen-laden cells. The plasma membrane forms strands from the centrally located nucleus out to the periphery—**spider cells**.

FIGURE 5–13. INTRAATRIAL LIPOMA (LIPOMATOUS HYPERTROPHY OF THE ATRIAL SEPTUM): An intraatrial lipoma is characterized by the presence of mature adipose tissue admixed with myocyte fibers.

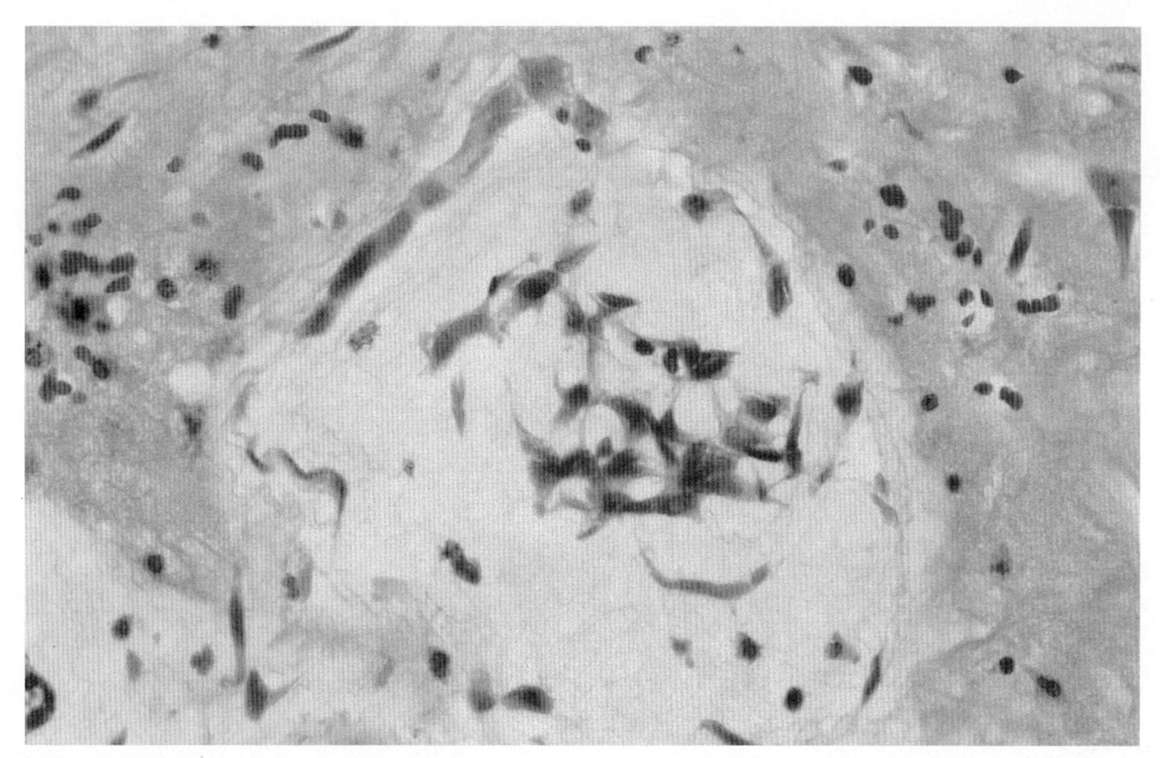

Figure 5–11

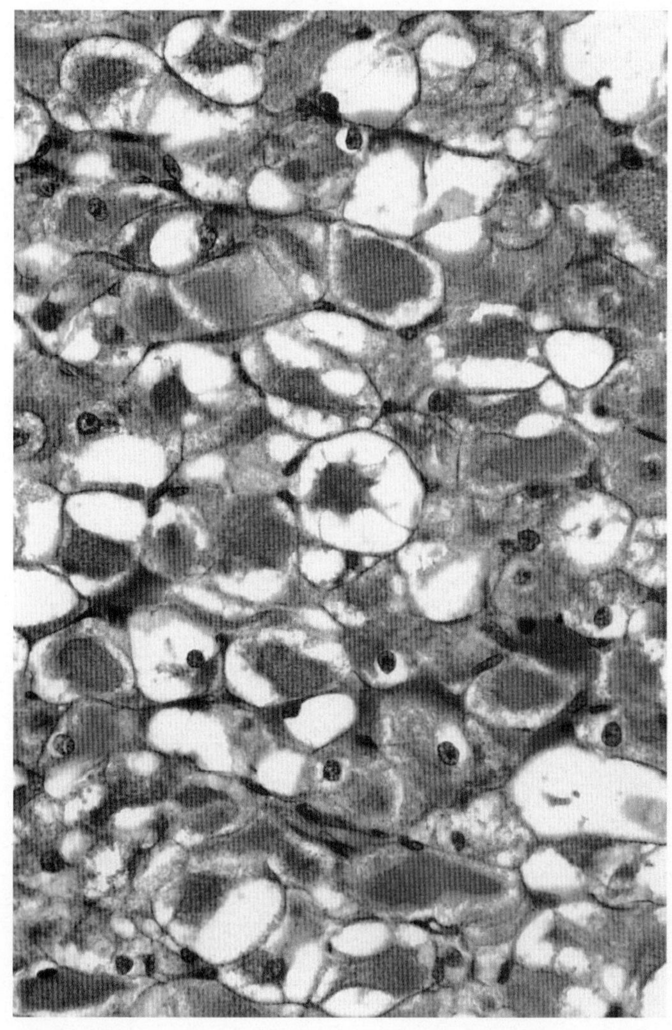

Figure 5–12

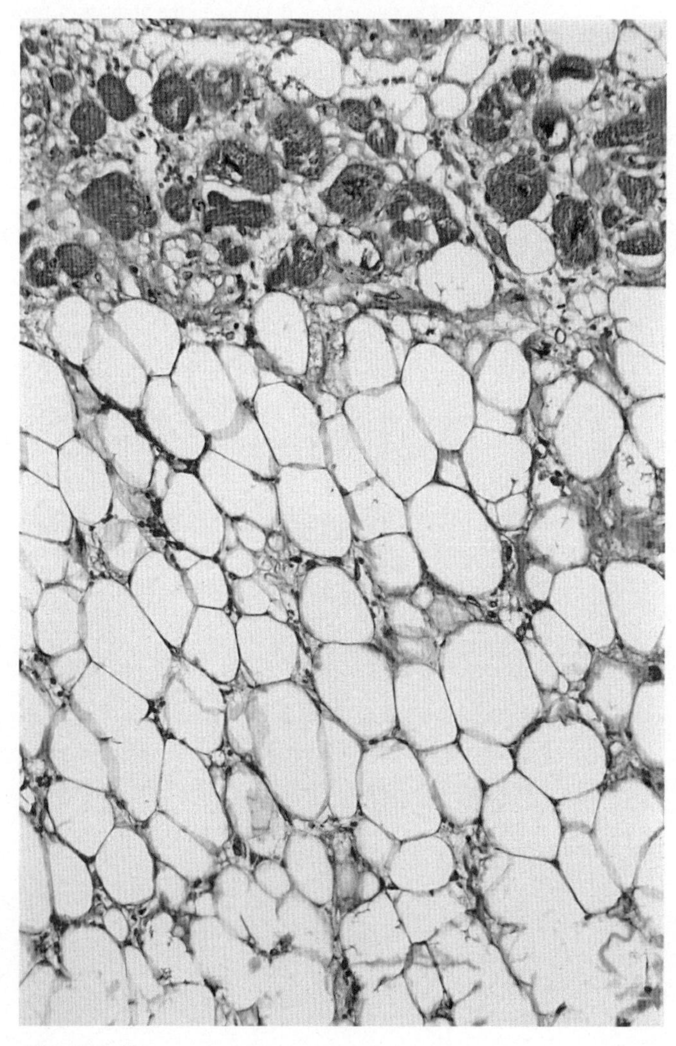

Figure 5–13

■ BLOOD VESSELS

ARTERIOSCLEROSIS

FIGURE 5–14. ATHEROSCLEROSIS: In atherosclerosis, the blood vessel lumen becomes narrowed because of an accumulation of intracellular and extracellular substances within the vessel wall. **A:** This is an example of minimal narrowing of the blood vessel lumen. **B:** The earliest change is a subendothelial accumulation of foamy macrophages. **C:** As the disease progresses, there is a proliferation of modified smooth muscle cells and a concomitant decrease in the number of foamy macrophages.

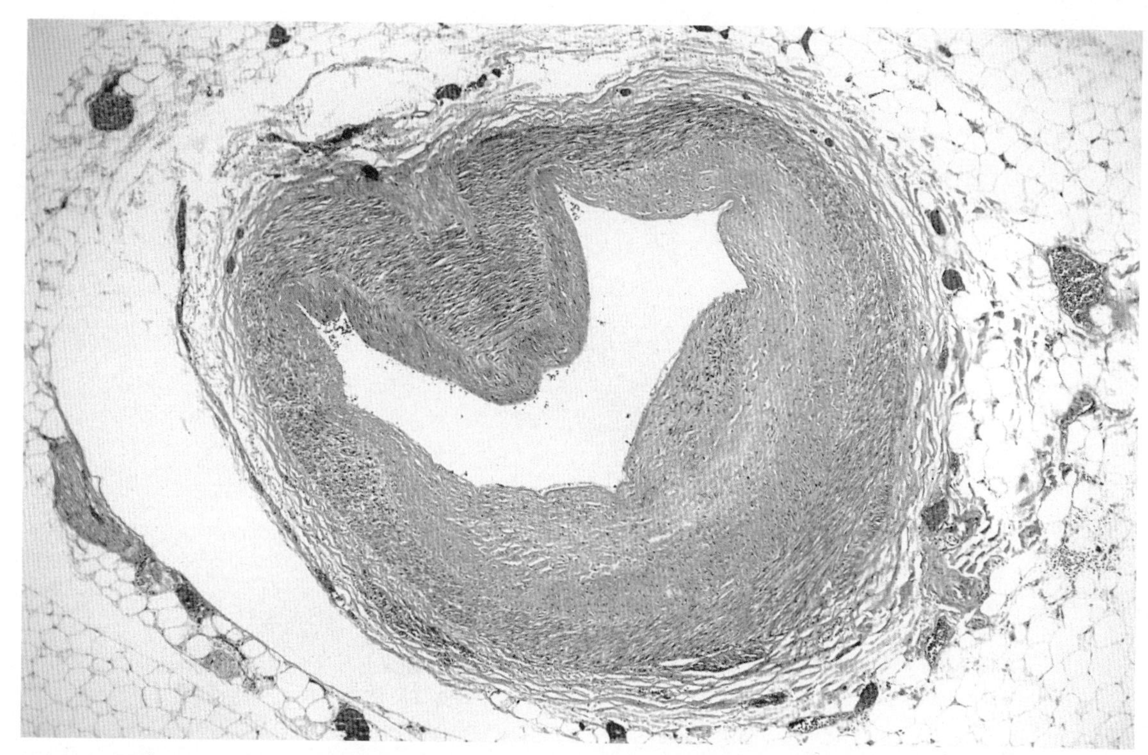

Figure 5–14 A

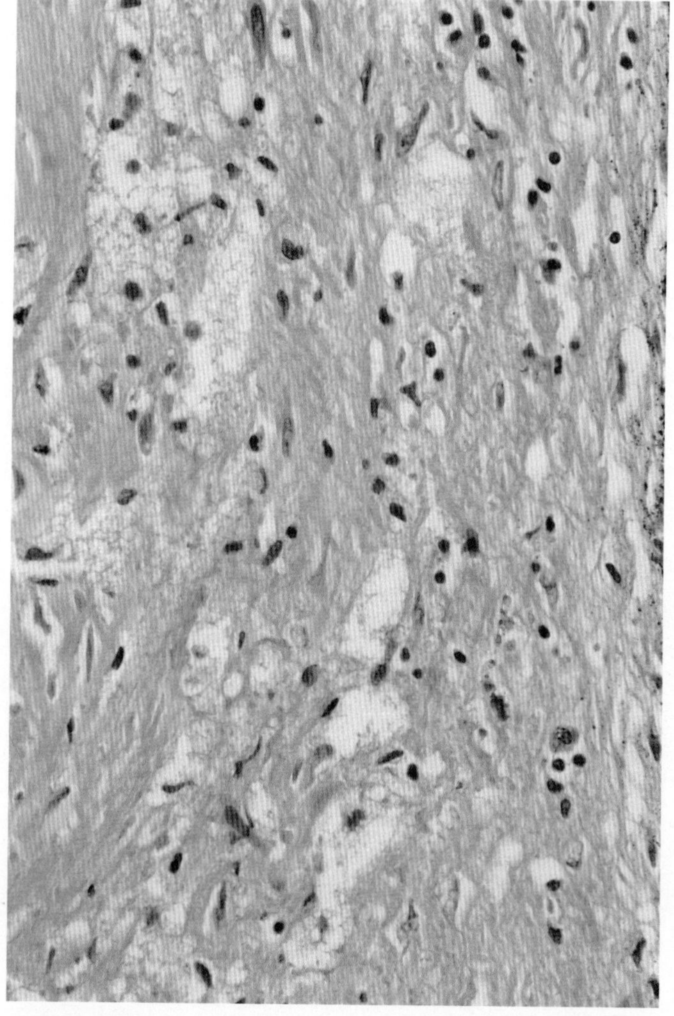

Figure 5–14 B

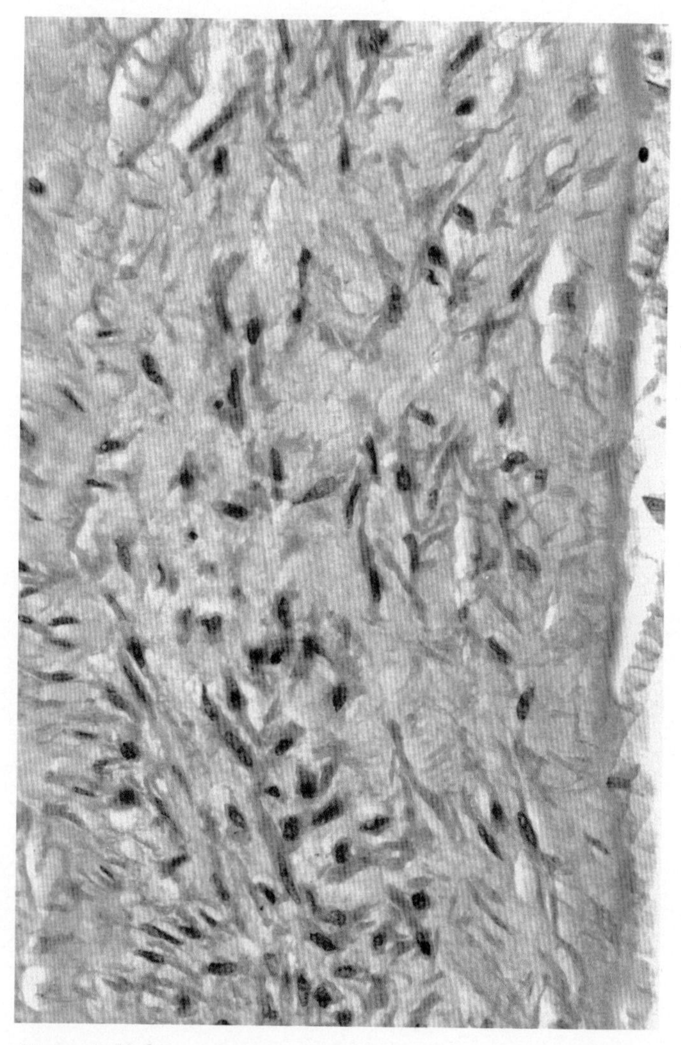

Figure 5–14 C

ARTERIOSCLEROSIS

FIGURE 5–14. ATHEROSCLEROSIS: D: This is an example of a **calcified plaque.** E: In a markedly narrowed vessel, the lumen is almost entirely occluded. F: Recanalization of the plaque has occurred.

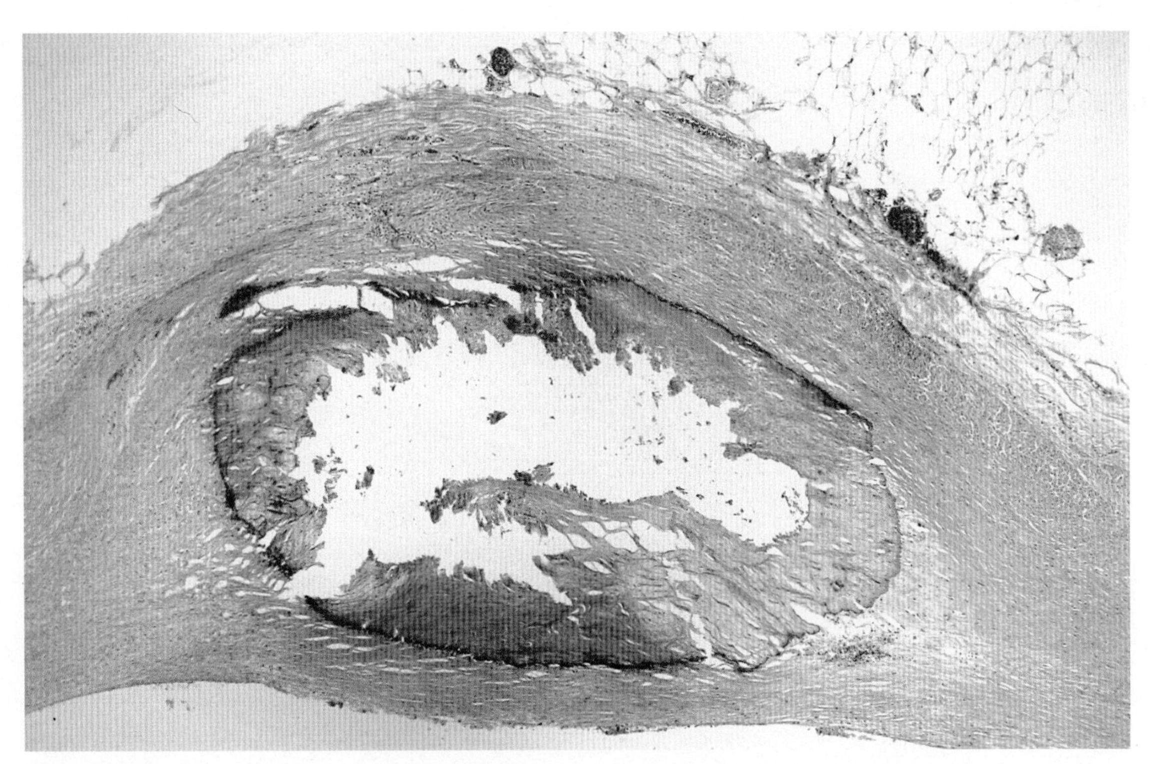

Figure 5–14 D

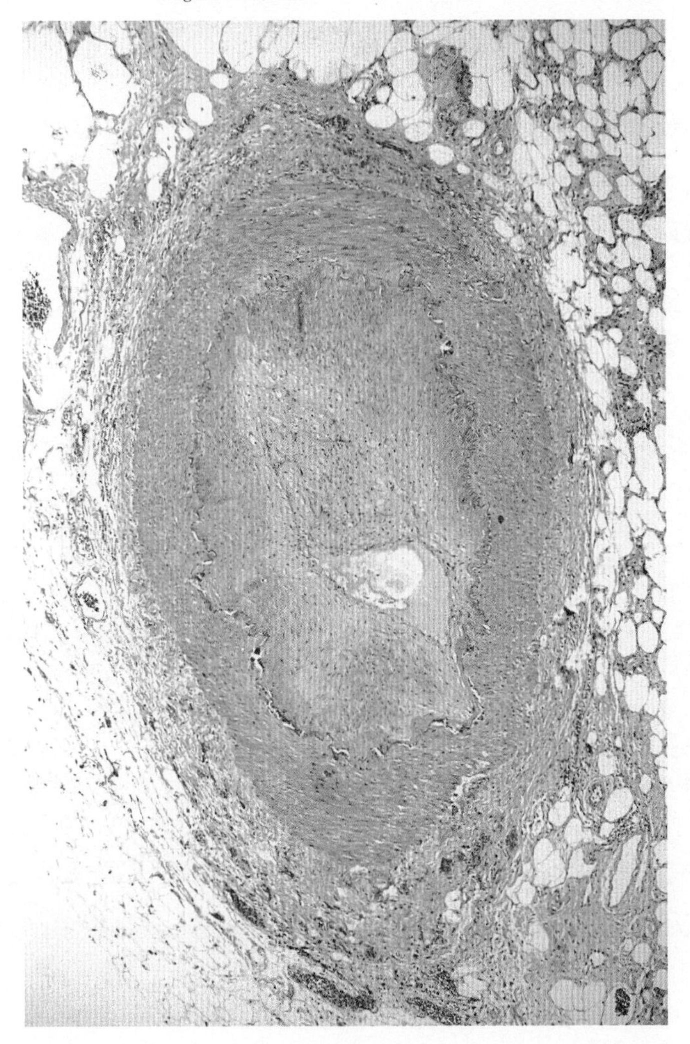

Figure 5–14 E

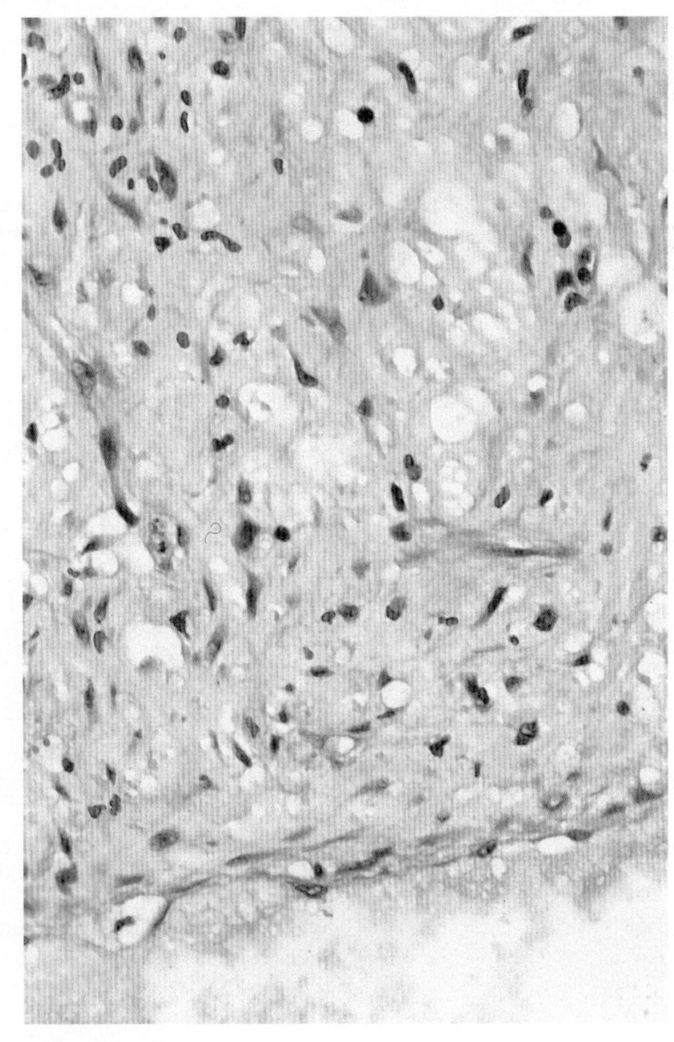

Figure 5–14 F

ARTERIOSCLEROSIS

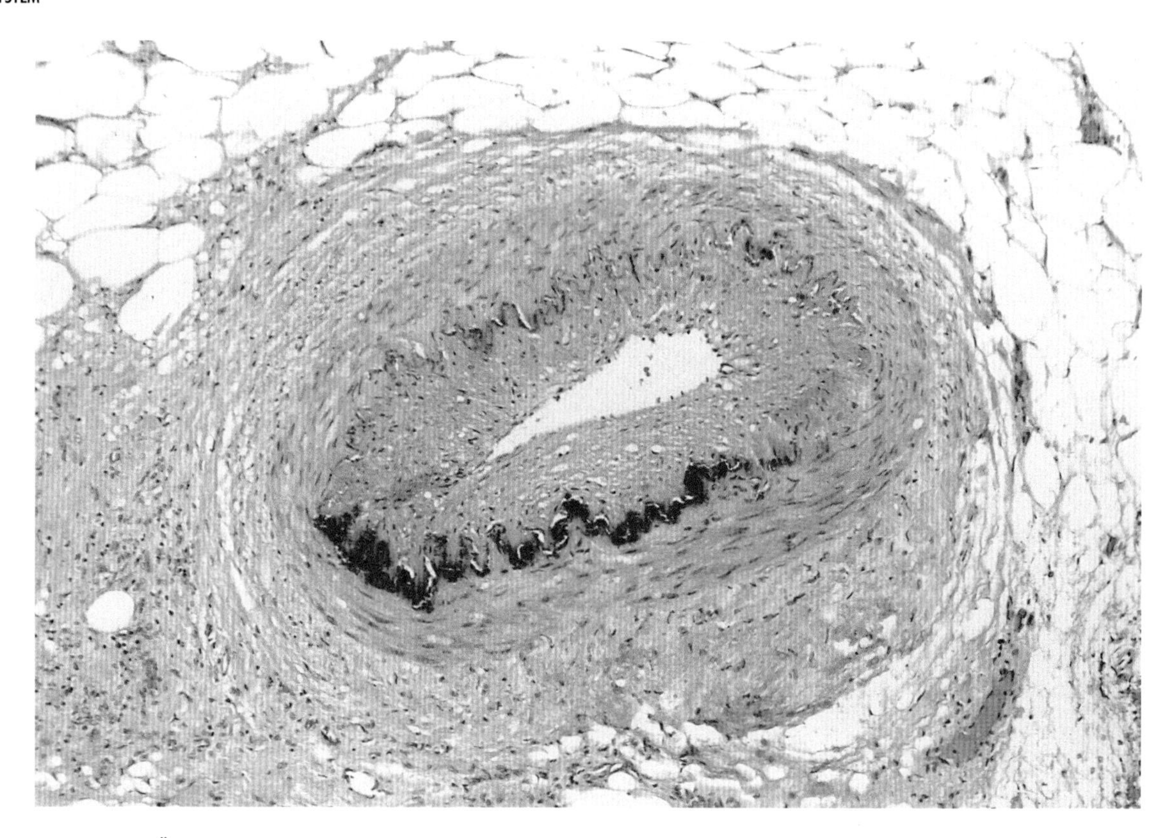

FIGURE 5–15. MÖNCKEBERG'S MEDIAL CALCIFIC SCLEROSIS: In Mönckeberg's medial calcific sclerosis, medium-sized muscular arteries have plates or rings of calcium deposited in the tunica media.

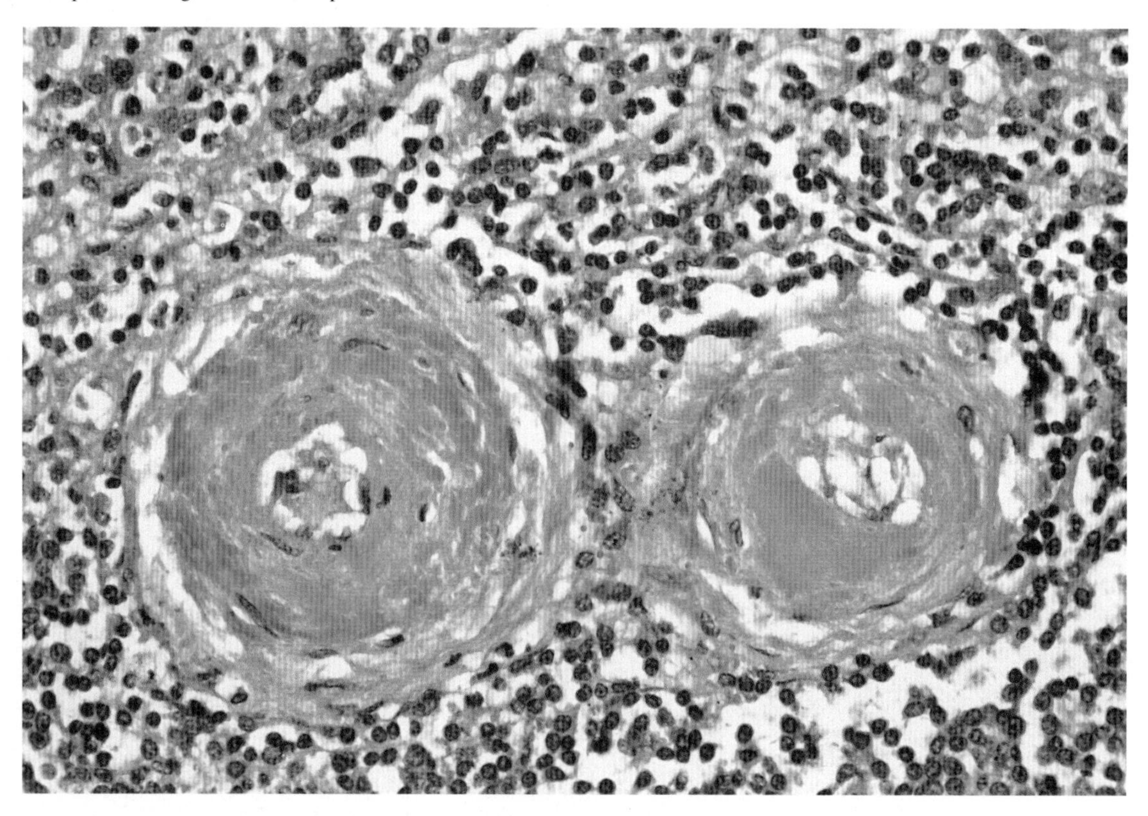

FIGURE 5–16. HYALINE ARTERIOLOSCLEROSIS: In hyaline arteriolosclerosis, the arteriolar walls are thickened by deposition of an amorphous eosinophilic staining material.

AGING

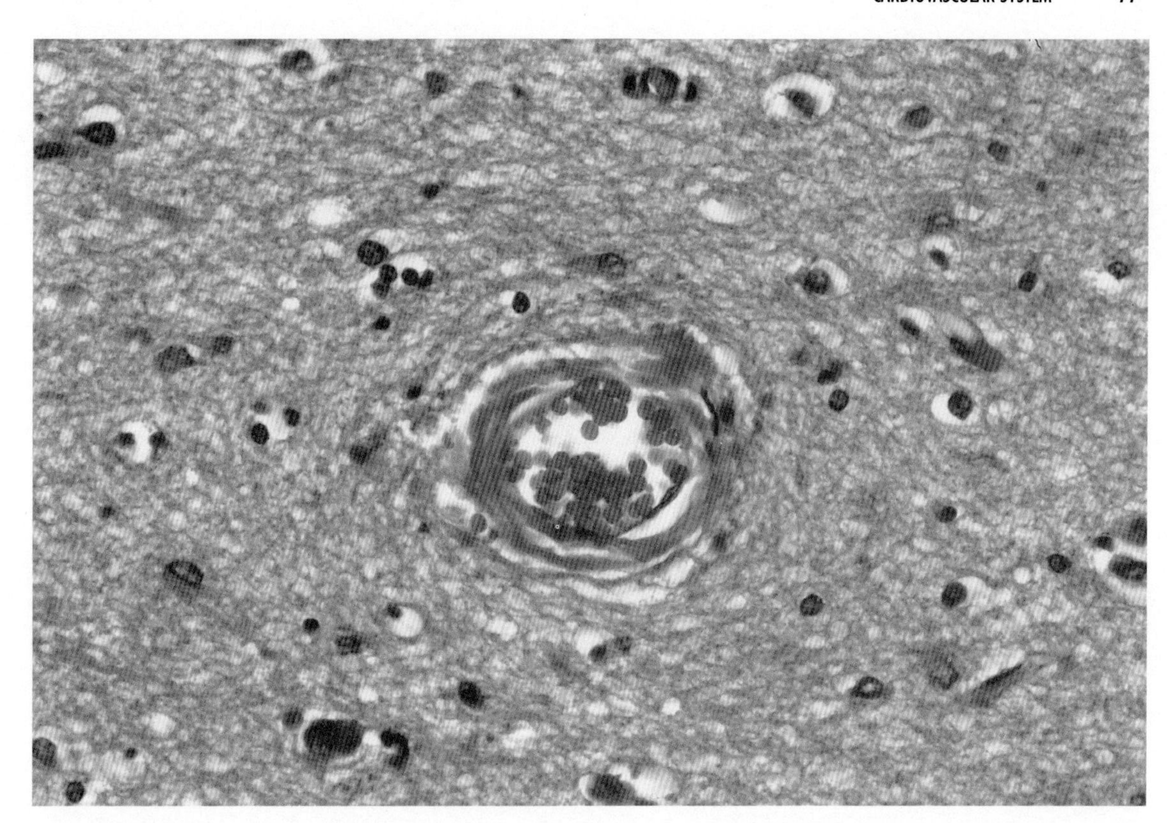

Figure 5–17 A

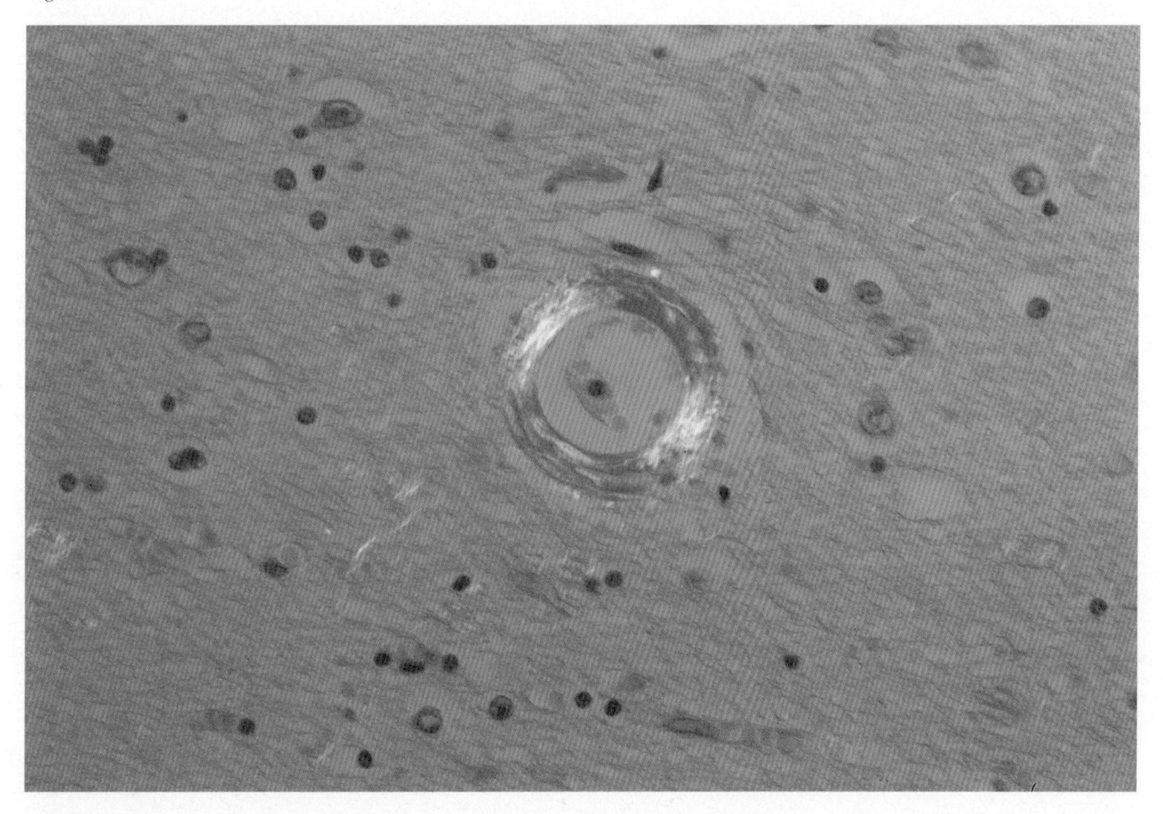

Figure 5–17 B

FIGURE 5–17. AMYLOID ANGIOPATHY: A: A change seen as part of the aging process is deposition of amyloid, an acellular eosinophilic staining material, within the vessel wall—similar to the deposition seen in hyaline arteriosclerosis (Figure 5–16). **B:** However, the amyloid nature of the material in amyloid angiopathy is demonstrated using a Congo red stain. Viewed with polarized light, the amyloid exhibits apple green birefringence.

INFLAMMATORY LESIONS

FIGURE 5–18. HYPERSENSITIVITY VASCULITIS (LEUKOCYTOCLASTIC VASCULITIS): Leukocytoclastic vasculitis is characterized by neutrophilic infiltration within and adjacent to the walls of arterioles, venules, or capillaries. **Leukocytoclasis,** fragmentation of neutrophilic nuclei, is a prominent feature.

FIGURE 5–19. WEGENER'S GRANULOMATOSIS: A: Wegener's granulomatosis is a necrotizing angiitis. **B:** Granulomatous inflammation with giant cells is present in an extravascular location.

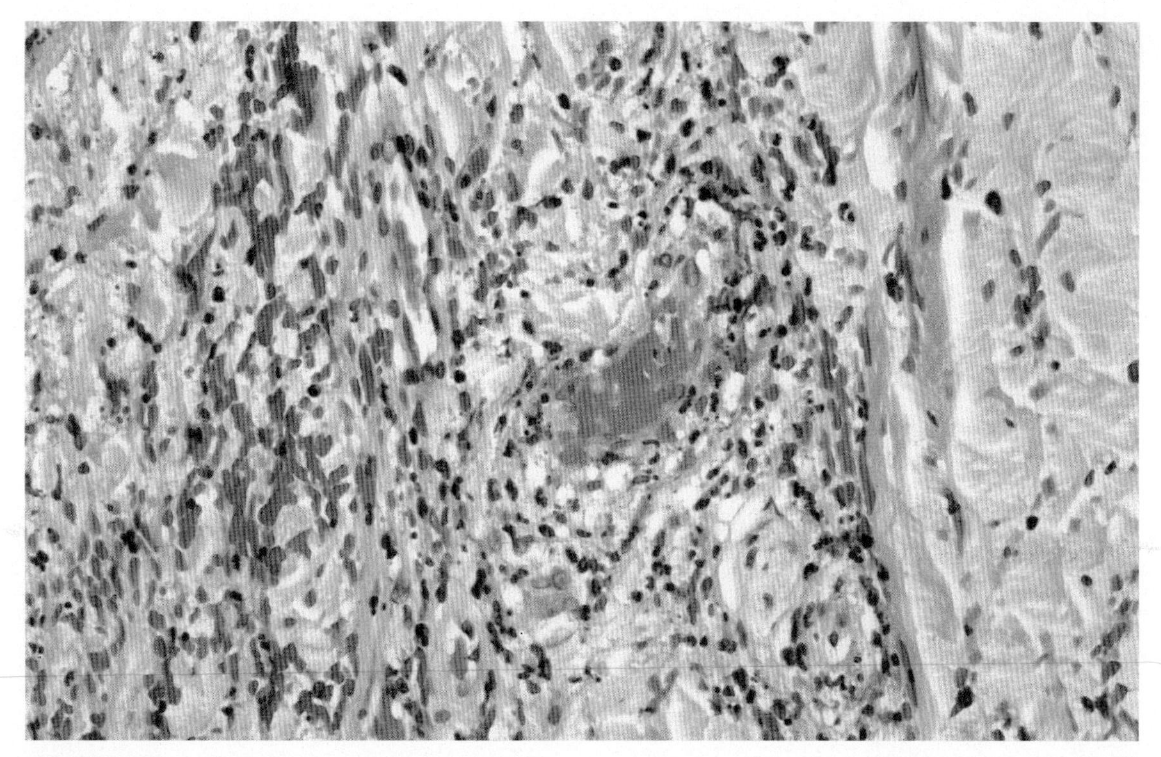

Figure 5–18

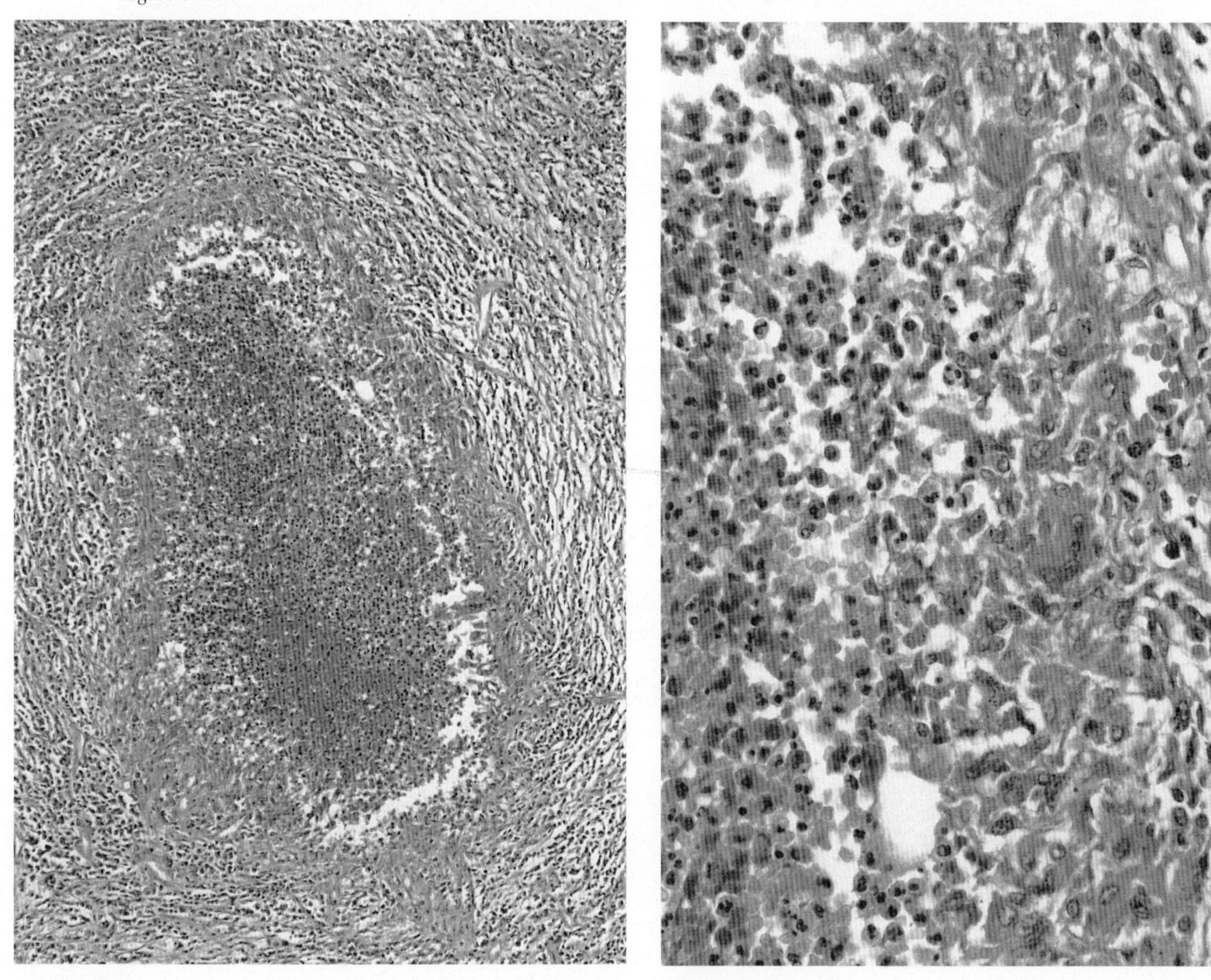

Figure 5–19 A

Figure 5–19 B

INFLAMMATORY LESIONS

FIGURE 5–20. CHURG-STRAUSS SYNDROME (ALLERGIC ANGIITIS AND GRANULOMATOSIS): In addition to the histologic findings of Wegener's granulomatosis (Figure 5–19 **A** and **B**), in Churg-Strauss syndrome, the inflammatory component is predominantly eosinophils.

FIGURE 5–21. TEMPORAL ARTERITIS (GIANT CELL ARTERITIS): A: In this photomicrograph of giant cell arteritis, the lumen of the blood vessel is completely obliterated. **B:** The inner half of the blood vessel wall is infiltrated by mononuclear cells and giant cells.

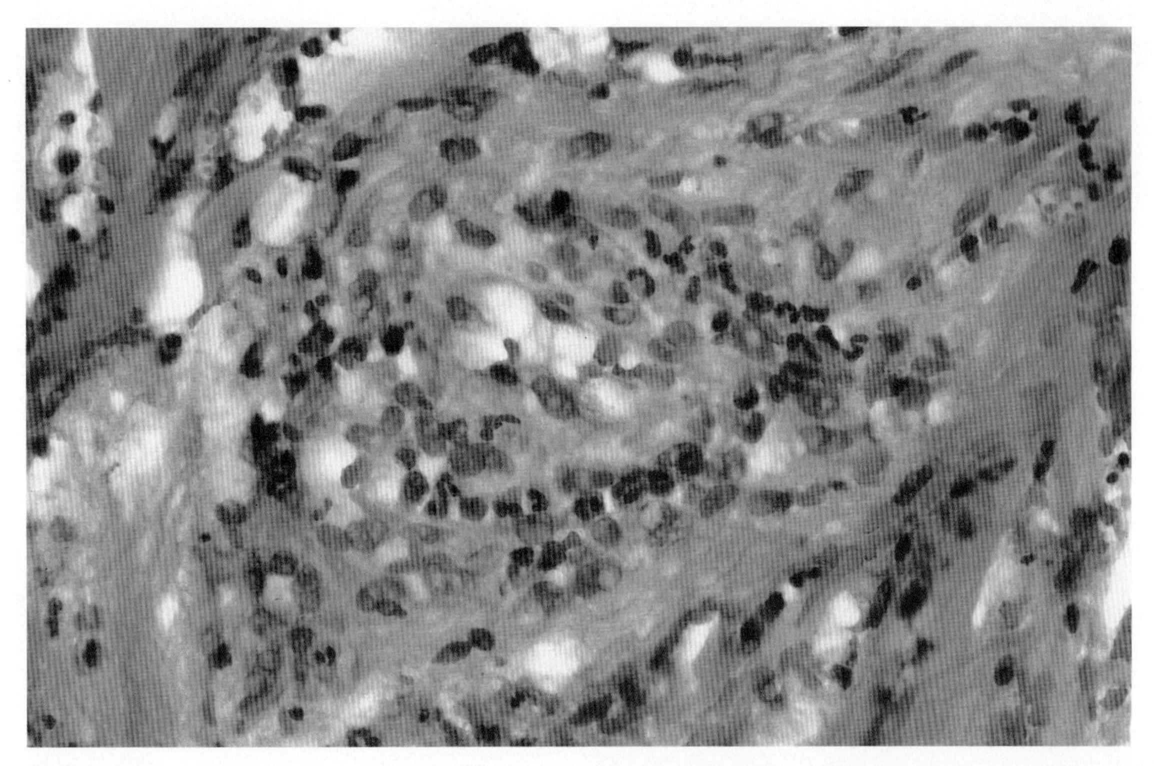

Figure 5–20

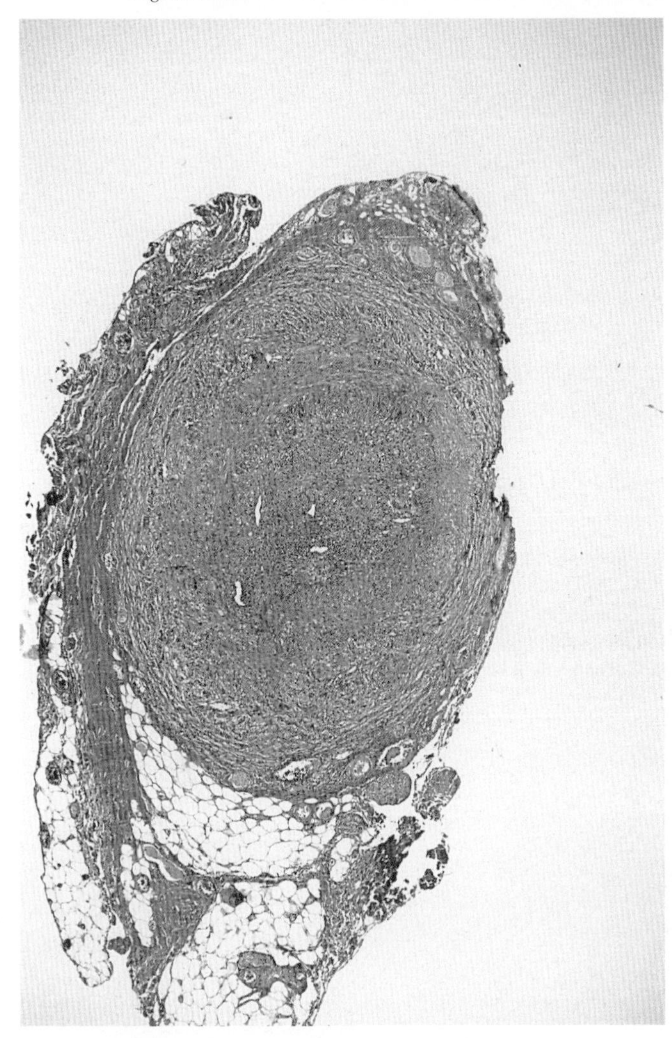

Figure 5–21 A

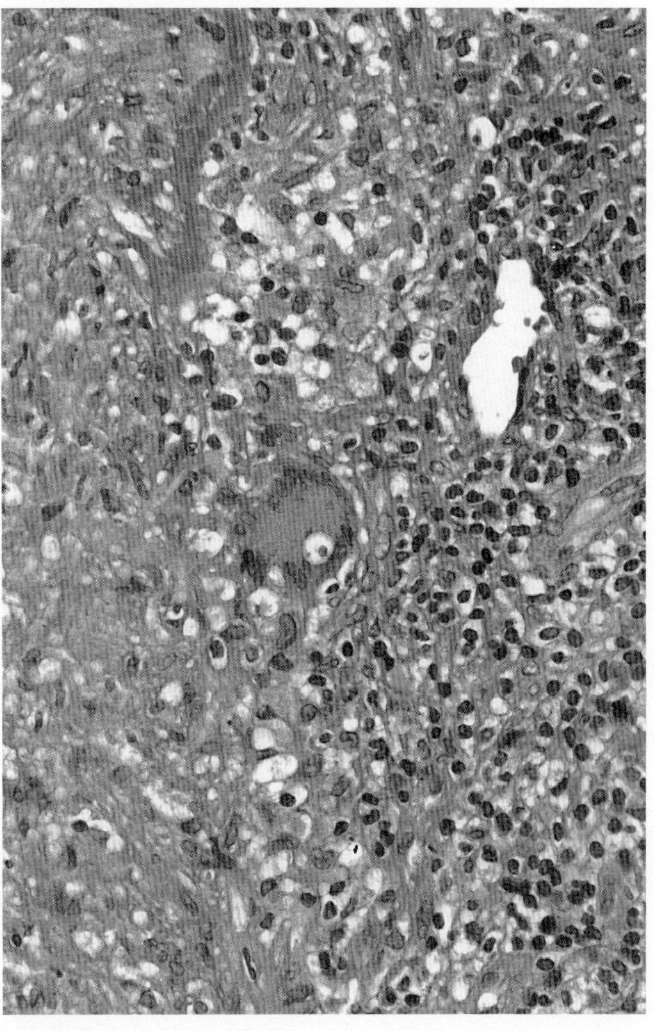

Figure 5–21 B

INFLAMMATORY LESIONS

FIGURE 5–22. POLYARTERITIS NODOSA: Polyarteritis nodosa is characterized by a transmural acute necrotizing arteritis with central fibrinoid necrosis and an infiltrate composed of neutrophils and eosinophils.

FIGURE 5–23. THROMBOANGIITIS OBLITERANS (BUERGER'S DISEASE): A: In the acute phase of Buerger's disease, the blood vessel lumen is obliterated by a thrombus and an associated abscess. **B:** A later stage shows organization and recanalization of the vessel lumen.

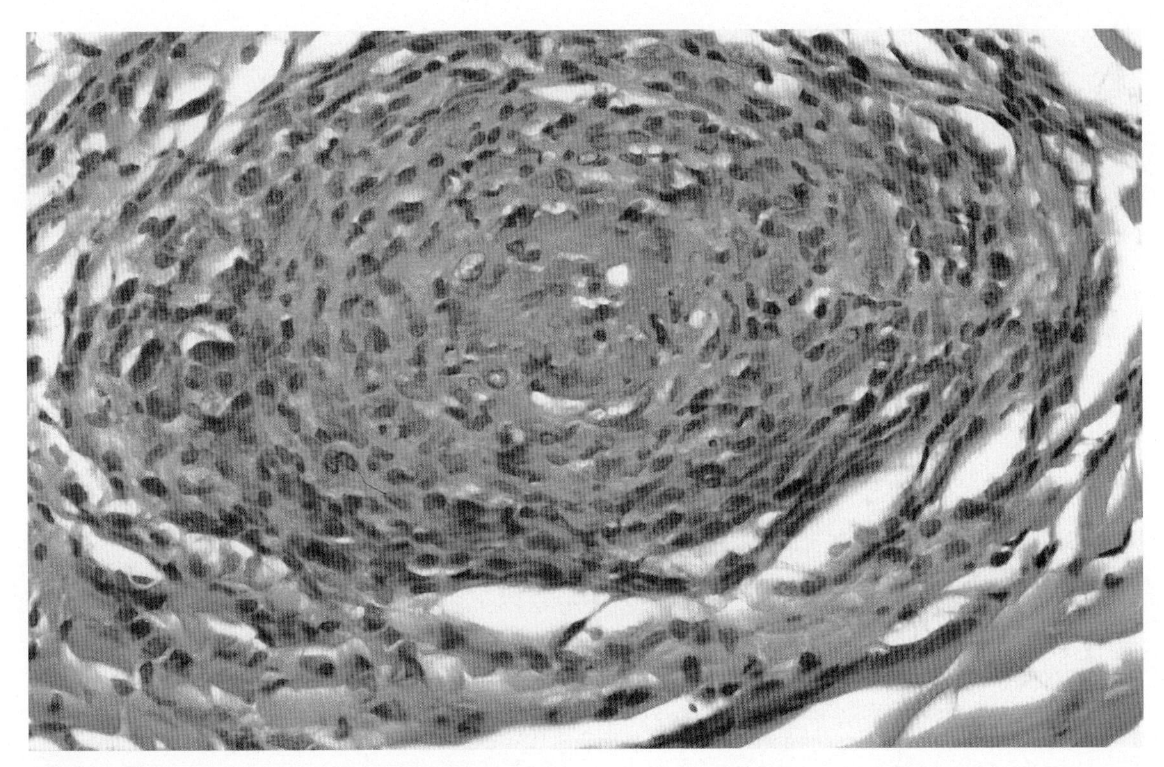

Figure 5–22

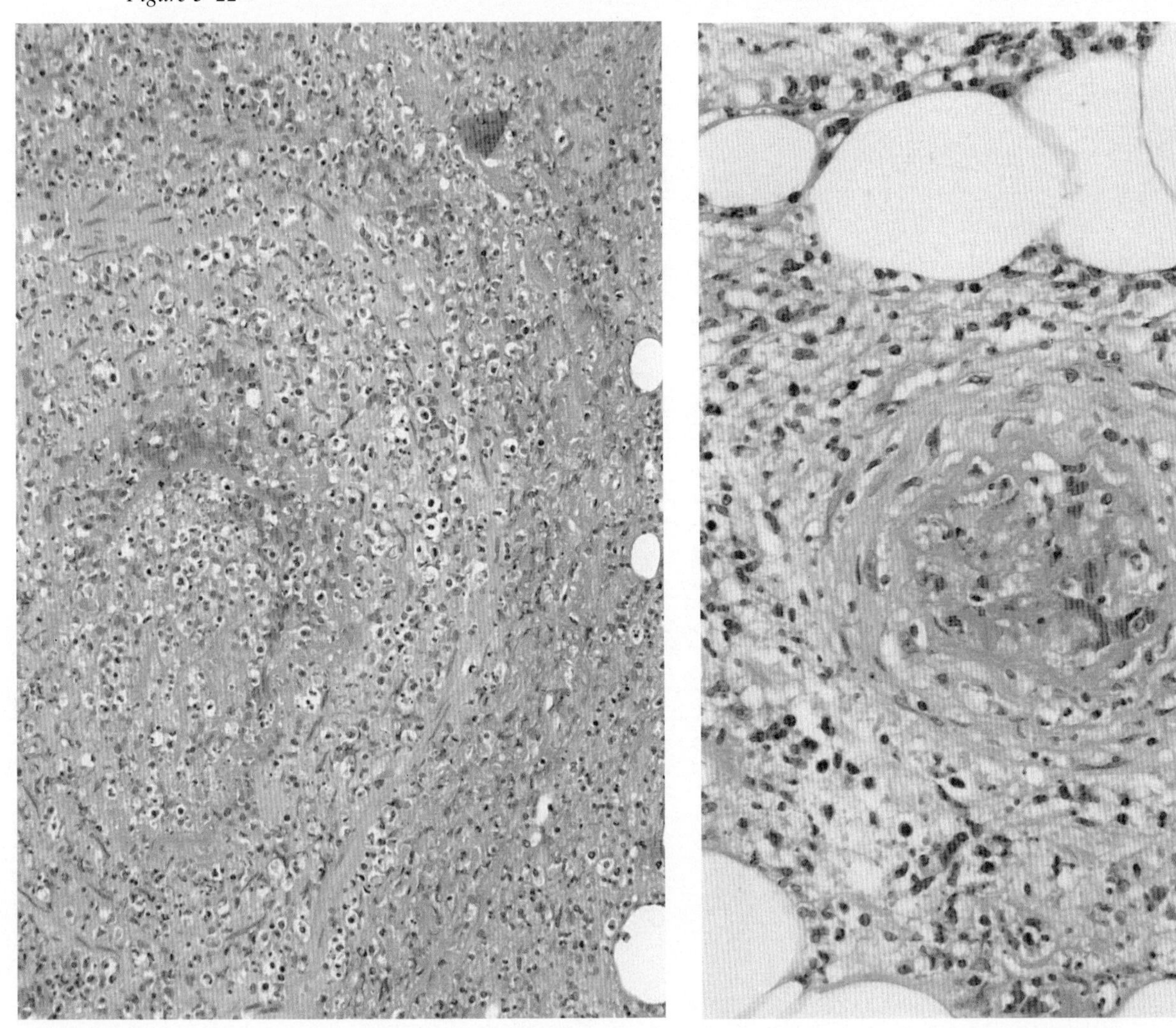

Figure 5–23 A

Figure 5–23 B

ANEURYSMS

FIGURE 5–24. MARFAN'S SYNDROME (CYSTIC MEDIAL DEGENERATION): A: In cystic medial degeneration, the tunica media is focally rarefied and filled with an amorphous basophilic substance. **B:** The basophilic material is an acid mucopolysaccharide, as shown with alcian blue staining. **C:** An elastic tissue stain shows the focal disruption of the elastic fibers in the tunica media.

Figure 5–24 A

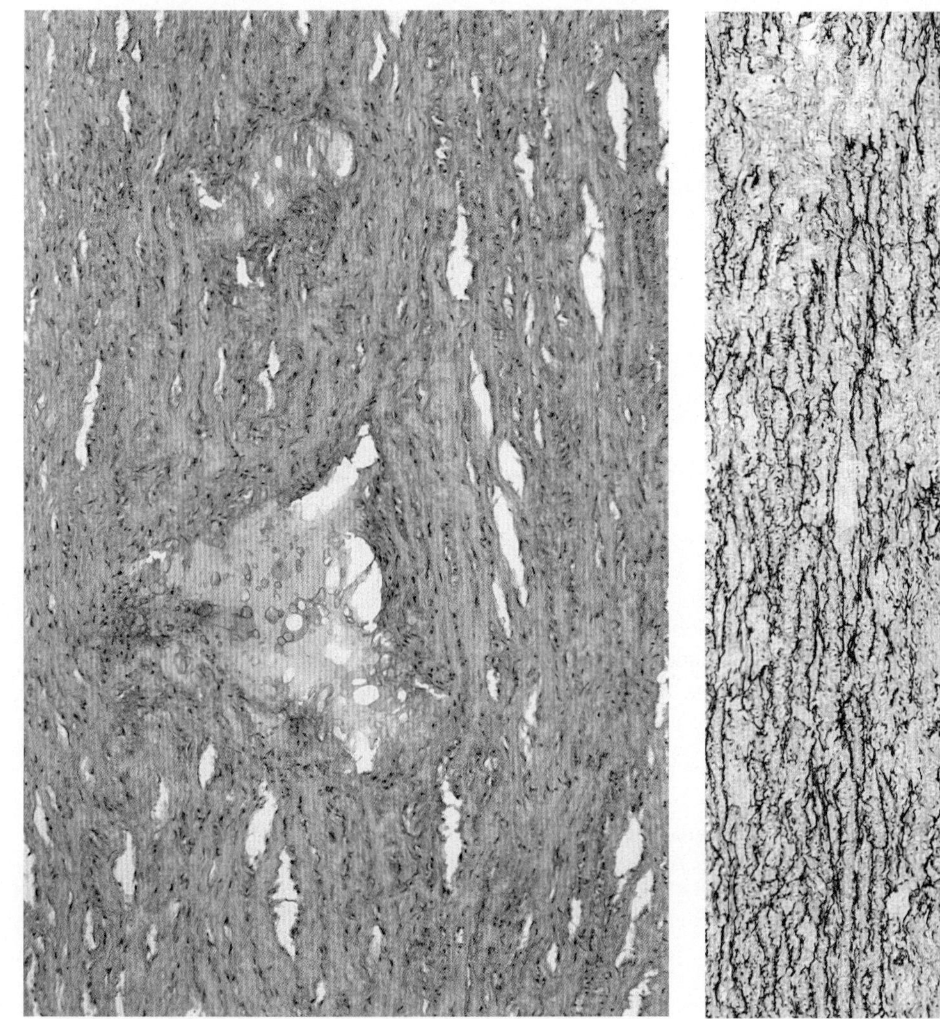

Figure 5–24 B

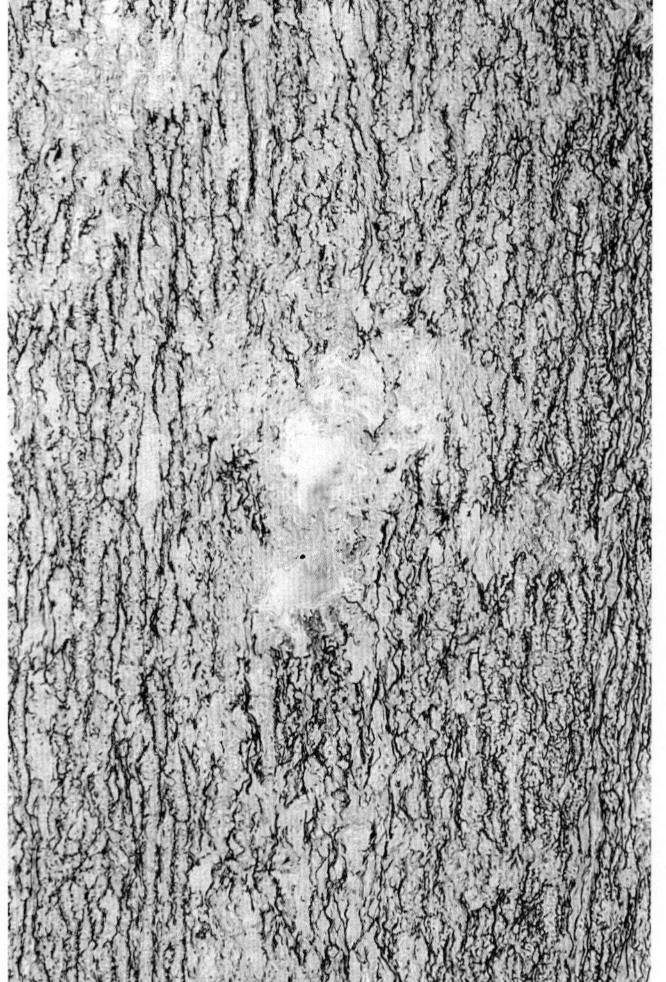

Figure 5–24 C

ANEURYSMS

FIGURE 5–25. DISSECTING ANEURYSM: Dissecting aneurysms can be seen in Marfan's syndrome, although, in most cases, the arteries and aorta appear normal. Blood enters the defect in the tunica media and splits it longitudinally.

FIGURE 5–26. SYPHILITIC AORTITIS (LUETIC AORTITIS): Syphilitic aortitis is characterized by a perivascular inflammation involving the vasa vasorum. The inflammatory infiltrate is lymphoplasmacytic in nature.

PHLEBOTHROMBOSIS

FIGURE 5–27. THROMBOPHLEBITIS: In this example of a venous thrombosis, the clot is adherent to the blood vessel wall. The thrombus is composed of alternating layers of red blood cells, platelets, and granulocytes.

Figure 5–25

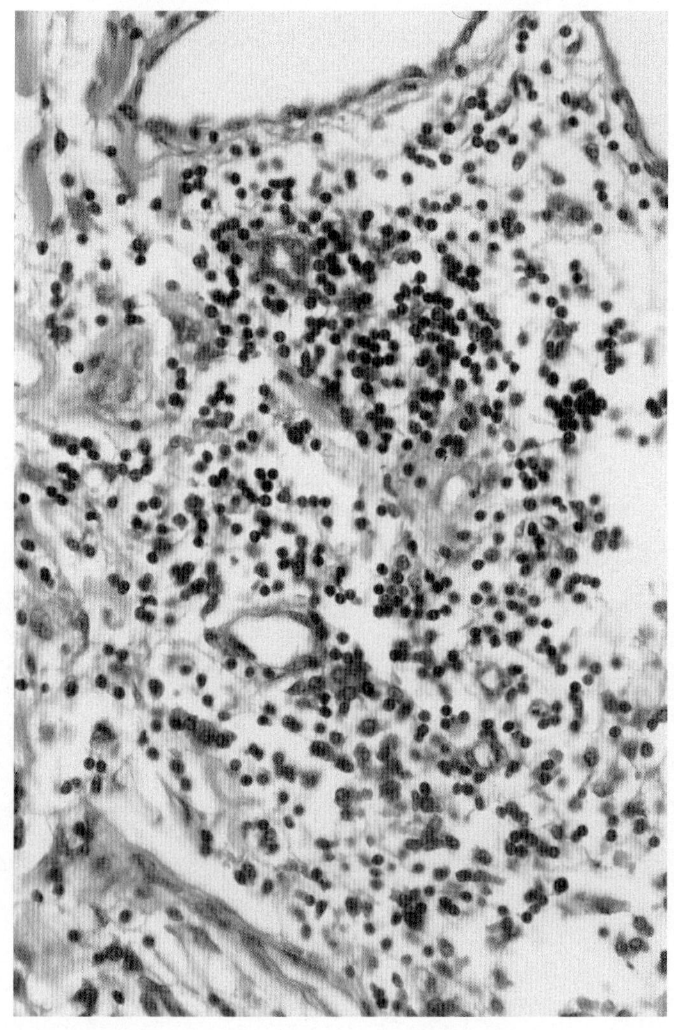

Figure 5–26

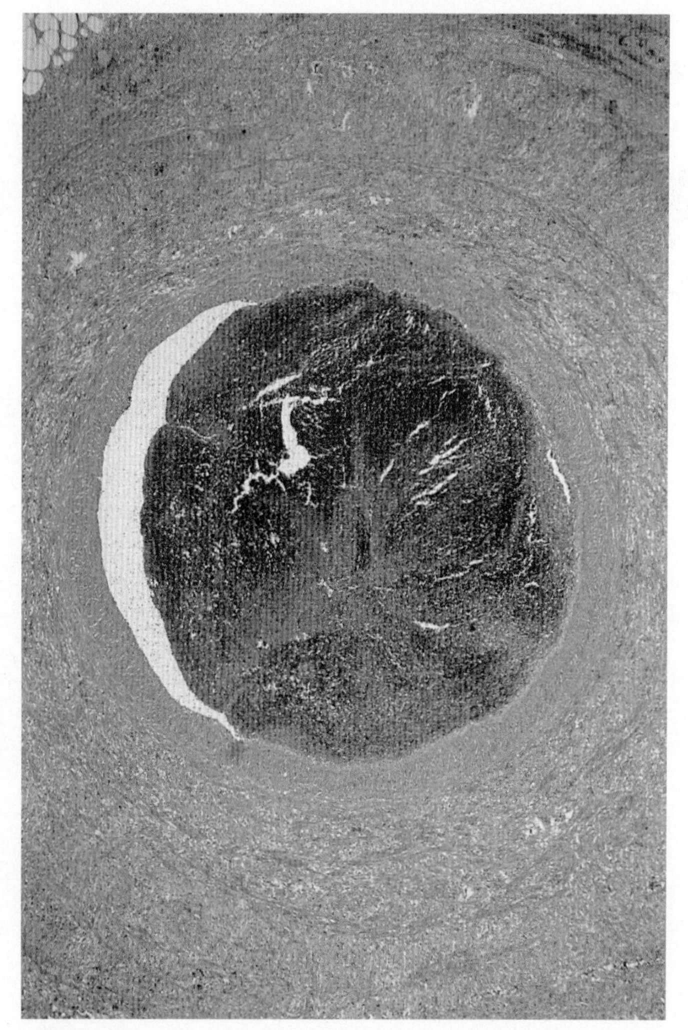

Figure 5–27

BENIGN VASCULAR TUMORS

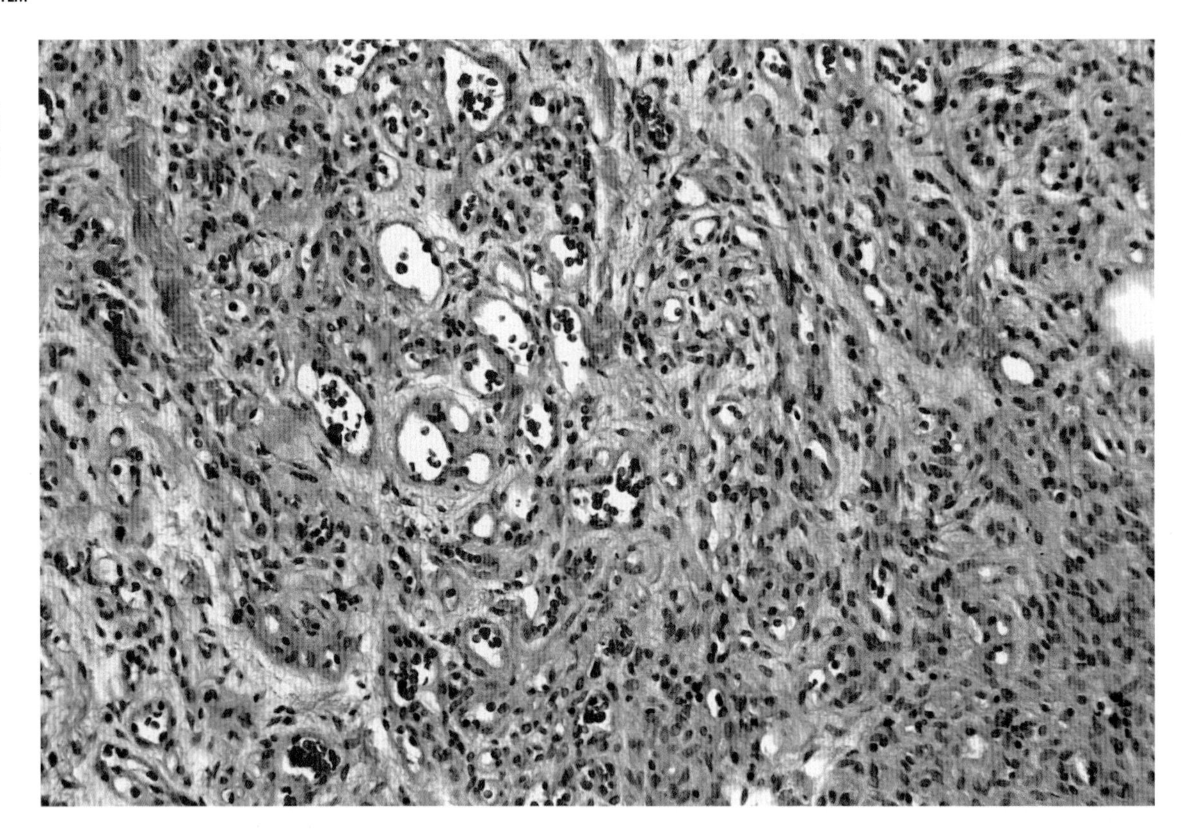

FIGURE 5–28. CAPILLARY HEMANGIOMA: A capillary hemangioma is characterized by a proliferation of capillaries with scant intervening stroma.

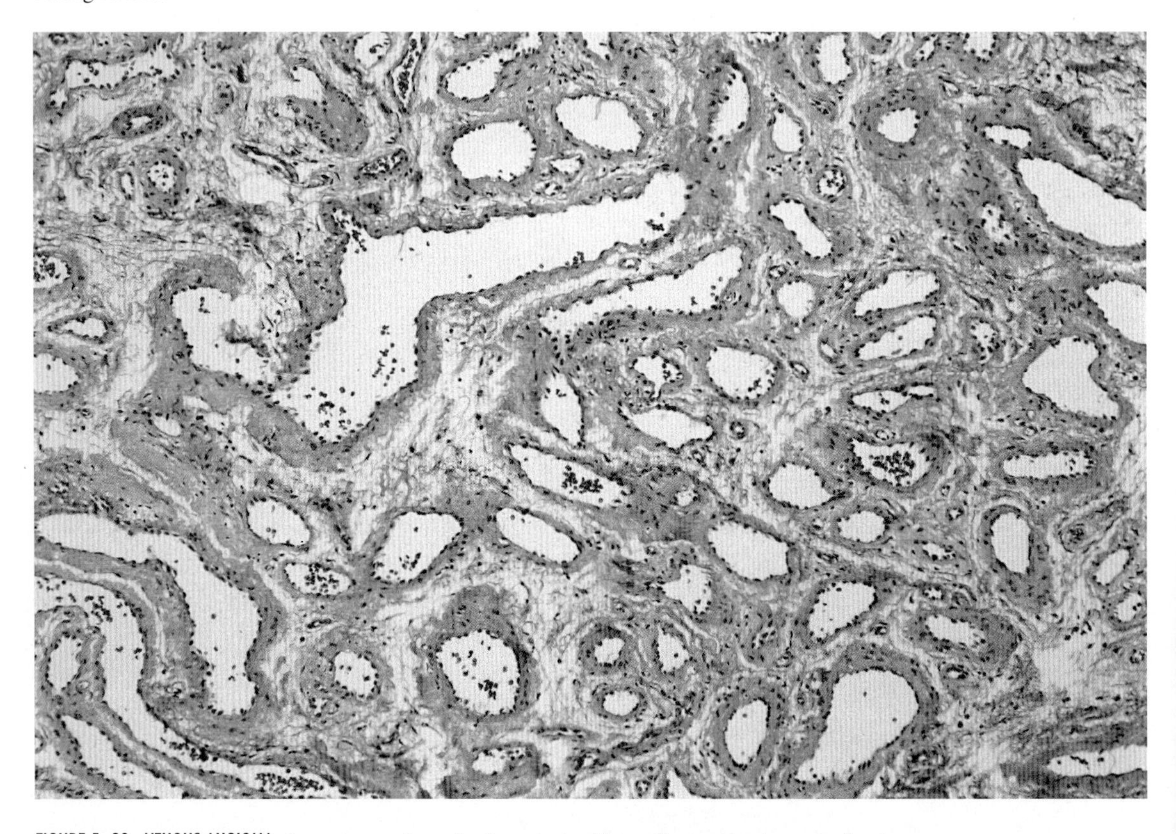

FIGURE 5–29. VENOUS ANGIOMA: A venous angioma is characterized by a disorganized group of veins.

BENIGN VASCULAR TUMORS

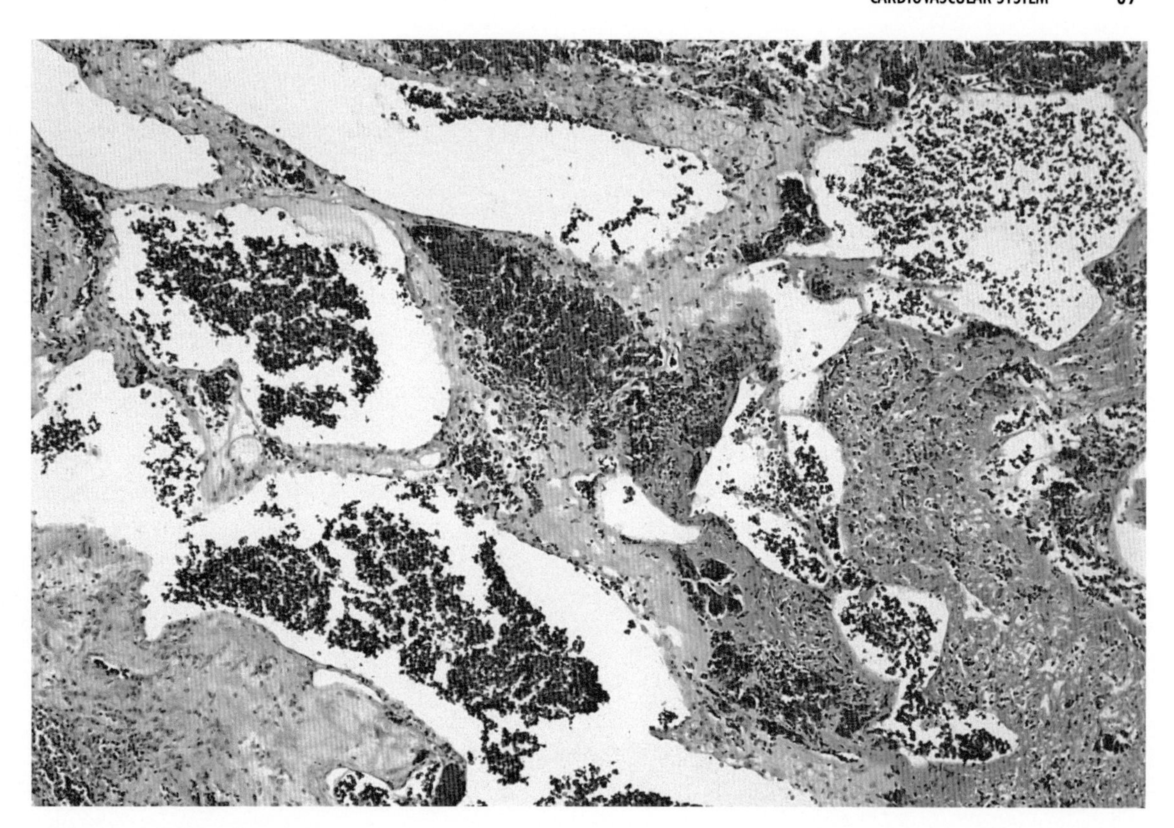

FIGURE 5–30. CAVERNOUS HEMANGIOMA: A cavernous hemangioma is characterized by large cavernous blood-filled spaces.

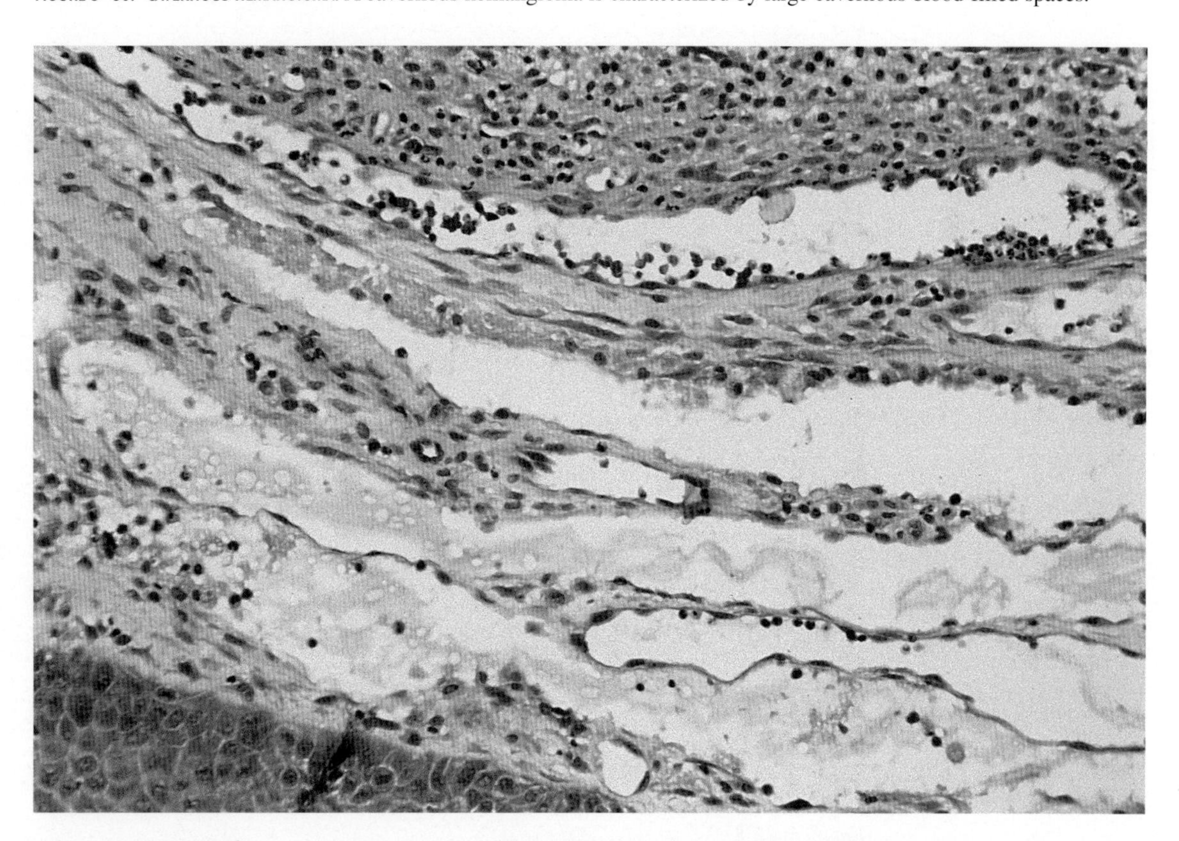

FIGURE 5–31. LYMPHANGIOMA: A lymphangioma is characterized by a proliferation of lymphatic vessels. The vessels are lined by an attenuated endothelium and filled with lymph, lymphocytes, and few, if any, red blood cells.

BENIGN VASCULAR TUMORS

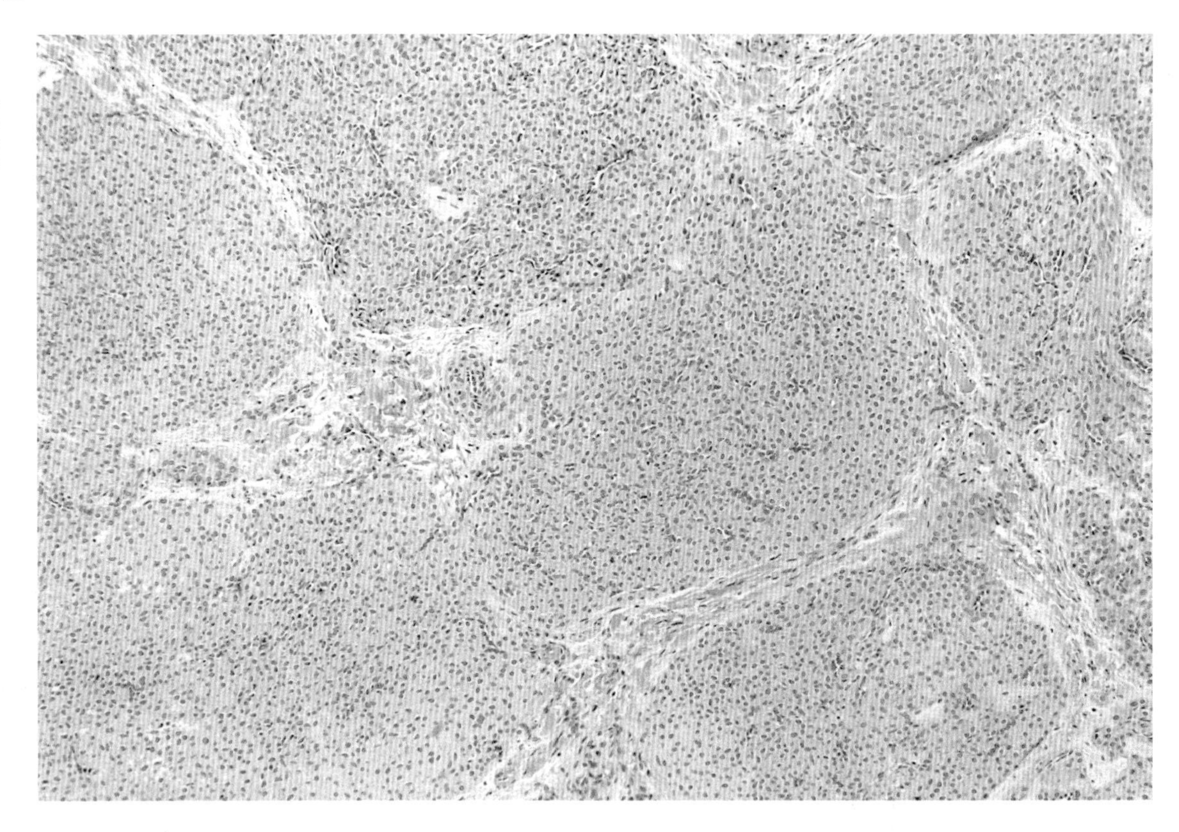

Figure 5–32 **A**

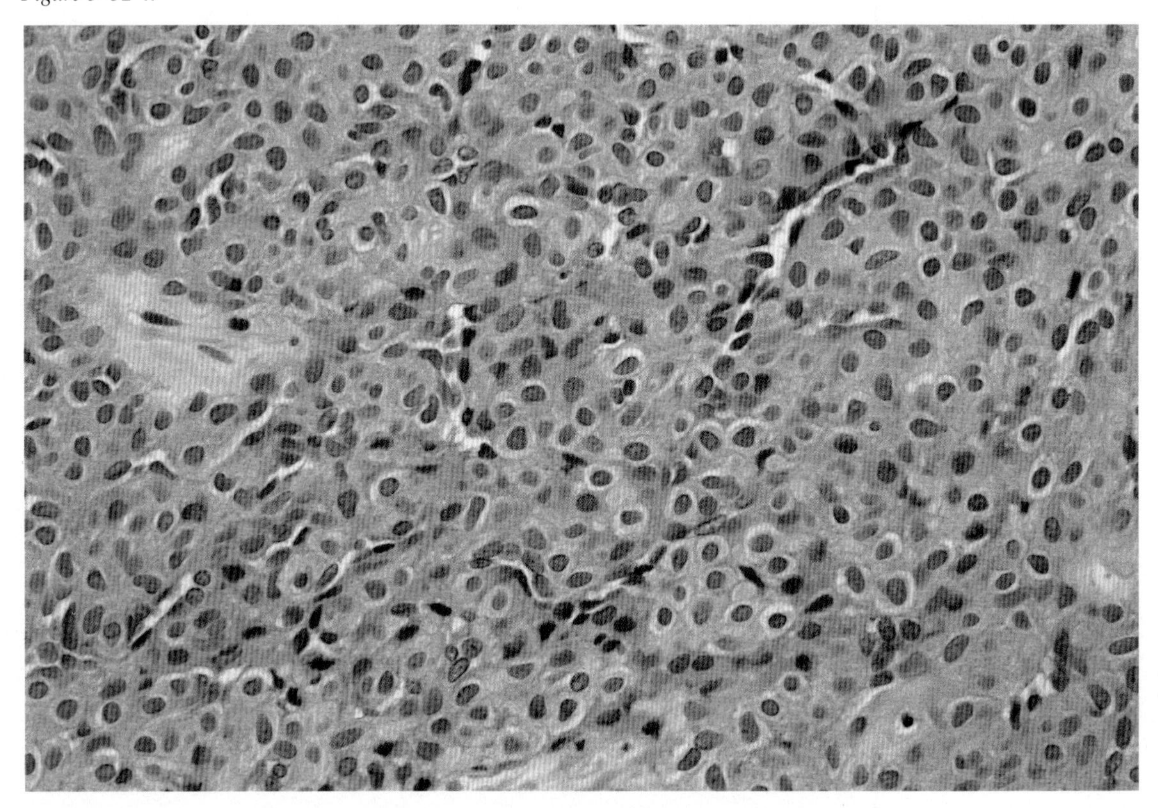

Figure 5–32 **B**

FIGURE 5–32. GLOMANGIOMA (GLOMUS TUMOR): A: A glomus tumor is derived from modified smooth muscle cells of a glomus body, with the cells arranged in nests. **B:** The glomus cells are uniform and polyhedral with centrally located nuclei, and the intervening stroma contains branching thick-walled vascular channels.

VASCULAR TUMORS OF INTERMEDIATE MALIGNANT POTENTIAL

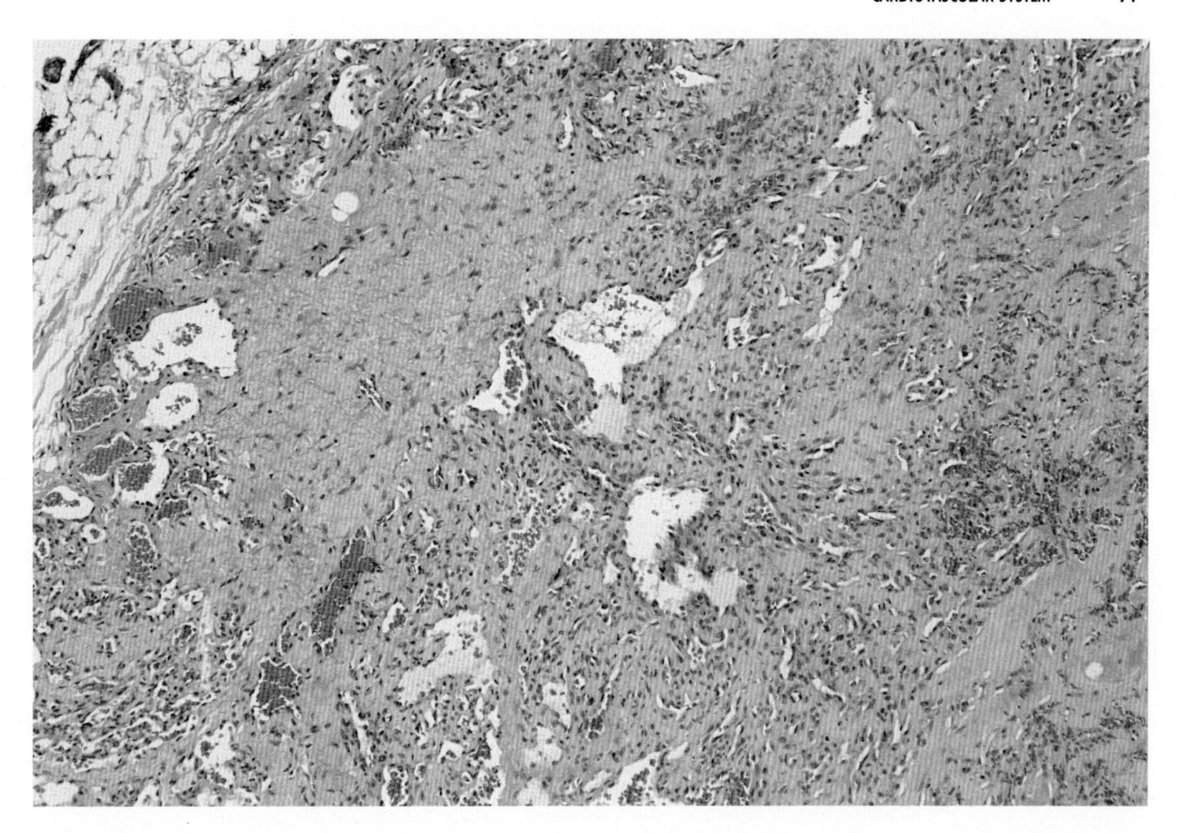

Figure 5–33 **A**

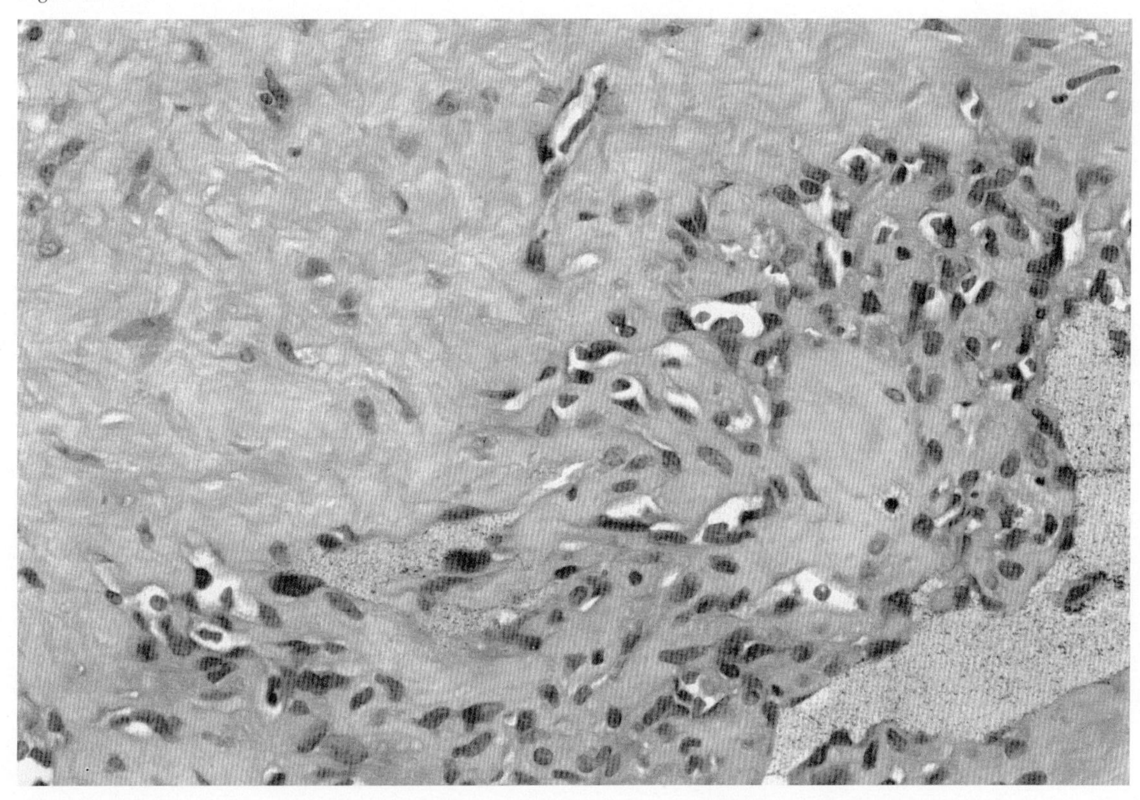

Figure 5–33 **B**

FIGURE 5–33. HEMANGIOENDOTHELIOMA: A hemangioendothelioma is a vascular tumor of intermediate malignant potential. **A:** It is characterized by vascular channels within a fibrocartilaginous matrix. **B:** The endothelial cells lining the vascular spaces appear normal.

MALIGNANT VASCULAR TUMORS

FIGURE 5–34. ANGIOSARCOMA: A: An angiosarcoma is characterized by a proliferation of anaplastic spindle cells within which are poorly formed vascular channels. **B:** The cells lining the vascular channels are enlarged with hyperchromatic nuclei that bulge into the lumina, giving it the characteristic **"hobnail" lining.** Compare these endothelial cells with those of the hemangioendothelioma (Figure 5–33). **C:** Individual cells forming vascular lumina in which red blood cells are present can also be seen.

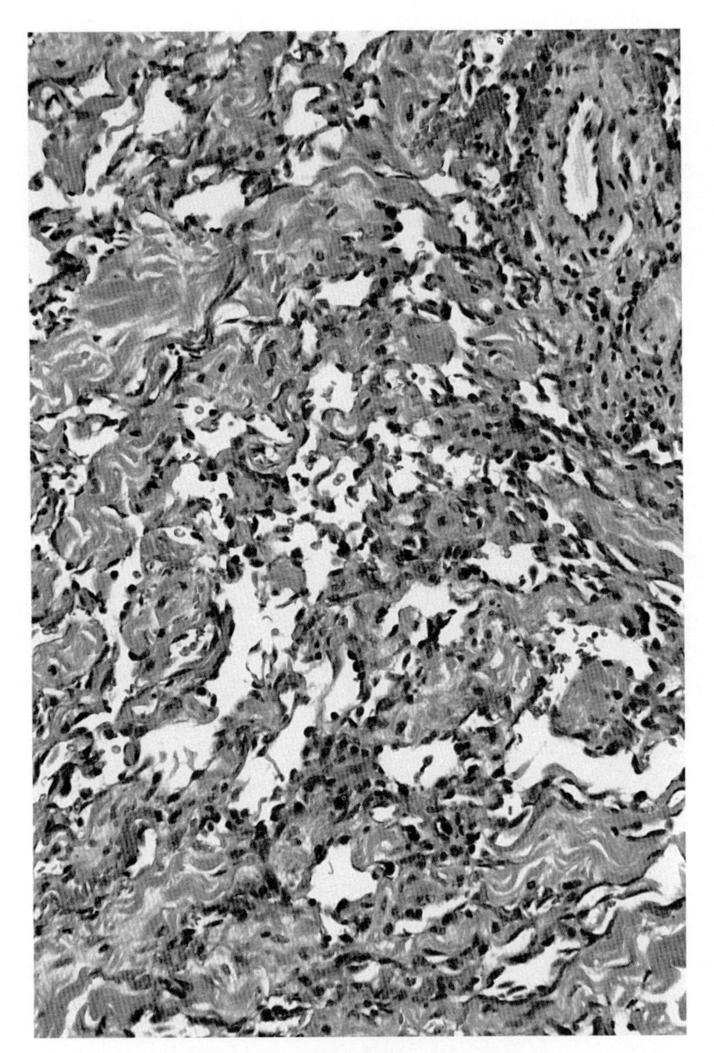

Figure 5–34 A

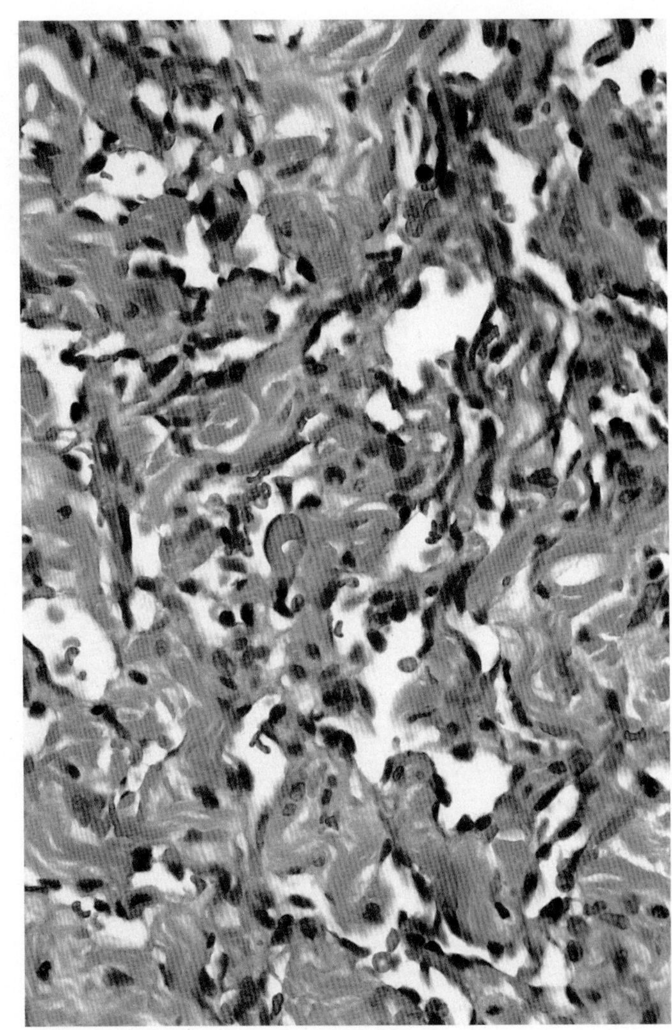

Figure 5–34 B

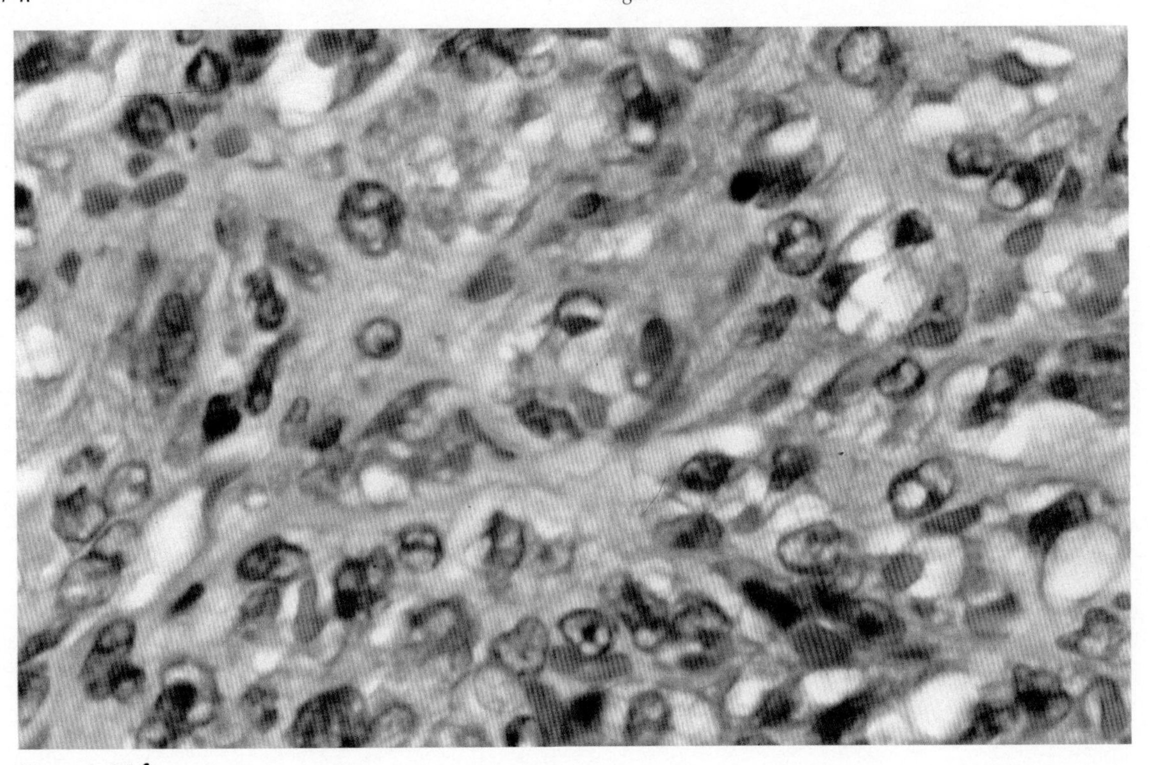

Figure 5–34 C

■ RED BLOOD CELL DISORDERS

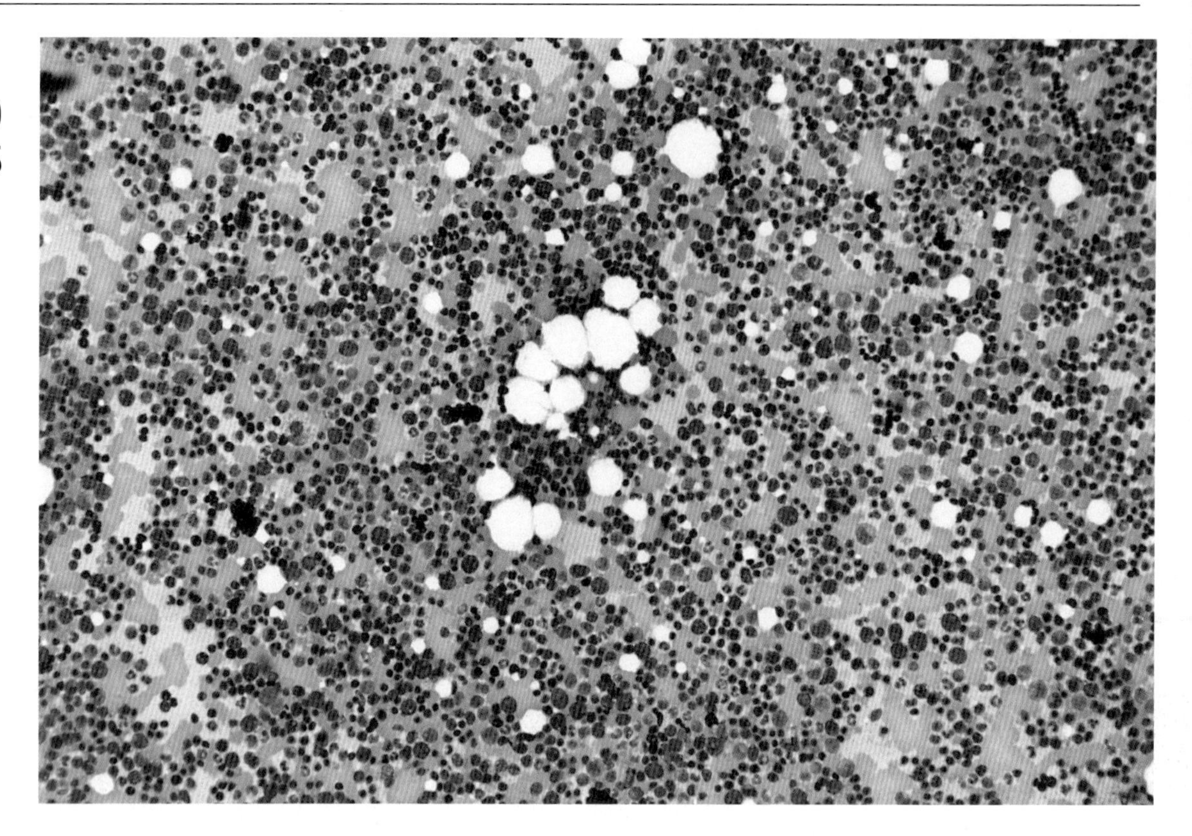

Figure 6–1 A

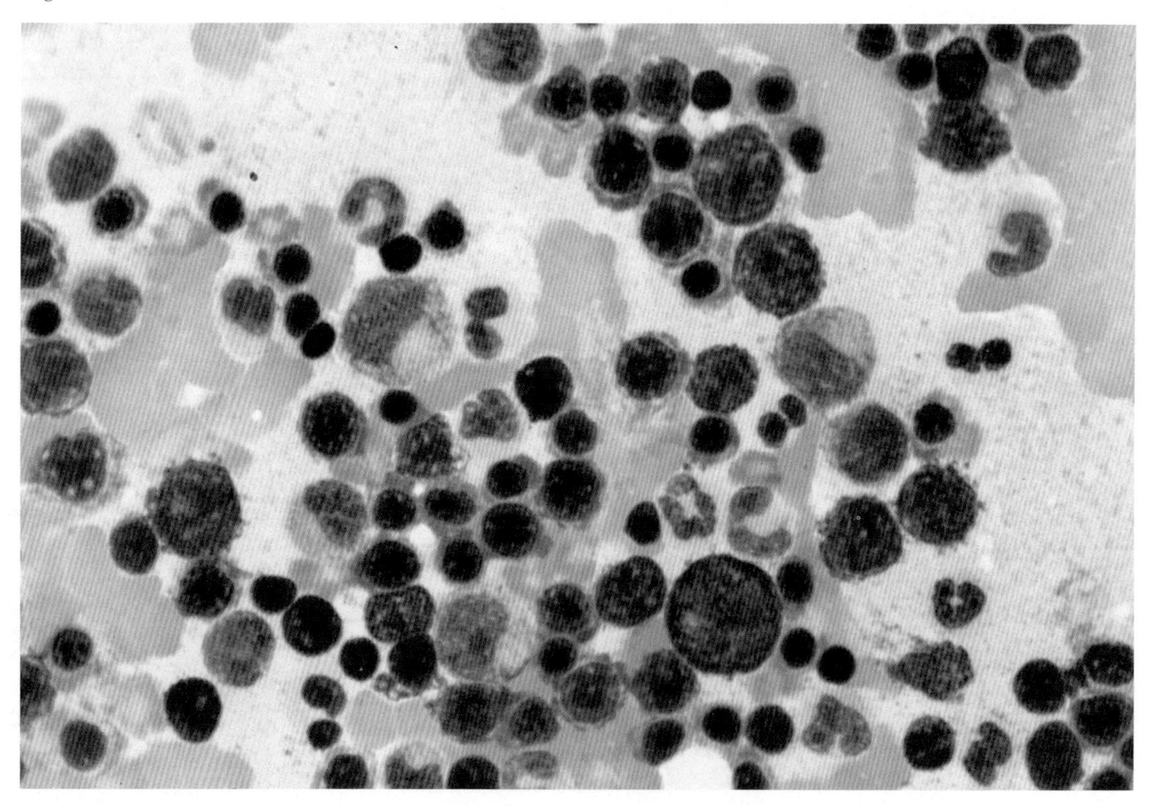

Figure 6–1 B

FIGURE 6–1. ANEMIA: In most cases of anemia, the bone marrow is hypercellular (**A**) and normoblasts are increased in number (**B**).

■ RED BLOOD CELL DISORDERS

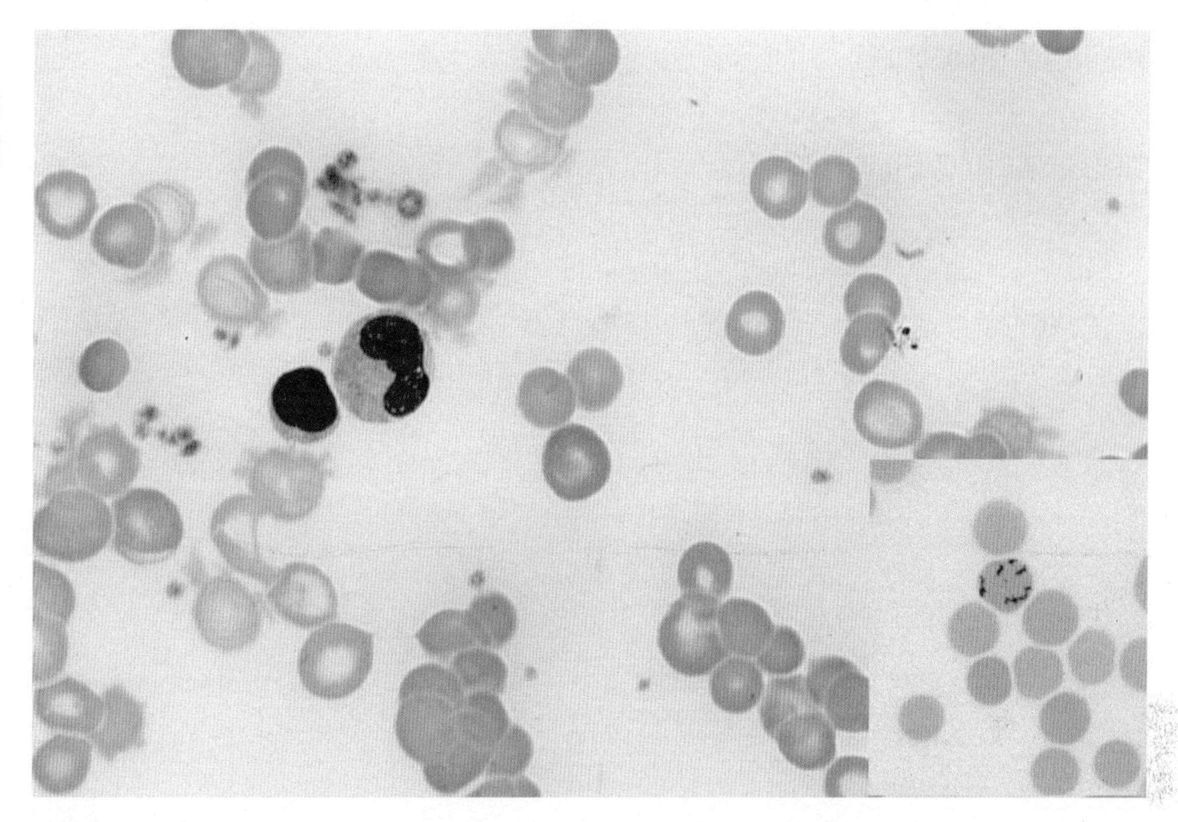

Figure 6–1 **C**

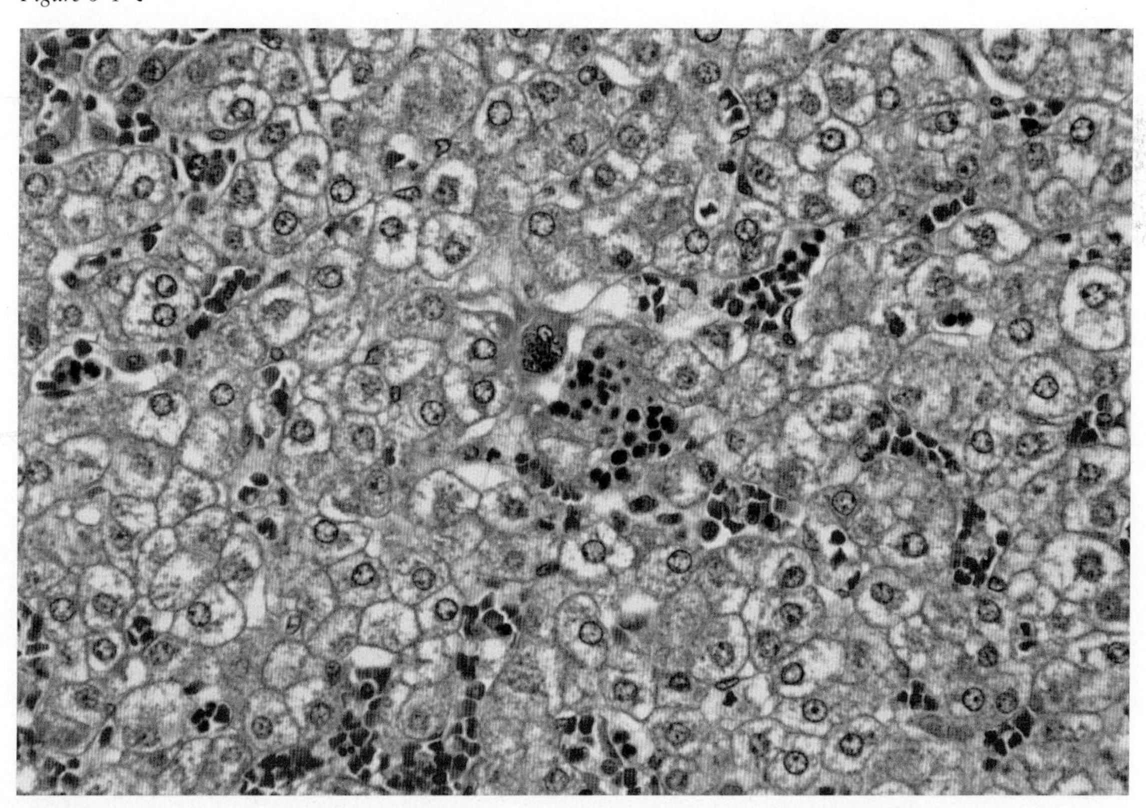

Figure 6–1 **D**

FIGURE 6–1. ANEMIA: C: In anemia, the peripheral blood smear shows an increase in **polychromatophils,** which are enlarged **macrocytic** cells that stain purple-gray with Wright stain and represent young red blood cells that have recently extruded their nuclei. When stained with a supravital dye (*inset*), such as new methylene blue B, residual fragments of cytoplasmic RNA stain purple-blue and the cells are called **reticulocytes. D:** When anemia is severe, compensatory extramedullary hematopoiesis occurs.

■ RED BLOOD CELL DISORDERS

INCREASED RED BLOOD CELL DESTRUCTION

FIGURE 6–2. HEREDITARY ELLIPTOCYTOSIS: This condition is characterized by red blood cells that have an elliptical configuration. The length of the red blood cell is twice its width.

FIGURE 6–3. SICKLE CELL DISEASE: Sickle cell disease is caused by an abnormal hemoglobin. In the homozygous state, **irreversibly sickled cells** are noted on the peripheral blood smear. These cells are canoe or sickle shaped. In addition, there is **anisocytosis,** variation in size, and **poikilocytosis,** variation in shape, of the other red blood cells. Nucleated red blood cells are also present.

FIGURE 6–4. THALASSEMIA: Thalassemia is caused by an excessive production of one of the globin chains. In peripheral blood smears, there is an increase in the number of **microcytic** red blood cells; **target cells,** which exhibit a dense zone of hemoglobin in the red blood cell center surrounded by a zone of pallor; **teardrop cells;** and **schistocytes,** fragmented red blood cells.

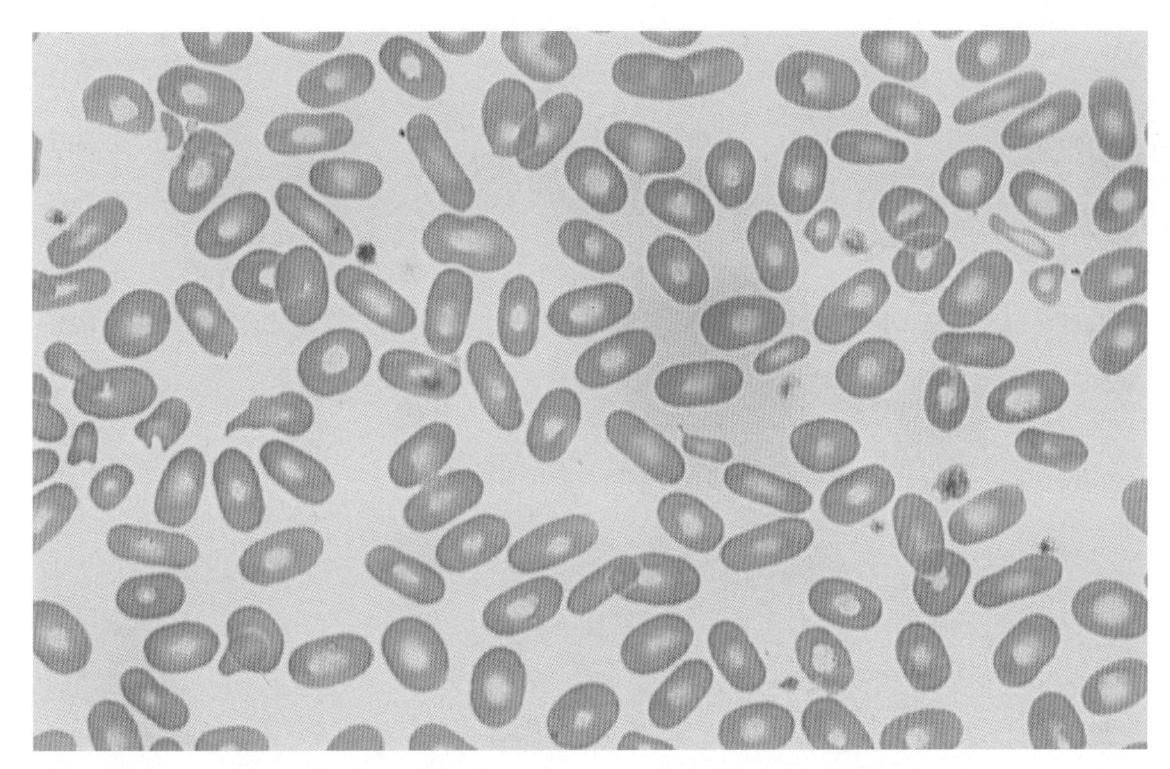

Figure 6–2

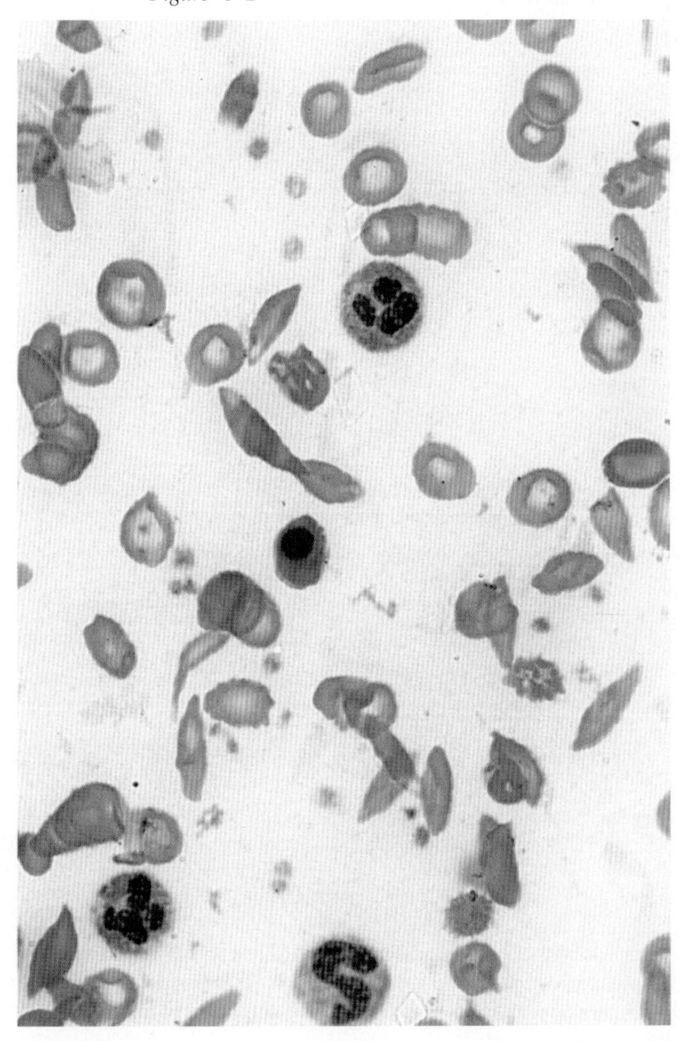

Figure 6–3

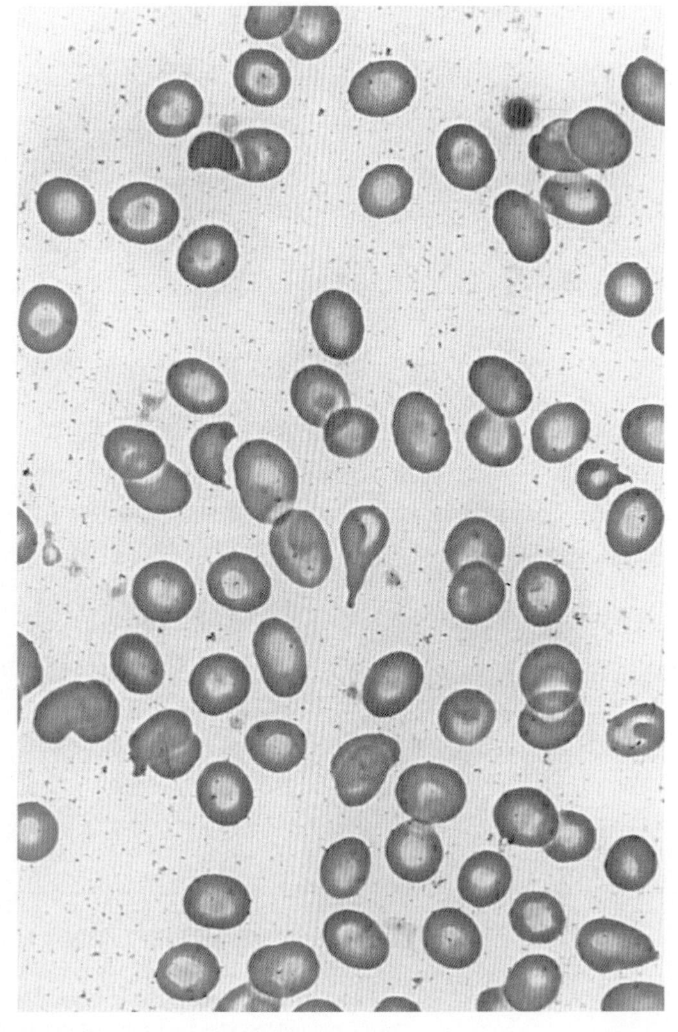

Figure 6–4

■ RED BLOOD CELL DISORDERS

INCREASED RED BLOOD CELL DESTRUCTION

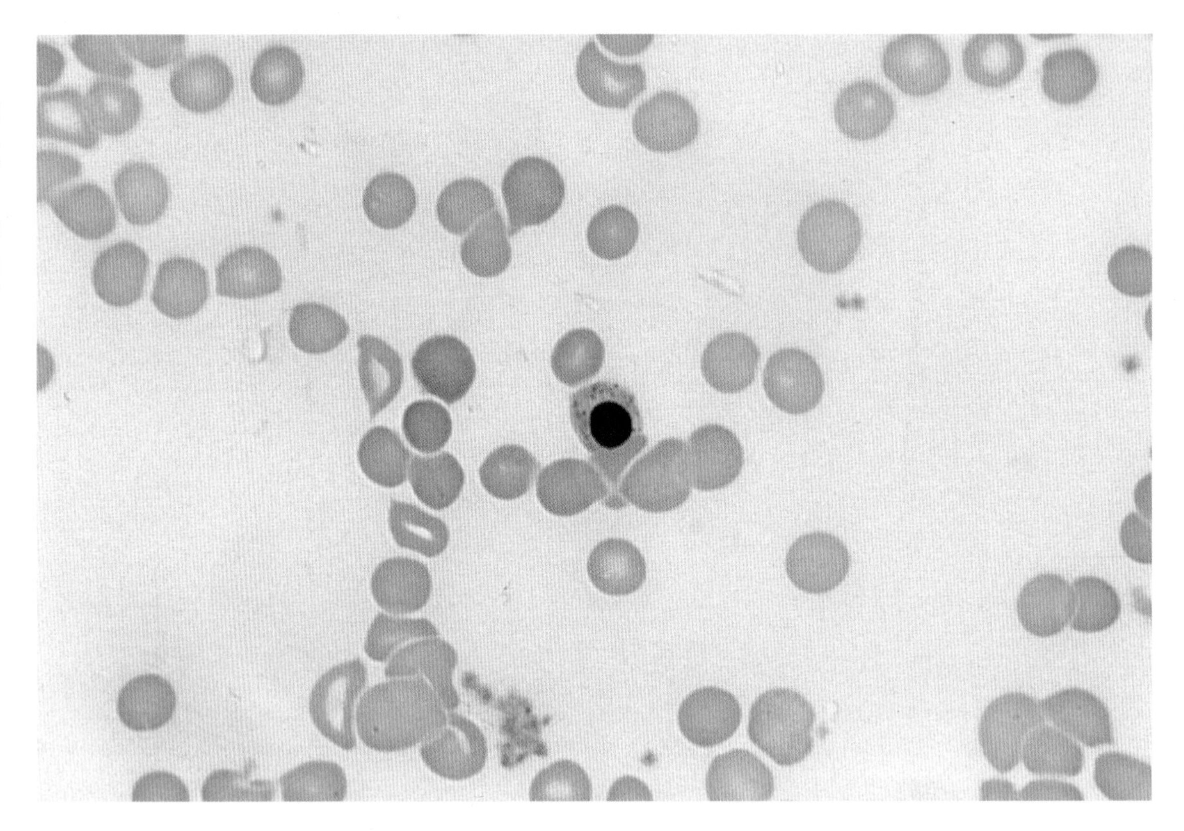

Figure 6–5 A

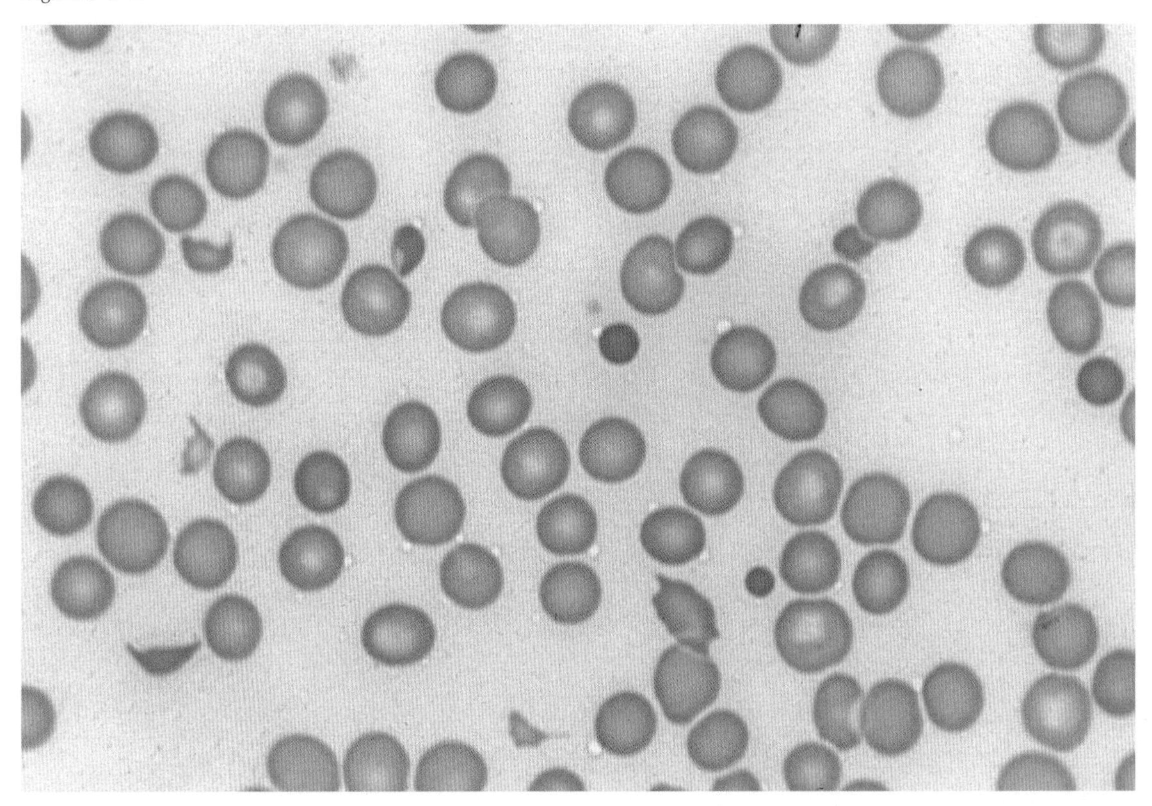

Figure 6–5 B

FIGURE 6–5. AUTOIMMUNE HEMOLYTIC ANEMIA: A: In autoimmune hemolytic anemia, the bone marrow reacts with an increase in the production of red blood cells, which results in the appearance of nucleated red blood cells in the circulation. The peripheral blood smear also shows polychromatophils. B: A dimorphic population of red blood cells composed of **microspherocytes** and **macrocytes** is also found in autoimmune hemolytic anemia.

■ RED BLOOD CELL DISORDERS

INCREASED RED BLOOD CELL DESTRUCTION

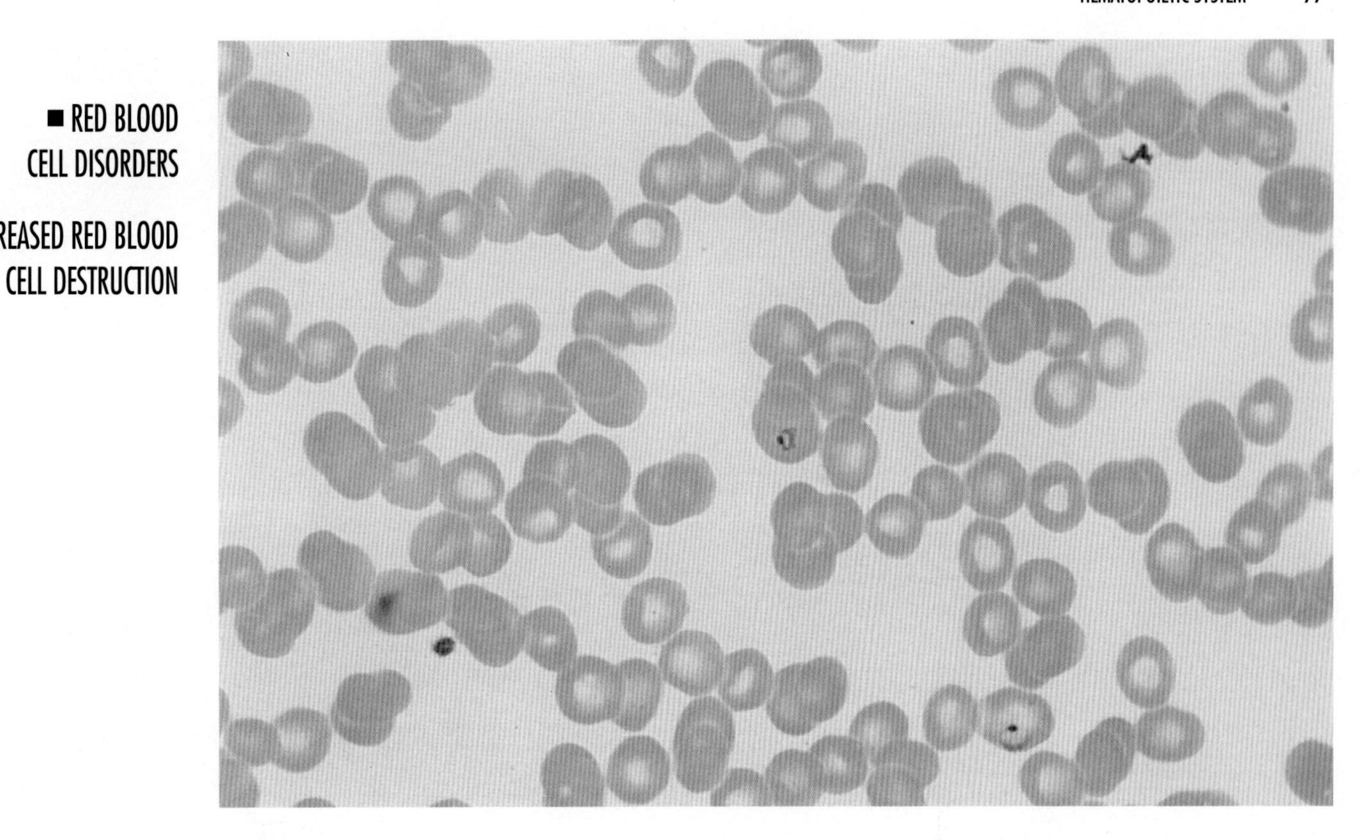

Figure 6–6 A

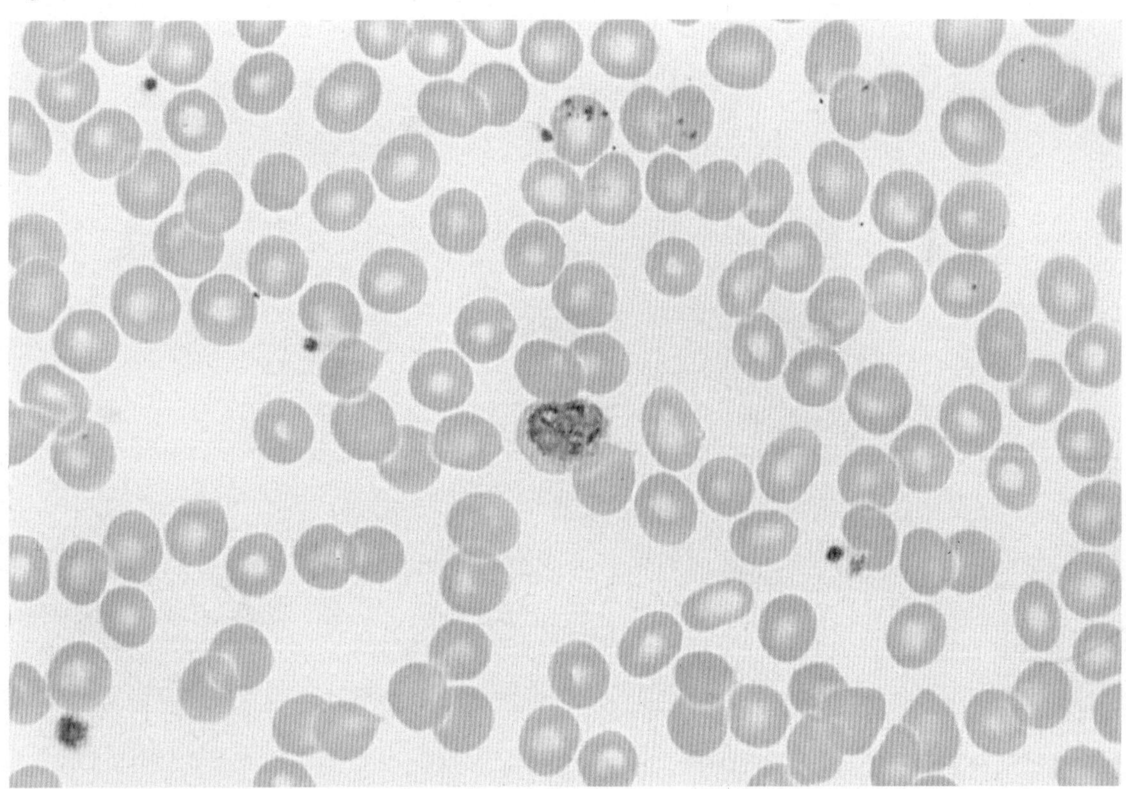

Figure 6–6 B

FIGURE 6–6. MALARIA: In malaria, the peripheral blood smear shows **ring forms** and intracellular **trophozoites** (A) and **gametocytes** (B).

■ RED BLOOD CELL DISORDERS

DECREASED RED BLOOD CELL PRODUCTION

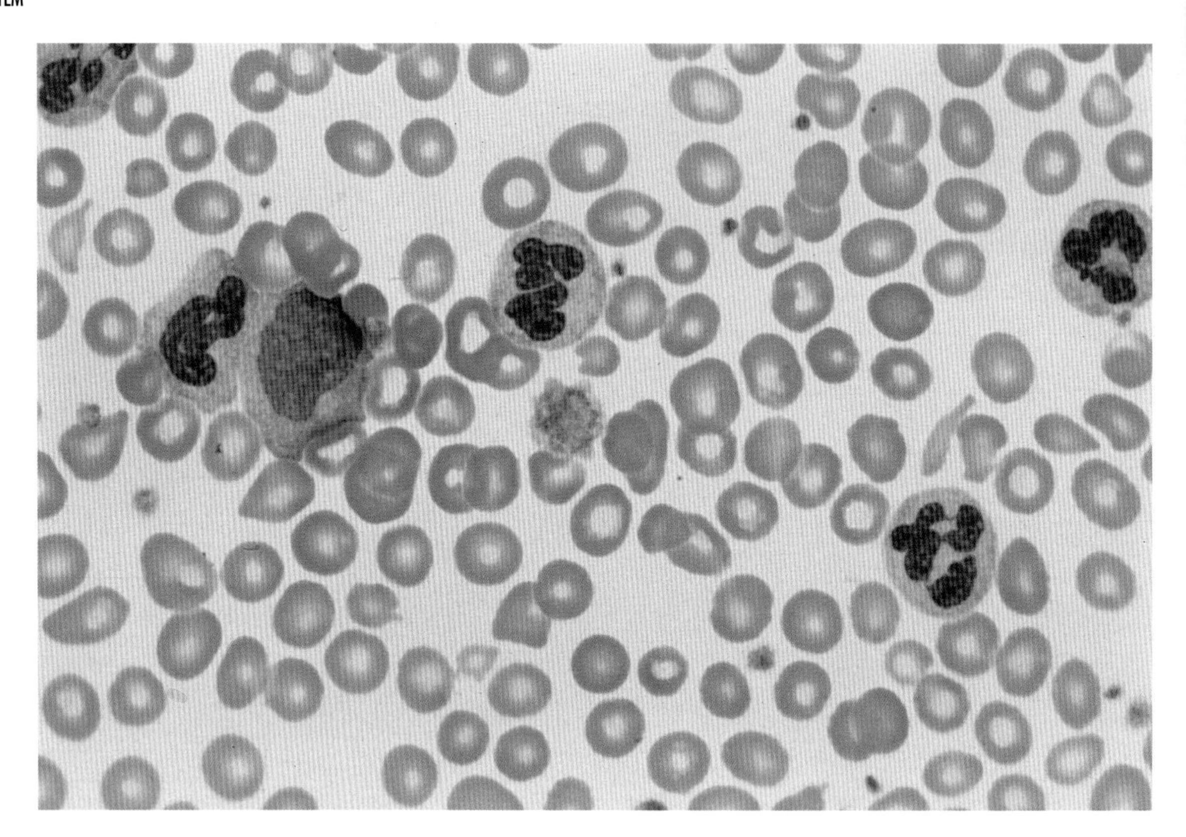

Figure 6–7 A

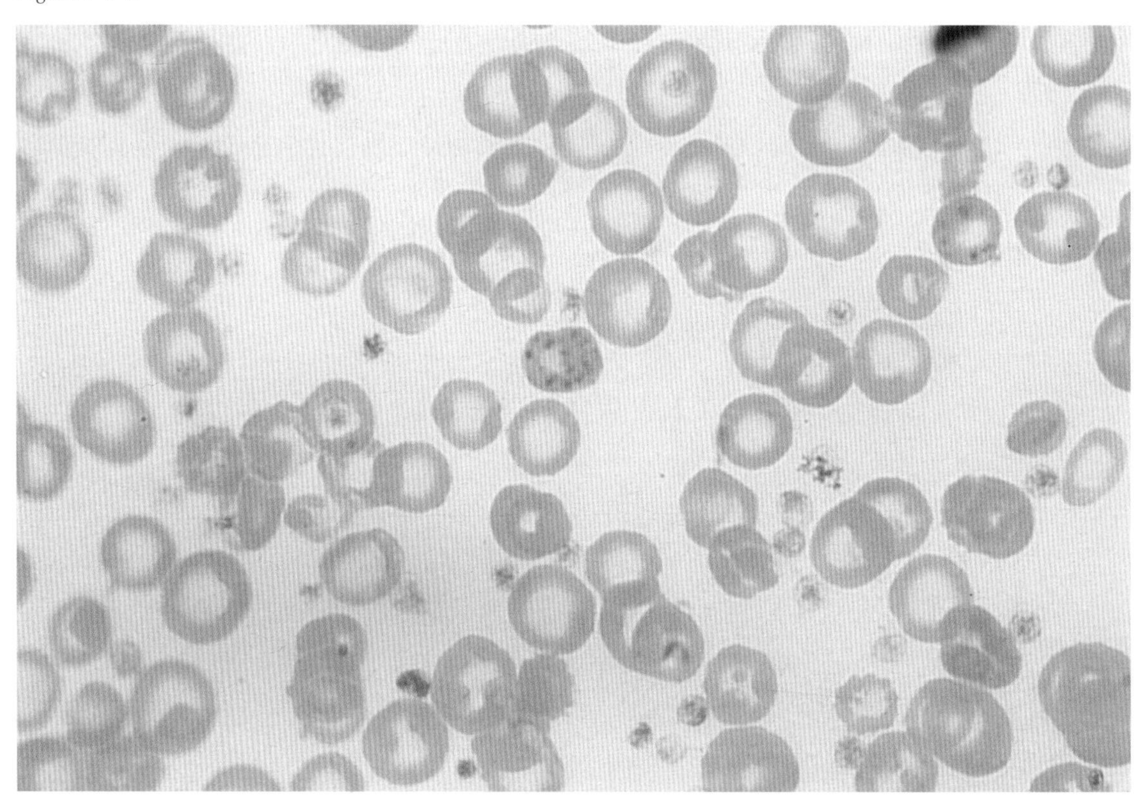

Figure 6–7 B

FIGURE 6–7. MEGALOBLASTIC ANEMIA: A: The characteristic findings in a peripheral blood smear in megaloblastic anemia are **hypersegmentation** of neutrophils, **macroovalocytes, schistocytes, teardrop cells,** and **giant platelets.** The hypersegmented neutrophils have six or more nuclear lobes. B: In severe cases, basophilic stippling is found.

■ RED BLOOD CELL DISORDERS

DECREASED RED BLOOD CELL PRODUCTION

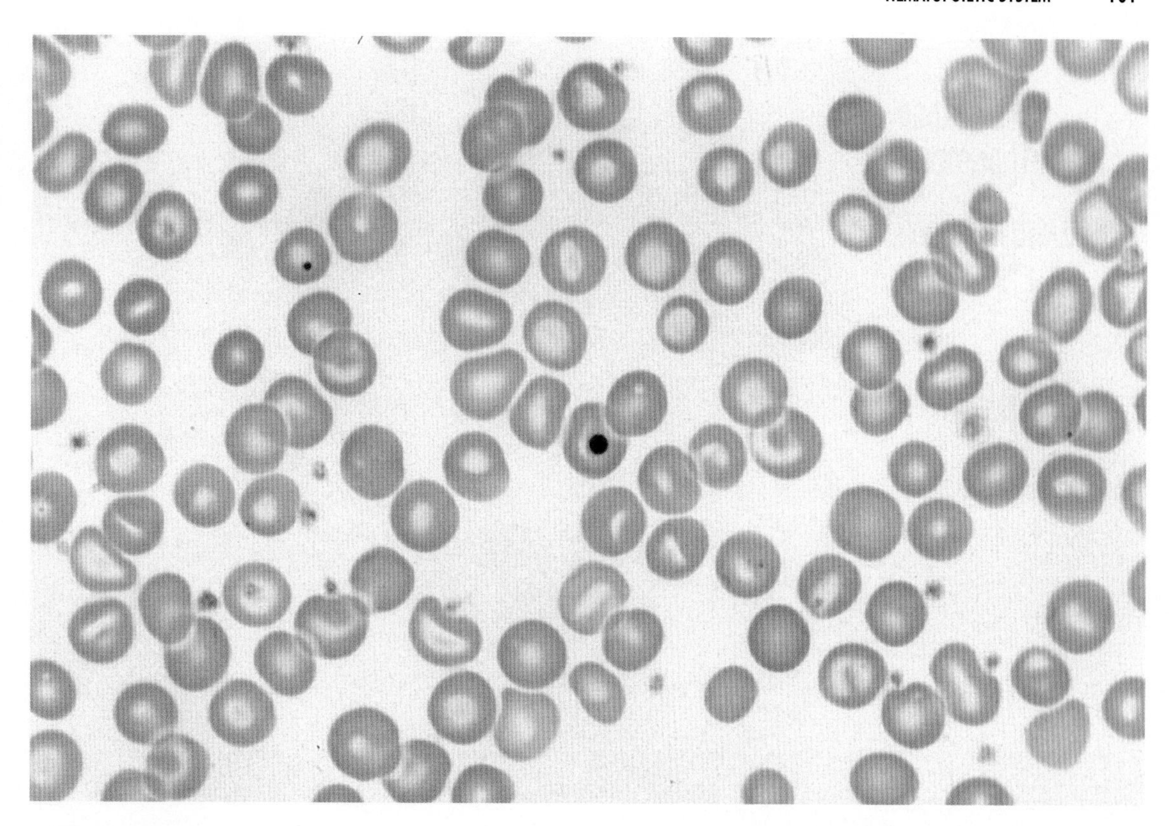

Figure 6–7 C

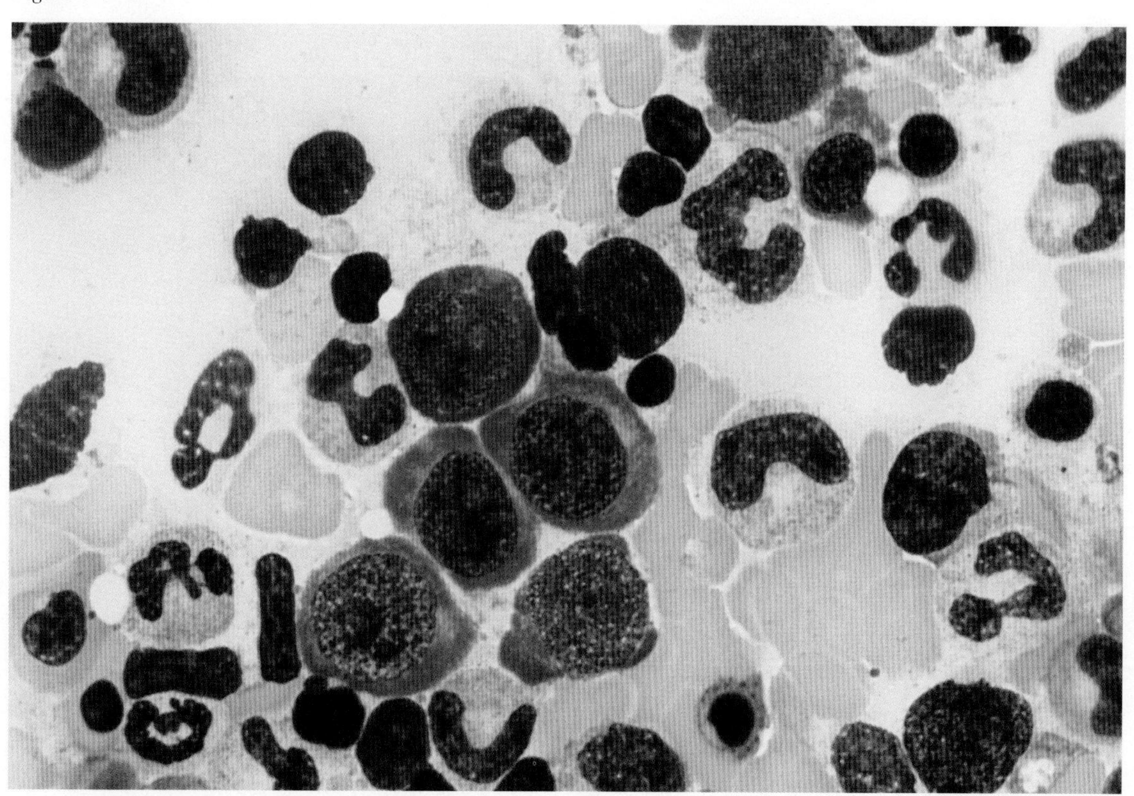

Figure 6–7 D

FIGURE 6–7. MEGALOBLASTIC ANEMIA: C: In megaloblastic anemia, nuclear fragments, **Howell-Jolly bodies,** are found within red blood cells. It is only in asplenic states that Howell-Jolly bodies are found in circulating red blood cells. D: The characteristic bone marrow changes seen in megaloblastic anemia are enlargement of cells of the erythroid, granulocytic, and megakaryocytic series. In this high-power photomicrograph, megaloblastic erythroblasts show a lag in nuclear maturation compared with cytoplasmic maturation. Giant metamyelocytes and giant bands are also present.

■ RED BLOOD CELL DISORDERS

DECREASED RED BLOOD CELL PRODUCTION

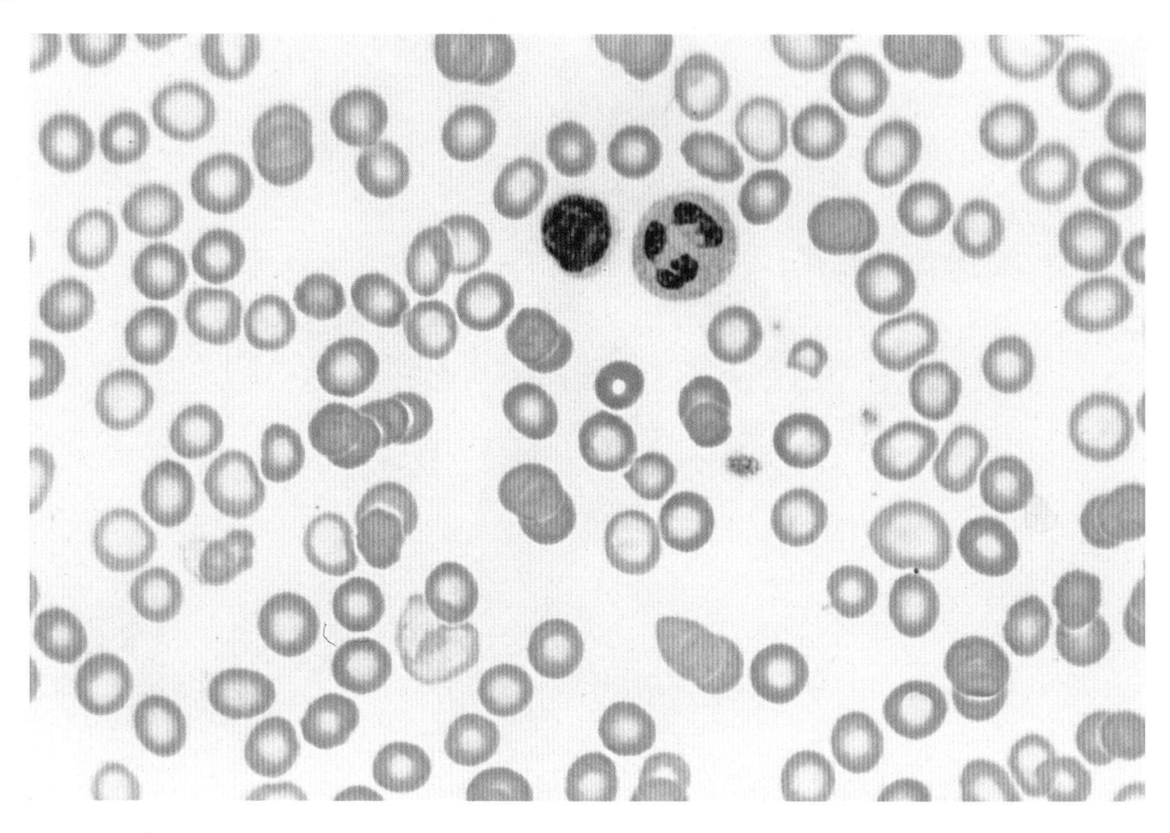

Figure 6–8 A

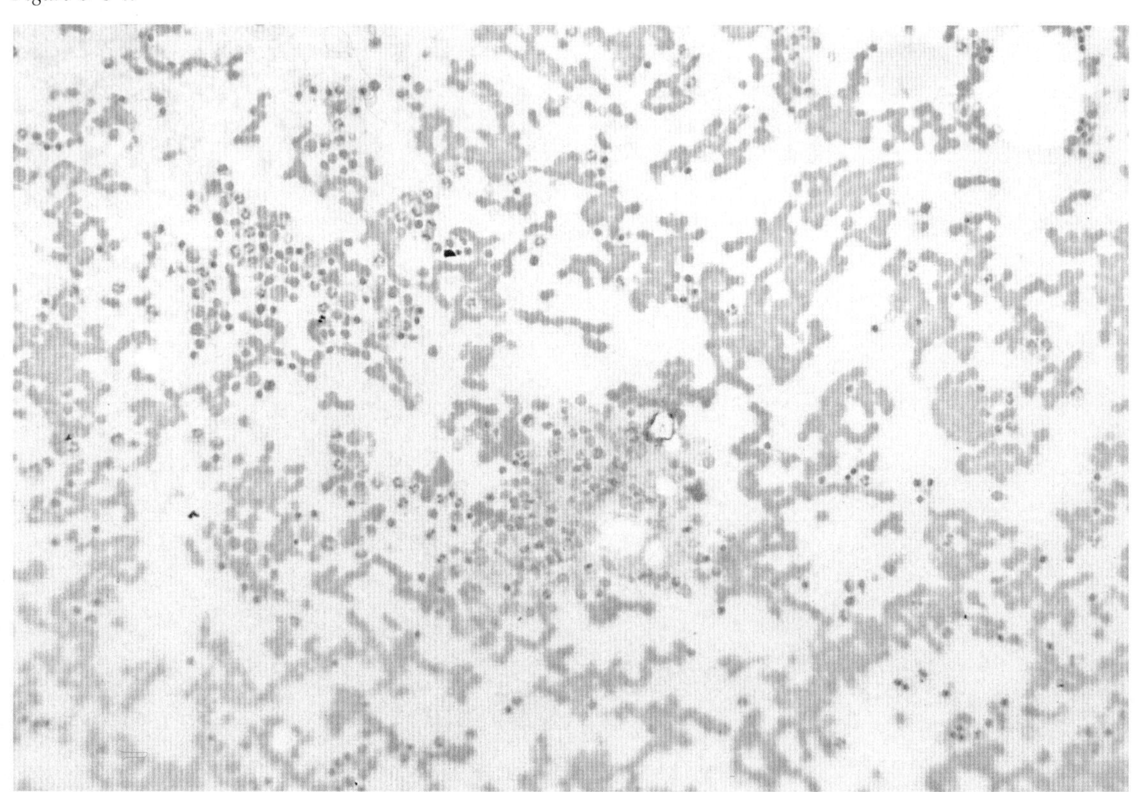

Figure 6–8 B

FIGURE 6–8. IRON DEFICIENCY ANEMIA: Iron deficiency anemia is the most prevalent type of anemia. A: In iron deficiency anemia, the peripheral smear shows **microcytic, hypochromic** red blood cells. Microcytic cells are smaller than the nucleus of a lymphocyte, and hypochromasia is characterized by an increased zone of central pallor. B: Prussian blue staining of the bone marrow is negative for iron (absence of blue-staining iron).

■ RED BLOOD CELL DISORDERS

DECREASED RED BLOOD CELL PRODUCTION

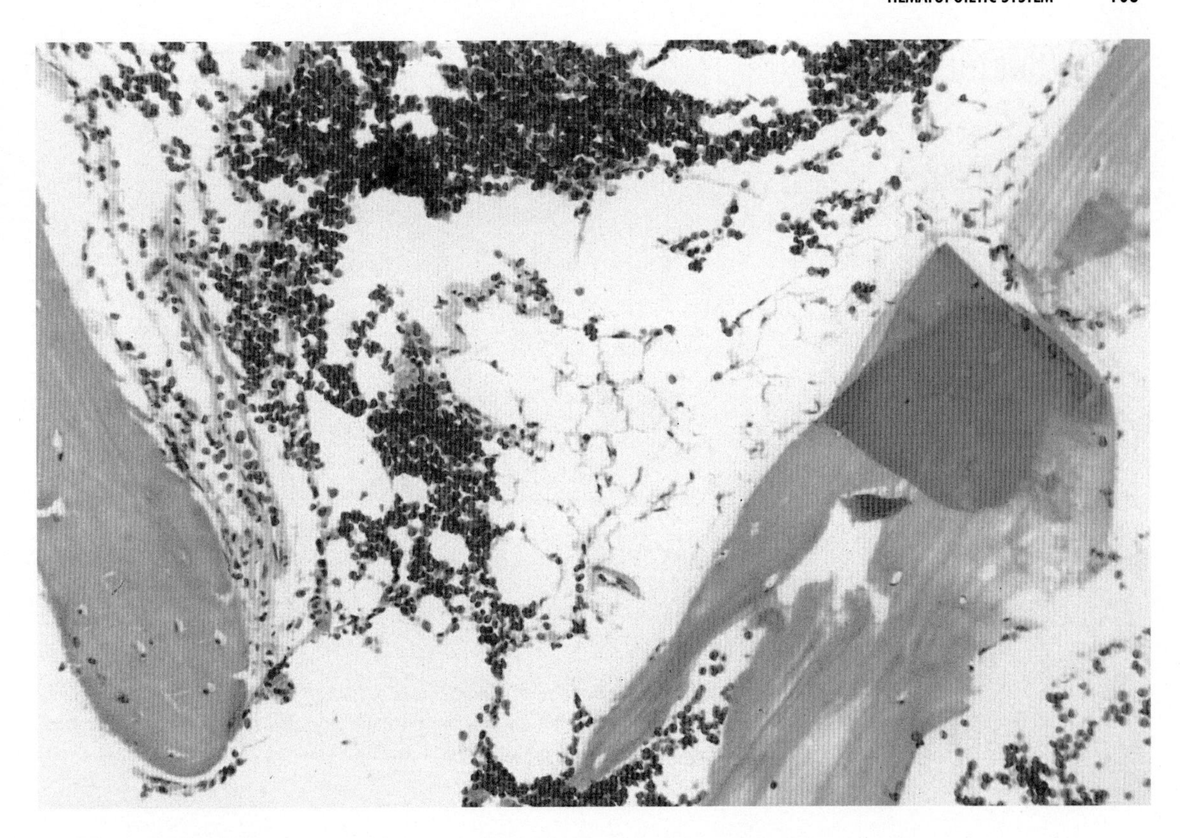

FIGURE 6–9. APLASTIC ANEMIA: In the most severe form of aplastic anemia, the bone marrow is replaced by fatty marrow and blood-forming elements are absent.

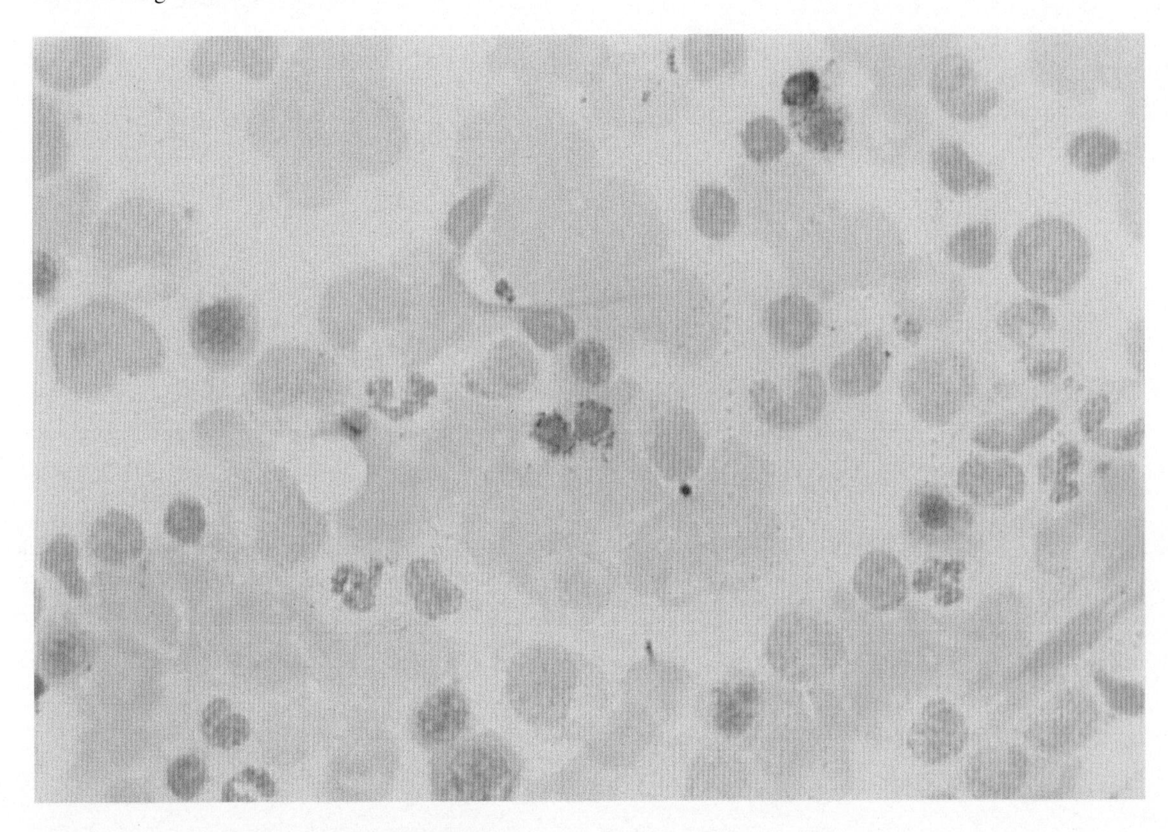

FIGURE 6–10. REFRACTORY ANEMIA WITH RINGED SIDEROBLASTS: Refractory anemia with ringed sideroblasts is one of the myelodysplastic syndromes. In the bone marrow, **ringed sideroblasts** have iron-containing granules that encircle erythroblast nuclei (Prussian blue stain).

■ WHITE BLOOD CELL DISORDERS

REACTIVE LEUKOCYTOSIS

FIGURE 6–11. BACTERIAL INFECTION: A: The peripheral blood smear in bacterial infections is characterized by **toxic granulations** and **Döhle bodies.** Toxic granulations are primary granules. B: Döhle bodies are blue-gray smudges that represent remnants of rough endoplasmic reticulum (*arrow*).

FIGURE 6–12. VIRAL INFECTIONS: In this example of infectious mononucleosis, **atypical lymphocytes** with abundant blue cytoplasm, enlarged nuclei, and prominent nucleoli are present. The cytoplasm of the atypical lymphocytes conforms to the shape of the red blood cells surrounding these lymphocytes.

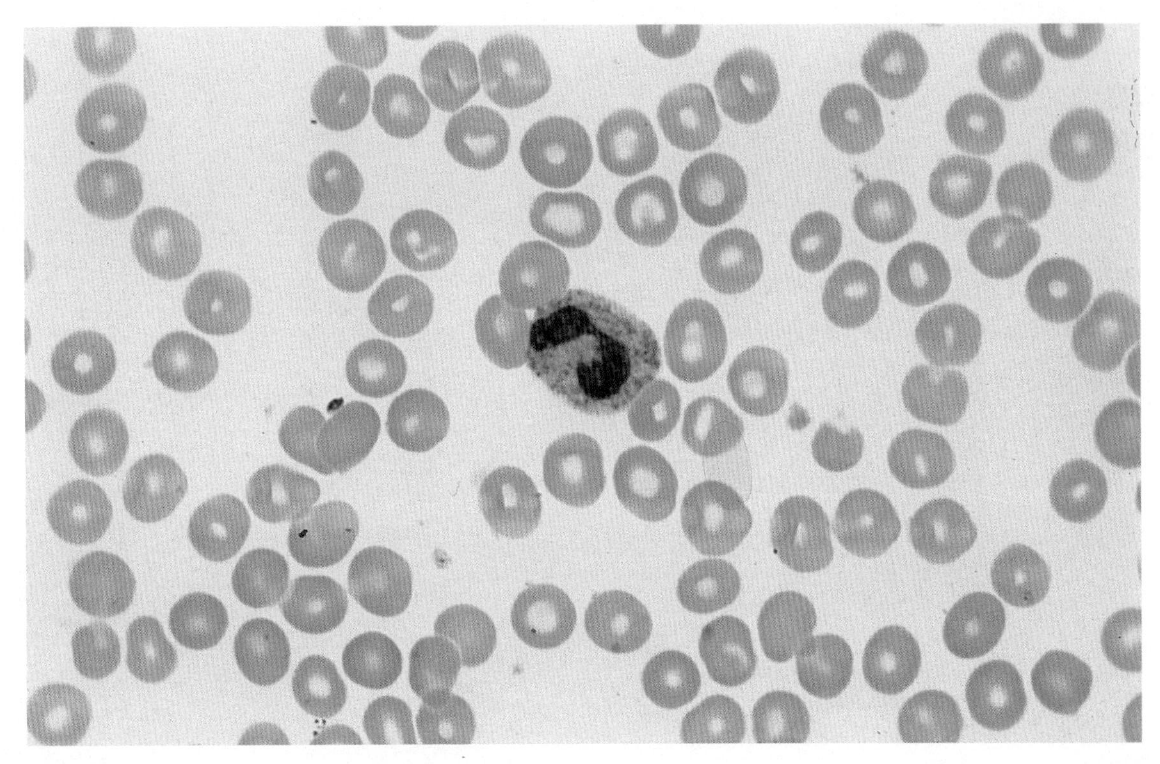

Figure 6–11 **A**

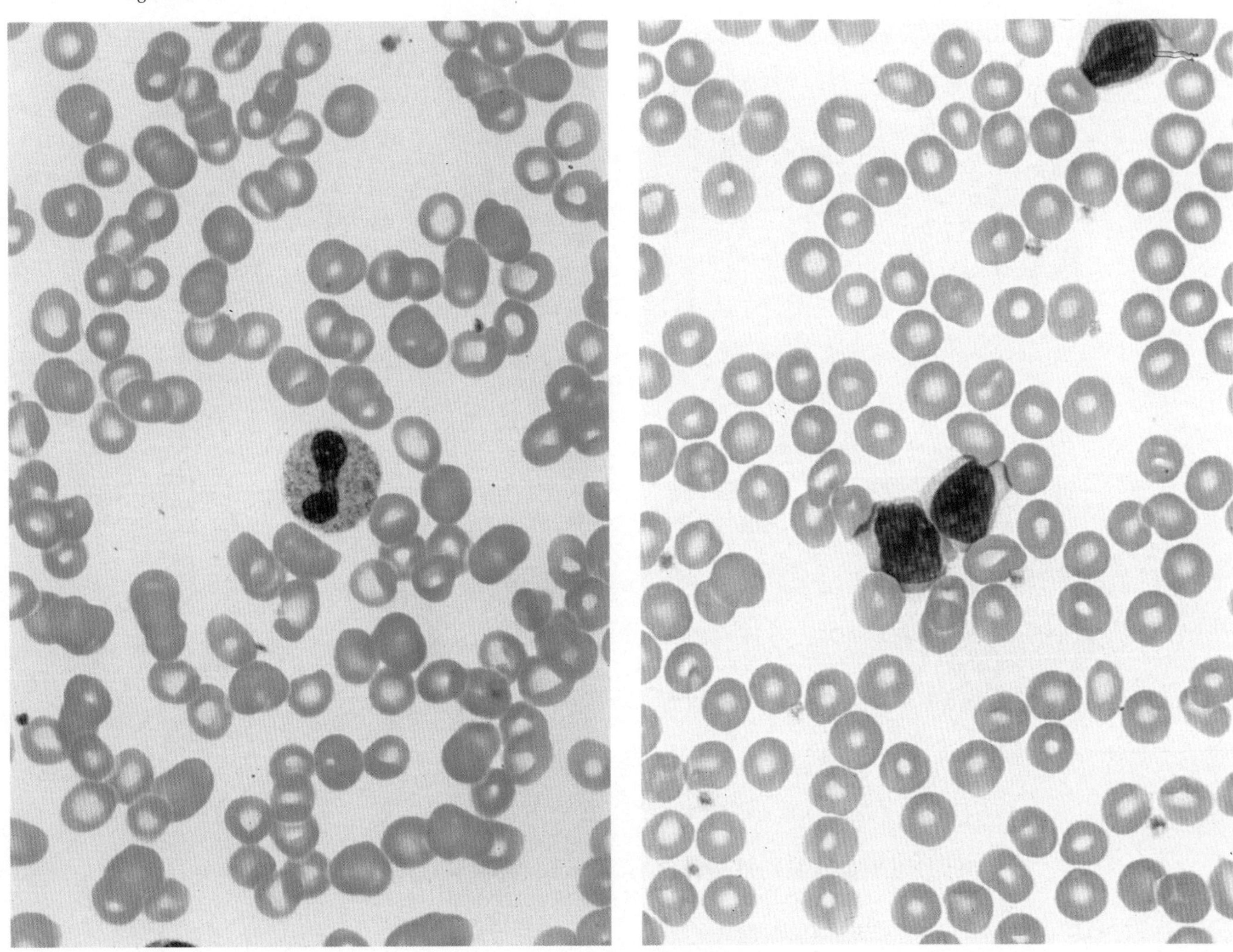

Figure 6–11 **B**

Figure 6–12

■ WHITE BLOOD CELL DISORDERS

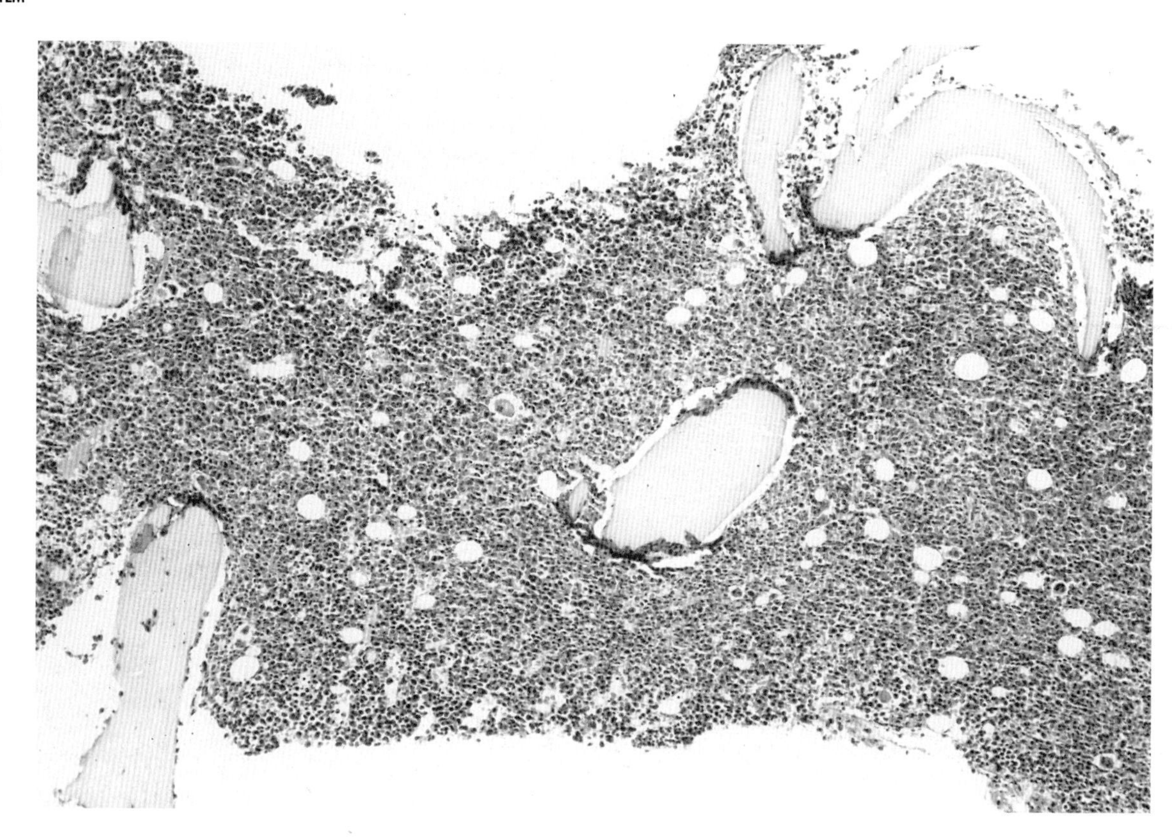

Figure 6–13 A

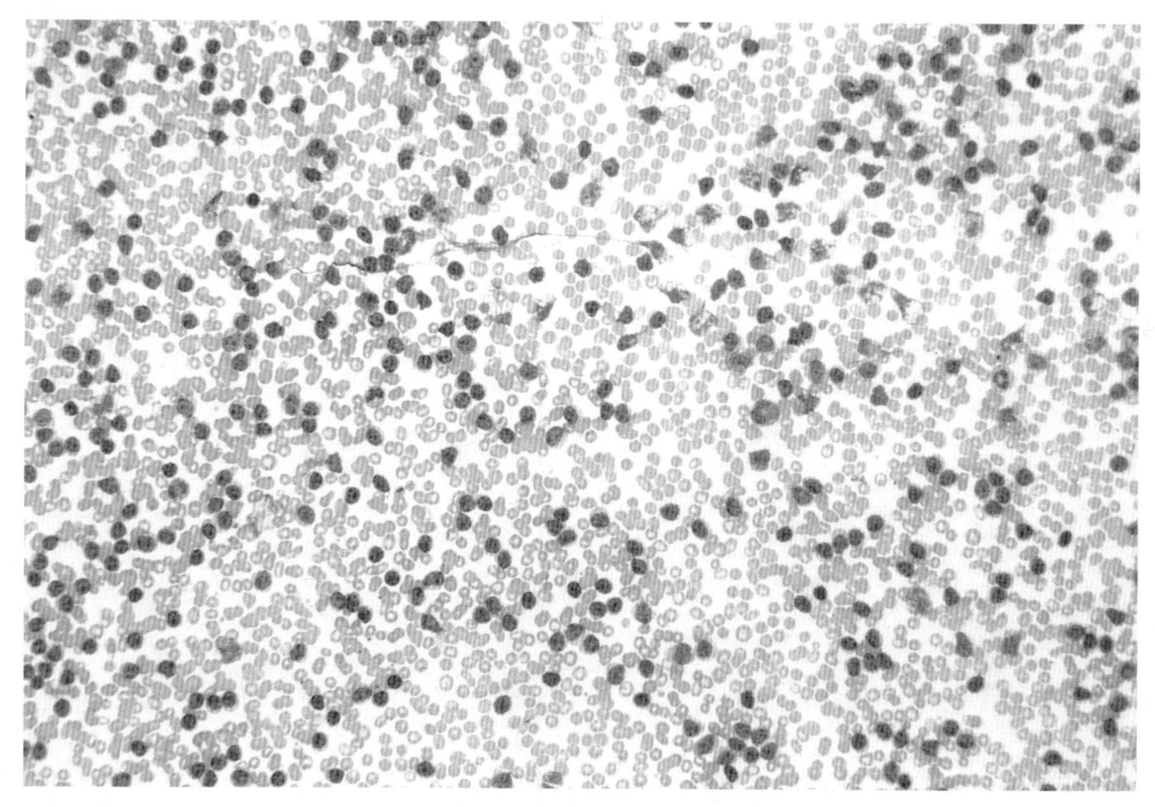

Figure 6–13 B

FIGURE 6–13. LEUKEMIAS: A: In both acute and chronic leukemia, the bone marrow is markedly hypercellular. **B:** The neoplastic cells spill out into the circulation, and there is marked leukocytosis.

■ WHITE BLOOD CELL DISORDERS

ACUTE LYMPHOCYTIC LEUKEMIA

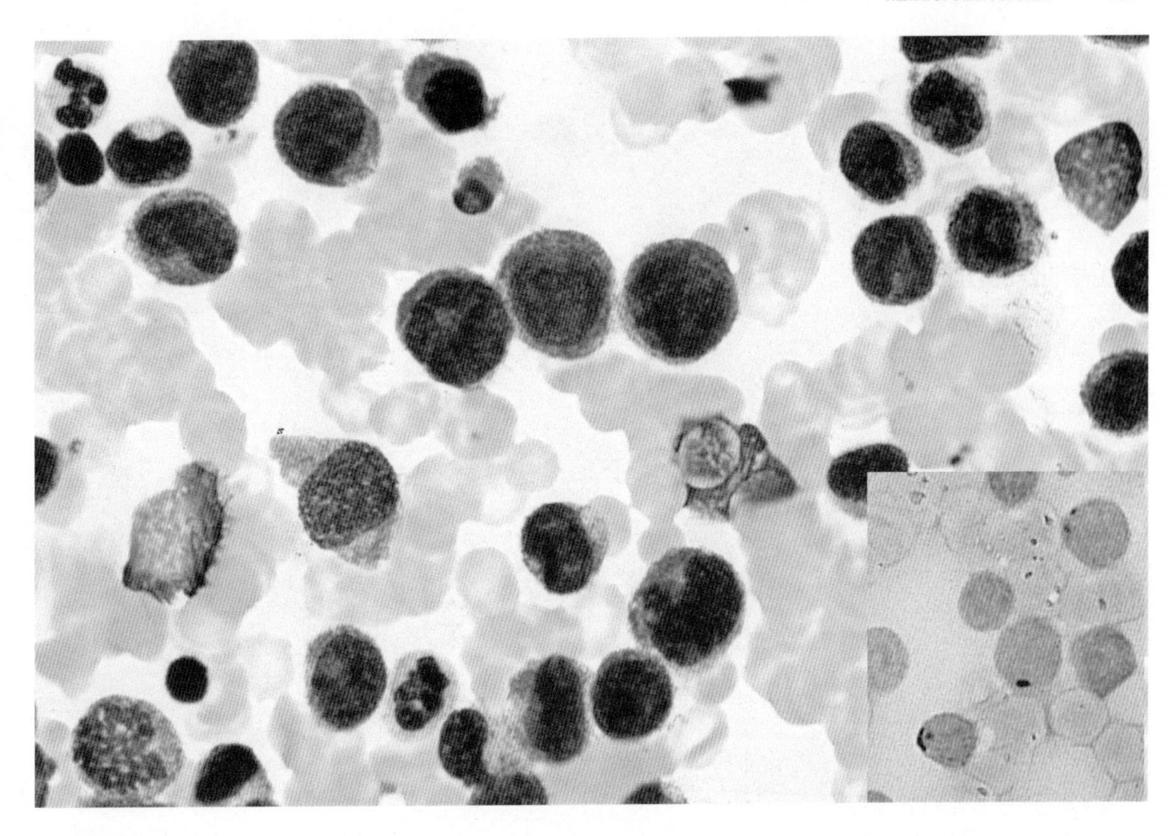

FIGURE 6–14. L1: The L1 variant of acute lymphocytic leukemia is the most common type and mostly affects children up to age 15 years. The lymphoblasts are uniform and have nuclei with finely granular nucleoplasm and faint nucleoli. When stained with periodic acid–Schiff (*inset*), lymphoblasts display red staining material in their cytoplasm.

FIGURE 6–15. L3 (BURKITT'S TYPE): The L3 variant of acute lymphocytic leukemia is extremely uncommon. The lymphoblasts are characterized by numerous intracytoplasmic vacuoles.

■ WHITE BLOOD CELL DISORDERS

CHRONIC LYMPHOPROLIFERATIVE DISORDERS

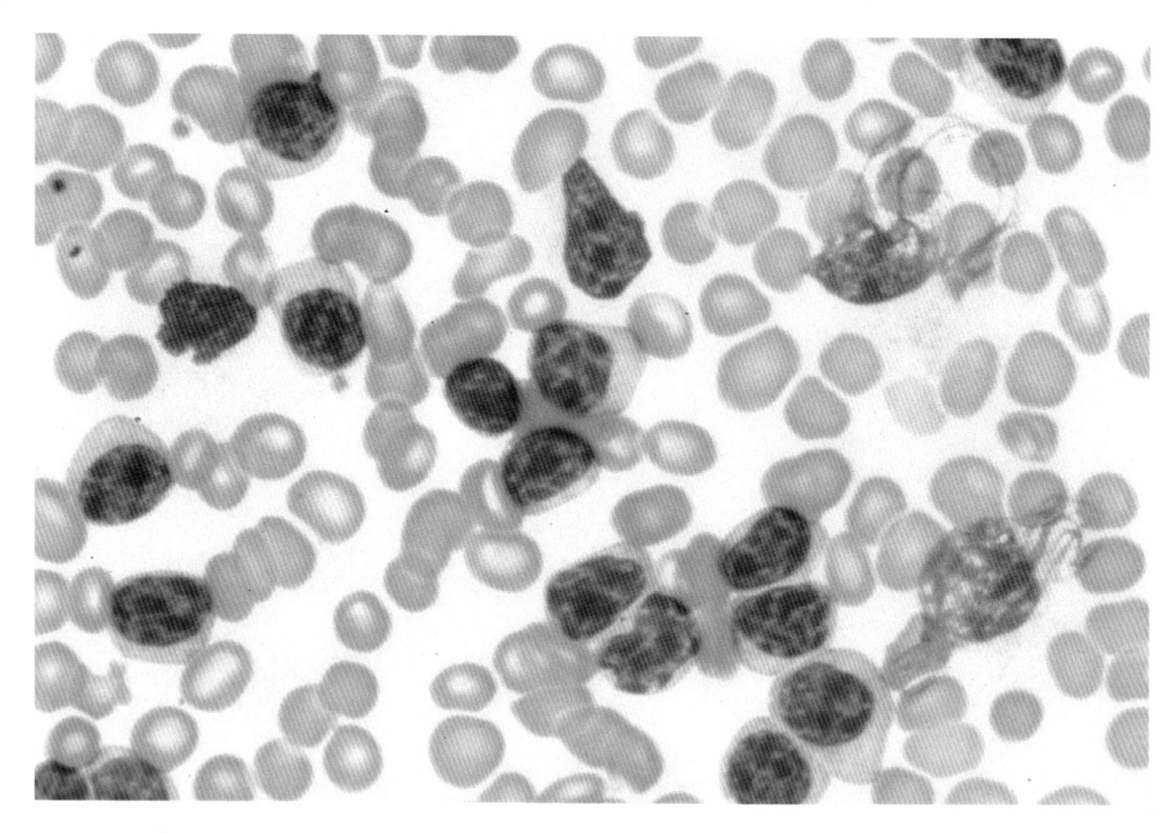

Figure 6–16 A

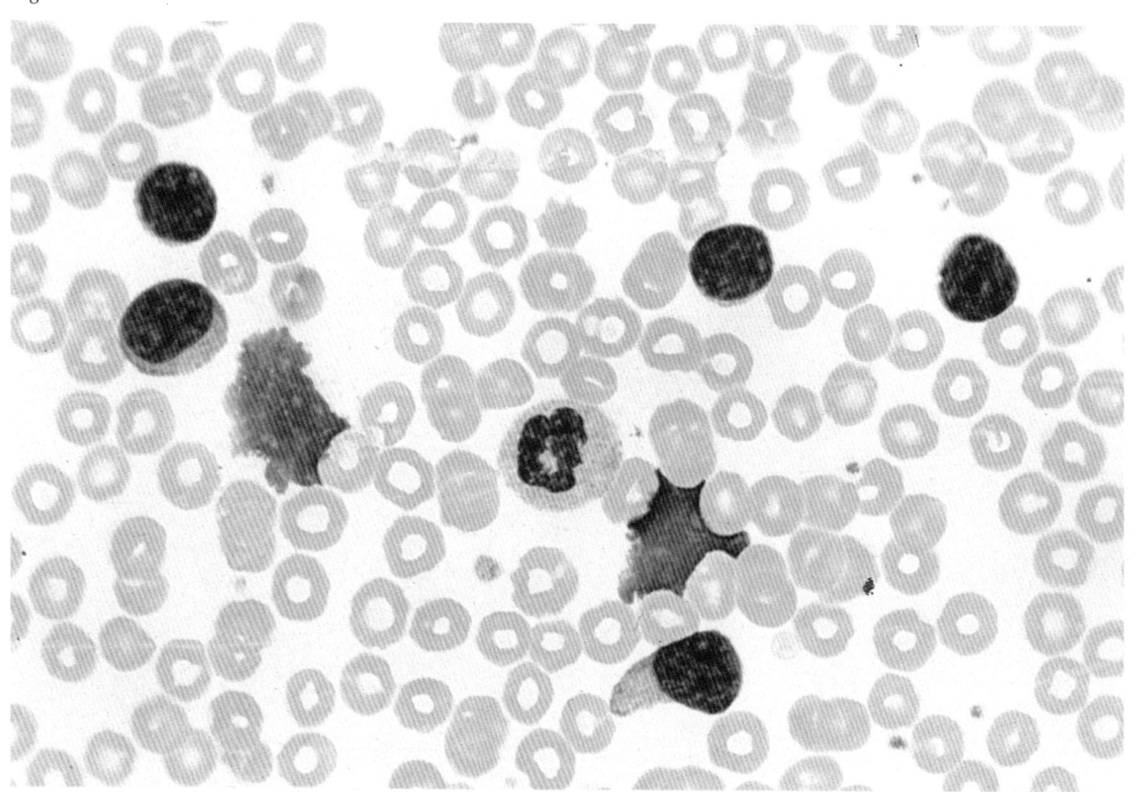

Figure 6–16 B

FIGURE 6–16. CHRONIC LYMPHOCYTIC LEUKEMIA: A: In chronic lymphocytic leukemia, the lymphocytes are hypermature and fragile. Lymphocyte cytoplasm is more abundant and chromatin is more coarsely condensed than in a normal mature lymphocyte. B: Because of the increased fragility of the lymphocytes, lymphocytes are disrupted when a blood smear is prepared and what remains are **smudge cells** or **basket cells**.

■ WHITE BLOOD CELL DISORDERS

CHRONIC LYMPHOPROLIFERATIVE DISORDERS

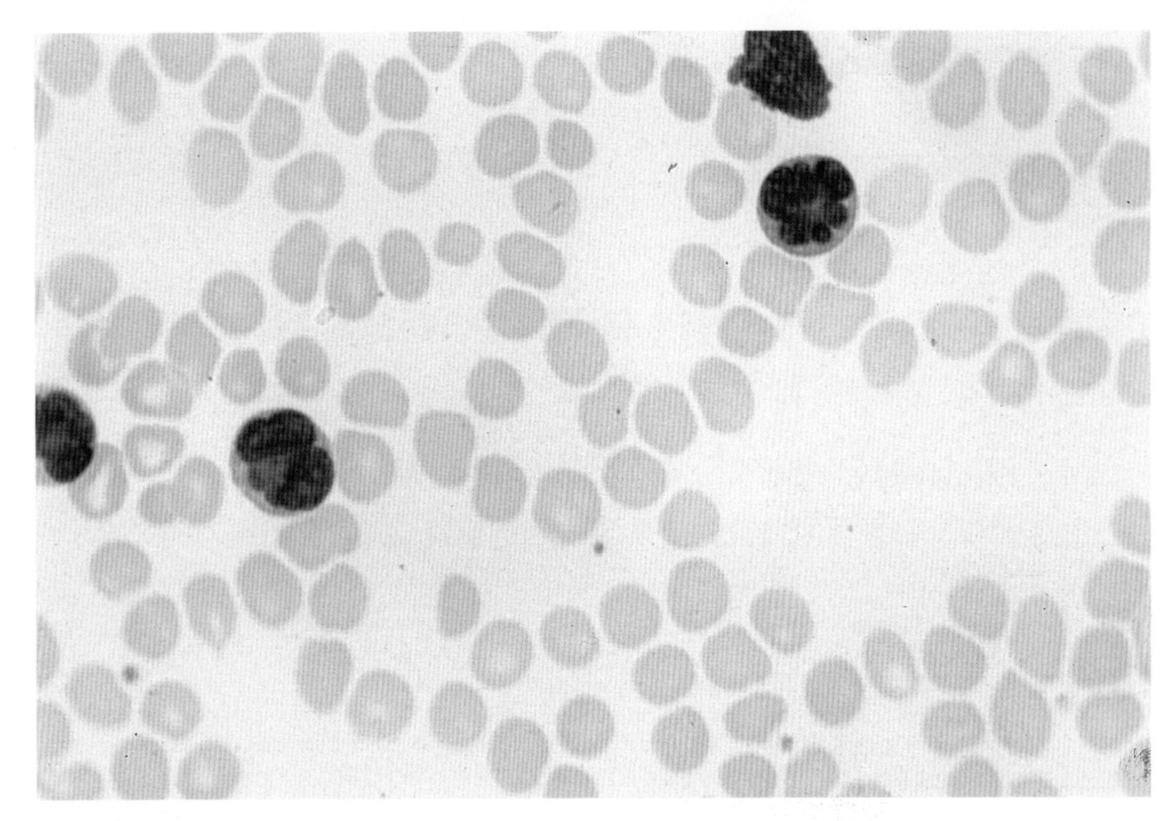

FIGURE 6–17. ADULT T-CELL LEUKEMIA-LYMPHOMA: This type of lymphoma is caused by human T-cell lymphotropic virus-I (HTLV-I). The leukemic phase of acute T-cell leukemia-lymphoma is characterized by the appearance of cerebriform nuclei in circulating neoplastic lymphocytes.

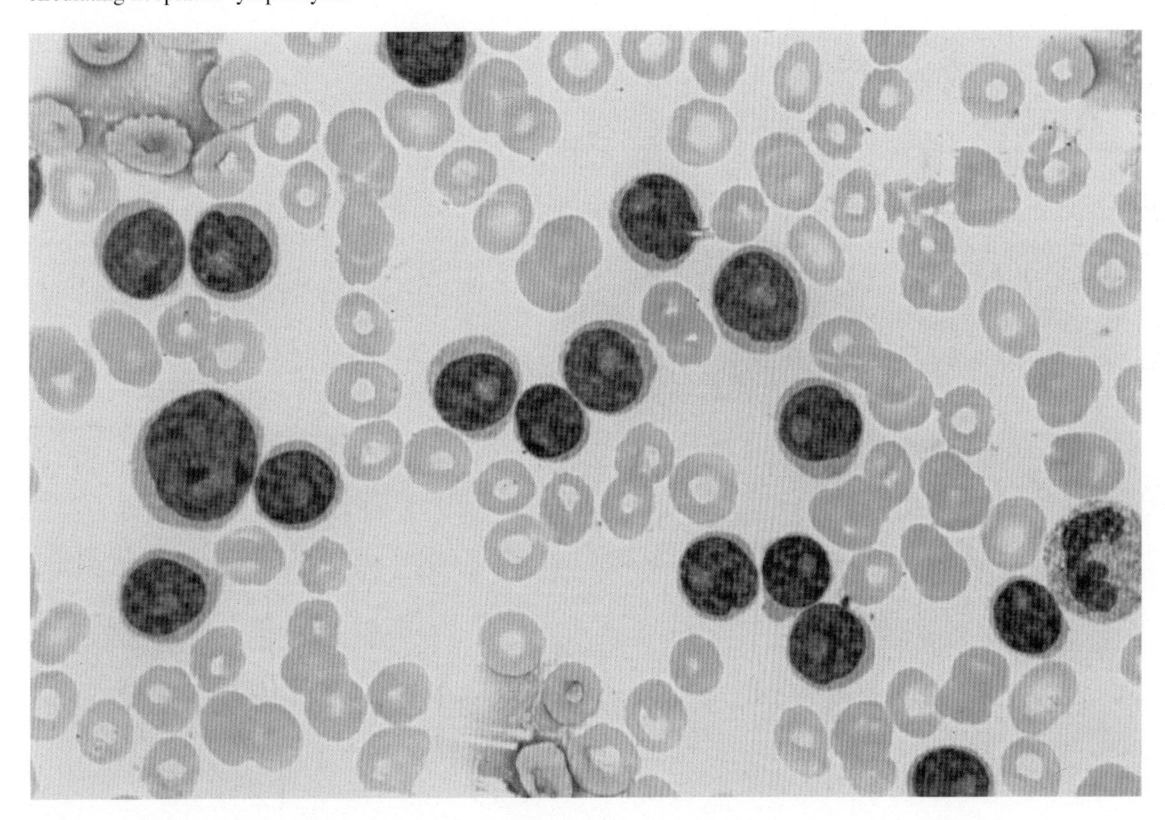

FIGURE 6–18. PROLYMPHOCYTIC LEUKEMIA: In prolymphocytic leukemia, the neoplastic lymphocytes are enlarged and the nuclei display prominent "punched out" nucleoli.

■ WHITE BLOOD CELL DISORDERS

CHRONIC LYMPHOPROLIFERATIVE DISORDERS

FIGURE 6–19. HAIRY CELL LEUKEMIA: A: In the peripheral smear in hairy cell leukemia, the neoplastic lymphocytes have hairlike cytoplasmic projections, the nucleus may be cleaved, and nucleoli are present. **B:** In bone marrow sections of hairy cell leukemia, the cells are evenly distributed and stand away from one another because of the hairlike cytoplasmic projections. **C:** The cells in hairy cell leukemia contain an acid phosphatase that is resistant to digestion by tartrate (tartrate-resistant acid phosphatase [TRAP] stain).

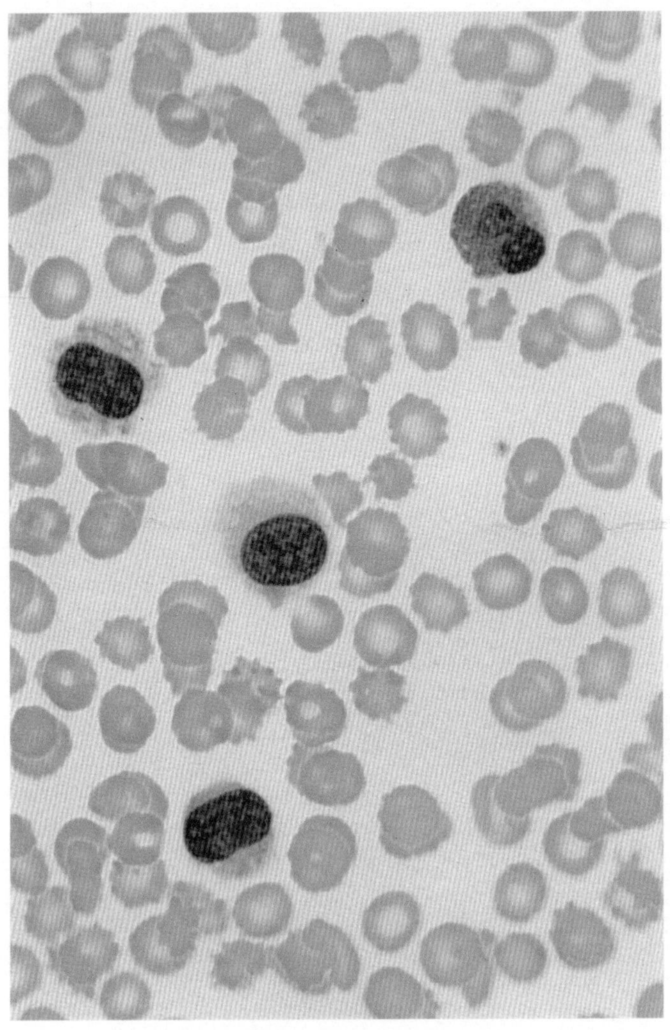

Figure 6–19 A

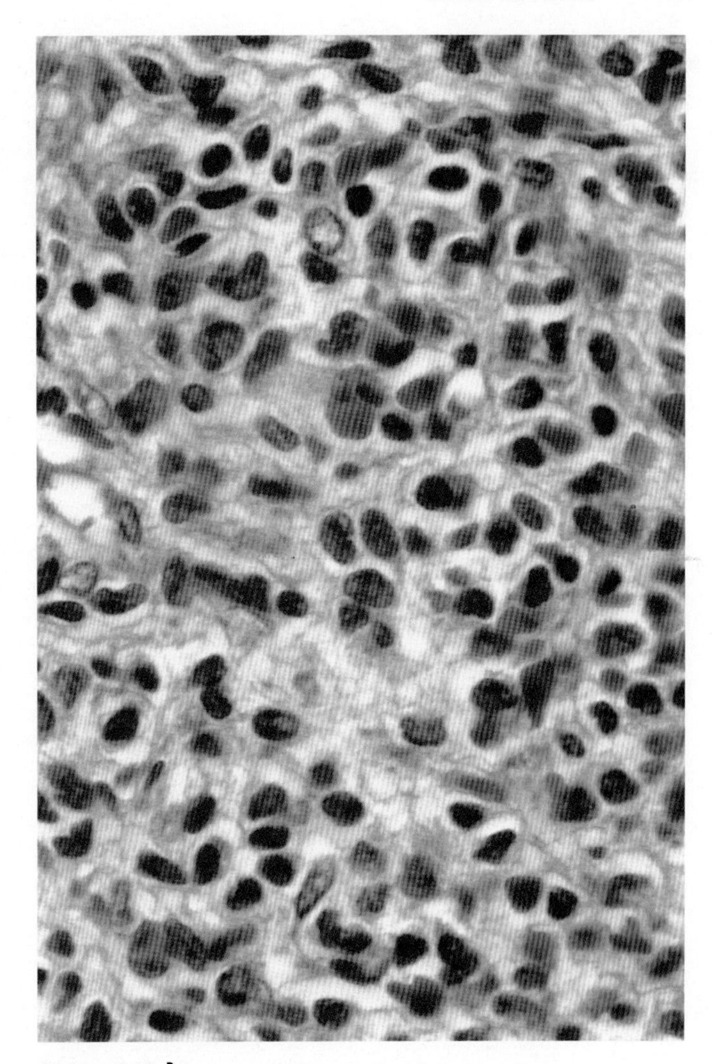

Figure 6–19 B

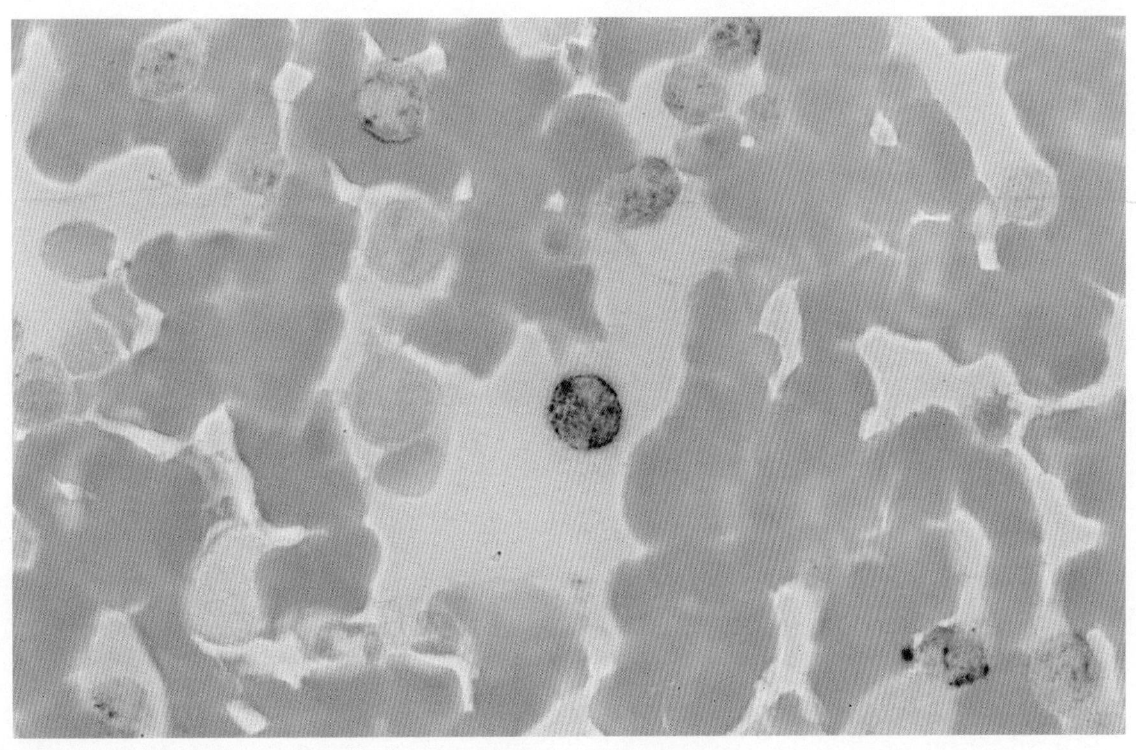

Figure 6–19 C

■ WHITE BLOOD CELL DISORDERS

ACUTE MYELOCYTIC LEUKEMIAS

FIGURE 6–20. ACUTE MYELOCYTIC LEUKEMIA: The diagnostic feature of the acute myelocytic leukemias is the presence of **Auer rods** in the cytoplasm of the neoplastic myelocytic cells. Auer rods are red rod-shaped structures consisting of condensed primary granules (*arrow*).

FIGURE 6–21. ACUTE MONOCYTIC LEUKEMIA: The diagnostic feature of acute monocytic leukemia is positive staining of monoblasts and promonocytes for nonspecific esterase.

MYELOPROLIFERATIVE DISORDERS

FIGURE 6–22. CHRONIC MYELOGENOUS LEUKEMIA: An important prognostic indicator seen in approximately 90% of patients with chronic myelogenous leukemia is the Philadelphia (Ph^1) chromosome. The peripheral smear is characterized by a "left shift" toward immature granulocytic cells, with myelocytes predominating. Basophils are increased in number.

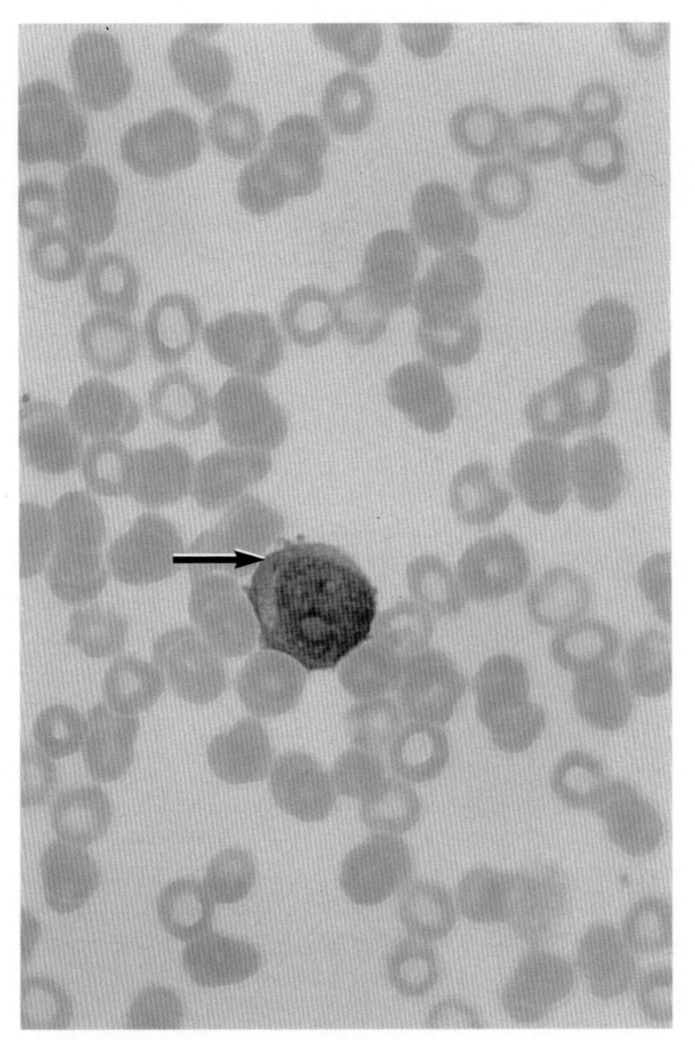

Figure 6–20

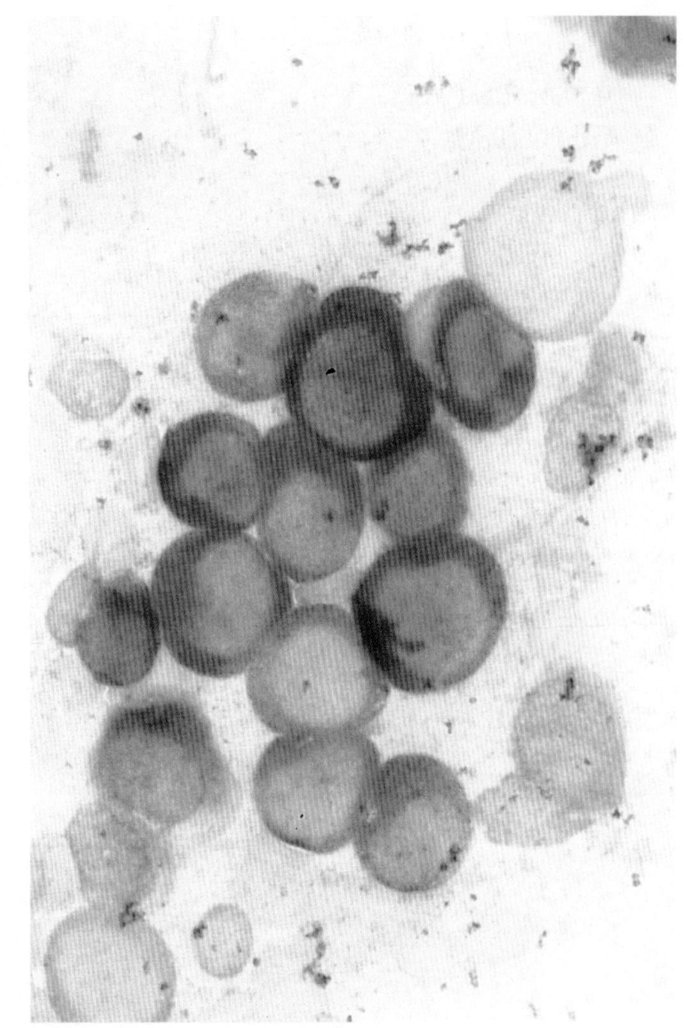

Figure 6–21

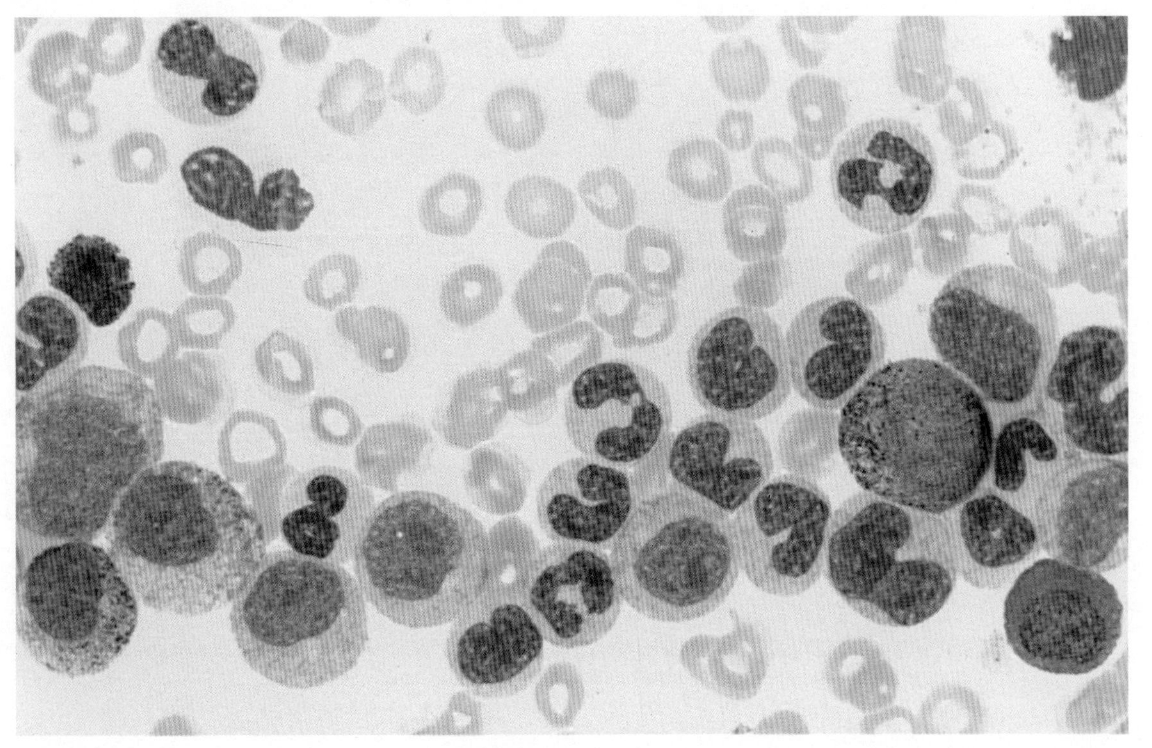

Figure 6–22

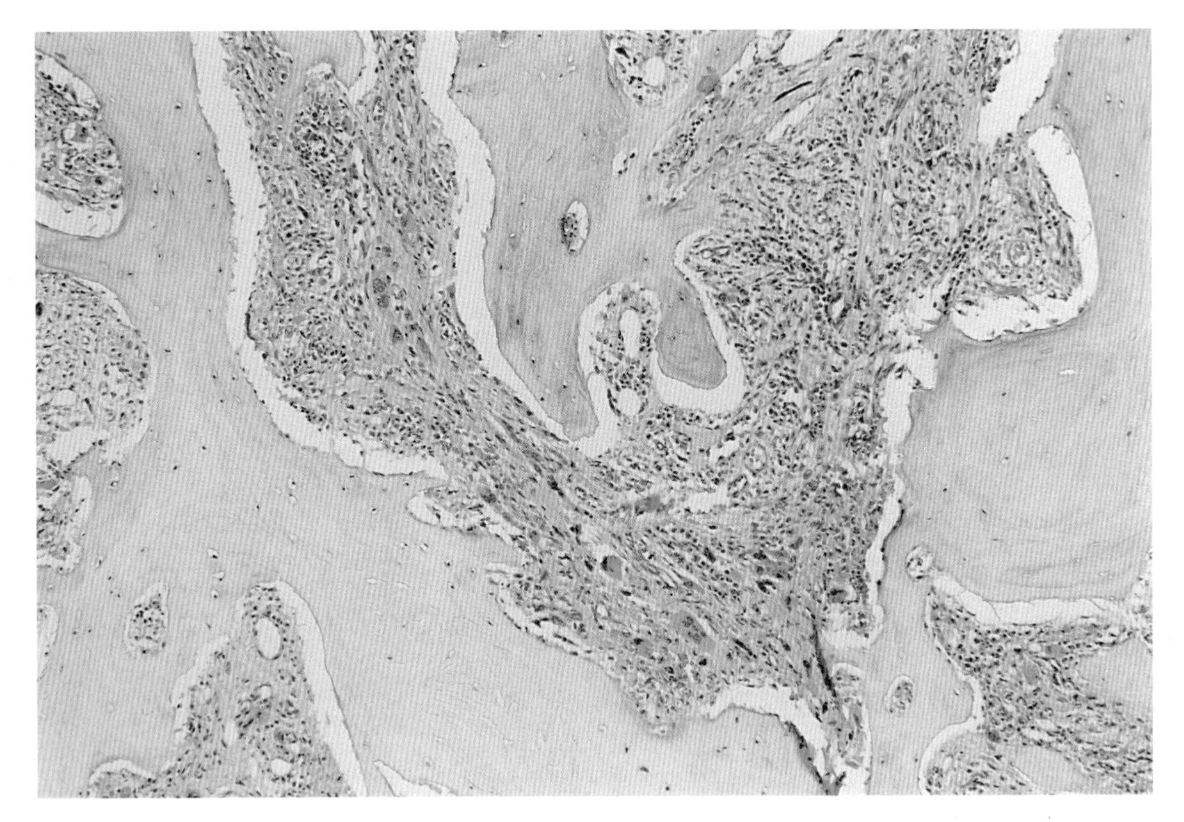

Figure 6–23 **A**

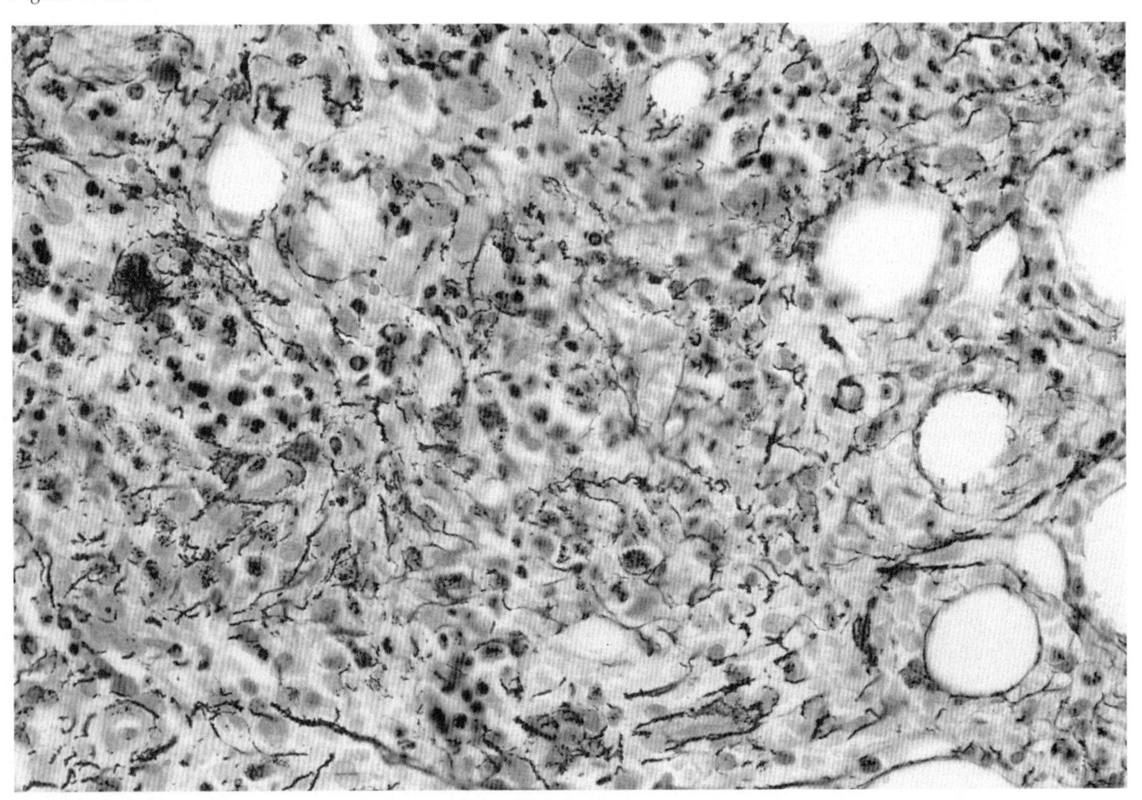

Figure 6–23 **B**

FIGURE 6–23. MYELOFIBROSIS WITH MYELOID METAPLASIA (AGNOGENIC MYELOID METAPLASIA): In this disorder, the bone marrow is fibrotic (**A**) and there is an increase in bone marrow collagen (**B**, reticulin stain).

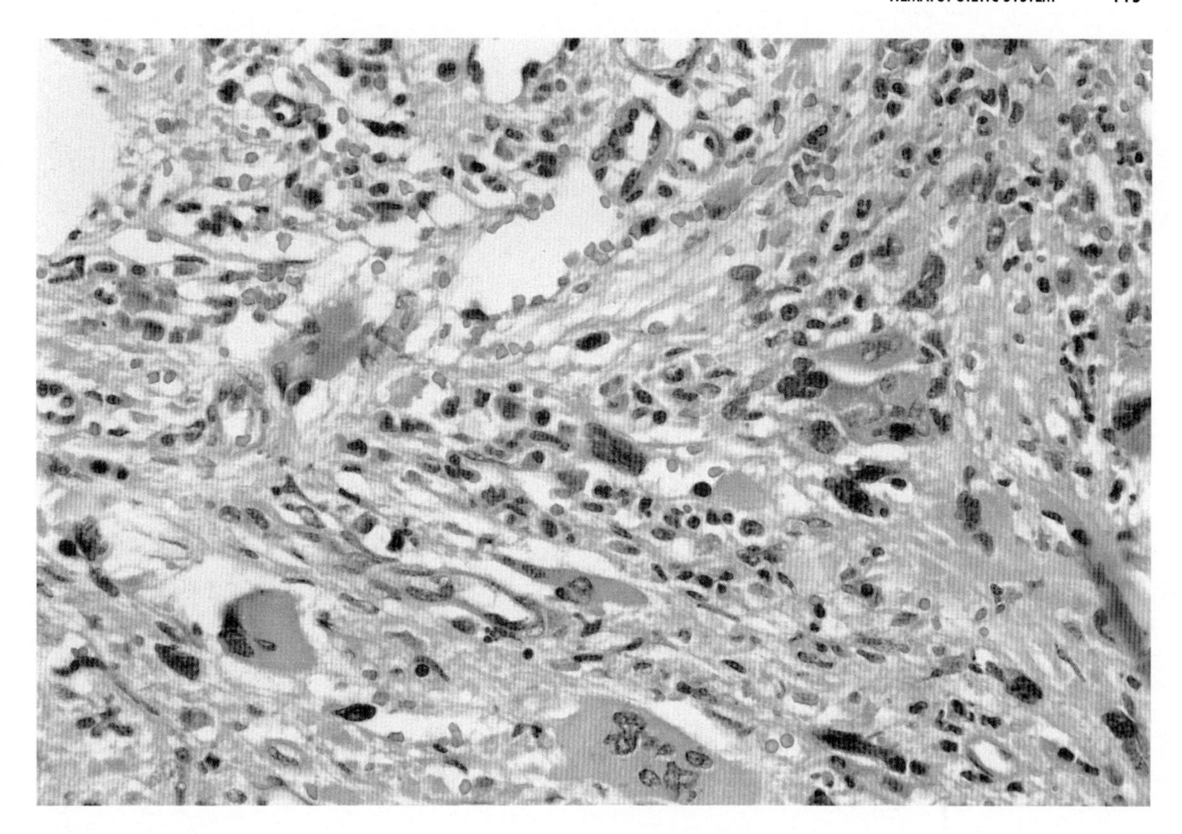

Figure 6–23 C

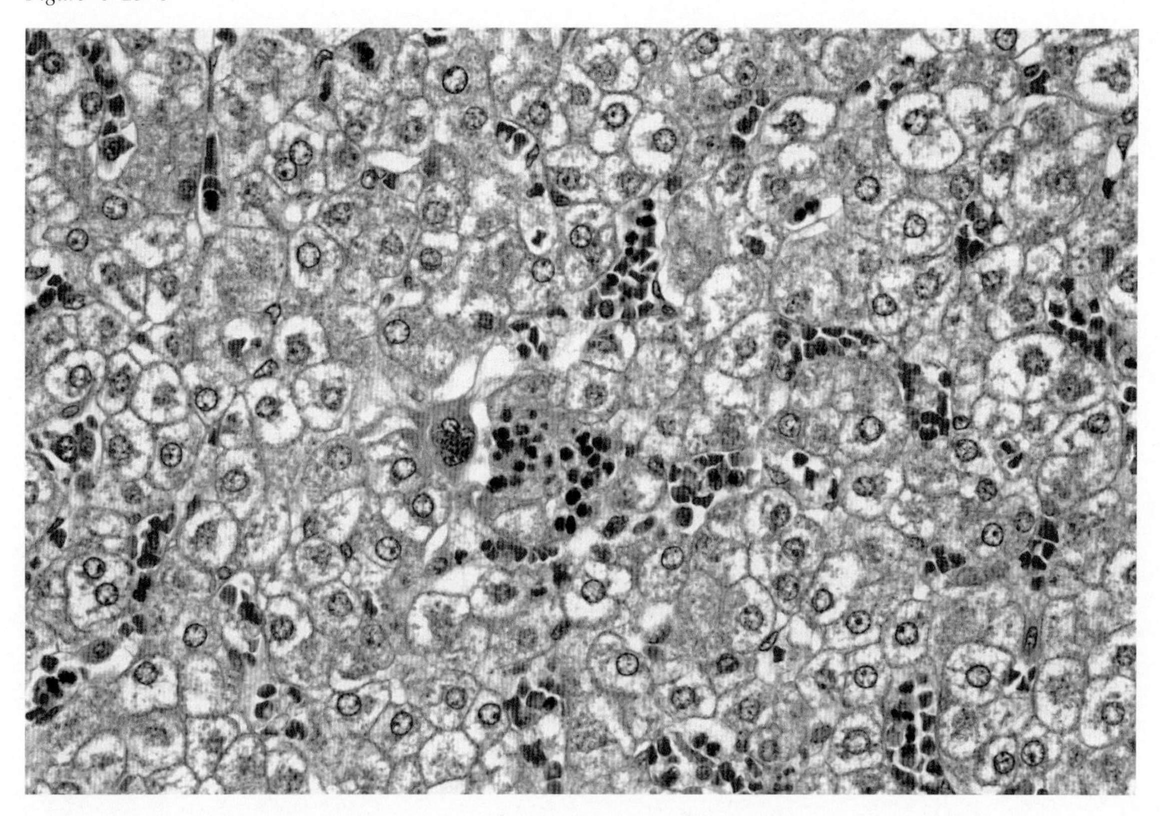

Figure 6–23 D

FIGURE 6–23. MYELOFIBROSIS WITH MYELOID METAPLASIA (AGNOGENIC MYELOID METAPLASIA): C: Also in myelofibrosis with myeloid metaplasia, groups of atypical megakaryocytes are entrapped within the fibrotic marrow. Compensatory "myeloid metaplasia" is present in the liver (**D**), spleen, and lymphoid organs.

7. LYMPHOID SYSTEM

FIGURE 7–1. ANTHRACOSIS: Intralymphatic macrophages that drain the lungs contain inhaled carbon pigment.

■ REACTIVE CHANGES

FIGURE 7–2. SINUS HISTIOCYTOSIS: A: In sinus histiocytosis, there is dilatation of the subcapsular and trabecular sinuses of lymph nodes. **B:** The sinuses are increased in size as a result of hypertrophy of the endothelial cells lining the blood vessels and an increase in the number of histiocytes.

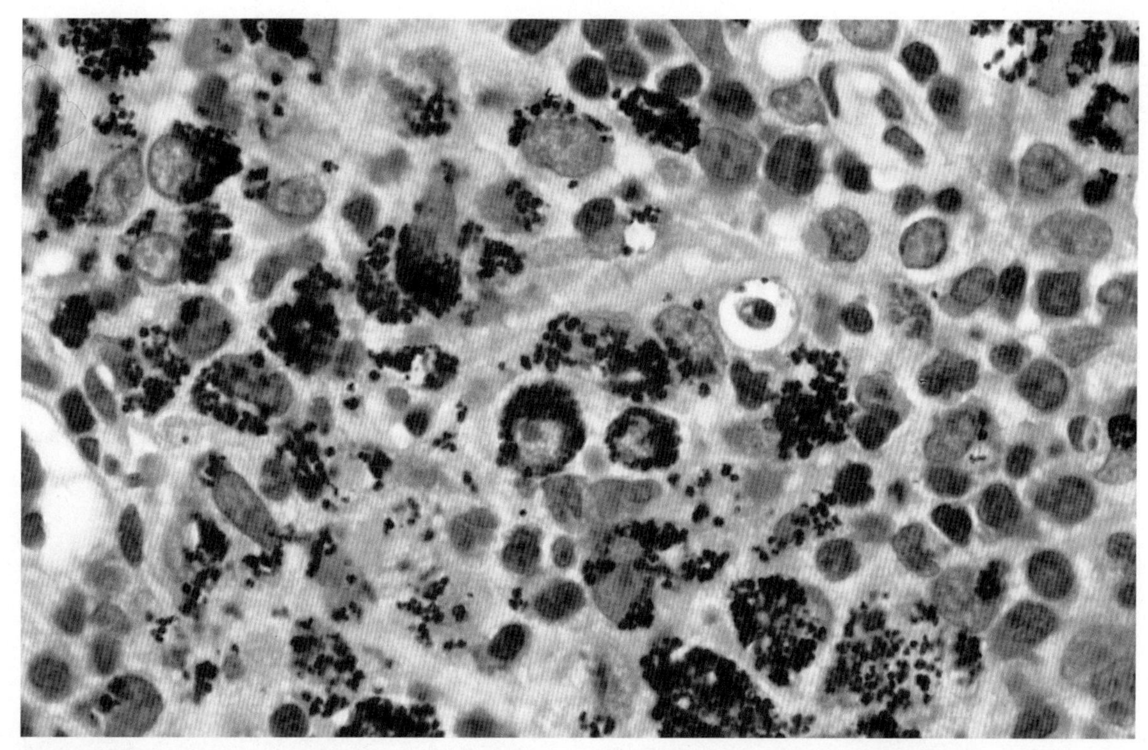

Figure 7–1

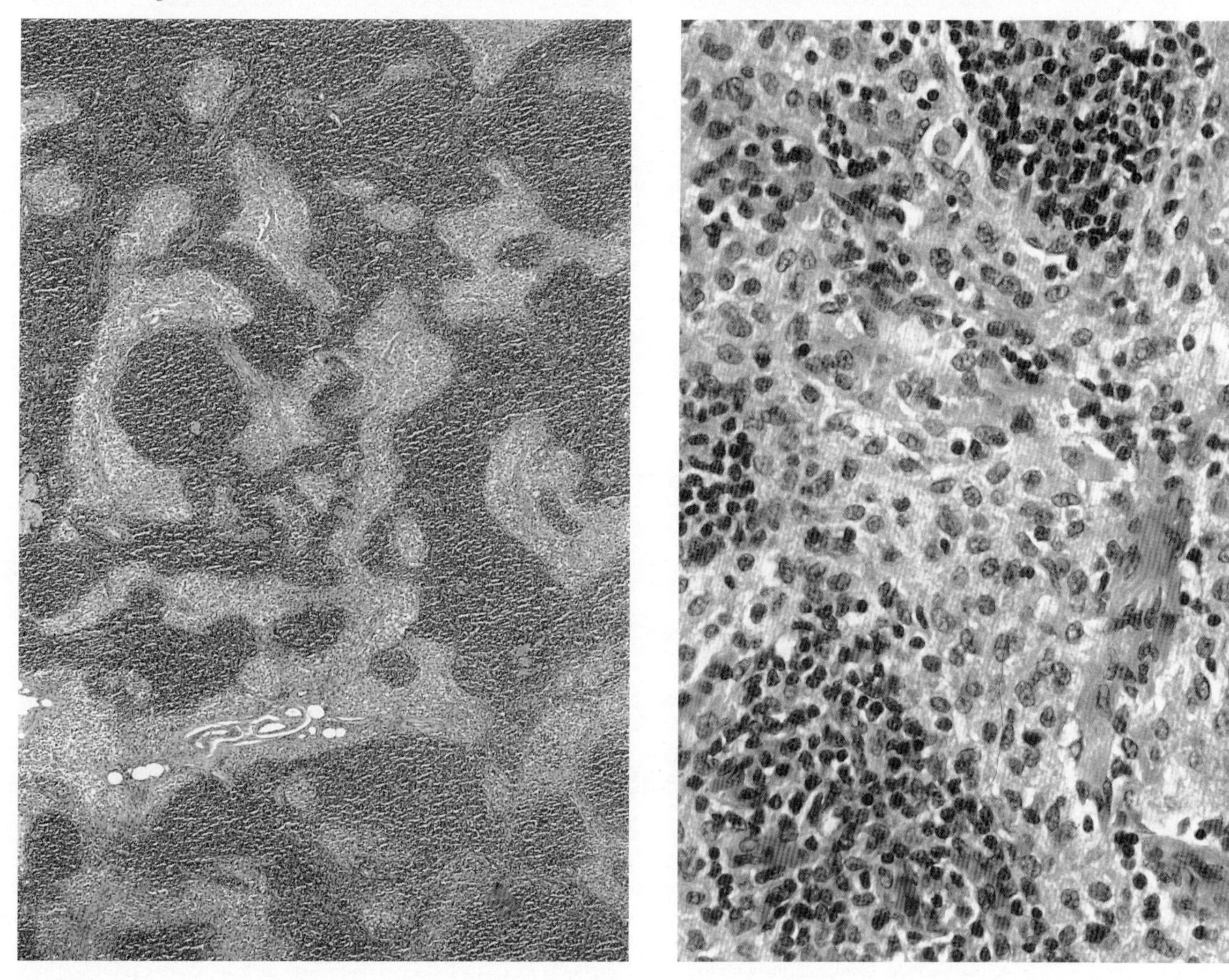

Figure 7–2 A

Figure 7–2 B

■ REACTIVE CHANGES

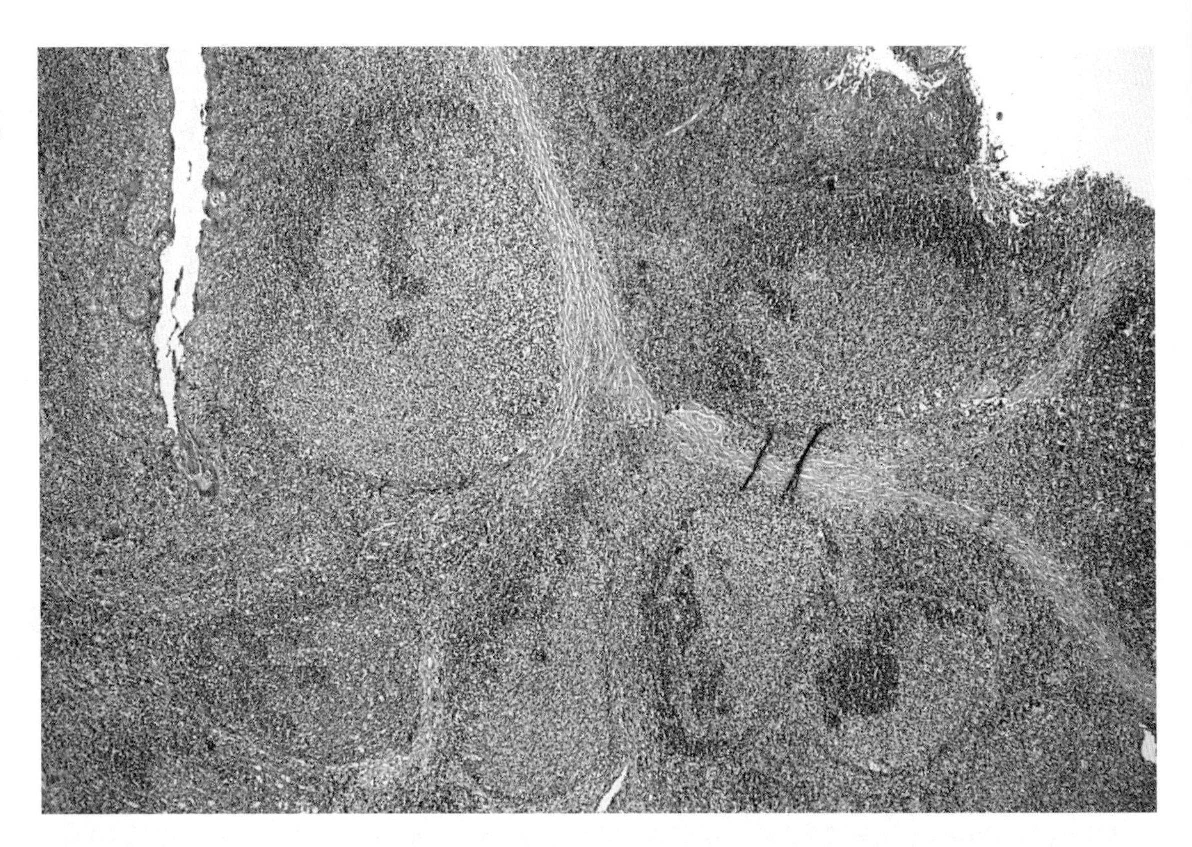

Figure 7–3 **A**

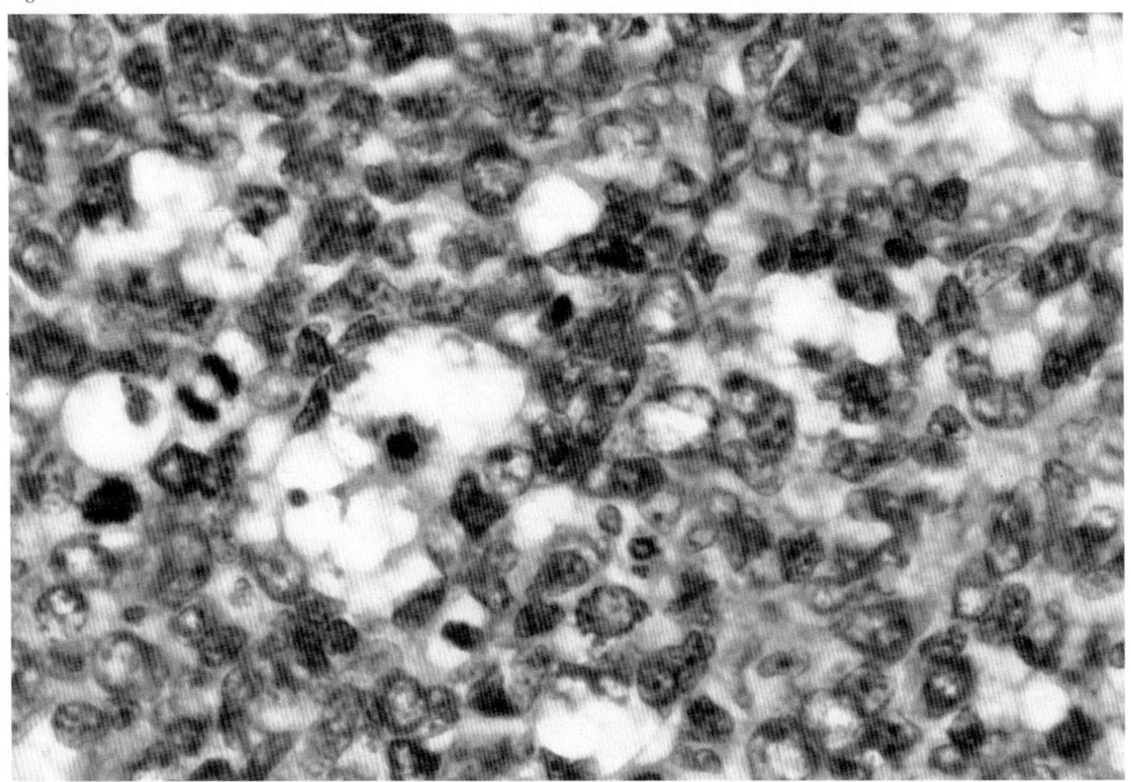

Figure 7–3 **B**

FIGURE 7–3. REACTIVE FOLLICULAR HYPERPLASIA: In reactive follicular hyperplasia involving lymphoid tissue, there is an increased number of follicles with prominent germinal centers. **A:** In this example of reactive follicular hyperplasia in the tonsil, follicles are increased in size and vary in size. The mantle zone surrounding the follicles is prominent, and the interfollicular zone is preserved. Compare this with follicular lymphoma in a lymph node (Figure 7–11 **B**). **B:** In reactive follicular hyperplasia, germinal centers are composed of a polymorphous population of activated cleaved and noncleaved lymphocytes and tingible body macrophages. Mitotic figures are also prominent.

■ REACTIVE CHANGES

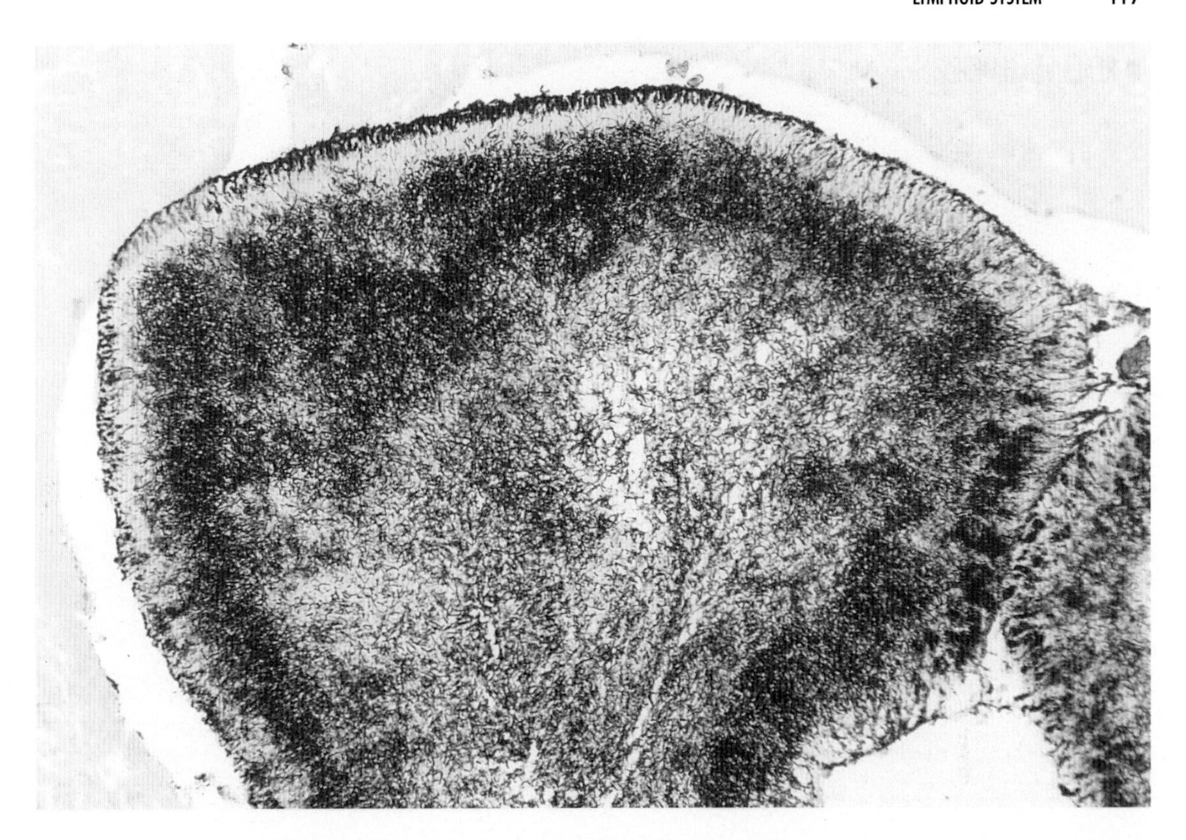

Figure 7–3 C

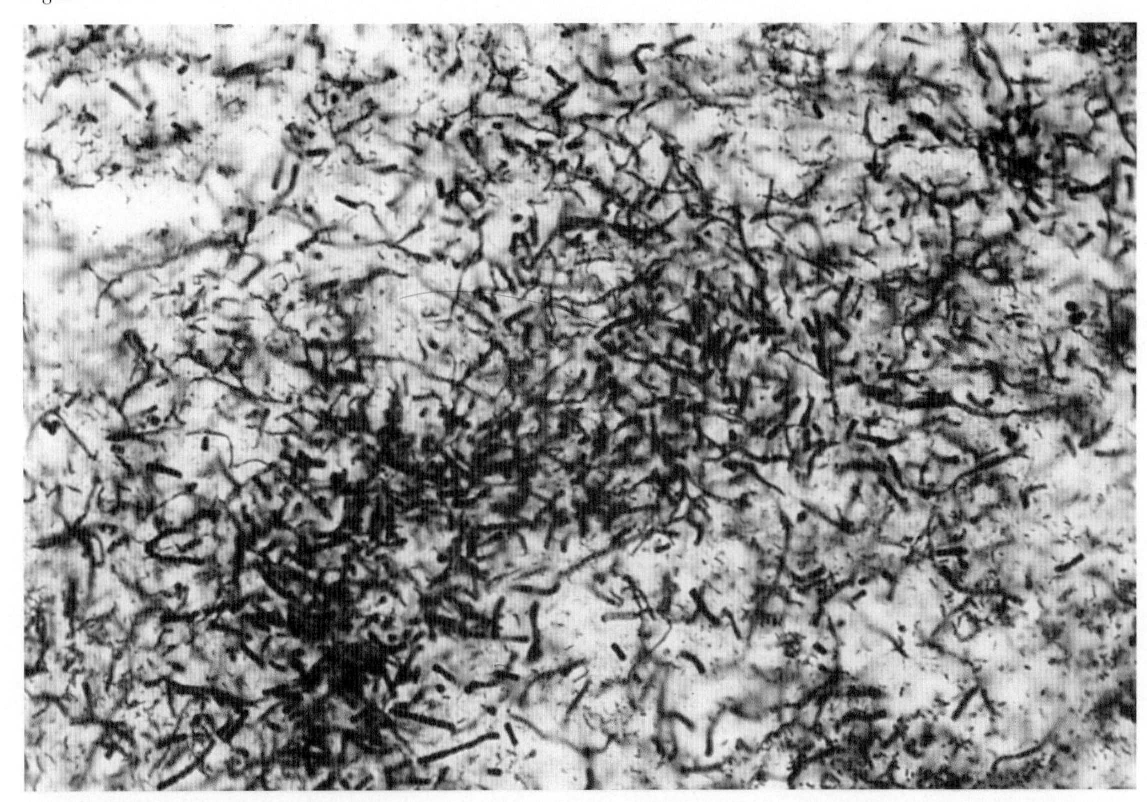

Figure 7–3 D

FIGURE 7–3. REACTIVE FOLLICULAR HYPERPLASIA: C: In this high-power photomicrograph taken from the crypts of the tonsils, a saprophytic actinomyces granule is noted. **D:** Silver stain shows the filamentous organisms that comprise the actinomyces granule.

■ REACTIVE CHANGES

FIGURE 7–4. SARCOIDOSIS: A: Sarcoidosis is characterized by noncaseating epithelioid granulomas with or without Langhans' giant cells. Laminated concretions—**Schaumann bodies** (**B**)—and stellate inclusions in giant cells—**asteroid bodies** (**C**)—are often seen in sarcoidosis.

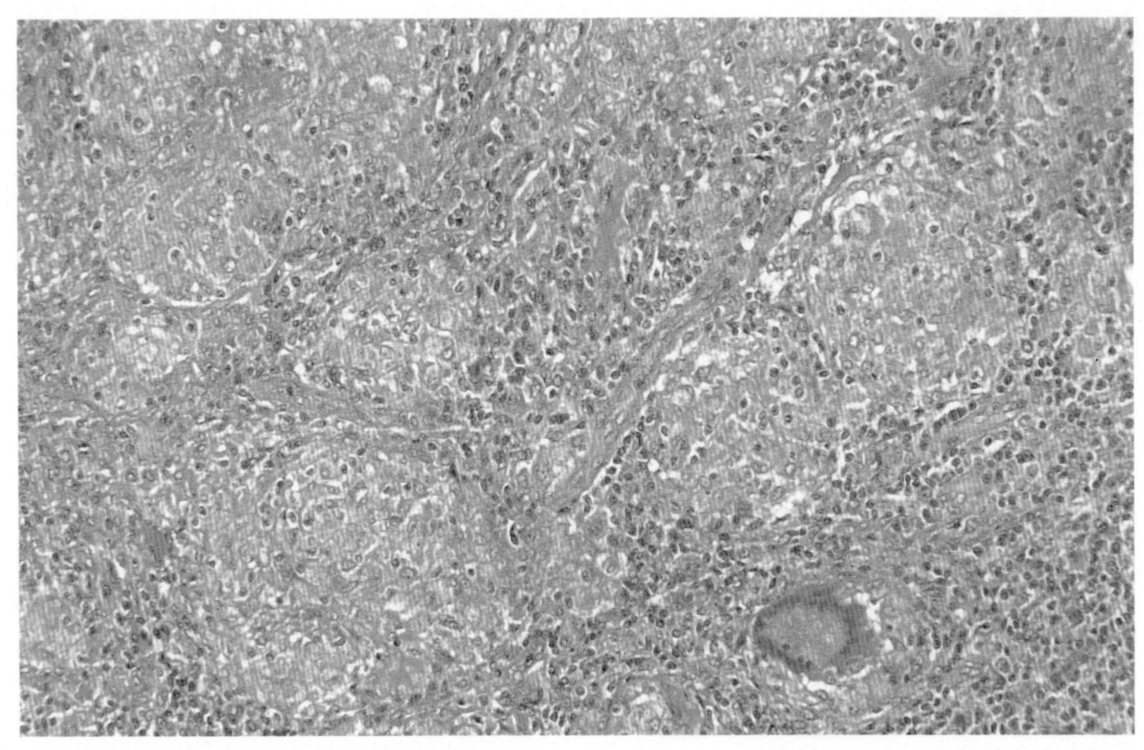

Figure 7–4 A

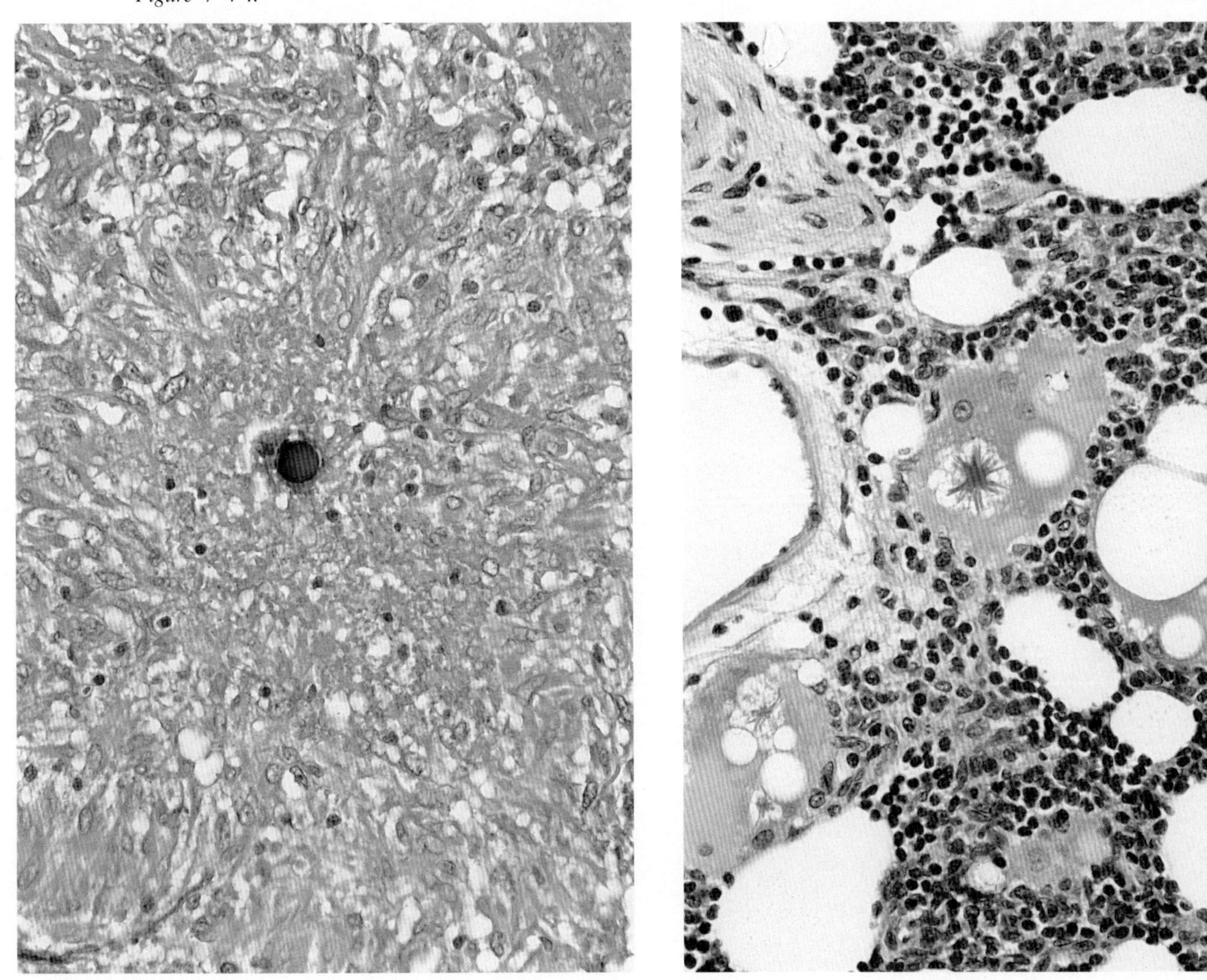

Figure 7–4 B

Figure 7–4 C

■ REACTIVE CHANGES

FIGURE 7–5. MYCOBACTERIAL GRANULOMATOUS INFLAMMATION: Infection with mycobacteria, tuberculosis, and atypical mycobacterial types is characterized by caseating granulomas (**A**), with Langhans' giant cells (**B**). **C**: The organisms stain red with an acid-fast stain.

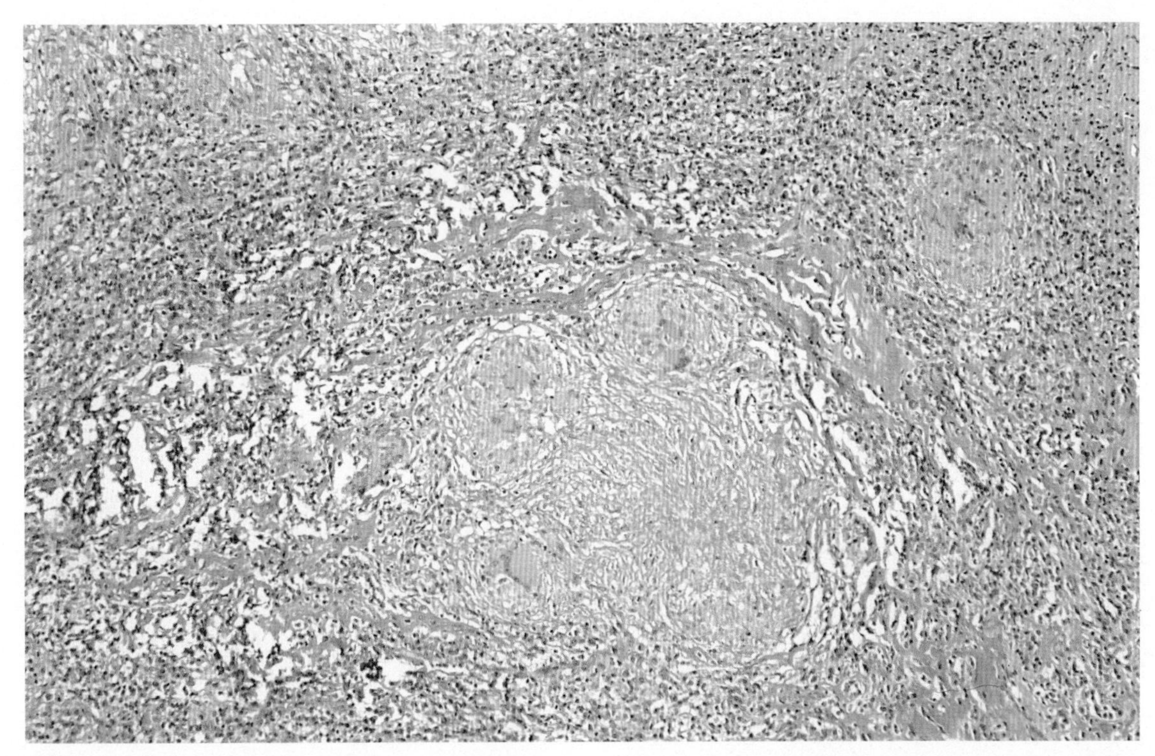

Figure 7–5 A

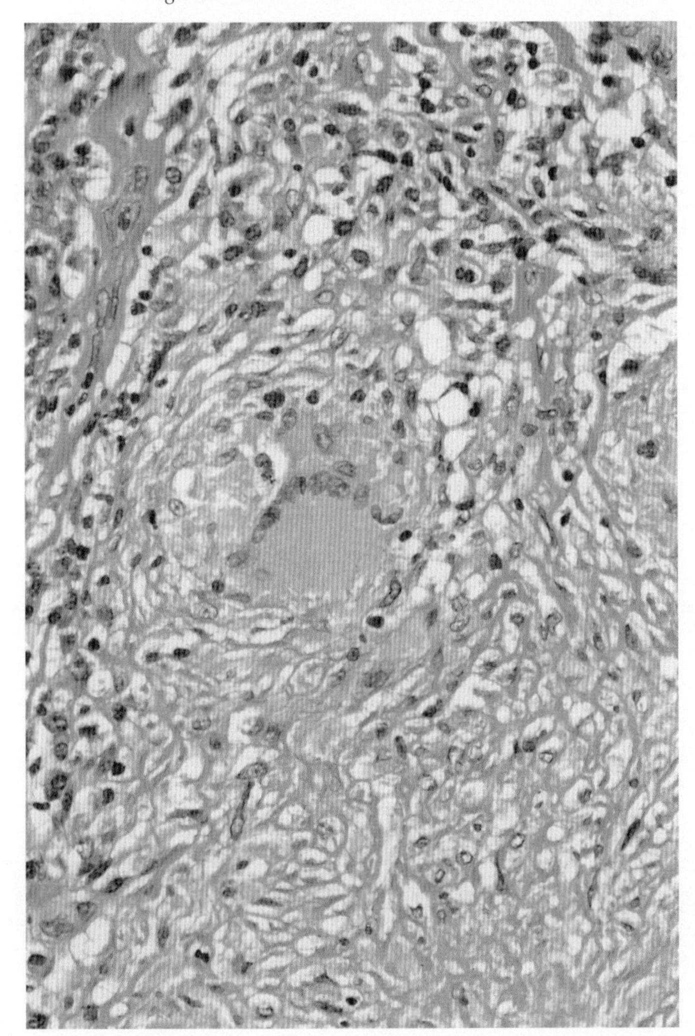

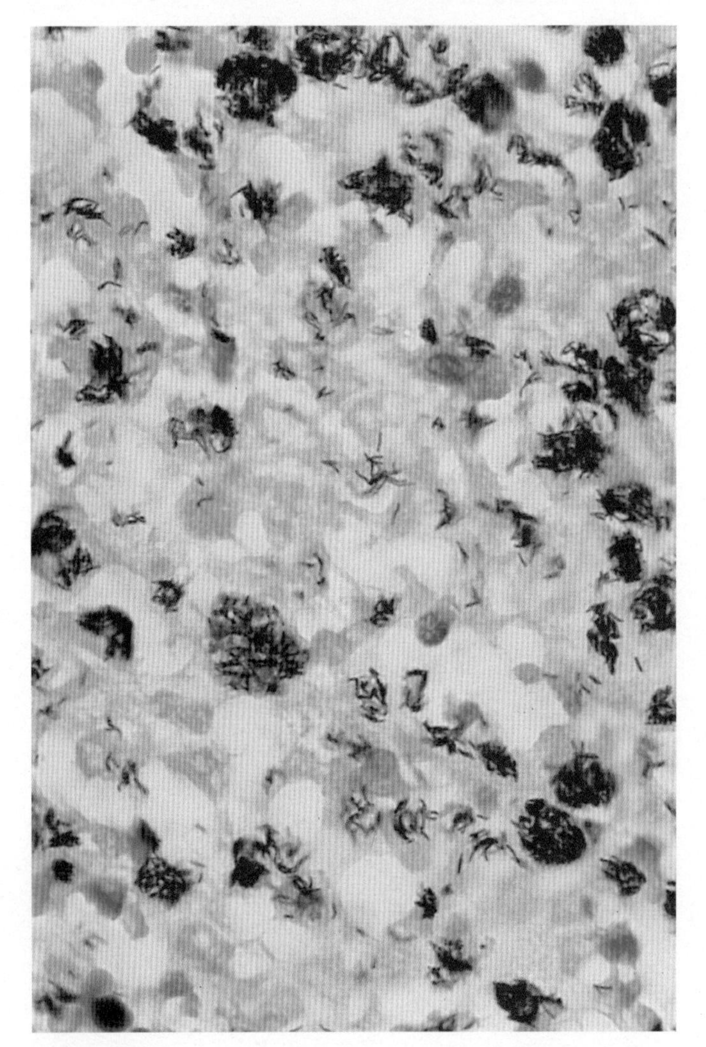

Figure 7–5 B

Figure 7–5 C

■ REACTIVE CHANGES

FIGURE 7–6. CAT-SCRATCH DISEASE: A: In the early stages of cat-scratch disease, follicular hyperplasia occurs. **B:** Suppurative granulomas are a characteristic feature. **C:** The coccobacilli that cause the disease are best seen with a Warthin-Starry silver stain.

Figure 7–6 A

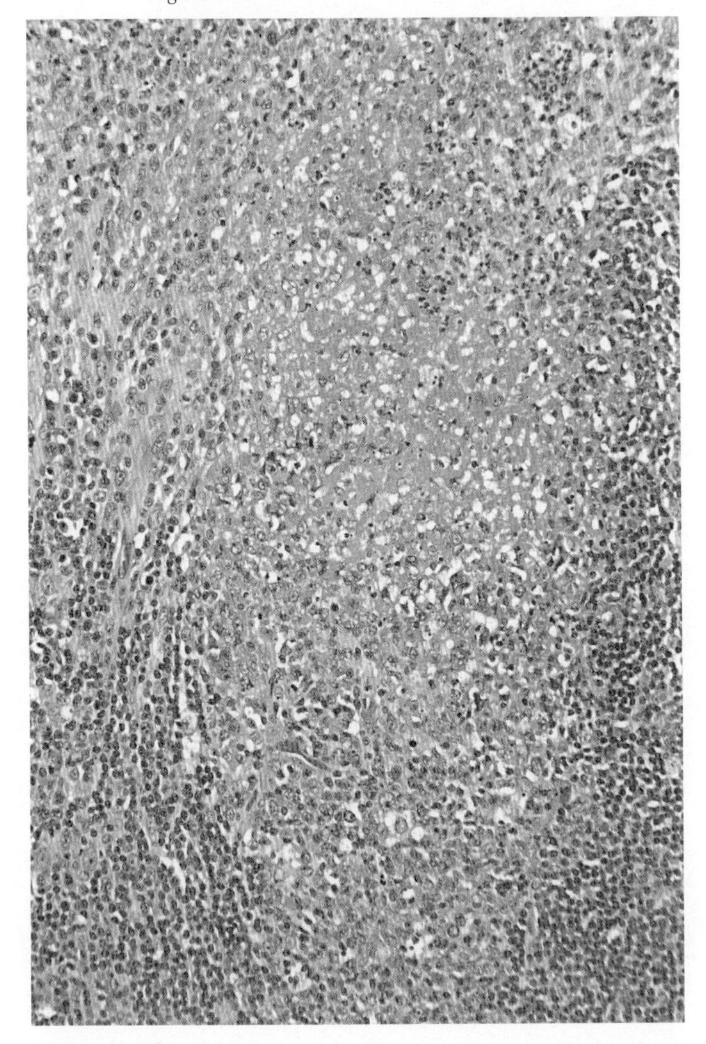

Figure 7–6 B

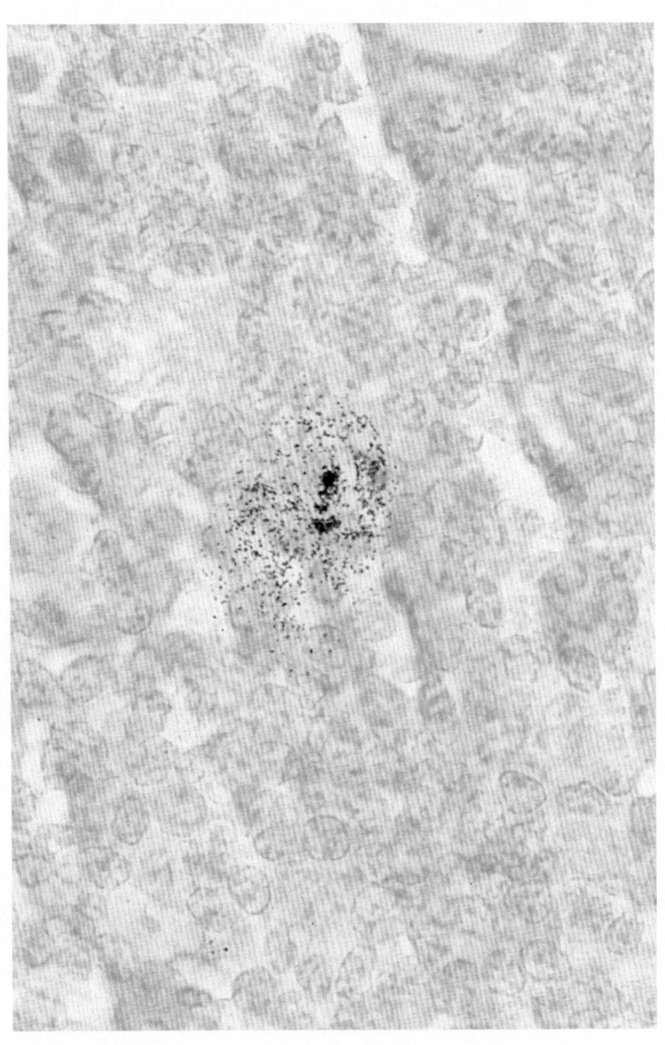

Figure 7–6 C

■ LYMPHOMAS

HODGKIN'S DISEASE

FIGURE 7–7. HODGKIN'S DISEASE: The diagnosis of Hodgkin's disease is made on the basis of finding a classic **Reed-Sternberg** (RS) cell. **A:** The classic **RS cell** is characterized as having abundant cytoplasm and binucleation, with prominent "owl-eye" inclusion-like nucleoli. In lymphocyte depletion Hodgkin's disease, RS cells predominate. **B:** **Mummified cells** are dying RS cells.

FIGURE 7–8. LYMPHOCYTE PREDOMINANCE HODGKIN'S DISEASE: Lymphocyte predominance Hodgkin's disease is uncommon. It is characterized by **"L/H" cells** that have polylobated folded nuclei with vesicular chromatin and indistinct nucleoli.

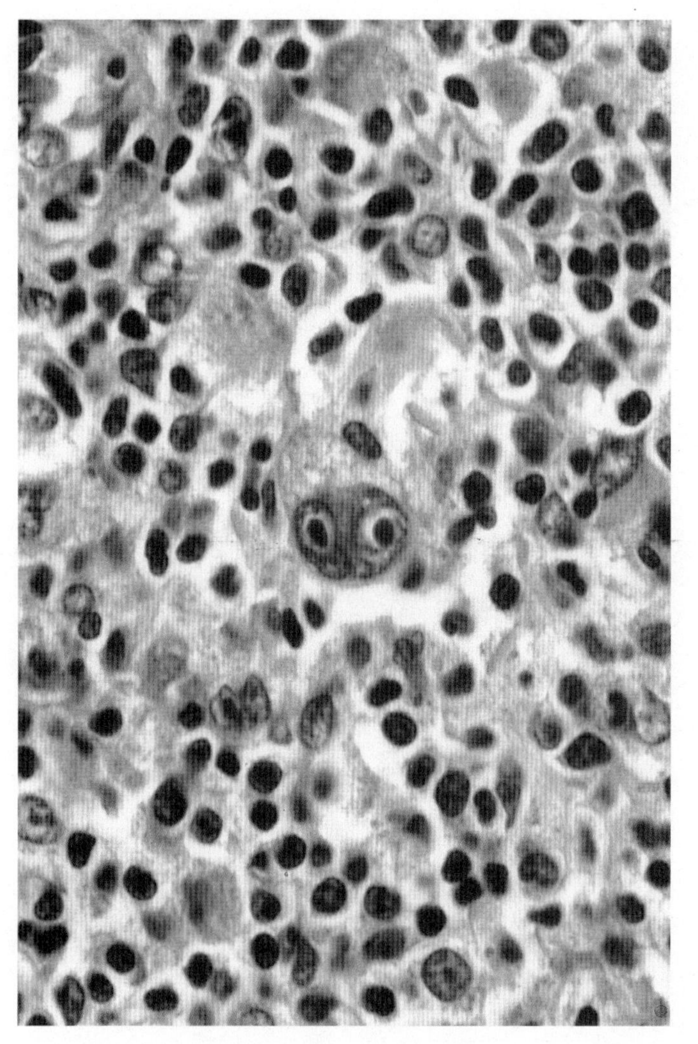

Figure 7–7 A

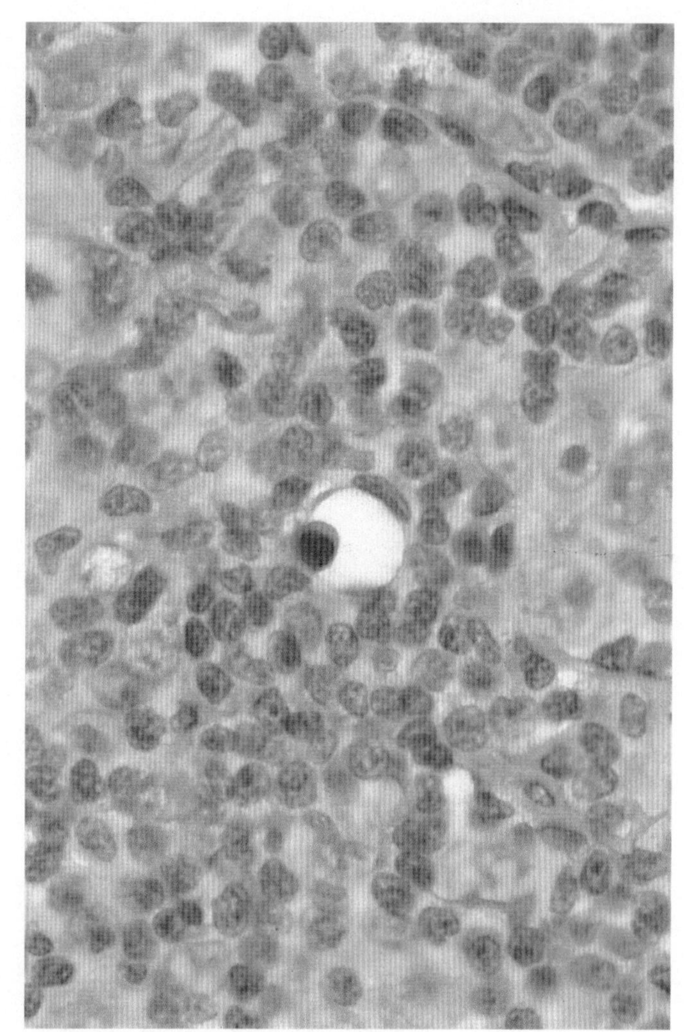

Figure 7–7 B

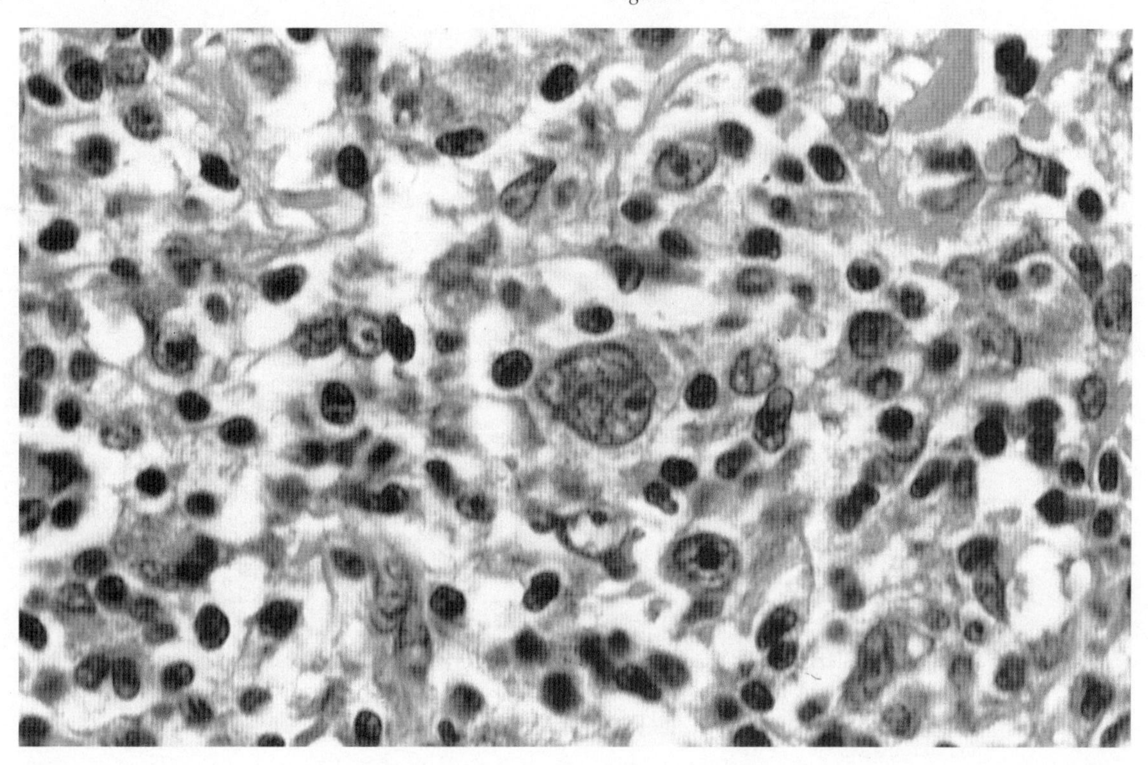

Figure 7–8

■ LYMPHOMAS

HODGKIN'S DISEASE

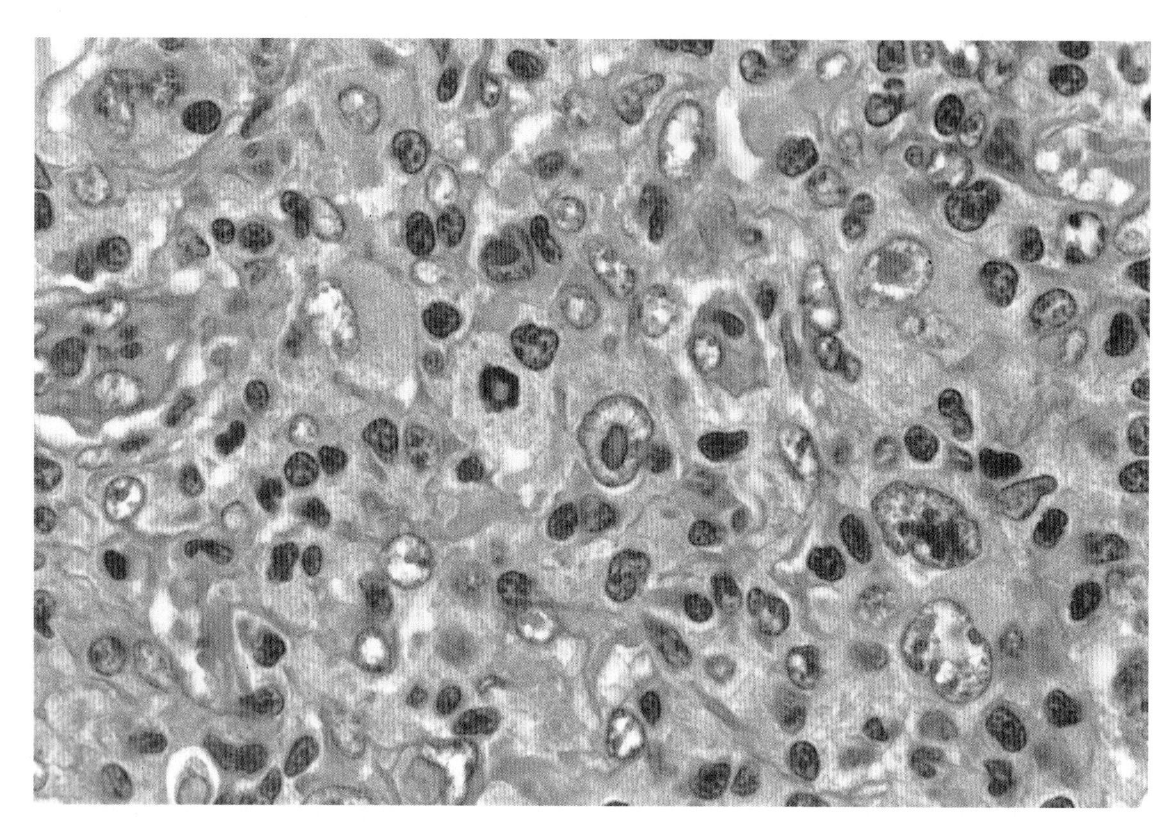

Figure 7–9 A

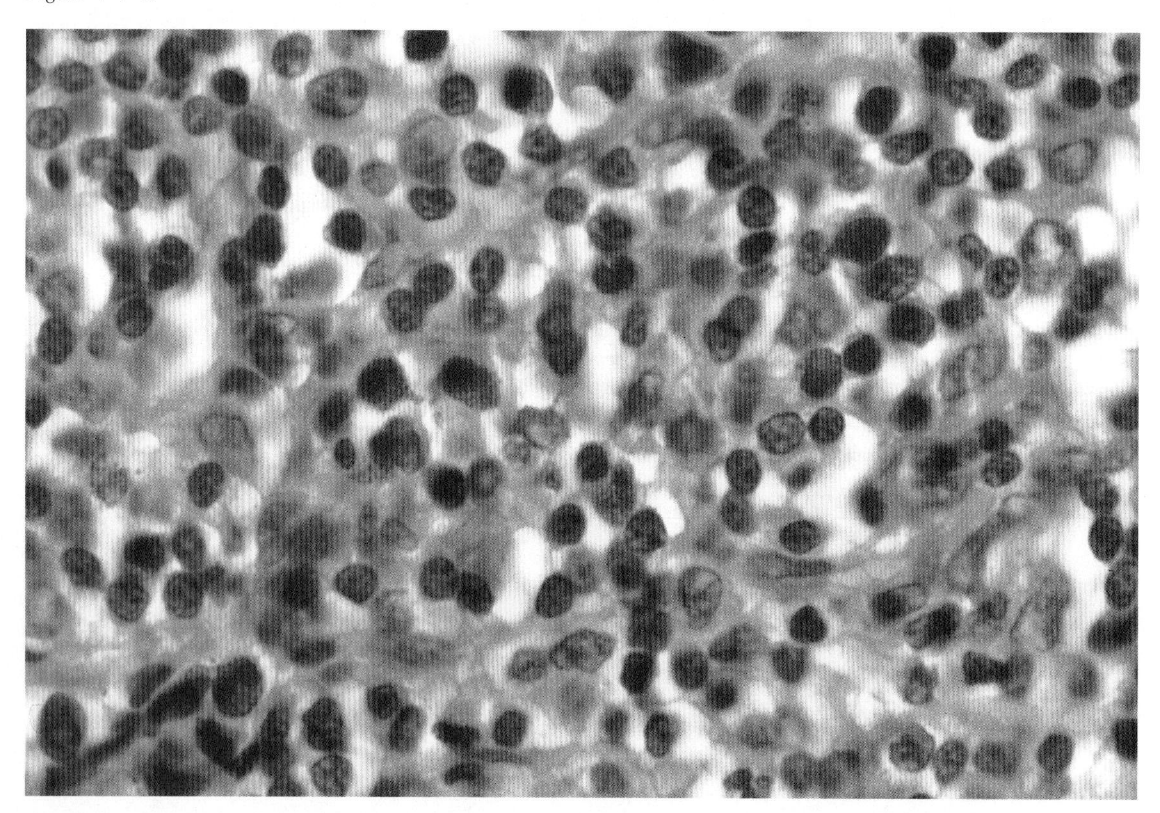

Figure 7–9 B

FIGURE 7–9. MIXED CELLULARITY HODGKIN'S DISEASE: Mixed cellularity Hodgkin's disease is the second most common variant of Hodgkin's disease. **A:** It is characterized by an abundance of RS cells and mononuclear Hodgkin's cells. **B:** The hallmark of this variant is the polymorphous background population composed of eosinophils, lymphocytes, histiocytes, plasma cells, and fibroblasts.

■ LYMPHOMAS

HODGKIN'S DISEASE

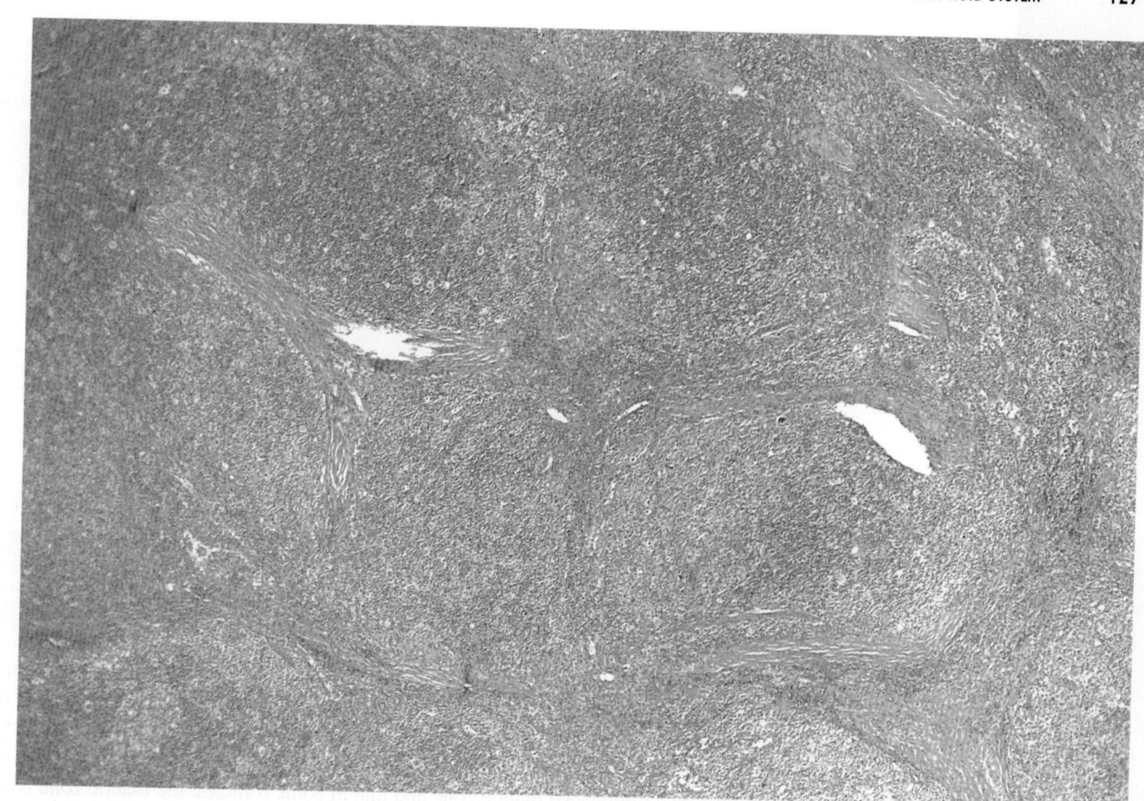

Figure 7–10 A

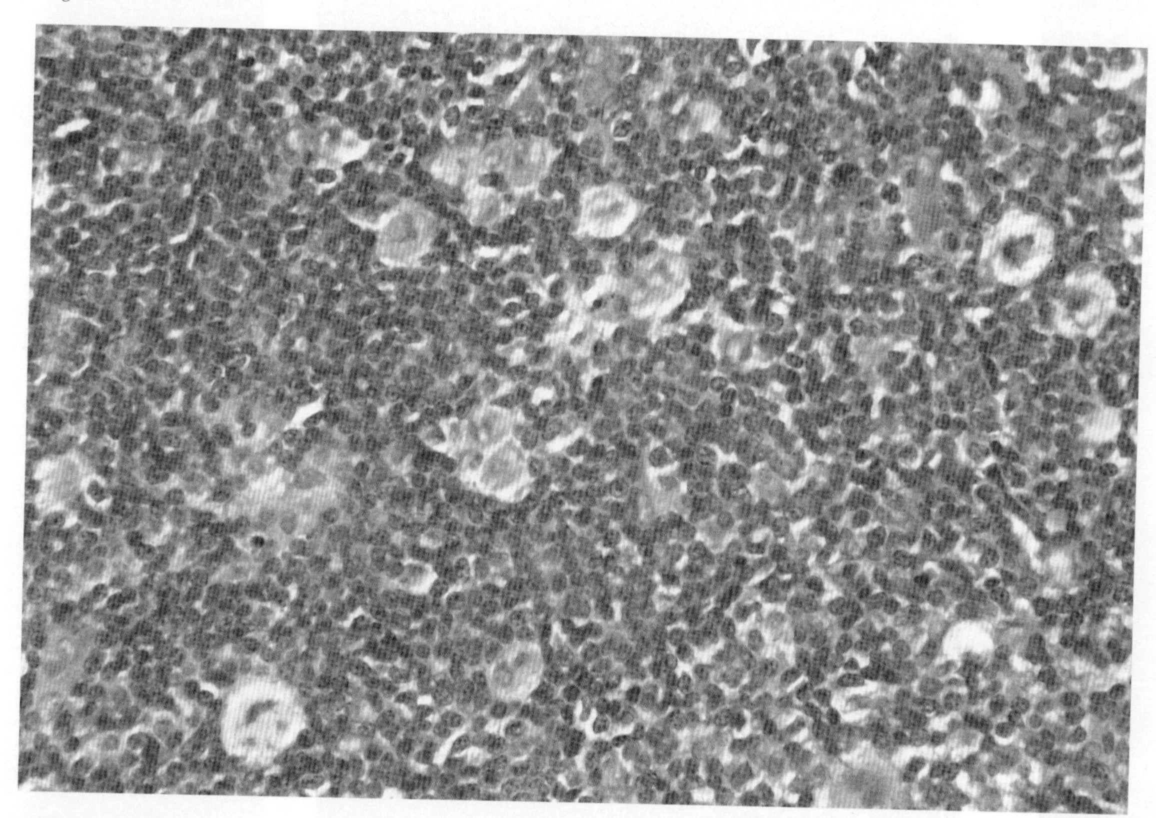

Figure 7–10 B

FIGURE 7–10. NODULAR SCLEROSING HODGKIN'S DISEASE: Nodular sclerosing Hodgkin's disease is the most common variant of Hodgkin's disease. A: Broad bands of collagen separate the lymph node into nodules. B: **Lacunar cells** are present only in formalin-fixed tissue and therefore are an artifact of fixation. Lacunar cells are mononucleated or multinucleated cells lying in a clear space. Classic RS cells are infrequent, and the diagnosis rests on the presence of the lacunar cells.

■ NON-HODGKIN'S LYMPHOMAS

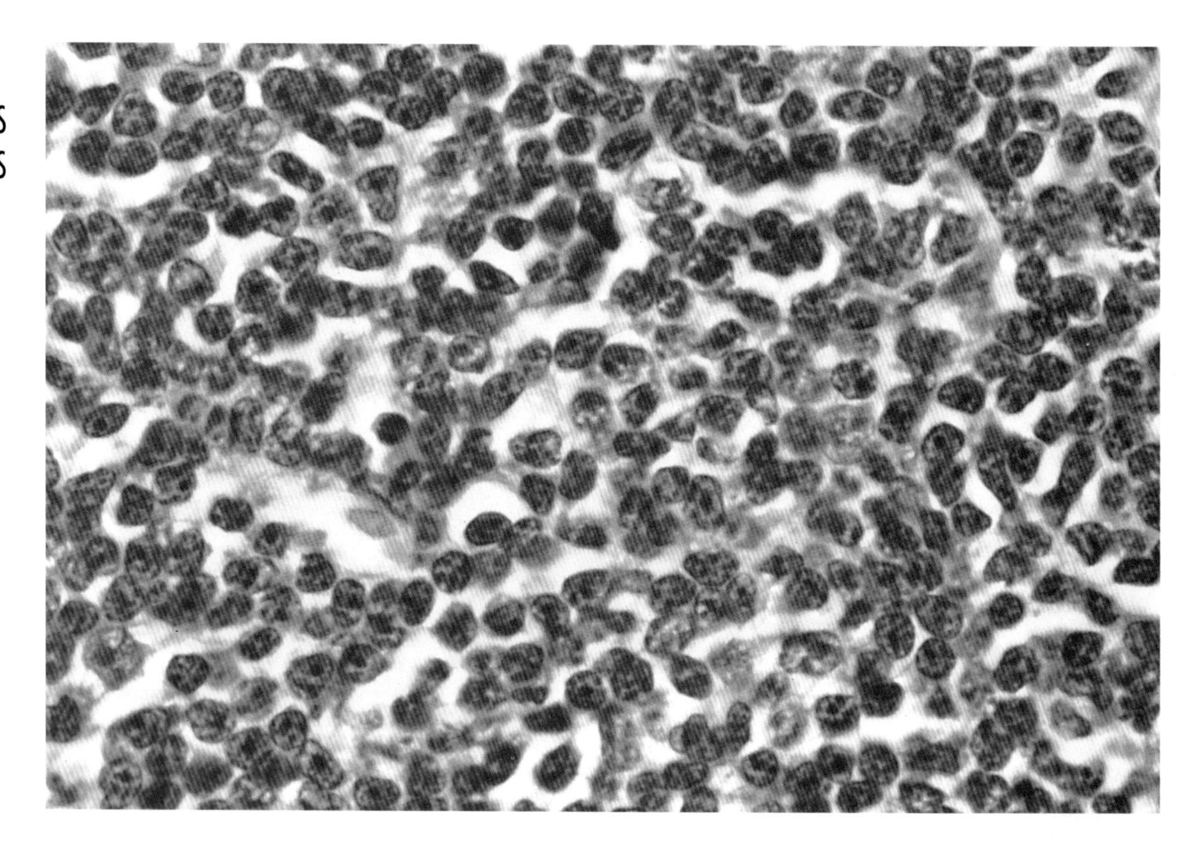

Figure 7–12 **A**

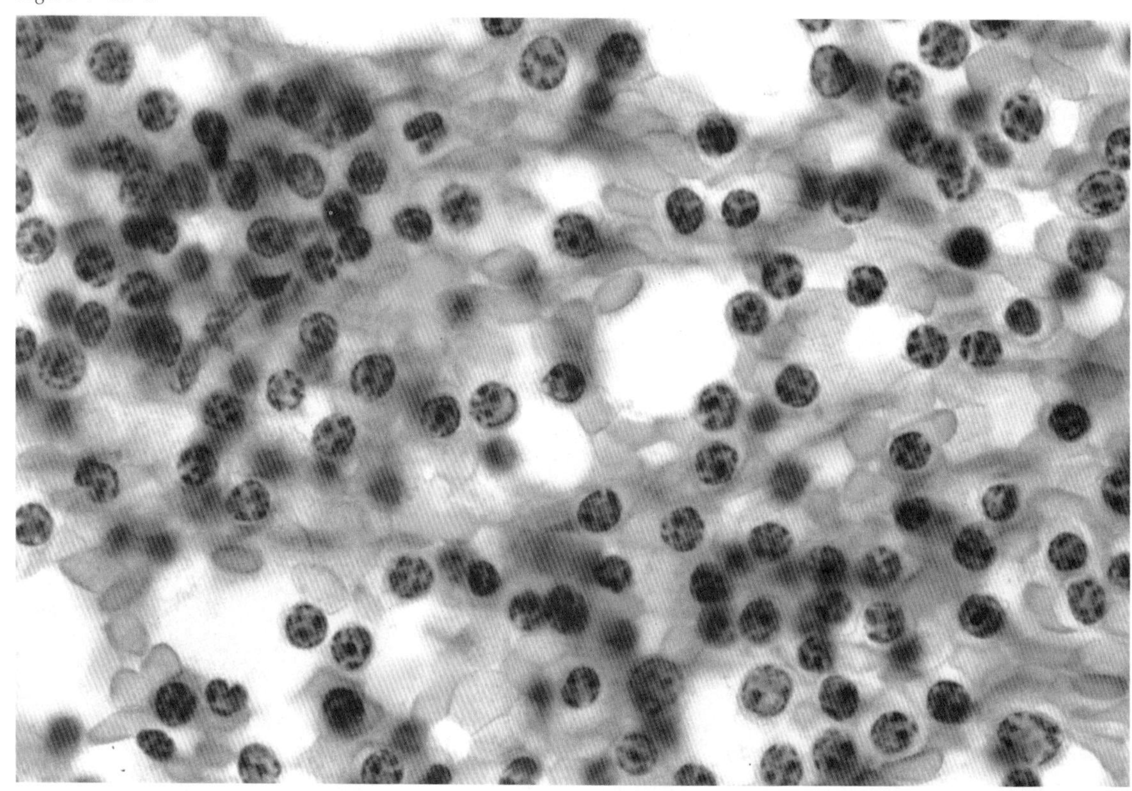

Figure 7–12 **B**

FIGURE 7–12. MALIGNANT LYMPHOMA, SMALL LYMPHOCYTIC TYPE (WELL-DIFFERENTIATED LYMPHOCYTIC LYMPHOMA): A: In malignant lymphoma, small lymphocytic type, there is diffuse effacement of the lymph node architecture. **B:** The cells are monotonous and small with scant cytoplasm, round regular nuclei, and clumped chromatin (touch preparation).

■ LYMPHOMAS

HODGKIN'S DISEASE

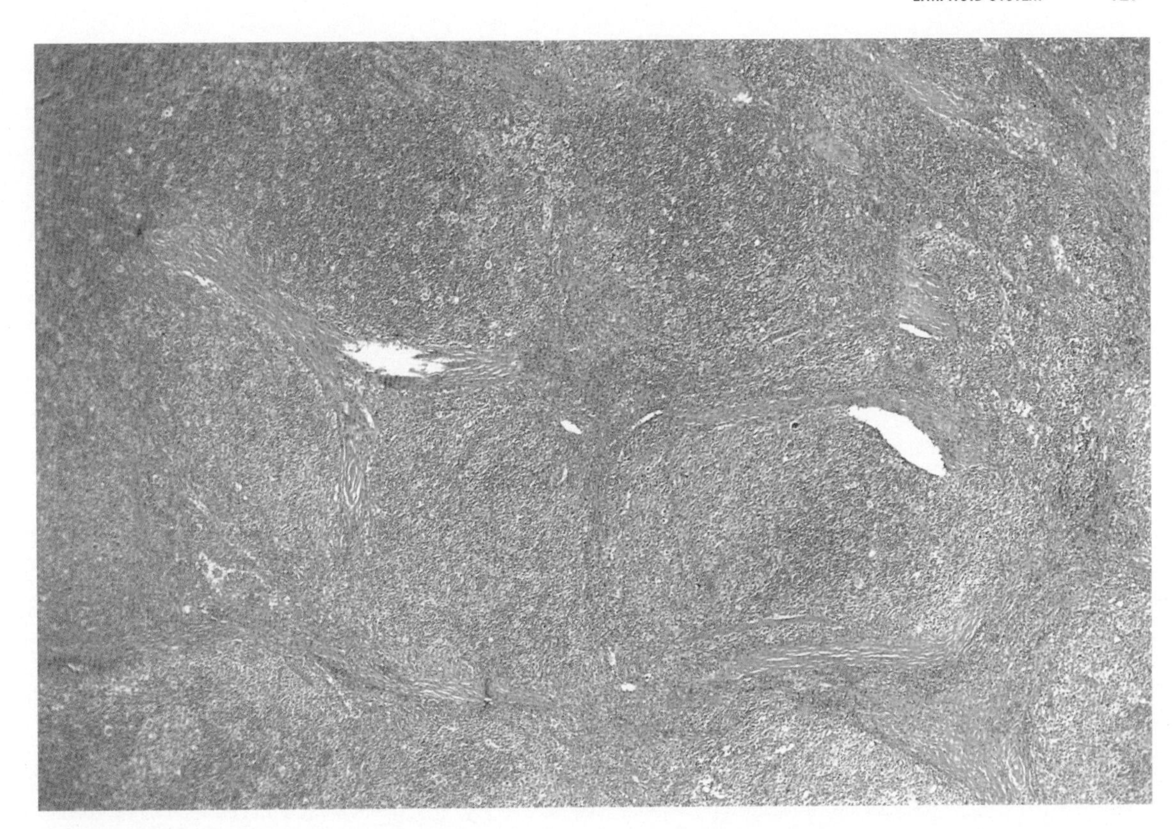

Figure 7–10 **A**

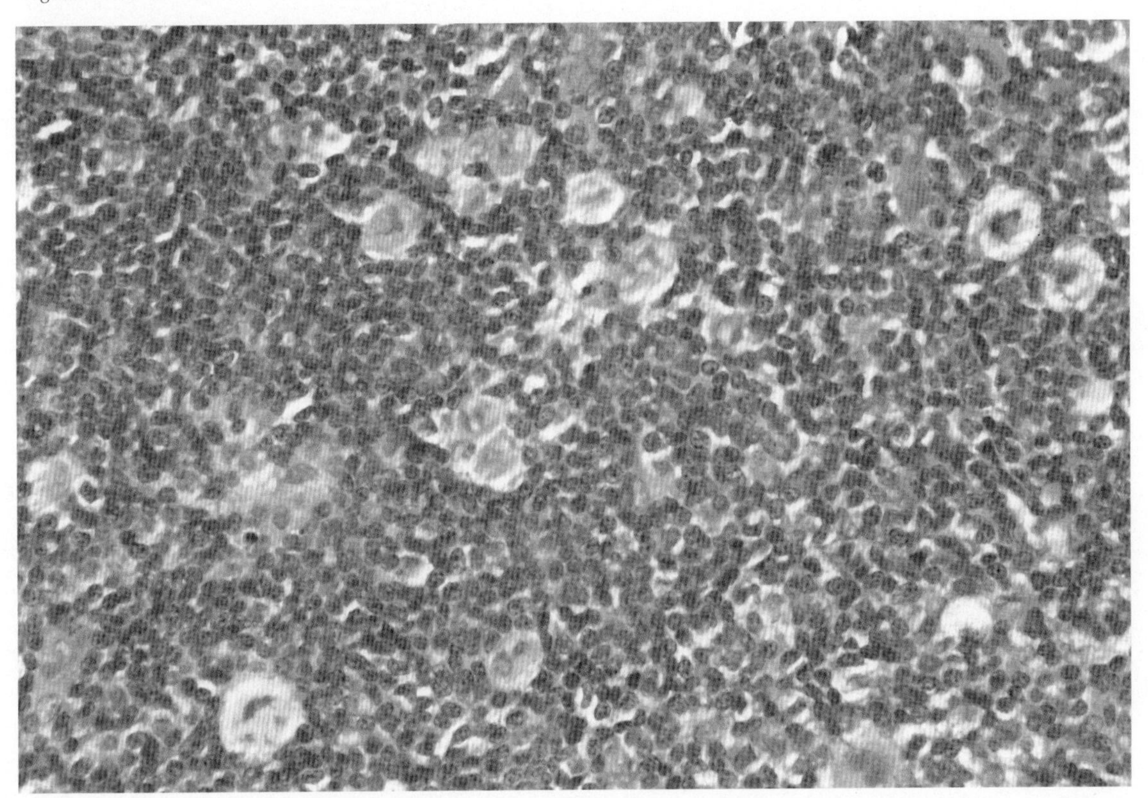

Figure 7–10 **B**

FIGURE 7–10. NODULAR SCLEROSING HODGKIN'S DISEASE: Nodular sclerosing Hodgkin's disease is the most common variant of Hodgkin's disease. **A:** Broad bands of collagen separate the lymph node into nodules. **B: Lacunar cells** are present only in formalin-fixed tissue and therefore are an artifact of fixation. Lacunar cells are mononucleated or multinucleated cells lying in a clear space. Classic RS cells are infrequent, and the diagnosis rests on the presence of the lacunar cells.

■ NON-HODGKIN'S LYMPHOMAS

FIGURE 7–11. MALIGNANT LYMPHOMA: A: Features supporting the diagnosis of malignant lymphoma are effacement of the normal lymph node architecture and extension of the malignant cells beyond the confines of the lymph node. The effacement can be in a follicular pattern (**B**) or a diffuse pattern (**C**). Contrast follicular lymphoma (Figure 7–11 **B**) with follicular hyperplasia (Figure 7–3 **A**). Note in follicular lymphoma that the follicles are surrounded by a diminished mantle zone and that in some follicles, the mantle zone is absent. Also note that the follicles appear to have fused with each other.

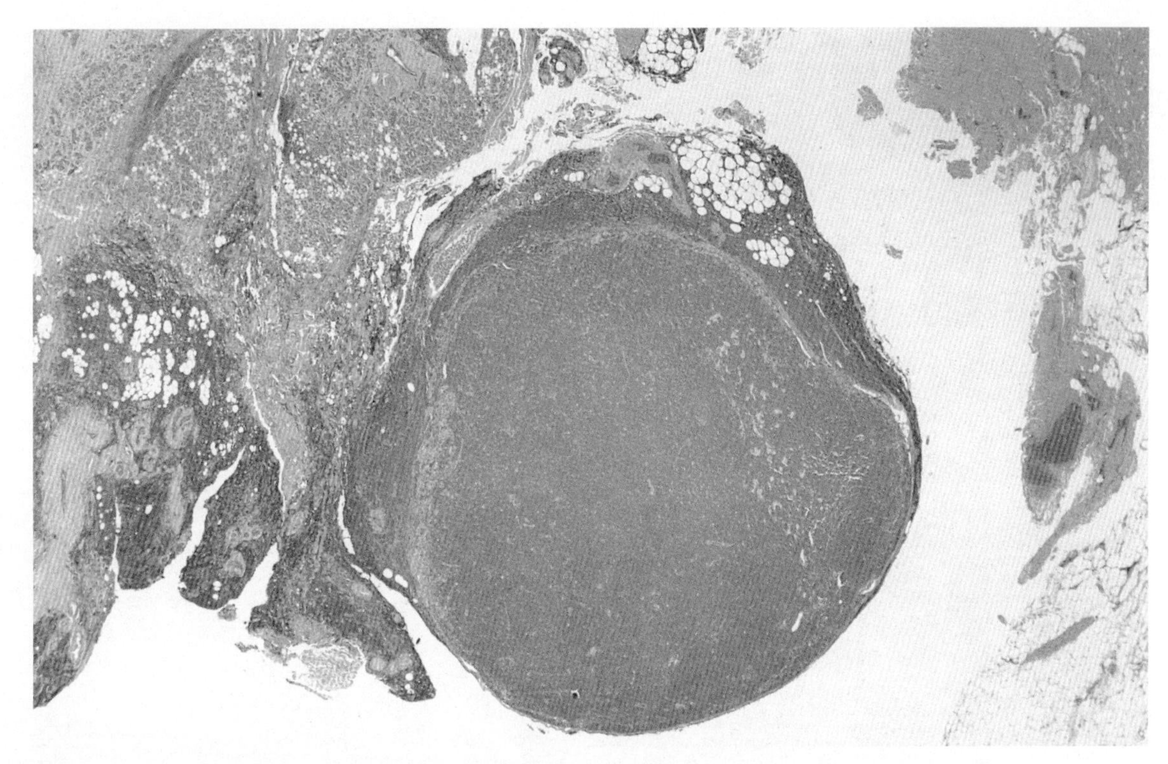

Figure 7–11 A

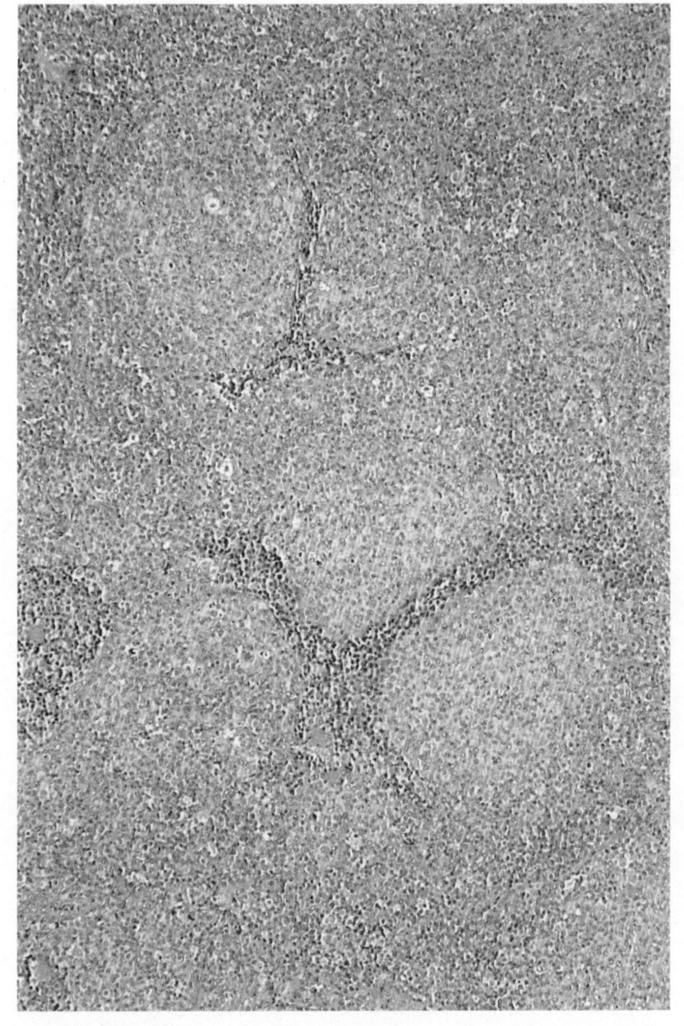

Figure 7–11 B

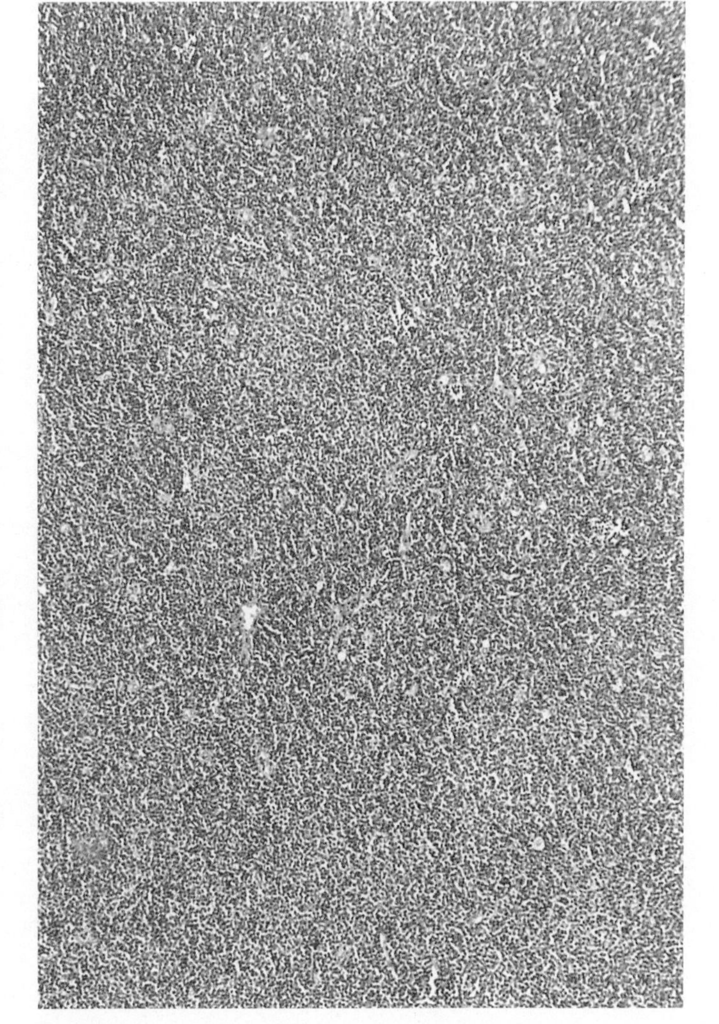

Figure 7–11 C

■ NON-HODGKIN'S LYMPHOMAS

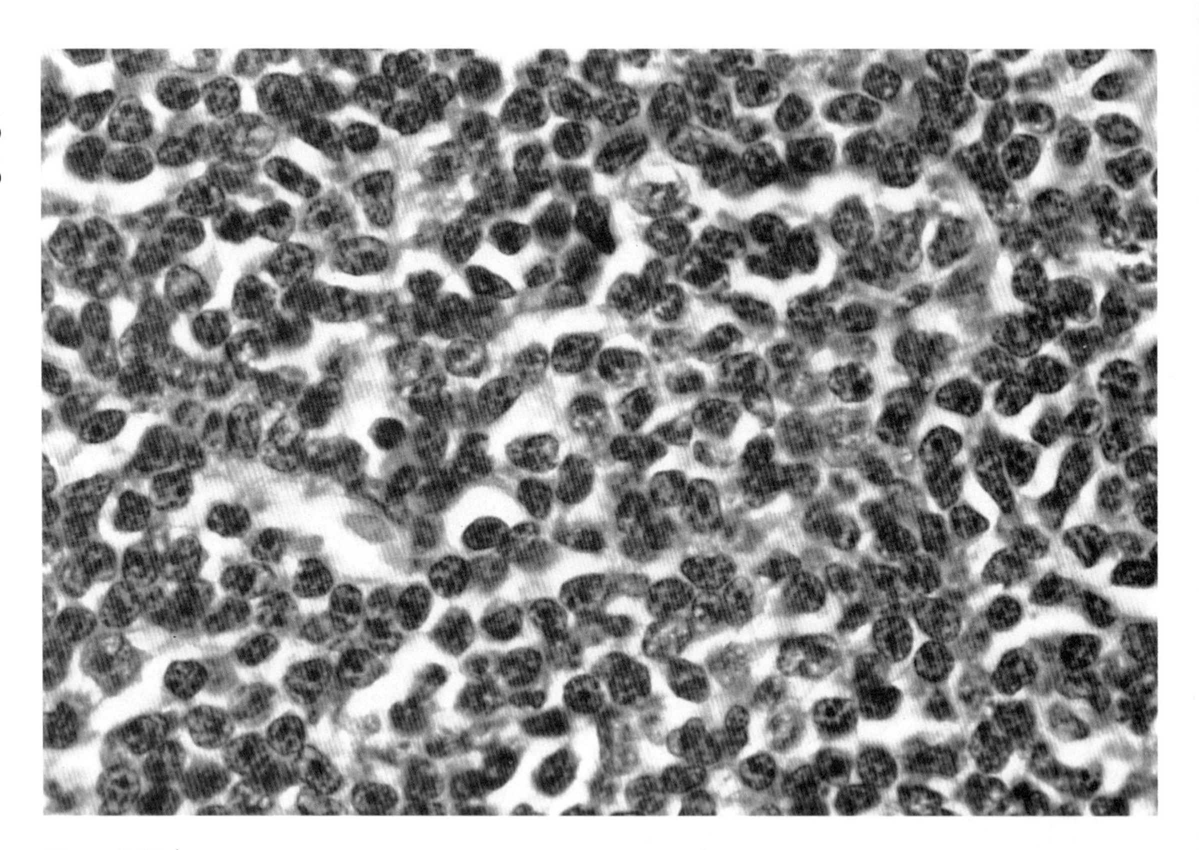

Figure 7–12 A

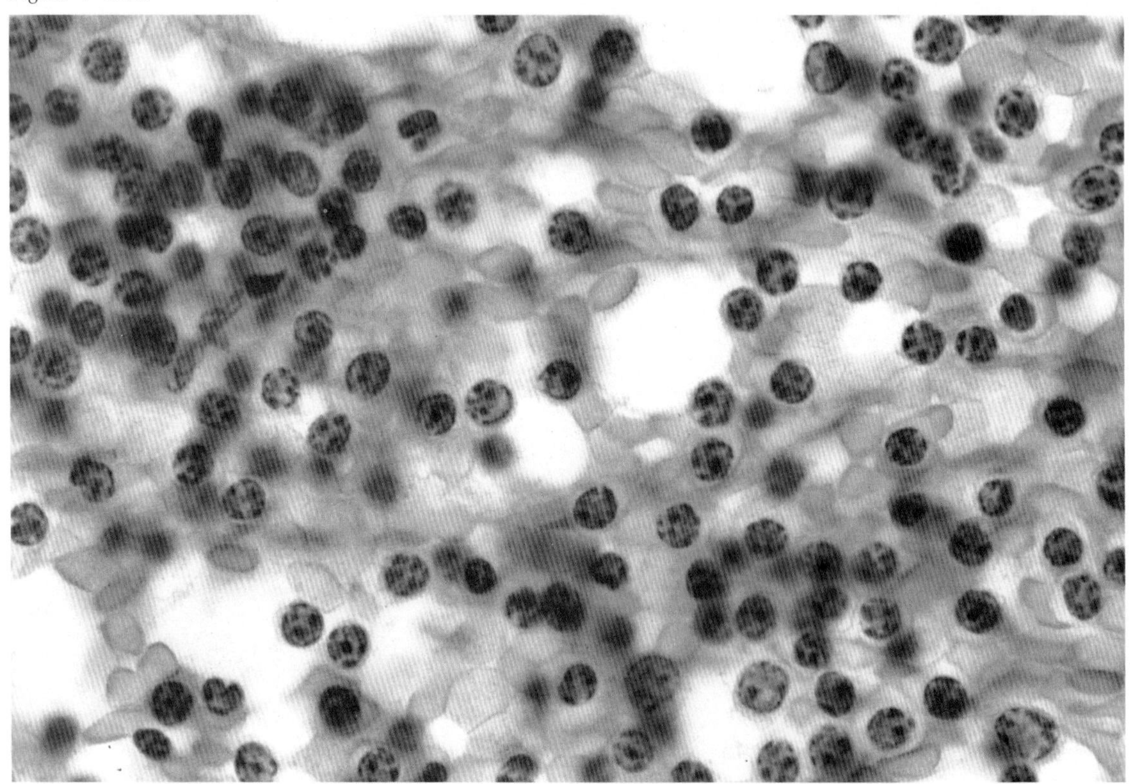

Figure 7–12 B

FIGURE 7–12. MALIGNANT LYMPHOMA, SMALL LYMPHOCYTIC TYPE (WELL-DIFFERENTIATED LYMPHOCYTIC LYMPHOMA): A: In malignant lymphoma, small lymphocytic type, there is diffuse effacement of the lymph node architecture. **B:** The cells are monotonous and small with scant cytoplasm, round regular nuclei, and clumped chromatin (touch preparation).

■ NON-HODGKIN'S LYMPHOMAS

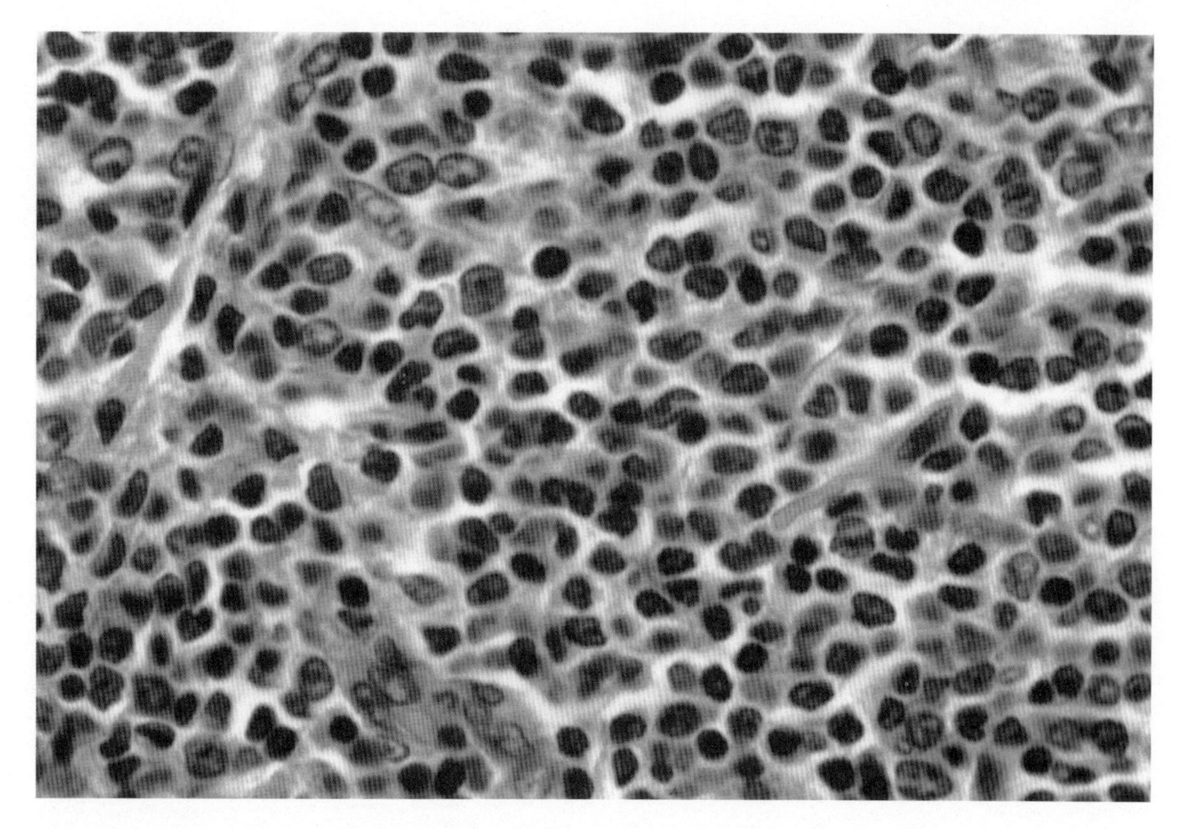

Figure 7–13 **A**

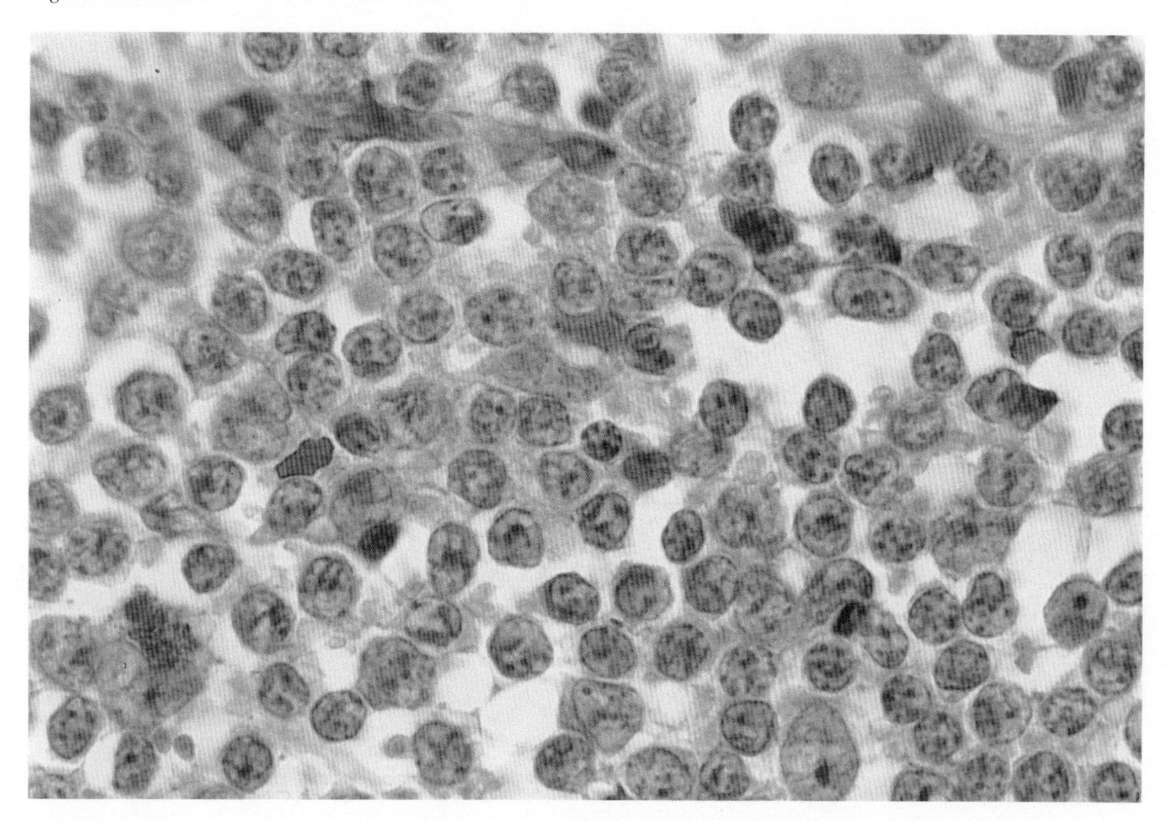

Figure 7–13 **B**

FIGURE 7–13. MALIGNANT LYMPHOMA, PREDOMINANTLY SMALL CLEAVED CELL TYPE (POORLY DIFFERENTIATED LYMPHOCYTIC LYMPHOMA): A: Small cleaved cells are slightly larger than their normal counterparts and, as the name implies, their nuclei are cleaved and chromatin is condensed (**B**, touch preparation).

■ NON-HODGKIN'S LYMPHOMAS

FIGURE 7–14. MALIGNANT LYMPHOMA, PREDOMINANTLY LARGE CELL TYPE (HISTIOCYTIC LYMPHOMA): A: The cells in large cell lymphoma are two to three times larger than their normal counterparts. Nuclei are larger than endothelial cell nuclei and are predominantly round to oval, with vesicular chromatin and small nucleoli (**B**, touch preparation).

FIGURE 7–15. MALIGNANT LYMPHOMA, MIXED, SMALL CLEAVED AND LARGE CELL TYPE (MIXED CELL LYMPHOMA): In small cleaved and large cell lymphoma, there is, as the name suggests, an admixture of small cleaved and large cells. The large cells have a diameter of two to three times that of a small lymphocyte. Endothelial cells lining a blood vessel are used as a size reference. Small cells are smaller than and large cells are larger than the endothelial cells.

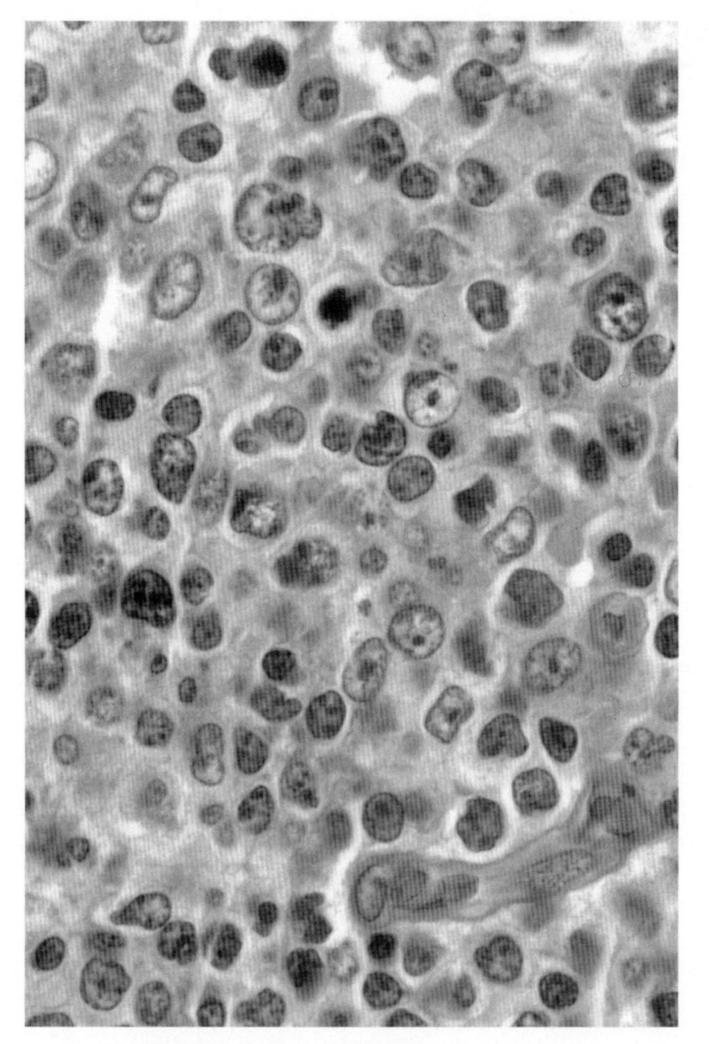

Figure 7–14 A

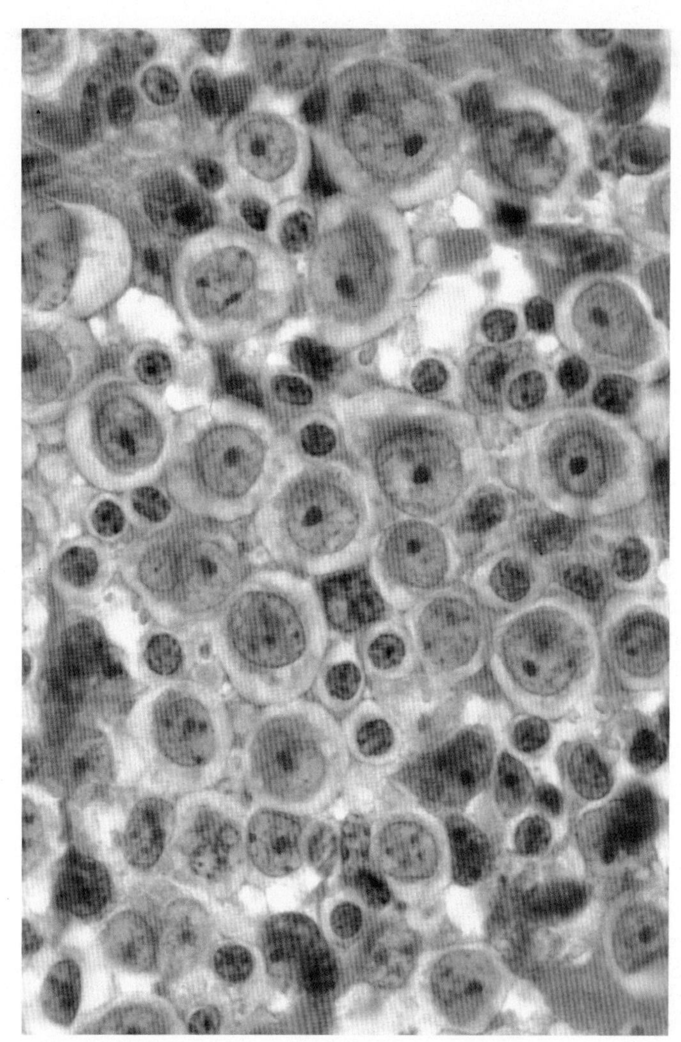

Figure 7–14 B

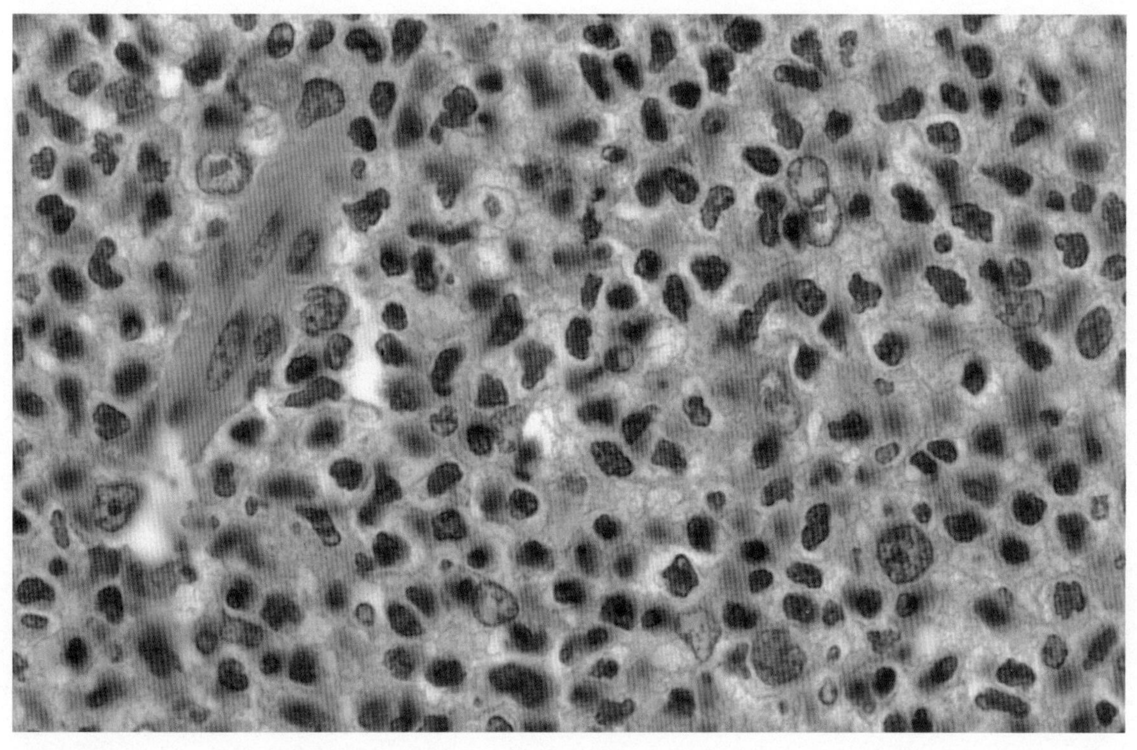

Figure 7–15

■ NON-HODGKIN'S LYMPHOMAS

FIGURE 7–16. MALIGNANT LYMPHOMA, SMALL NON-CLEAVED CELL TYPE (BURKITT'S LYMPHOMA): A: Burkitt's lymphoma classically has a **"starry sky"** appearance at low magnification. **B:** The background is of medium-size dark blue neoplastic lymphocytes punctuated with clear spaces occupied by nonneoplastic macrophages. **C:** The malignant cells are round to oval with scant cytoplasm and round or oval nuclei with multiple nucleoli (touch preparation).

Figure 7–16 A

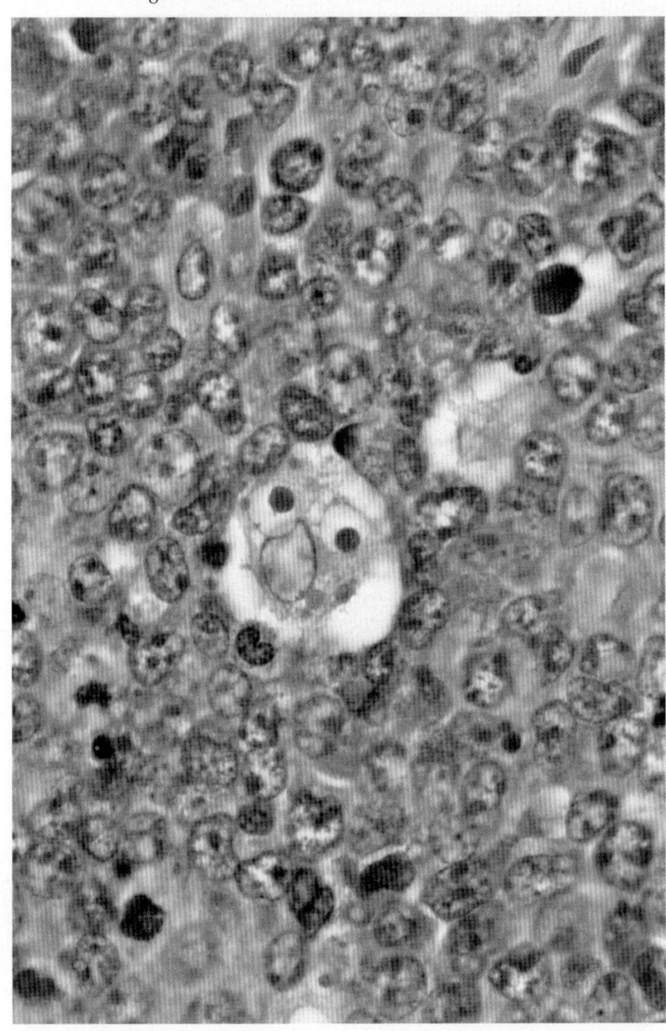

Figure 7–16 B

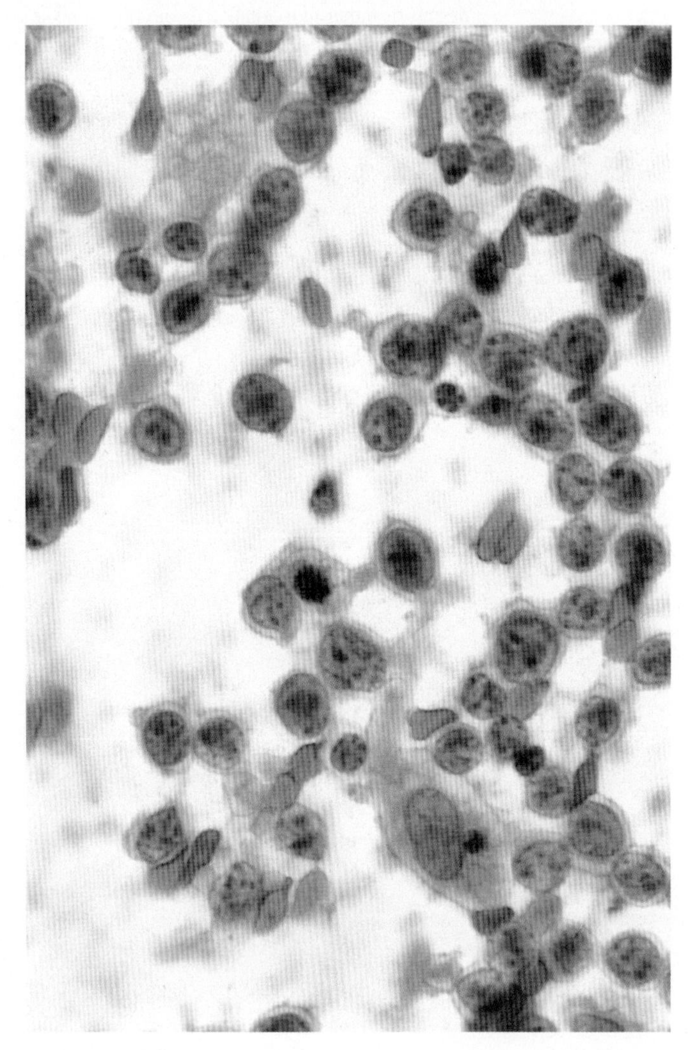

Figure 7–16 C

■ NON-HODGKIN'S LYMPHOMAS

FIGURE 7–17. MYCOSIS FUNGOIDES/SÉZARY SYNDROME: This type of lymphoma is a neoplasm composed of mature T helper cells. Mycosis fungoides and Sézary syndrome affect the skin and lymph nodes. **A:** The neoplastic lymphocytes are present in a lichenoid (bandlike) infiltrate in the superficial dermis. **B:** The epidermis is acanthotic, with elongation of rete ridges and focal parakeratosis. Neoplastic cells infiltrate the epidermis as either individual cells or collections known as **Pautrier's microabscesses.** **C:** Cytologically, the cells are convoluted and "cerebriform." In Sézary syndrome, neoplastic lymphocytes spill out into the peripheral blood.

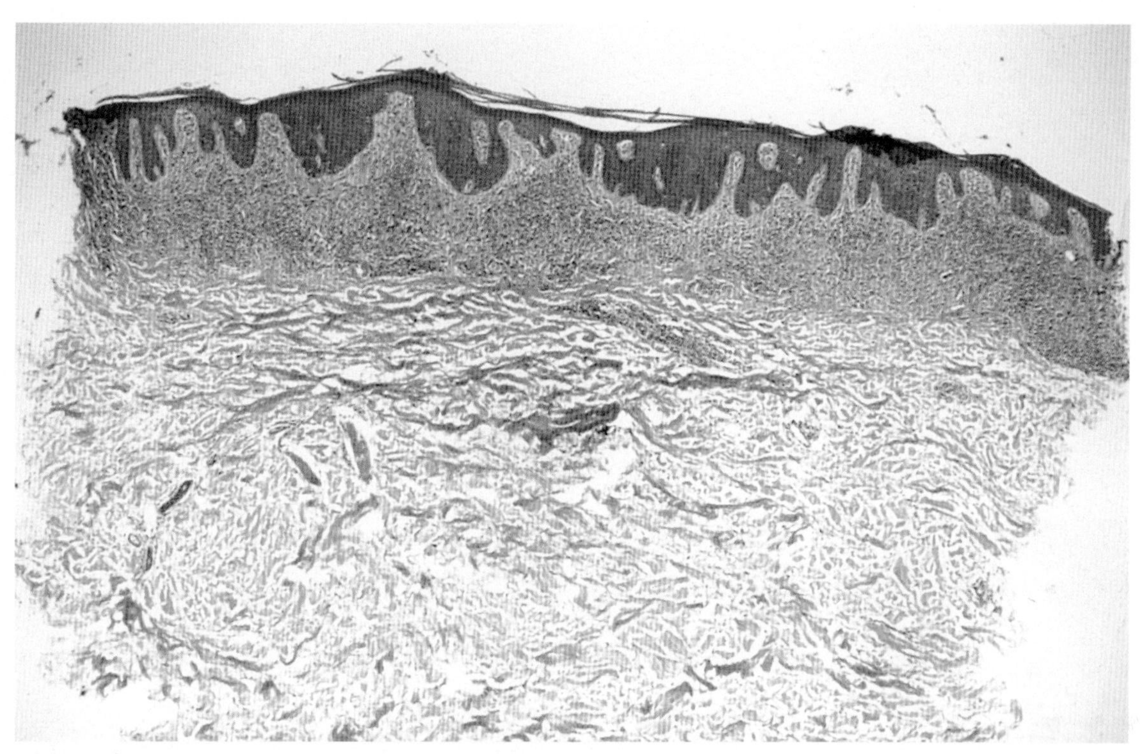

Figure 7–17 A

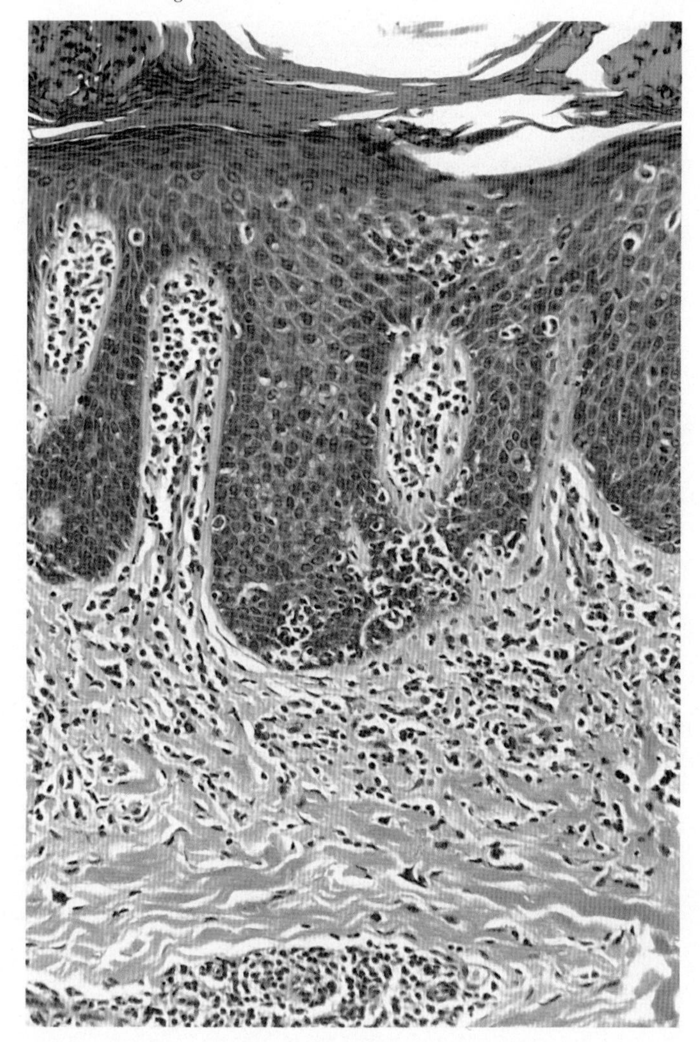

Figure 7–17 B

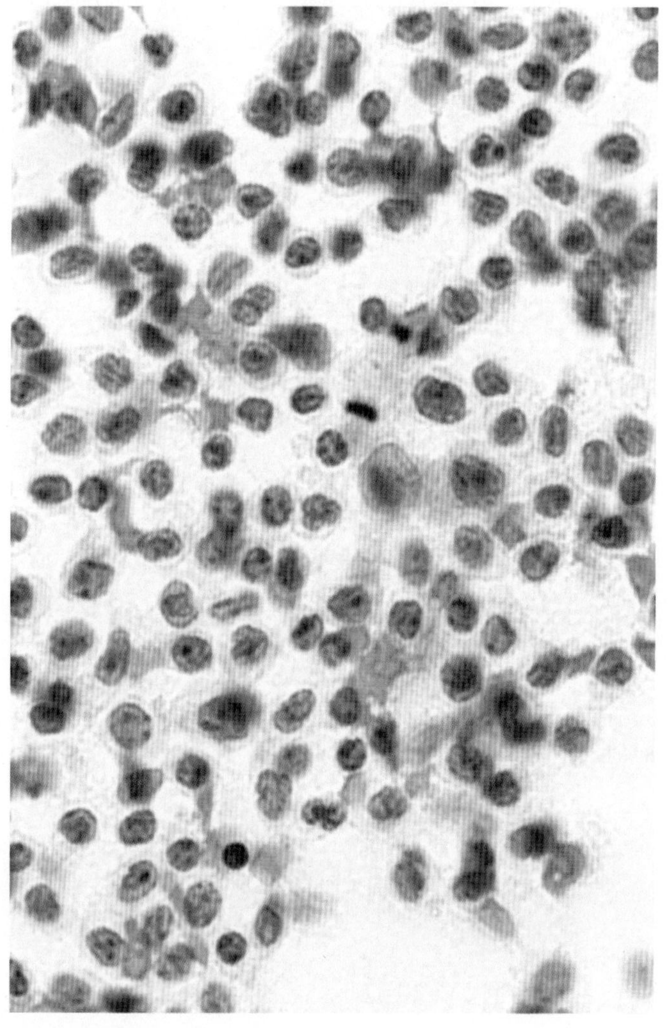

Figure 7–17 C

■ MULTIPLE MYELOMA

FIGURE 7–18. PLASMA CELL MYELOMA: A: Plasma cell myelomas are composed of sheets of neoplastic plasma cells representing greater than 10% of the cell population. **B:** An excess of immunoglobulin is noted as eosinophilic intracytoplasmic droplets within the plasma cells. An excess of abnormal immunoglobulins may enter the serum and produce an **"M spike"** on electrophoresis. **C:** In this immunofixation electrophoretogram, there is evidence of immunoglobulin G (IgG) kappa monoclonal gammopathy.

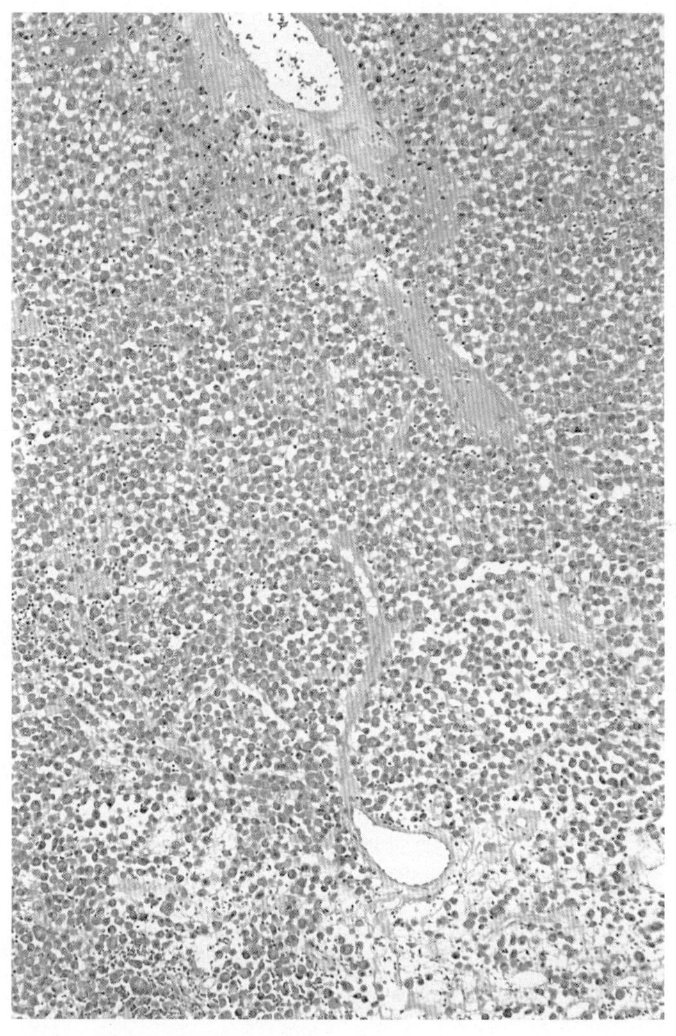

Figure 7–18 A

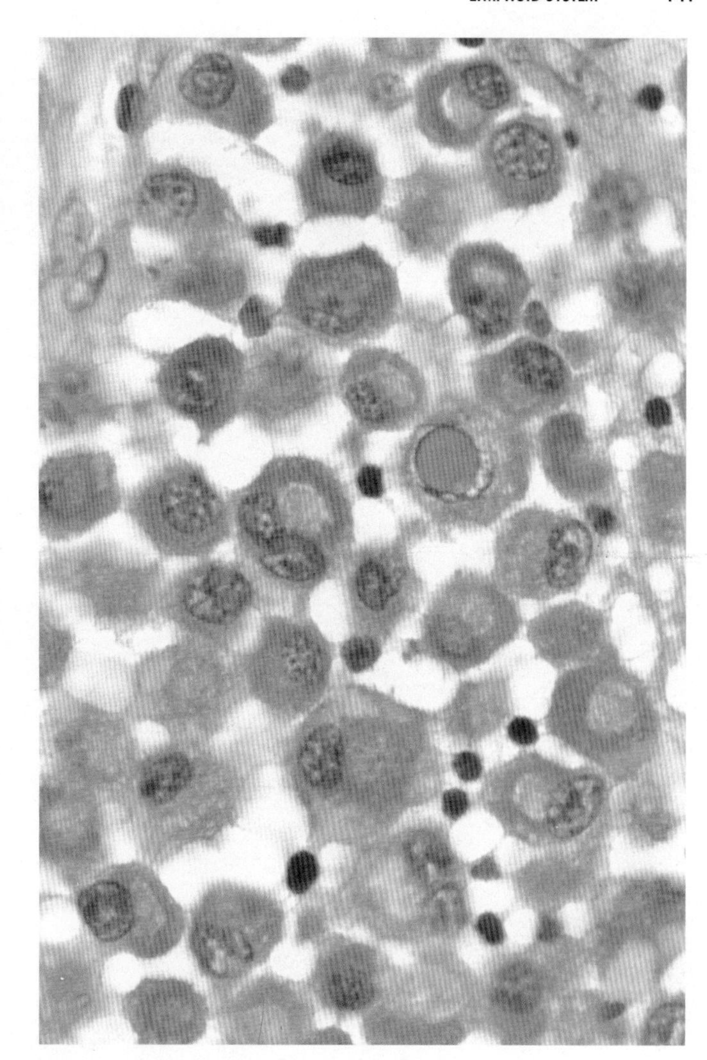

Figure 7–18 B

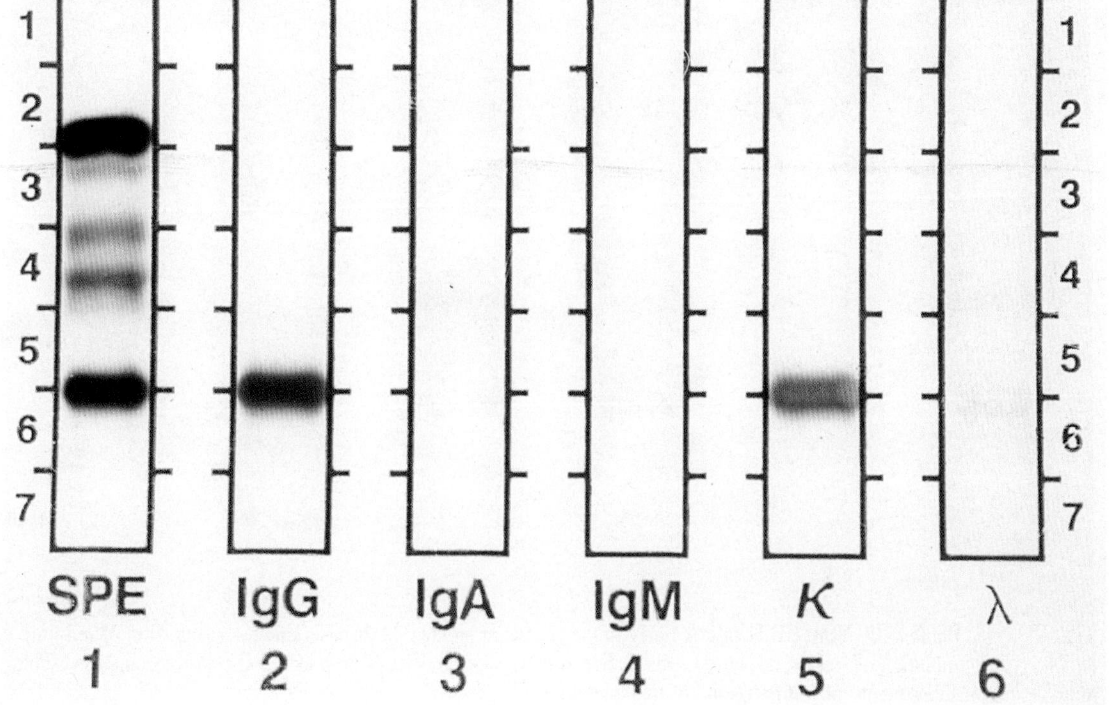

Figure 7–18 C

■ SPLEEN

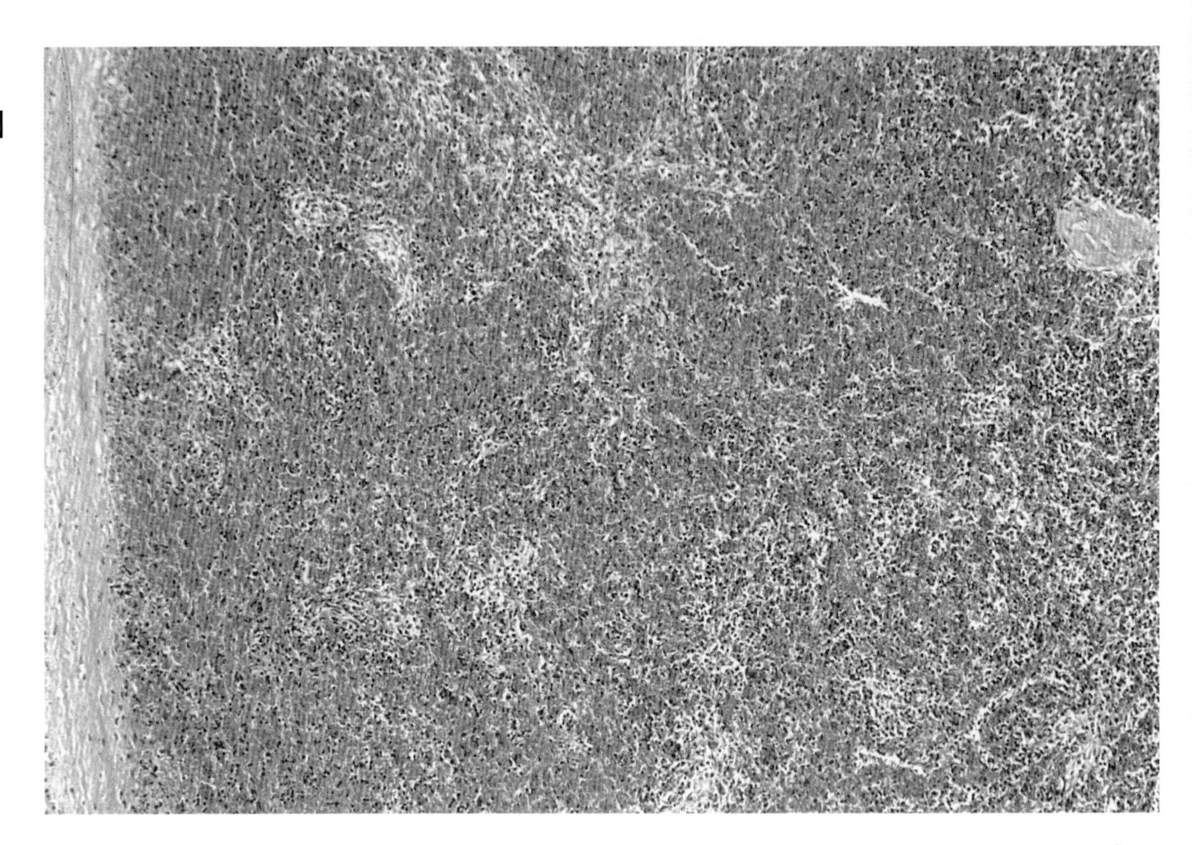

Figure 7–19 **A**

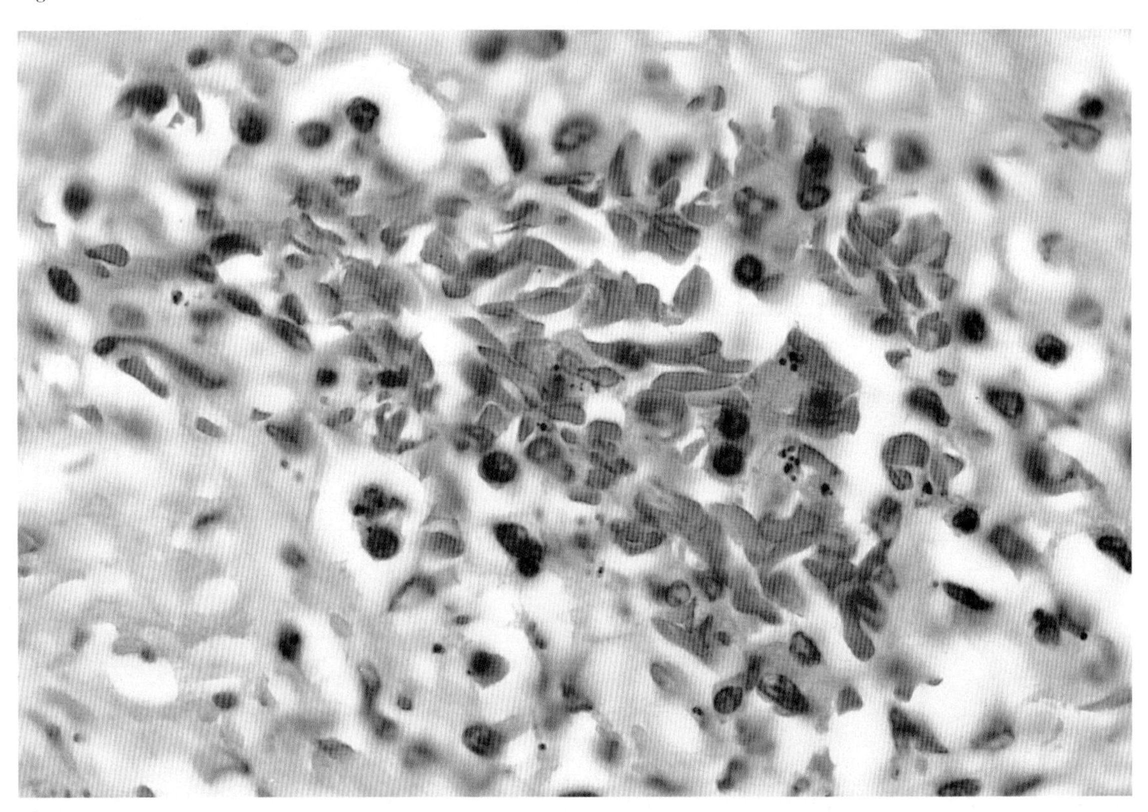

Figure 7–19 **B**

FIGURE 7–19. SICKLE CELL DISEASE: A: Early in the course of sickle cell disease, there is congestion of red pulp because of accumulation of sickled red blood cells. **B:** Irreversibly sickled cells pile up in blood vessels, causing vascular occlusion and subsequent **"autoinfarction"** of the spleen.

■ SPLEEN

Figure 7–19 C

Figure 7–19 D

FIGURE 7–19. SICKLE CELL DISEASE: C: Healed infarcts—**Gamna-Gandy bodies**—are areas of organized fibrous tissue. D: Calcium is deposited in the Gamna-Gandy bodies and shows up as red-stained regions on Alizarin red staining.

■ TRANSPLANT REJECTION

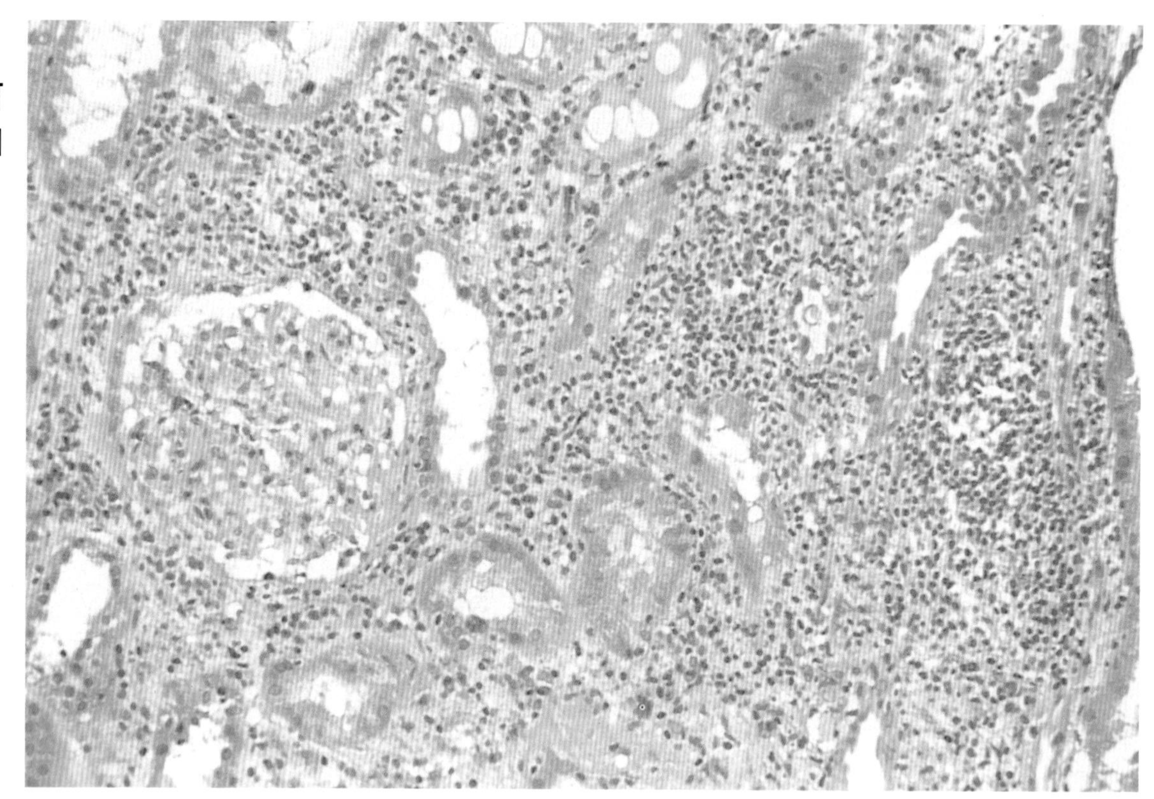

Figure 8–1 A

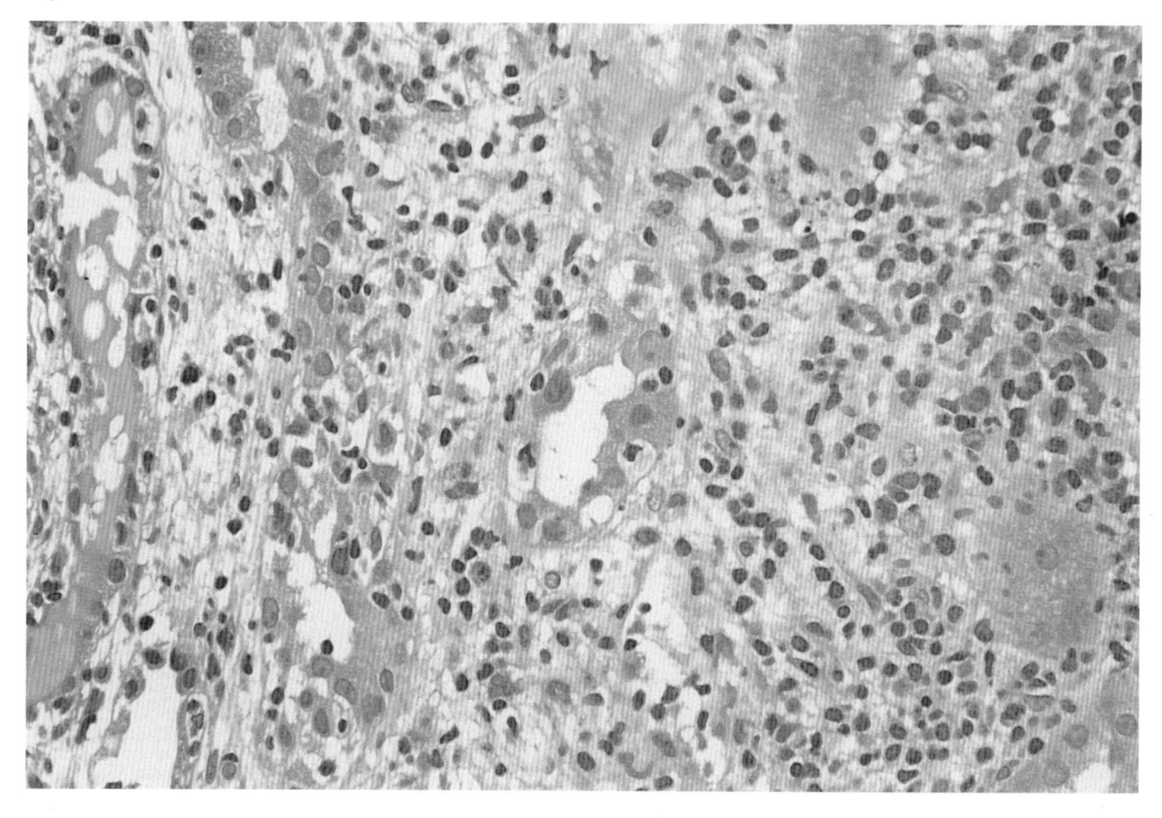

Figure 8–1 B

FIGURE 8–1. ACUTE TUBULOINTERSTITIAL REJECTION: A: In acute tubulointerstitial rejection there is an interstitial infiltrate composed of lymphocytes. B: The tubules are infiltrated by lymphocytes, **"tubulitis,"** and focally tubular epithelial cells are necrotic.

■ TRANSPLANT REJECTION

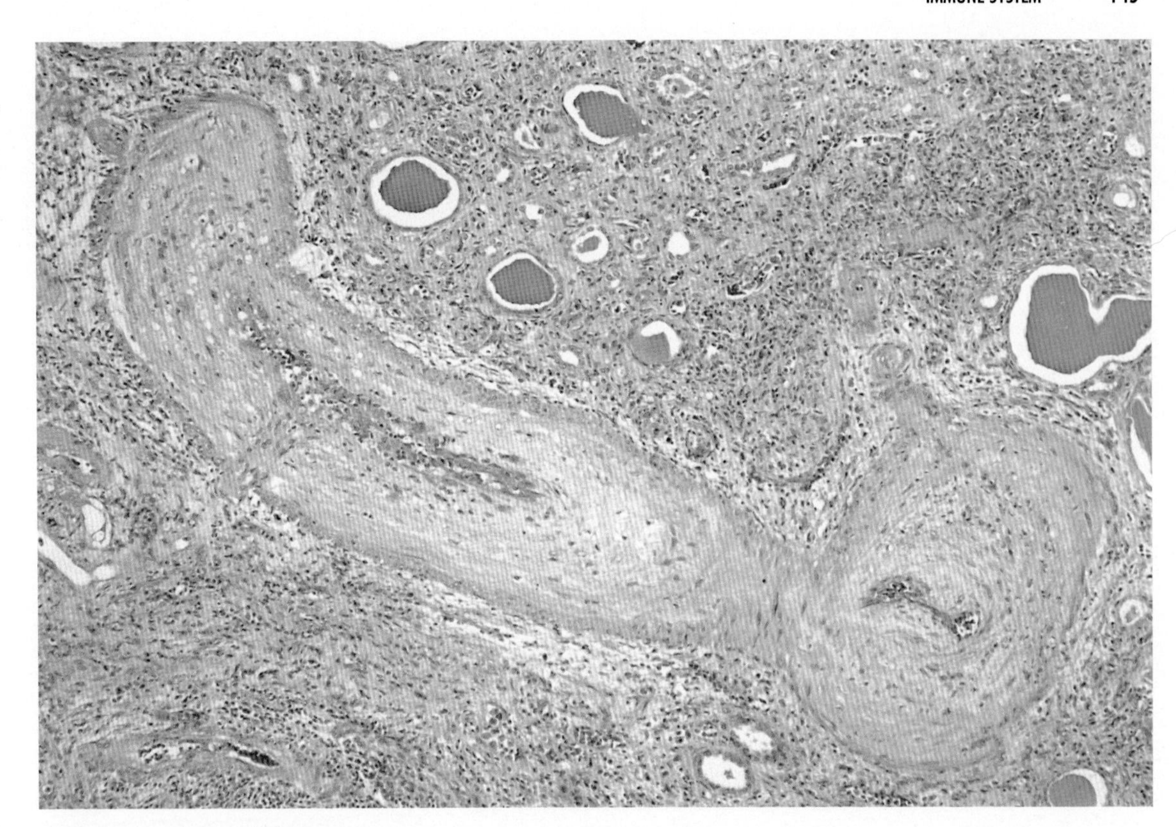

FIGURE 8–2. CHRONIC VASCULAR REJECTION: In chronic vascular rejection, there is intimal fibrosis with narrowing of the lumen.

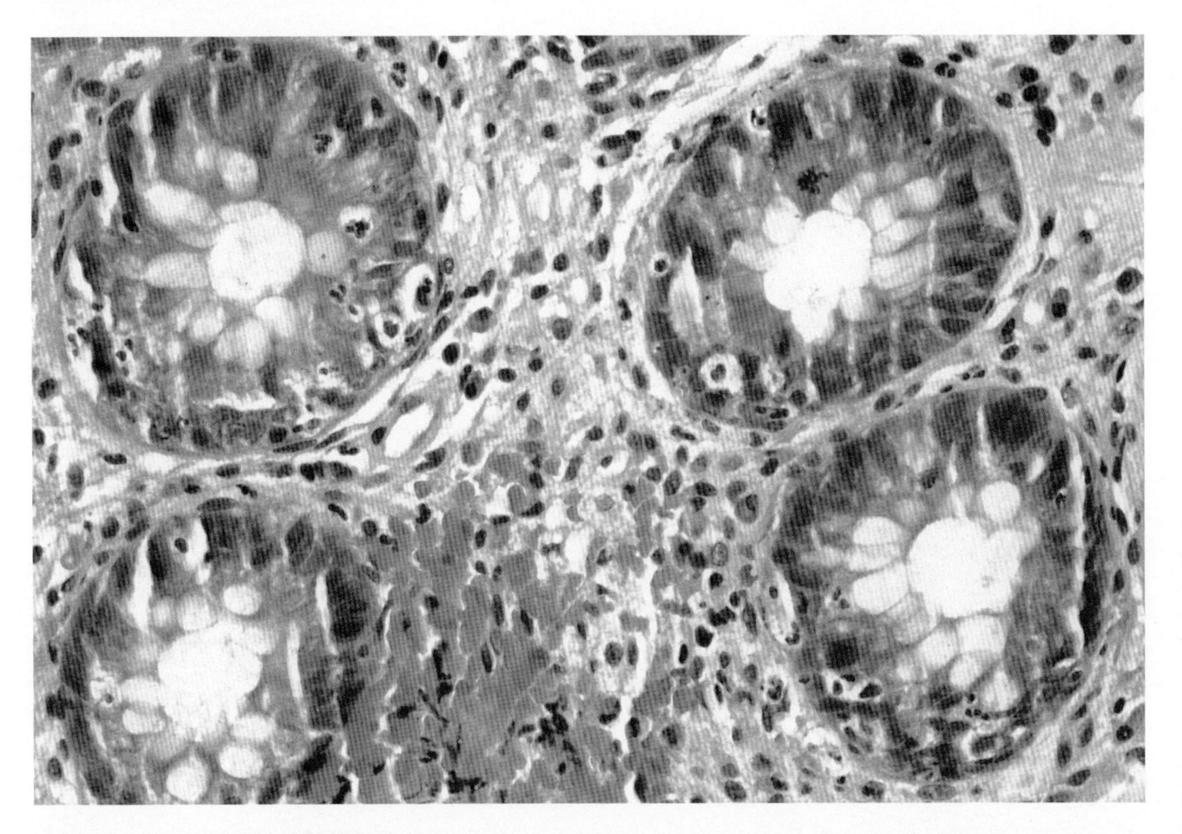

FIGURE 8–3. GRAFT VERSUS HOST DISEASE (GVHD): In this photomicrograph of GVHD in a specimen obtained from the colon, there is necrosis of individual cells. The necrosis is characterized by vacuolization of epithelial cells at the base of the crypts.

■ THYMUS

FIGURE 8–4. THYMIC DEPLETION: Thymic depletion can be seen in acquired immunodeficiency syndrome (AIDS) and other immune-deficient diseases. **A:** In thymic depletion, thymocytes are depleted and there is a reversal of the normal corticomedullary architecture in which cortical thymocytes are preferentially depleted. **B:** At high power, a crowding of Hassall's corpuscles can be observed.

FIGURE 8–5. THYMIC HYPERPLASIA: Thymic hyperplasia can be seen in myasthenia gravis and other autoimmune diseases and is characterized by the appearance of lymphoid follicles in the medulla.

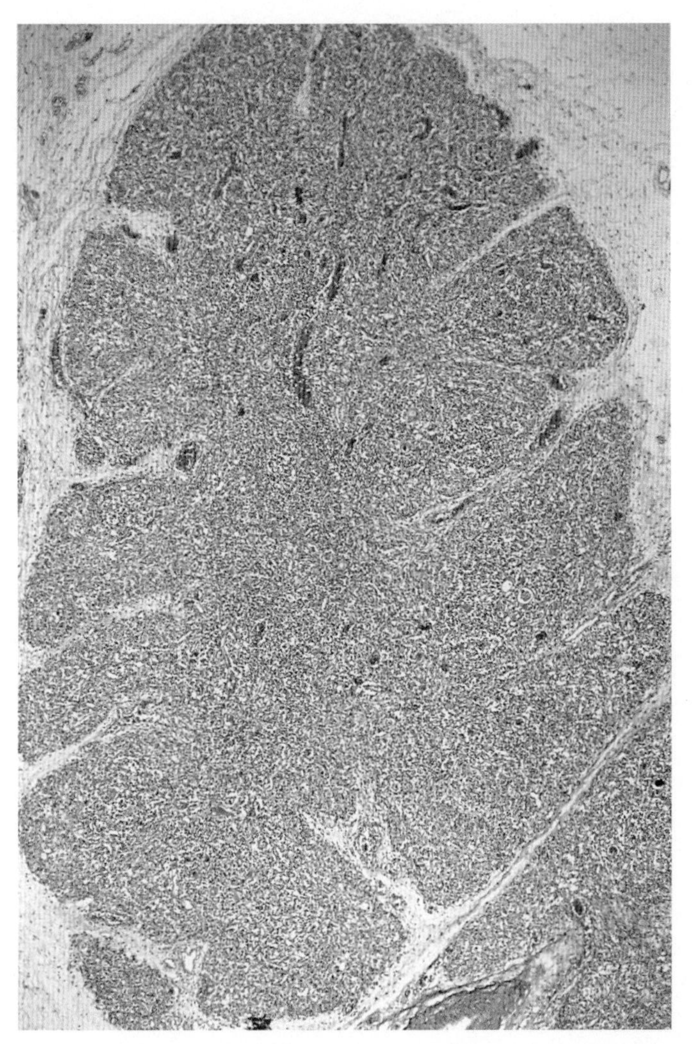

Figure 8–4 **A**

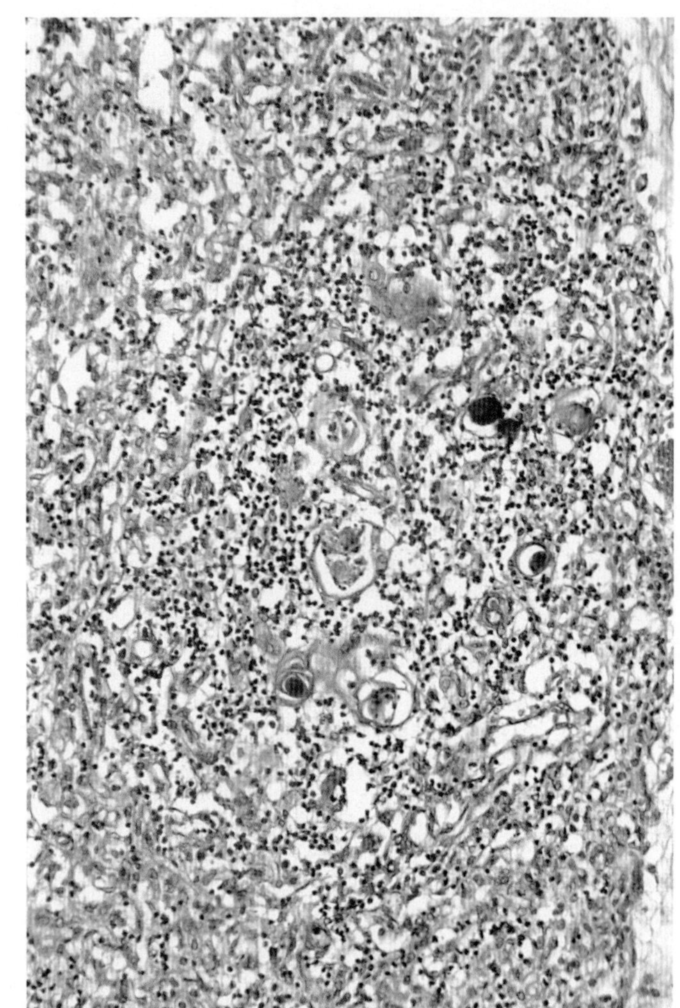

Figure 8–4 **B**

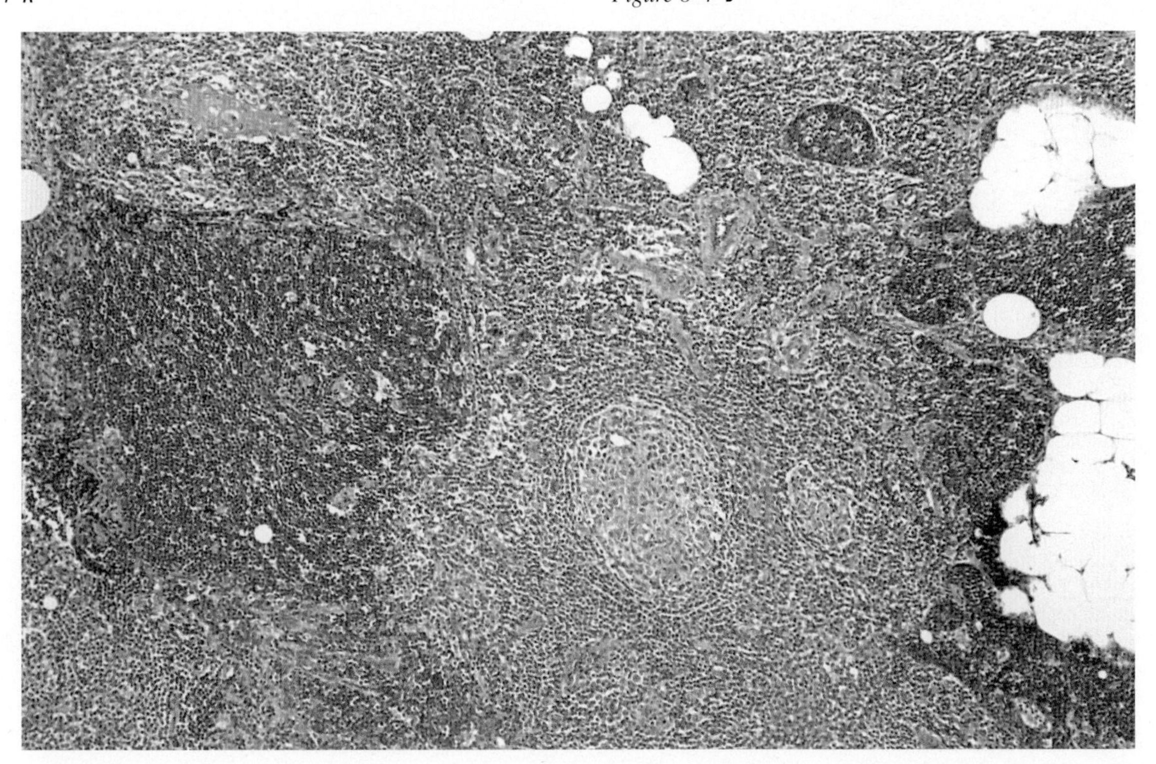

Figure 8–5

■ THYMUS

FIGURE 8–6. THYMOMA: A: Thymomas have a thick fibrous capsule from which fibrous bands emanate and divide the tumor into lobules. **B:** Foci of cystic degeneration are present. **C:** The tumor is composed of an admixture of malignant epithelial cells and benign lymphocytes.

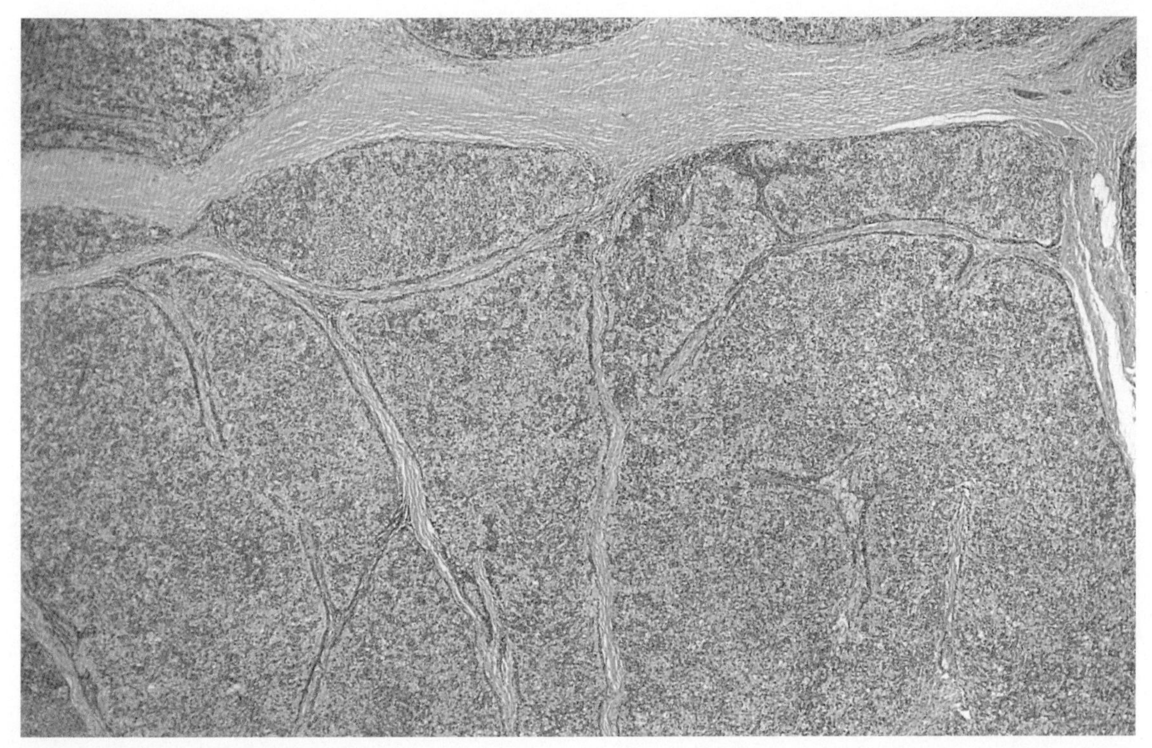

Figure 8–6 A

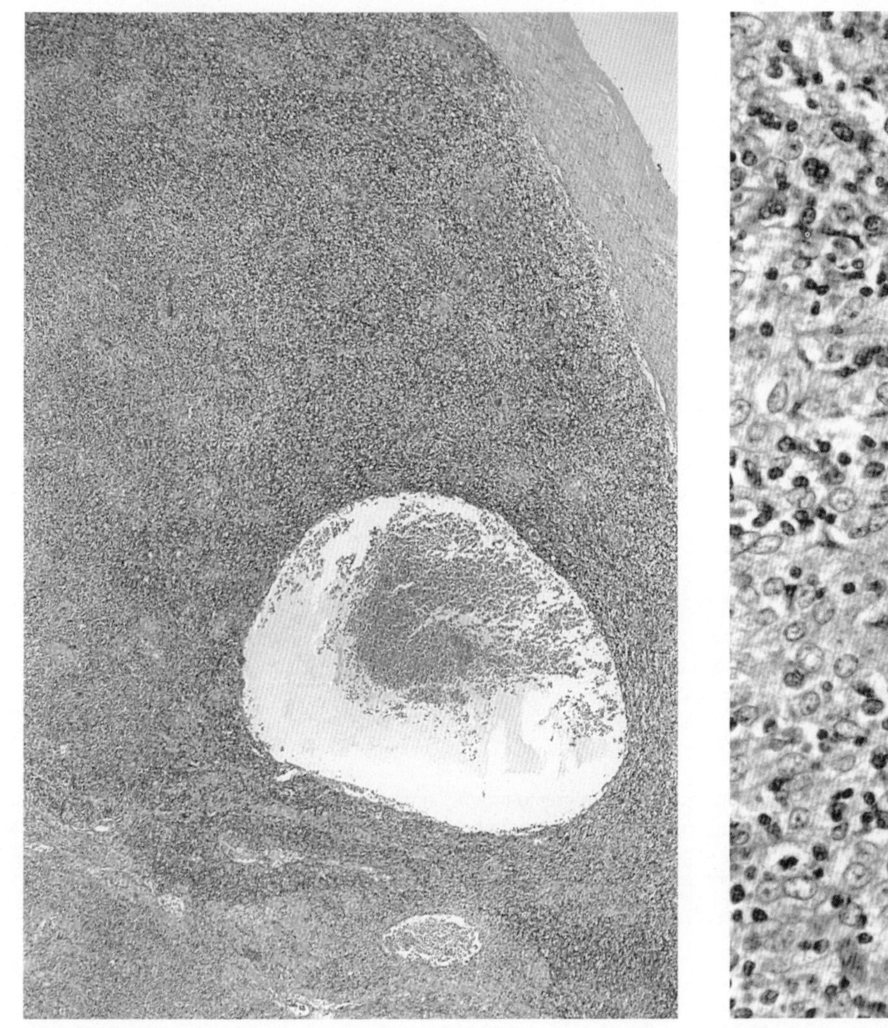

Figure 8–6 B

Figure 8–6 C

■ ACQUIRED IMMUNODEFICIENCY SYNDROME (AIDS)

LYMPHOID SYSTEM

FIGURE 8–7. LYMPHOID DEPLETION: Late-stage AIDS is characterized by a depletion of lymphocytes in the immune system. Lymphoid depletion in a lymph node can be seen in this photomicrograph. Note the absence of lymphoid follicles.

RESPIRATORY SYSTEM

FIGURE 8–8. PULMONARY LYMPHOID HYPERPLASIA: Pulmonary lymphoid hyperplasia is more commonly found in the pediatric population than in the adult population. It is characterized by peribronchial and peribronchiolar nodular lymphoid aggregates with or without germinal centers.

FIGURE 8–9. LYMPHOID INTERSTITIAL PNEUMONITIS: This condition also is more commonly found in the pediatric population than in the adult population. It is characterized by diffuse infiltration of alveolar septa and peribronchial tissue by small irregular lymphocytes.

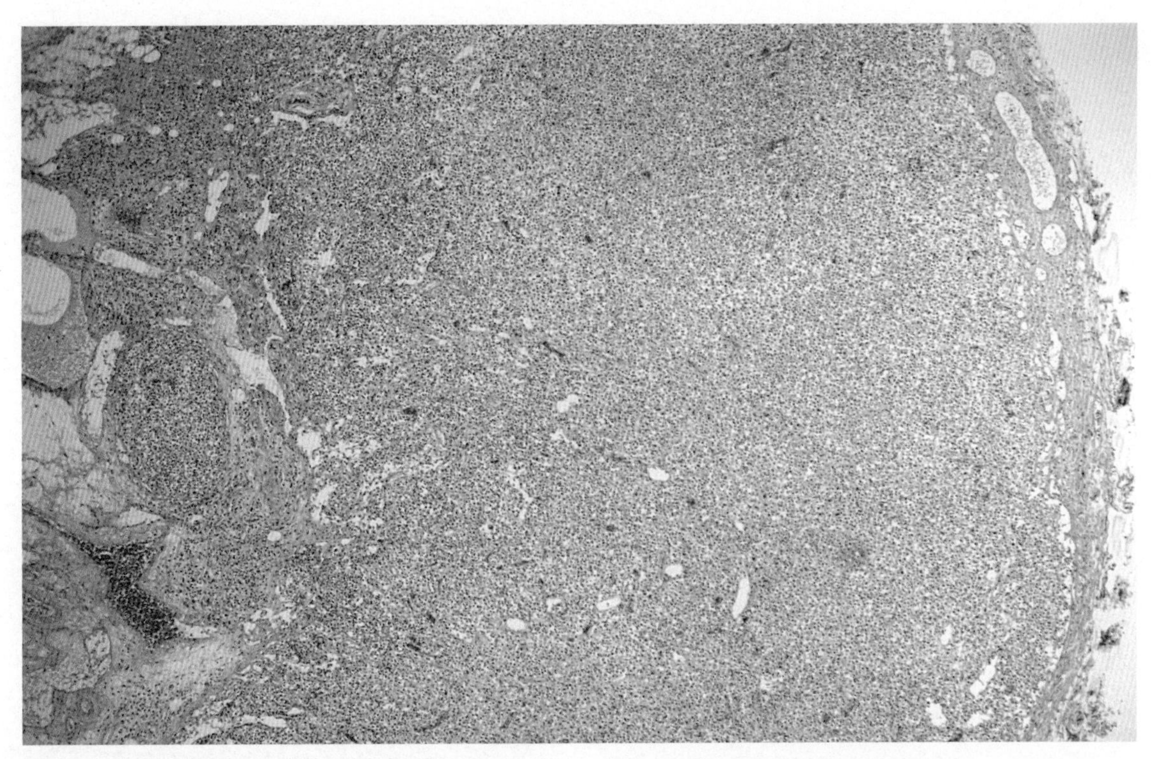

Figure 8–7

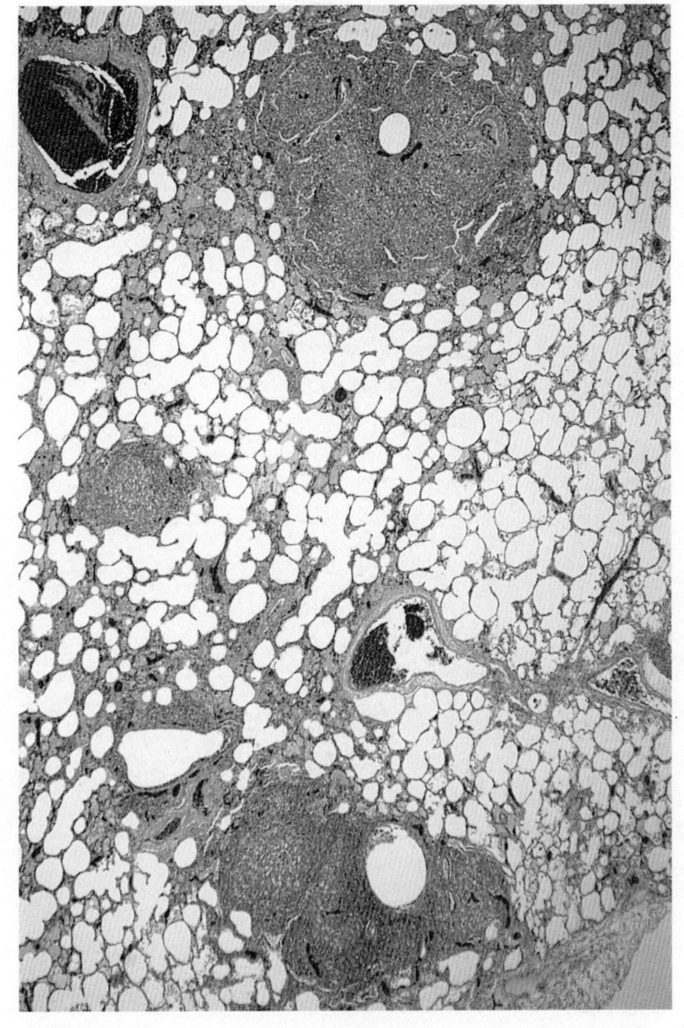

Figure 8–8

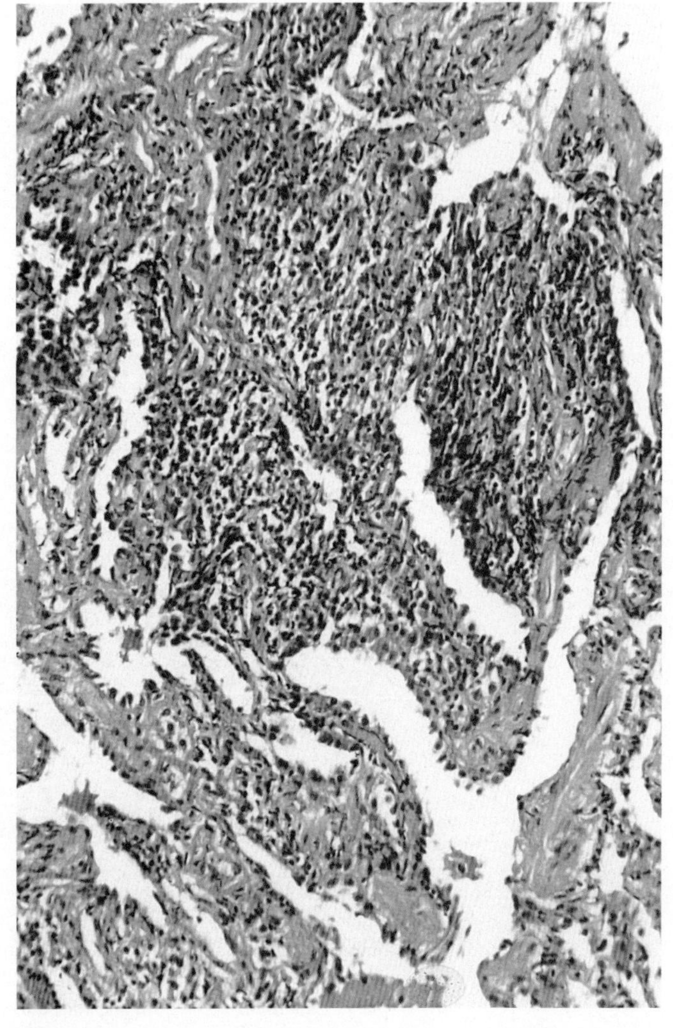

Figure 8–9

■ AIDS

SALIVARY GLANDS

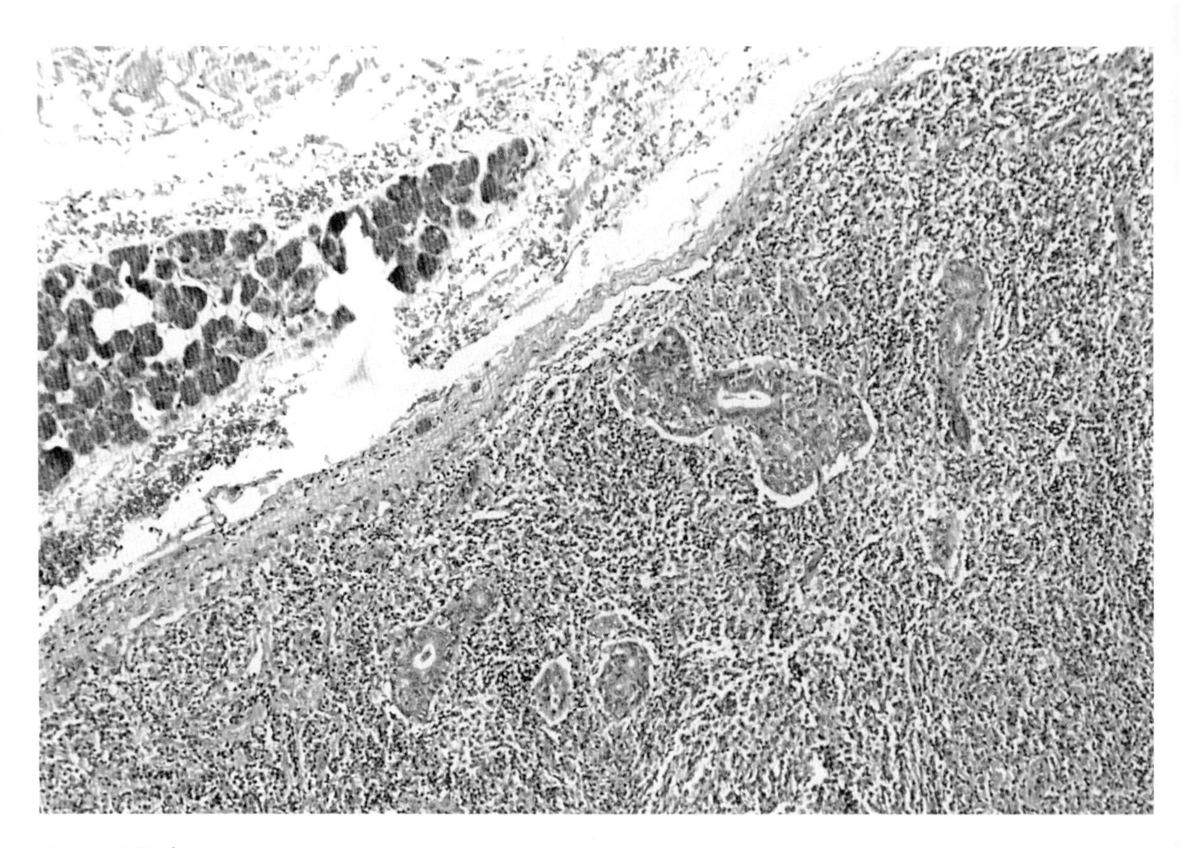

Figure 8–10 **A**

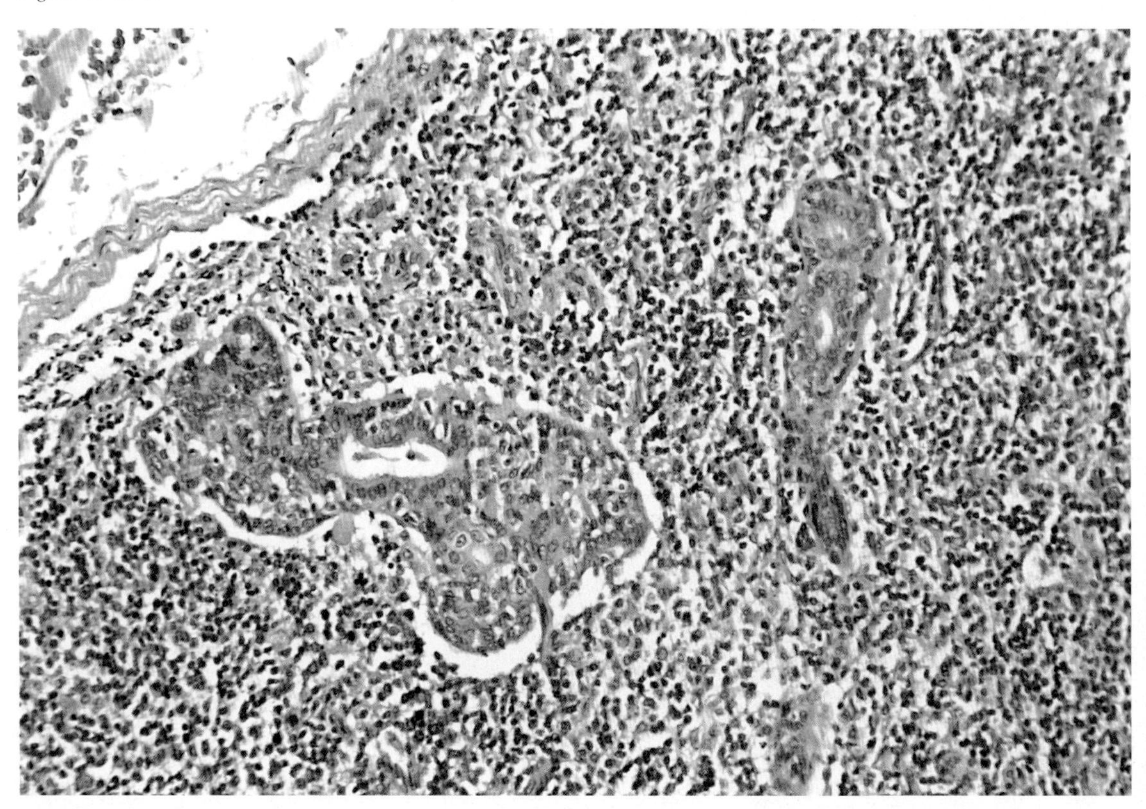

Figure 8–10 **B**

FIGURE 8–10. BENIGN LYMPHOEPITHELIAL LESION: This lesion affects major salivary glands. **A:** It is composed of masses of lymphoid tissue with embedded epithelial islands. **B:** The epithelial islands are presumed to be dilated ducts in which there is metaplastic squamous epithelium and formation of microcysts.

■ AIDS

INFLAMMATORY OR INFECTIOUS LESIONS

BACTERIAL INFECTIONS

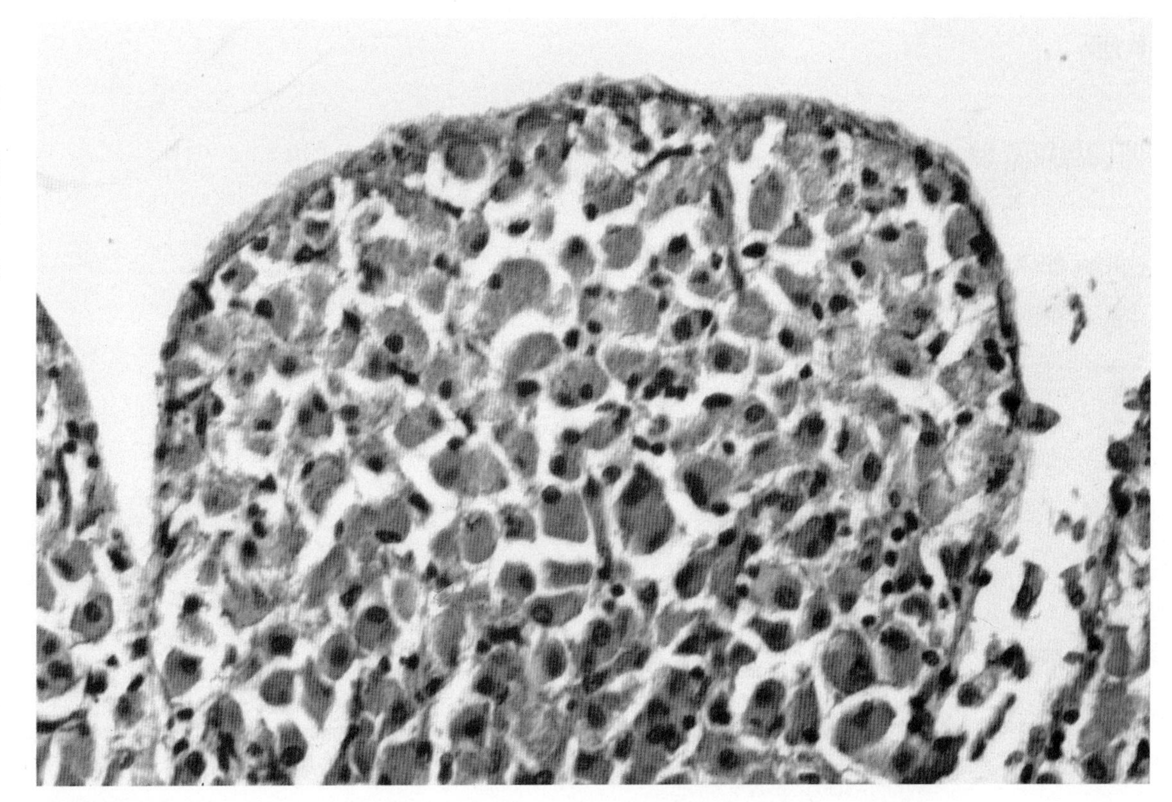

Figure 8–11 **A**

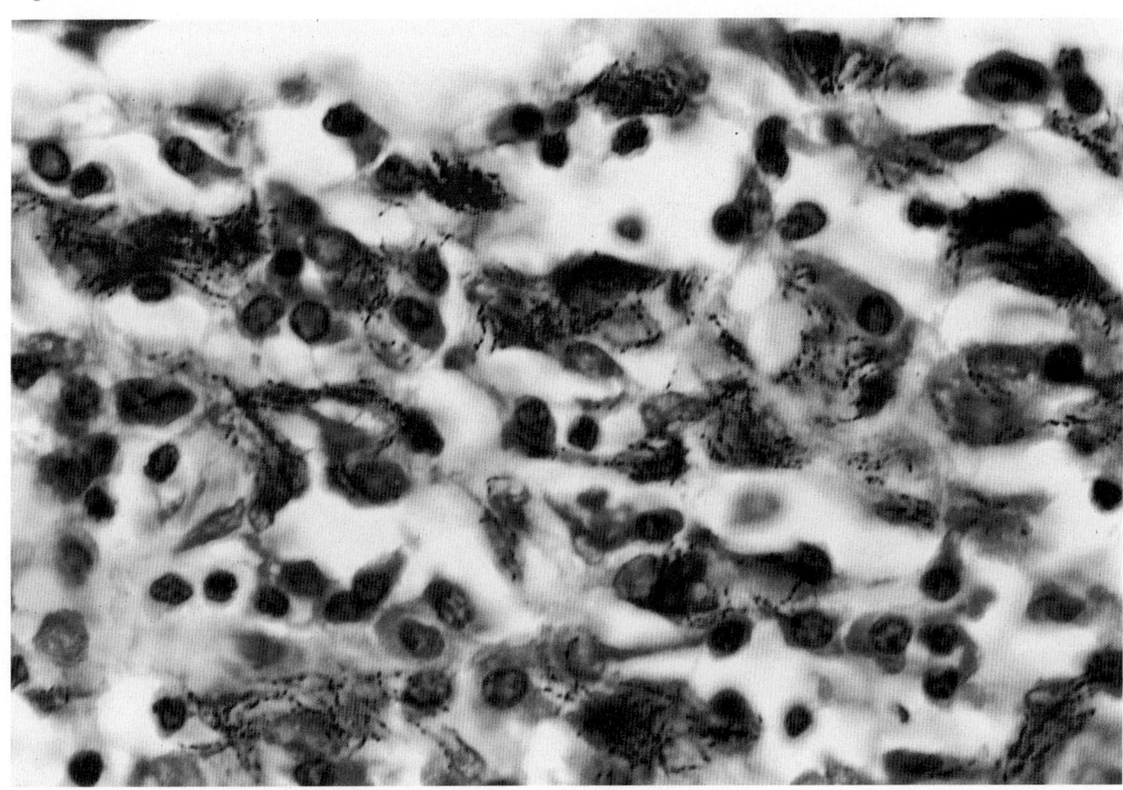

Figure 8–11 **B**

FIGURE 8–11. ***MYCOBACTERIUM AVIUM INTRACELLULARE:*** In AIDS, *Mycobacterium avium intracellulare* is the most common cause of disseminated bacterial infection. **A:** In this sample taken from the small intestine, there is widening and blunting of the villous architecture. The villi are filled with foamy macrophages. **B:** The rod-shaped mycobacteria, which are present in the macrophages, stain red with Ziehl-Neelsen stain.

■ AIDS

INFLAMMATORY OR INFECTIOUS LESIONS

BACTERIAL INFECTIONS

FIGURE 8–12. BACILLARY EPITHELIOID ANGIOMATOSIS: A: Bacillary epithelioid angiomatosis is an exophytic nodule with an epidermal collarette at the base. **B:** There is a proliferation of blood vessels lined by plump endothelial cells. Present in the stroma are large histiocytes containing eosinophilic material that represents the organisms. **C:** The small bacillary bacterium that causes this lesion is *Rochalimaea henselae,* best seen with Warthin-Starry silver staining.

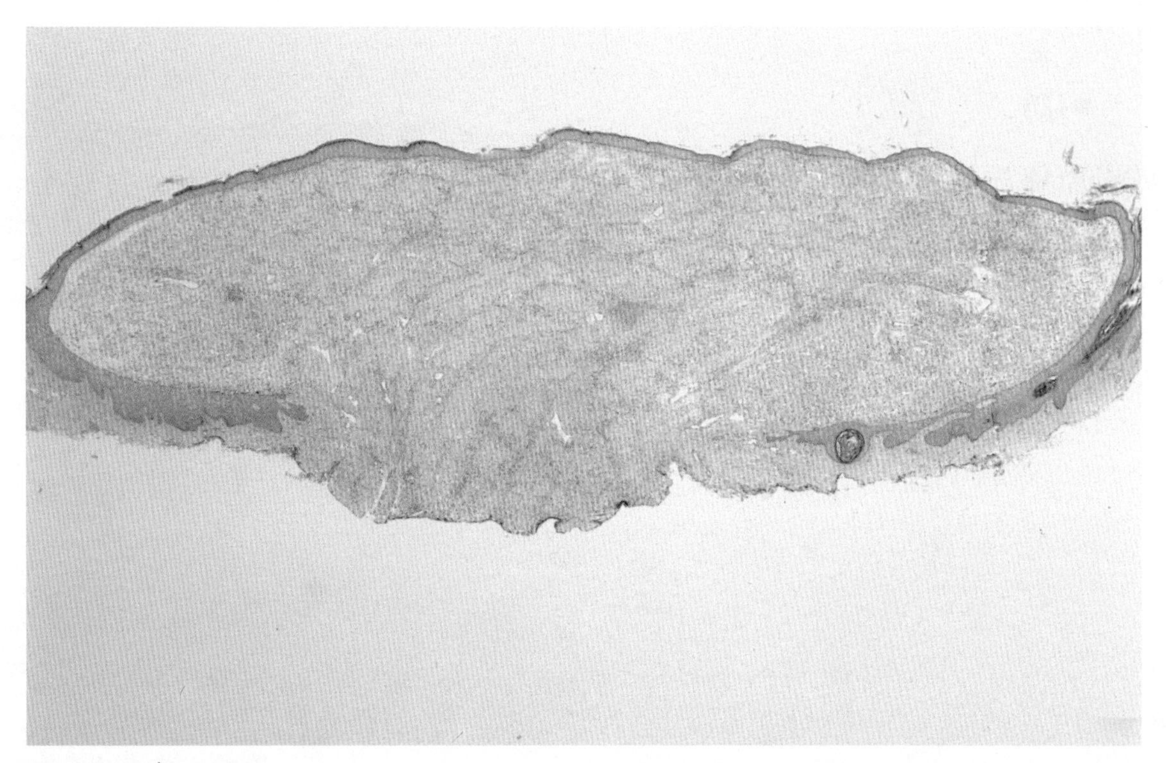

Figure 8–12 A

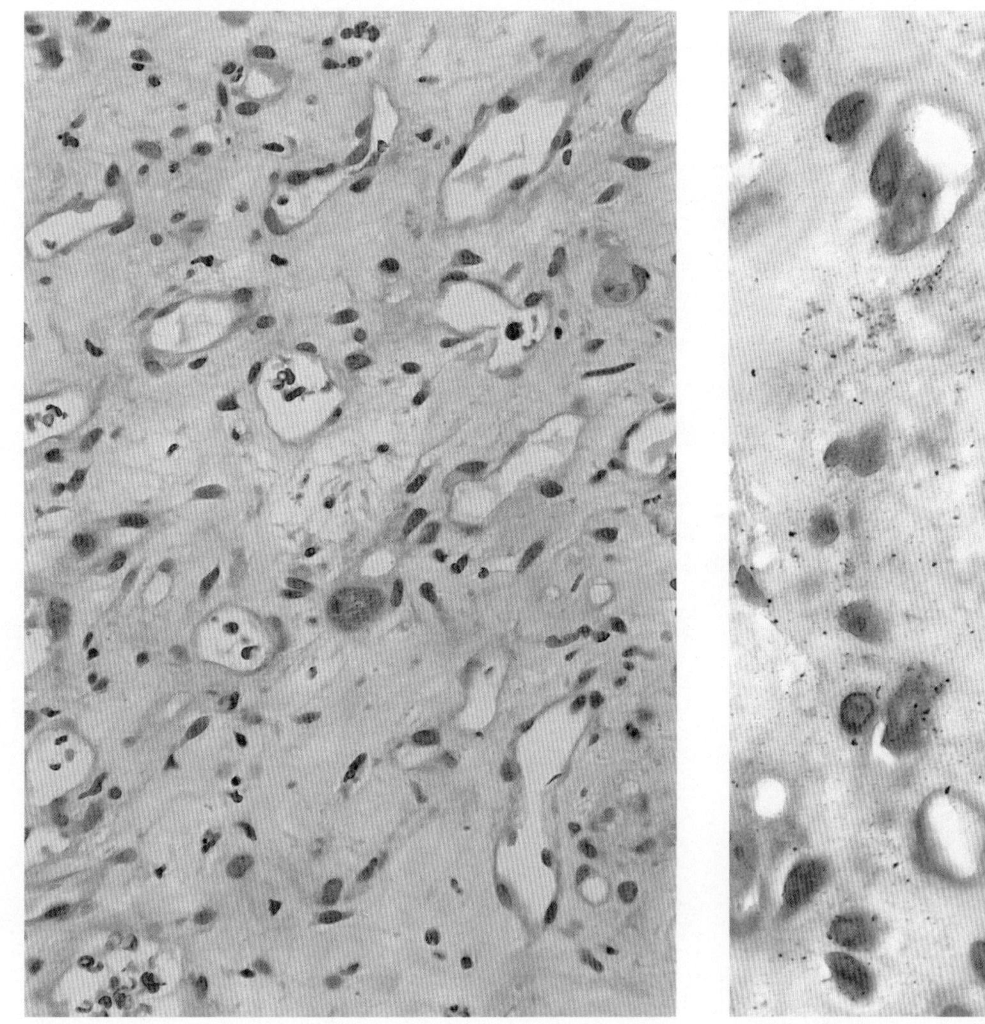

Figure 8–12 B

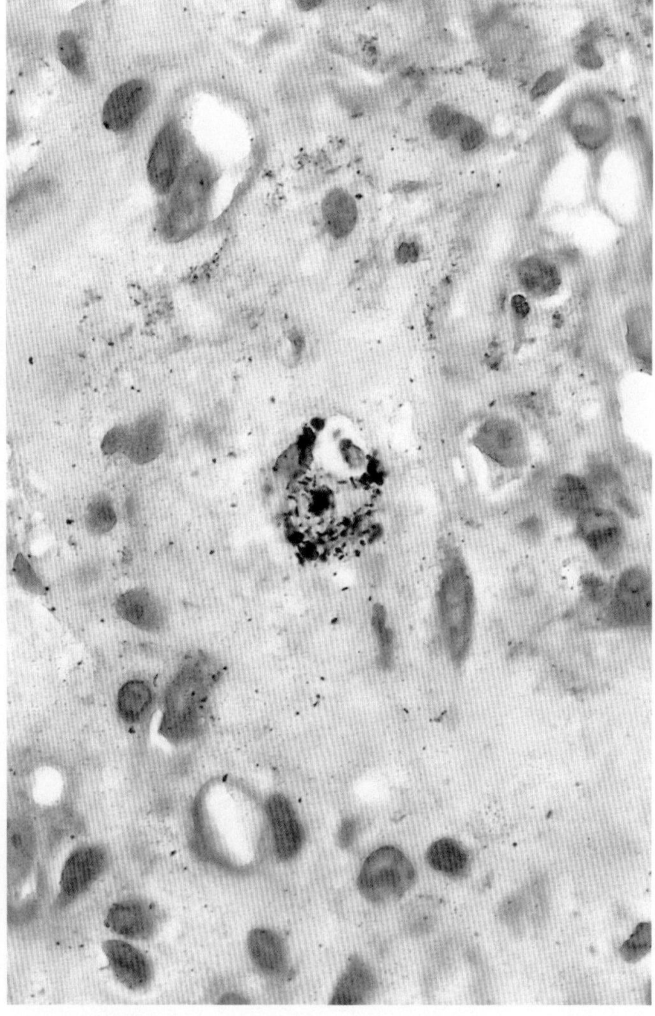

Figure 8–12 C

■ AIDS

INFLAMMATORY OR INFECTIOUS LESIONS

VIRAL INFECTIONS

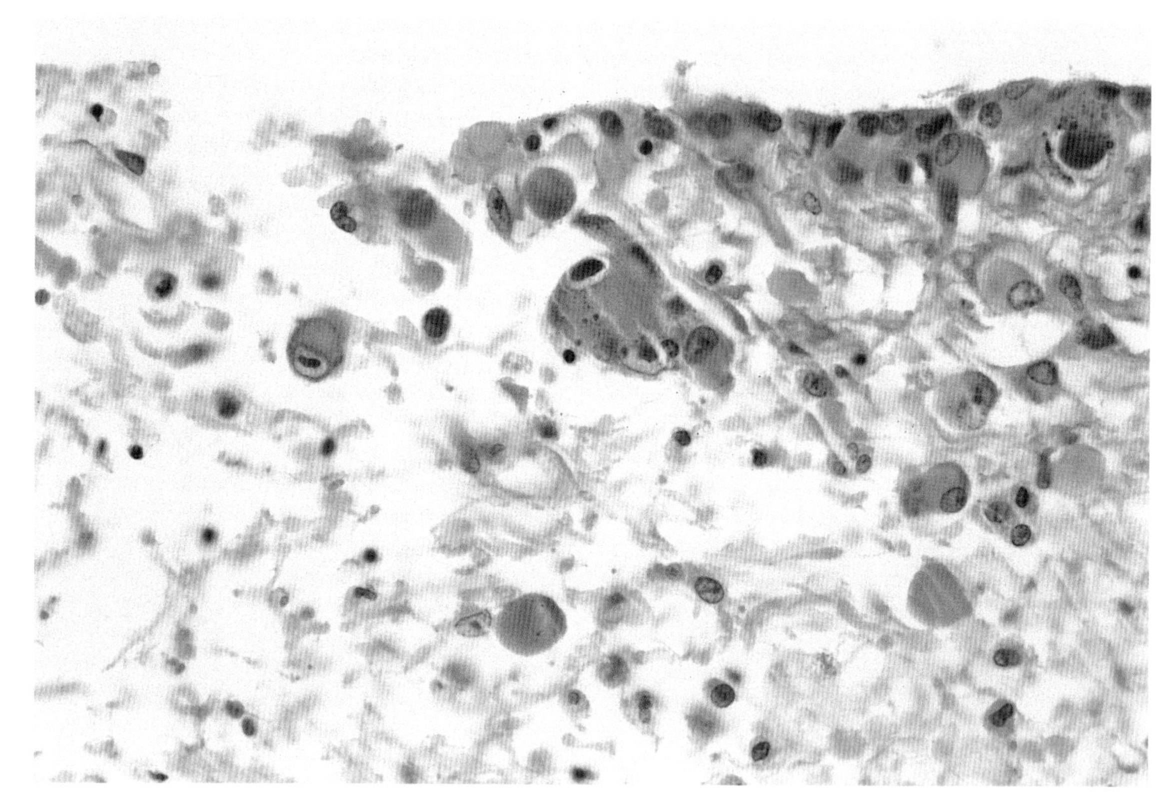

FIGURE 8–13. CYTOMEGALOVIRUS: In immunocompromised hosts, cytomegalovirus is the most common opportunistic infection from any cause. Characteristics of cytomegalovirus-infected cells include having basophilic intranuclear inclusions surrounded by a clear halo, giving the classic **"owl eye"** appearance of the nucleus.

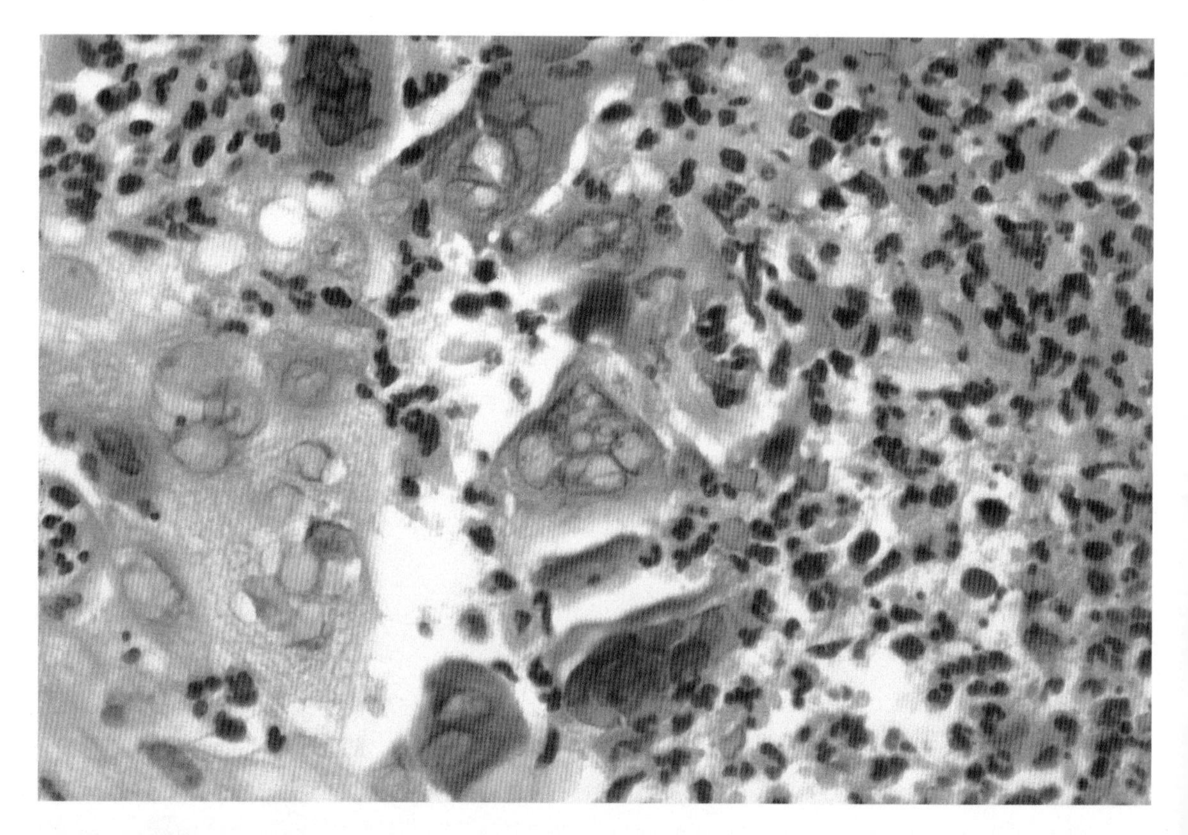

FIGURE 8–14. HERPES SIMPLEX VIRUS: Herpes simplex virus is the most common cause of nongonococcal proctitis in sexually active homosexual men. This perianal lesion shows the characteristic multinucleated giant cells with **"ground glass"** nuclei. Neutrophilic infiltration is prominent.

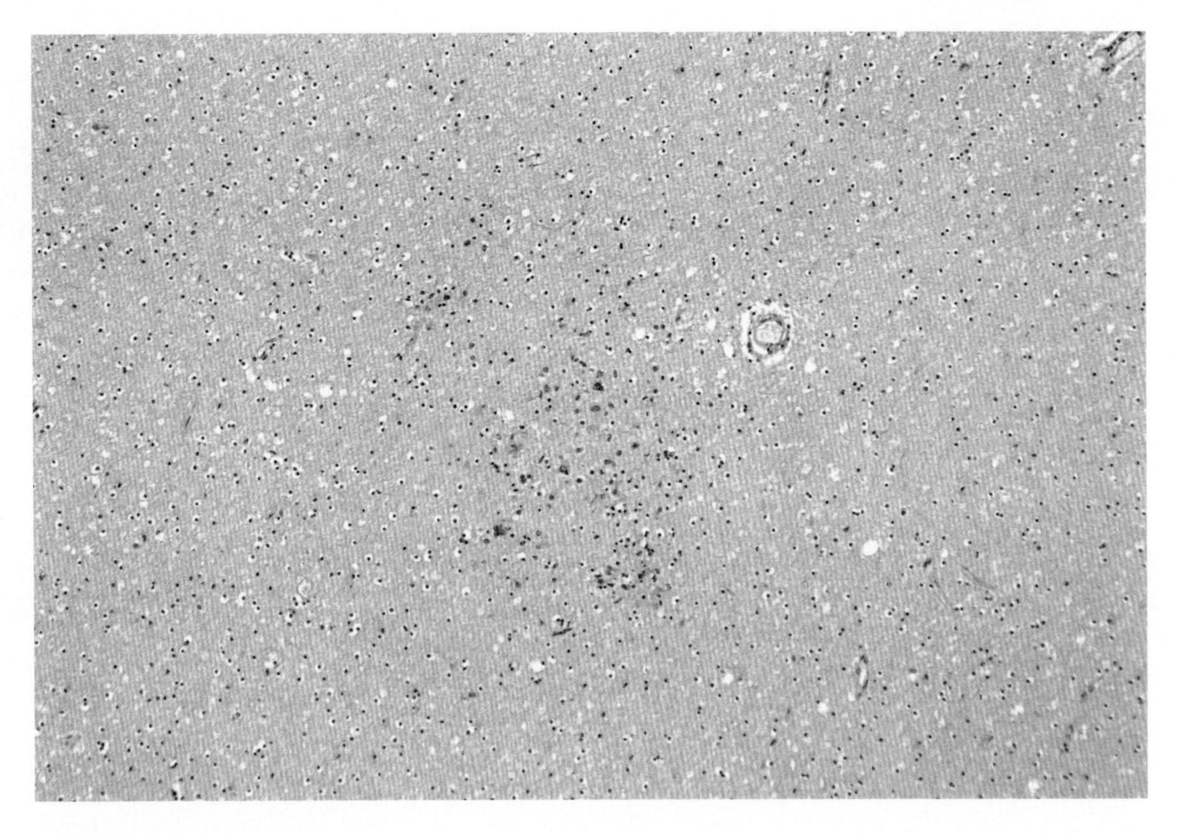

Figure 8–15 A

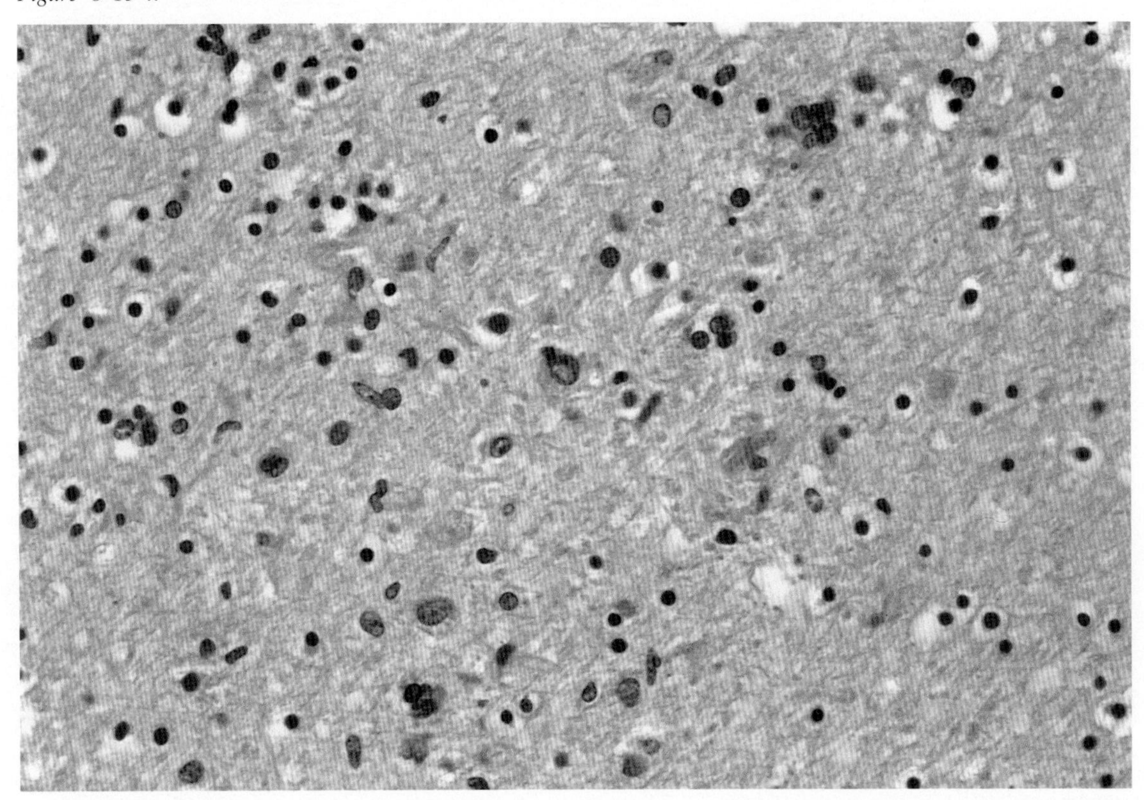

Figure 8–15 B

FIGURE 8–15. HIV-1 ENCEPHALITIS: A: HIV-1 encephalitis is characterized by microglial nodules in the vicinity of small blood vessels. **B:** Within the microglial nodules are multinucleated giant cells.

■ AIDS

INFLAMMATORY OR INFECTIOUS LESIONS

VIRAL INFECTIONS

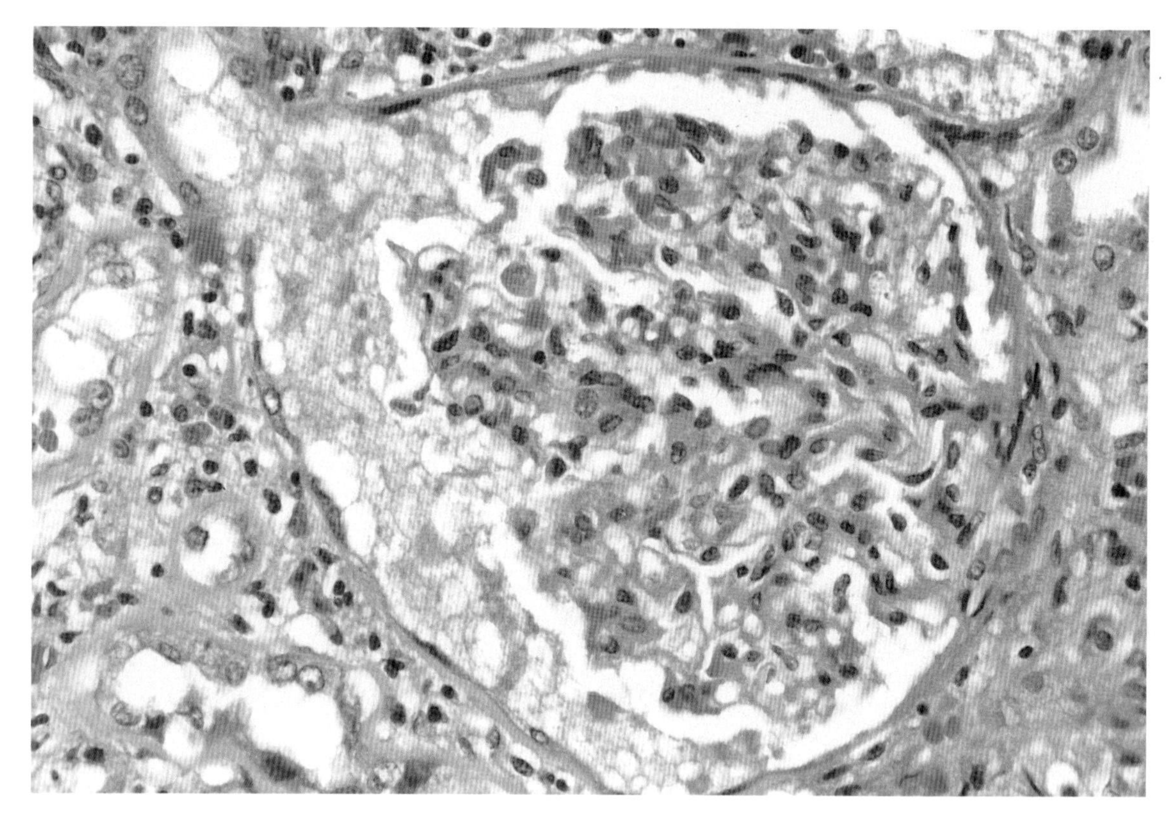

Figure 8–16 A

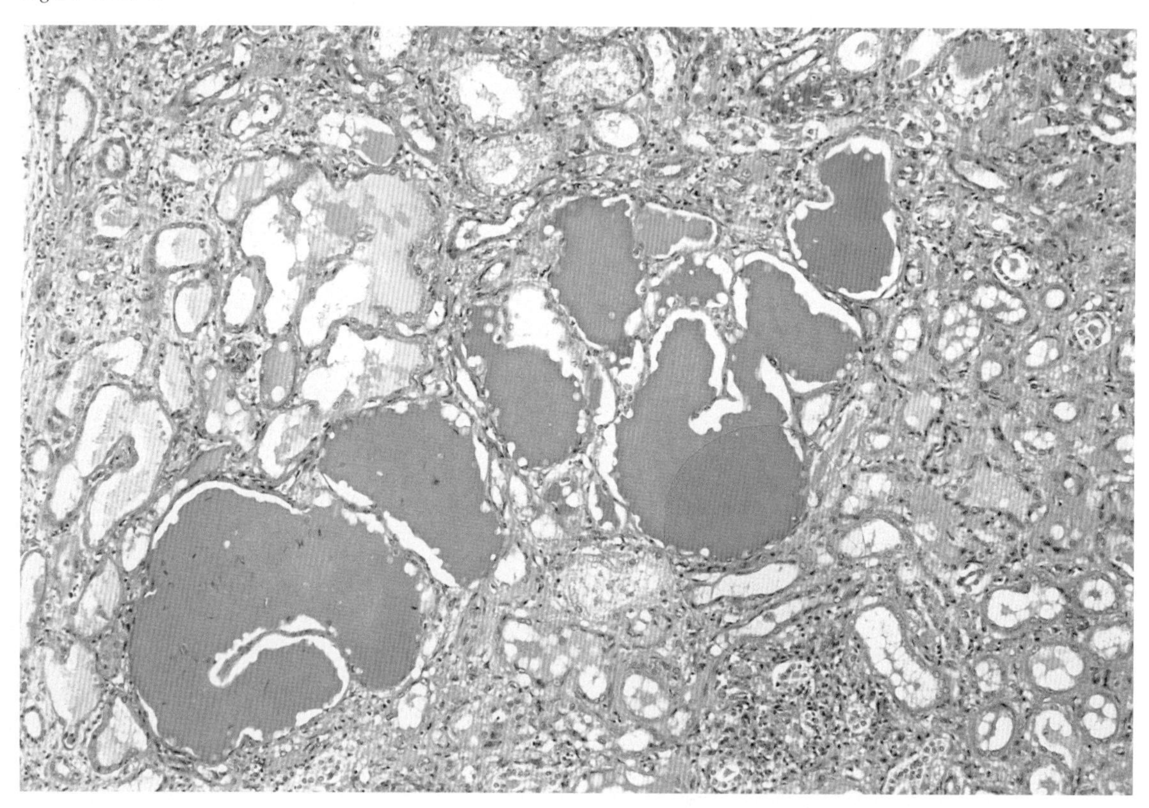

Figure 8–16 B

FIGURE 8–16. HIV-1–ASSOCIATED NEPHROPATHY: A: HIV-1–associated nephropathy is characterized by dilatation of Bowman's spaces. The glomeruli are hypercellular and focally sclerotic. B: Clusters of renal tubules are dilated, **"microcysts,"** which contain protein casts.

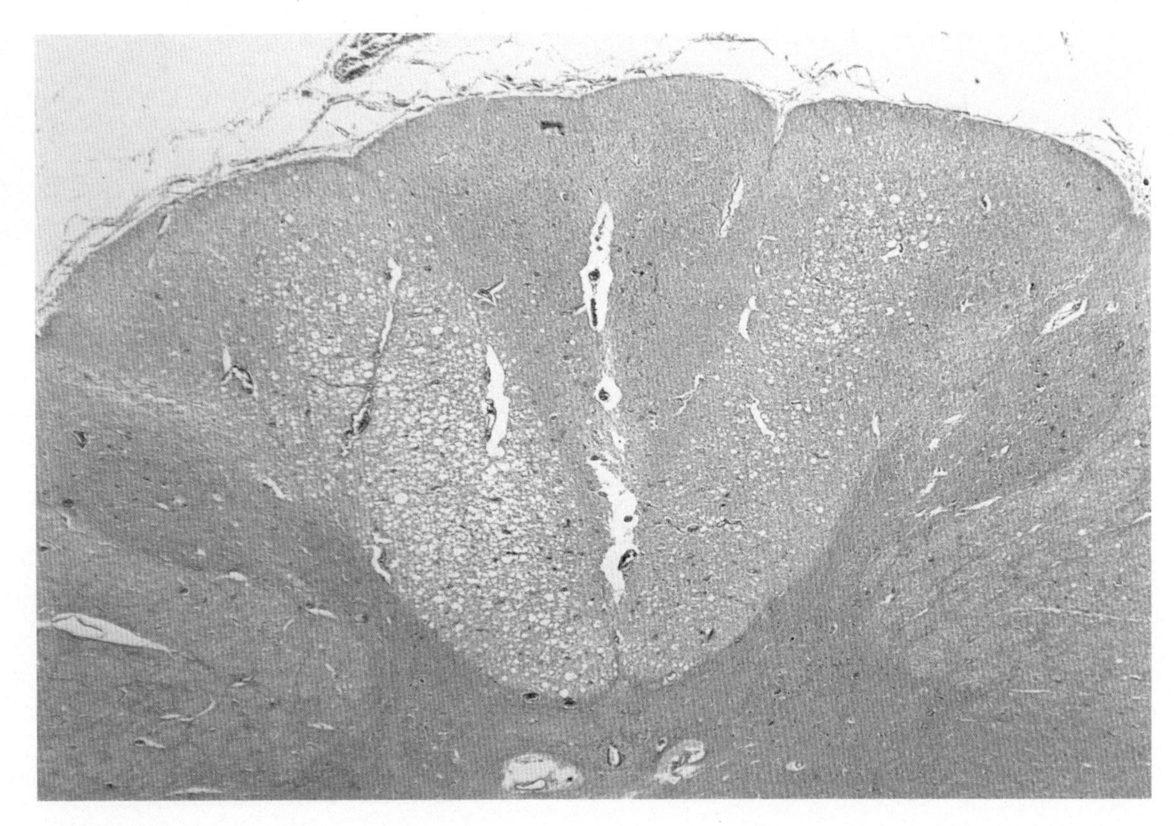

Figure 8–17 **A**

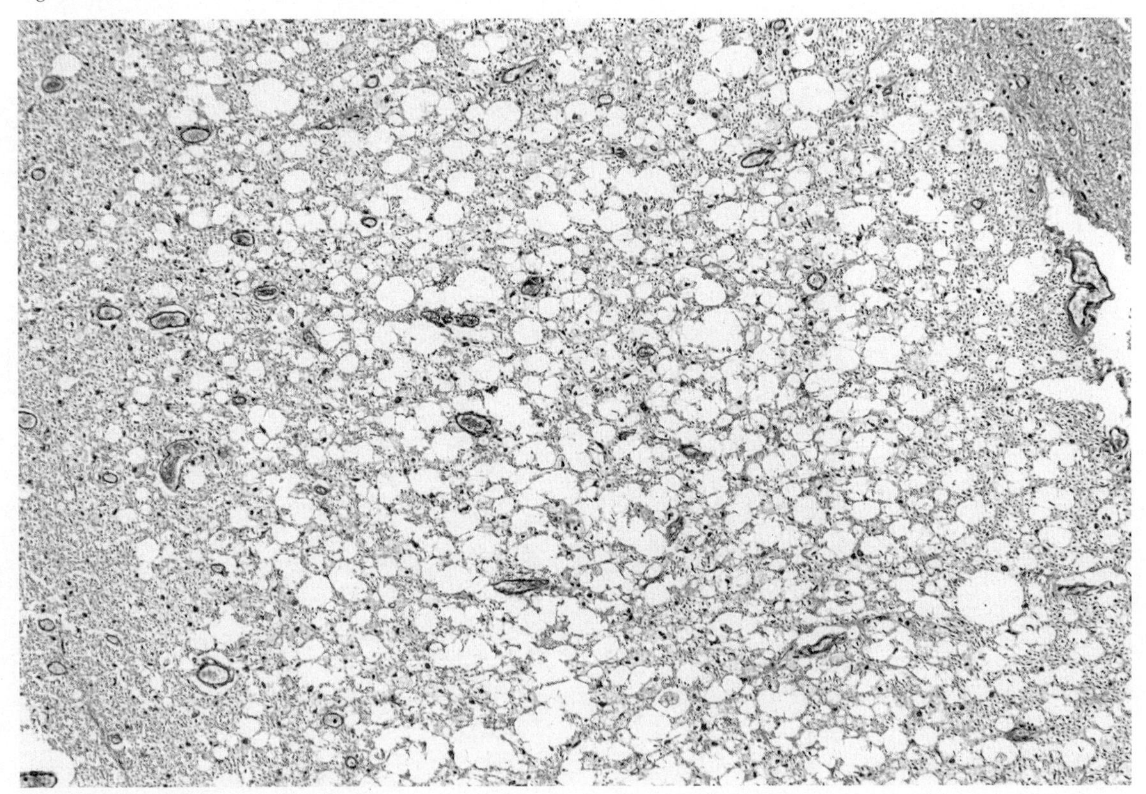

Figure 8–17 **B**

FIGURE 8–17. VACUOLAR MYELOPATHY: A: Vacuolar myelopathy affects the posterior and lateral columns of the lower thoracic spinal cord. **B:** Many vacuoles, ranging from 10 to 100 μm, are present in these regions.

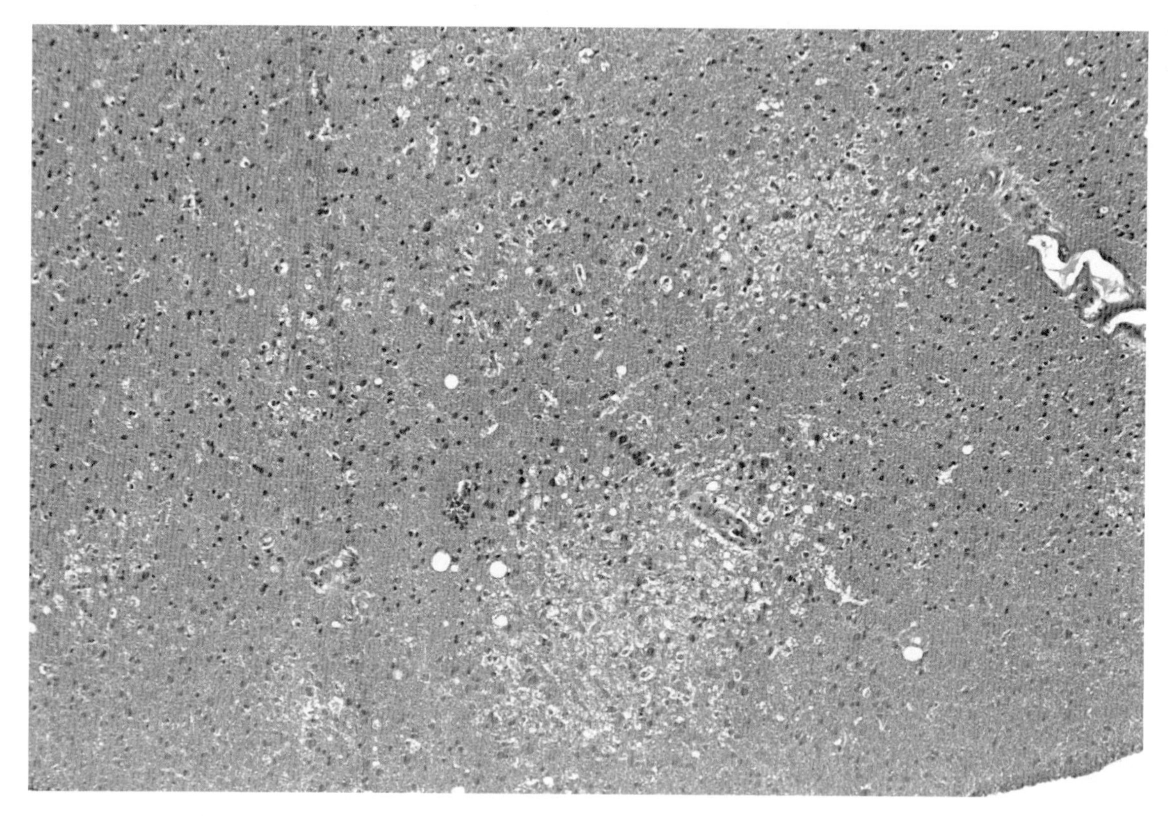

Figure 8–18 **A**

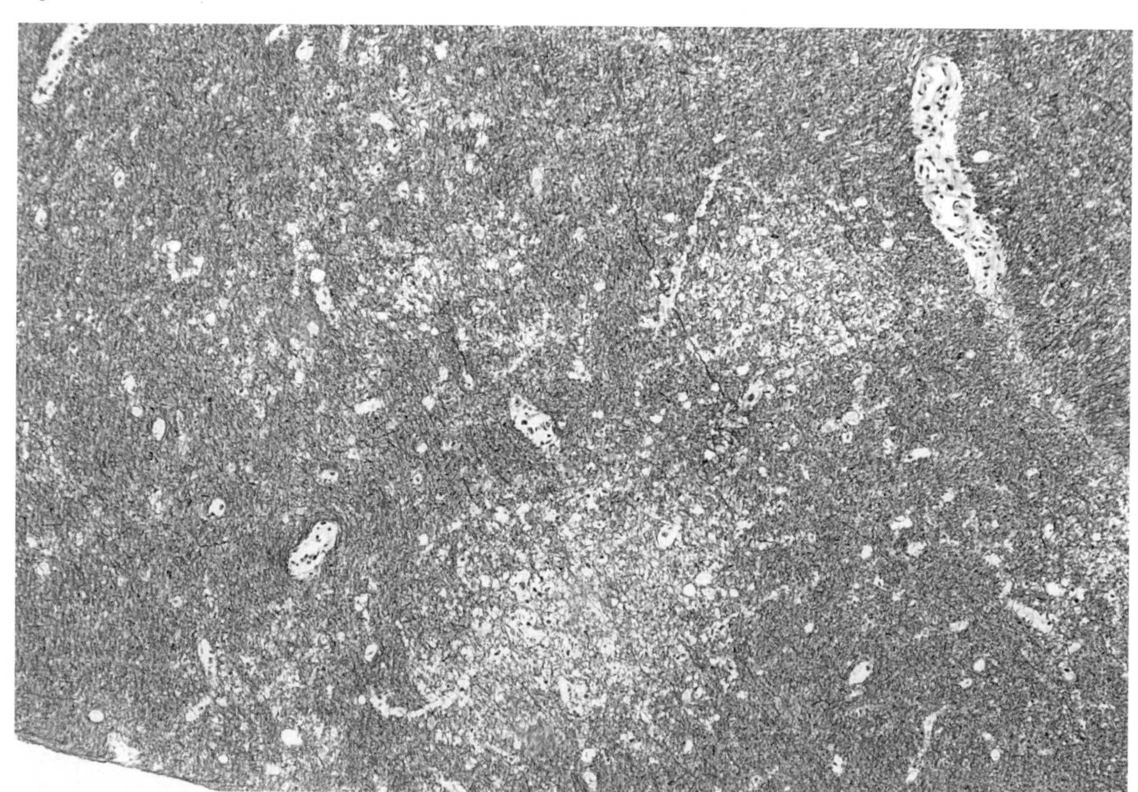

Figure 8–18 **B**

FIGURE 8–18. PROGRESSIVE MULTIFOCAL LEUKOENCEPHALOPATHY: This condition is characterized by stellate areas of demyelination (**A**), better seen as pale areas on Luxol fast blue staining (**B**).

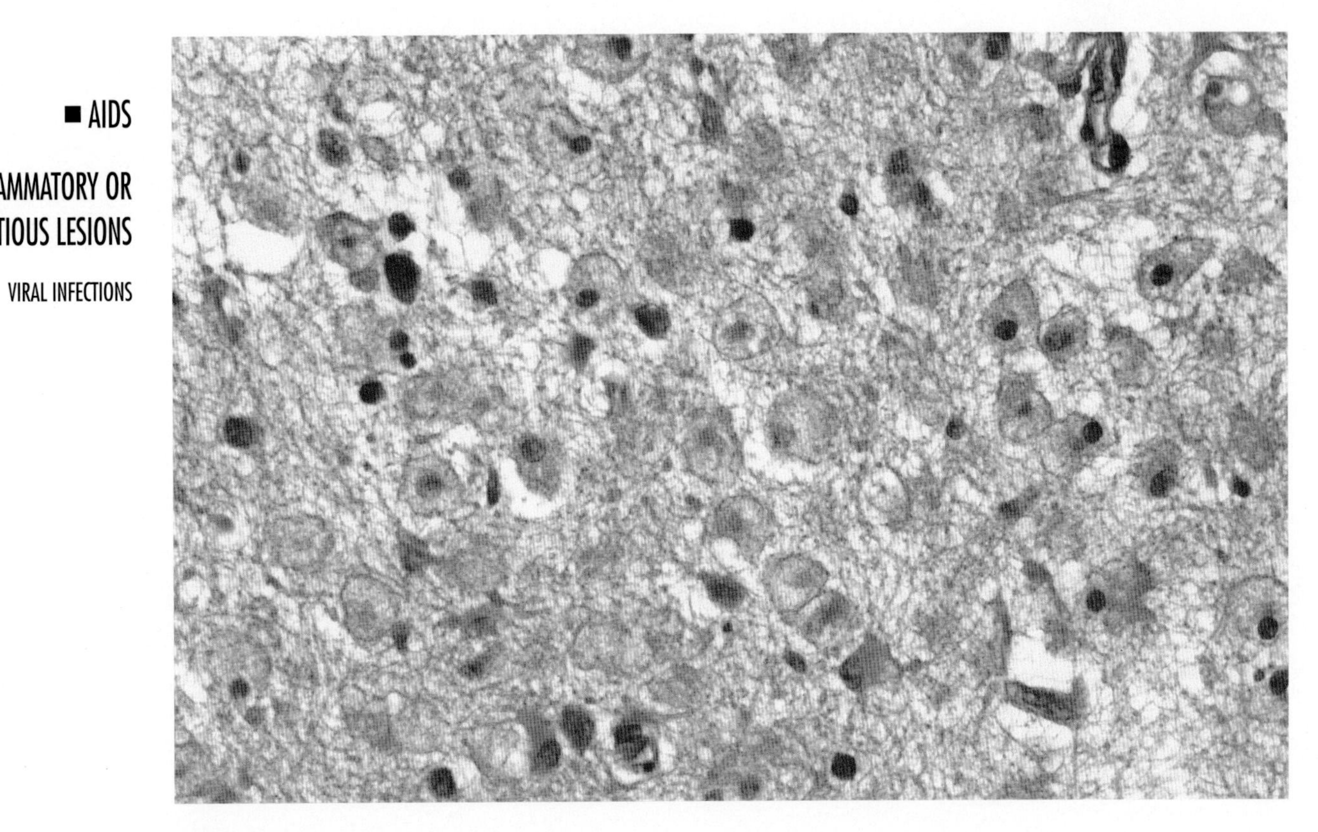

Figure 8–18 C

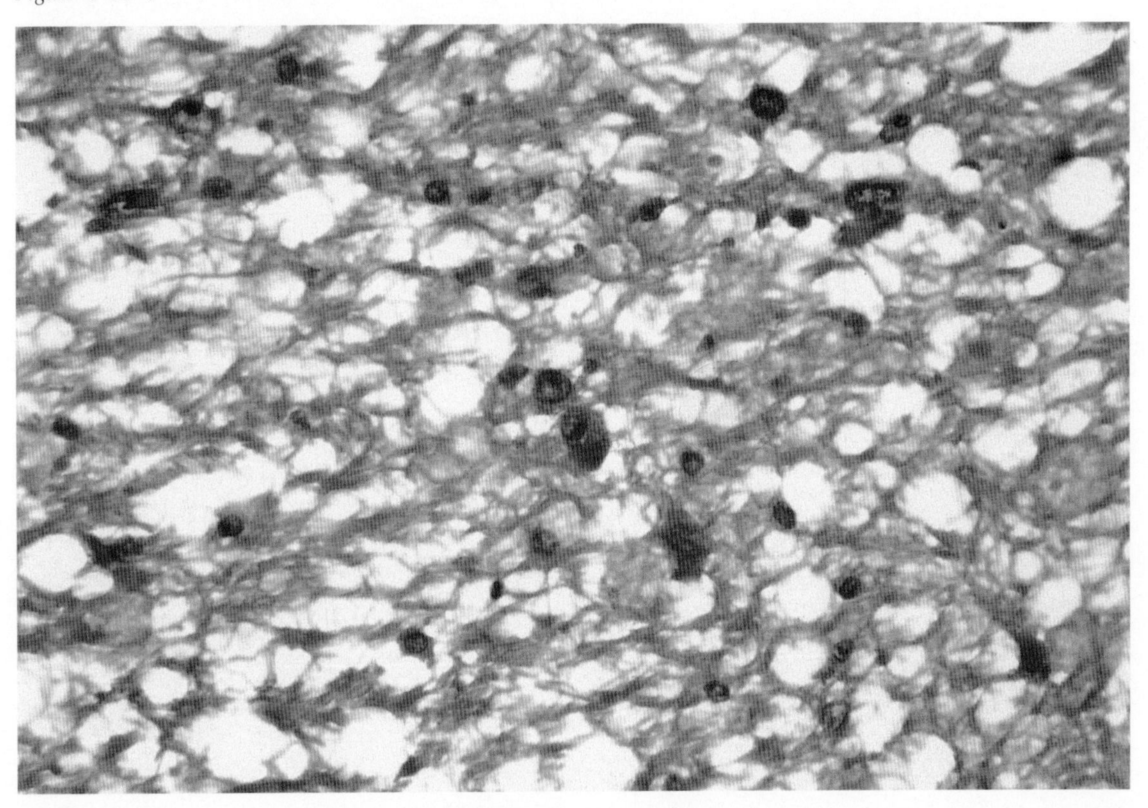

Figure 8–18 D

FIGURE 8–18. PROGRESSIVE MULTIFOCAL LEUKOENCEPHALOPATHY: C: Foamy macrophages are abundant in demyelinated areas. **D:** Oligodendroglia are enlarged and have violet smudges in the nuclei.

■ AIDS

INFECTIOUS AND INFLAMMATORY LESIONS

FUNGAL INFECTIONS

FIGURE 8–19. ***CANDIDA*** **SPECIES:** *Candida* is the most commonly identified fungal infection in immunocompromised persons. The oral cavity and esophagus are most often affected. **A:** In this example of fungal esophagitis, there is ulceration of the epithelium. Gomori methenamine-silver stain (**B**) and periodic acid–Schiff stain (**C**) clearly show the presence of pseudohyphal and spore forms of *Candida* species.

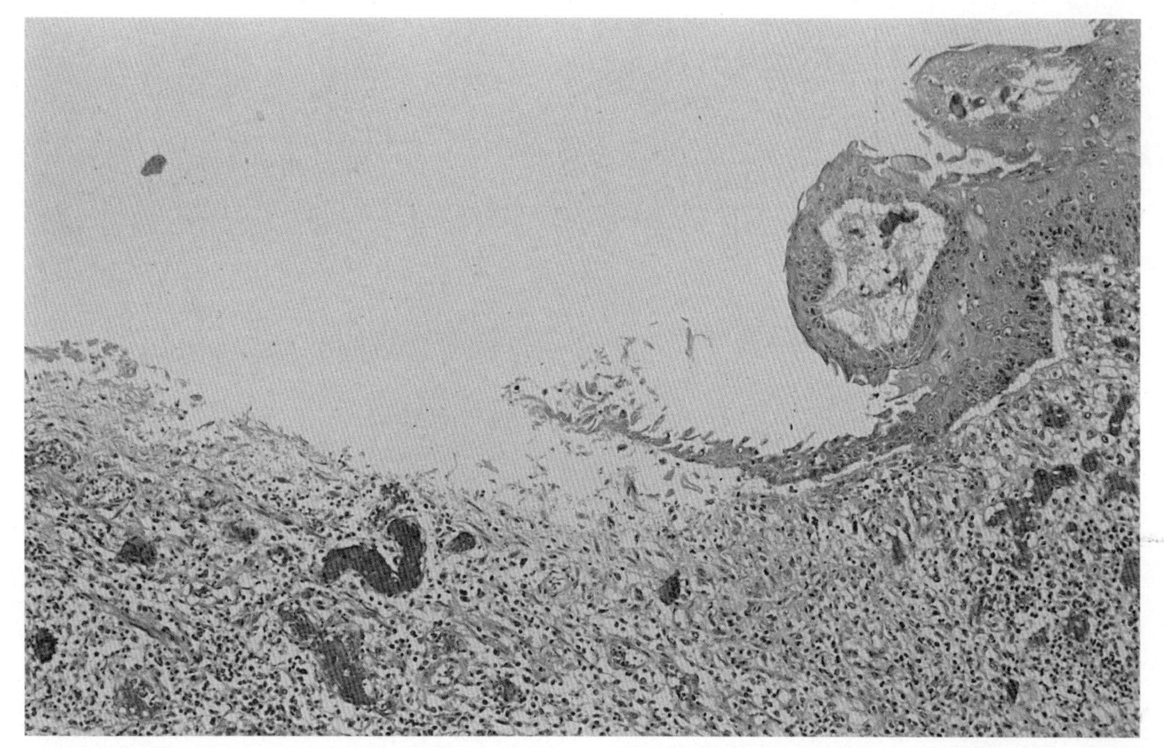

Figure 8–19 **A**

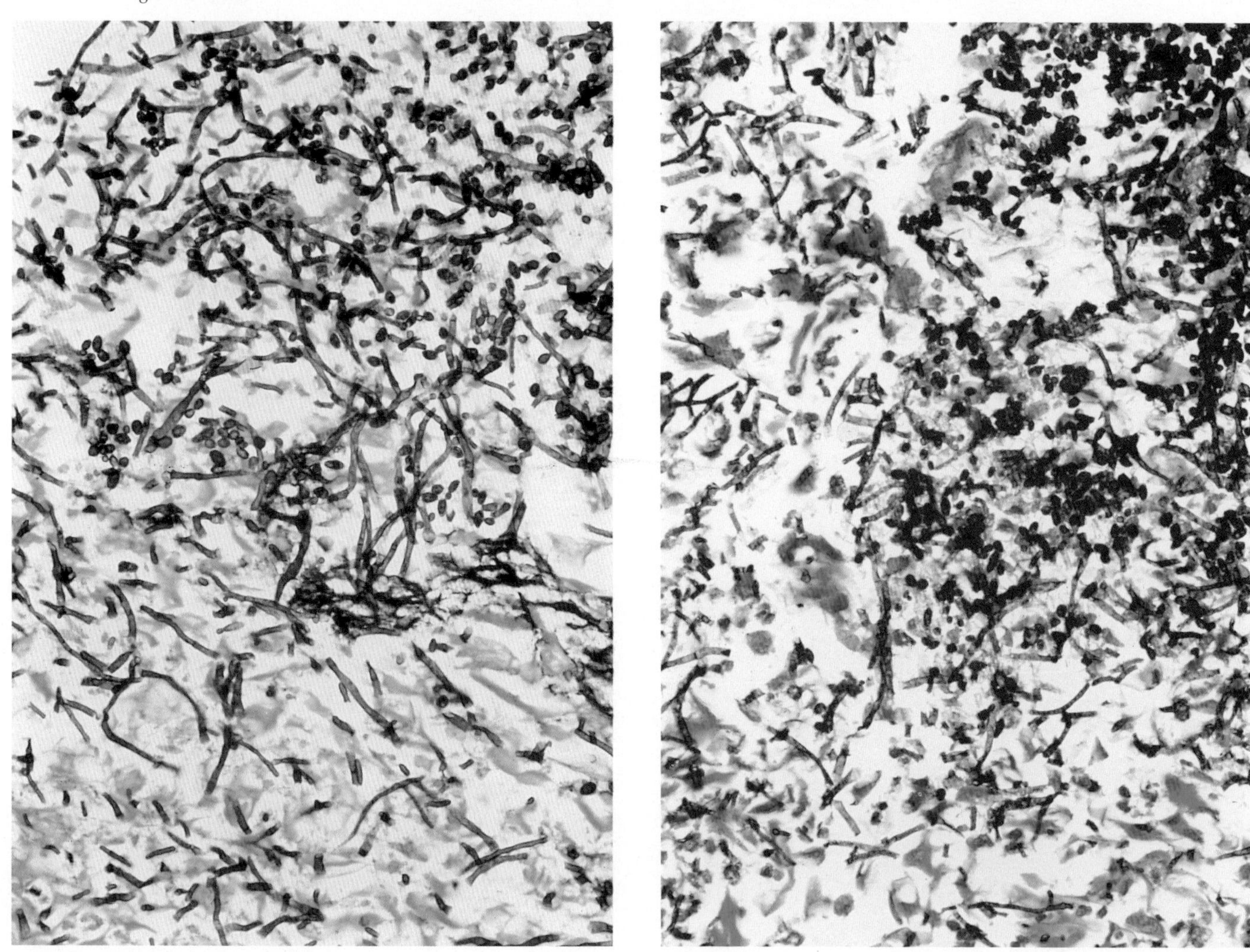

Figure 8–19 **B**

Figure 8–19 **C**

■ AIDS

INFECTIOUS AND INFLAMMATORY LESIONS

FUNGAL INFECTIONS

FIGURE 8–20. *CRYPTOCOCCUS NEOFORMANS:* A: In this example of cryptococcal lymphadenitis, large clear confluent areas are present. B: The clear areas are caused by a retraction artifact of the thick capsule that surrounds the pale blue-gray organisms. C: Mucicarmine stains the capsule red and confirms the mucopolysaccharide nature of the capsule. The organisms are round and have a single bud.

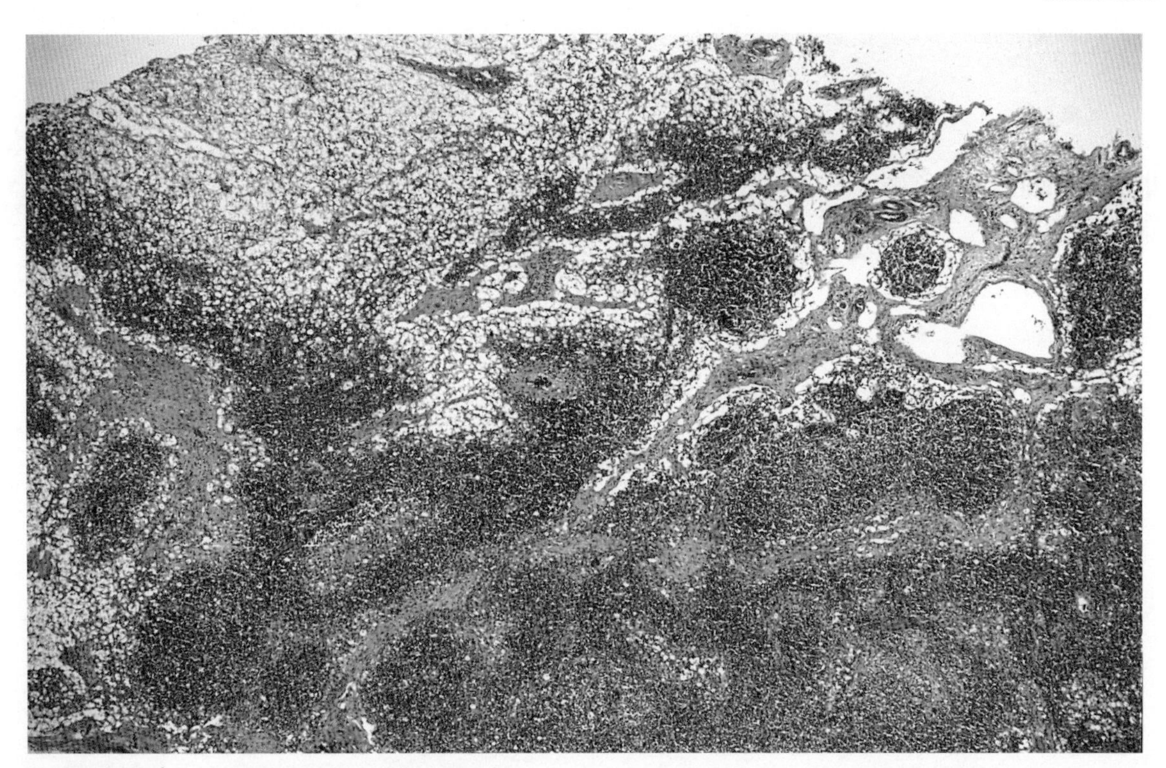

Figure 8–20 A

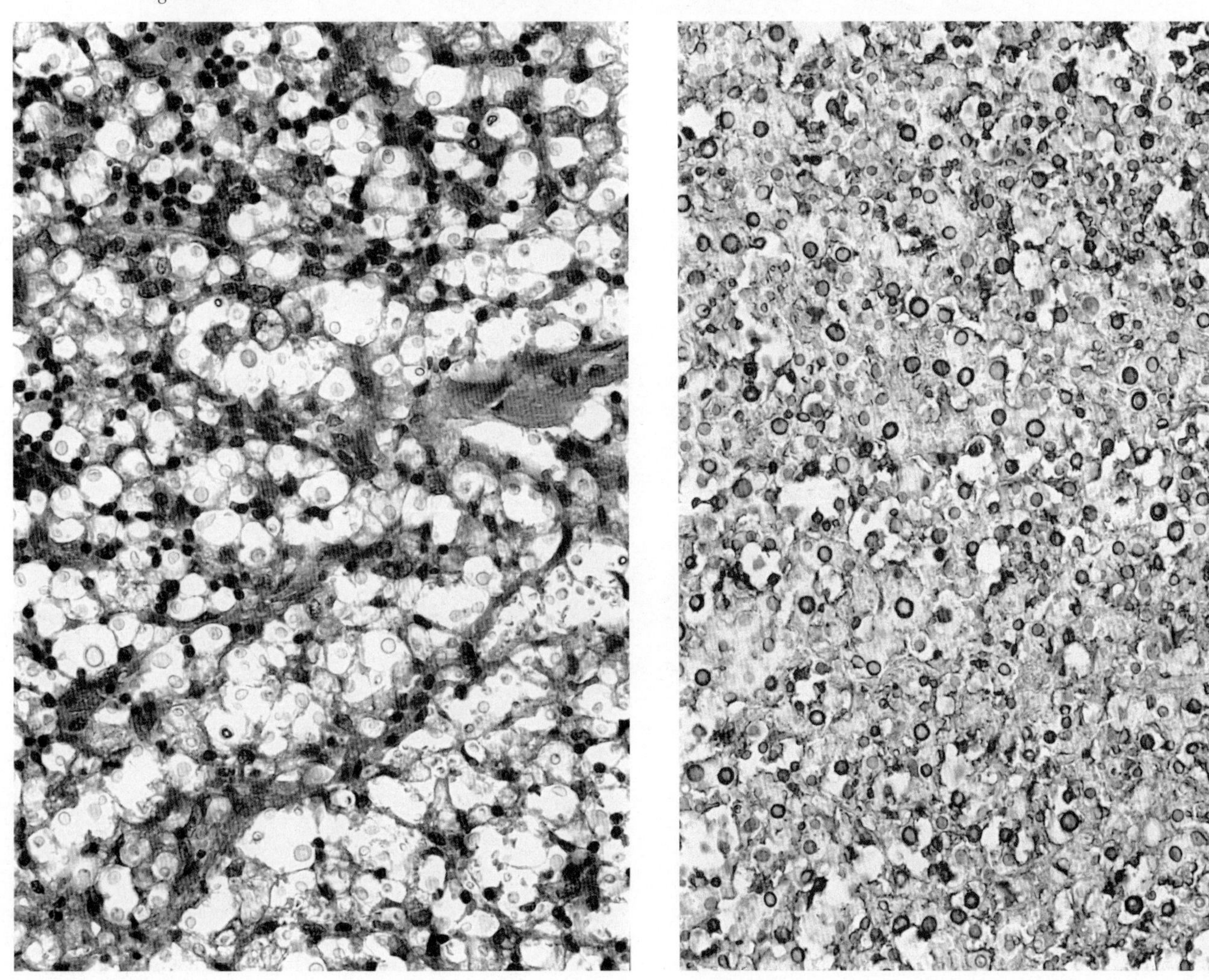

Figure 8–20 B

Figure 8–20 C

■ AIDS

INFECTIOUS AND INFLAMMATORY LESIONS

FUNGAL INFECTIONS

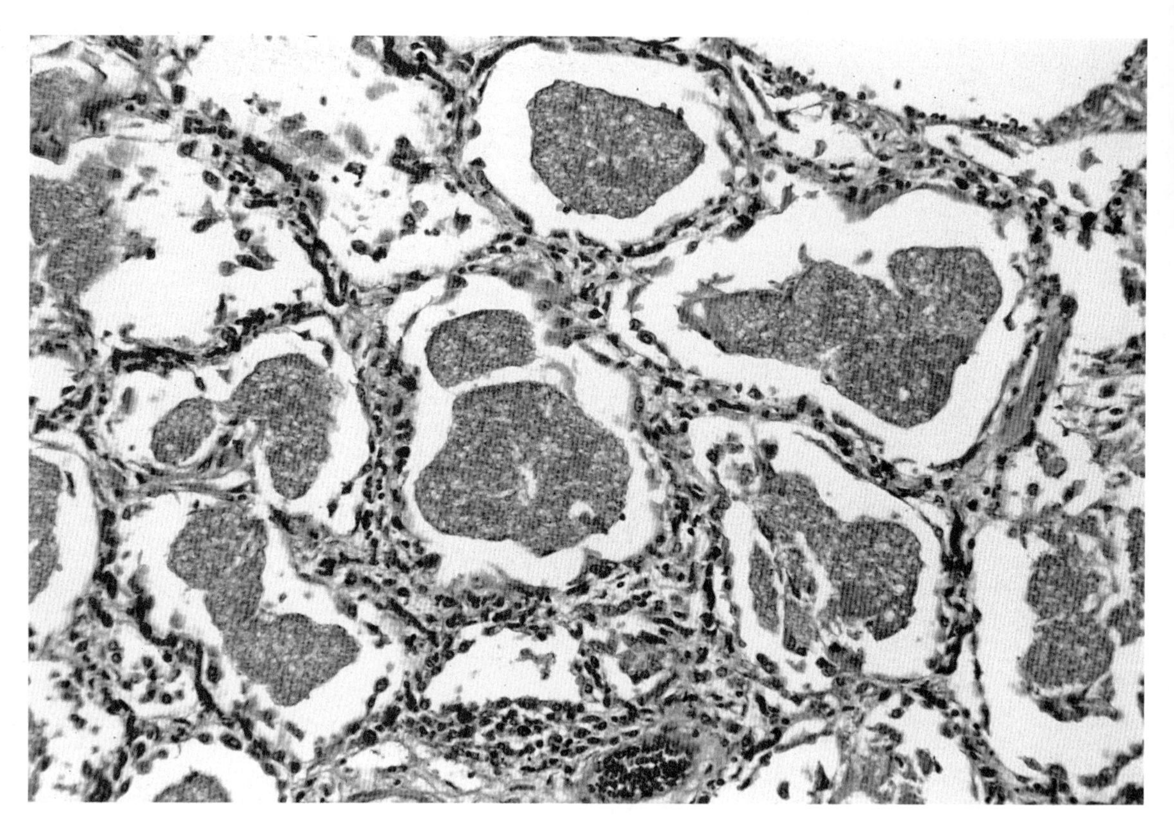

Figure 8–21 A

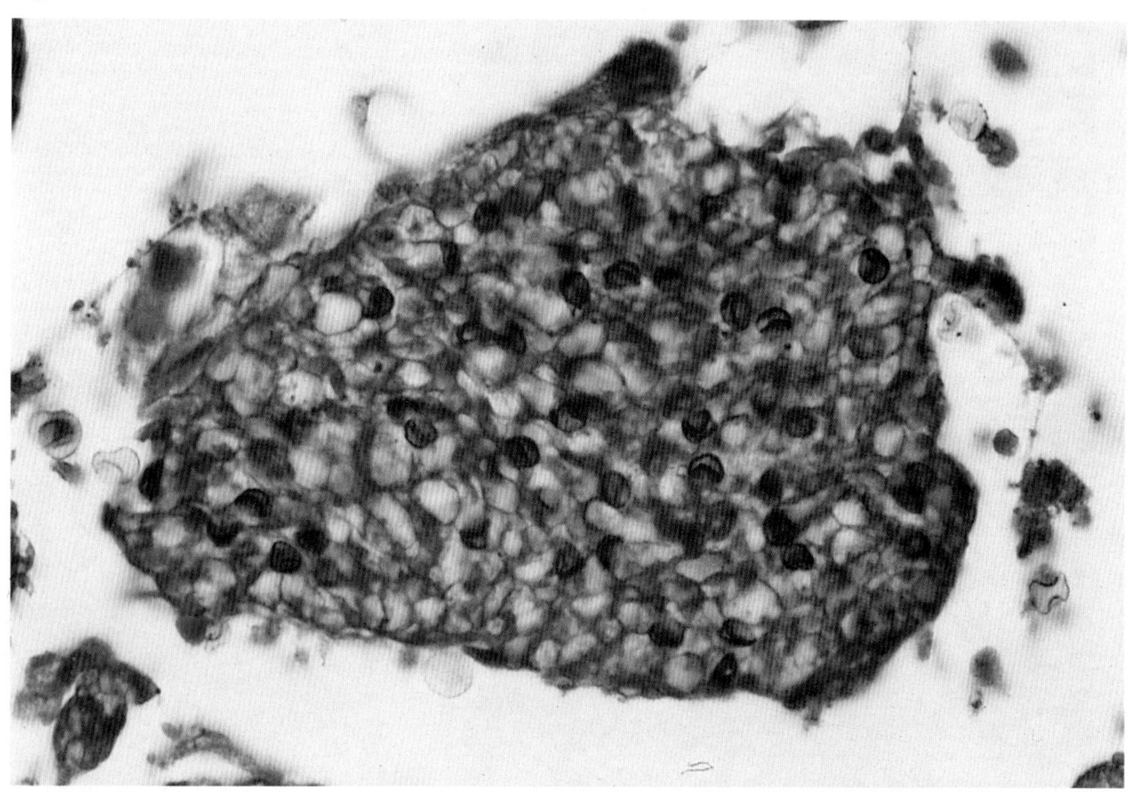

Figure 8–21 B

FIGURE 8–21. ***PNEUMOCYSTIS CARINII:*** *Pneumocystis carinii* was originally classified as a protozoan but has been reclassified as a fungus. It is the most common pulmonary disease and opportunistic infection in AIDS. **A:** Characteristic of *Pneumocystis carinii*–related pneumonia are intra-alveolar collections of eosinophilic foamy material composed of serum exudate and enmeshed organisms. The alveolar septa are widened secondary to interstitial edema and inflammation. **B:** Gomori methenamine-silver stain stains the oval organisms gray-black.

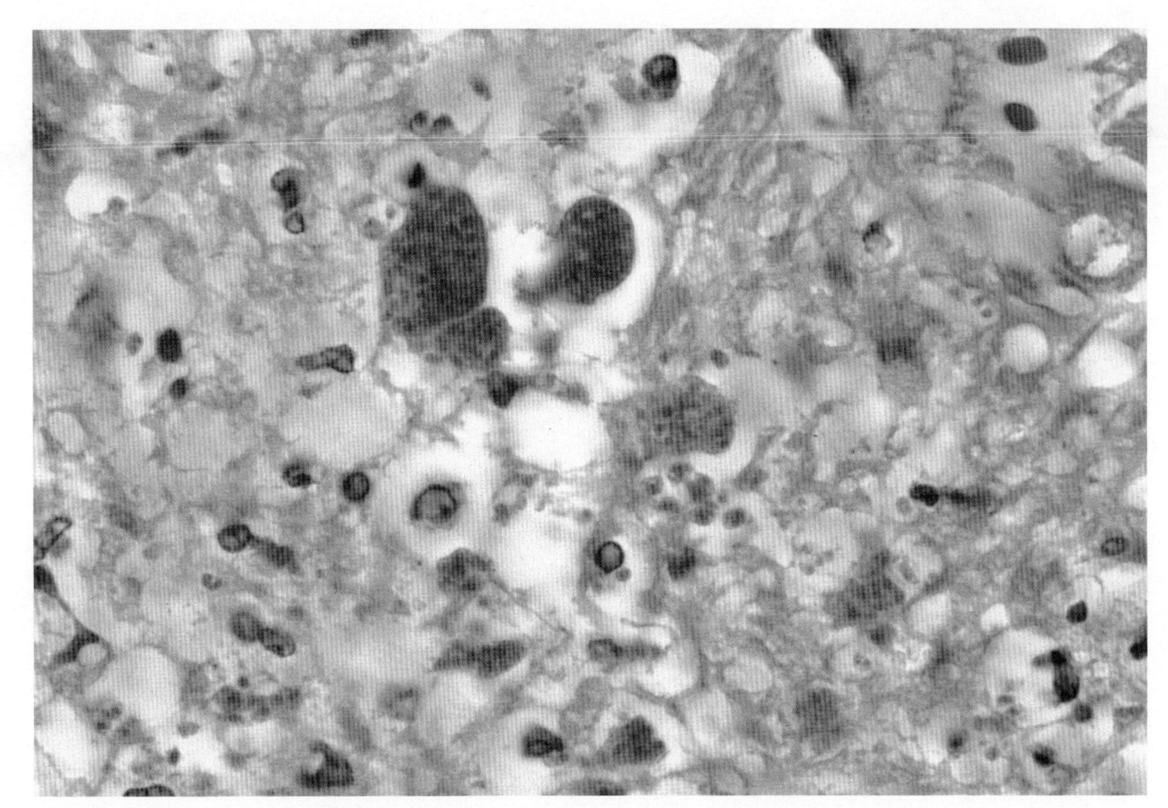

Figure 8–22 **A**

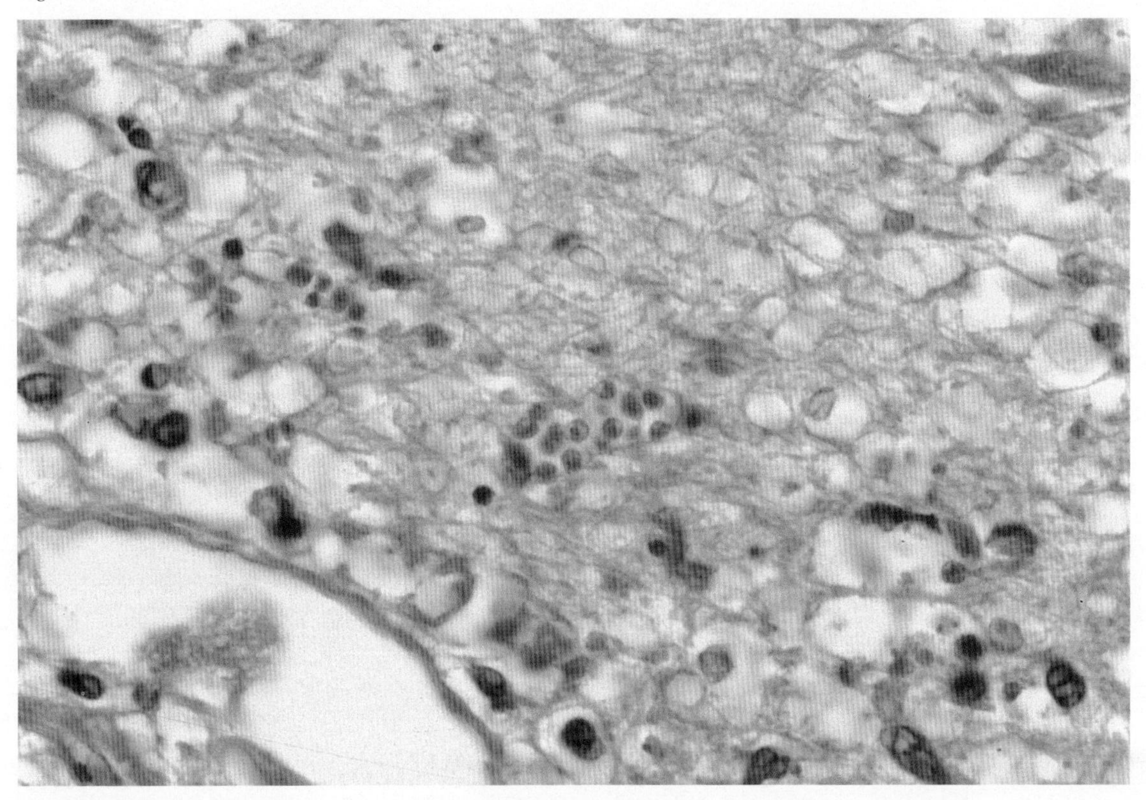

Figure 8–22 **B**

FIGURE 8–22. ***TOXOPLASMA GONDII:*** **A**: *Toxoplasma gondii* is one of the most common causes of neural symptoms and is the most common opportunistic infection involving the central nervous system in AIDS. In this example of necrotizing encephalitis caused by *Toxoplasma gondii,* cysts filled with bradyzoites (**A**) and free tachyzoites (**B**) are seen peripheral to the necrotizing areas.

■ AIDS

INFECTIOUS AND INFLAMMATORY LESIONS

PARASITIC INFECTIONS

FIGURE 8–23. CRYPTOSPORIDIA: In this photomicrograph of cryptosporidiosis, in a sample taken from the colon, spherical basophilic organisms are present along the luminal border of the crypt. The surrounding parenchyma is infiltrated with lymphocytes, plasma cells, neutrophils, and, in this photomicrograph, eosinophils as well.

SKIN

FIGURE 8–24. PRURITIC PAPULAR ERUPTION OF AIDS: A: Perivascular inflammation in the dermis characterizes pruritic papular eruption of AIDS. **B:** The inflammatory exudate is composed of lymphocytes, plasma cells, eosinophils, and neutrophils.

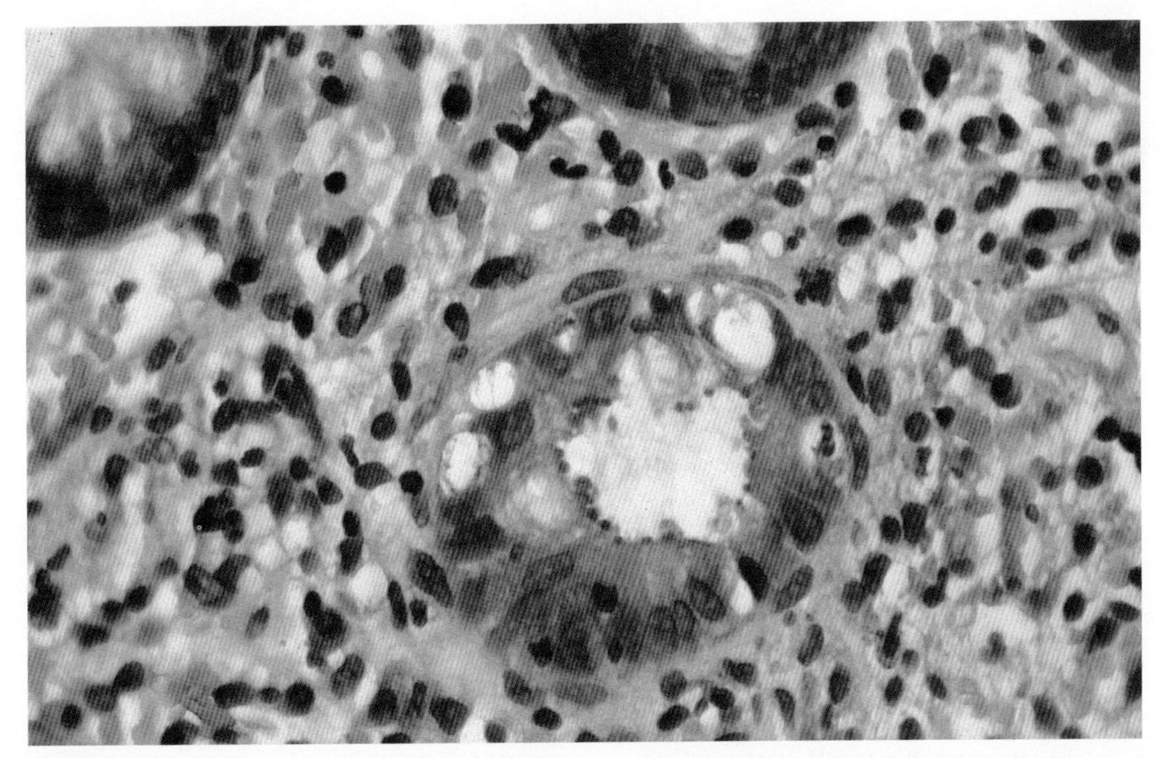

Figure 8–23

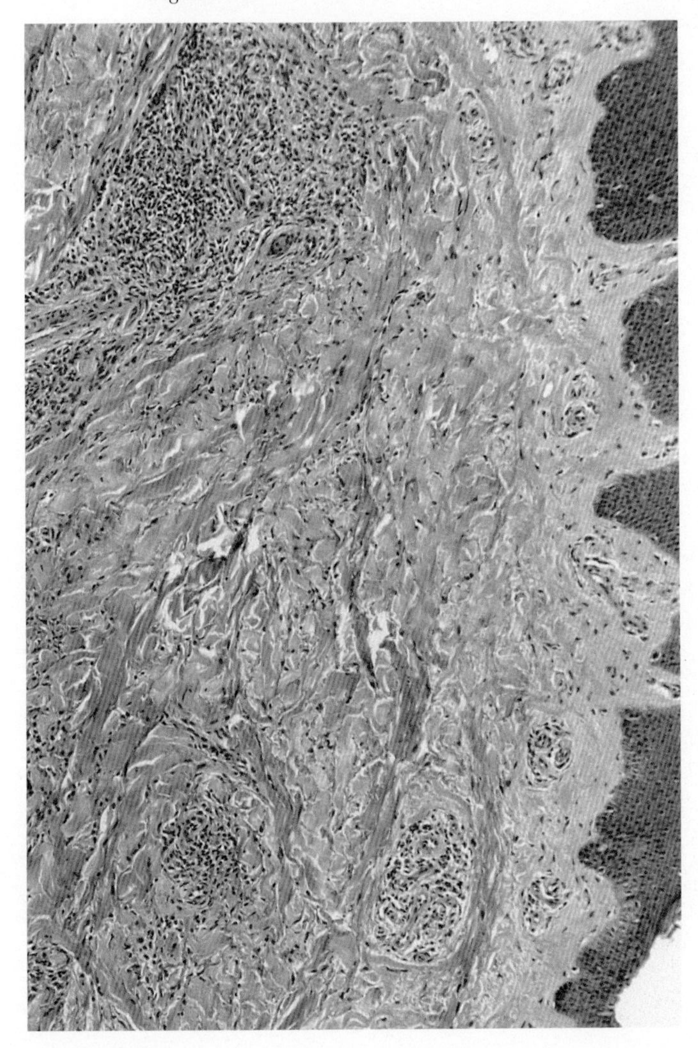

Figure 8–24 A

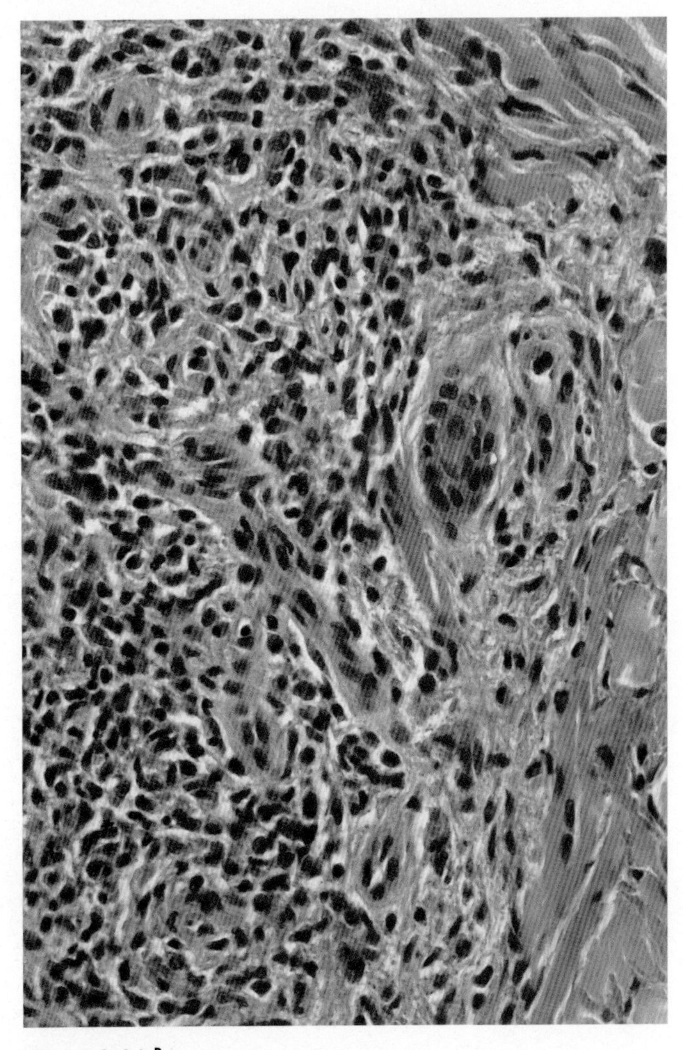

Figure 8–24 B

■ AIDS

MALIGNANT TUMORS

FIGURE 8–25. KAPOSI'S SARCOMA: Kaposi's sarcoma is the most common neoplasm in AIDS. **A:** In this example involving the skin, there is a dermal nodule. The nodule is composed of a proliferation of atypical spindle cells lining slitlike vascular channels. **B:** There is red blood cell extravasation, and eosinophilic globules are present extracellularly, in macrophages, and in tumor cells. **C:** In older lesions of Kaposi's sarcoma, hemosiderin is also present.

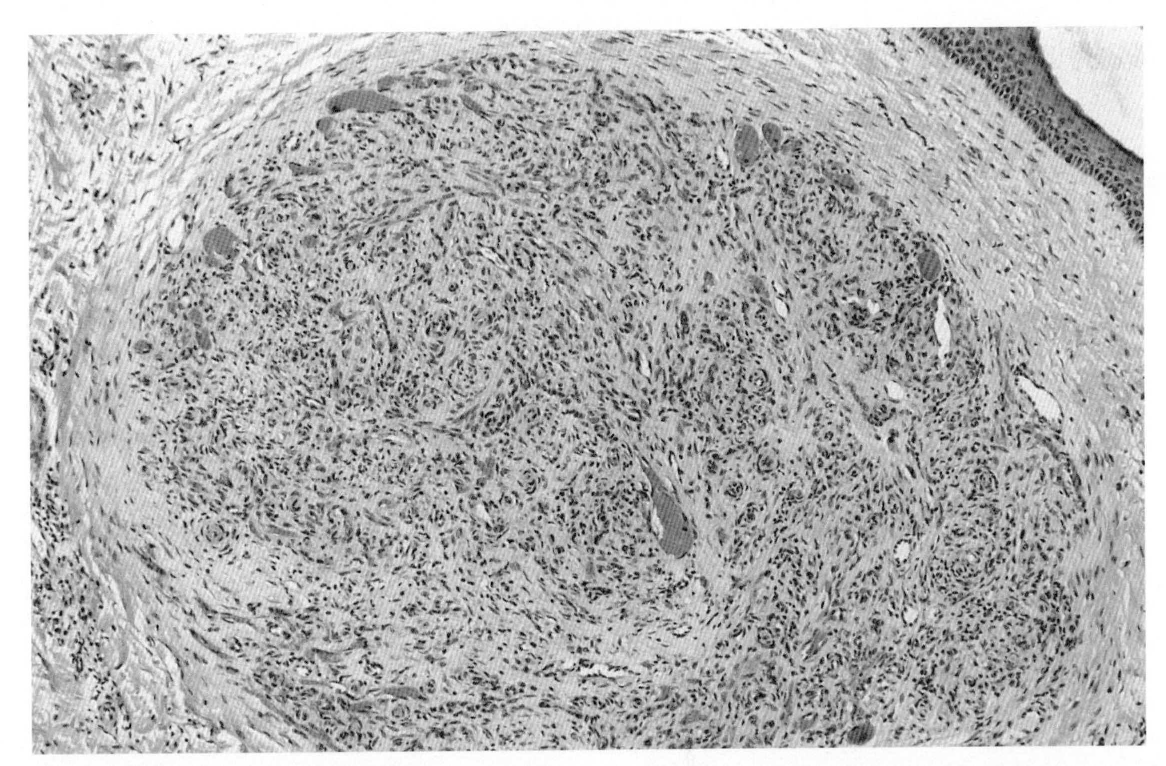

Figure 8–25 A

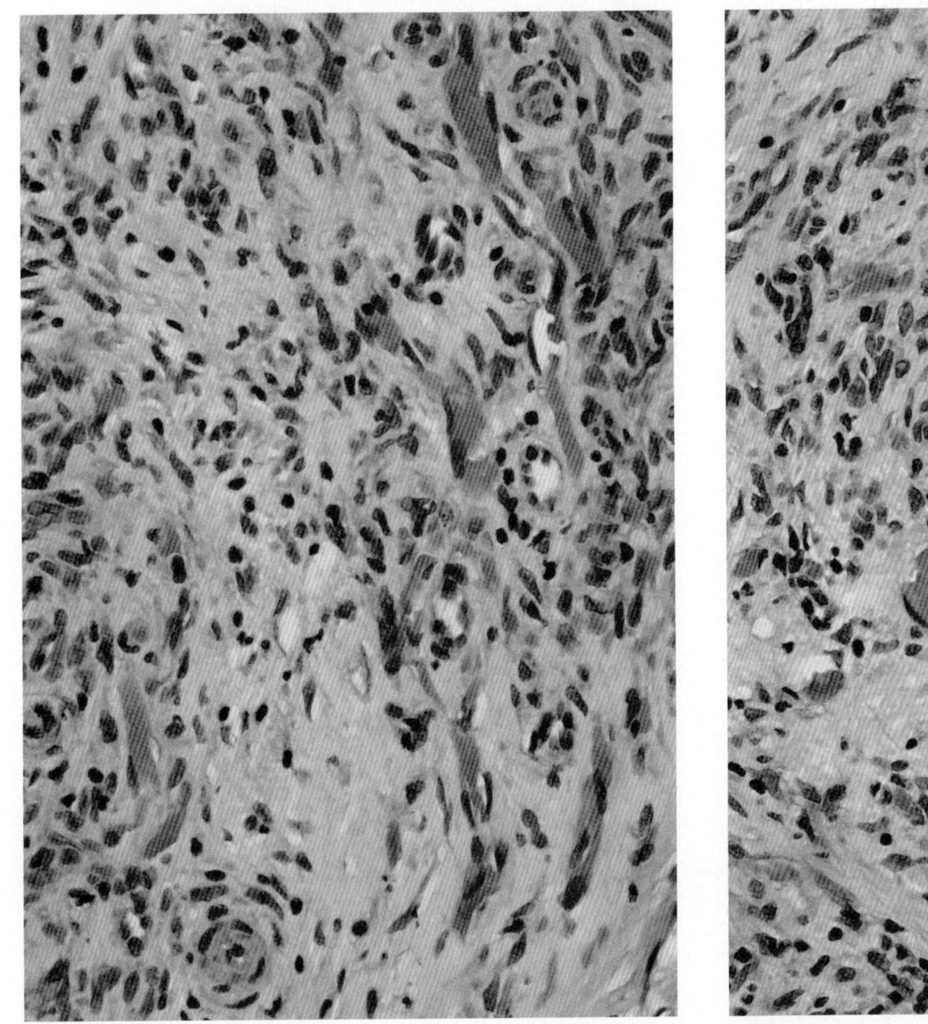

Figure 8–25 B

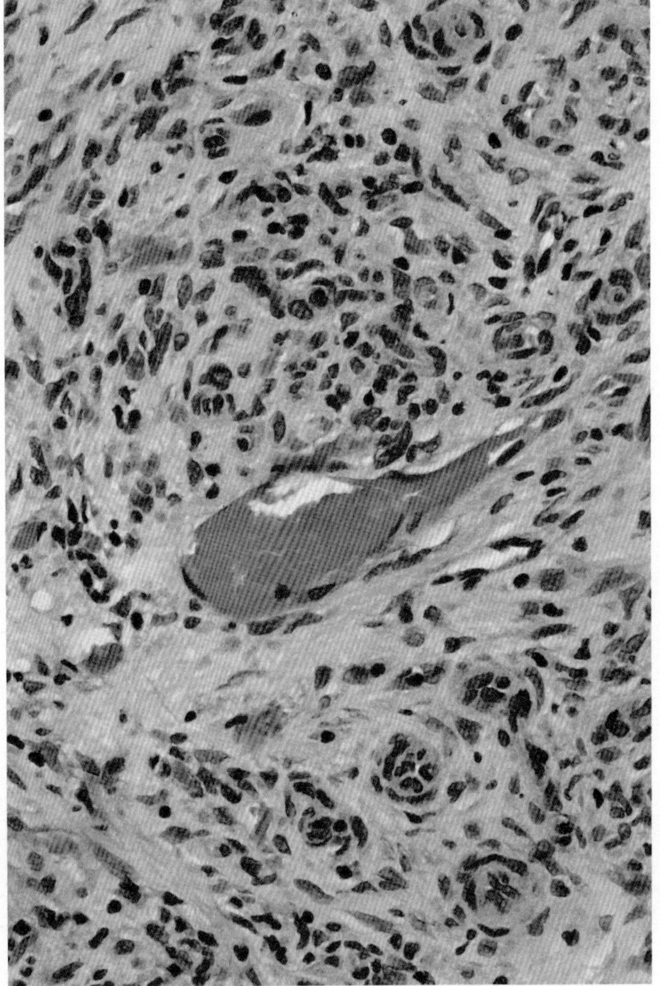

Figure 8–25 C

9. RESPIRATORY SYSTEM

■ OBSTRUCTIVE LUNG DISEASES

FIGURE 9–1. ASTHMA: A: In asthma, there is mucus plugging of bronchi and compensatory hypertrophy of the smooth muscle in the bronchial wall. The same process affects bronchioles. **B:** In addition, there is goblet cell hypertrophy, submucosal inflammation, and an increase in fibrous tissue beneath the basement membrane, giving the appearance of a thickened basement membrane. **C:** The inflammatory infiltrate is composed of eosinophils, plasma cells, and histiocytes. Eosinophils are also found within the mucus plugs.

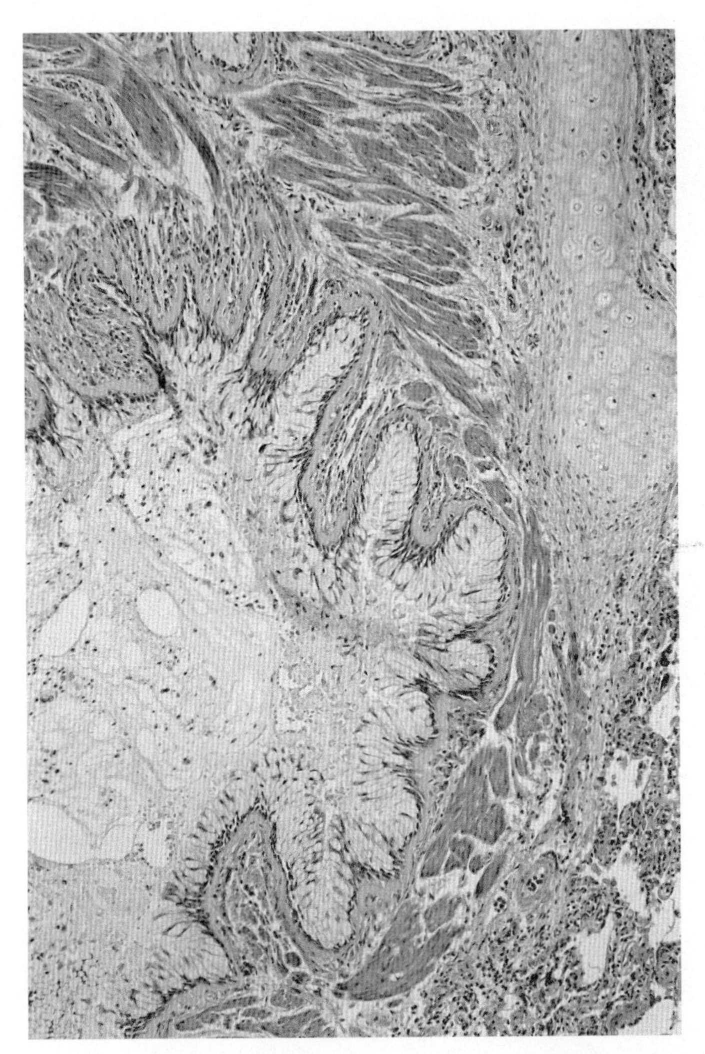

Figure 9–1 A

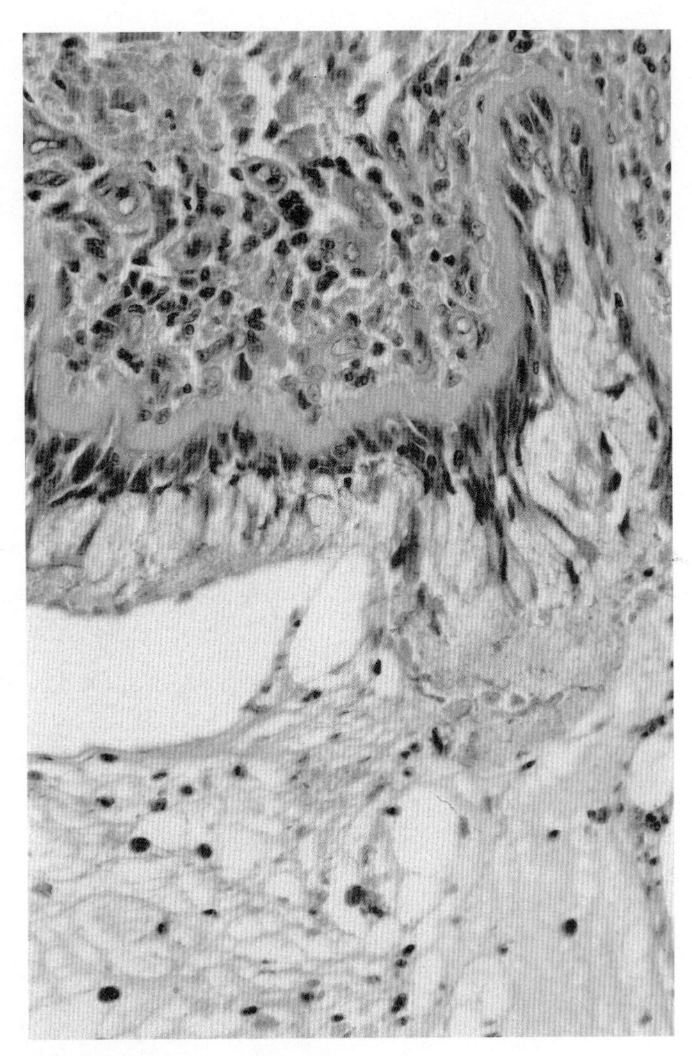

Figure 9–1 B

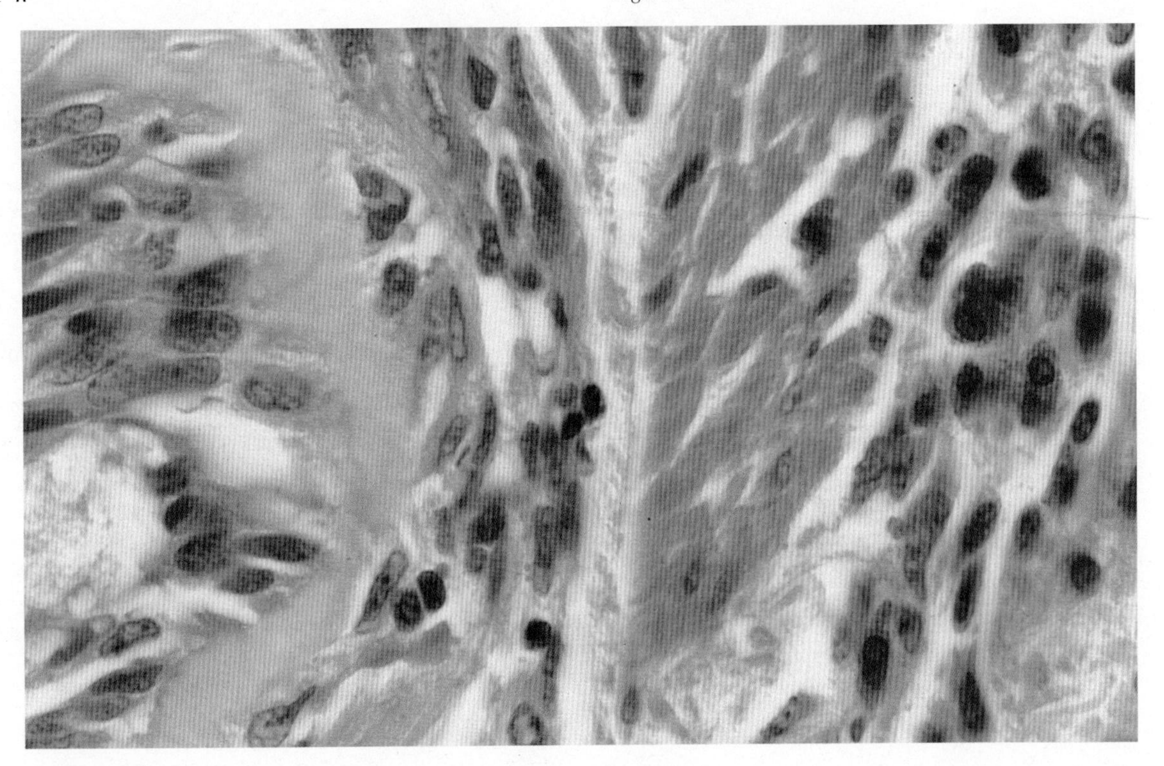

Figure 9–1 C

■ OBSTRUCTIVE LUNG DISEASES

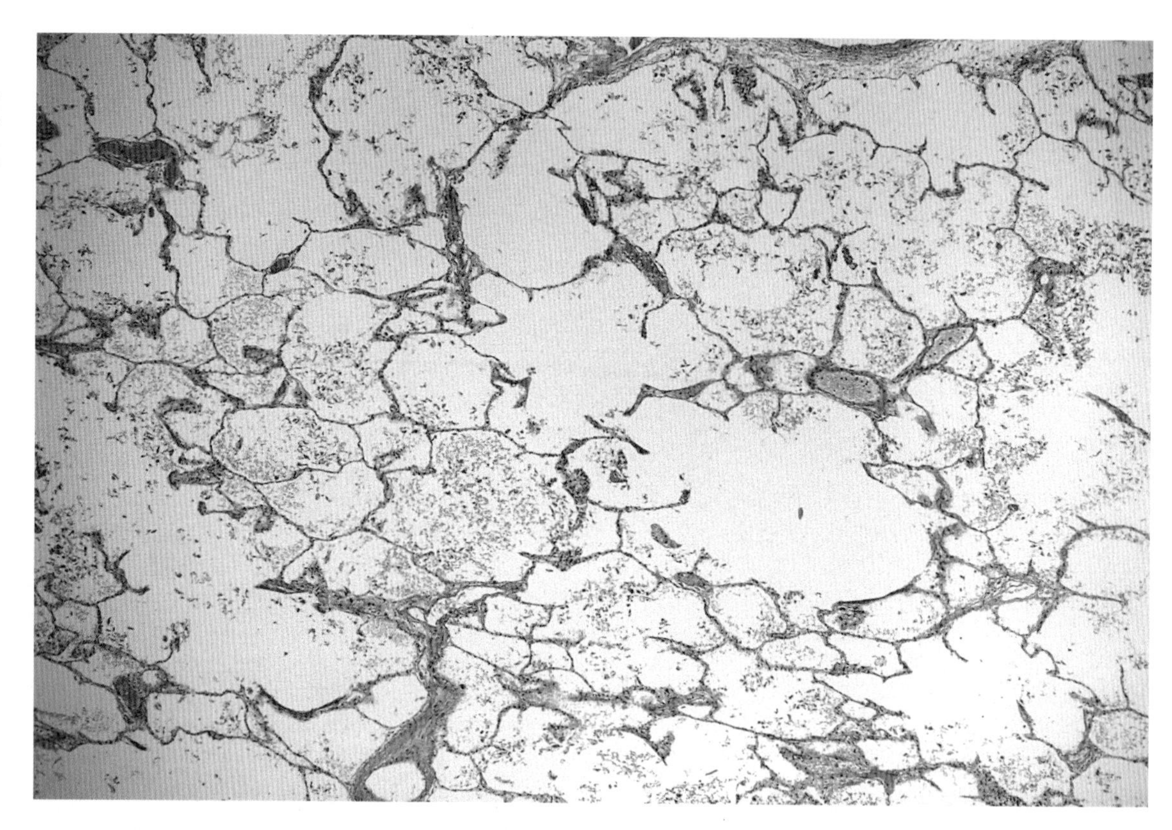

FIGURE 9–2. EMPHYSEMA: In emphysema, alveolar and bronchiolar walls are destroyed, which results in large air spaces.

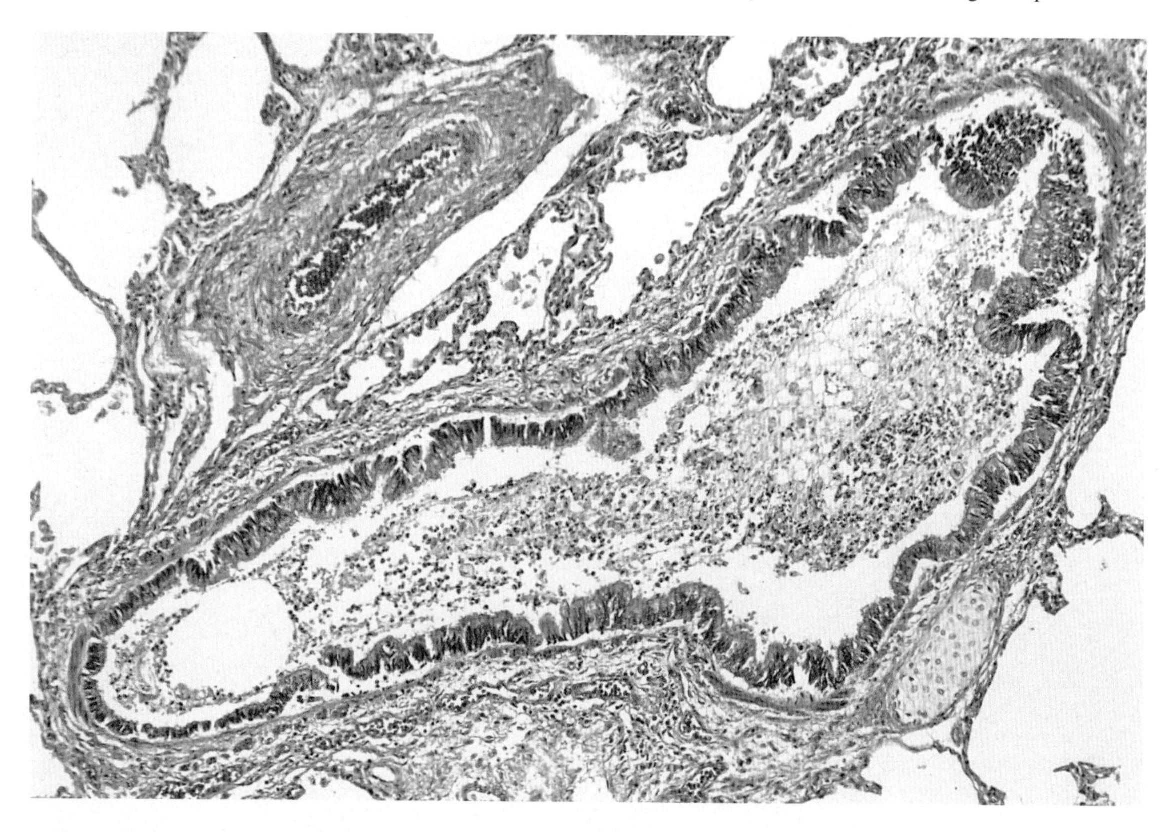

FIGURE 9–3. CYSTIC FIBROSIS: In early-stage cystic fibrosis, dilatation of airways with inspissated mucus secretions is noted.

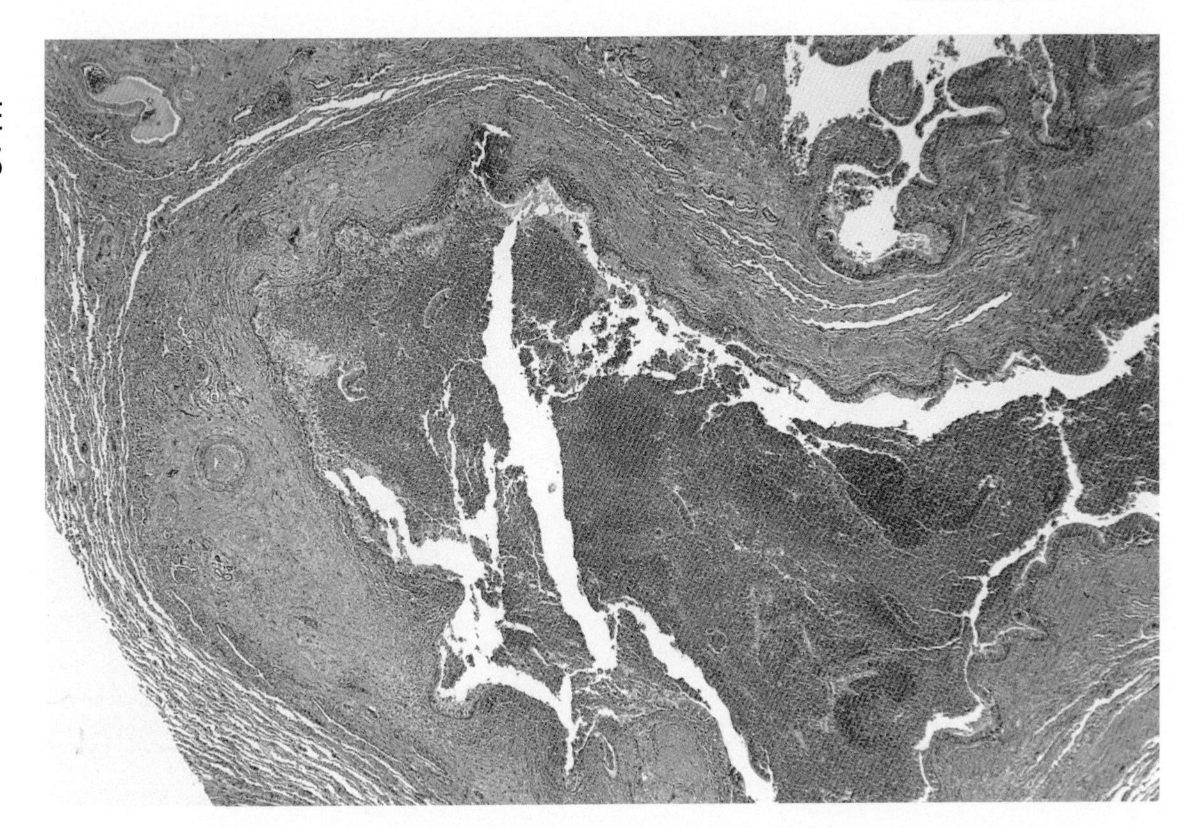

Figure 9–4 **A**

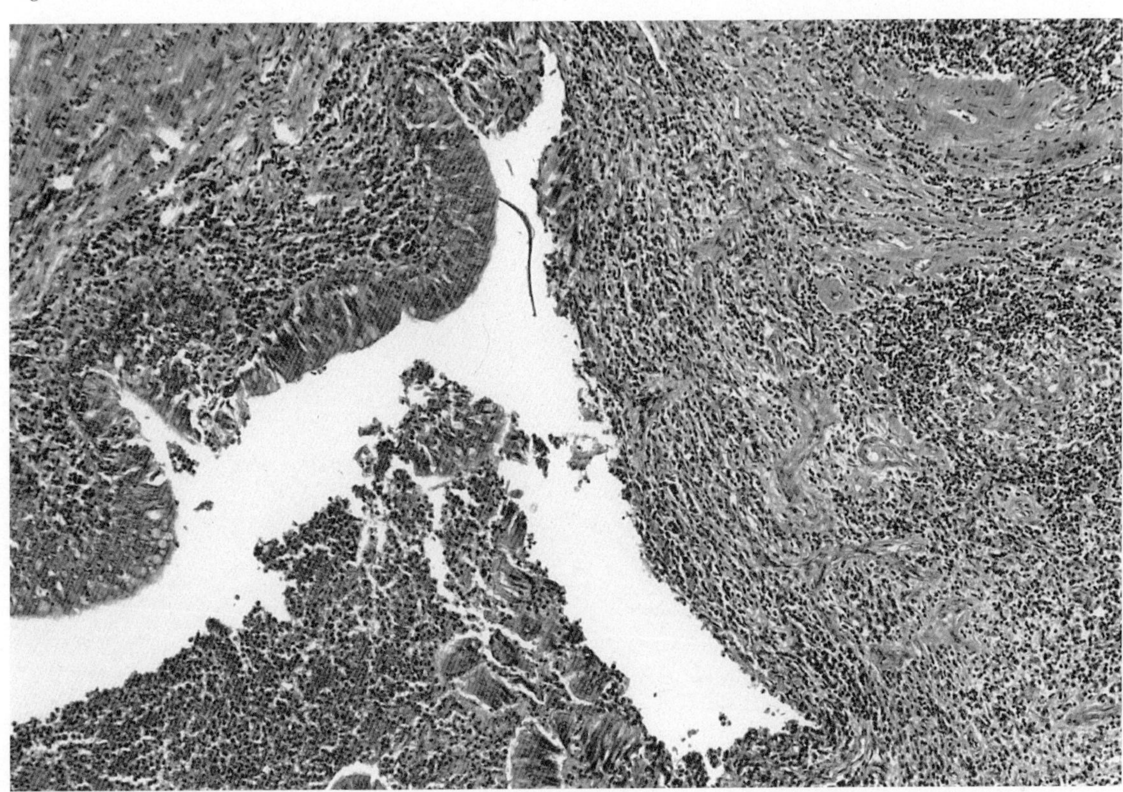

Figure 9–4 **B**

FIGURE 9–4. BRONCHIECTASIS: A: In bronchiectasis, airways are markedly dilated and filled with a dense inflammatory exudate. Focally there is epithelial ulceration and eventual replacement by metaplastic squamous epithelium. **B:** The ulcerated areas are associated with granulation tissue, and the adjacent mucosa is infiltrated with acute and chronic inflammatory cells.

■ OBSTRUCTIVE LUNG DISEASES

FIGURE 9–5. BRONCHITIS: A: Bronchitis is characterized by hyperplasia of the submucosal gland layer in bronchial walls. **B:** A sparse mononuclear cell infiltrate accompanies the smooth muscle hyperplasia in the mucosa. **C:** Goblet cell metaplasia of the lining epithelium is present.

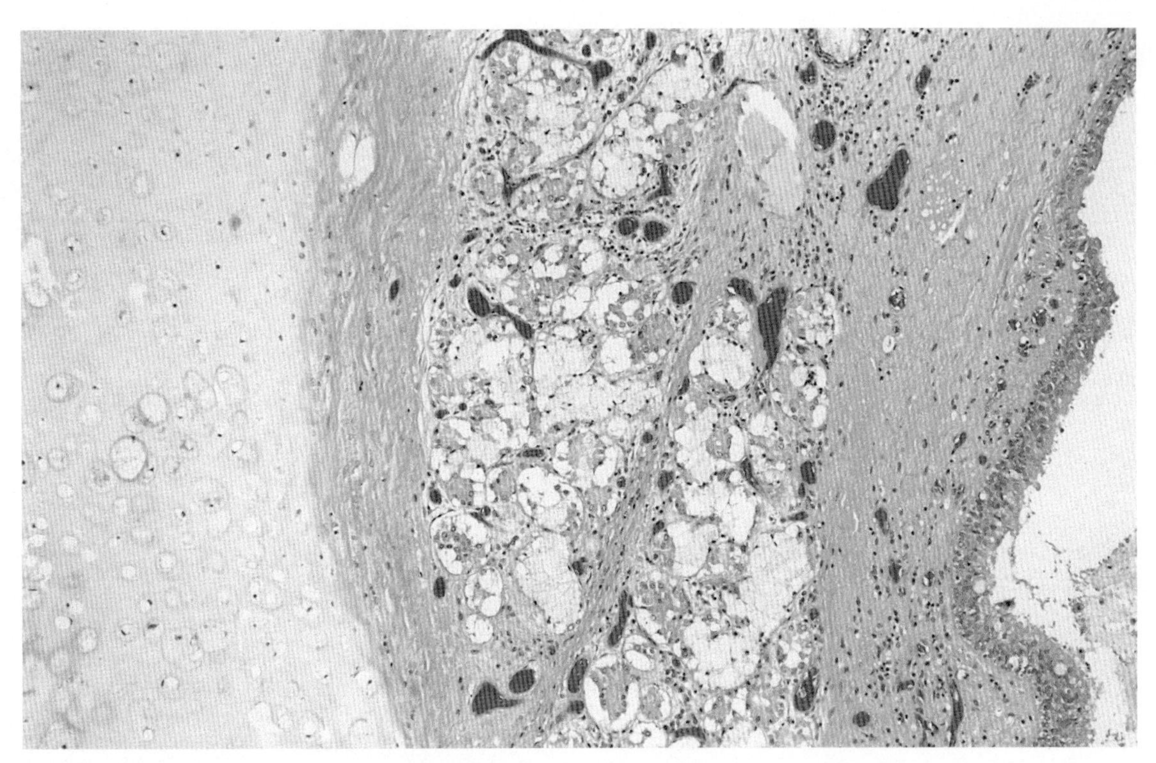

Figure 9–5 A

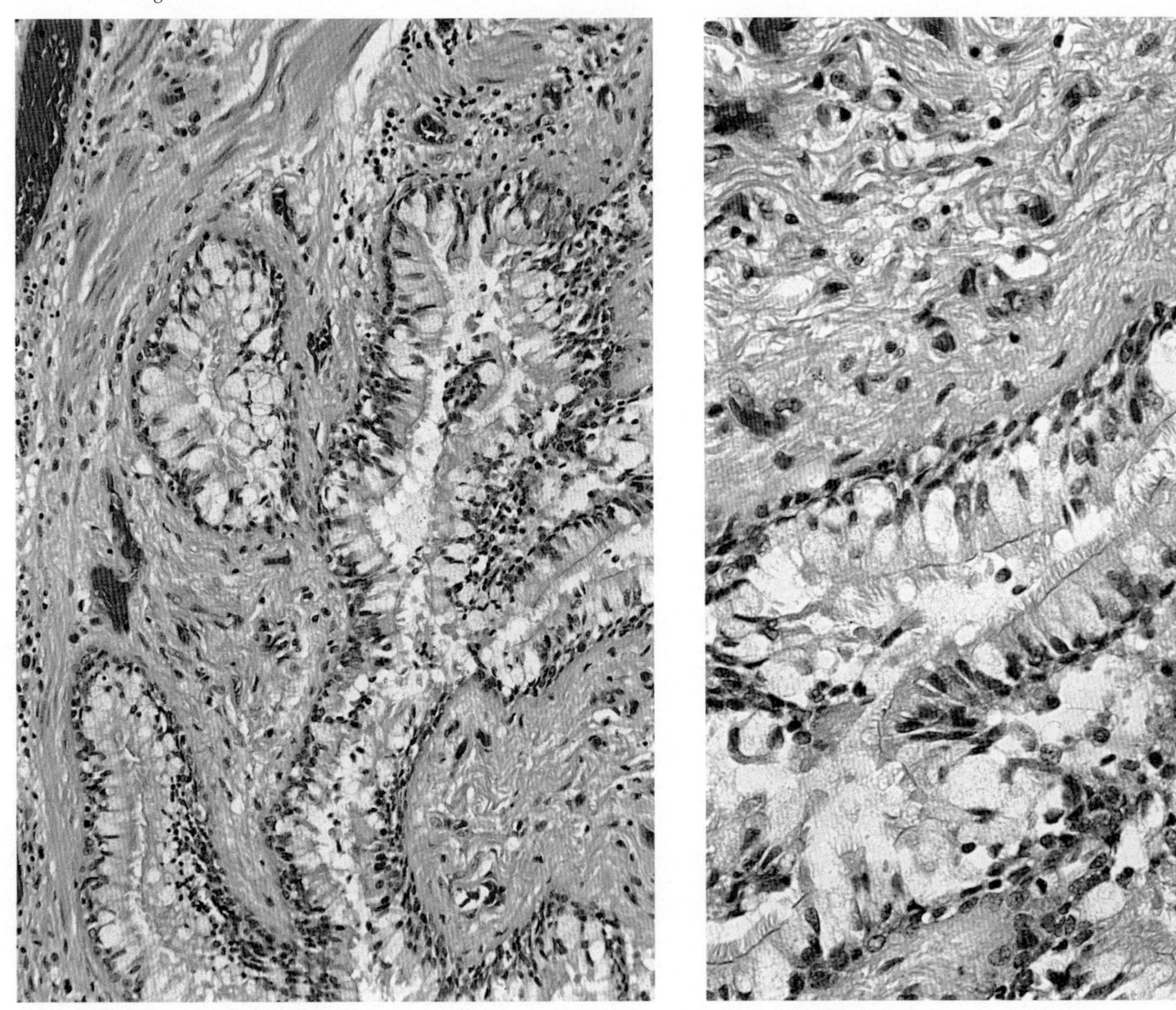

Figure 9–5 B

Figure 9–5 C

■ RESTRICTIVE LUNG DISEASES

FIGURE 9–6. DIFFUSE ALVEOLAR DAMAGE (ADULT RESPIRATORY DISTRESS SYNDROME [ARDS]): A: In the acute, **exudative phase** of diffuse alveolar damage, there is intra-alveolar edema and hemorrhage. B: Eosinophilic-staining hyaline membranes composed of protein-rich edema fluid mixed with necrotic epithelial cells and fibrin are prominent features in dilated alveolar ducts. C: In the **fibrotic, organizing phase,** there is fibroblastic proliferation involving the exudates and alveolar septa, with eventual fibrosis.

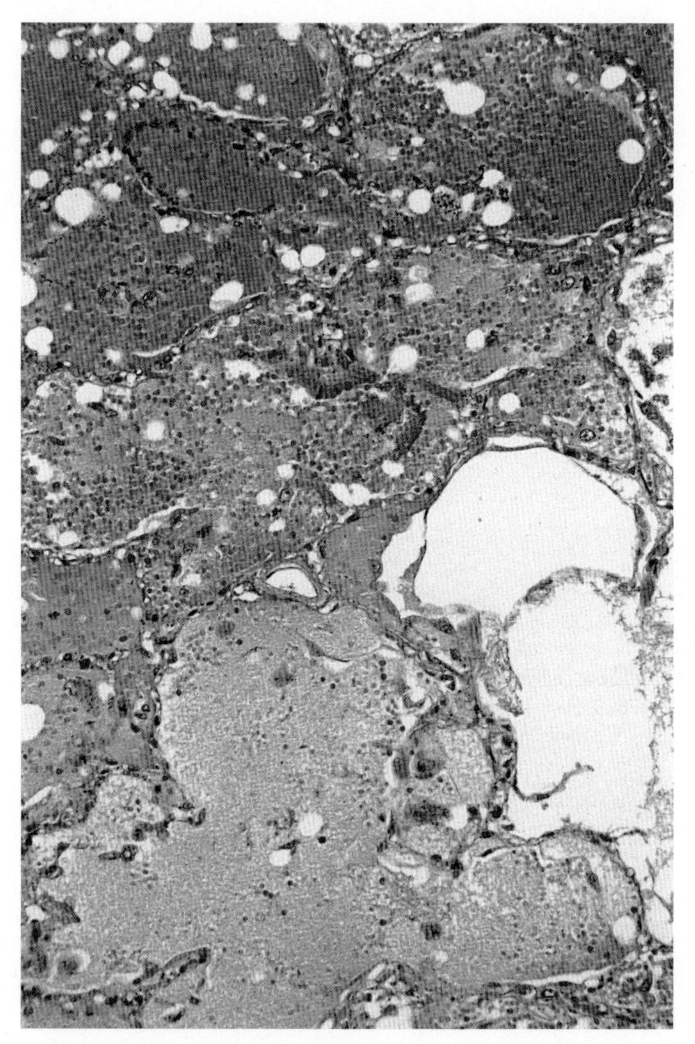

Figure 9–6 A

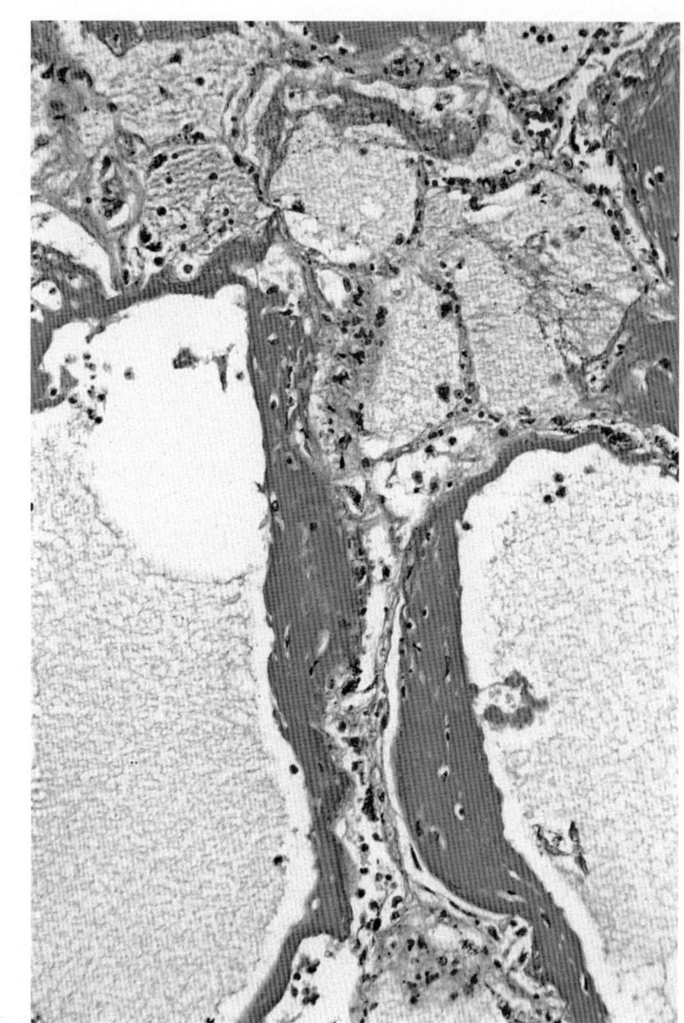

Figure 9–6 B

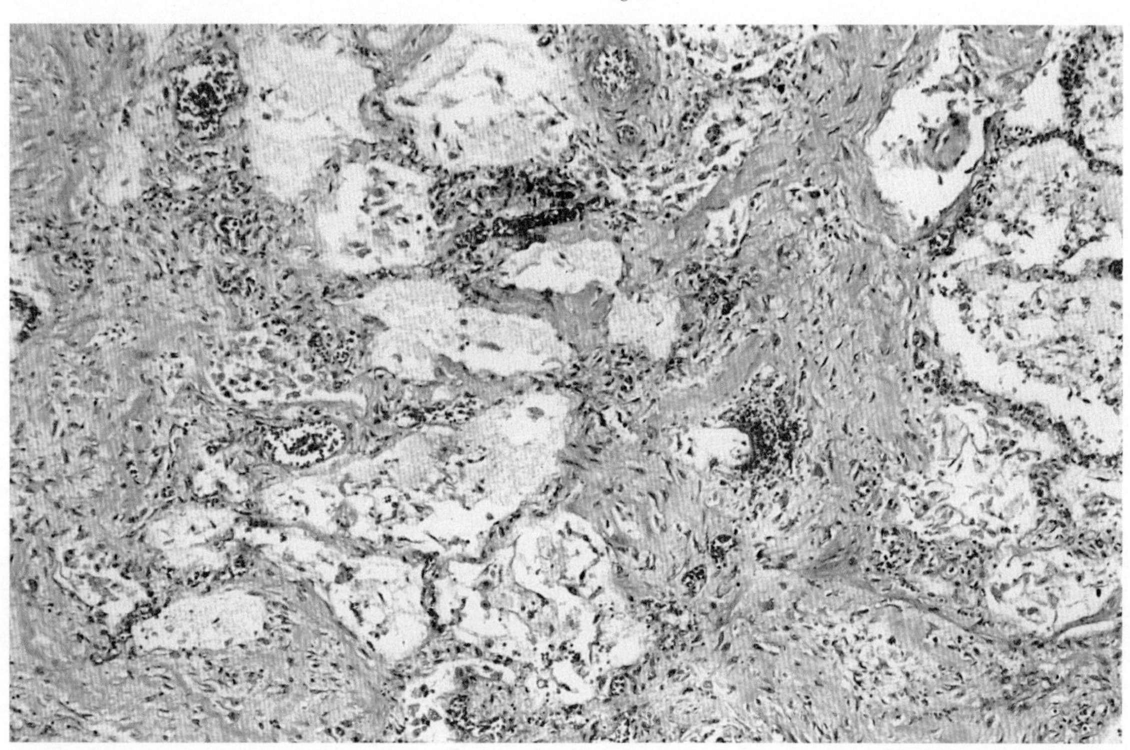

Figure 9–6 C

■ RESTRICTIVE LUNG DISEASES

FIGURE 9–7. HYPERSENSITIVITY PNEUMONITIS (EXTRINSIC ALLERGIC ALVEOLITIS): A: In hypersensitivity pneumonitis, there is patchy interstitial pneumonia. **B:** The interstitial infiltrate is composed of epithelioid granulomas accompanied by lymphocytes, plasma cells, and histiocytes. **C:** Foamy macrophages are present in the alveolar spaces.

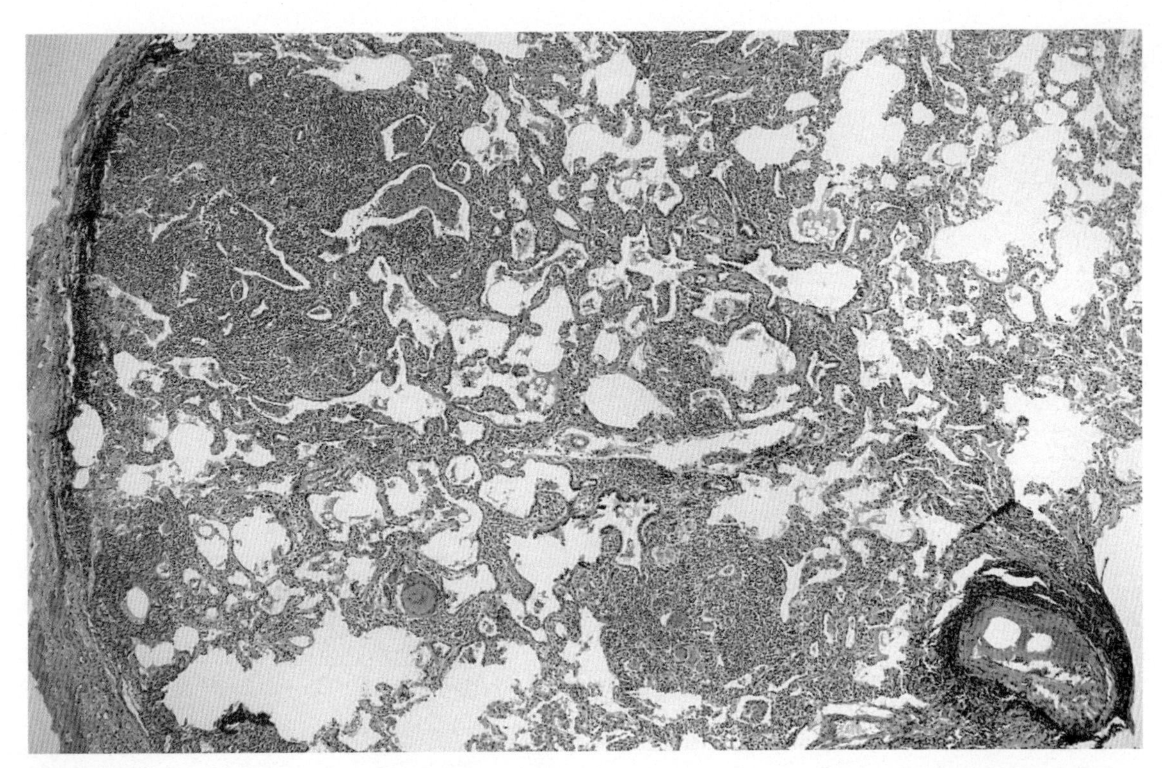

Figure 9–7 A

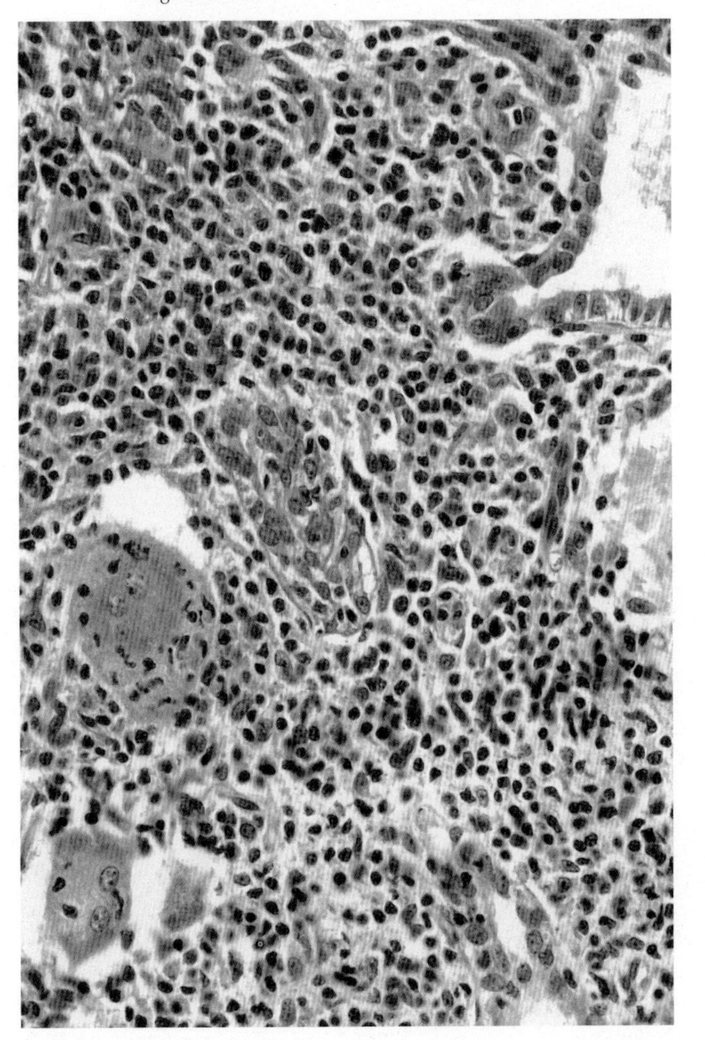

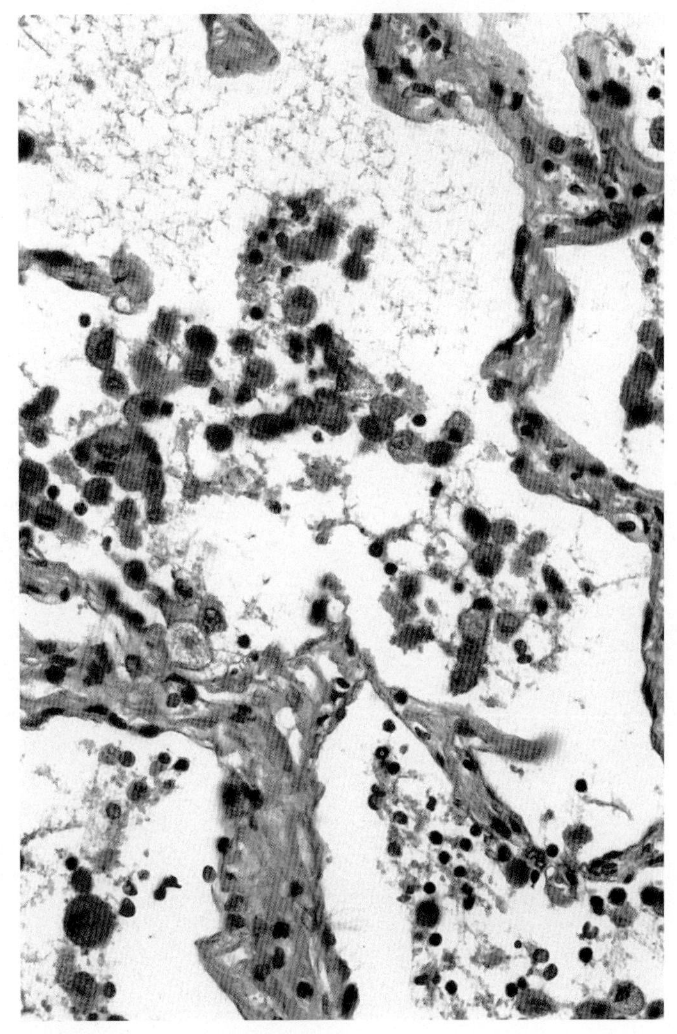

Figure 9–7 B

Figure 9–7 C

■ RESTRICTIVE LUNG DISEASES

FIGURE 9–8. SILICOSIS: In silicosis, nodules are well delimited (**A**) and fibrotic (**B**). **C:** In polarized light, **silica** is seen as bright particles in the parenchyma of the lung.

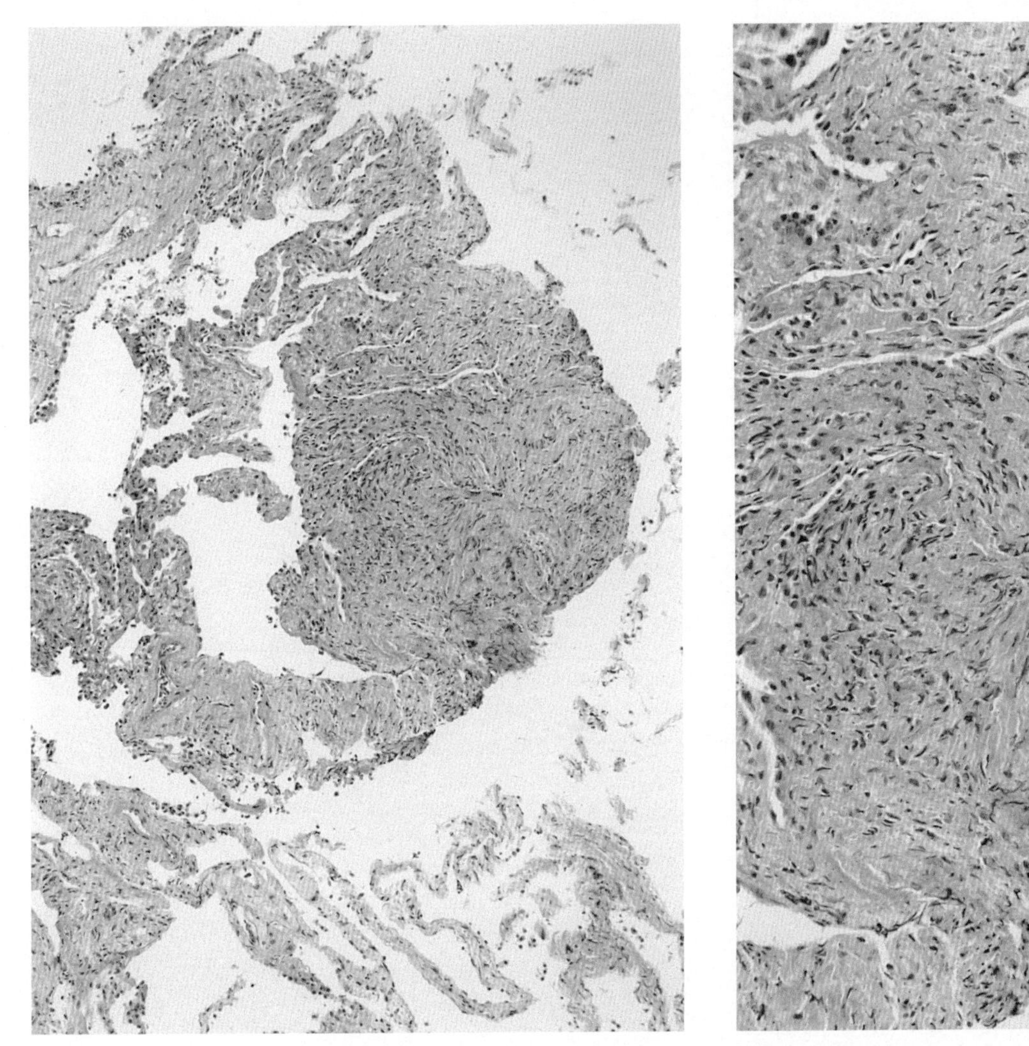

Figure 9–8 A

Figure 9–8 B

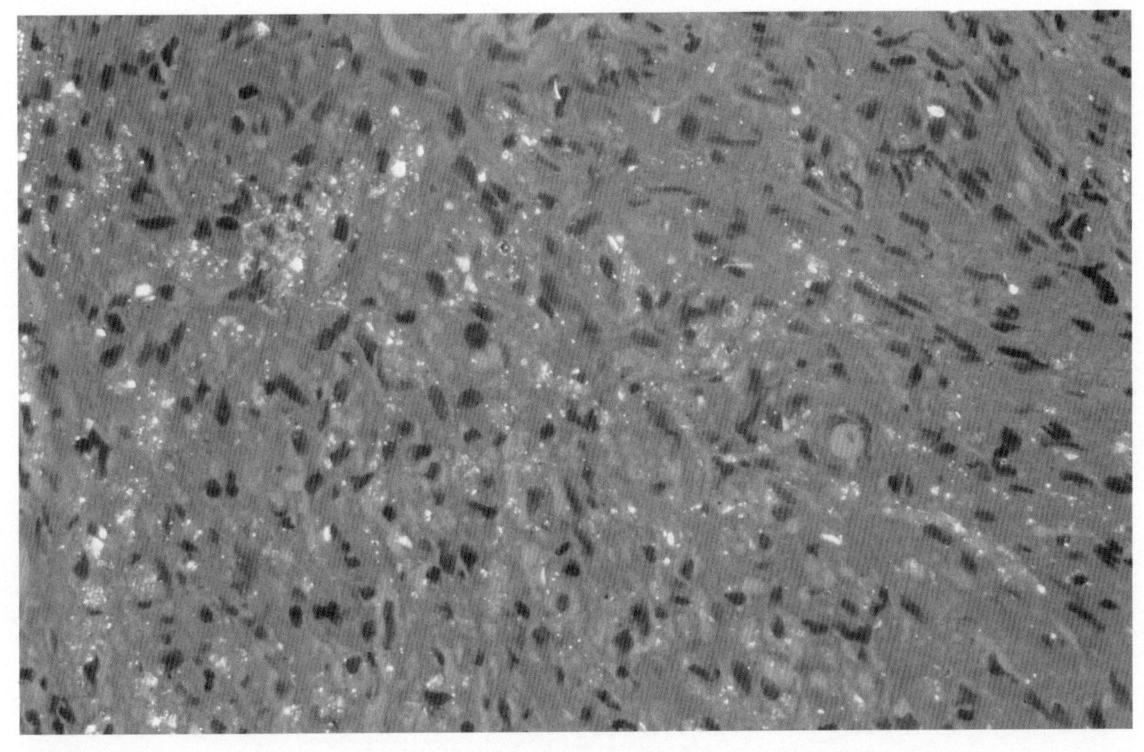

Figure 9–8 C

■ RESTRICTIVE LUNG DISEASES

FIGURE 9–9. ASBESTOSIS: A: In this photomicrograph of late-stage asbestosis, there is diffuse interstitial fibrosis. B: Needlelike **asbestos fibers** are present in macrophages. C: **Asbestos bodies** are fibers coated with iron that imparts a golden hue. The asbestos bodies are beaded with terminal bulbs.

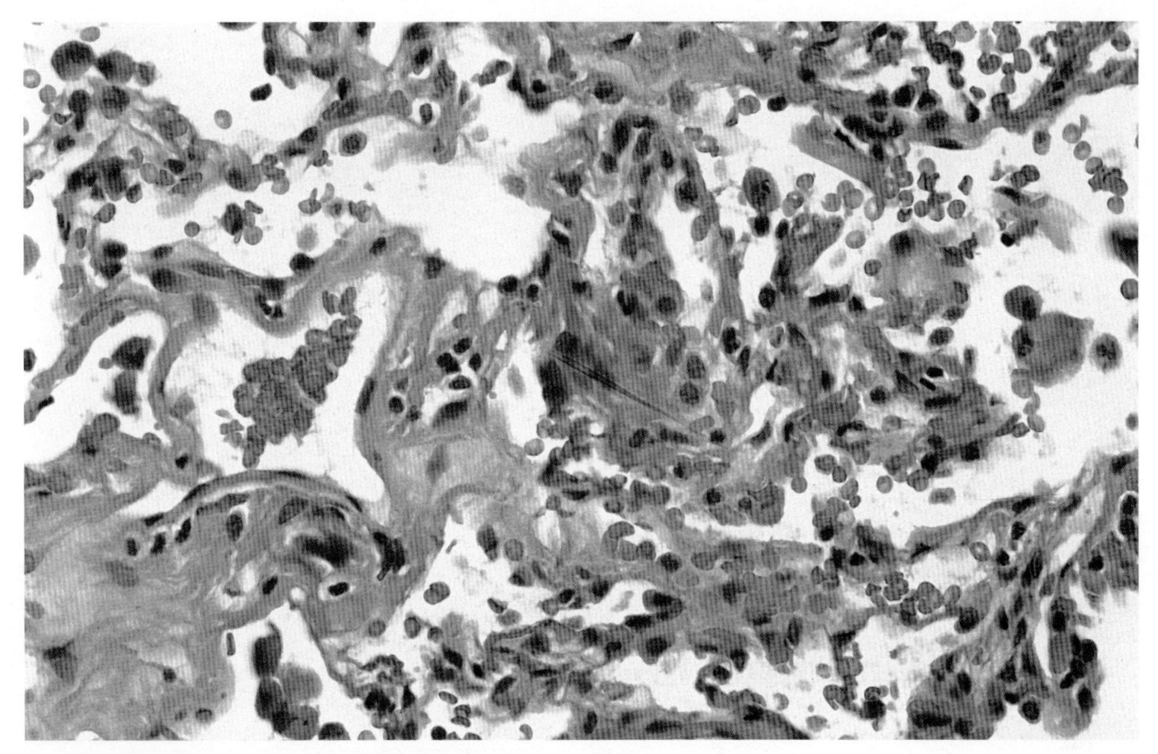

Figure 9–9 A

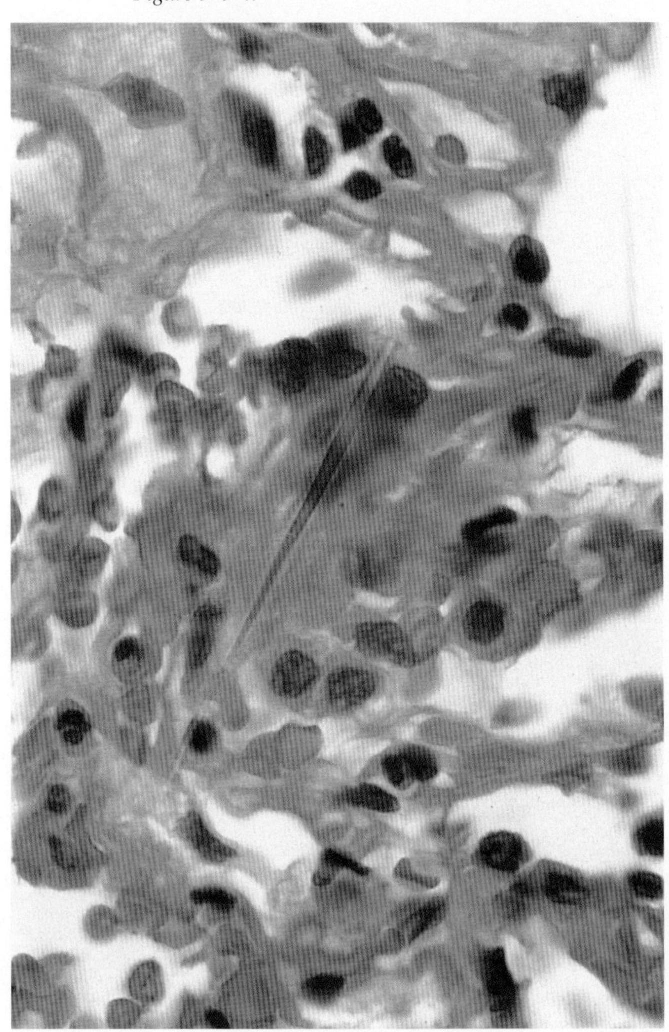

Figure 9–9 B

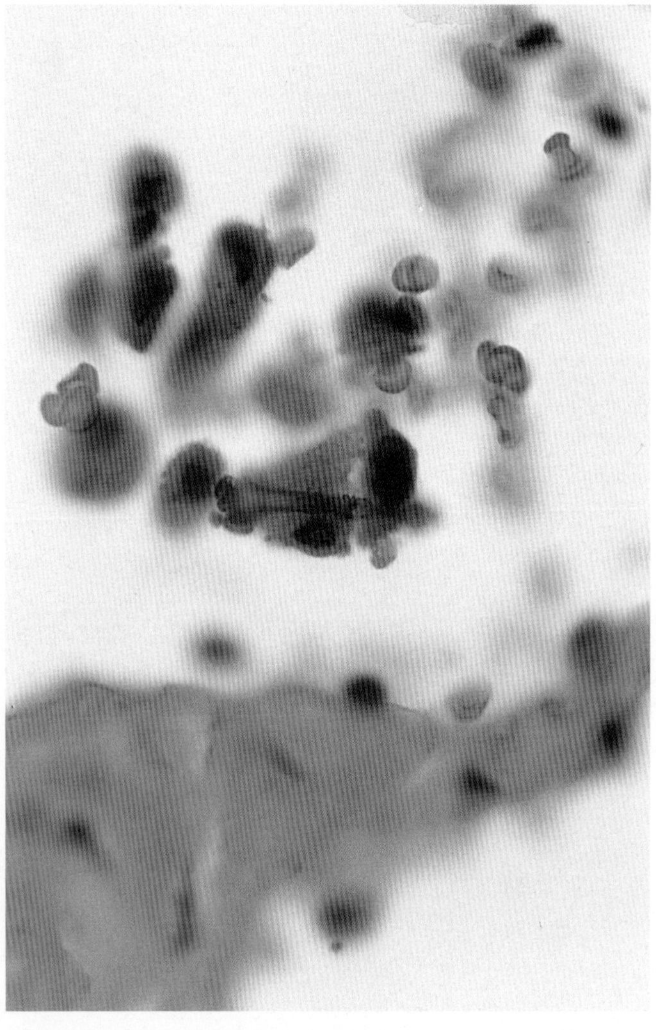

Figure 9–9 C

FIGURE 9–10. WEGENER'S GRANULOMATOSIS: A: Wegener's granulomatosis is characterized by large irregular islands of necrosis. The two hallmarks of Wegener's granulomatosis are granulomatous inflammation and vasculitis. **B:** The vasculitis is transmural and necrotizing. **C:** The granulomatous inflammation is present in the pulmonary parenchyma and is composed of lymphocytes, plasma cells, and occasional multinucleated giant cells.

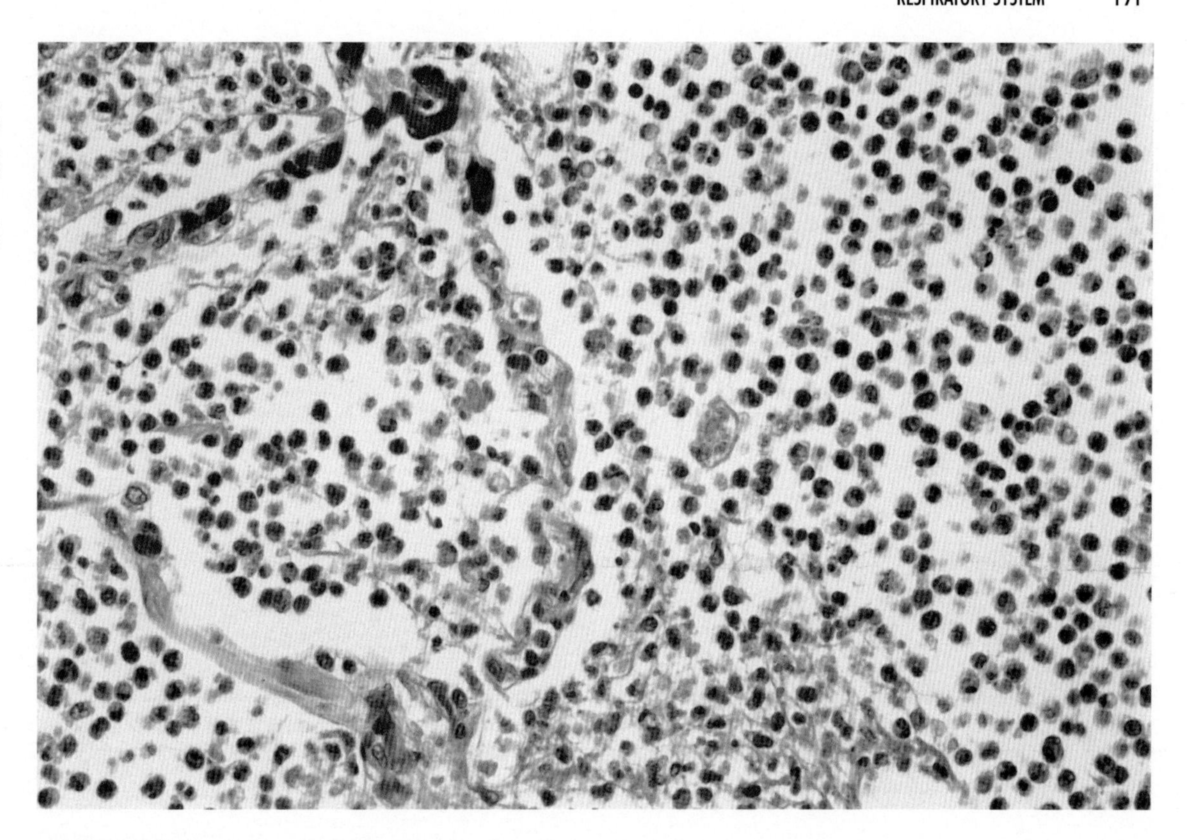

Figure 9–15 A

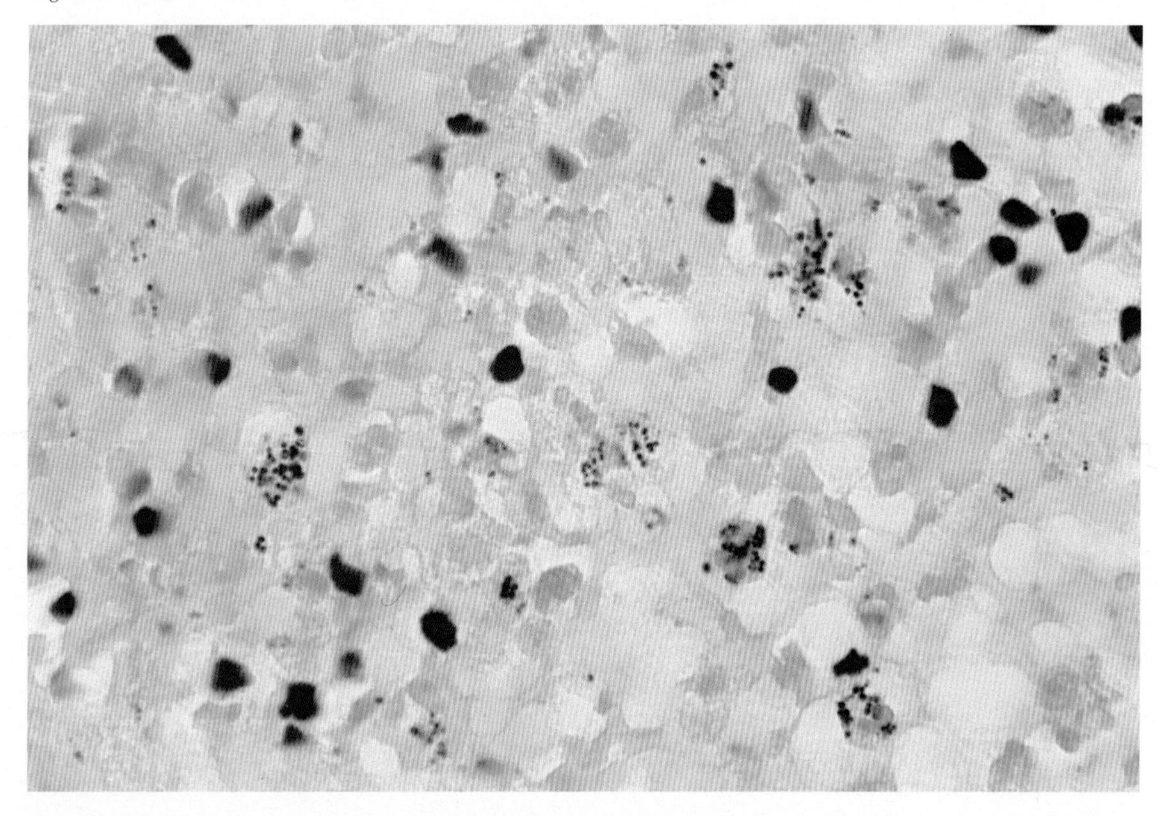

Figure 9–15 B

FIGURE 9–15. ACUTE BACTERIAL PNEUMONIA: A: In acute bacterial pneumonia, there is vascular congestion and an acute inflammatory exudate filling the airways. **B:** Brown-Brenn staining shows gram-positive cocci (blue) within the neutrophils of the inflammatory exudate.

■ INFECTIOUS DISEASES

BACTERIAL INFECTIONS

FIGURE 9–16. MYCOBACTERIUM TUBERCULOSIS: A: Tuberculosis is characterized by caseating granulomatous inflammation. **B:** Rimming the necrotic areas are epithelioid histiocytes and few lymphocytes. Langhan's giant cells, with nuclei in a horseshoe pattern, are characteristically present. **C:** Mycobacteria stain red with an acid-fast stain.

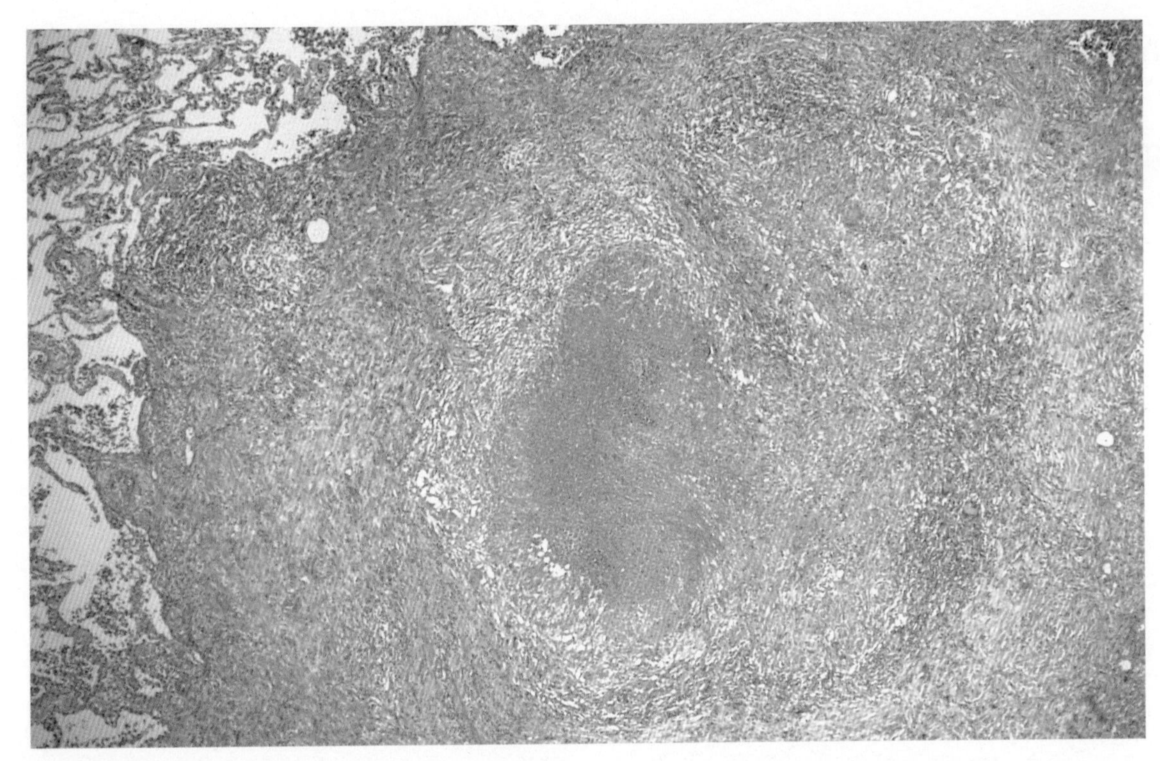

Figure 9–16 A

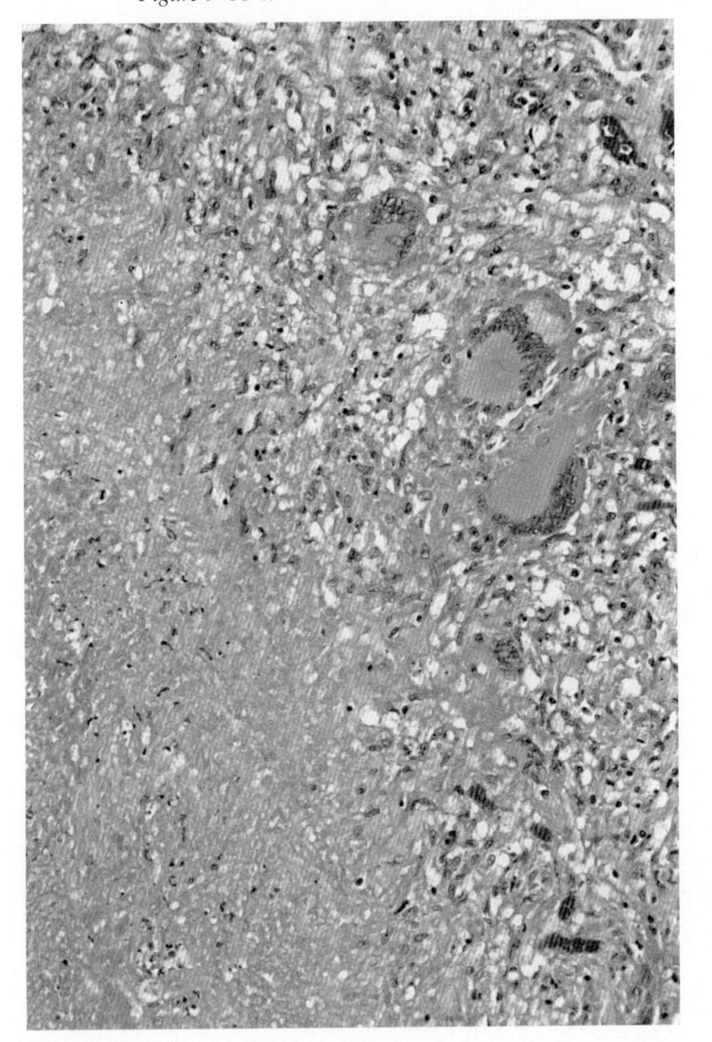

Figure 9–16 B

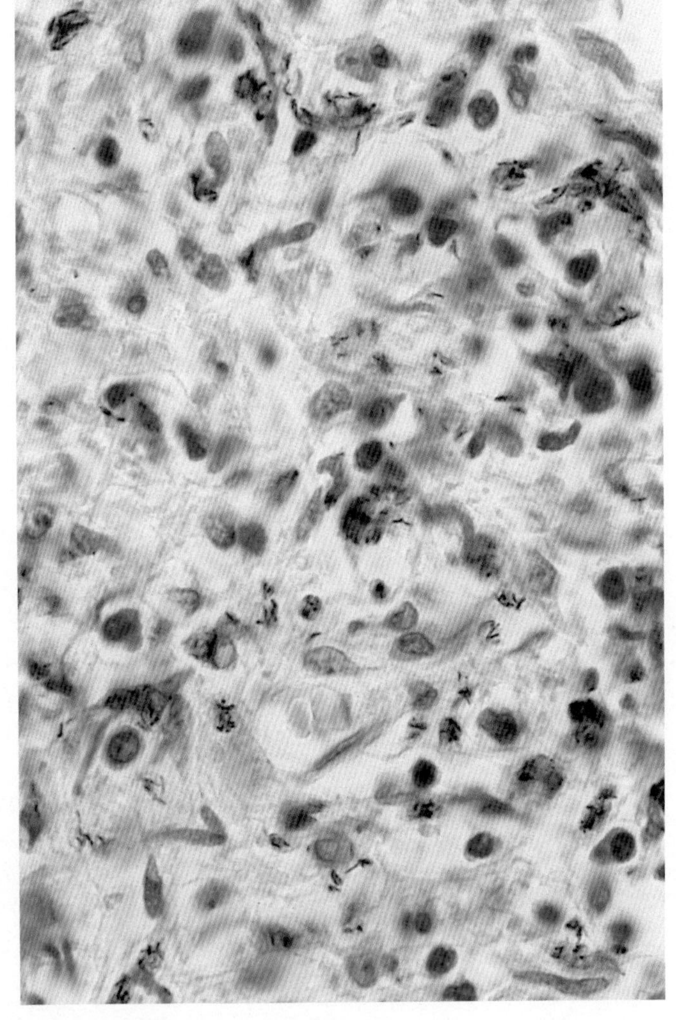

Figure 9–16 C

■ INFECTIOUS DISEASES

FUNGAL INFECTIONS

FIGURE 9–17. HISTOPLASMOSIS: A: Histoplasmosis is characterized by noncaseating granulomatous inflammation. **B:** The granuloma is rimmed by parallel bundles of collagen containing Langhan's giant cells. **C:** This is an example of a Gomori methenamine-silver stain in which oval yeast forms of *Histoplasma capsulatum* are seen in the alveoli.

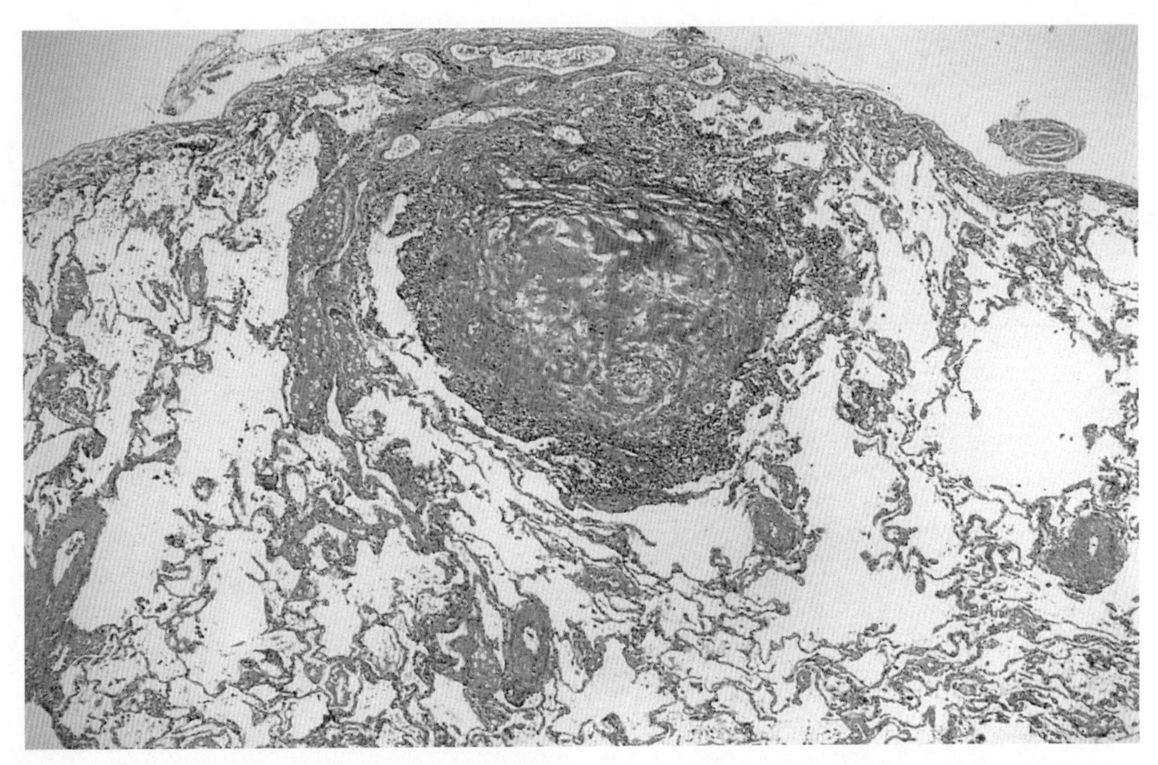

Figure 9–17 A

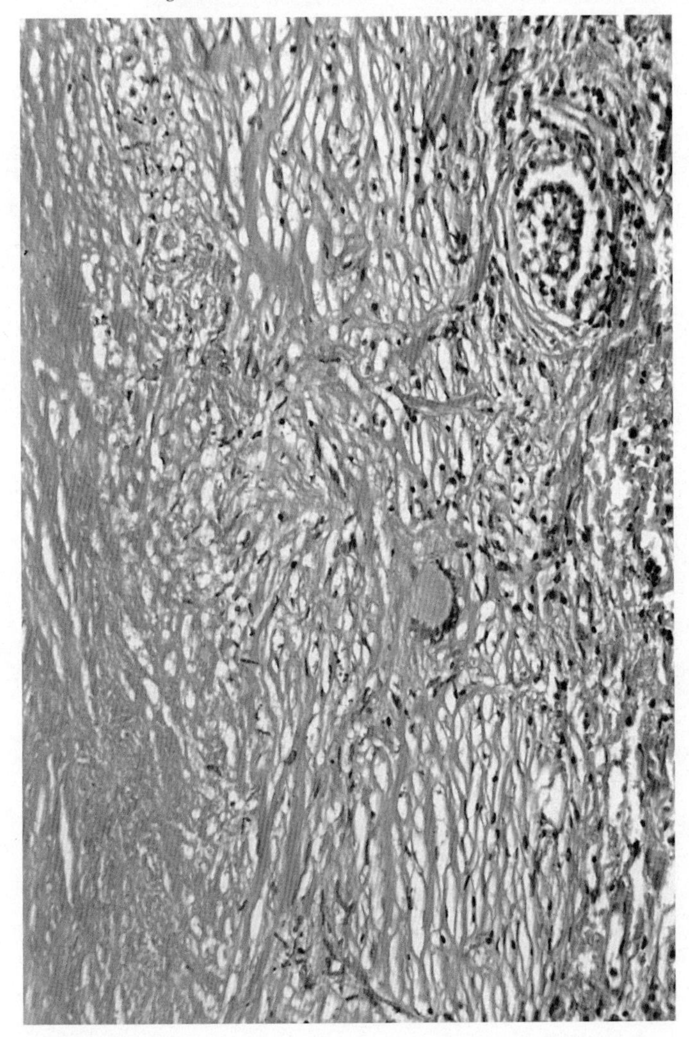

Figure 9–17 B

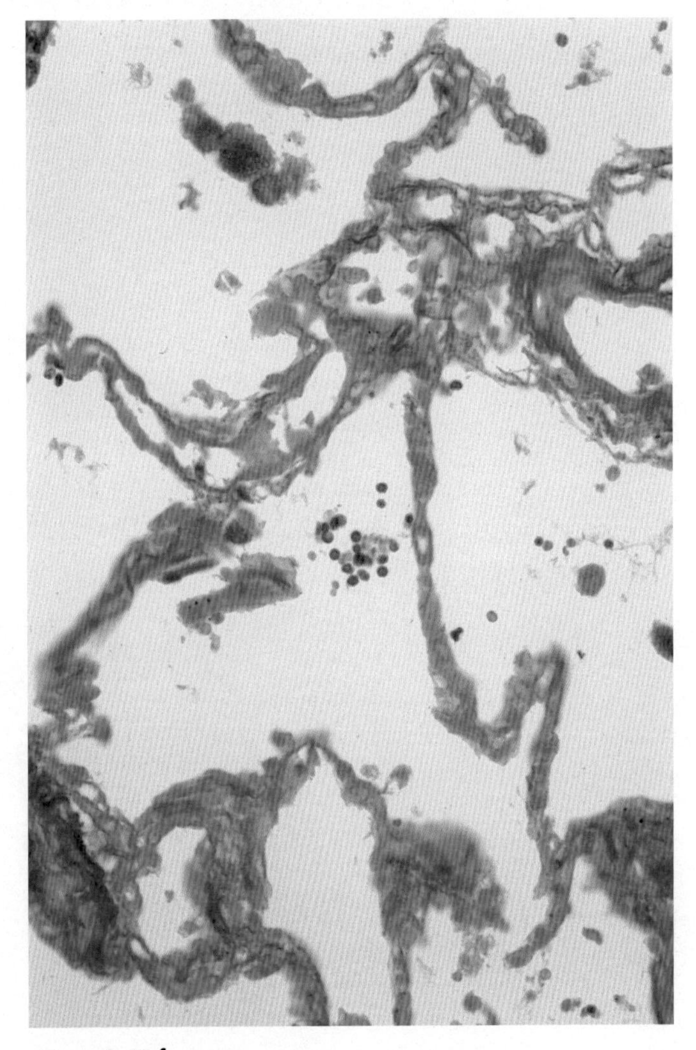

Figure 9–17 C

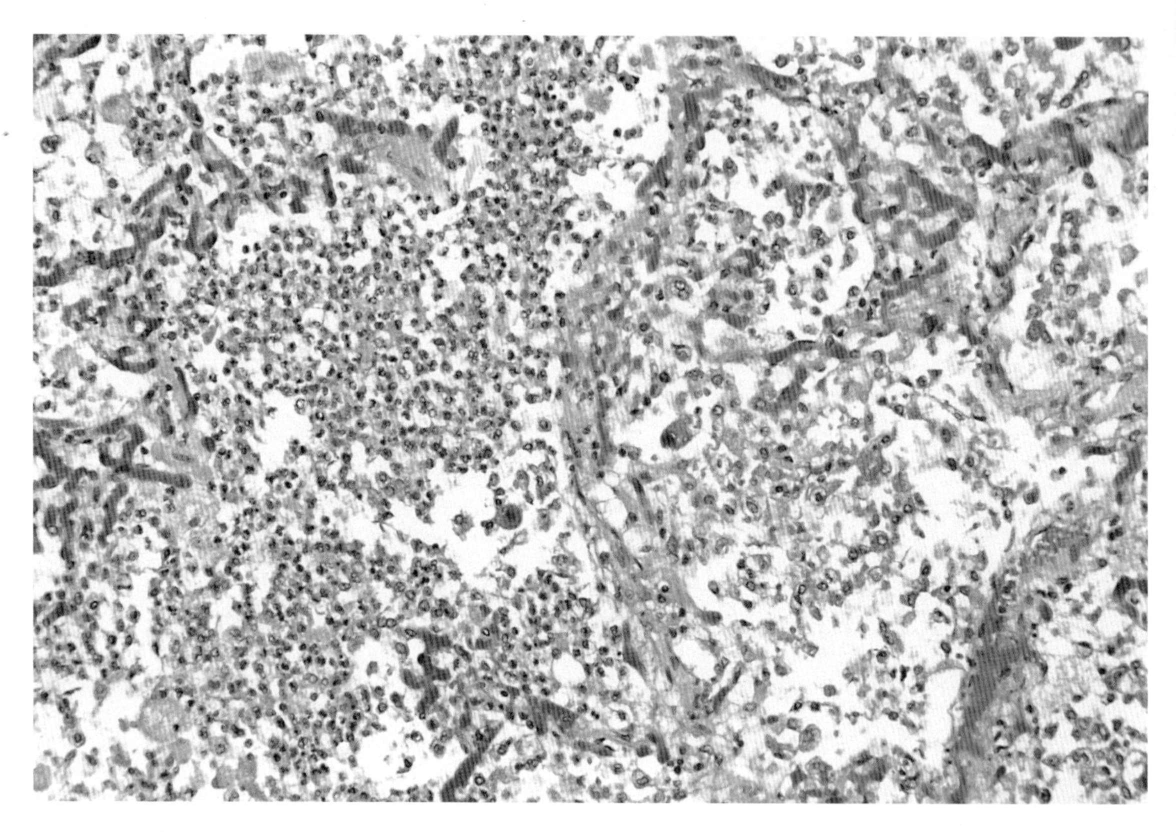

Figure 9–18 A

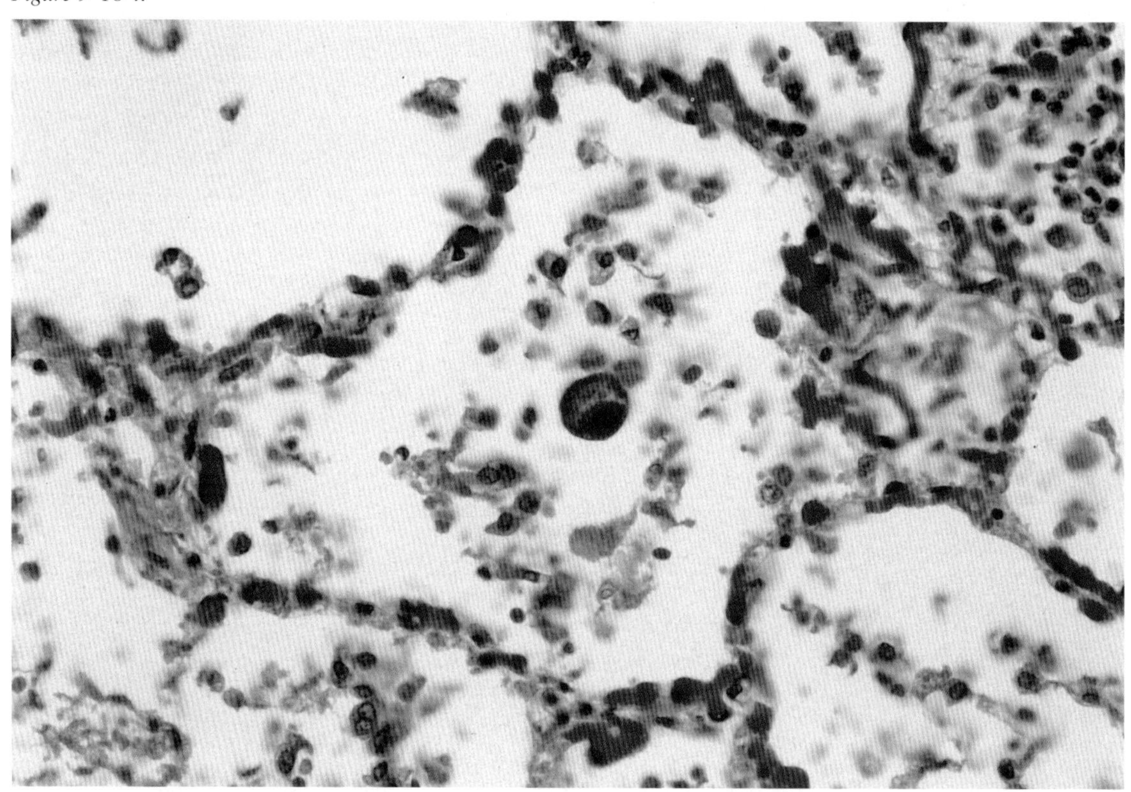

Figure 9–18 B

FIGURE 9–18. CYTOMEGALOVIRUS: A: In cytomegalovirus pneumonitis, there is a marked inflammatory exudate in the alveoli. **B:** Cytomegalovirus-infected cells are enlarged, with a dark intranuclear inclusion surrounded by a halo, giving the classic "owl eye" appearance.

■ PEDIATRIC PULMONARY DISEASES

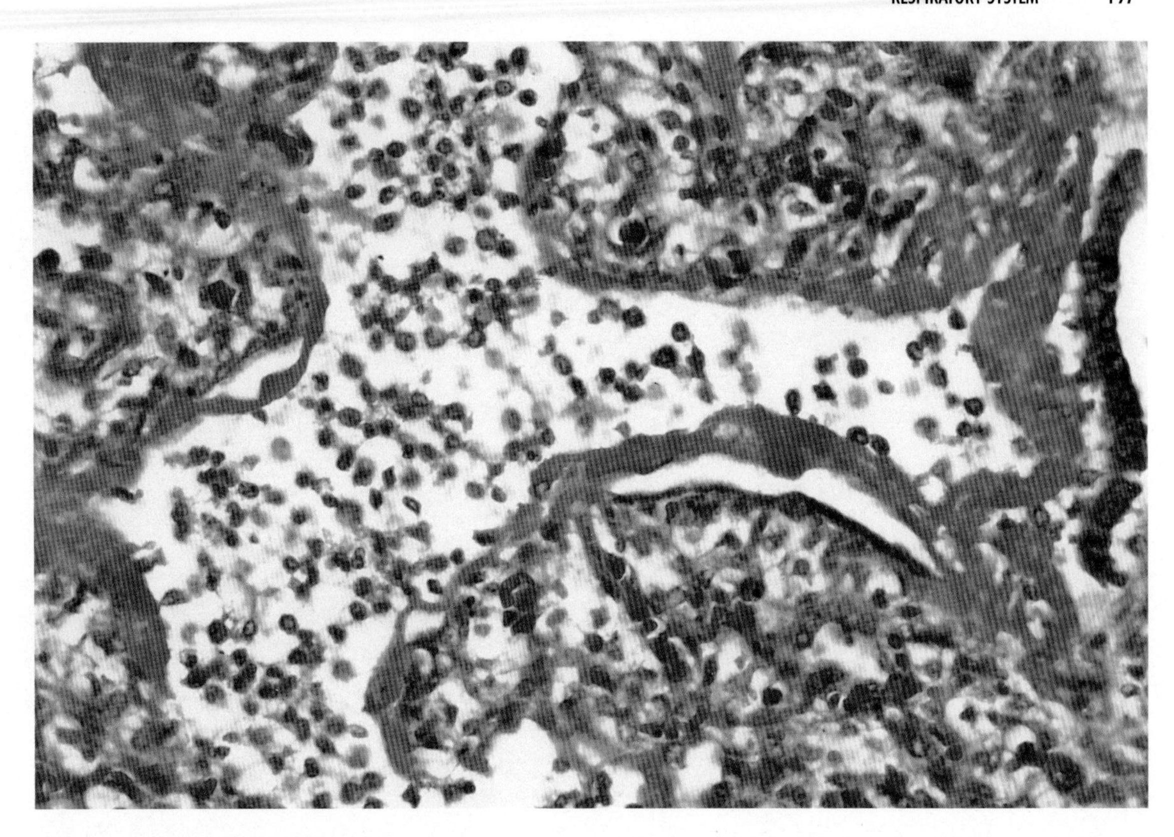

FIGURE 9–19. HYALINE MEMBRANE DISEASE (HMD): Hyaline membrane disease occurs in the perinatal period. Hyaline membranes, which are eosinophilic proteinaceous lamellar deposits, are present in the respiratory bronchioles, alveolar ducts, and alveoli. In this example of a bronchiole with a hyaline membrane, there is a superimposed pneumonic process.

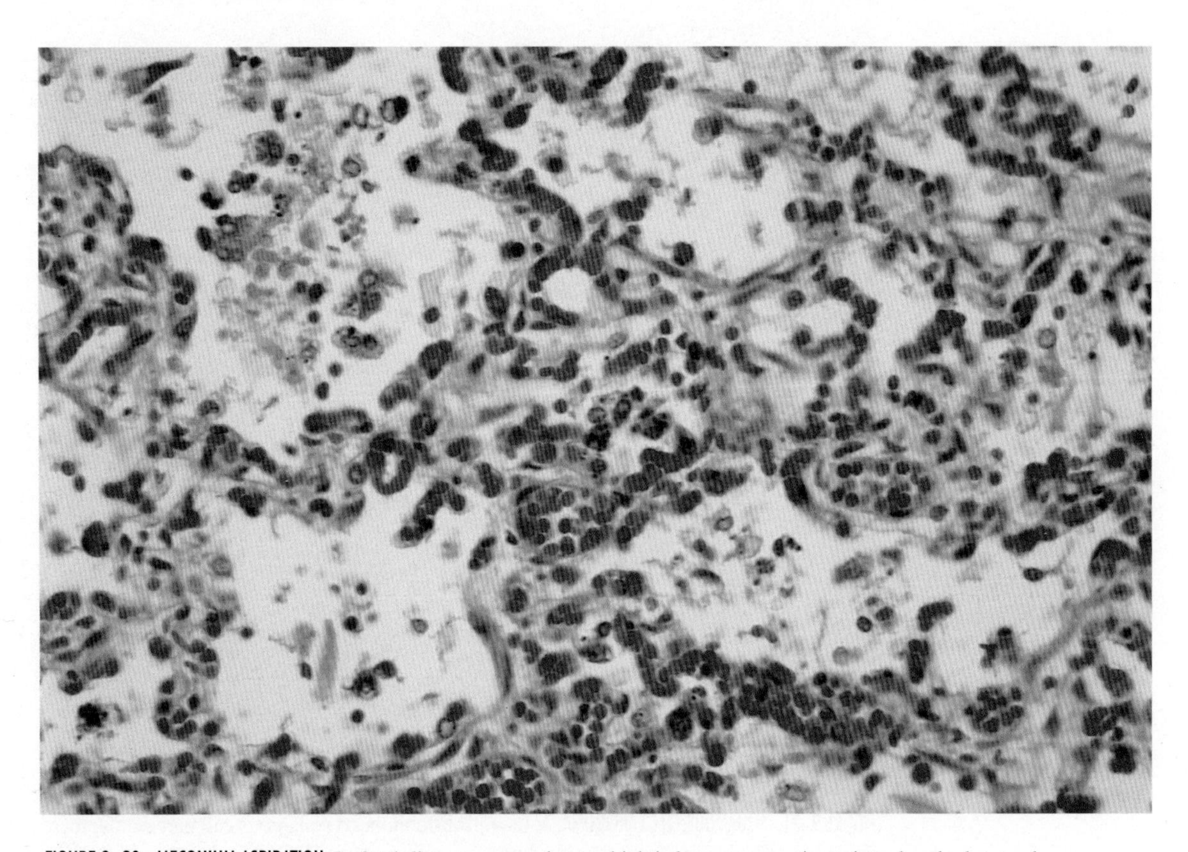

FIGURE 9–20. MECONIUM ASPIRATION: In fetal distress, meconium, which is brown-green, is aspirated and taken up by macrophages. Brown-green pigmented macrophages, squamous cells, and amniotic fluid are present in the alveoli.

■ BRONCHOGENIC CARCINOMAS

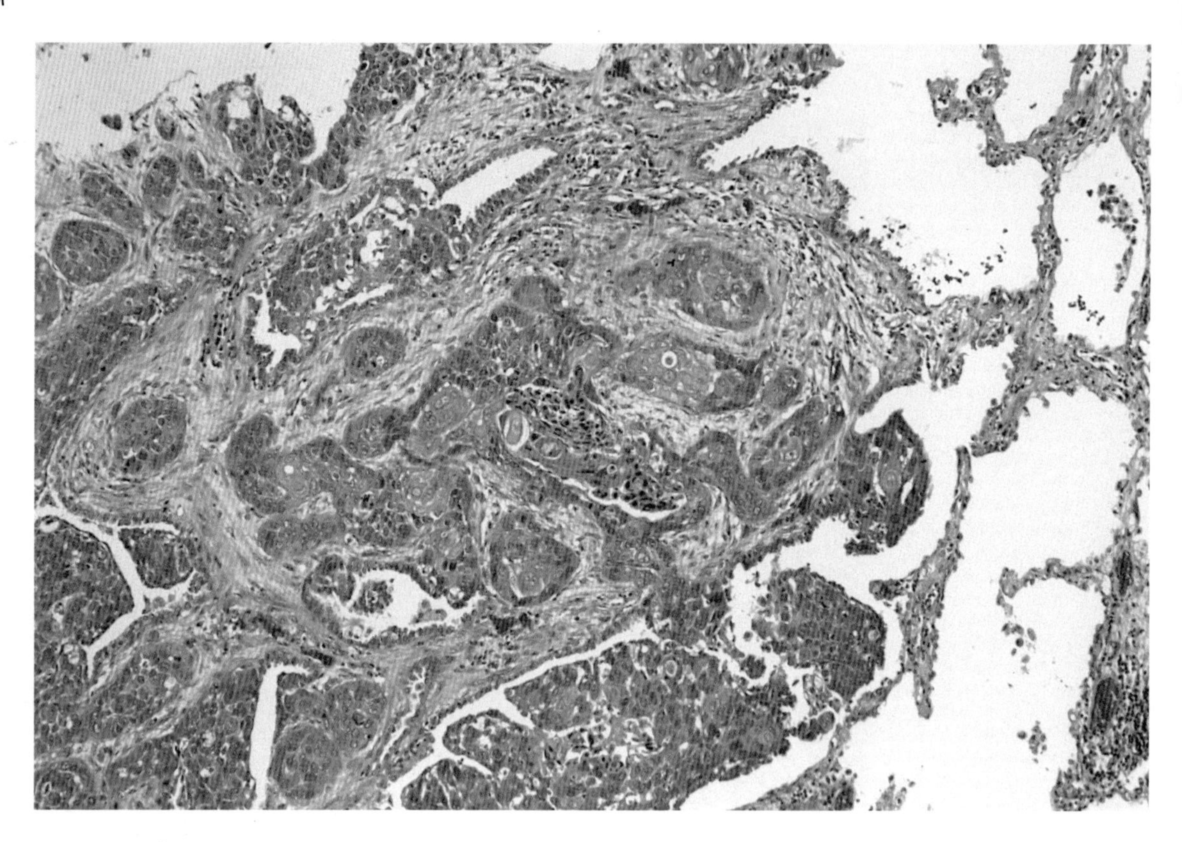

Figure 9–21 A

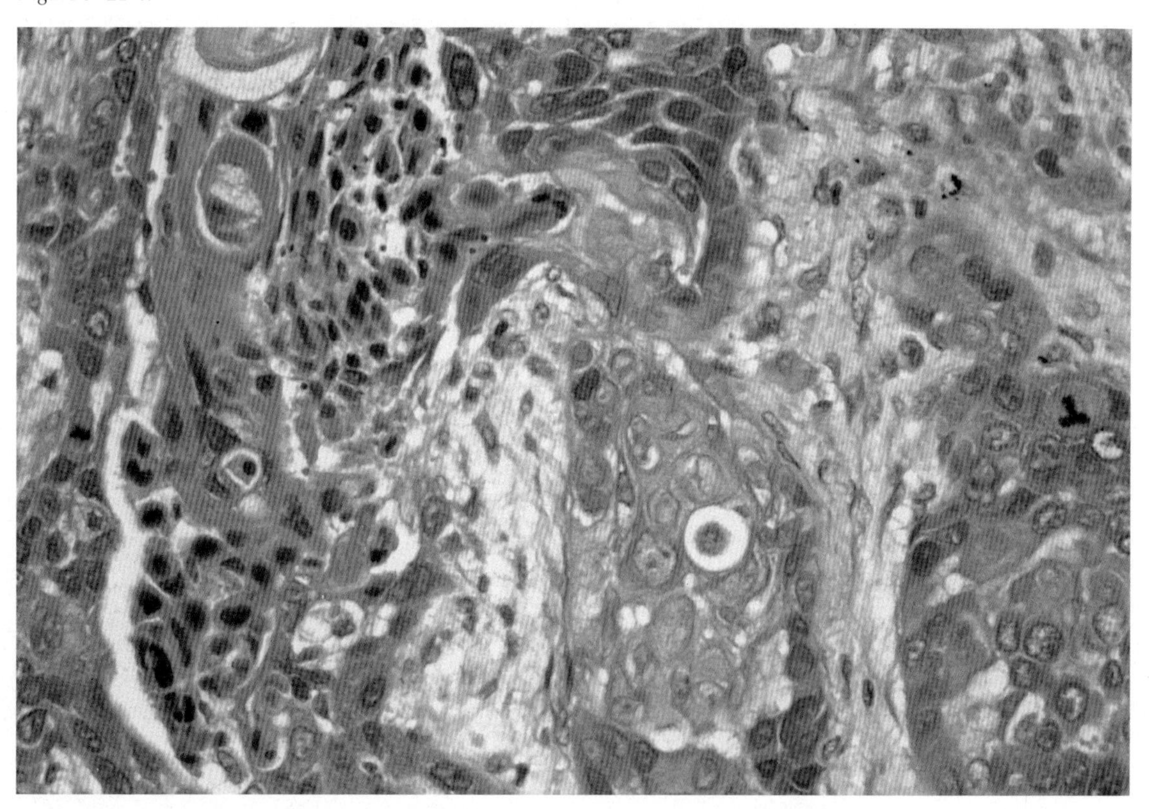

Figure 9–21 B

FIGURE 9–21. SQUAMOUS CELL CARCINOMA: A: In well-differentiated squamous cell carcinoma, there are islands of neoplastic epithelial cells that produce a desmoplastic stromal response. **B:** Confirming the squamous cell origin of the tumor are keratin pearls, keratinization of individual cells, and intercellular bridges. Mitotic figures are also noted.

■ BRONCHOGENIC CARCINOMAS

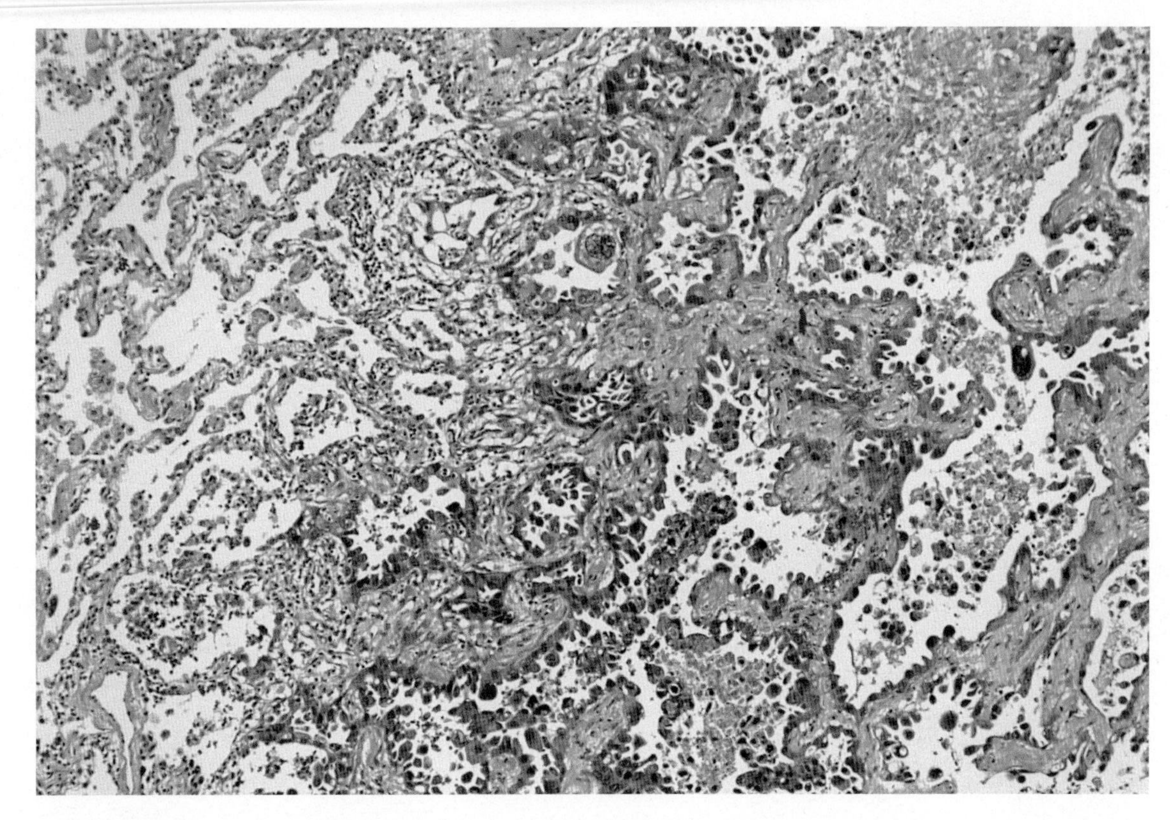

Figure 9–22 A

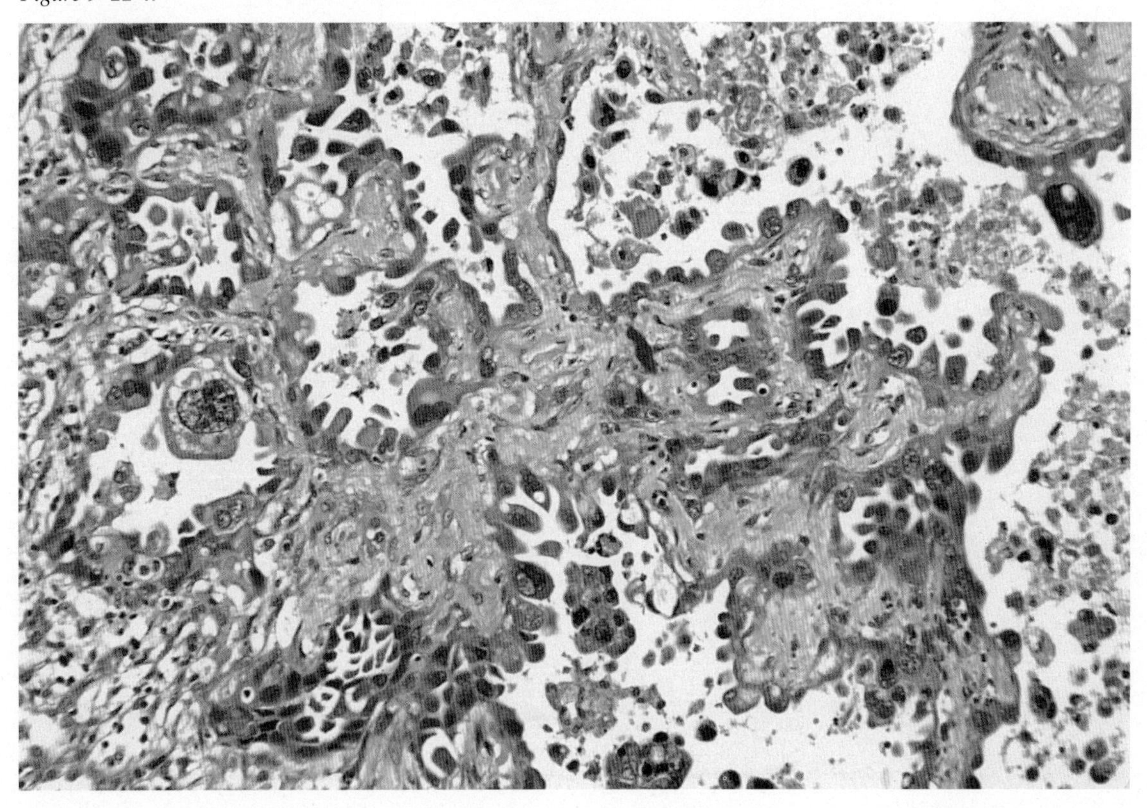

Figure 9–22 B

FIGURE 9–22. ADENOCARCINOMA: A: In adenocarcinoma, cells are arranged in glandular configurations. **B:** In this poorly differentiated adenocarcinoma, the cells are pleomorphic and cuboidal to columnar, with anaplastic nuclei.

■ BRONCHOGENIC CARCINOMAS

FIGURE 9–23. BRONCHIOLOALVEOLAR CARCINOMA: A: In bronchioloalveolar carcinoma, the neoplastic cells use the alveolar septa as a scaffold along which they grow. **B:** The cells are bland, columnar, and mucin-producing, with basally located nuclei. **C:** Mucicarmine staining accentuates the intracellular and extracellular mucin.

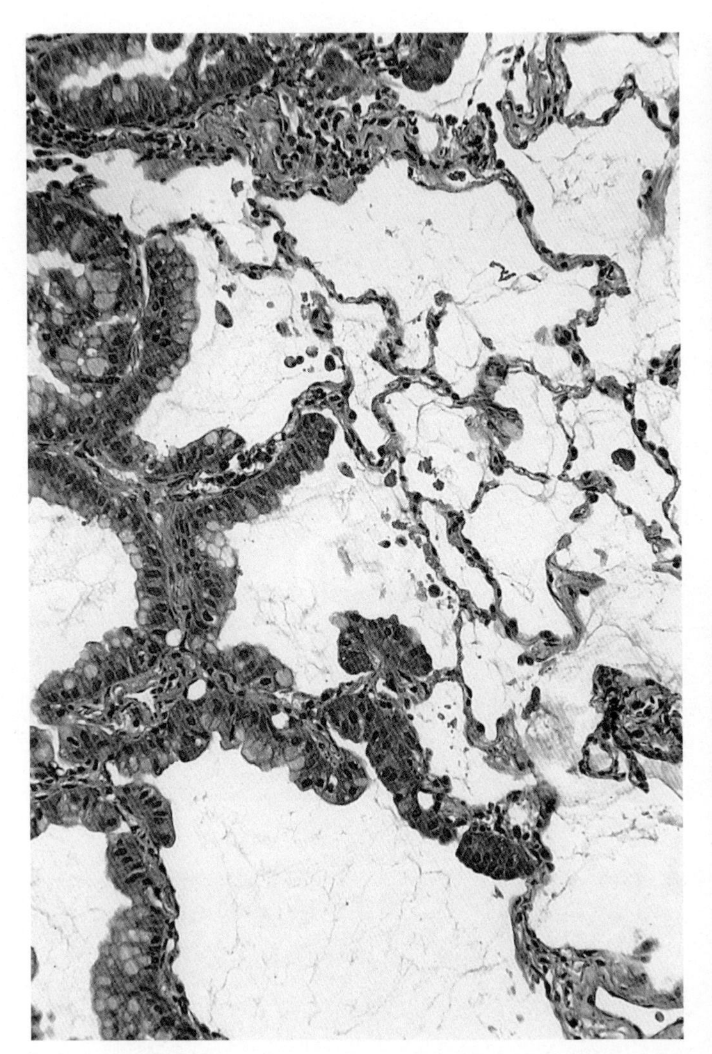

Figure 9–23 A

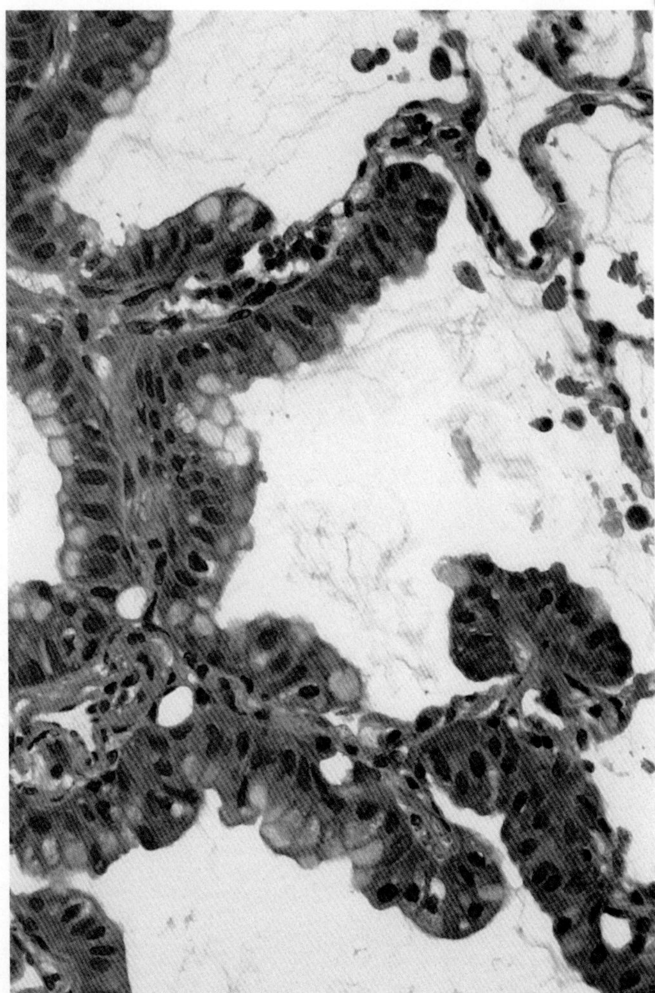

Figure 9–23 B

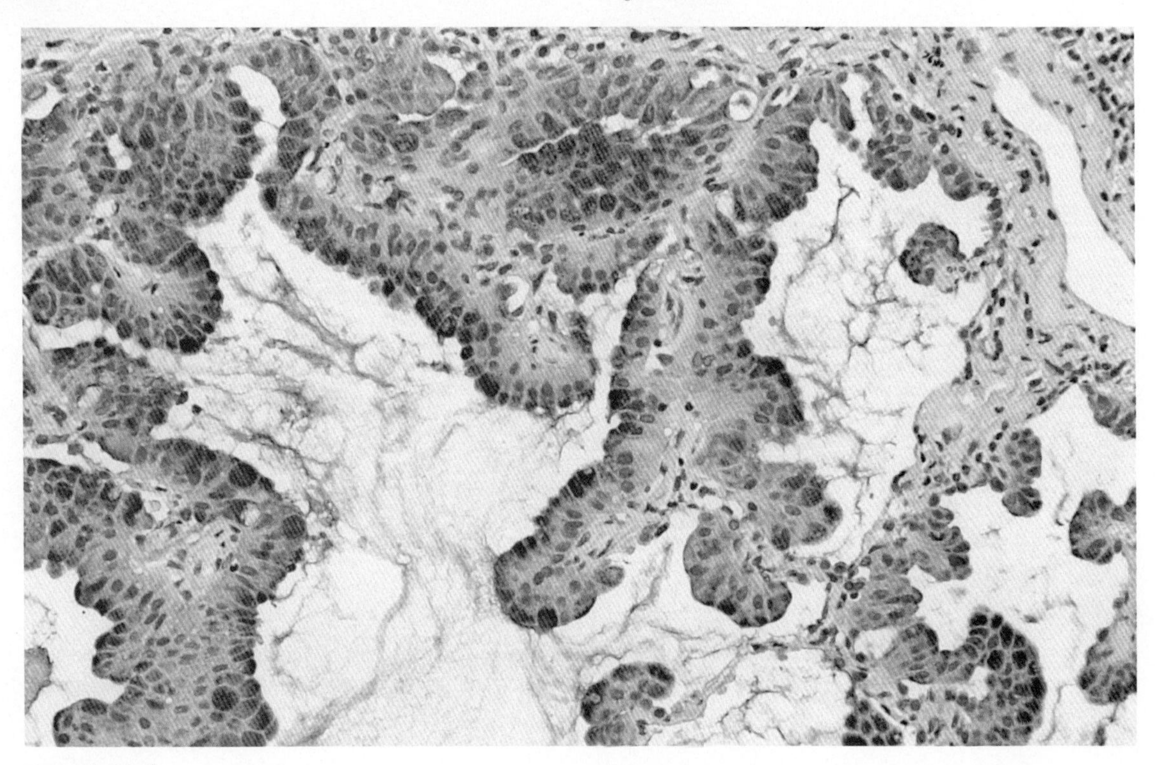

Figure 9–23 C

■ BRONCHOGENIC CARCINOMAS

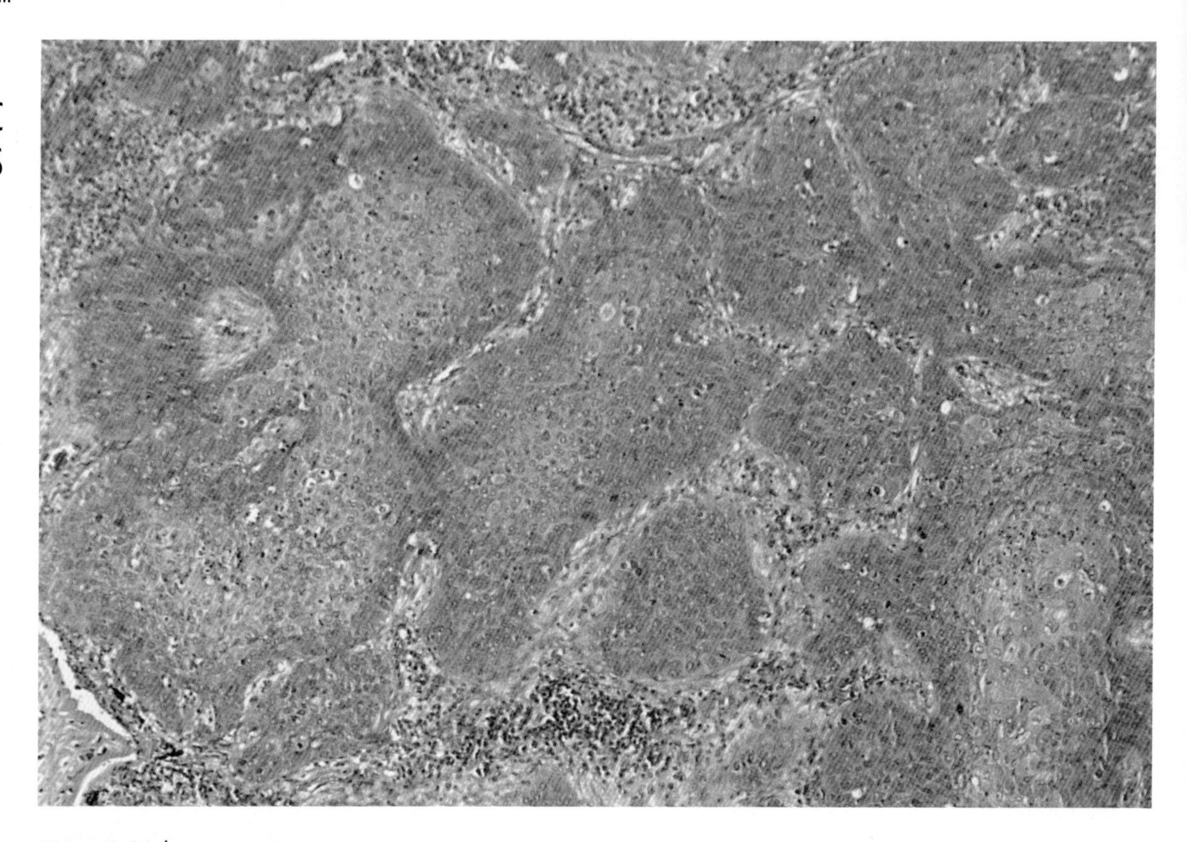

Figure 9–24 A

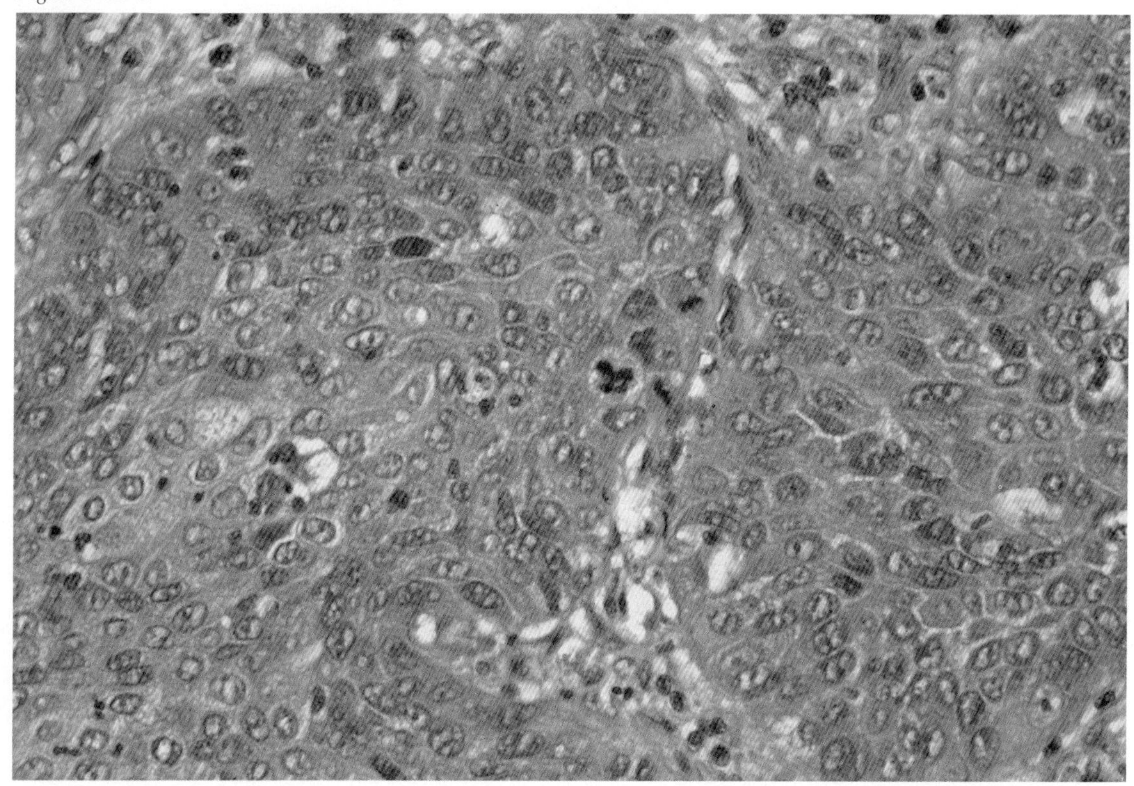

Figure 9–24 B

FIGURE 9–24. LARGE CELL CARCINOMA: A: Large cell carcinoma is characterized by large islands of anaplastic cells. **B:** The cells are polyhedral, with abundant cytoplasm, vesicular nuclei, and prominent nucleoli. Numerous mitotic figures are present.

■ BRONCHOGENIC CARCINOMAS

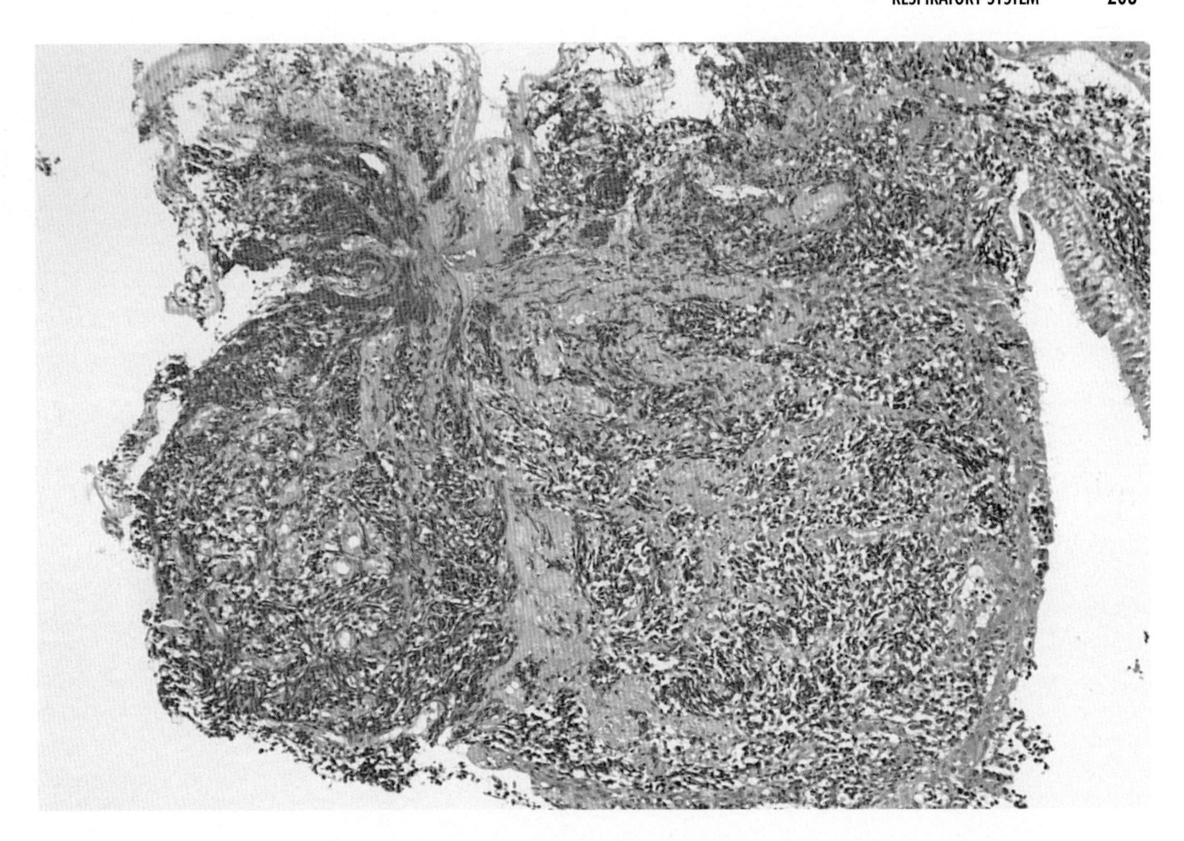

Figure 9–25 A

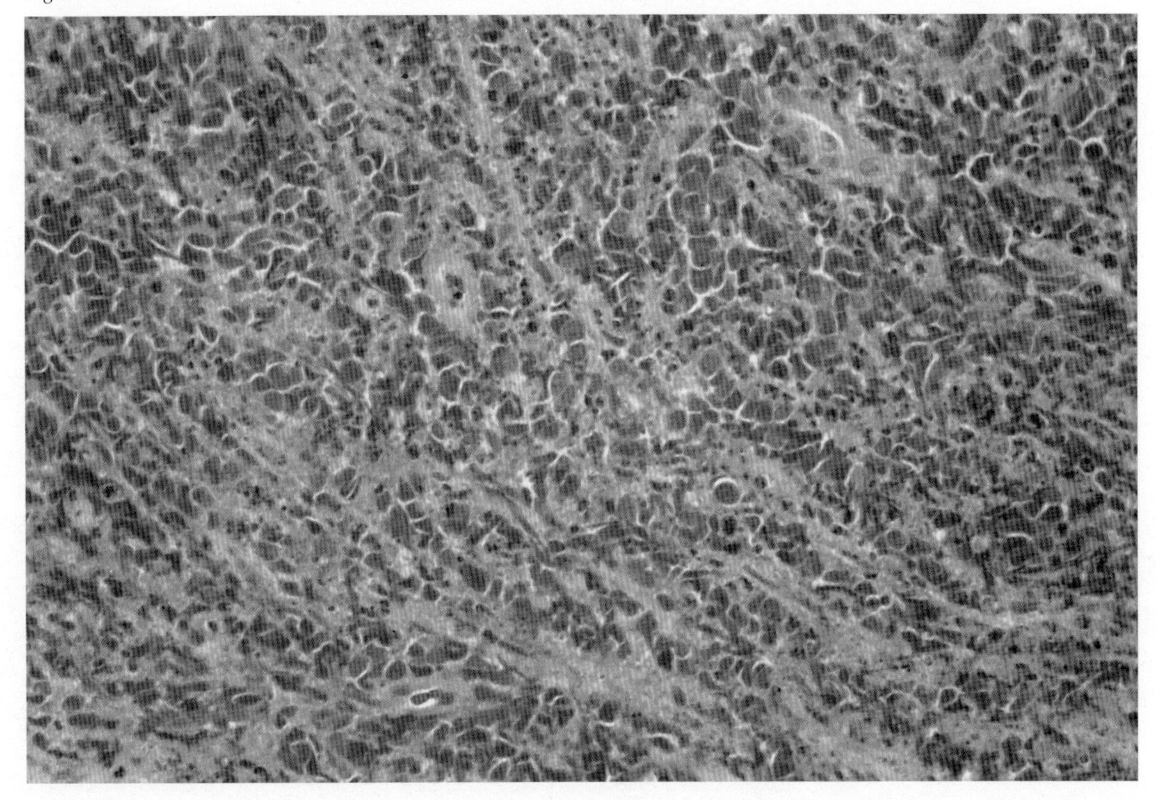

Figure 9–25 B

FIGURE 9–25. SMALL CELL CARCINOMA (OAT CELL CARCINOMA): A: Small cell carcinoma is characterized by nests and sheets of small lymphocyte-like cells with areas of necrosis. Because of the fragility of the nuclear chromatin, "crush artifact" is a prominent feature. **B:** Cells have scant cytoplasm and hyperchromatic nuclei that mold to one another.

■ BRONCHIAL CARCINOID

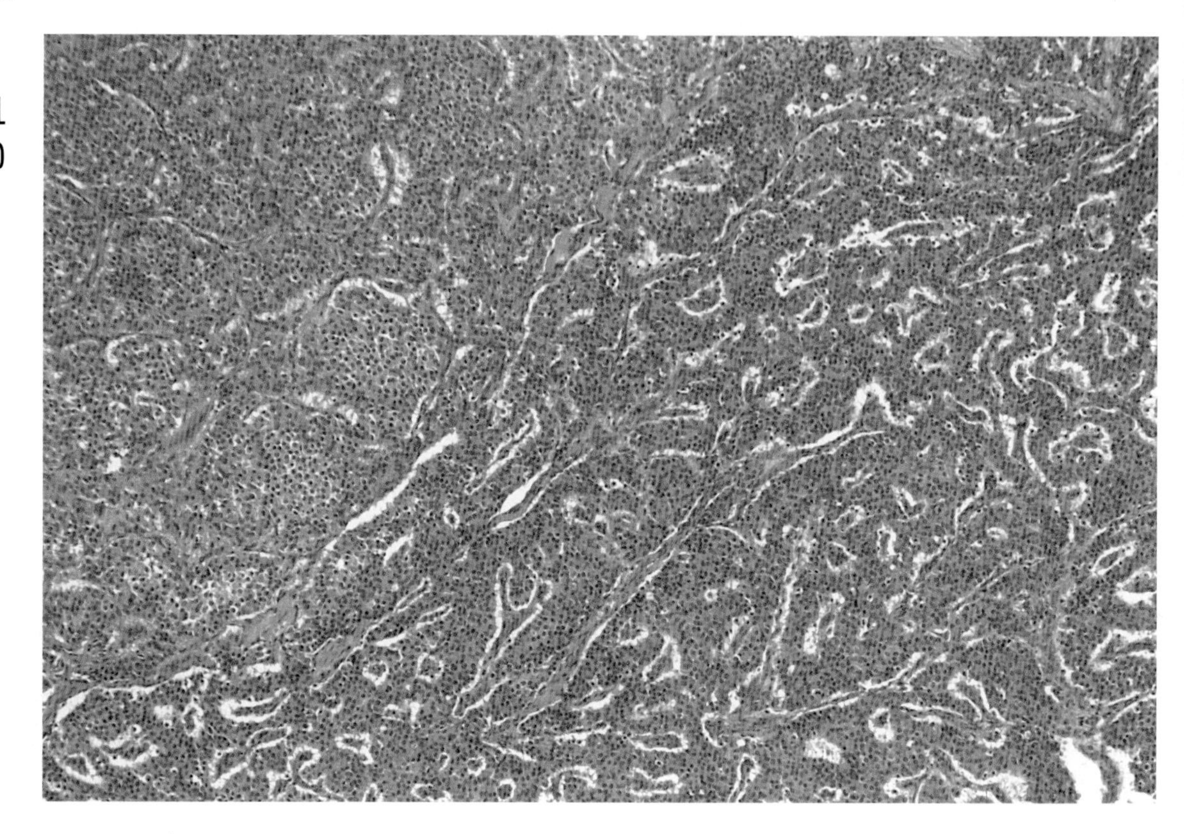

Figure 9–26 A

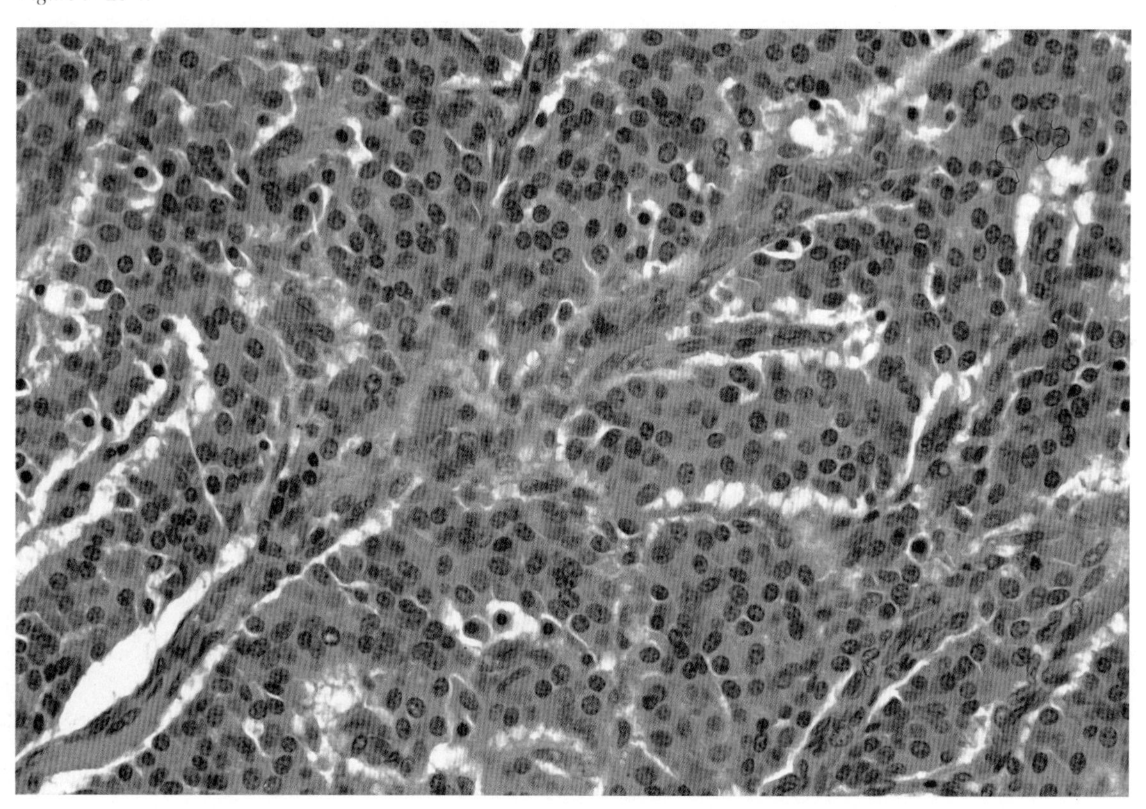

Figure 9–26 B

FIGURE 9–26. BRONCHIAL CARCINOID: A: Bronchial carcinoids are composed of nests and trabeculae of cells in a fibrovascular stroma. **B:** The cells are uniform and cuboidal, with regular round salt-and-pepper nuclei.

■ PLEURA

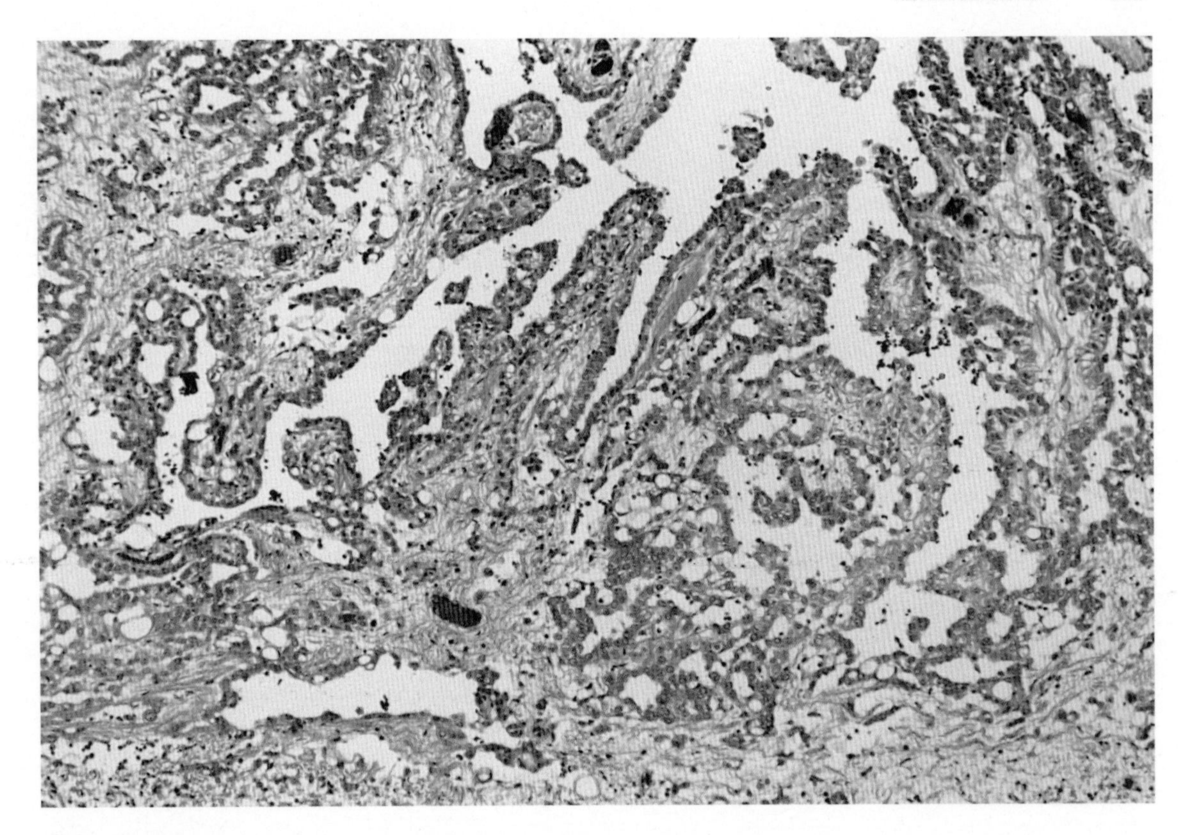

Figure 9–27 A

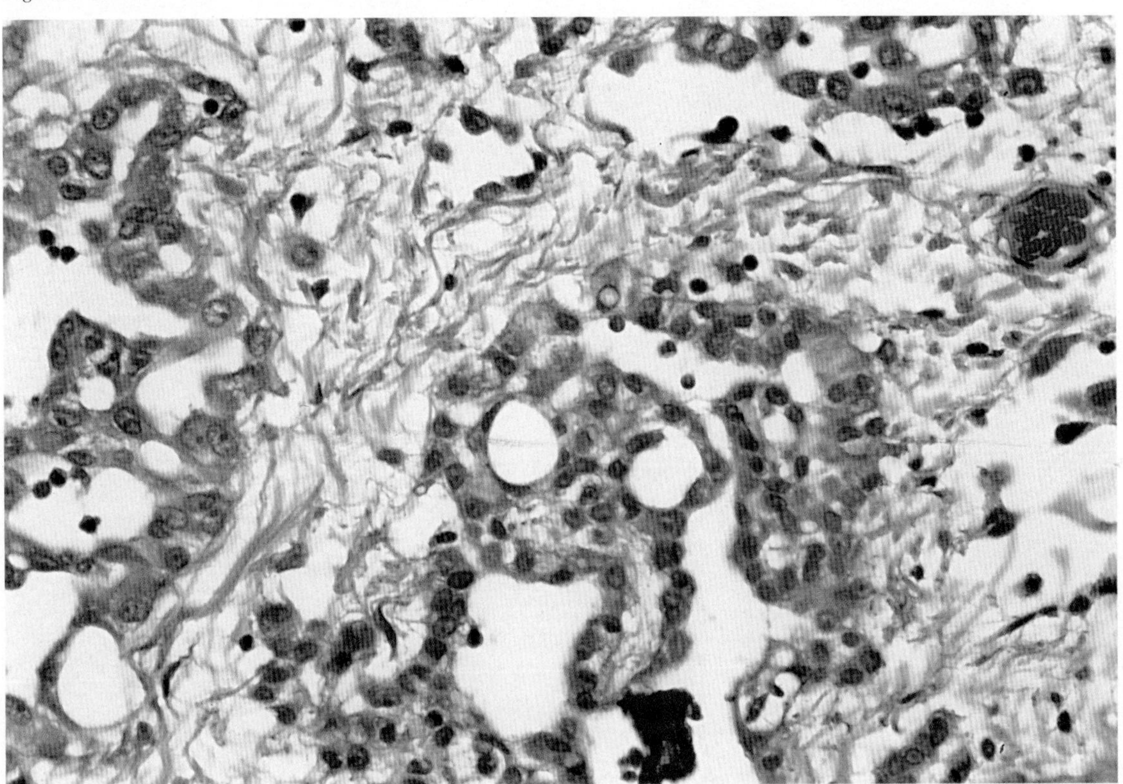

Figure 9–27 B

FIGURE 9–27. MESOTHELIOMA: A: This is a photomicrograph of an epithelial-type mesothelioma that consists of tubules and papillary buds that protrude into microcystic spaces. **B**: The cells lining the tubules and microcysts are cuboidal.

■ LARYNX

FIGURE 9–28. VOCAL CORD POLYP (SINGER'S NODULE): A vocal cord polyp is characterized by an edematous mucosa containing thin-walled blood vessels. The epithelium is stratified squamous.

FIGURE 9–29. SQUAMOUS CELL CARCINOMA: A: In squamous cell carcinoma involving the larynx, islands of neoplastic nonkeratinizing squamous epithelium infiltrate the stroma. **B:** At high power, the features of malignancy can be seen: loss of cell polarity, pleomorphism of cells and nuclei, nuclear hyperchromasia, an increased nuclear-to-cytoplasmic ratio, and mitotic figures, some of which are abnormal.

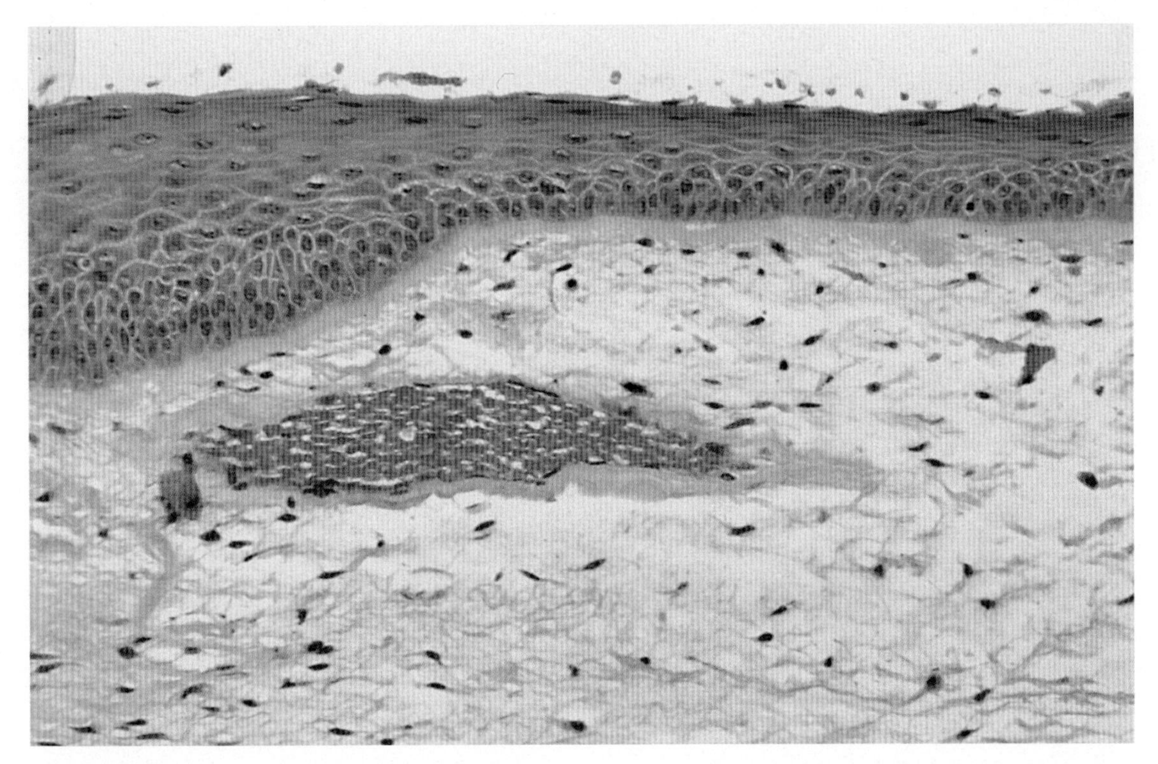

Figure 9–28

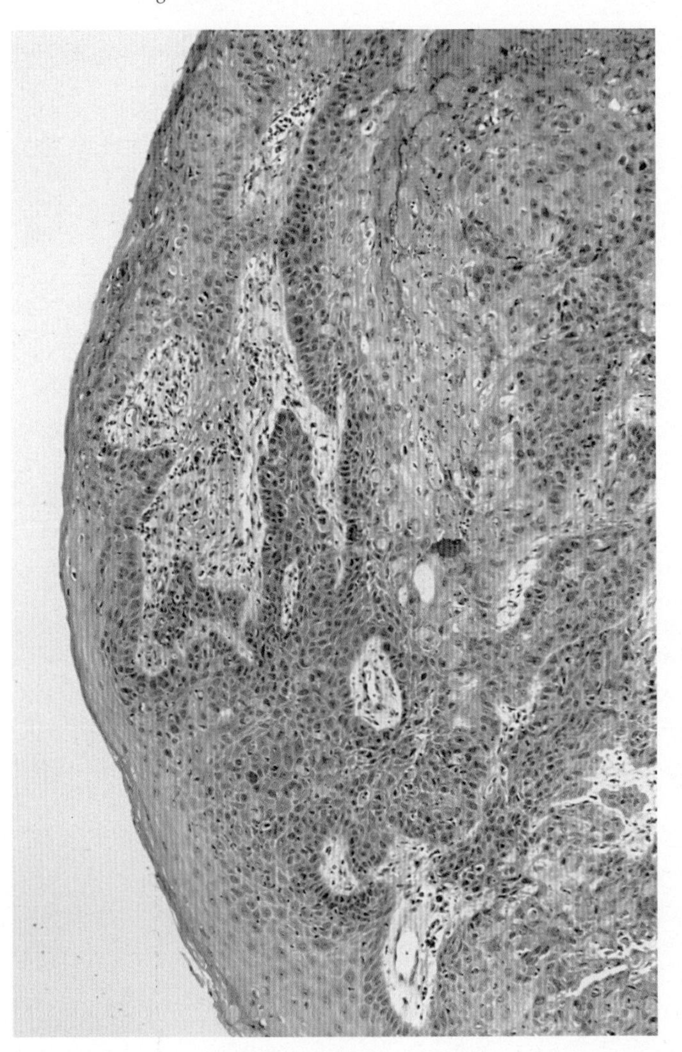

Figure 9–29 A

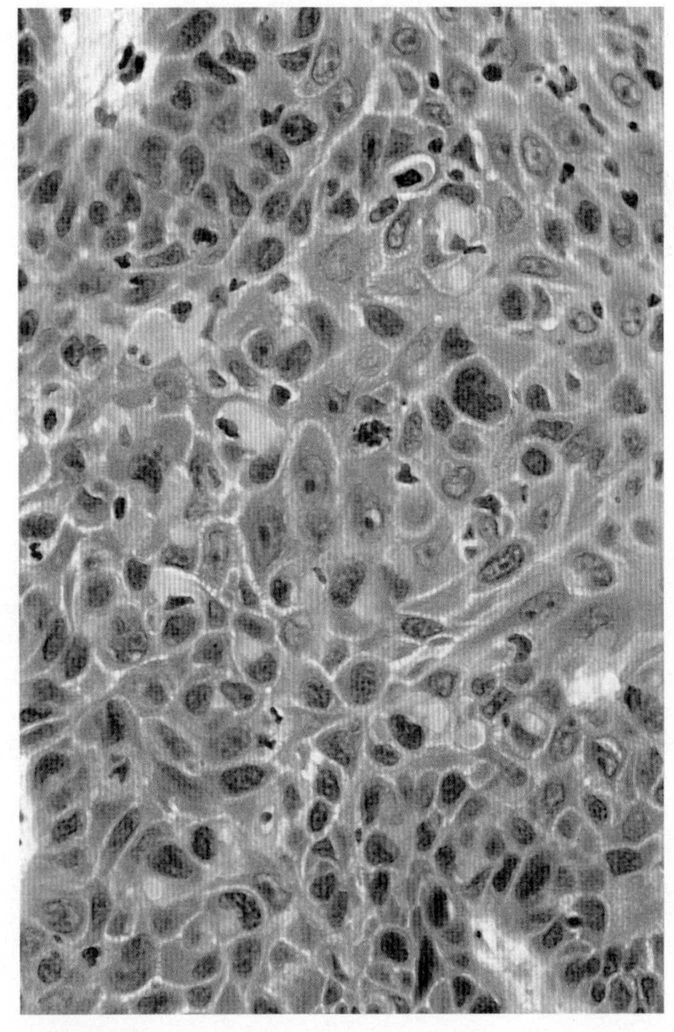

Figure 9–29 B

10. ORAL CAVITY AND SALIVARY GLANDS

■ ORAL CAVITY

LESIONS OF ODONTOGENIC ORIGIN

FIGURE 10–1. ODONTOGENIC KERATOCYST: The odontogenic keratocyst is a developmental cyst. The main histologic features are a thin stratified squamous epithelium composed of six to eight layers of cells and a parakeratin lining.

FIGURE 10–2. DENTIGEROUS CYST: A: The dentigerous cyst is characterized by a cyst lined by a thin stratified squamous epithelium. **B:** There is an intense chronic inflammatory infiltrate beneath and involving the epithelial lining.

Figure 10–1

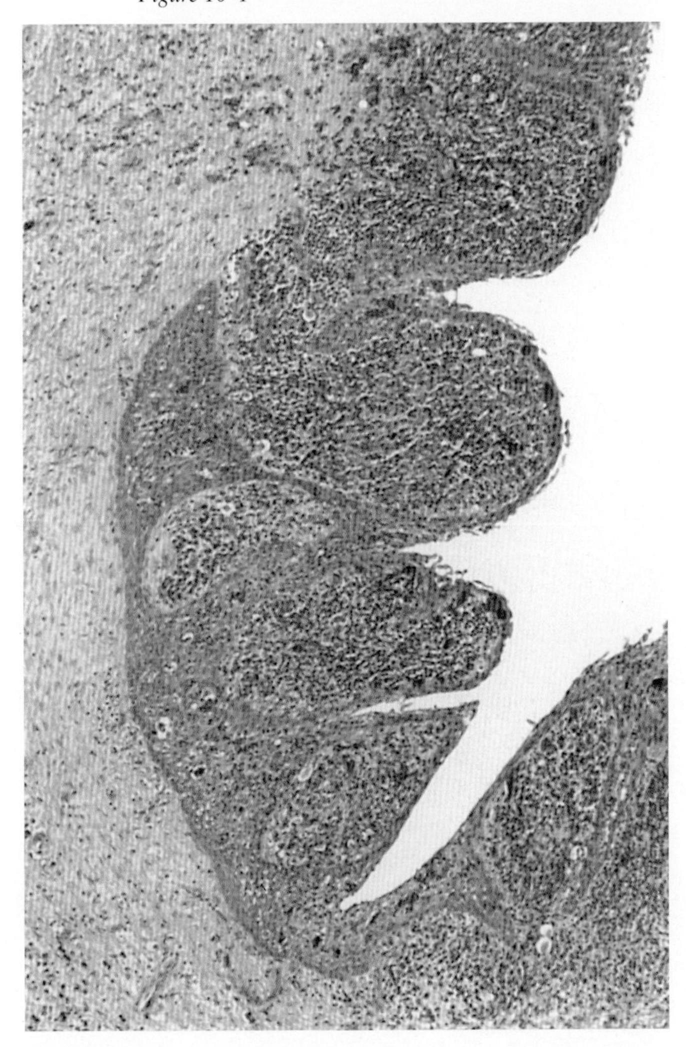

Figure 10–2 A

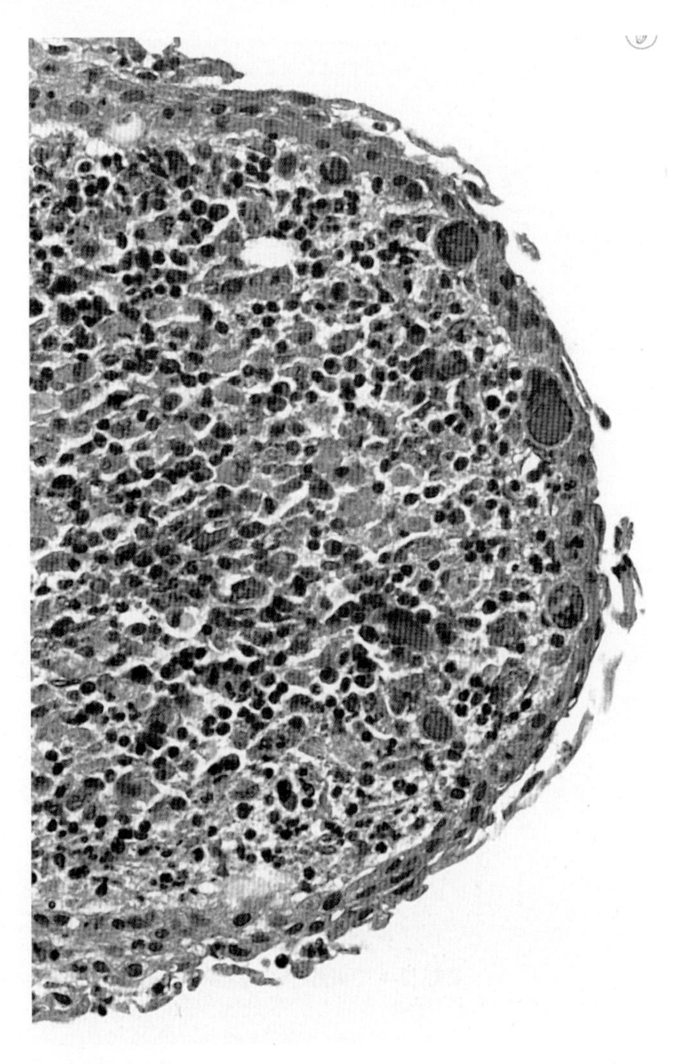

Figure 10–2 B

LESIONS OF ODONTOGENIC ORIGIN

FIGURE 10–3. AMELOBLASTOMA (ADAMANTINOMA): A: In ameloblastomas, the tumor is composed of sheets, nests, and anastomosing epithelial islands within a fibrovascular connective tissue stroma. **B:** The islands are rimmed by a row of cuboidal or columnar cells. Centrally, the cells are reminiscent of a primitive stellate reticulum.

FIGURE 10–4. CEMENTIFYING FIBROMA: Cementifying fibroma is characterized by rounded masses of calcified cementum within a fibroblastic stroma.

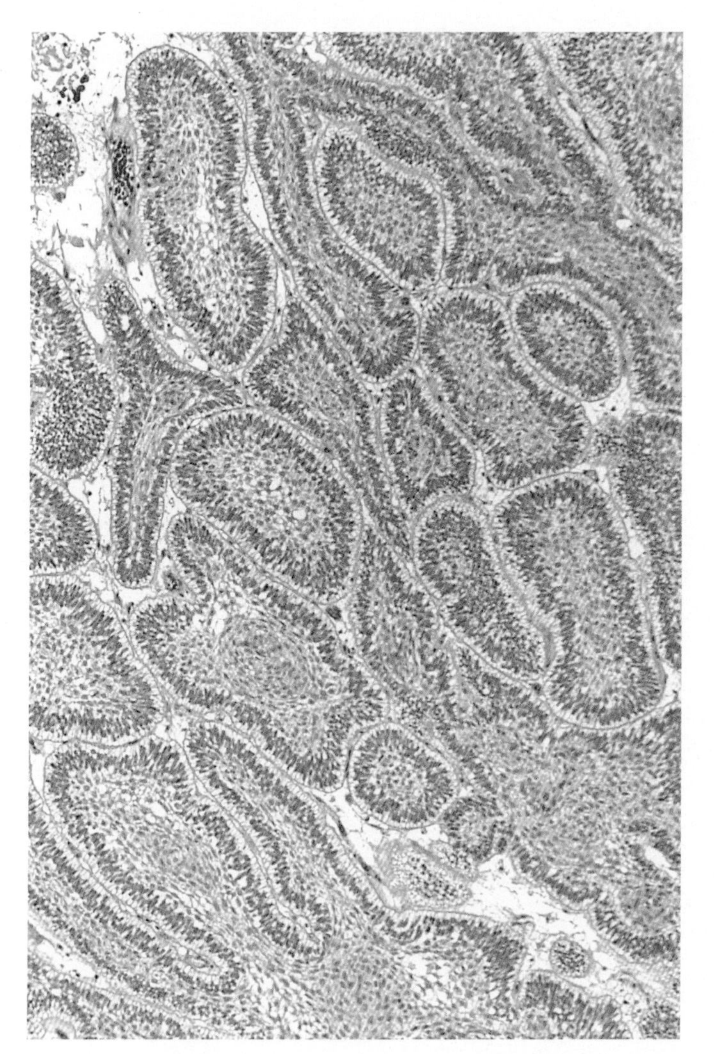

Figure 10–3 **A**

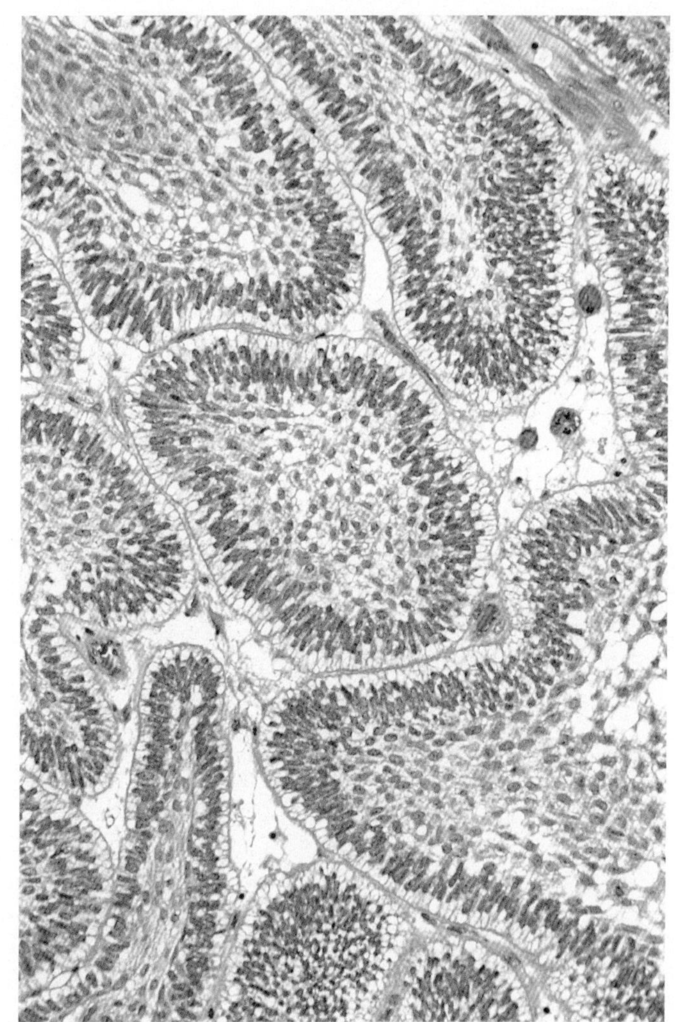

Figure 10–3 **B**

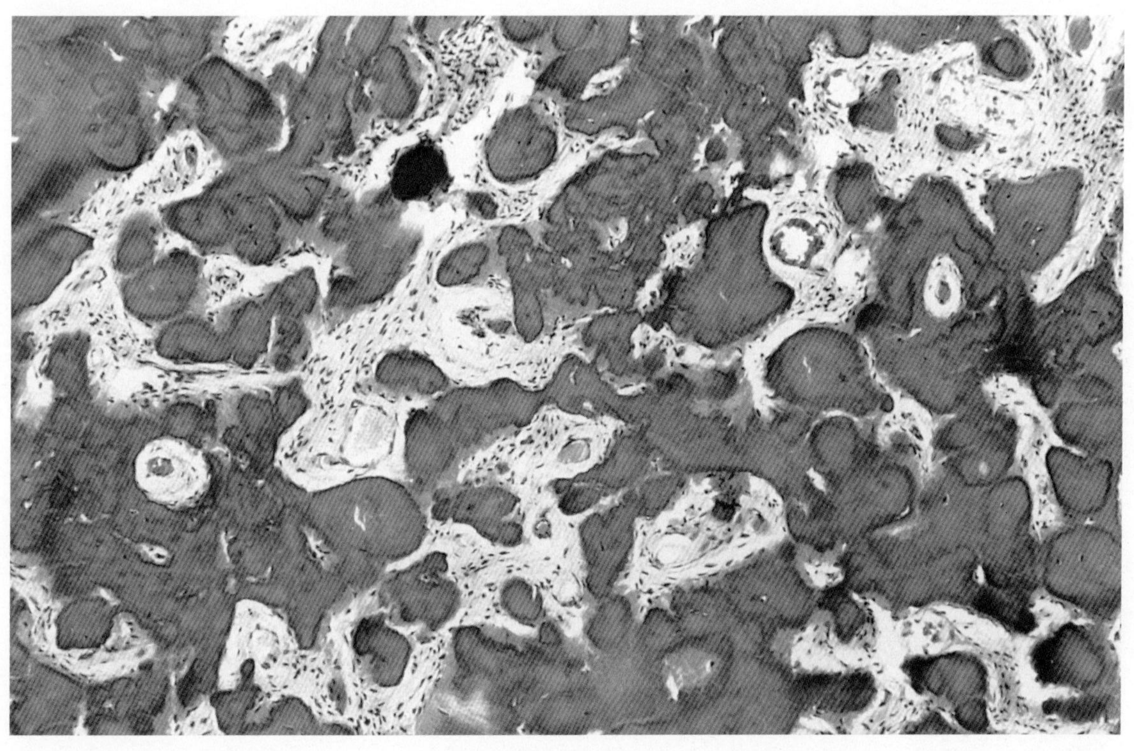

Figure 10–4

INFLAMMATORY LESIONS

FIGURE 10–5. RADICULAR CYST: The radicular cyst is an inflammatory cyst. It is the most common cyst of the jaws. Fragments of squamous epithelium, lining the cyst and penetrating the wall (**A**), are supported by an intensely and chronically inflamed connective tissue wall (**B**). .

FIGURE 10–6. MUCOCELE: In this example of mucocele, the dilated cyst is lined by an attenuated epithelium. The cyst contains mucinous material, debris, and inflammatory cells. A chronic inflammatory infiltrate surrounds the cyst.

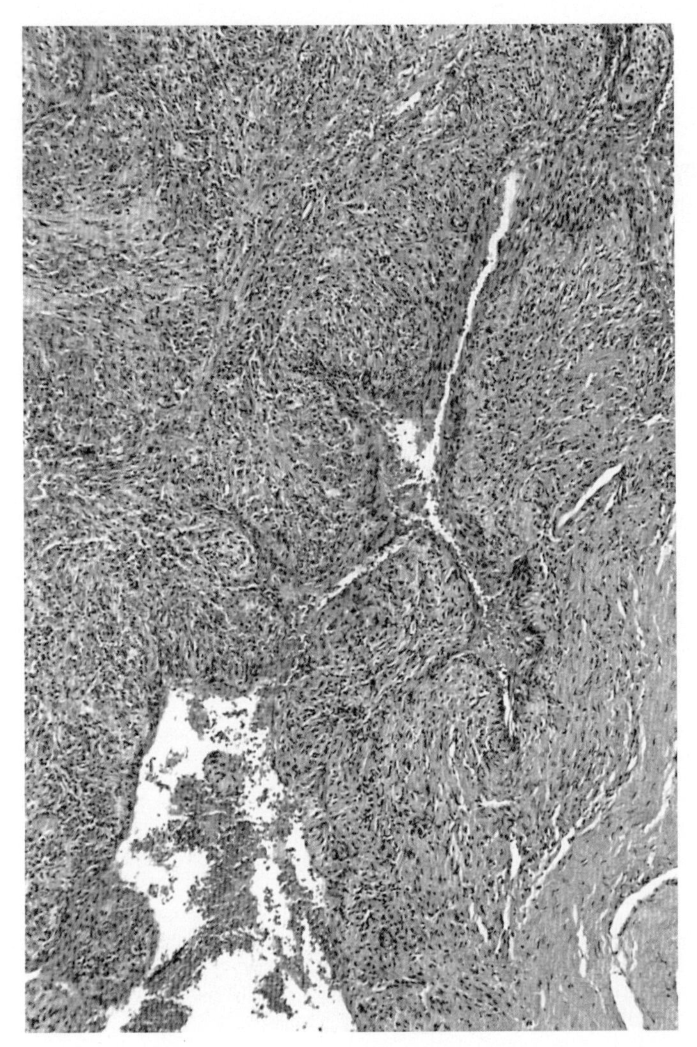

Figure 10–5 A

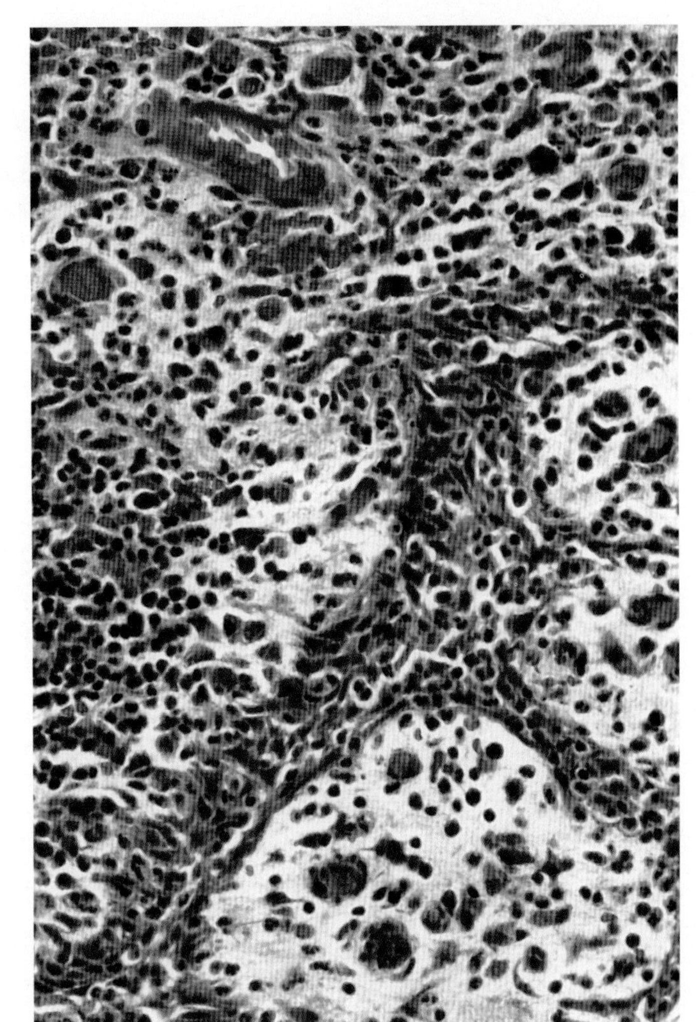

Figure 10–5 B

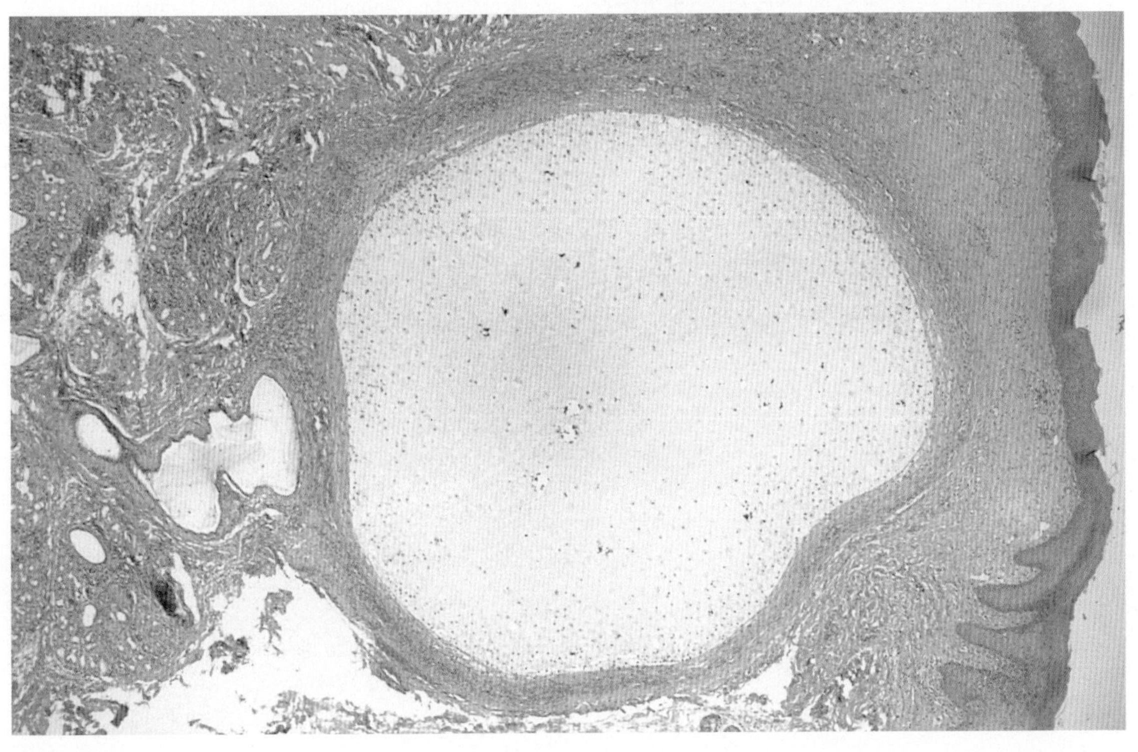

Figure 10–6

NONODONTOGENIC TUMORS

Leukoplakia is a clinical term meaning white patch. The histology of this lesion may include simple hyperkeratosis, pseudoepitheliomatous hyperplasia, in situ squamous cell carcinoma, or infiltrating squamous cell carcinoma.

FIGURE 10–7. PSEUDOEPITHELIOMATOUS HYPERPLASIA: In pseudoepitheliomatous hyperplasia, the rete ridges that penetrate the underlying stroma are elongated and "razor sharp."

FIGURE 10–8. SQUAMOUS CELL CARCINOMA IN SITU: In squamous cell carcinoma in situ, the entire thickness of the epithelium is replaced by neoplastic cells. The cells are pleomorphic, with a high nuclear-to-cytoplasmic ratio. Mitotic figures are present at all levels of the epithelium. The neoplastic cells are confined to the epithelial layer and do not extend beyond the basement membrane.

FIGURE 10–9. SQUAMOUS CELL CARCINOMA: In squamous cell carcinoma, in addition to the cytologic features of in situ carcinoma (Figure 10–8), the neoplastic cells penetrate the underlying stroma.

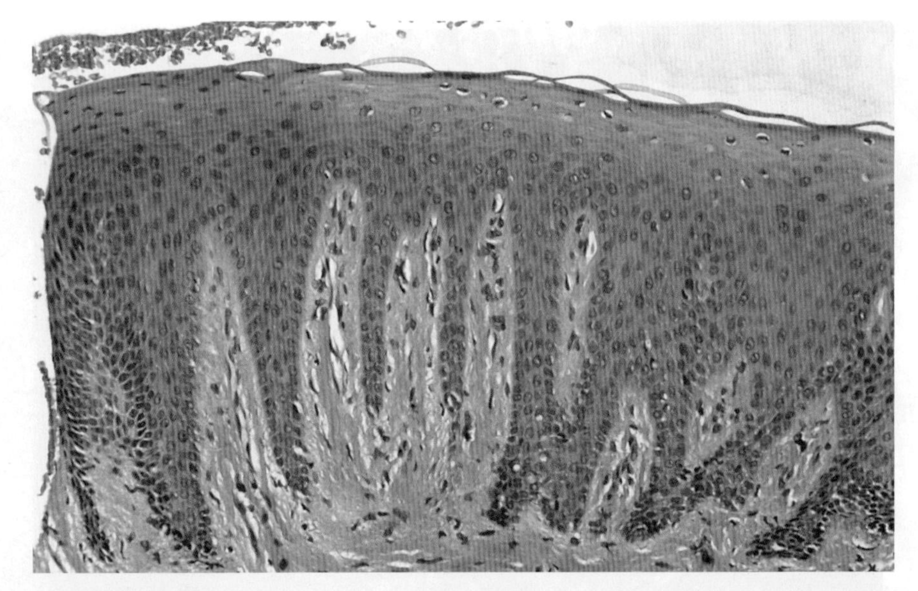

Figure 10–7

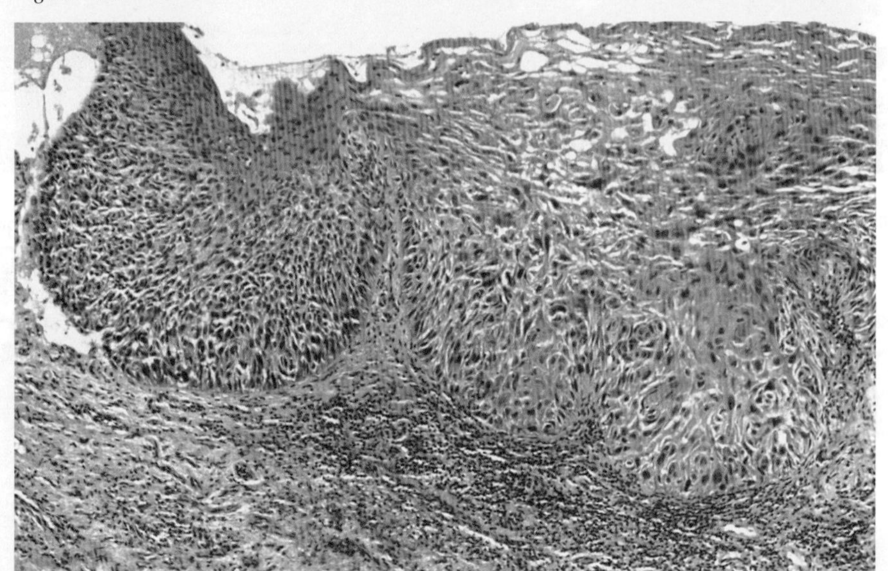

Figure 10–8

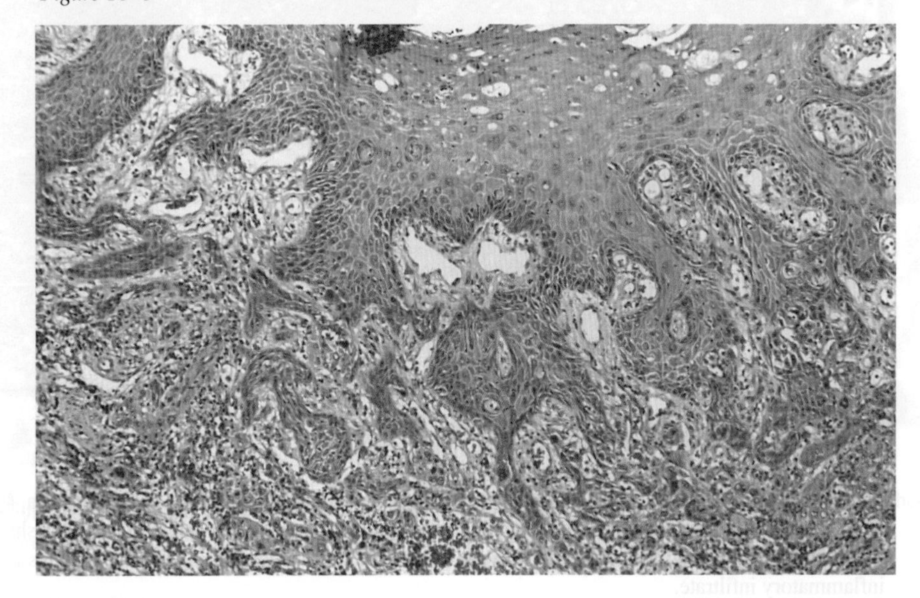

Figure 10–9

BENIGN SALIVARY GLAND TUMORS

FIGURE 10–16. PLEOMORPHIC ADENOMA (BENIGN MIXED TUMOR): Pleomorphic adenoma is the most common salivary gland tumor. As the name suggests, pleomorphic adenomas have a variable histologic appearance in both the amount and arrangement of the epithelial and the mesenchymal components. The epithelial components can be arranged in sheets (**A**), cords (**B**), and ducts and tubules (**C**). The mesenchymal component is chondromyxoid (**B** and **D**) and is produced by myoepithelial cells. The tumor may undergo metaplastic changes. **D:** The epithelial component may undergo squamous metaplasia with keratin cyst formation, and the background may undergo osseous metaplasia (**B**).

MESENCHYMAL LESIONS

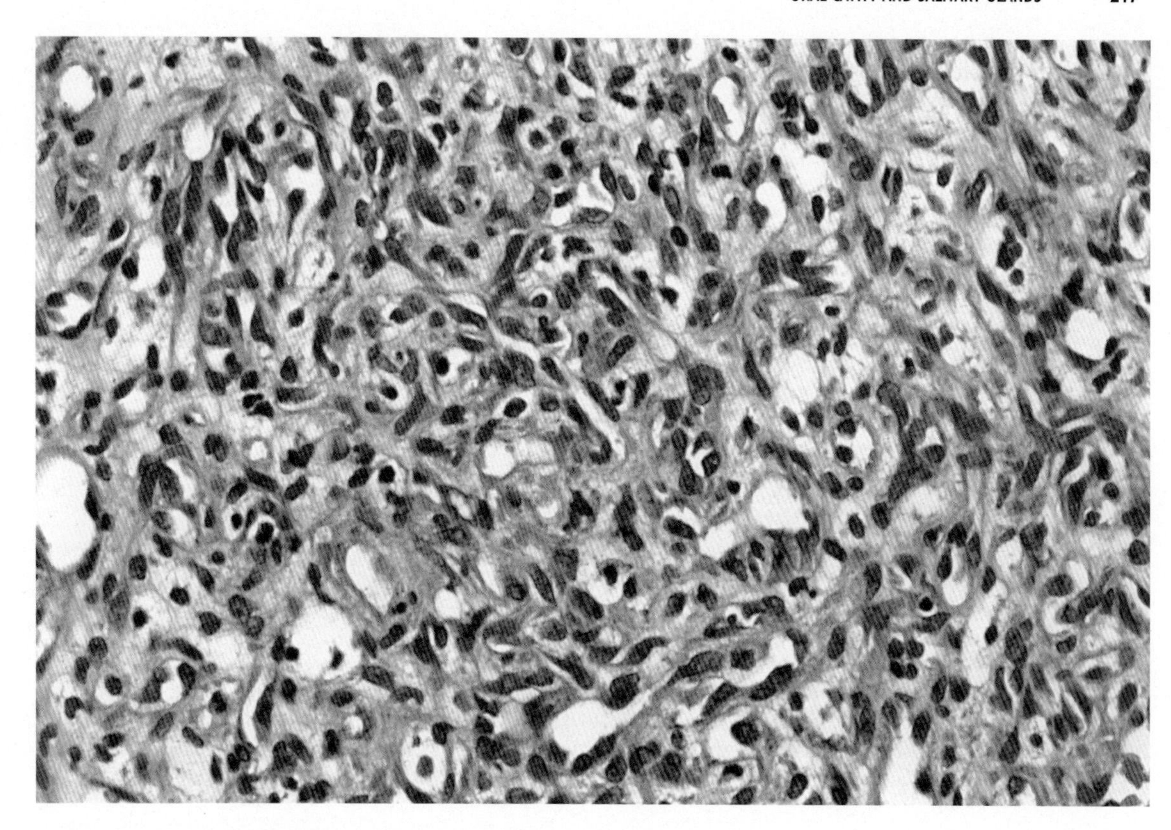

FIGURE 10–11. PYOGENIC GRANULOMA: In pyogenic granuloma, the subepithelial connective tissue contains a proliferation of vascular channels, fibroblasts, and acute and chronic inflammatory cells.

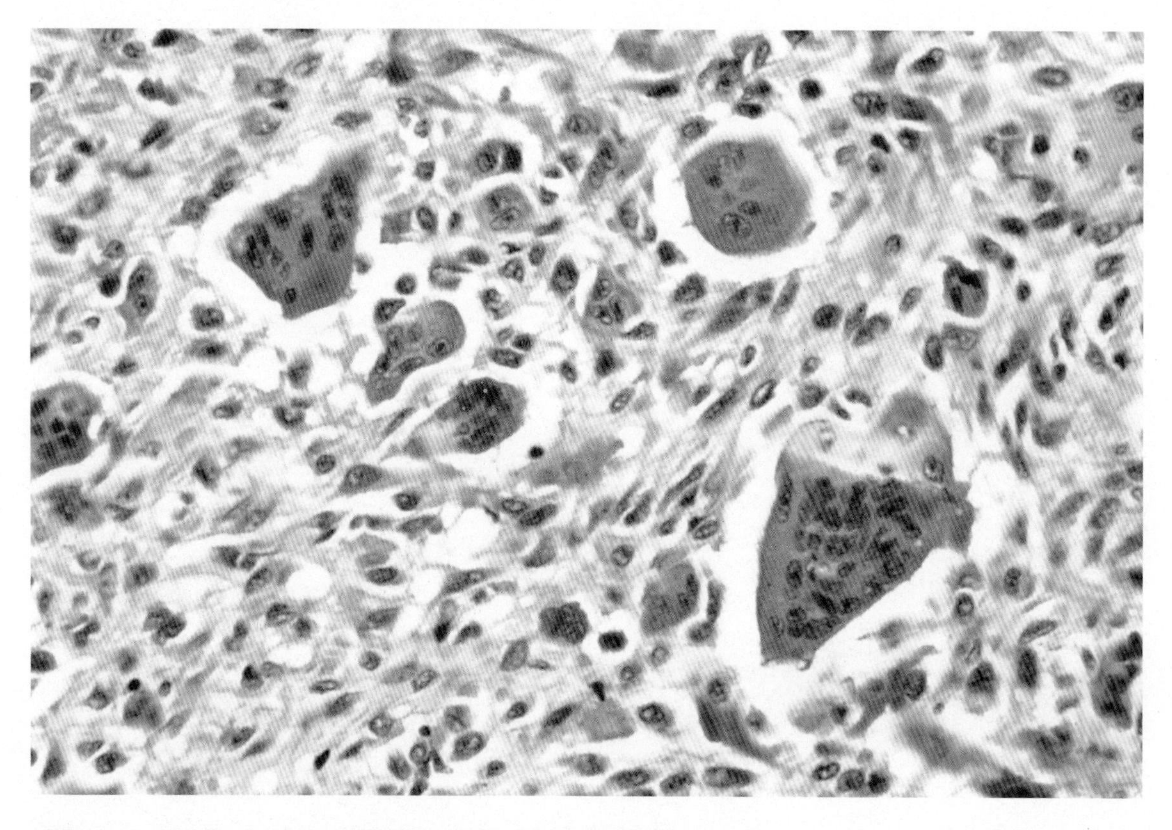

FIGURE 10–12. GIANT CELL GRANULOMA (PERIPHERAL GIANT CELL REPARATIVE GRANULOMA): In giant cell granuloma, there is proliferation of fibroblasts, capillaries, and multinucleated giant cells. Mitoses may be abundant.

MESENCHYMAL LESIONS

FIGURE 10–13. GRANULAR CELL TUMOR (GRANULAR CELL MYOBLASTOMA): A: In granular cell tumor, the overlying epithelium shows pseudoepitheliomatous hyperplasia. **B:** Within the connective tissue is a tumor mass composed of sheets and nests of large eosinophilic granular cells. Nuclei are small, and mitoses are infrequent. **C:** Immunoperoxidase staining with S-100 protein confirms the neural origin of this tumor.

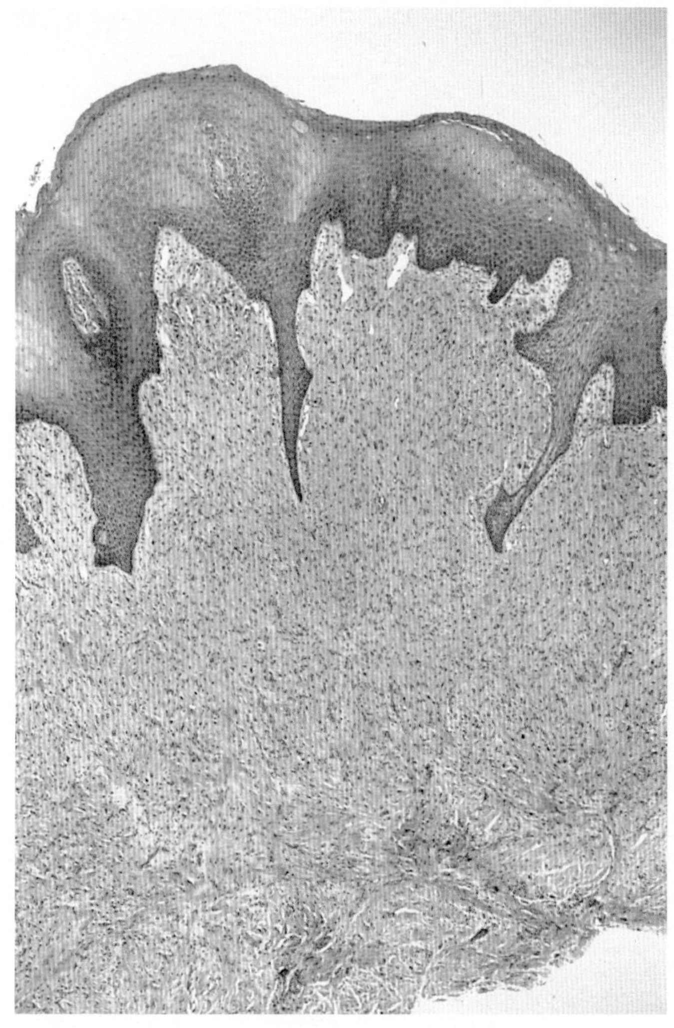

Figure 10–13 A

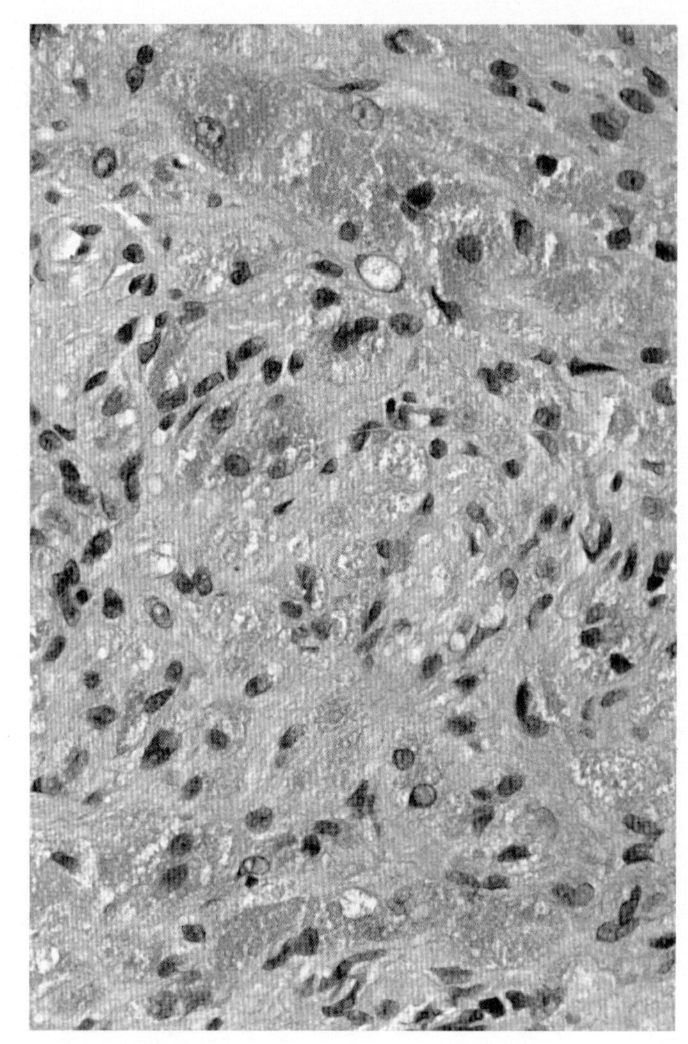

Figure 10–13 B

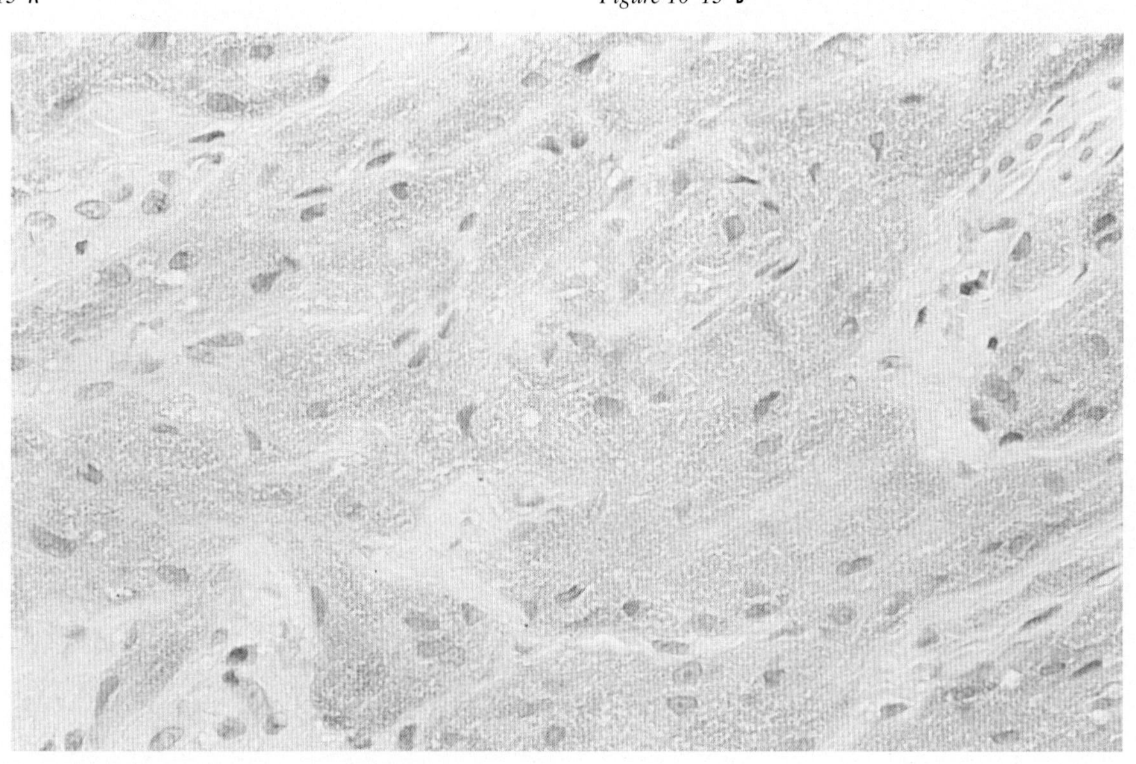

Figure 10–13 C

■ SALIVARY GLANDS

INFLAMMATORY LESIONS

FIGURE 10–14. CHRONIC SIALADENITIS: In chronic sialadenitis, there is duct ectasia, atrophy of the glandular acini, a chronic inflammatory infiltrate, and fibrosis.

FIGURE 10–15. LYMPHOEPITHELIAL LESION (SJÖGREN'S SYNDROME): A and **B:** In this example of an early-stage lymphoepithelial lesion of a salivary gland, there is destruction of the glandular epithelium and acinar structures associated with a dense lymphoplasmacytic infiltrate.

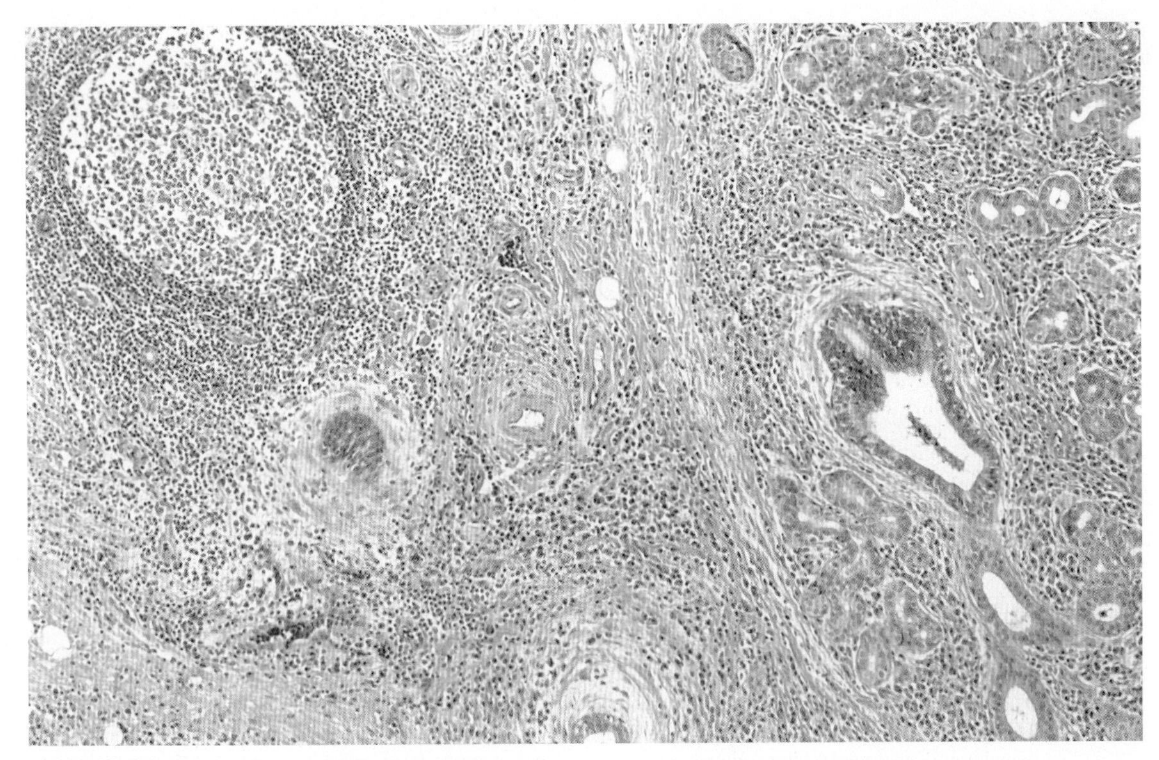

Figure 10–14

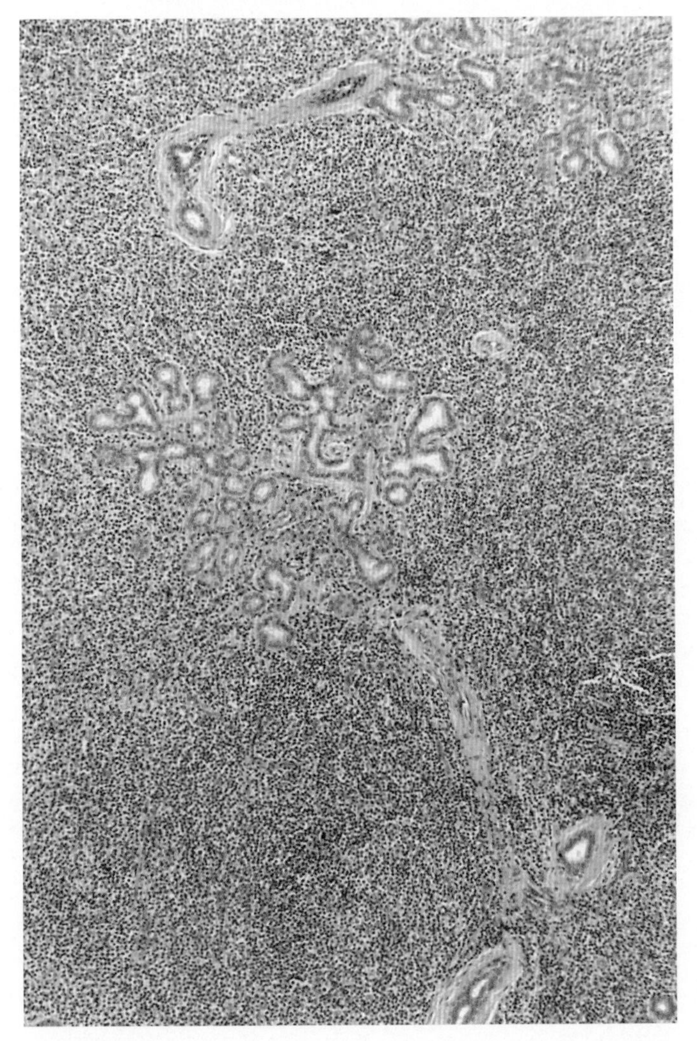

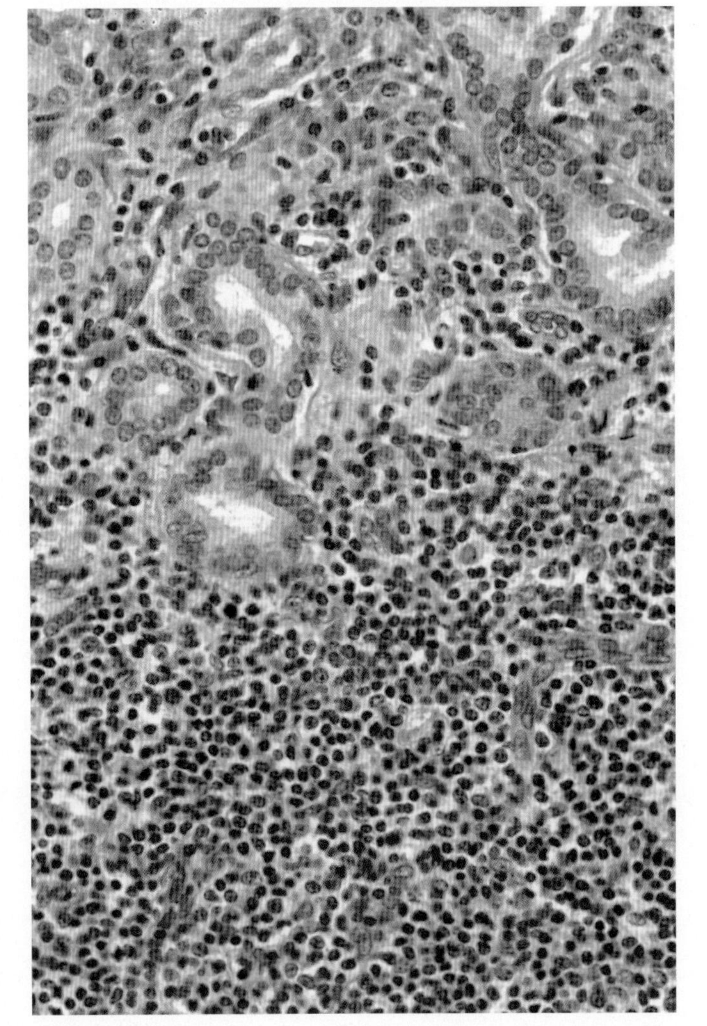

Figure 10–15 A

Figure 10–15 B

BENIGN SALIVARY GLAND TUMORS

FIGURE 10–16. PLEOMORPHIC ADENOMA (BENIGN MIXED TUMOR): Pleomorphic adenoma is the most common salivary gland tumor. As the name suggests, pleomorphic adenomas have a variable histologic appearance in both the amount and arrangement of the epithelial and the mesenchymal components. The epithelial components can be arranged in sheets (**A**), cords (**B**), and ducts and tubules (**C**). The mesenchymal component is chondromyxoid (**B** and **D**) and is produced by myoepithelial cells. The tumor may undergo metaplastic changes. **D:** The epithelial component may undergo squamous metaplasia with keratin cyst formation, and the background may undergo osseous metaplasia (**B**).

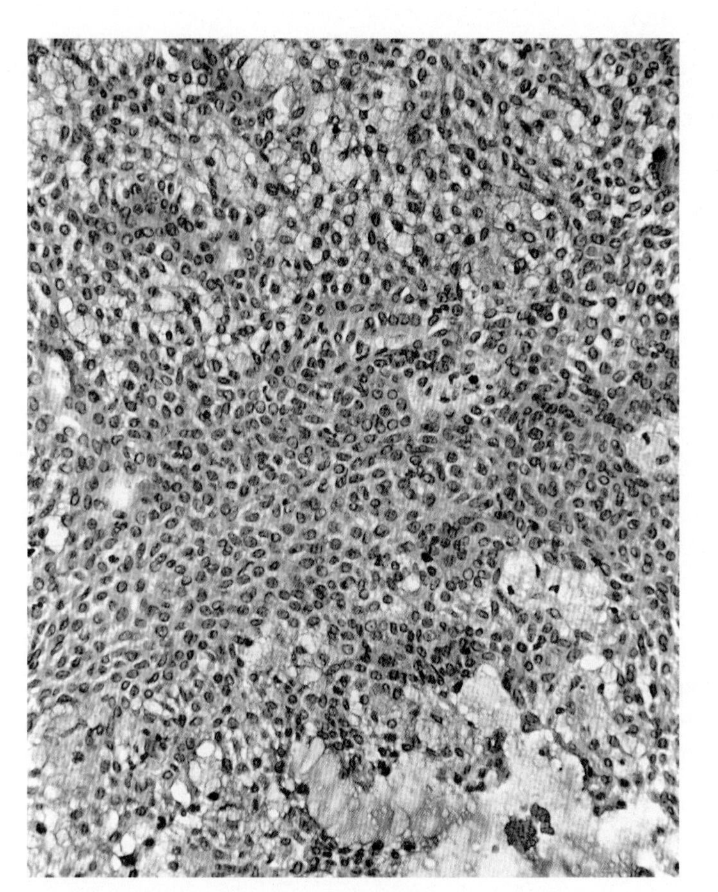

Figure 10–16 A

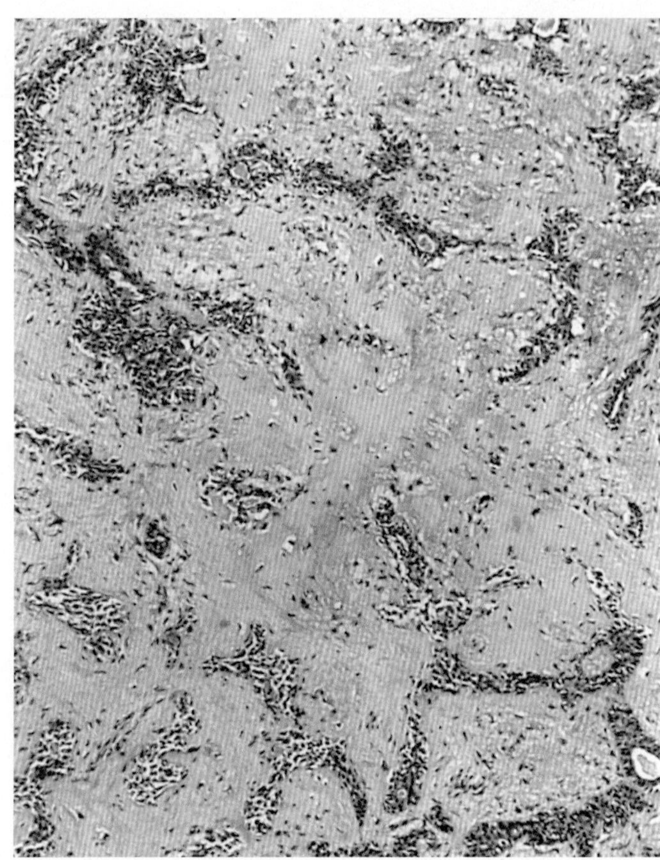

Figure 10–16 B

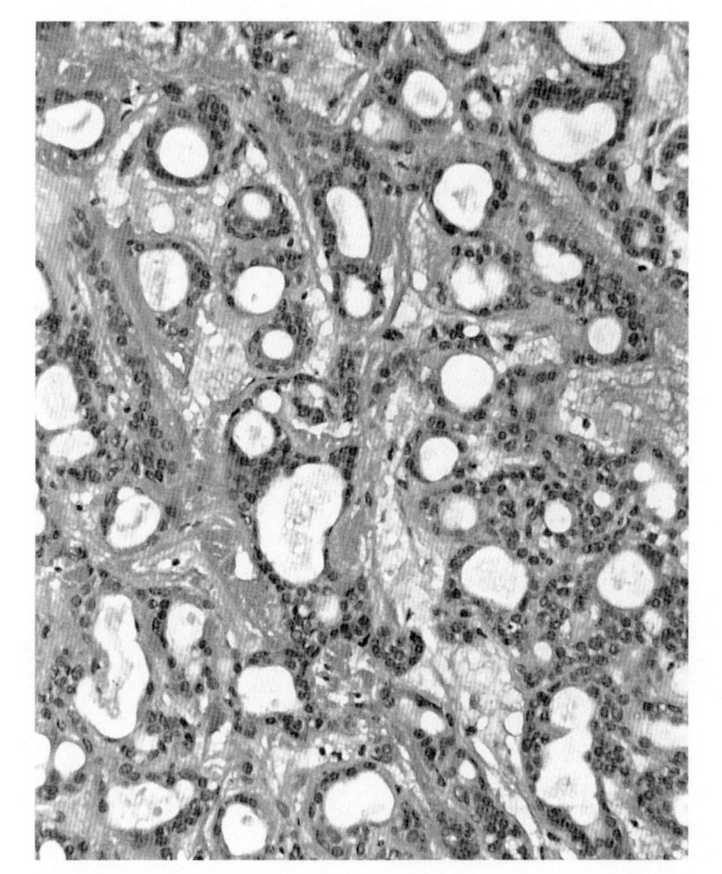

Figure 10–16 C

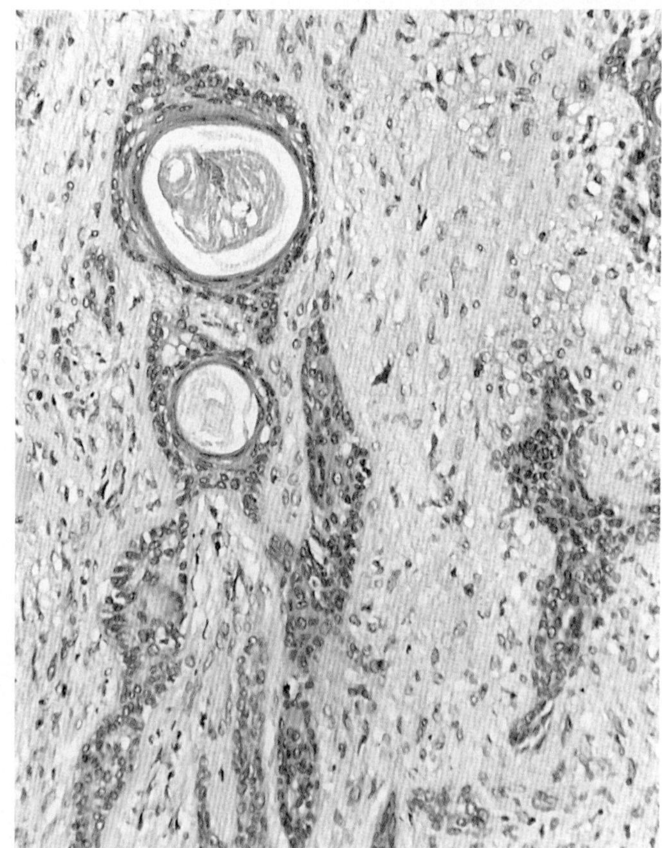

Figure 10–16 D

BENIGN SALIVARY GLAND TUMORS

FIGURE 10–17. WARTHIN'S TUMOR (PAPILLARY CYSTADENOMA LYMPHOMATOSA, ADENOLYMPHOMA): A: Warthin's tumors are papillary lesions that grow within a cystic space. **B:** The epithelium lining the papillae is oncocytic and has two layers of cells, a basal cuboidal layer and an apical columnar layer. The cores of the papillary projections are composed of reactive lymphoid tissue.

FIGURE 10–18. ONCOCYTOMA (OXYPHIL ADENOMA): Oncocytomas are composed of solid round acinar groups. The cells are uniform and polyhedral, with eosinophilic granular cytoplasm. The nuclei are small and centrally located.

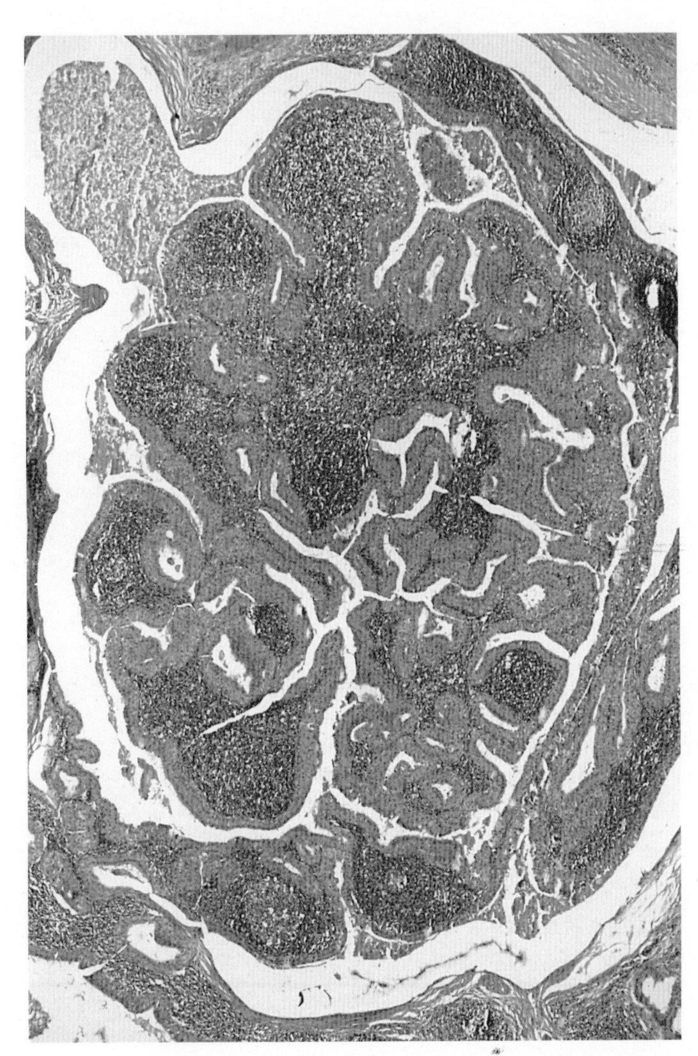

Figure 10–17 **A**

Figure 10–17 **B**

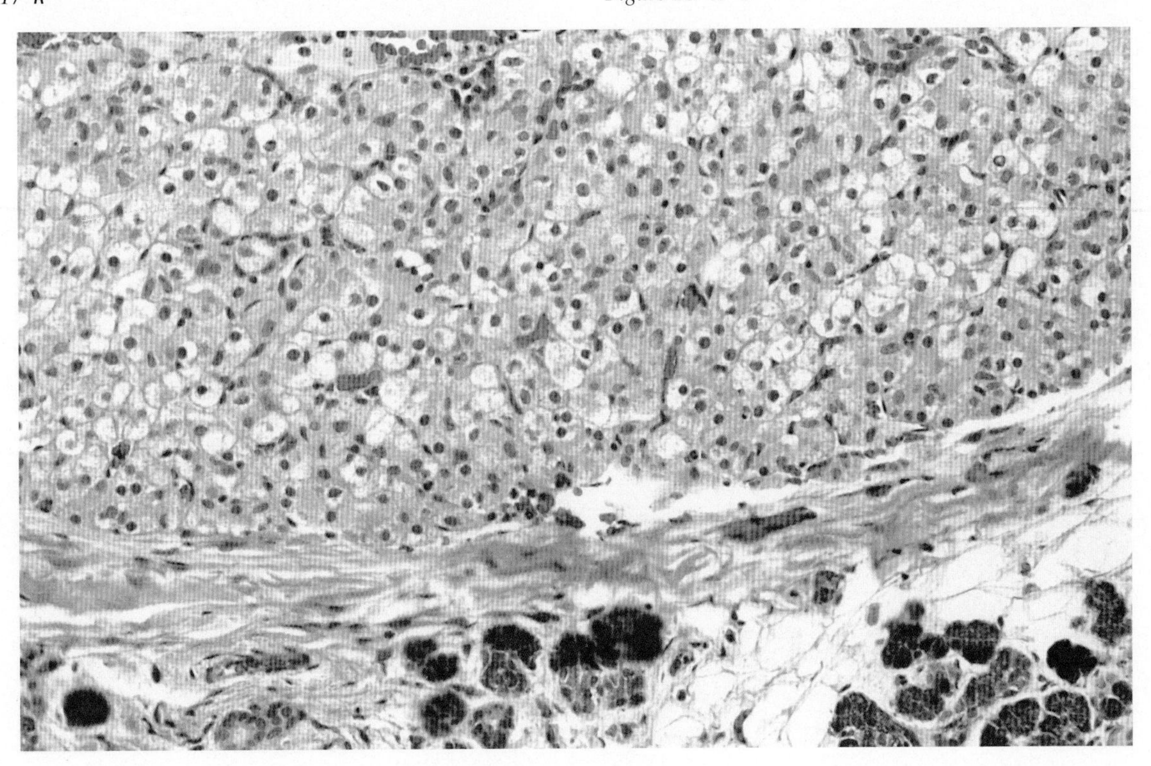

Figure 10–18

MALIGNANT SALIVARY GLAND TUMORS

FIGURE 10–19. MUCOEPIDERMOID CARCINOMA: In mucoepidermoid carcinomas, the presence of two types of cells—mucus and squamous—is essential for diagnosis. An undifferentiated cell that is difficult to identify has the potential to differentiate into either of these cell types. In **low-grade mucoepidermoid carcinoma,** the tumor contains cysts (A) lined by mucus and squamous cells (B). C: **High-grade mucoepidermoid carcinoma** is characterized by a solid pattern almost completely composed of squamous cells. Mitotic figures and cellular crowding are present.

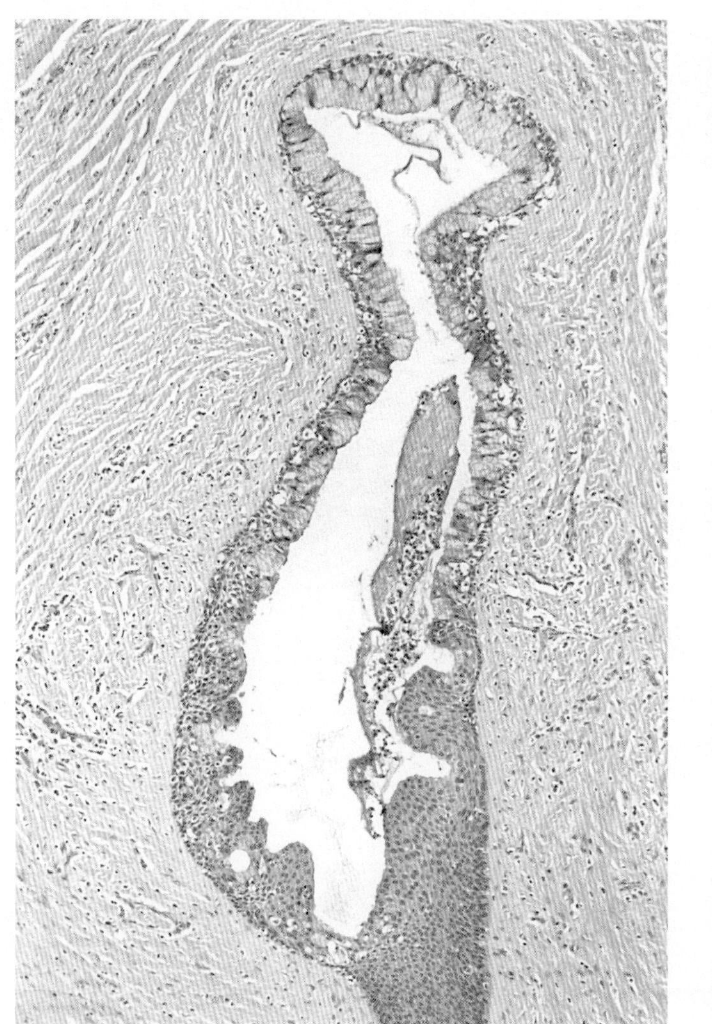

Figure 10–19 A

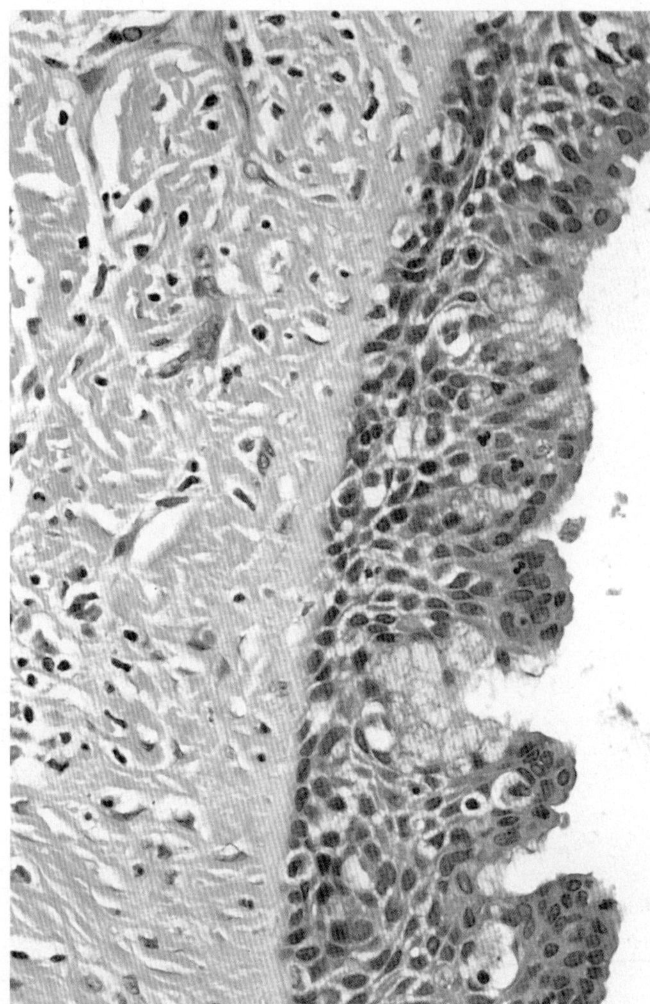

Figure 10–19 B

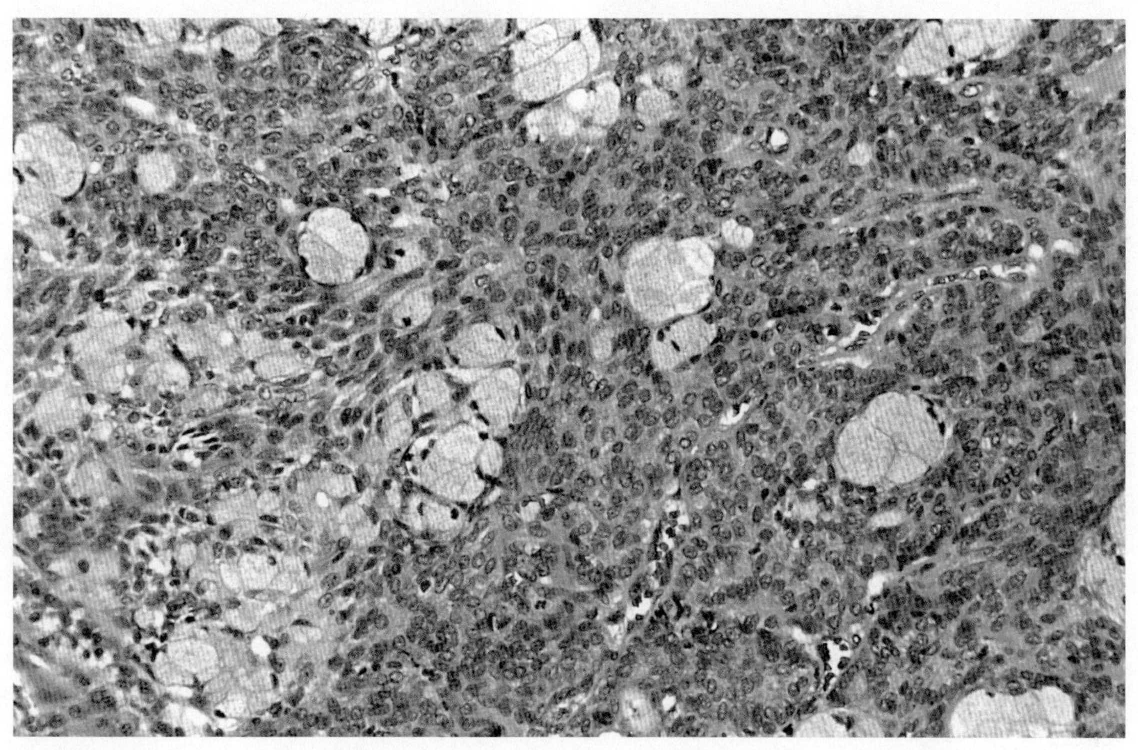

Figure 10–19 C

MALIGNANT SALIVARY GLAND TUMORS

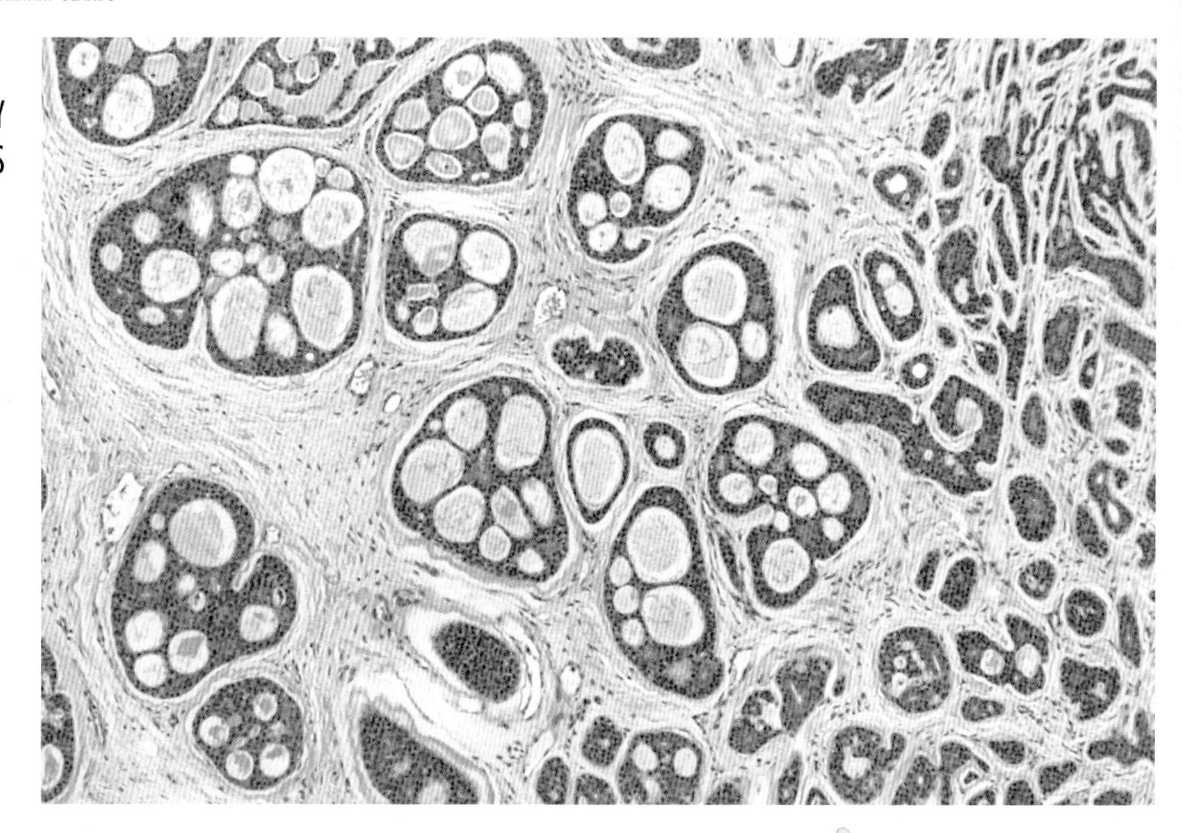

Figure 10–20 A

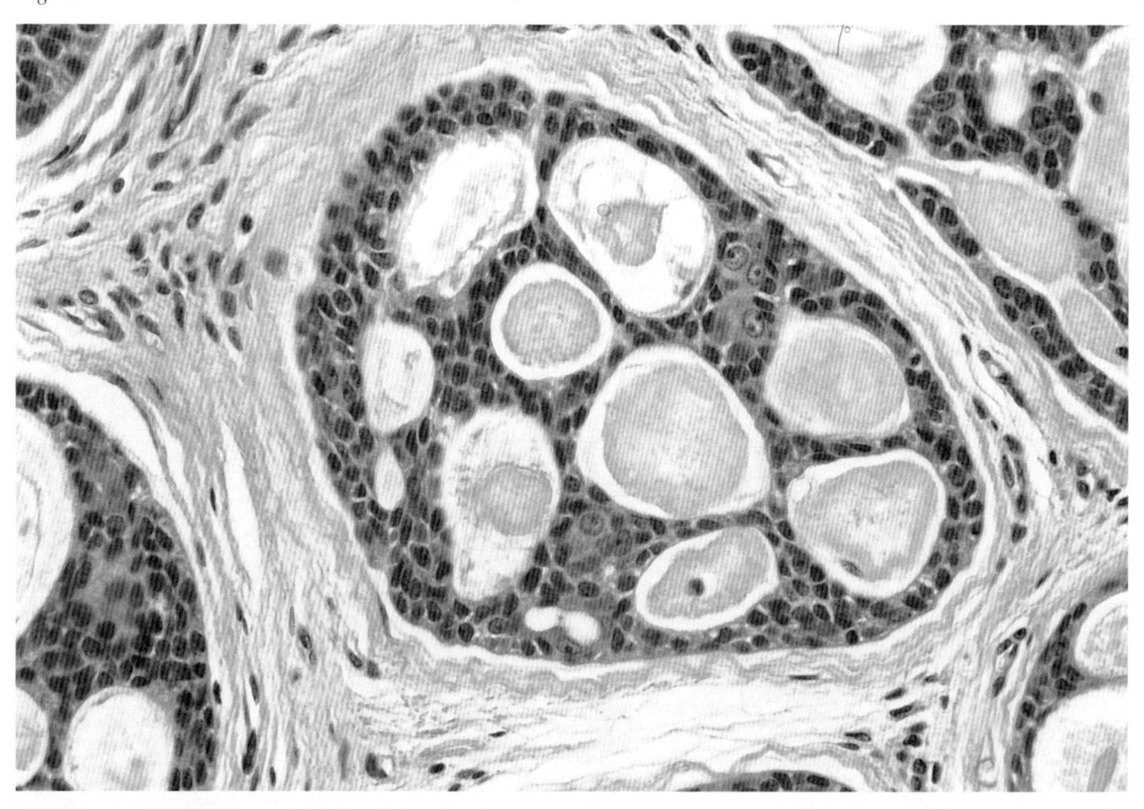

Figure 10–20 B

FIGURE 10–20. ADENOID CYSTIC CARCINOMA: A and **B:** In adenoid cystic carcinoma, the tumor is composed of islands of small uniform cells arranged in a cribriform pattern. Periodic acid–Schiff–positive eosinophilic material surrounds and fills the cystic spaces of the islands.

■ ESOPHAGUS

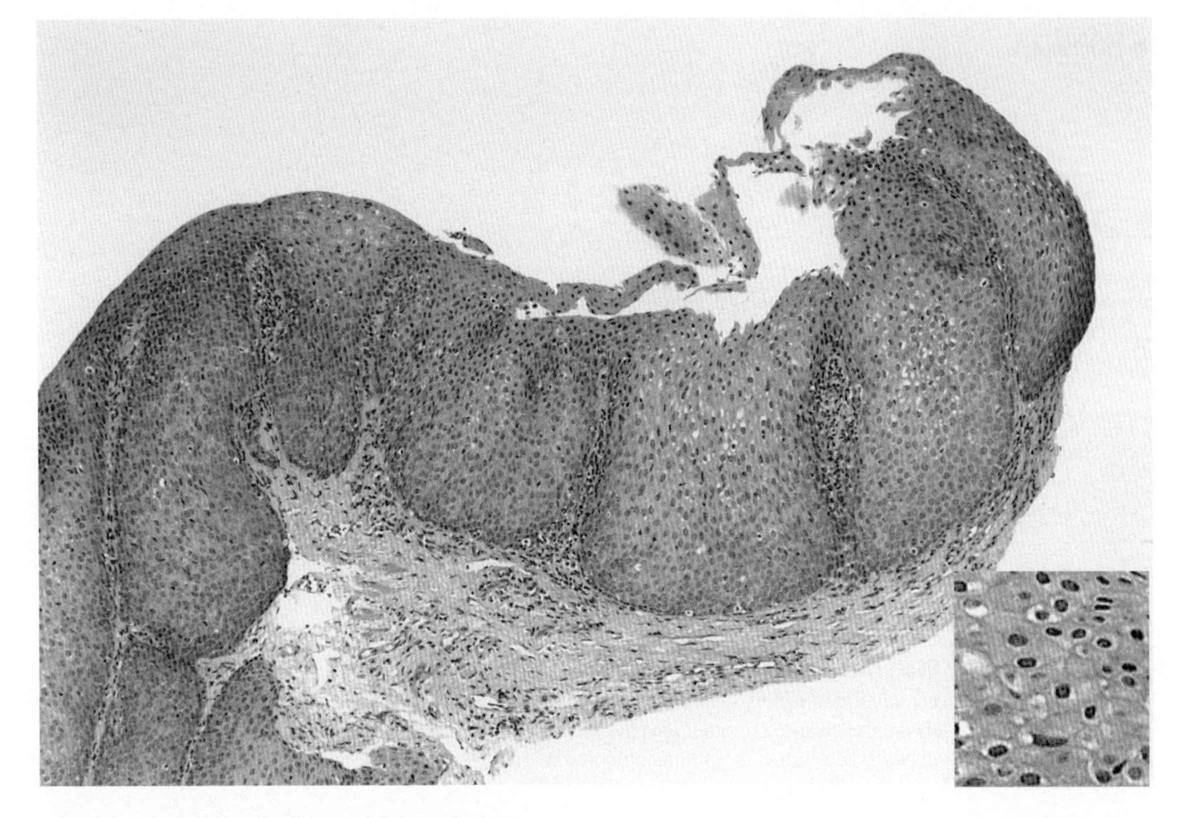

FIGURE 11–1. GASTROESOPHAGEAL REFLUX (REFLUX ESOPHAGITIS): Gastroesophageal reflux is characterized by basal cell hyperplasia and elongation of the papillae. Eosinophils are present in the epithelium and lamina propria (inset).

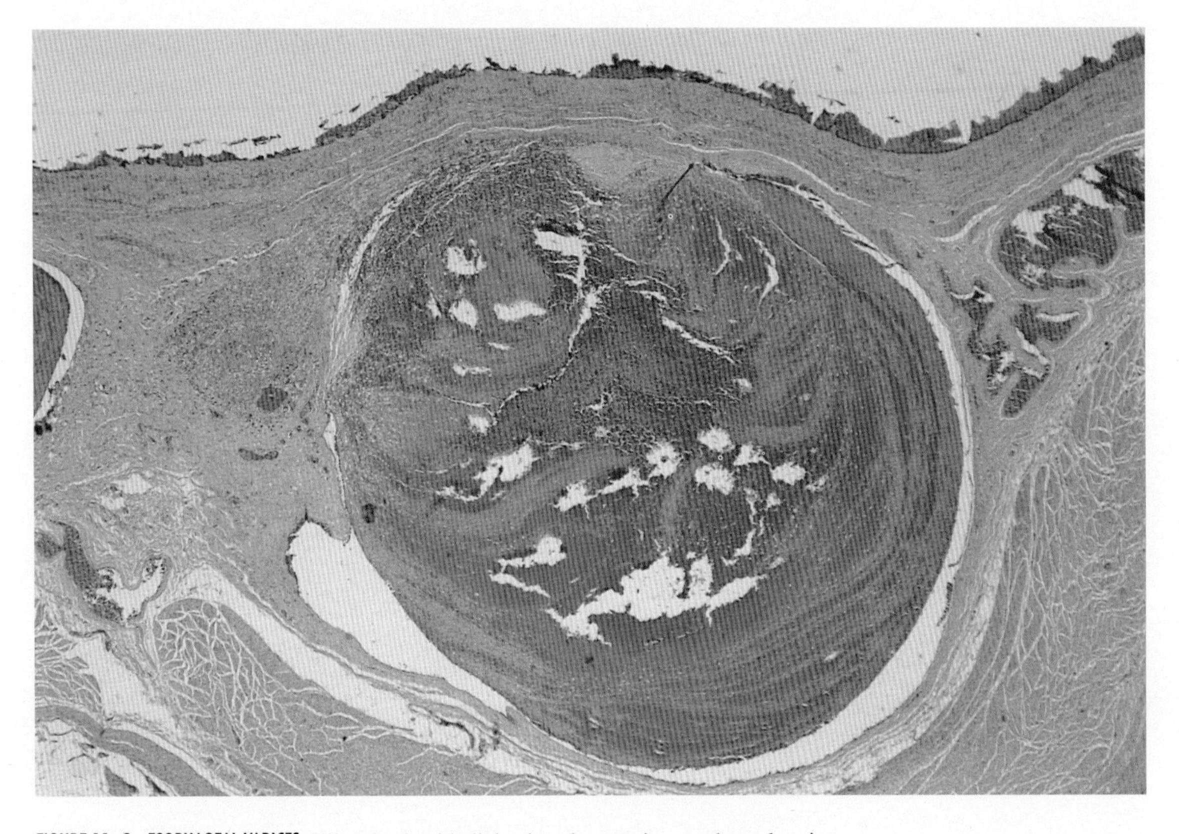

FIGURE 11–2. ESOPHAGEAL VARICES: Dilated subepithelial veins characterize esophageal varices.

■ ESOPHAGUS

FIGURE 11–3. FUNGAL ESOPHAGITIS: A: In fungal esophagitis, there is ulceration of the epithelium. Acute and chronic inflammatory cells are present in the lamina propria. **B:** Gomori methenamine-silver stain clearly shows the presence of pseudohyphal and spore forms of *Candida.* **C:** A Brown-Brenn stain shows a superimposed bacterial colonization composed of gram-positive (blue) and gram-negative (red) bacilli and cocci.

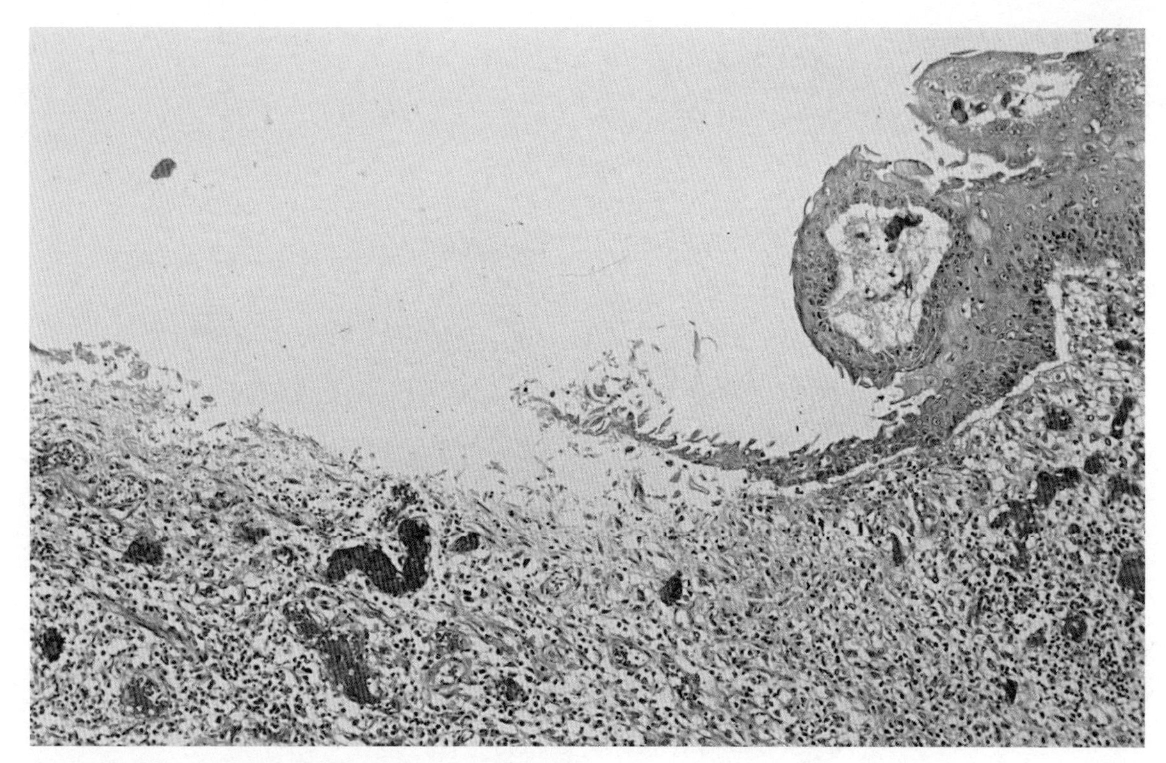

Figure 11–3 **A**

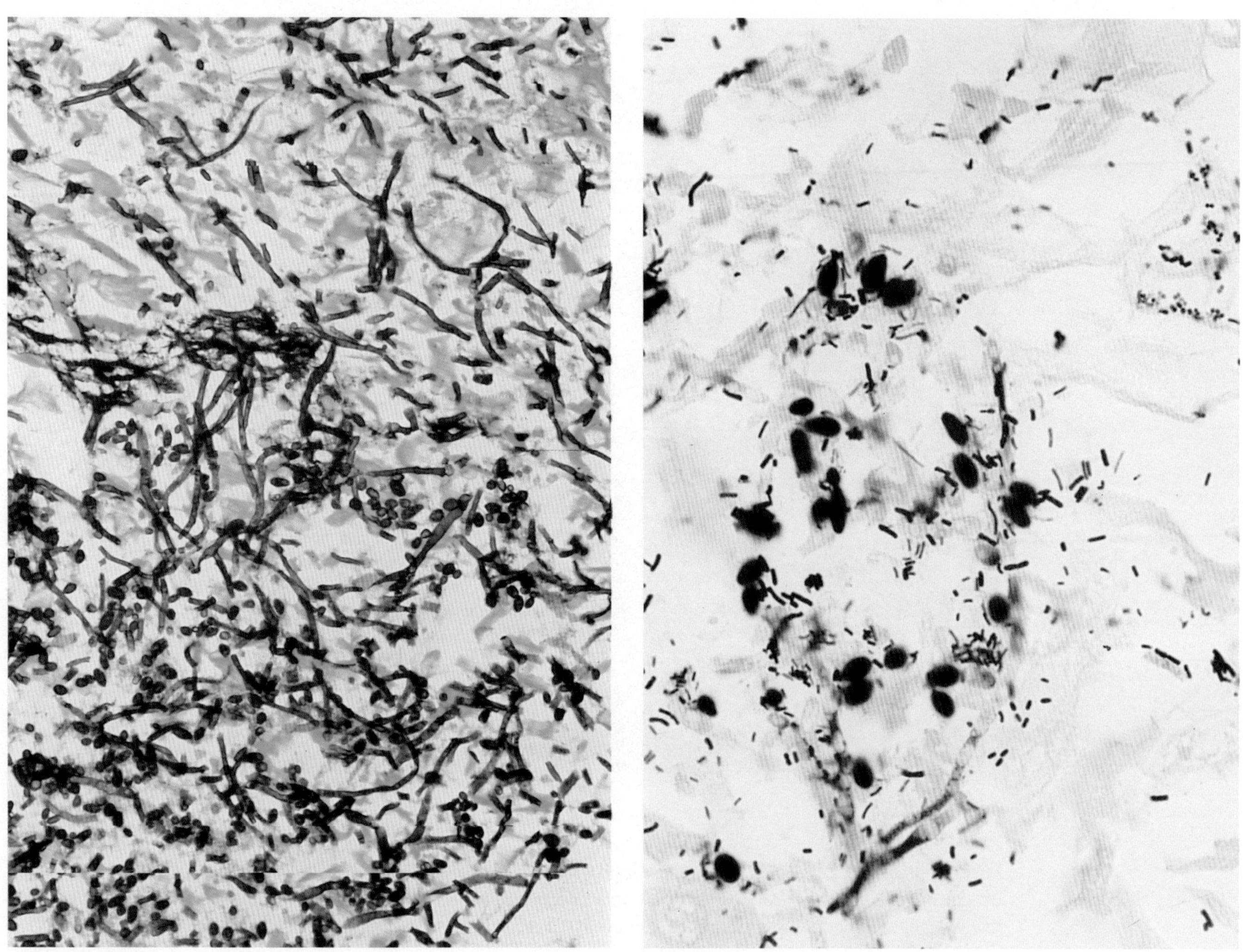

Figure 11–3 **B**

Figure 11–3 **C**

■ ESOPHAGUS

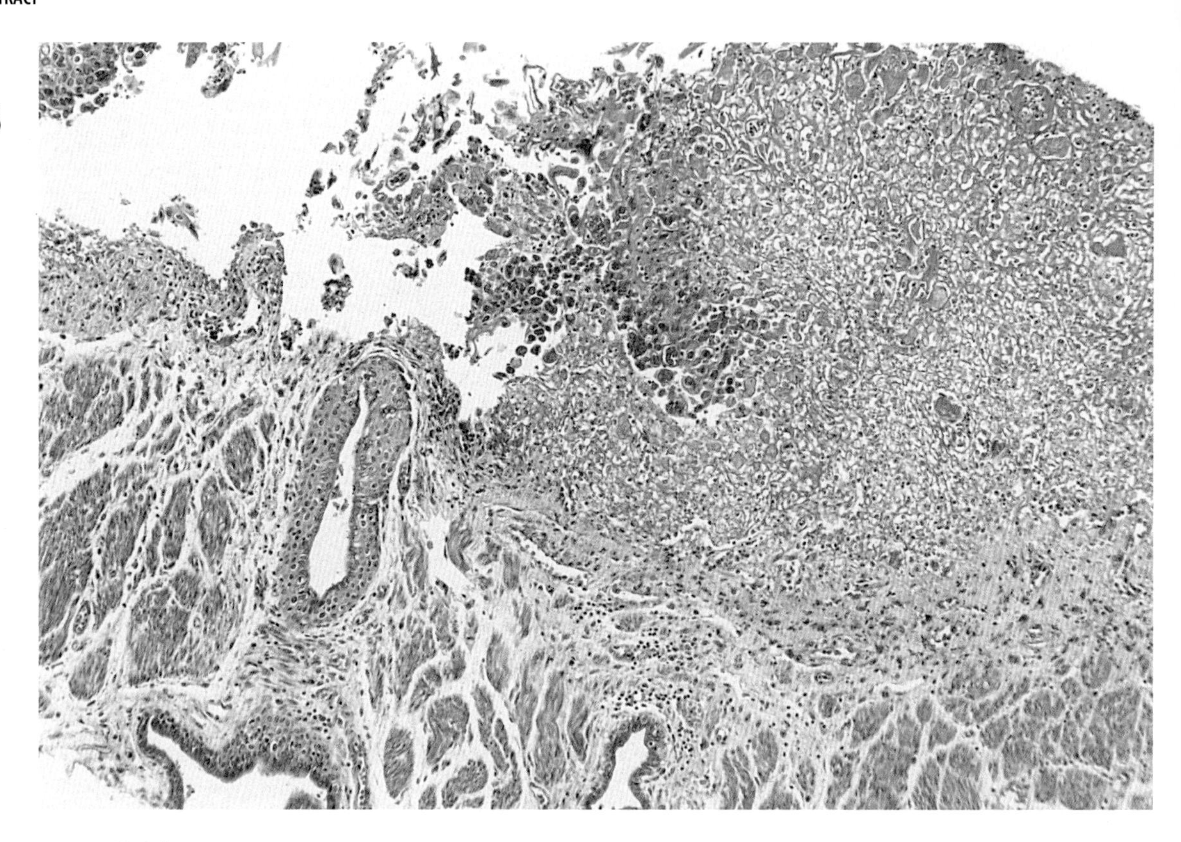

Figure 11–4 A

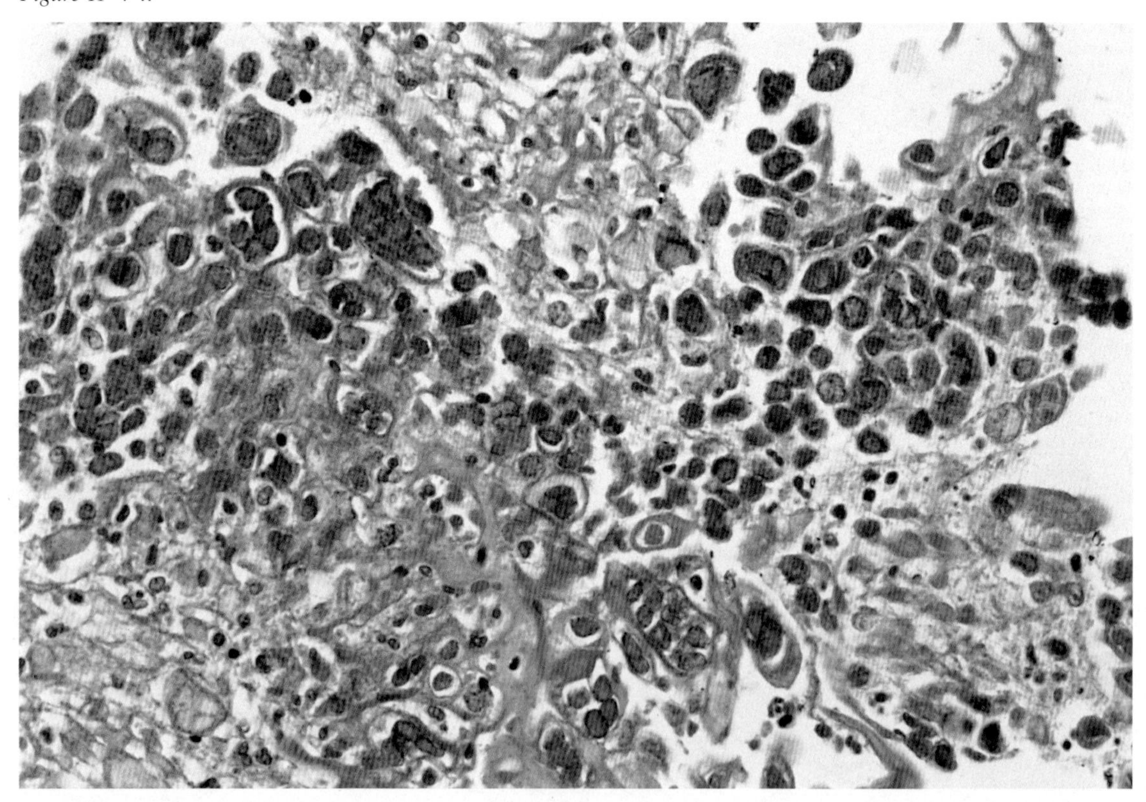

Figure 11–4 B

FIGURE 11–4. HERPETIC ESOPHAGITIS: A: Herpetic esophagitis is characterized by mucosal ulceration and an acute inflammatory exudate. **B:** Characteristic features of herpesvirus infection are ballooning degeneration of cells, Cowdry's type A intranuclear inclusion bodies, and multinucleated syncytial cells with ground-glass nuclei.

■ ESOPHAGUS

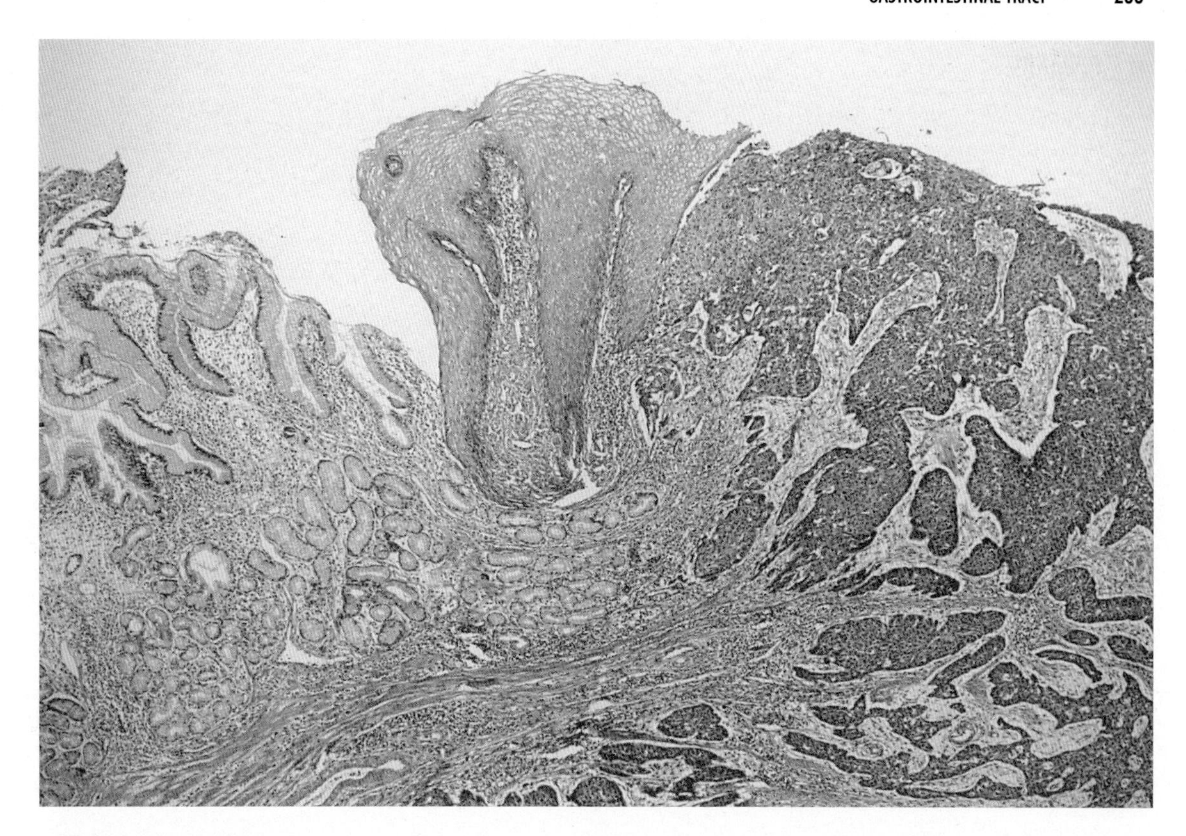

Figure 11–5 **A**

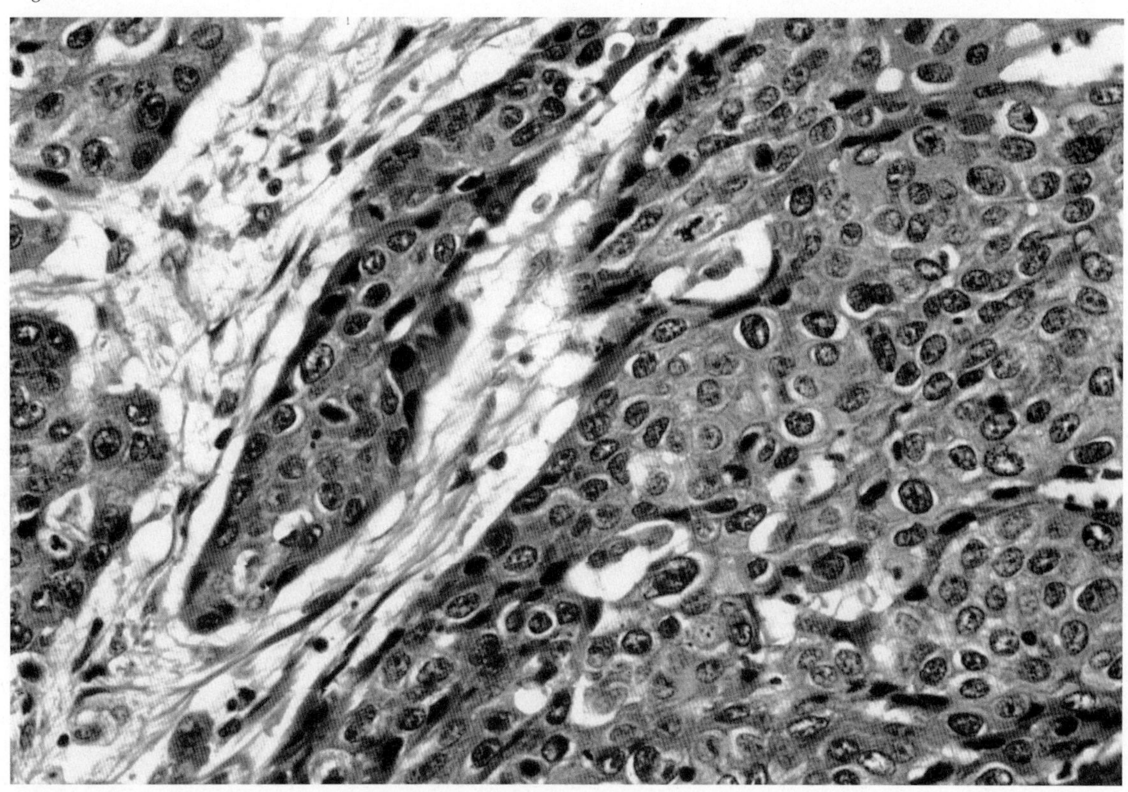

Figure 11–5 **B**

FIGURE 11–5. SQUAMOUS CELL CARCINOMA: A: This is a photomicrograph of poorly differentiated infiltrating squamous cell carcinoma arising at the gastroesophageal junction. **B:** There is cellular and nuclear pleomorphism, loss of polarity, and increased numbers of mitotic figures. The stroma shows a desmoplastic response.

■ ESOPHAGUS

FIGURE 11–6. BARRETT'S ESOPHAGUS: A: In Barrett's esophagus, the normal squamous epithelium is replaced by glandular epithelium. **B:** The most characteristic type, shown here, has a small intestinal-type epithelium with villus formation, goblet cells, and mucus cells.

FIGURE 11–7. BARRETT'S ESOPHAGUS WITH DYSPLASIA: In addition to the features of Barrett's esophagus (Figure 11–6 **A** and **B**), dysplasia is characterized by stratification of the glandular epithelial cells with a loss of polarity. Nuclei are hyperchromatic and enlarged, and mitotic figures are present at levels other than the basal layer.

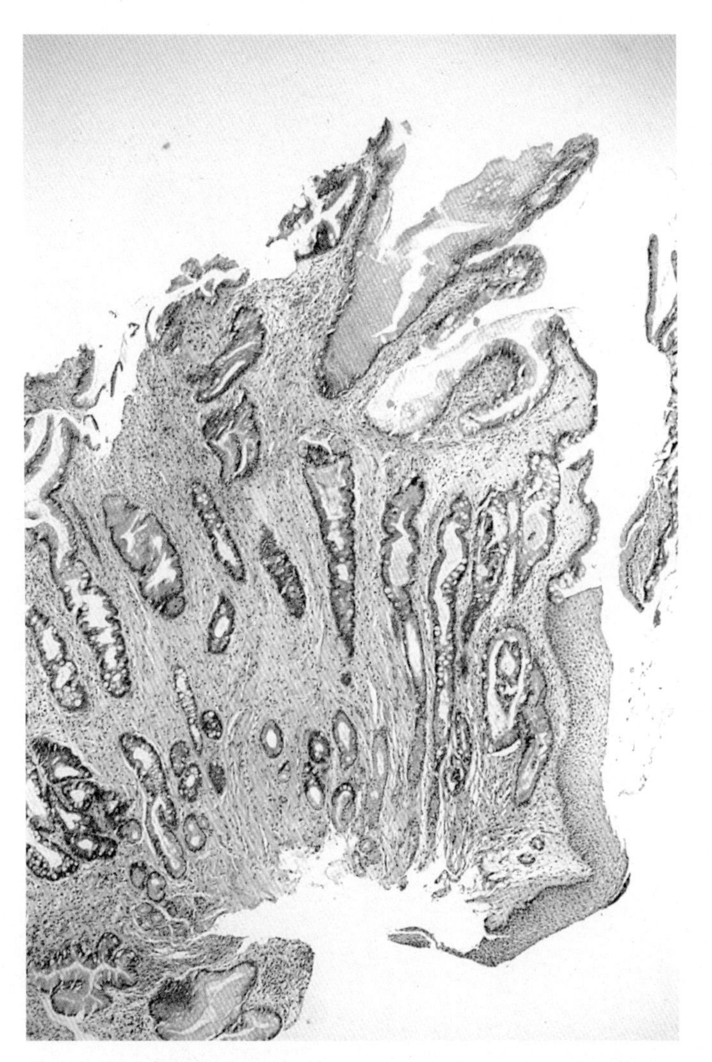

Figure 11–6 A

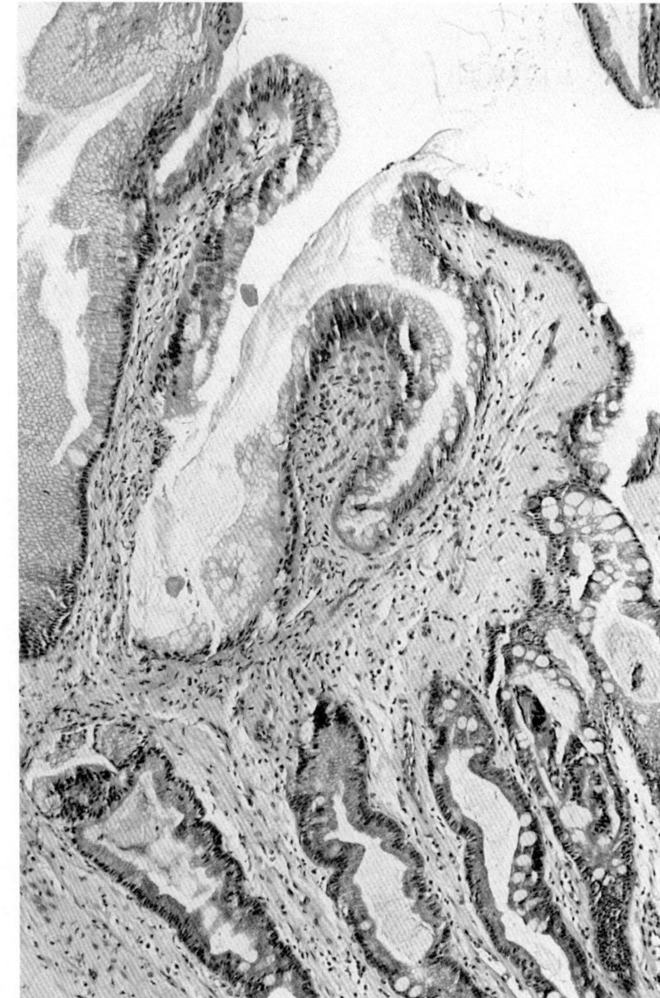

Figure 11–6 B

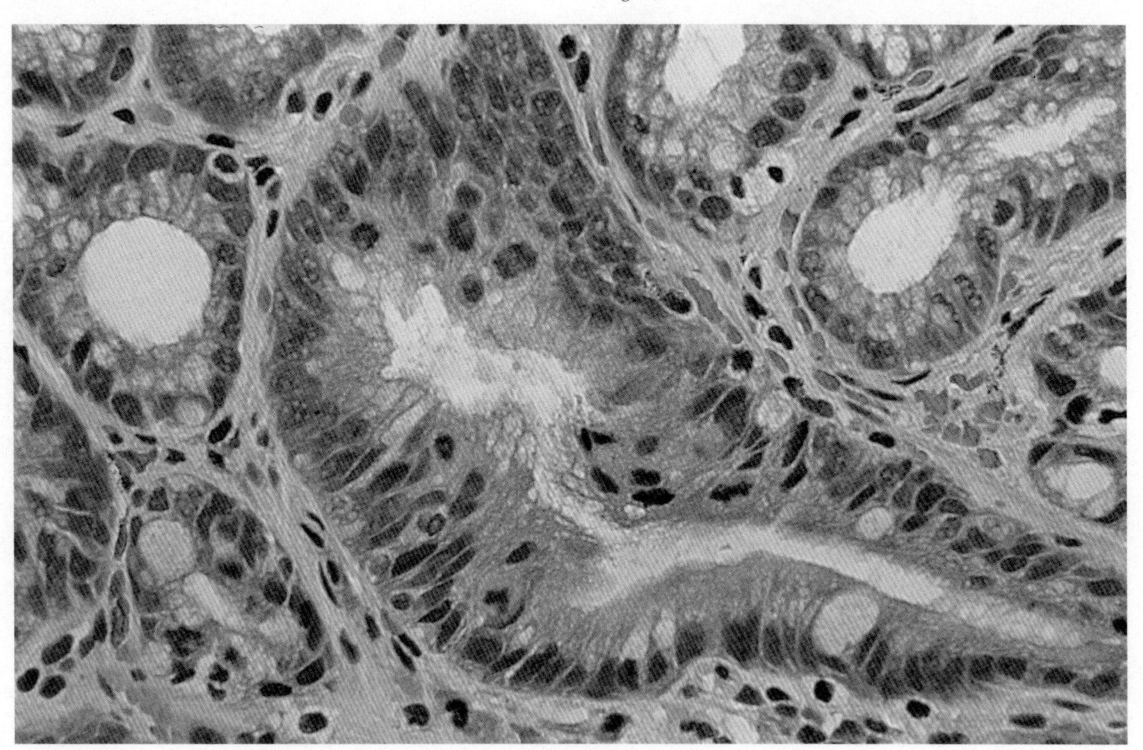

Figure 11–7

■ ESOPHAGUS

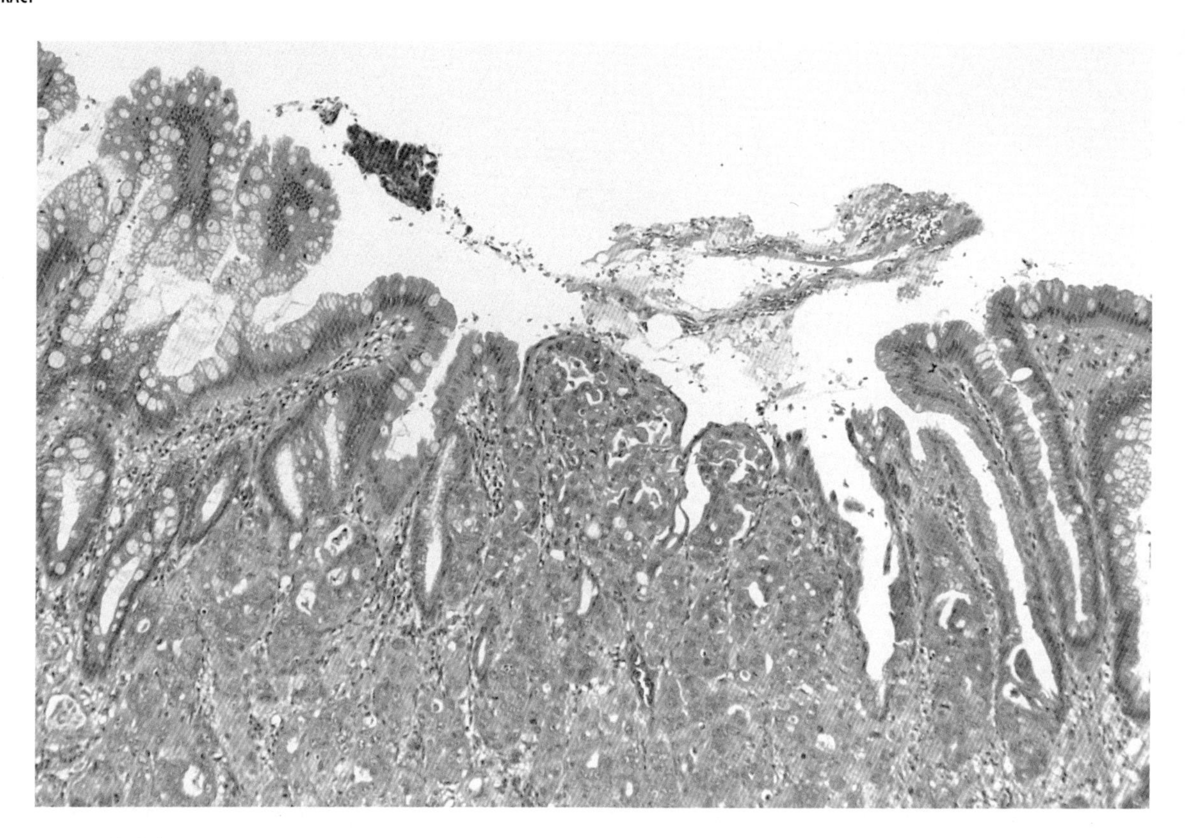

Figure 11–8 A

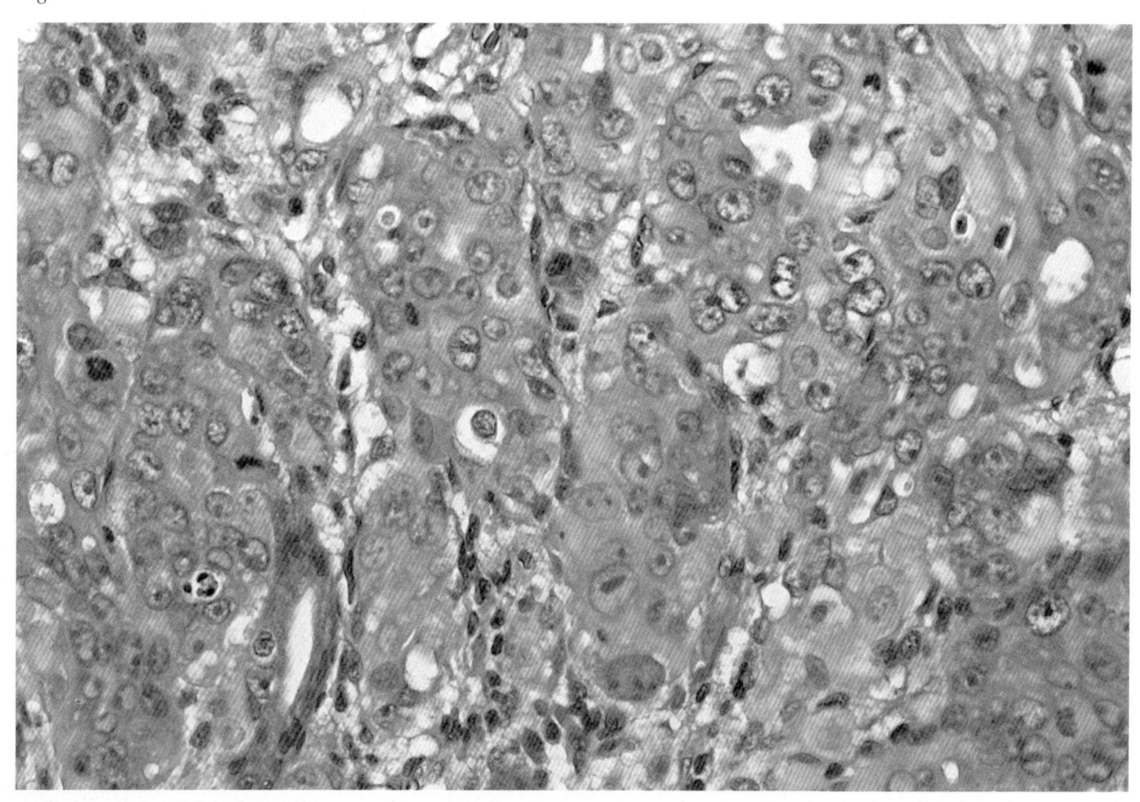

Figure 11–8 B

FIGURE 11–8. ADENOCARCINOMA: A: This is a photomicrograph of an infiltrating adenocarcinoma arising in Barrett's esophagus. **B:** The adenocarcinoma is moderately to poorly differentiated and retains a glandular pattern. There is cellular pleomorphism, nuclear pleomorphism and hyperchromasia, prominent nucleoli, and increased numbers of mitotic figures.

■ STOMACH

Figure 11–9 A

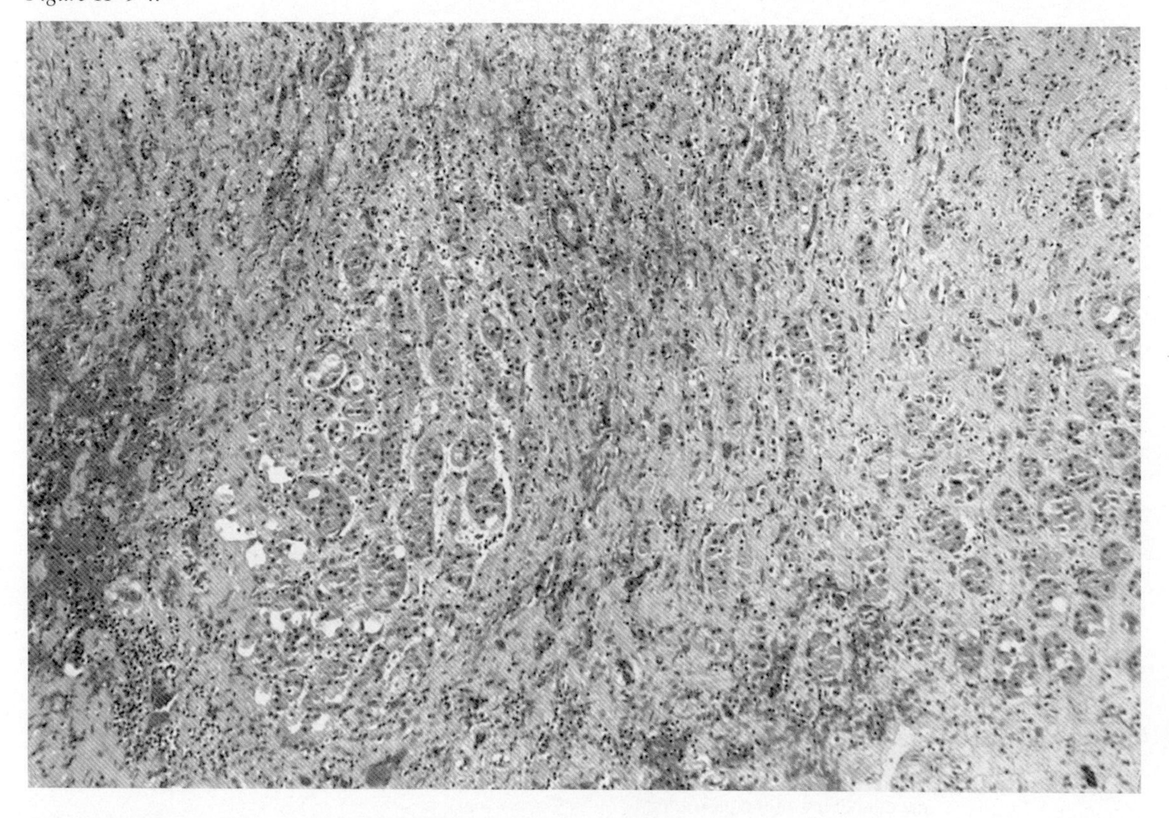

Figure 11–9 B

FIGURE 11–9. ACUTE HEMORRHAGIC GASTRITIS (STRESS ULCERS): A and **B**: In acute hemorrhagic gastritis, there is loss of the superficial epithelium that extends into the gastric pits, with associated hemorrhage into the mucosa.

■ STOMACH

FIGURE 11–10. *HELICOBACTER PYLORI: Helicobacter pylori* are seagull-shaped bacilli. This is a Steiner-stained photomicrograph showing the bacilli in the mucus overlying the epithelium of the stomach.

There are several histopathologic patterns that occur secondary to infection by *Helicobacter pylori.* Two examples are shown here.

FIGURE 11–11. FOLLICULAR GASTRITIS: Follicular gastritis is characterized by prominent lymphoid follicles in the mucosa and a chronic inflammatory infiltrate in the lamina propria.

FIGURE 11–12. CHRONIC ACTIVE GASTRITIS: Chronic active gastritis is characterized by an intense inflammatory infiltrate in the lamina propria and the epithelium of the gastric pits. It can be seen that the infiltrate in the lamina propria is composed of lymphocytes, plasma cells, and few eosinophils. Neutrophils are present in the epithelium of the gastric pits.

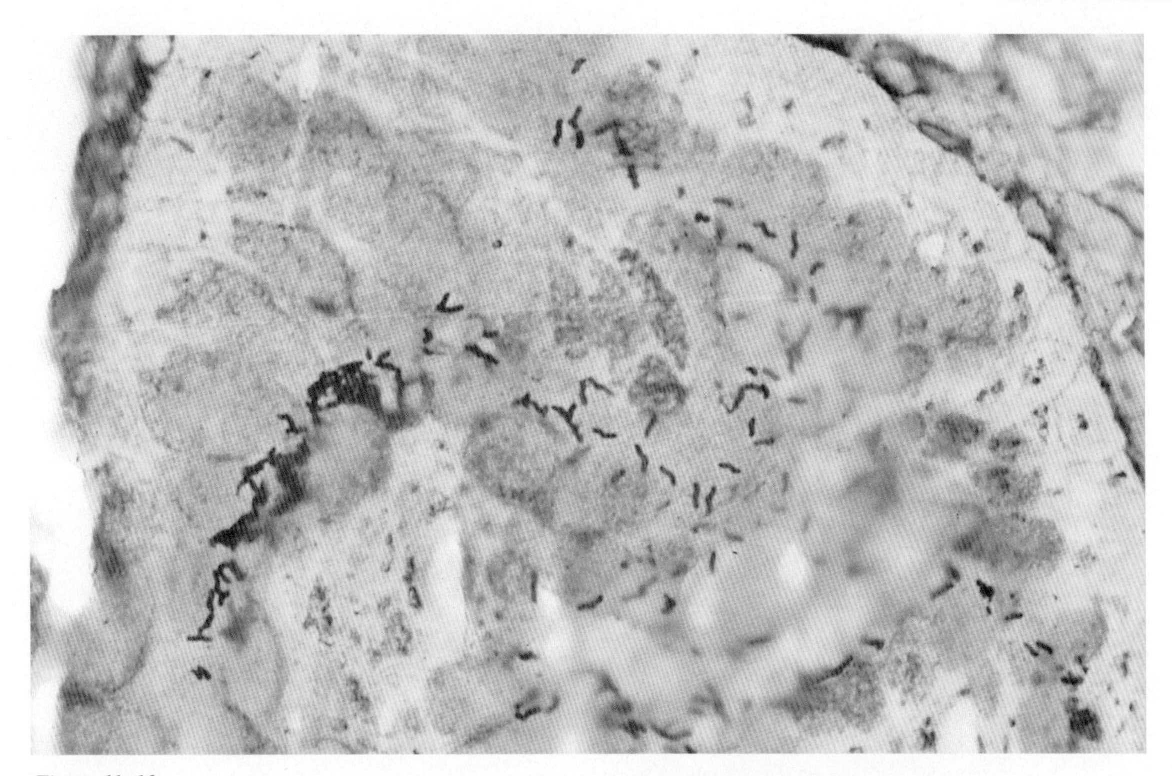

Figure 11–10

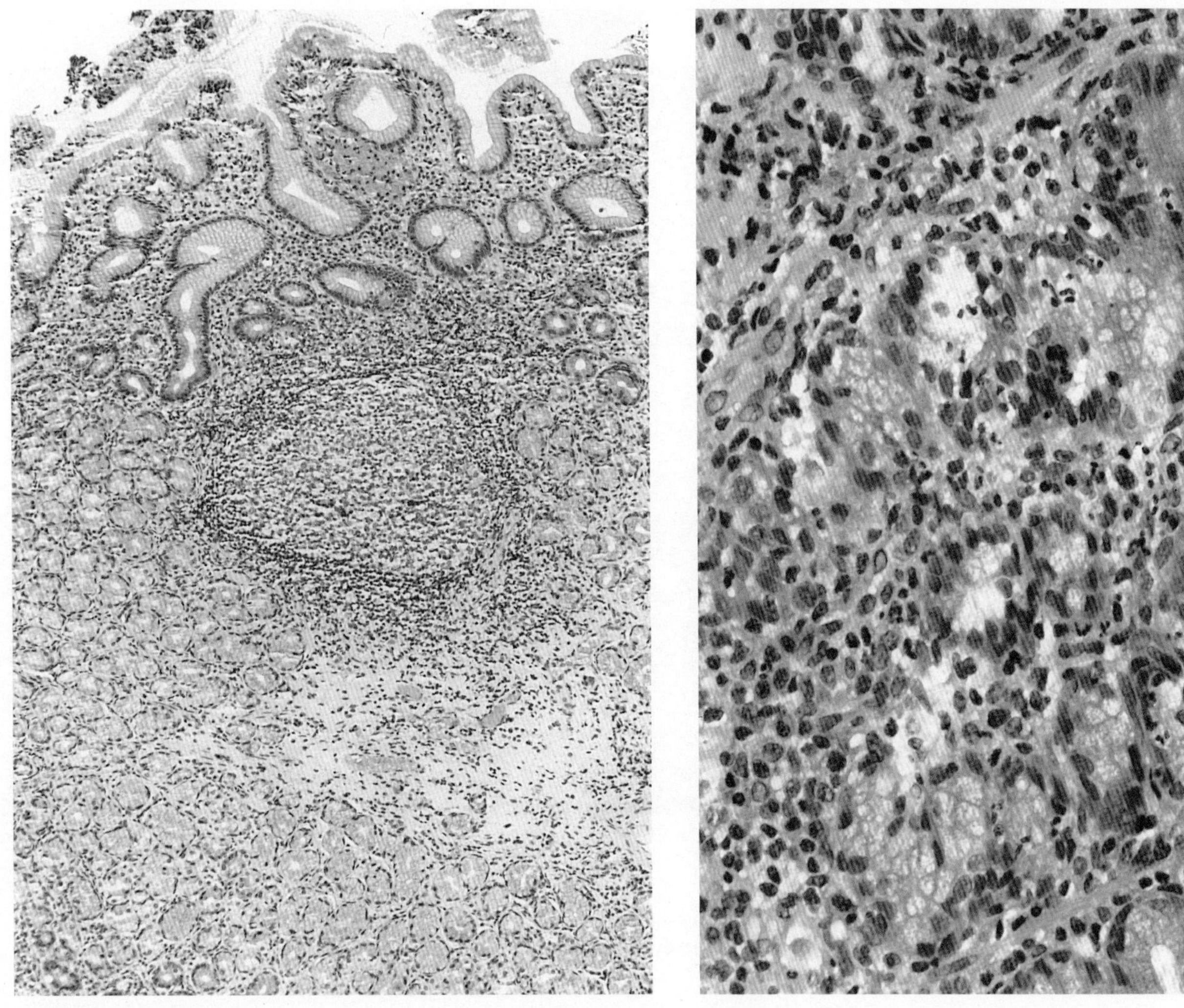

Figure 11–11

Figure 11–12

■ STOMACH

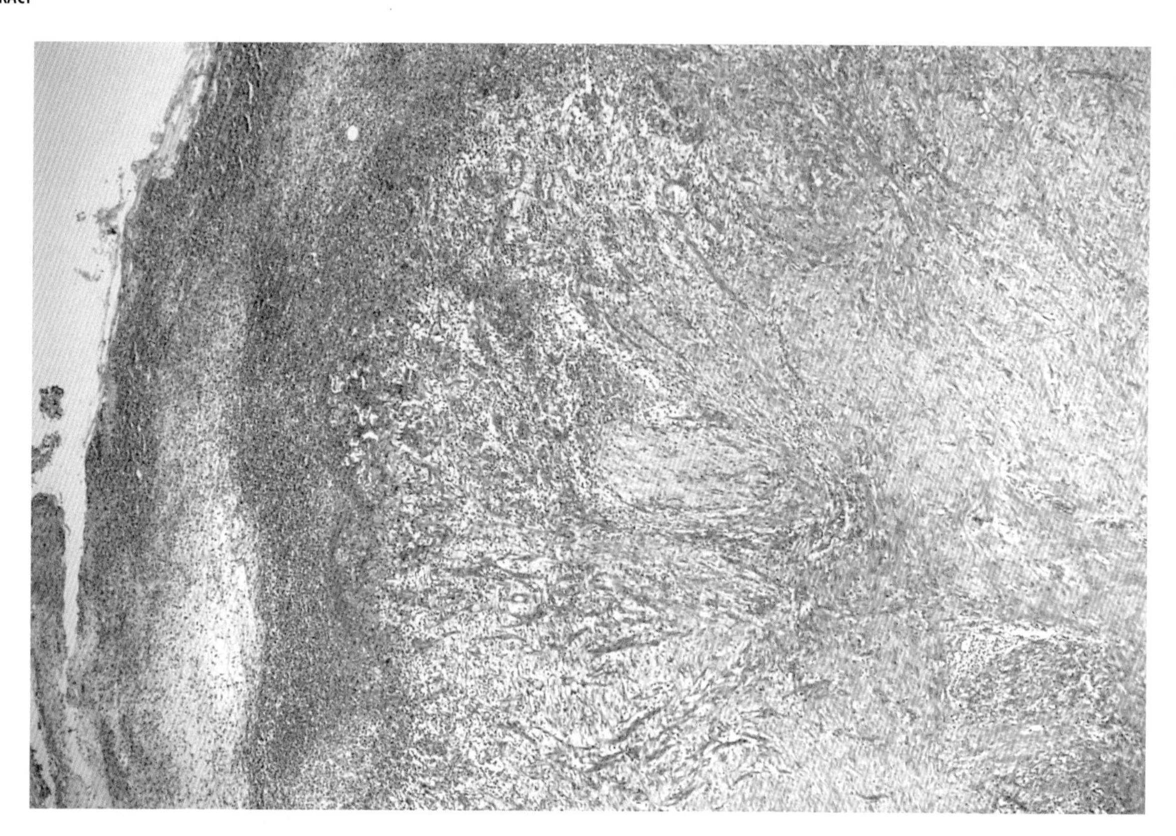

Figure 11–13 **A**

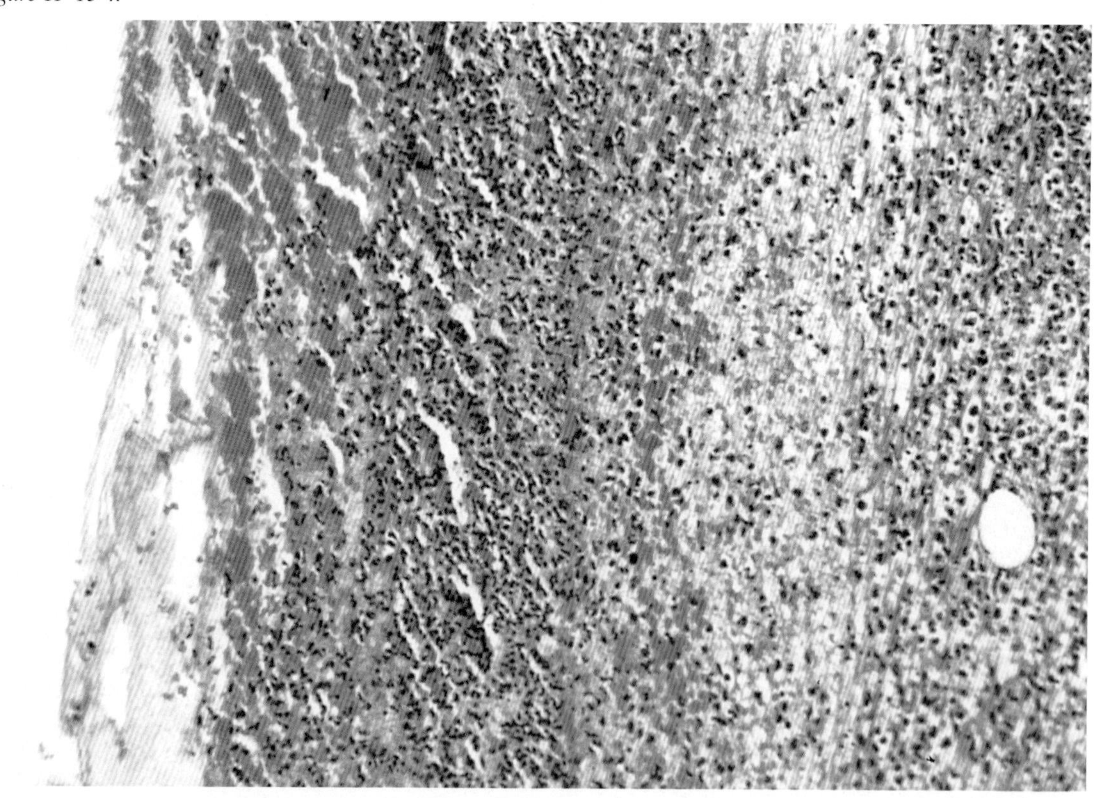

Figure 11–13 **B**

FIGURE 11–13. GASTRIC ULCER (PEPTIC ULCER): A: Chronic active ulcers are characterized by four distinct zones. **B:** The first two zones can be seen. The ulcer base is covered by fibrinopurulent debris (zone 1), beneath which is an acute suppurative inflammatory infiltrate (zone 2).

■ STOMACH

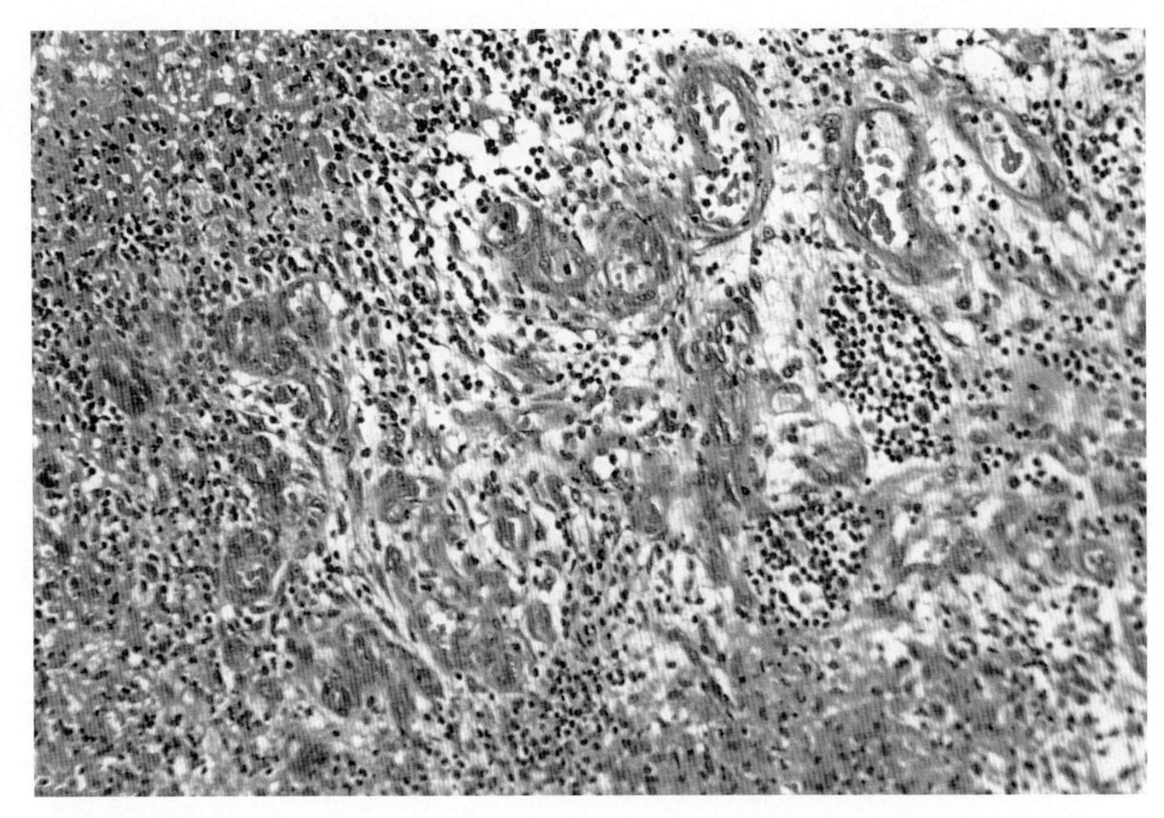

Figure 11–13 C

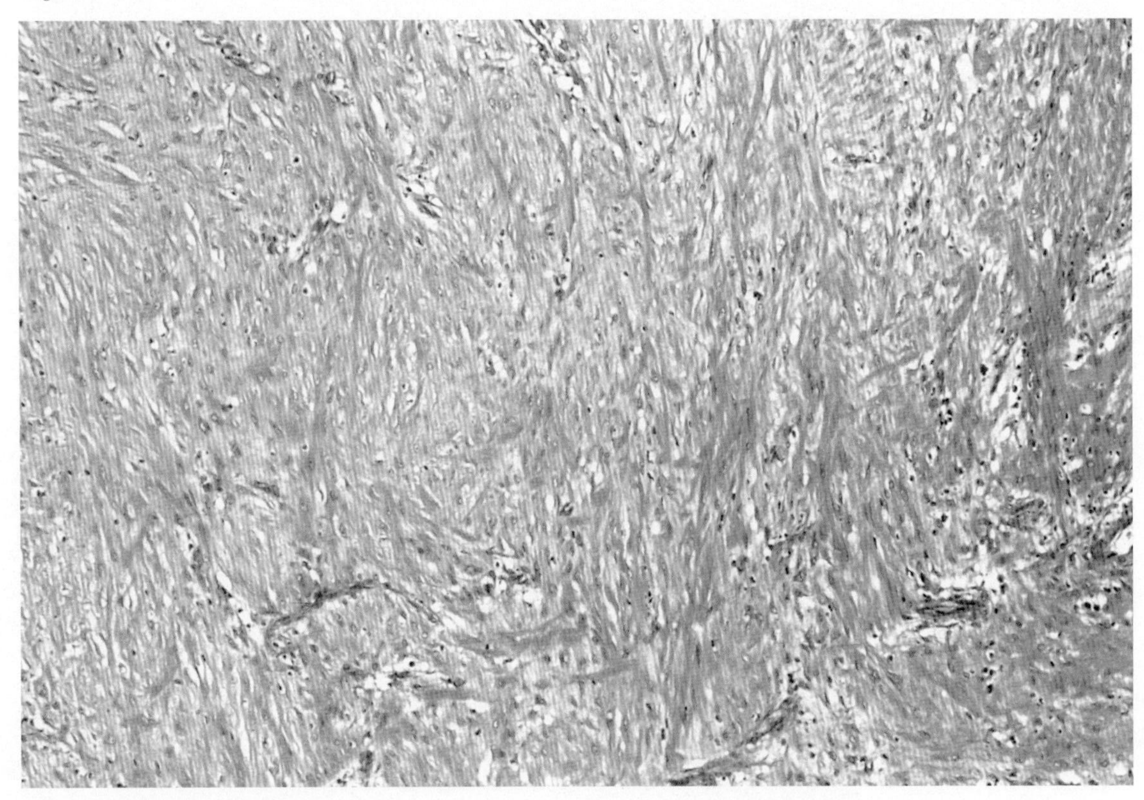

Figure 11–13 D

FIGURE 11–13. GASTRIC ULCER (PEPTIC ULCER): The third zone of a chronic active ulcer is composed of granulation tissue (**C**), and the fourth zone is composed of dense fibrous tissue (**D**).

■ STOMACH

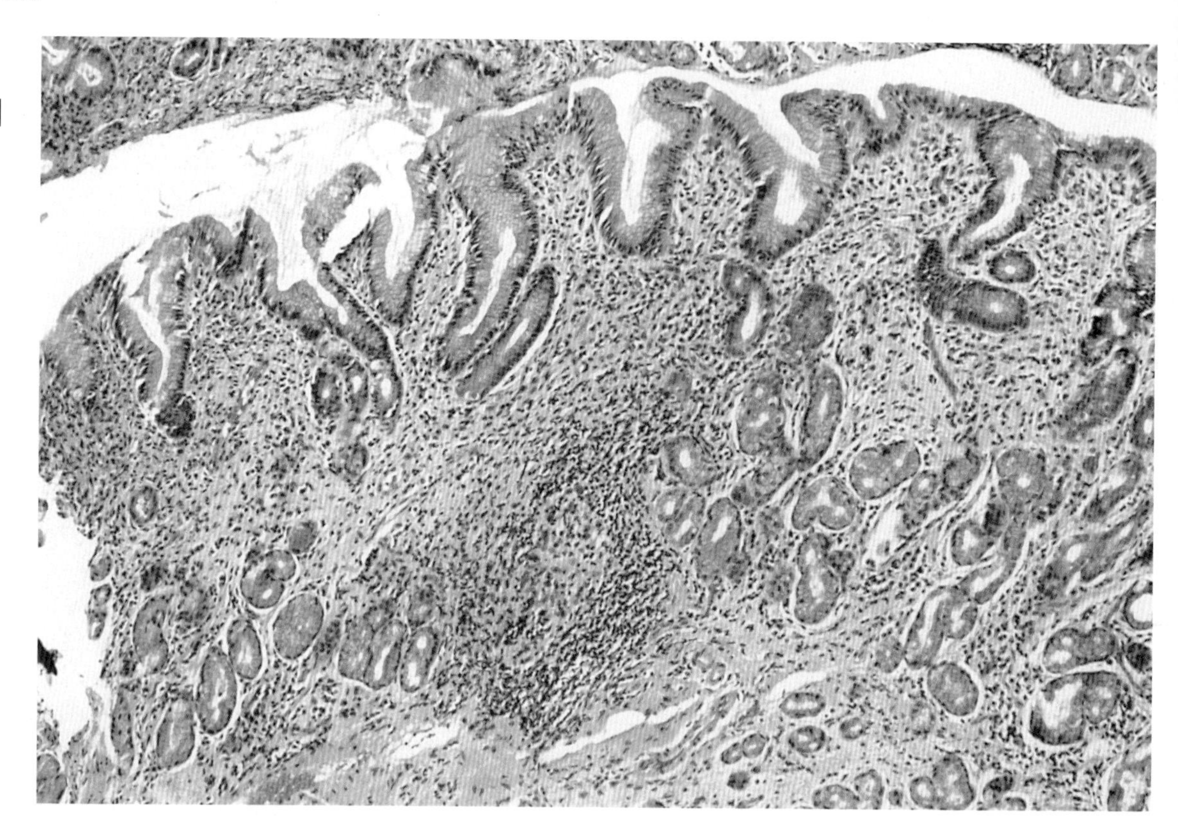

FIGURE 11–14. CHRONIC ATROPHIC GASTRITIS: Chronic atrophic gastritis is characterized by a decreased number of gastric glands associated with an interglandular inflammatory infiltrate.[1]

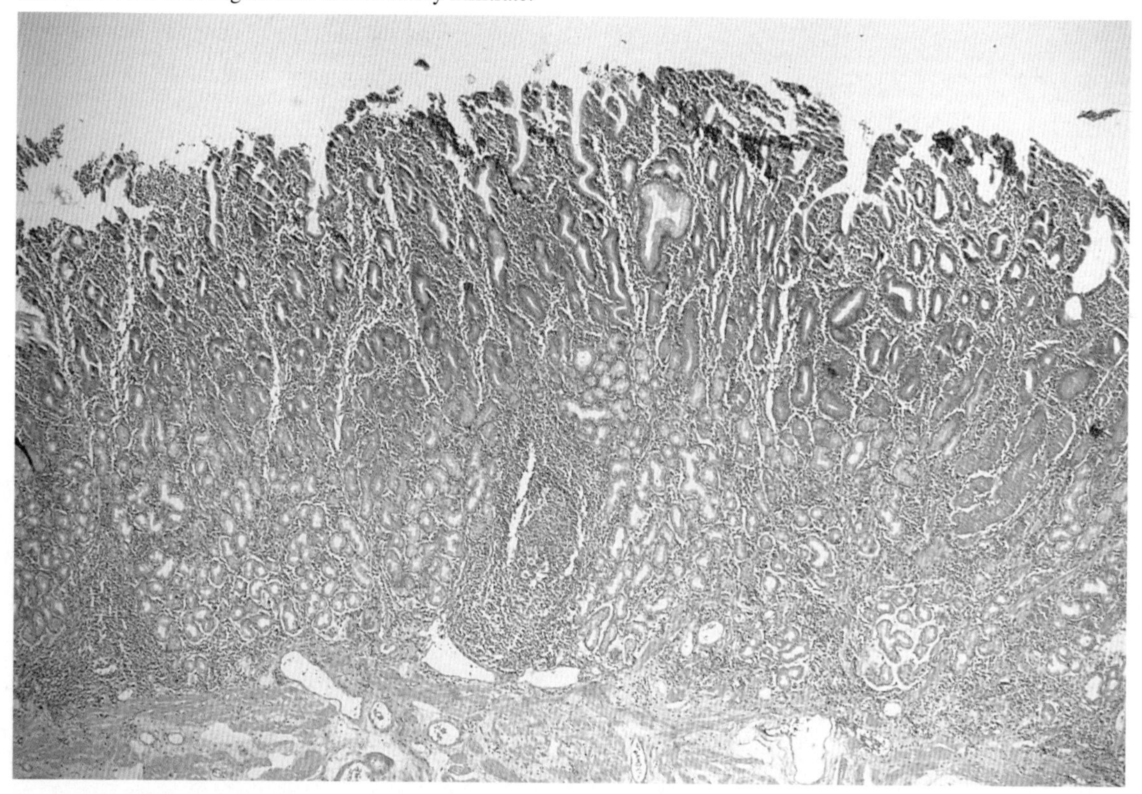

FIGURE 11–15. HYPERTROPHIC GASTRITIS: In this photomicrograph of hypertrophic gastritis, there is hyperplasia of the gastric glands with an associated mixed inflammatory infiltrate.[1]

[1]*Note: Chronic atrophic gastritis, final magnification 107.5 × versus hypertrophic gastritis, final magnification 43 ×.*

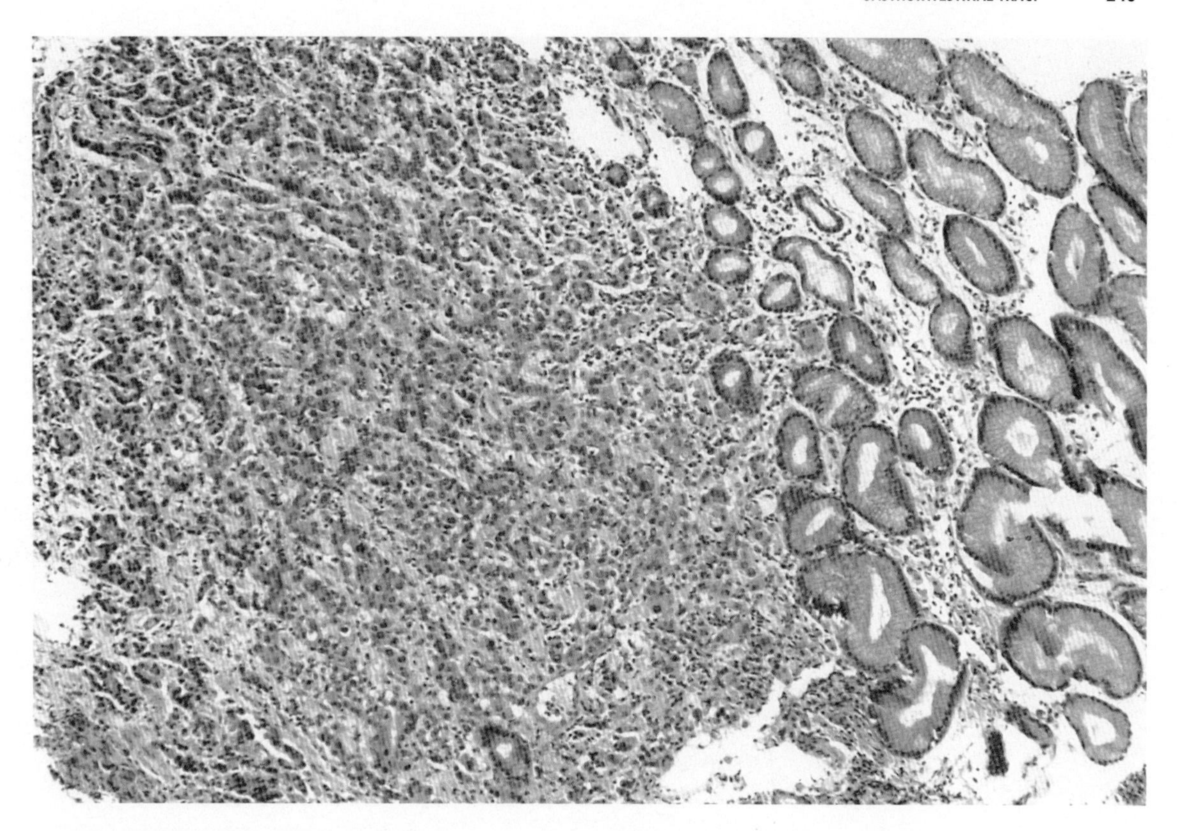

Figure 11–16 A

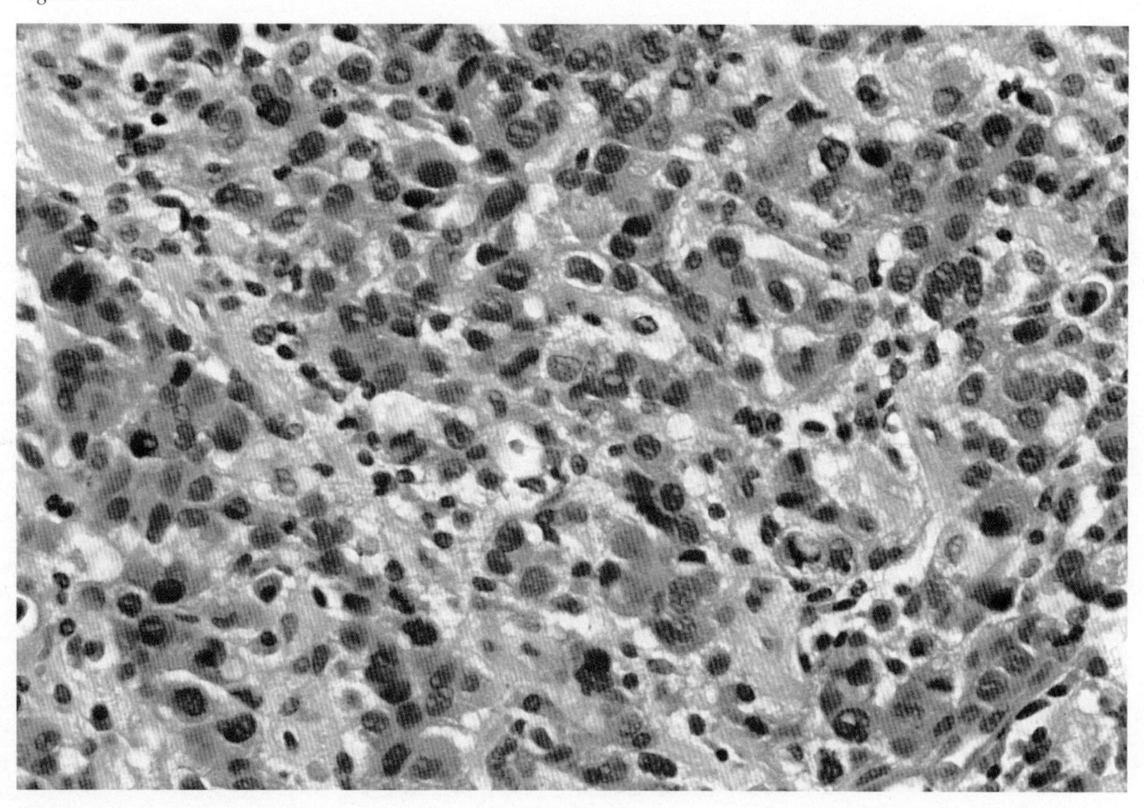

Figure 11–16 B

FIGURE 11–16. ADENOCARCINOMA: Adenocarcinoma of the stomach can grow in a polypoid, ulcerative, or infiltrative pattern. A: This is an example of an infiltrative **linitis plastica**–type adenocarcinoma that produces a stiff **"water-bottle"** stomach and no discernible intraluminal mass. B: This tumor is poorly differentiated with few ill-defined glands. The cells are anaplastic with extensive pleomorphism. Nuclei are enlarged and hyperchromatic. Mitotic figures are abundant.

■ STOMACH

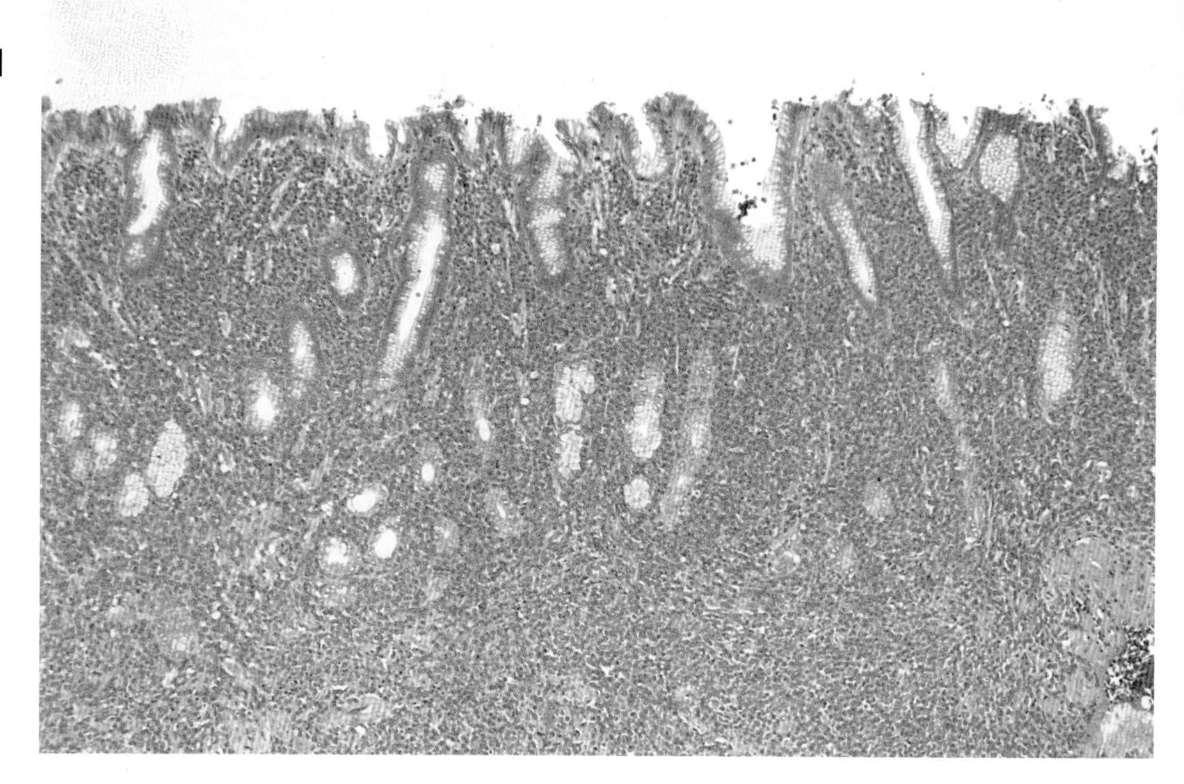

Figure 11–17 A

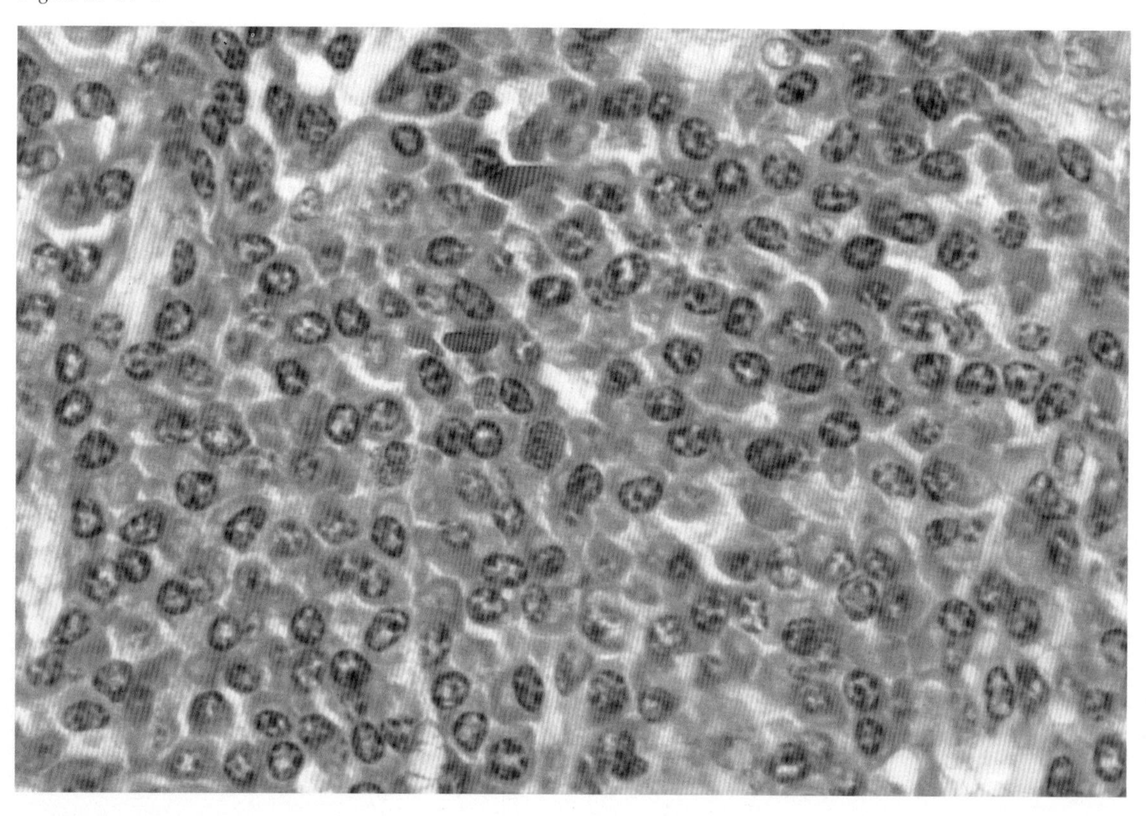

Figure 11–17 B

FIGURE 11–17. LYMPHOMA: The stomach is the most common site for gastrointestinal lymphomas. Most are high-grade non-Hodgkin's lymphomas. **A** and **B**: These are examples of a well-differentiated lymphocytic lymphoma, plasmacytoid variant, in which the mucosa and submucosa are infiltrated by neoplastic lymphocytes.

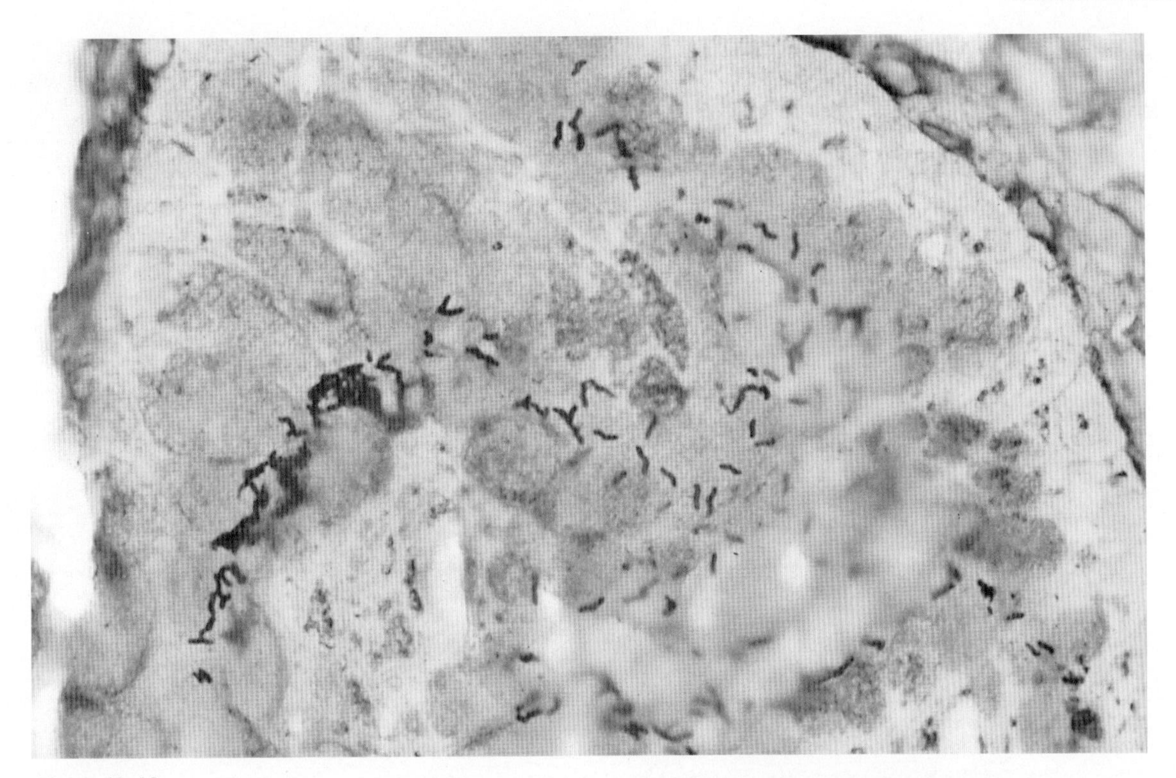

Figure 11–10

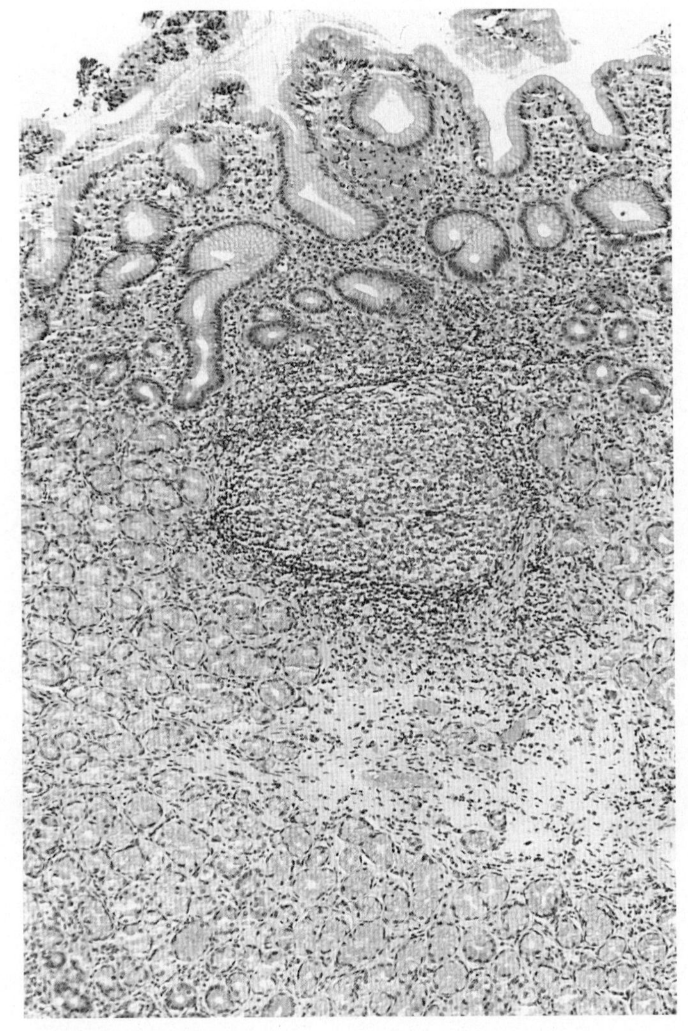

Figure 11–11

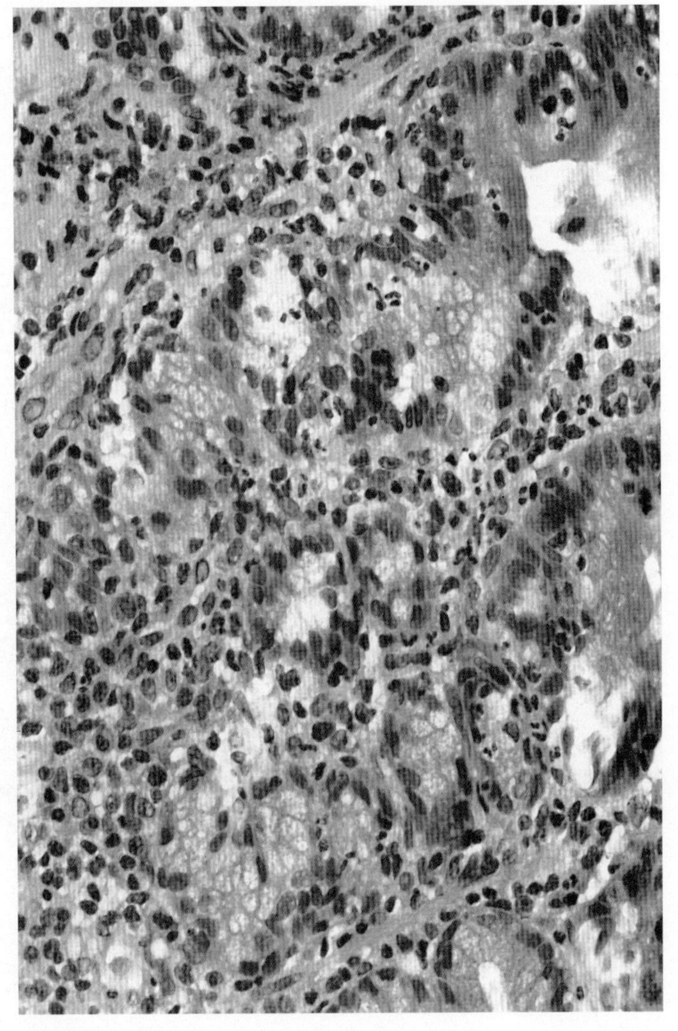

Figure 11–12

■ STOMACH

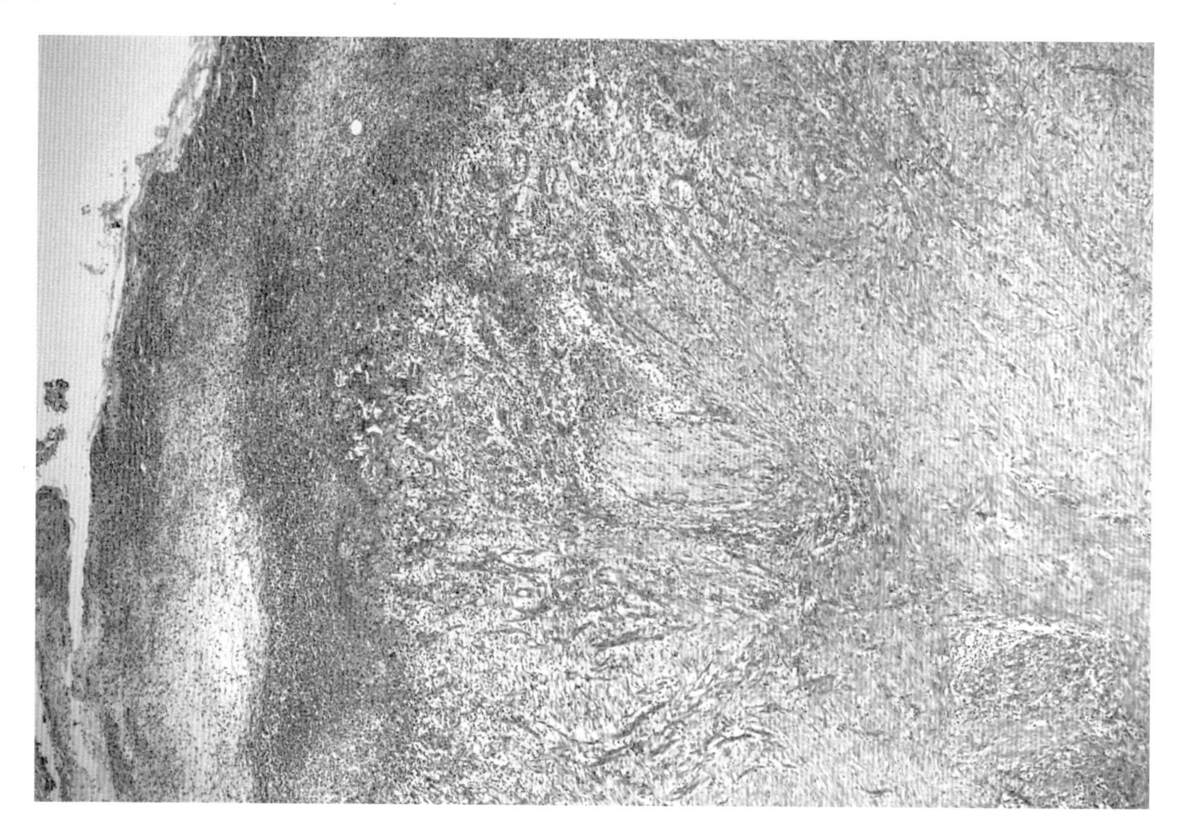

Figure 11–13 **A**

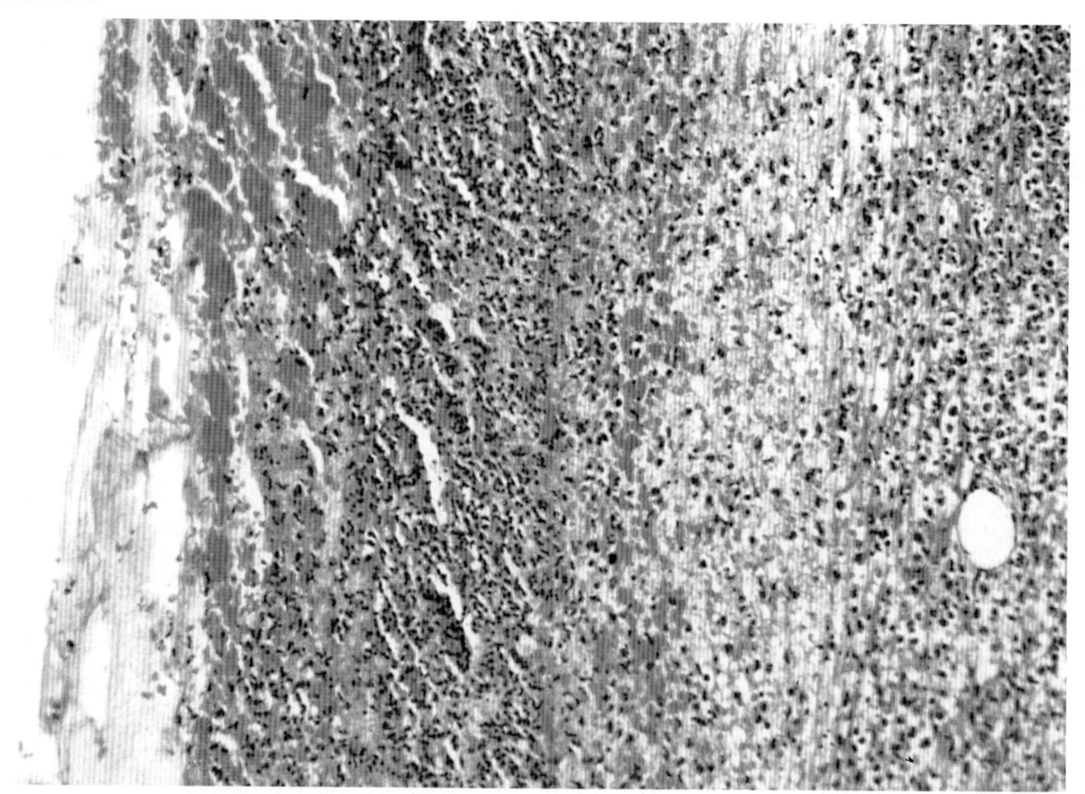

Figure 11–13 **B**

FIGURE 11–13. GASTRIC ULCER (PEPTIC ULCER): A: Chronic active ulcers are characterized by four distinct zones. **B:** The first two zones can be seen. The ulcer base is covered by fibrinopurulent debris (zone 1), beneath which is an acute suppurative inflammatory infiltrate (zone 2).

■ STOMACH

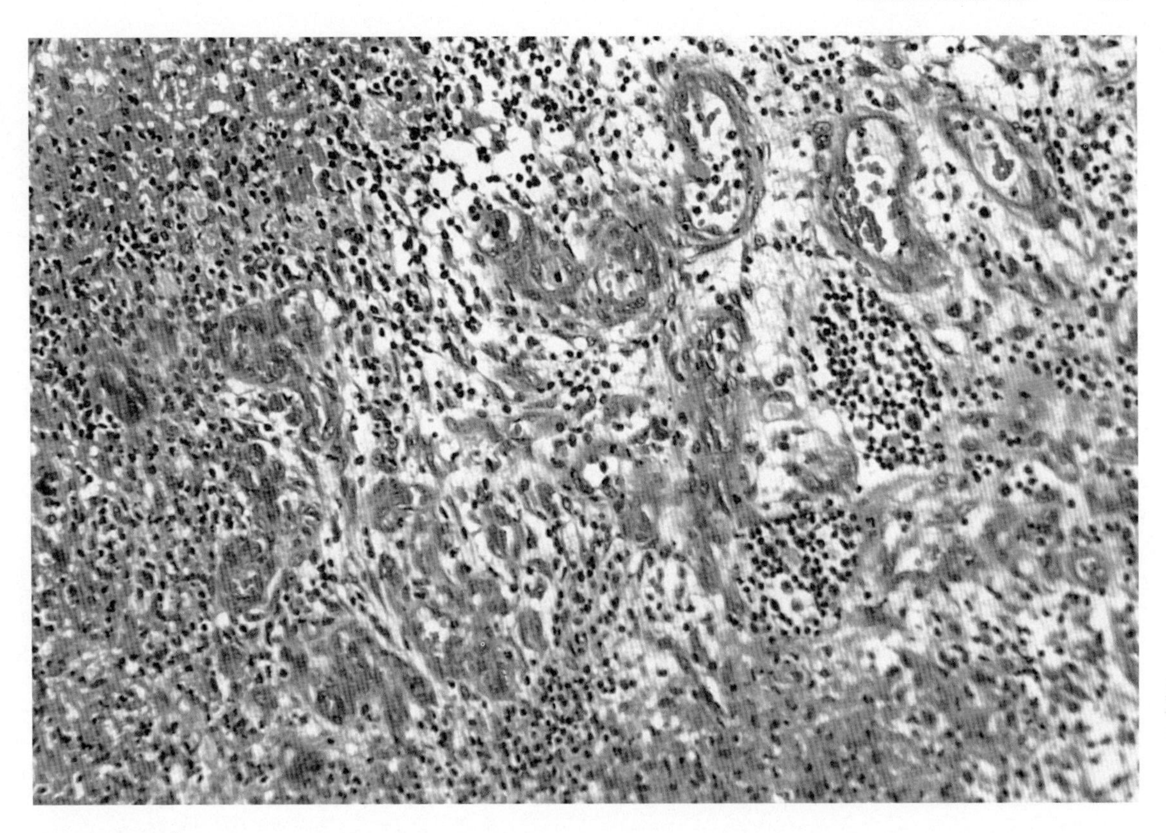

Figure 11–13 **C**

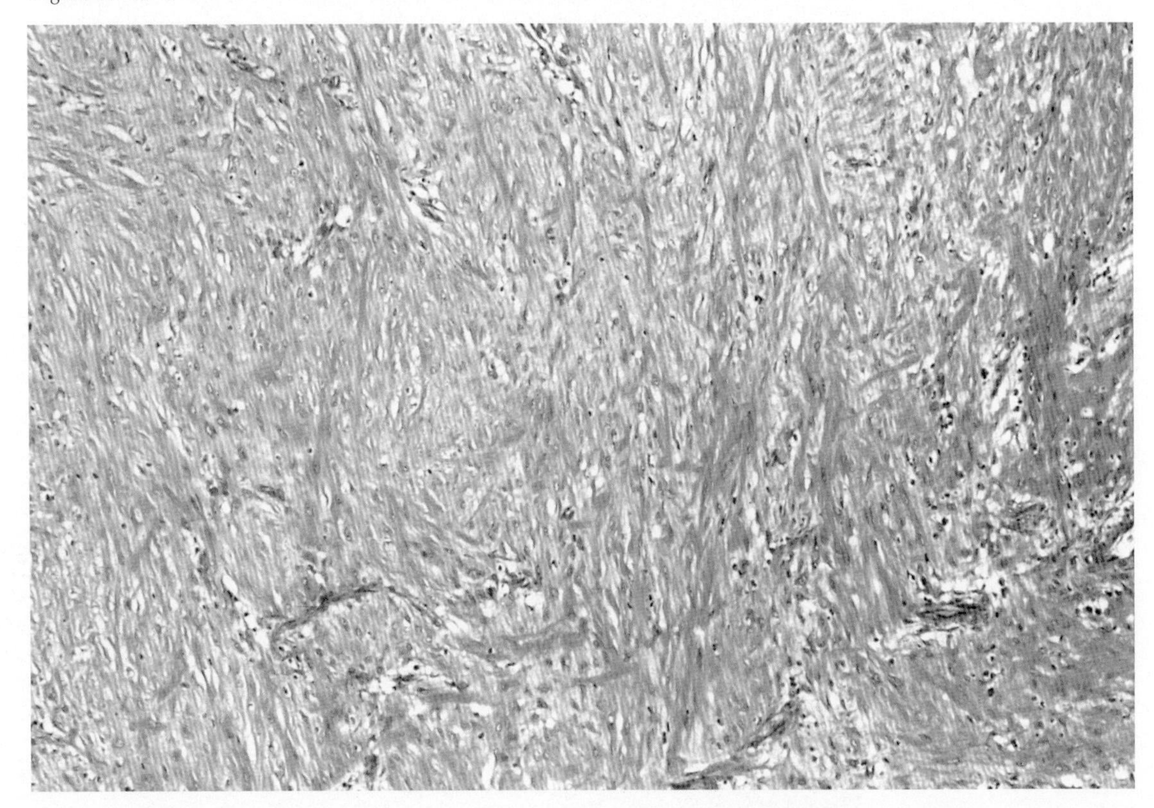

Figure 11–13 **D**

FIGURE 11–13. GASTRIC ULCER (PEPTIC ULCER): The third zone of a chronic active ulcer is composed of granulation tissue (**C**), and the fourth zone is composed of dense fibrous tissue (**D**).

■ STOMACH

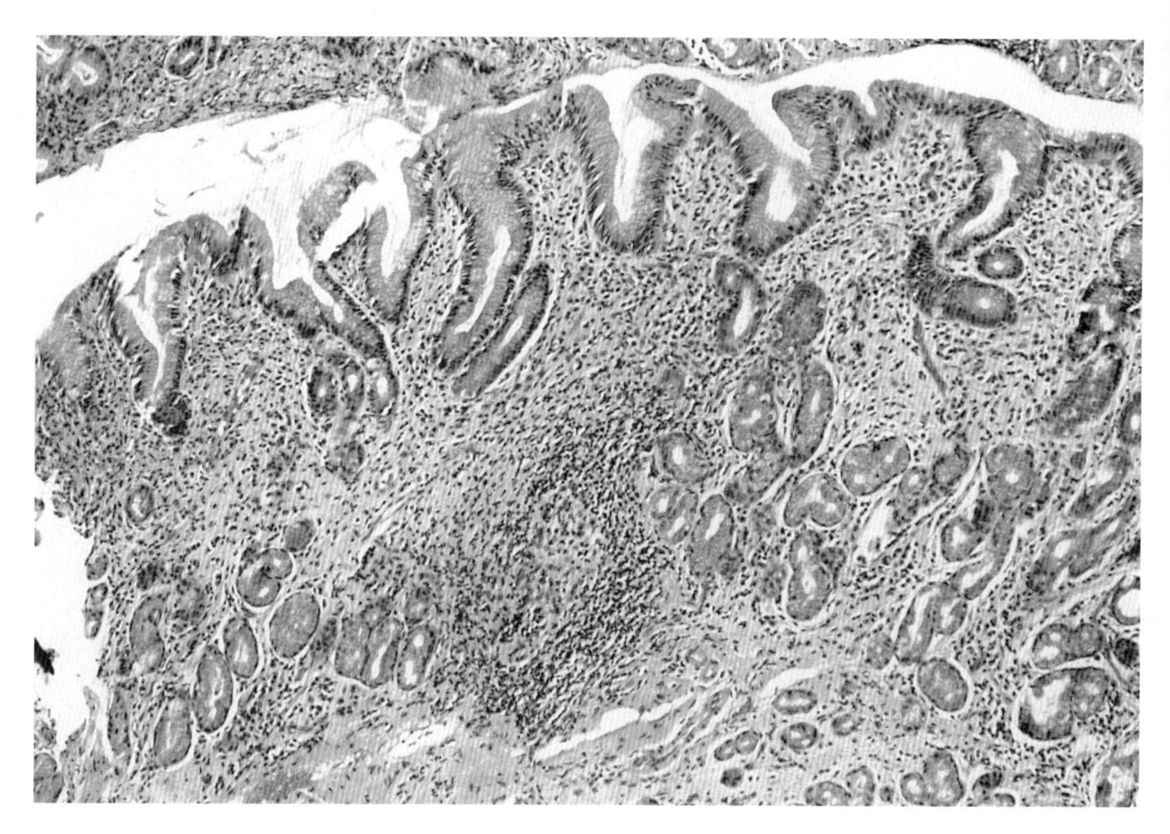

FIGURE 11–14. CHRONIC ATROPHIC GASTRITIS: Chronic atrophic gastritis is characterized by a decreased number of gastric glands associated with an interglandular inflammatory infiltrate.[1]

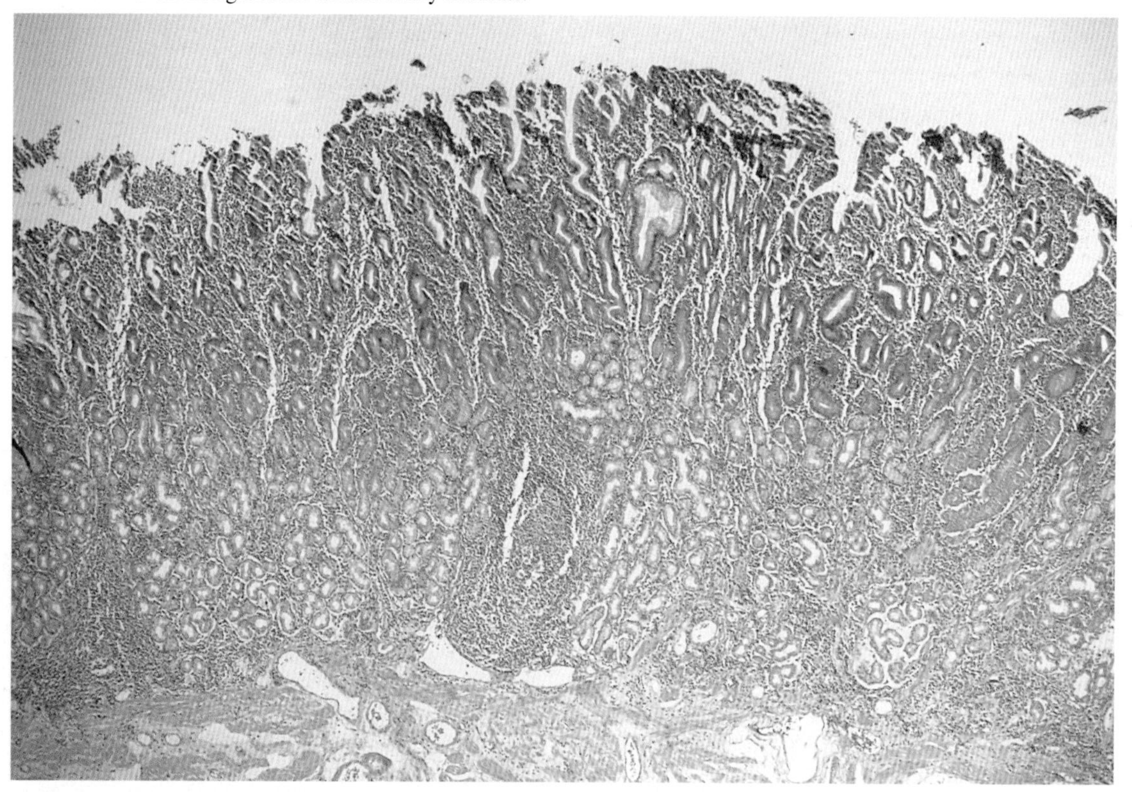

FIGURE 11–15. HYPERTROPHIC GASTRITIS: In this photomicrograph of hypertrophic gastritis, there is hyperplasia of the gastric glands with an associated mixed inflammatory infiltrate.[1]

[1] *Note: Chronic atrophic gastritis, final magnification 107.5 × versus hypertrophic gastritis, final magnification 43 ×.*

■ STOMACH

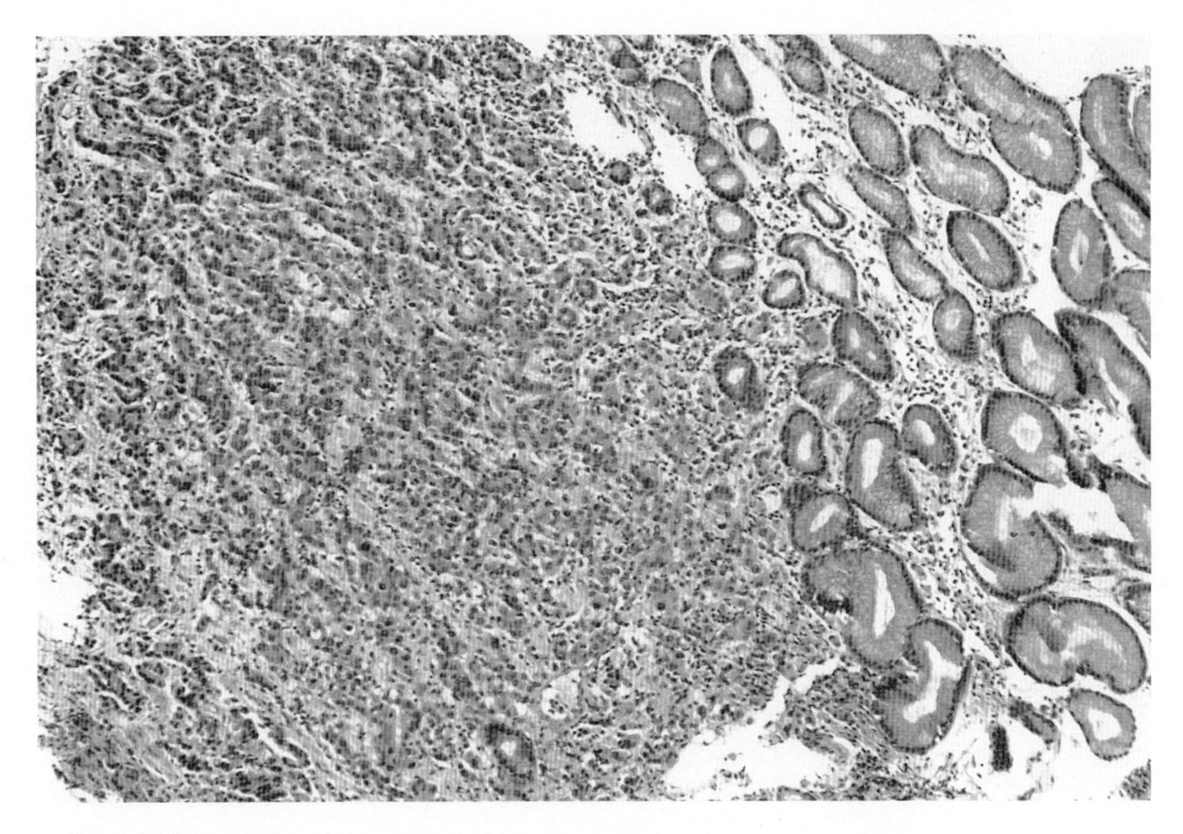

Figure 11–16 A

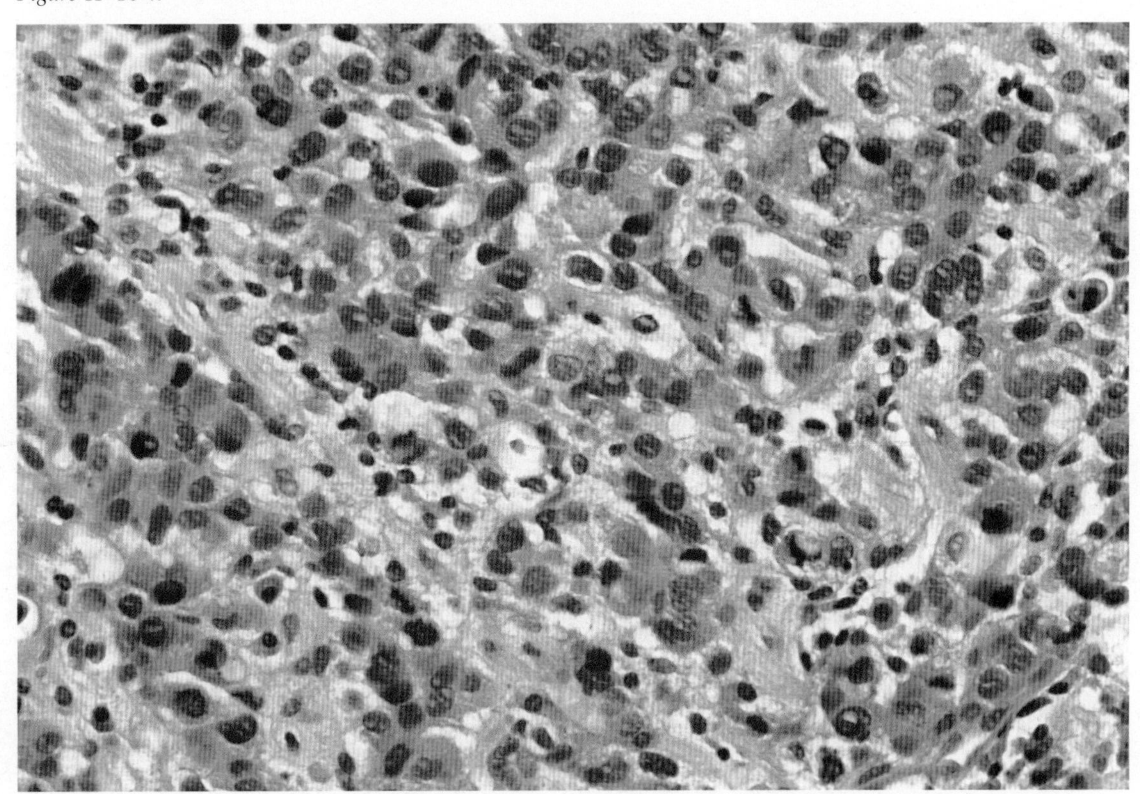

Figure 11–16 B

FIGURE 11–16. ADENOCARCINOMA: Adenocarcinoma of the stomach can grow in a polypoid, ulcerative, or infiltrative pattern. A: This is an example of an infiltrative **linitis plastica**–type adenocarcinoma that produces a stiff **"water-bottle"** stomach and no discernible intraluminal mass. B: This tumor is poorly differentiated with few ill-defined glands. The cells are anaplastic with extensive pleomorphism. Nuclei are enlarged and hyperchromatic. Mitotic figures are abundant.

■ STOMACH

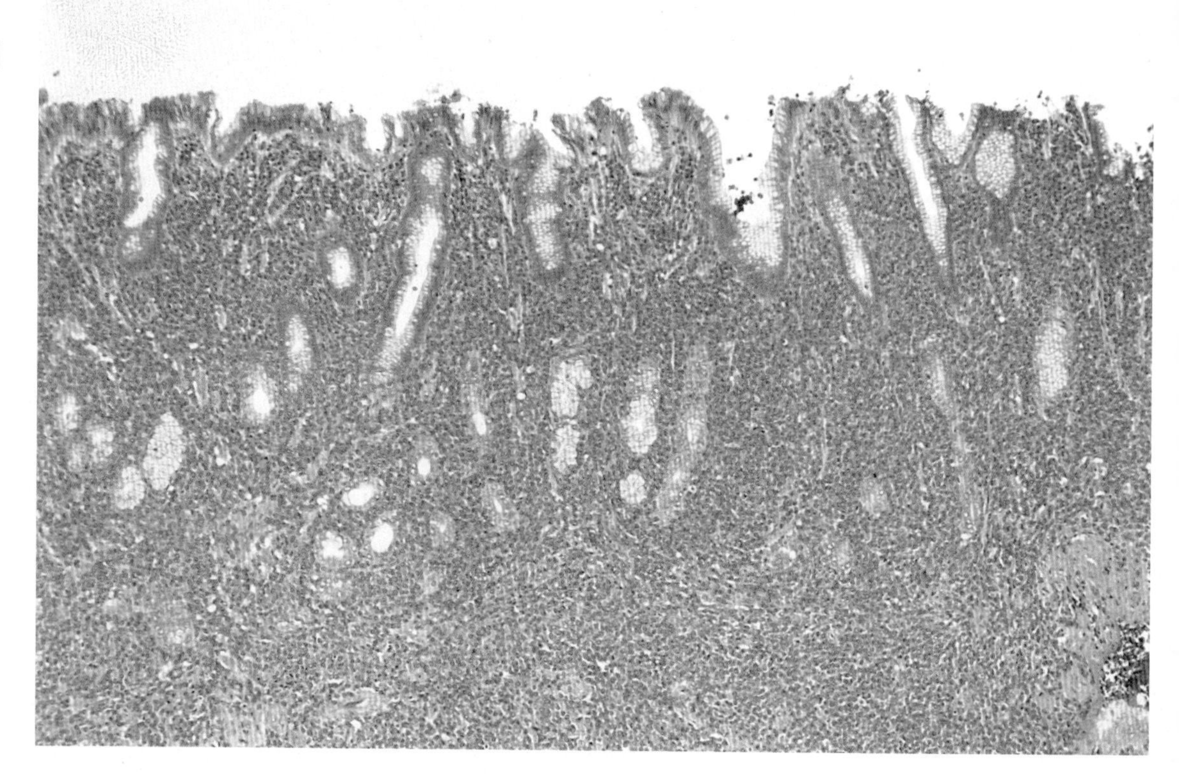

Figure 11–17 A

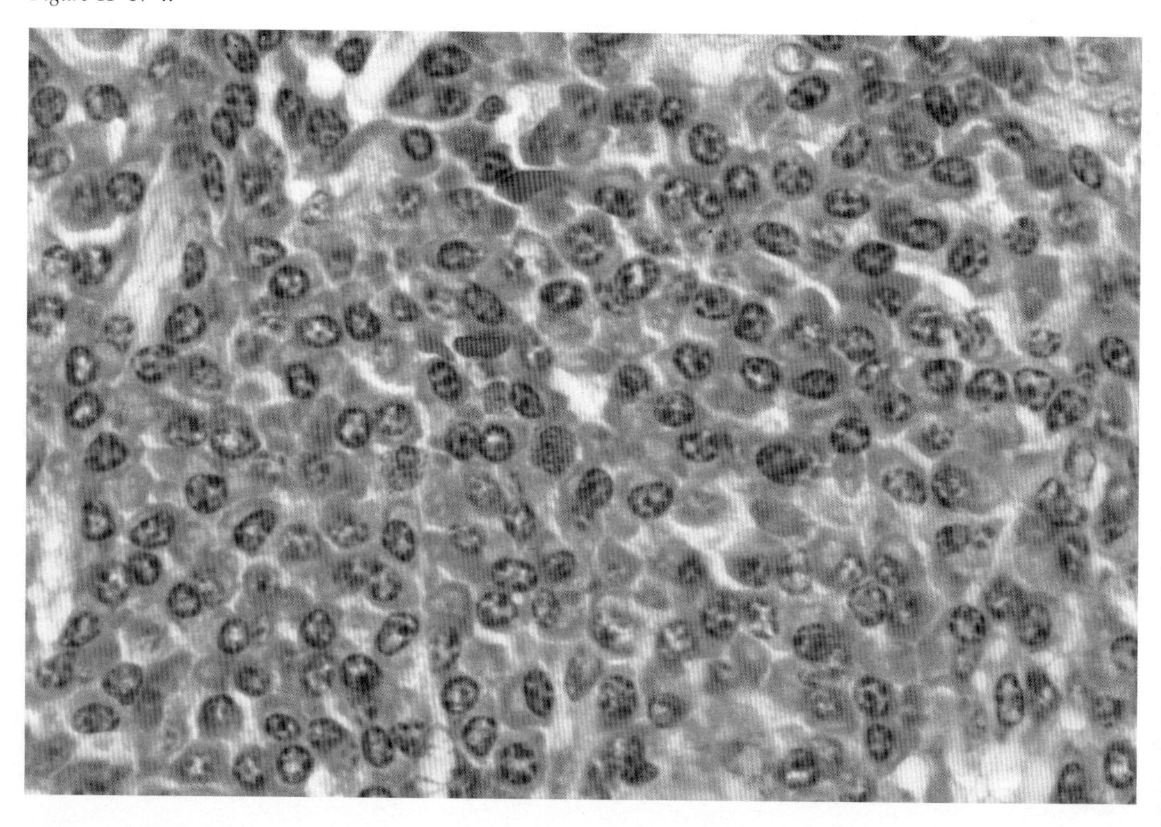

Figure 11–17 B

FIGURE 11–17. LYMPHOMA: The stomach is the most common site for gastrointestinal lymphomas. Most are high-grade non-Hodgkin's lymphomas. **A** and **B**: These are examples of a well-differentiated lymphocytic lymphoma, plasmacytoid variant, in which the mucosa and submucosa are infiltrated by neoplastic lymphocytes.

■ STOMACH

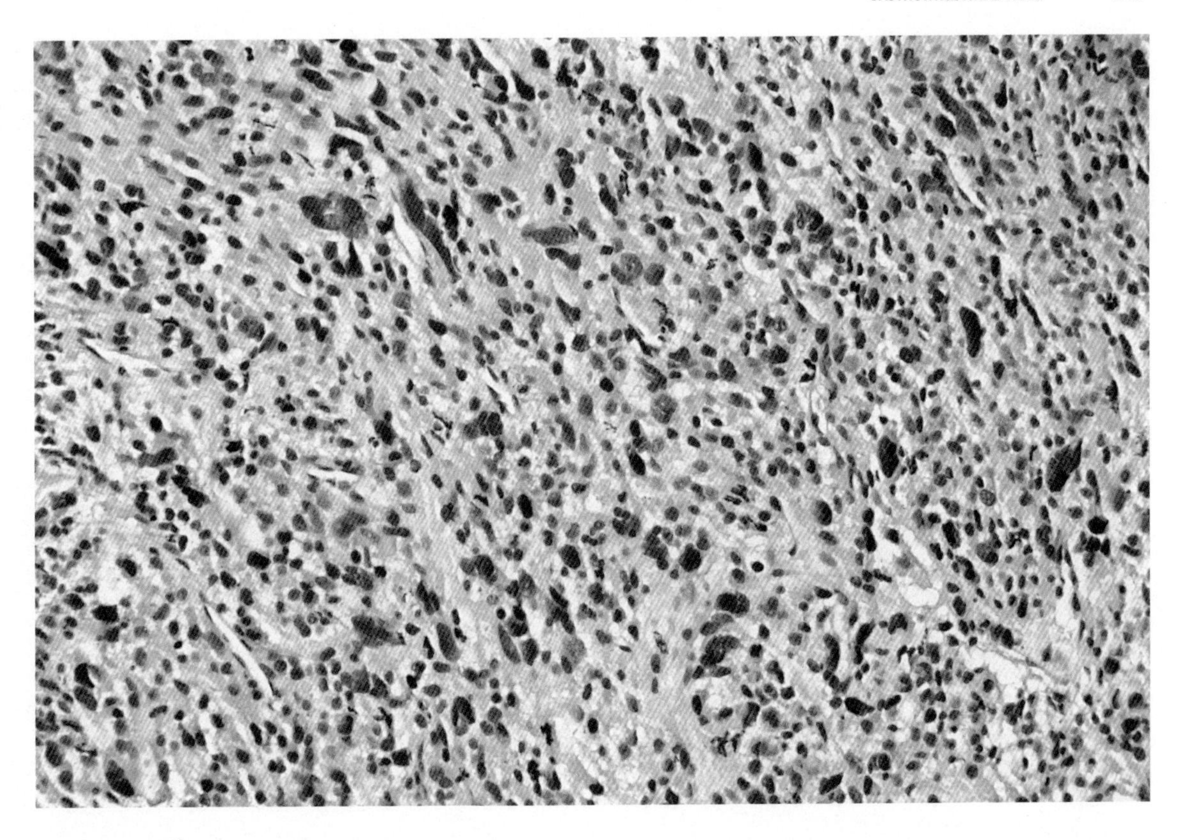

Figure 11–18 A

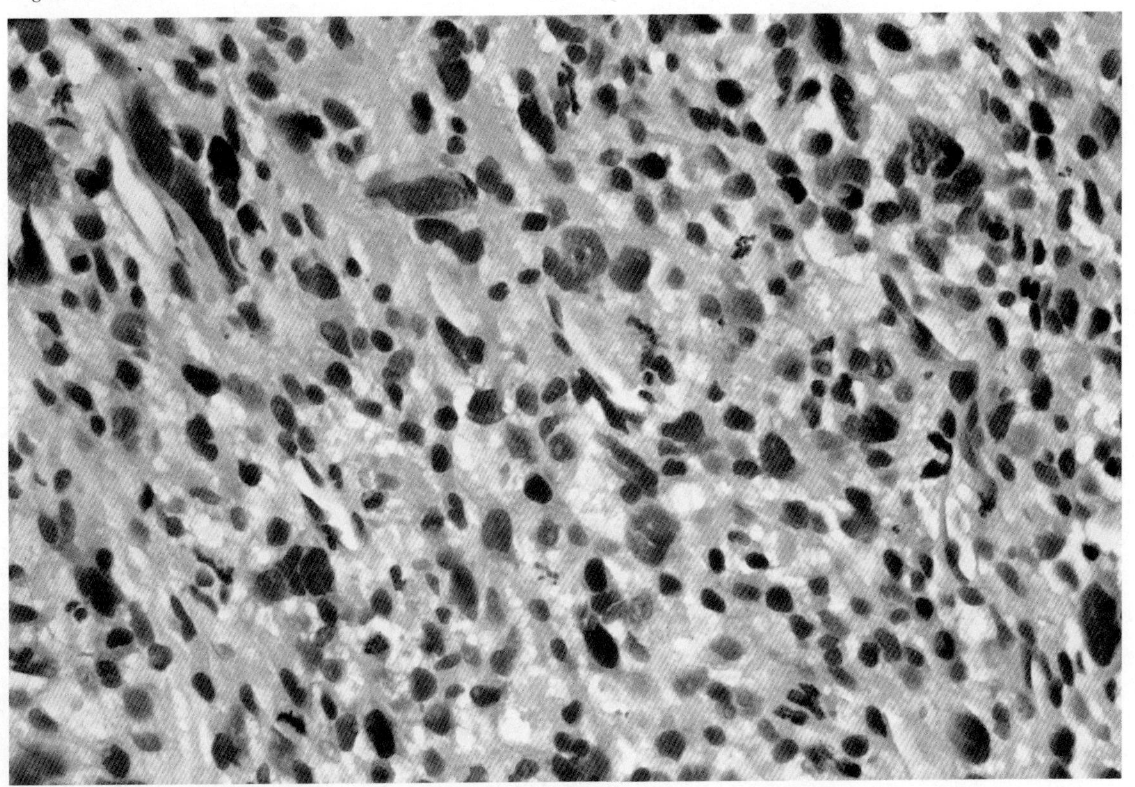

Figure 11–18 B

FIGURE 11–18. SARCOMAS: A high-grade leiomyosarcoma involving the stomach can be seen in **A** and **B**. **A:** The tumor is cellular and shows cellular and nuclear pleomorphism. **B:** Cells are anaplastic and haphazardly arranged, and some are multinucleated. Nuclei are enlarged, pleomorphic, and hyperchromatic. Nucleoli and mitotic figures are present.

■ SMALL INTESTINE

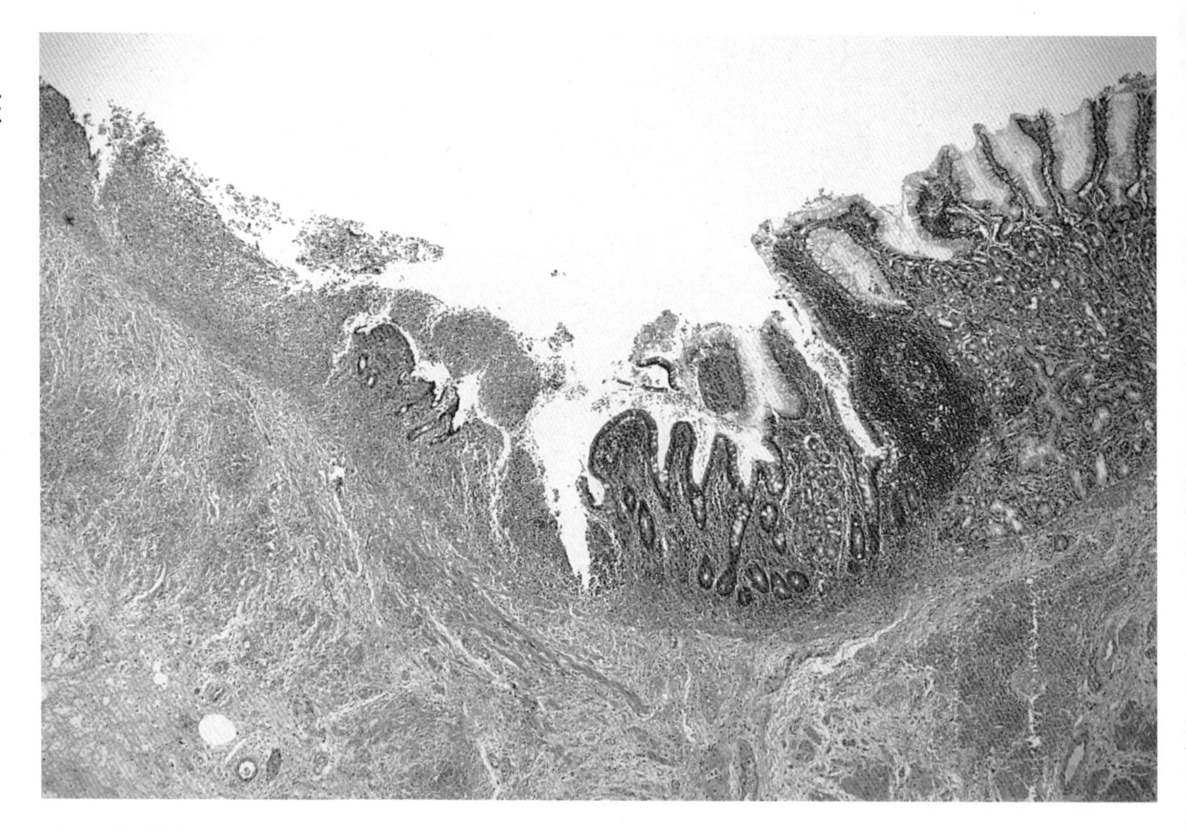

Figure 11–19 **A**

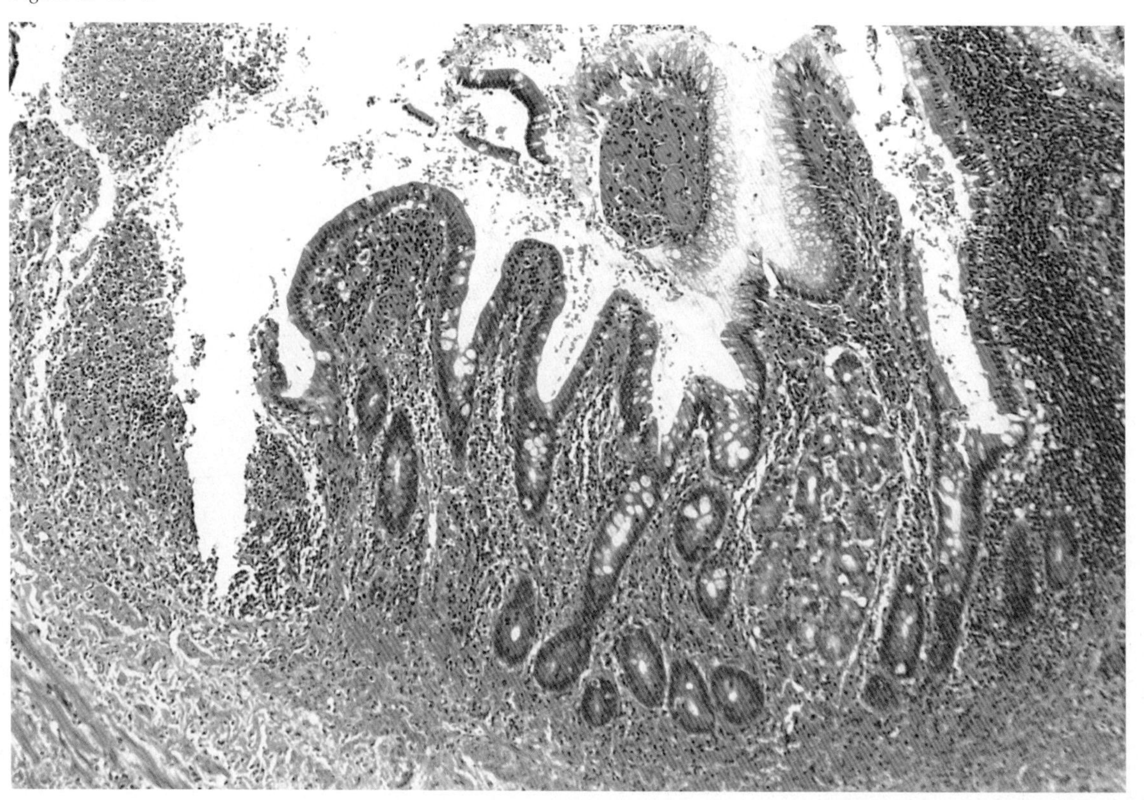

Figure 11–19 **B**

FIGURE 11–19. MECKEL'S DIVERTICULUM: Meckel's diverticulum is a remnant of the embryologically derived omphalomesenteric duct. The diverticulum is lined by small intestinal epithelium. In approximately 50% of cases, as in this example (**A** and **B**), there is heterotopic gastric epithelium. Peptic ulceration is also present.

■ SMALL INTESTINE

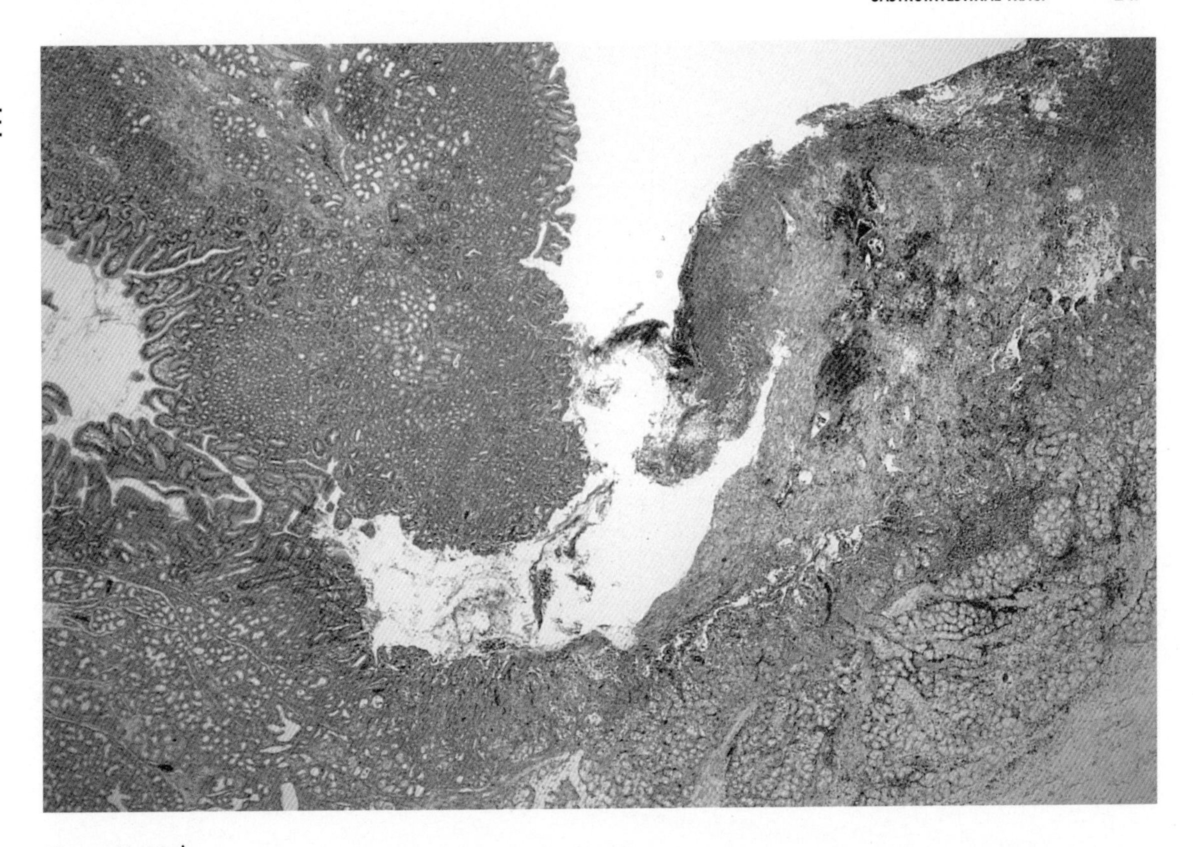

Figure 11–20 **A**

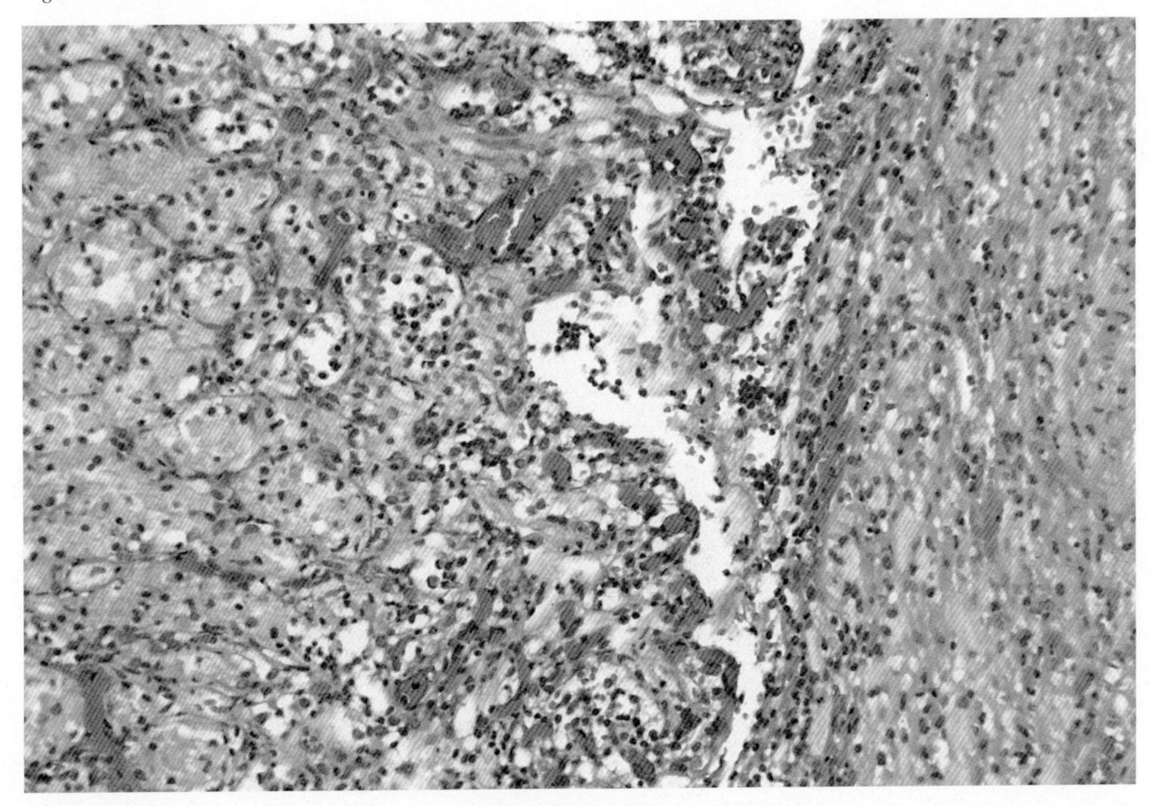

Figure 11–20 **B**

FIGURE 11–20. PEPTIC DUODENITIS: A: In this example of severe peptic duodenitis, there is ulceration of the epithelium. **B:** The lamina propria is infiltrated by acute and chronic inflammatory cells. Fibrinopurulent material admixed with necrotic cellular debris fills the ulcer crater (right side of slide).

■ SMALL INTESTINE

FIGURE 11–21. BRUNNER GLAND HYPERPLASIA: Brunner gland hyperplasia can occur in association with mild duodenitis.

FIGURE 11–22. ISCHEMIC BOWEL DISEASE: A: In this example of ischemic bowel disease, there is mucosal and submucosal infarction with sparing of the muscular layers. **B:** The mucosa is necrotic and hemorrhagic.

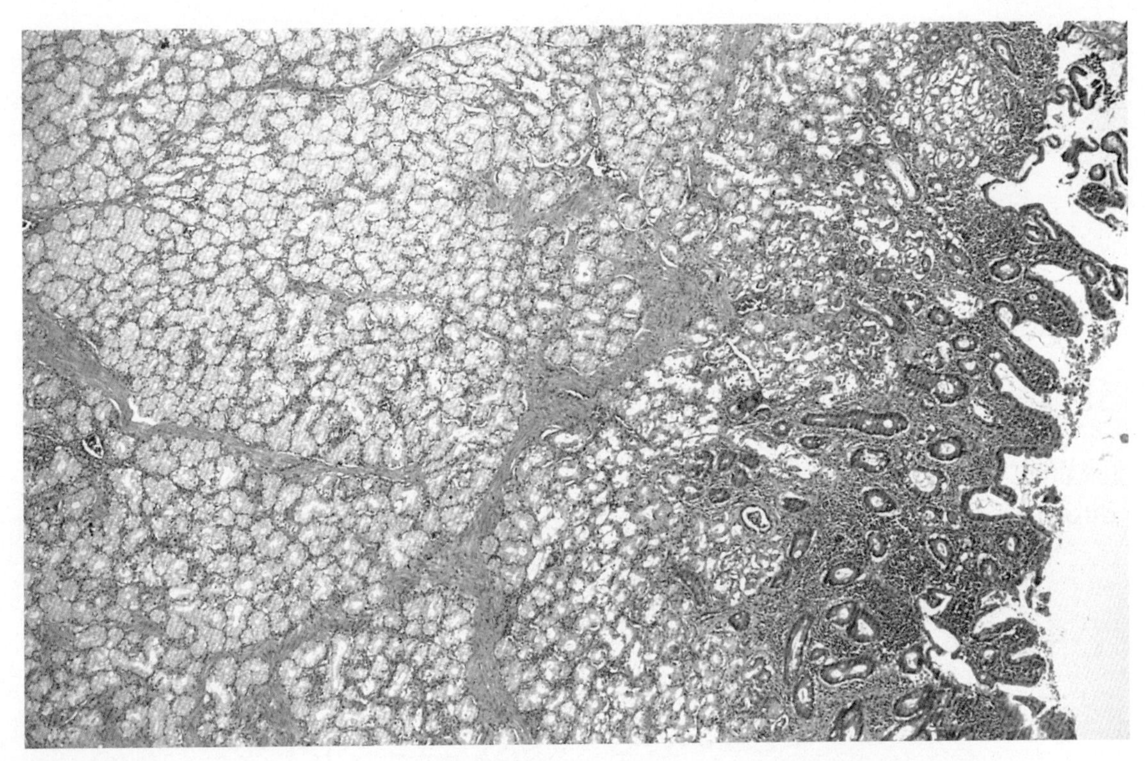

Figure 11–21

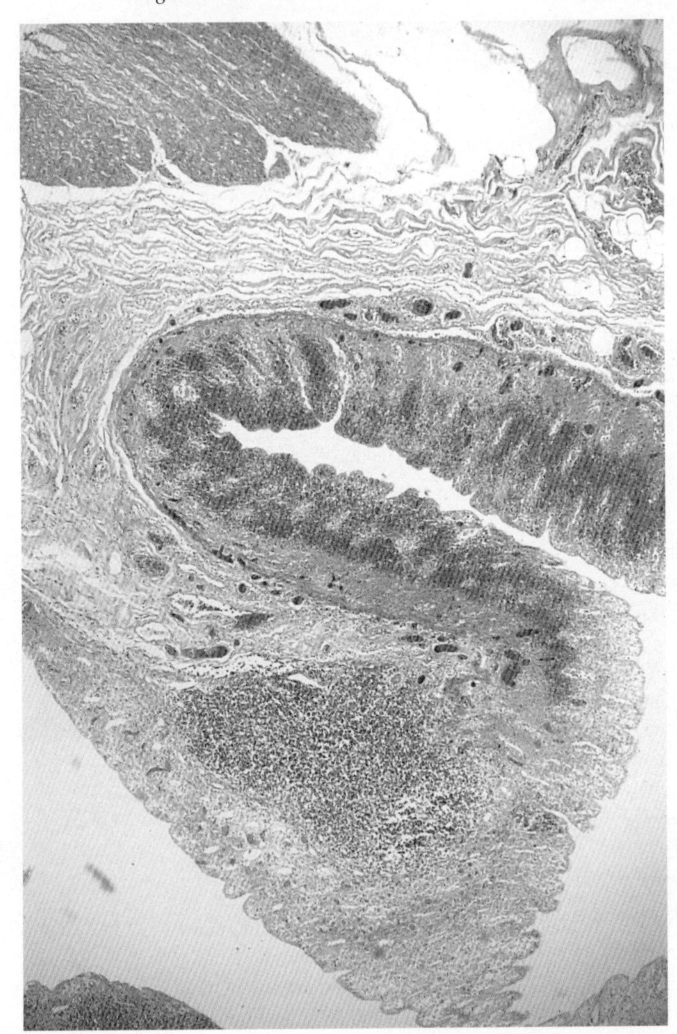

Figure 11–22 A

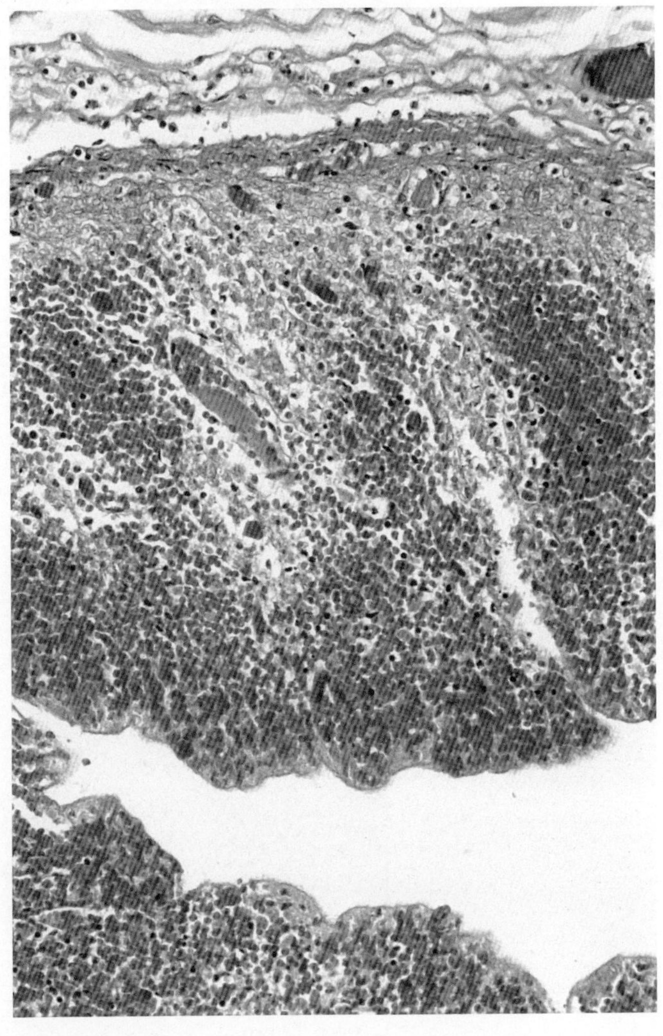

Figure 11–22 B

Crohn's disease is one of the inflammatory bowel diseases. Its characteristic features are

1. noncaseating granulomata
2. lymphoid aggregates
3. aphthous ulceration
4. fissures and fistulas
5. transmural inflammation and fibrosis
6. neural hyperplasia.

■ CROHN'S DISEASE (REGIONAL ENTERITIS, GRANULOMATOUS ENTERITIS)

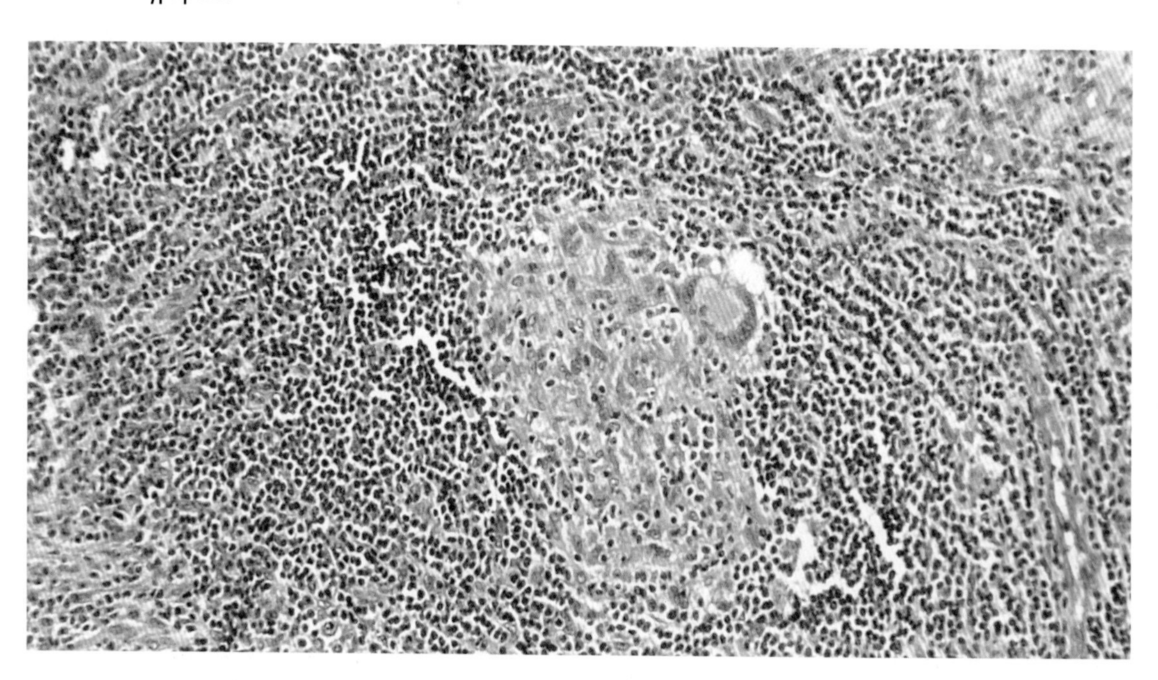

FIGURE 11–23. NONCASEATING GRANULOMATA: A: Noncaseating granulomata with giant cells are found in 40–60% of cases.

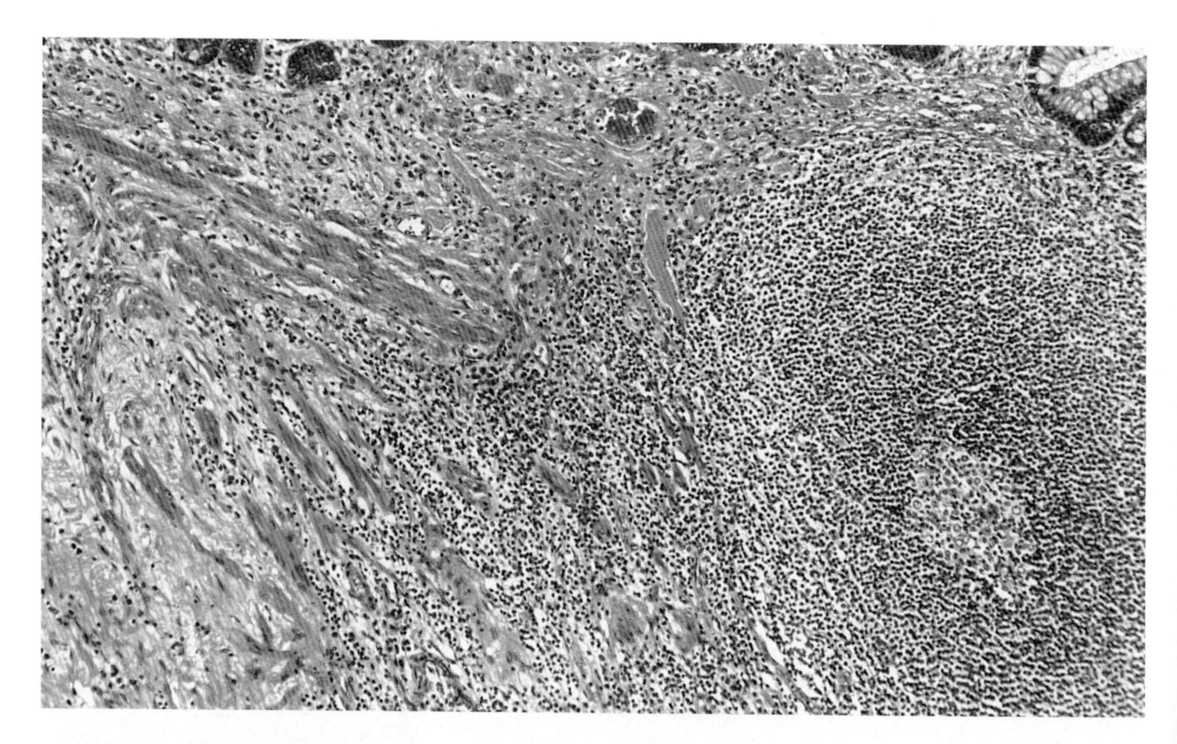

FIGURE 11–23. LYMPHOID AGGREGATES: B: Lymphoid aggregates, some with germinal centers, are found more frequently than are granulomata. These aggregates are present at the junction of the mucosa and submucosa and splay the muscular fibers between the two.

■ CROHN'S DISEASE

FIGURE 11–23. TRANSMURAL INFLAMMATION AND FIBROSIS: The transmural inflammatory infiltra
phocytes, plasma cells, and histiocytes (**E** and **F**). **F:** The fibrosis causes notable thicke

FIGURE 11–23. NEURAL HYPERPLASIA: G: Neural hyperplasia in the myenteric and submu
prominent feature of Crohn's disease.

■ SMALL INTESTINE

MALABSORPTION SYNDROMES

FIGURE 11–24. WHIPPLE'S DISEASE: Whipple's disease is a systemic disorder affecting joints, serosal surfaces, lymphoid organs, heart, liver, lung, skin, and nervous system. **A:** In Whipple's disease, there is widening and blunting of the villi. **B:** The lamina propria is stuffed with foamy macrophages. **C:** The foamy macrophages stain intense pink with periodic acid–Schiff stain. Note the built-in positive control of goblet cells lining the villi.

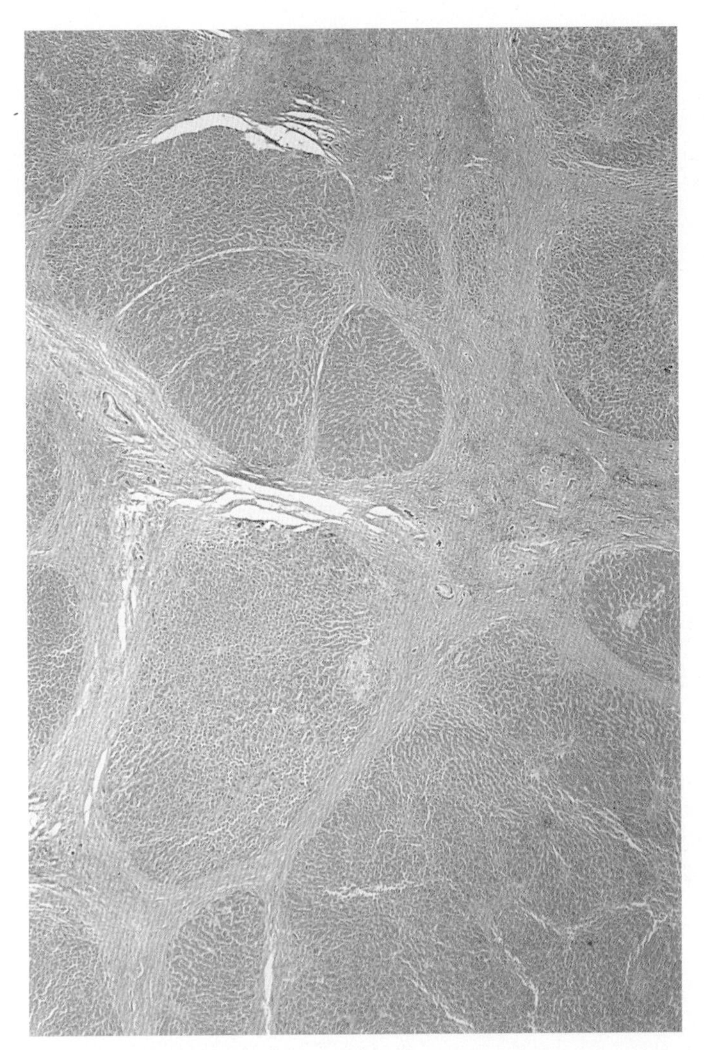

Figure 12–7 A

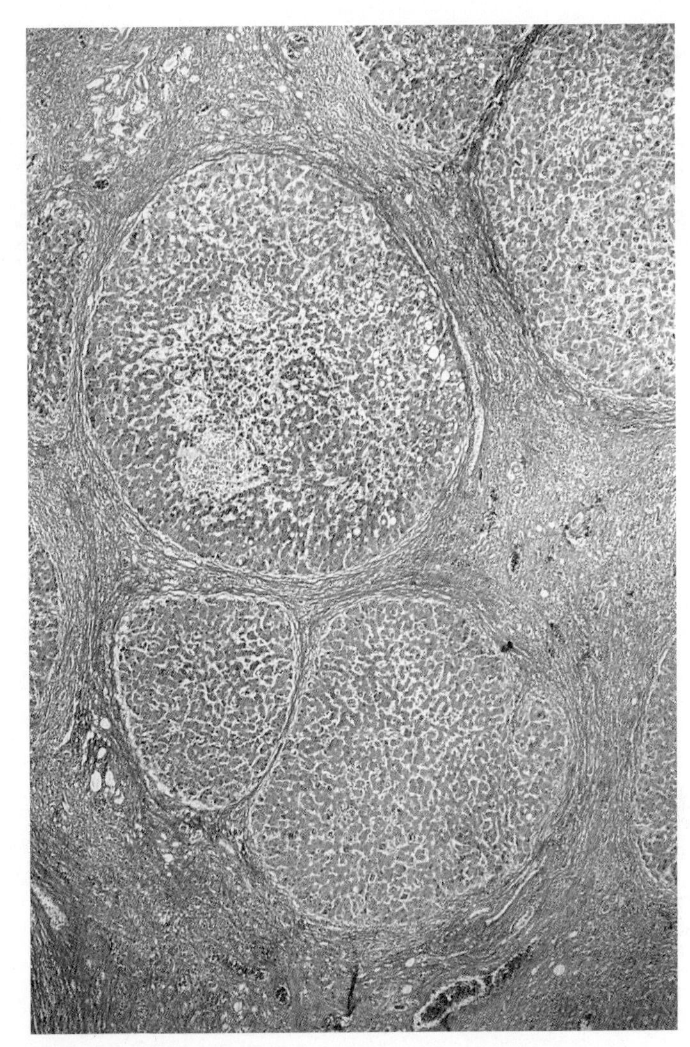

Figure 12–7 B

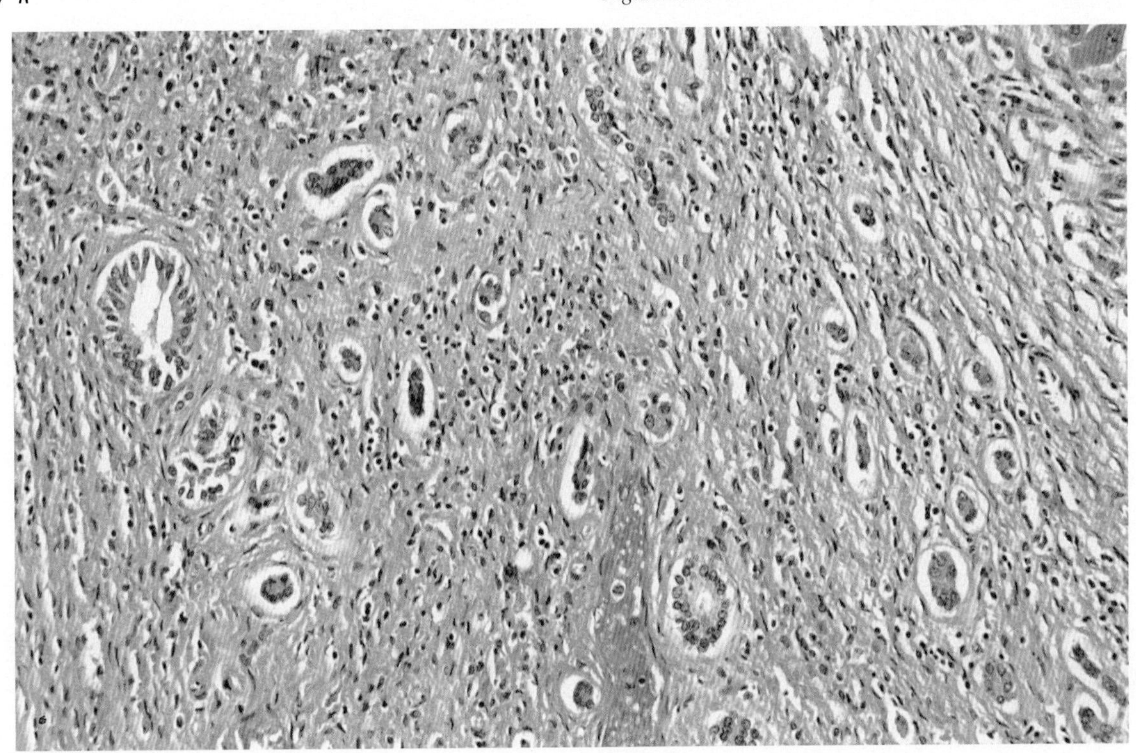

Figure 12–7 C

■ LIVER

CIRRHOSIS AND CIRRHOSIS-ASSOCIATED DISORDERS

ALCOHOLIC LIVER DISEASE

FIGURE 12–8. FATTY LIVER (STEATOSIS): The earliest change in alcoholic liver disease is microvesicular and macrovesicular steatosis.

FIGURE 12–9. ALCOHOLIC HEPATITIS (STEATOHEPATITIS): In alcoholic hepatitis, the hepatocytes undergo ballooning degeneration; neutrophils are attracted to these degenerating cells. Condensed ropelike eosinophilic material known as **Mallory hyaline** can be found in the degenerated cells.

FIGURE 12–10. ALCOHOLIC CIRRHOSIS (LAËNNEC'S CIRRHOSIS): In early alcoholic cirrhosis, bridging fibrosis with entrapped portal tracts is noted. Hepatocytes degenerate, and central veins and portal tracts become trapped in the fibrous scar. Bile plugs and fatty change is noted in the remaining hepatocytic parenchyma.

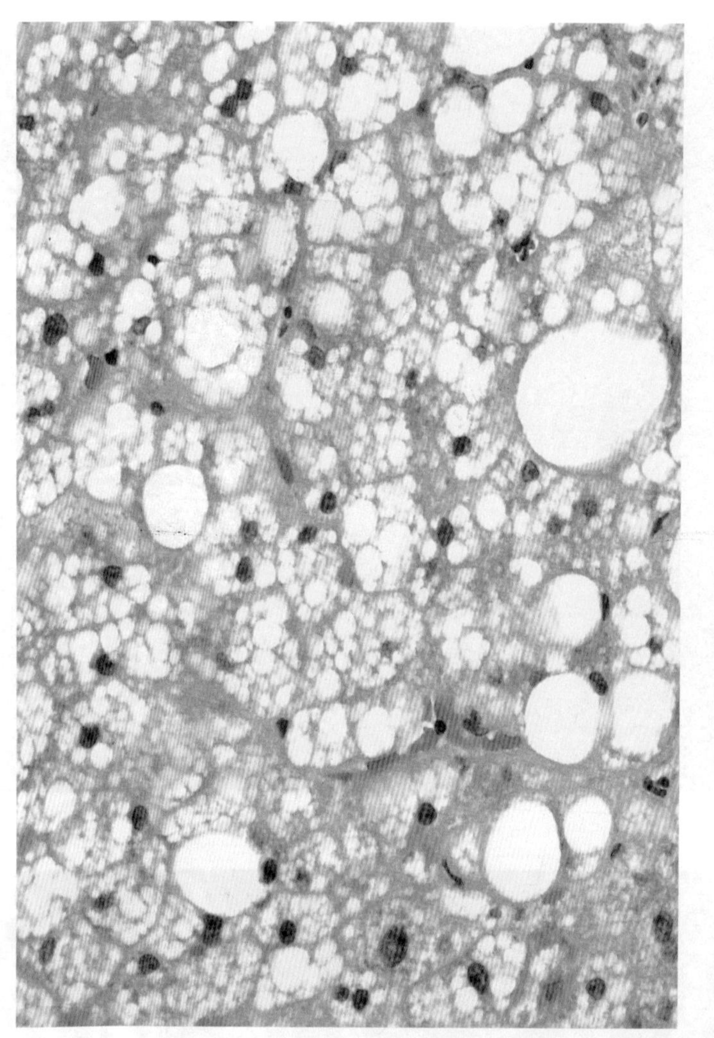

Figure 12–8

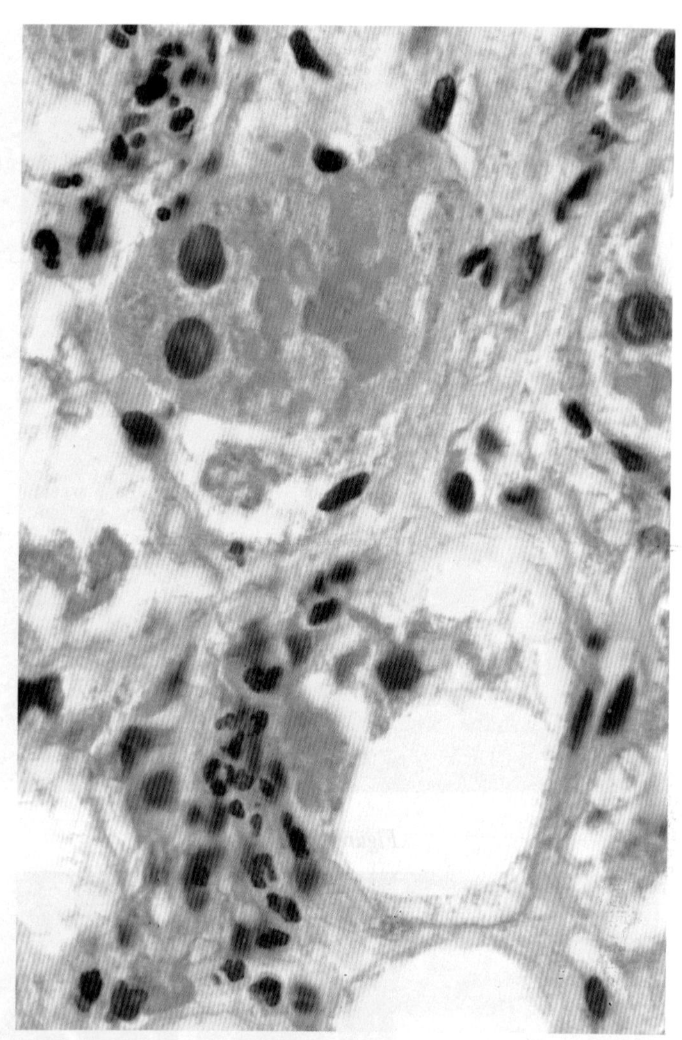

Figure 12–9

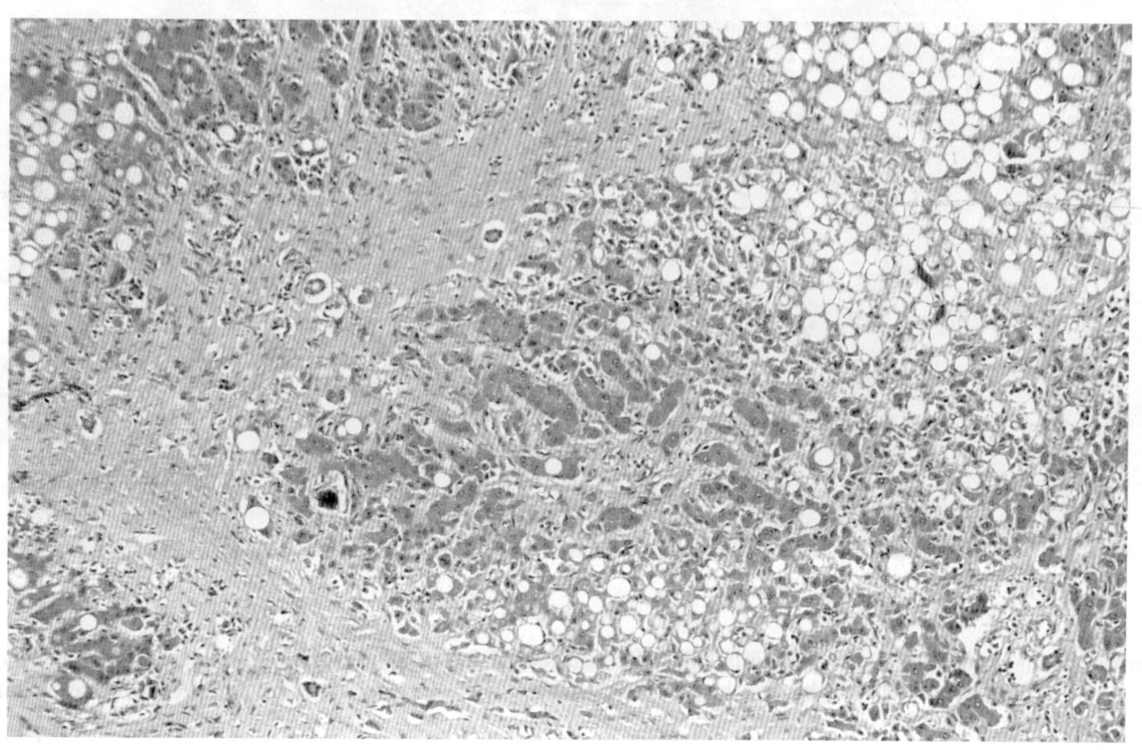

Figure 12–10

■ LIVER

MALIGNANT TUMORS

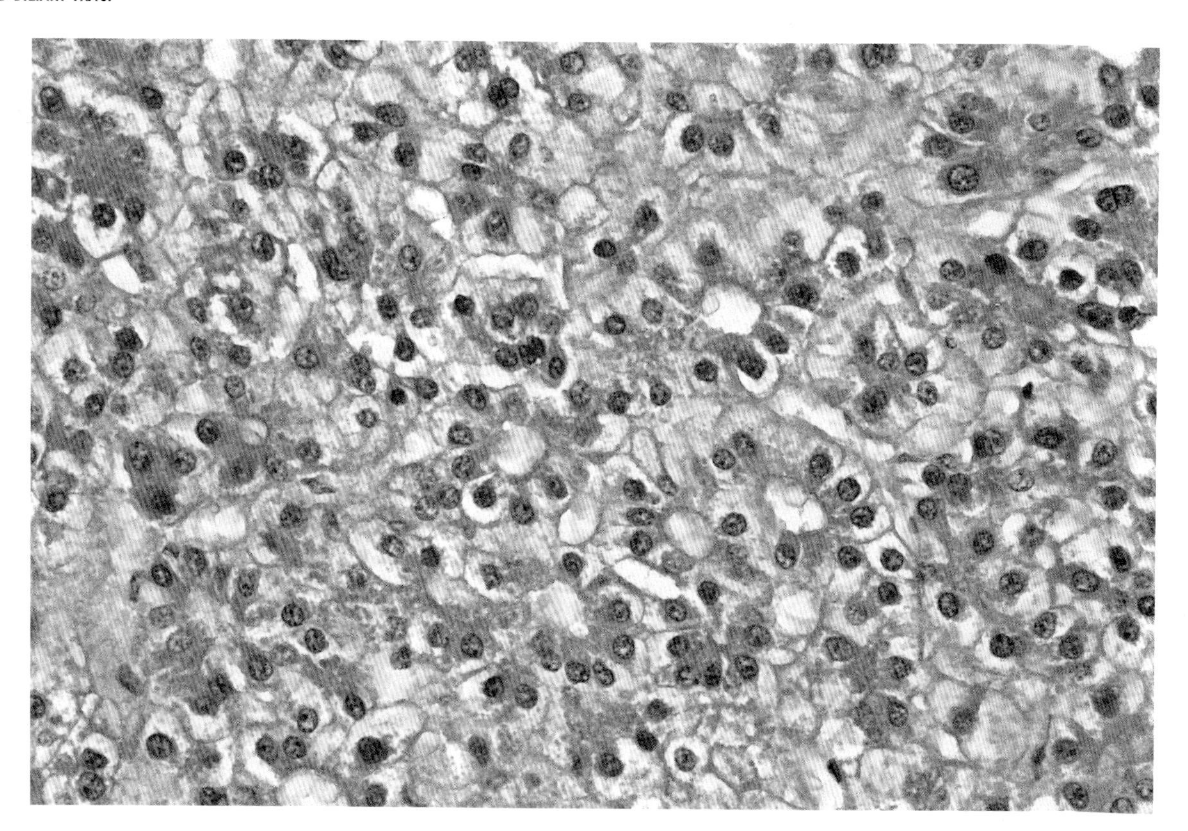

Figure 12–21 A

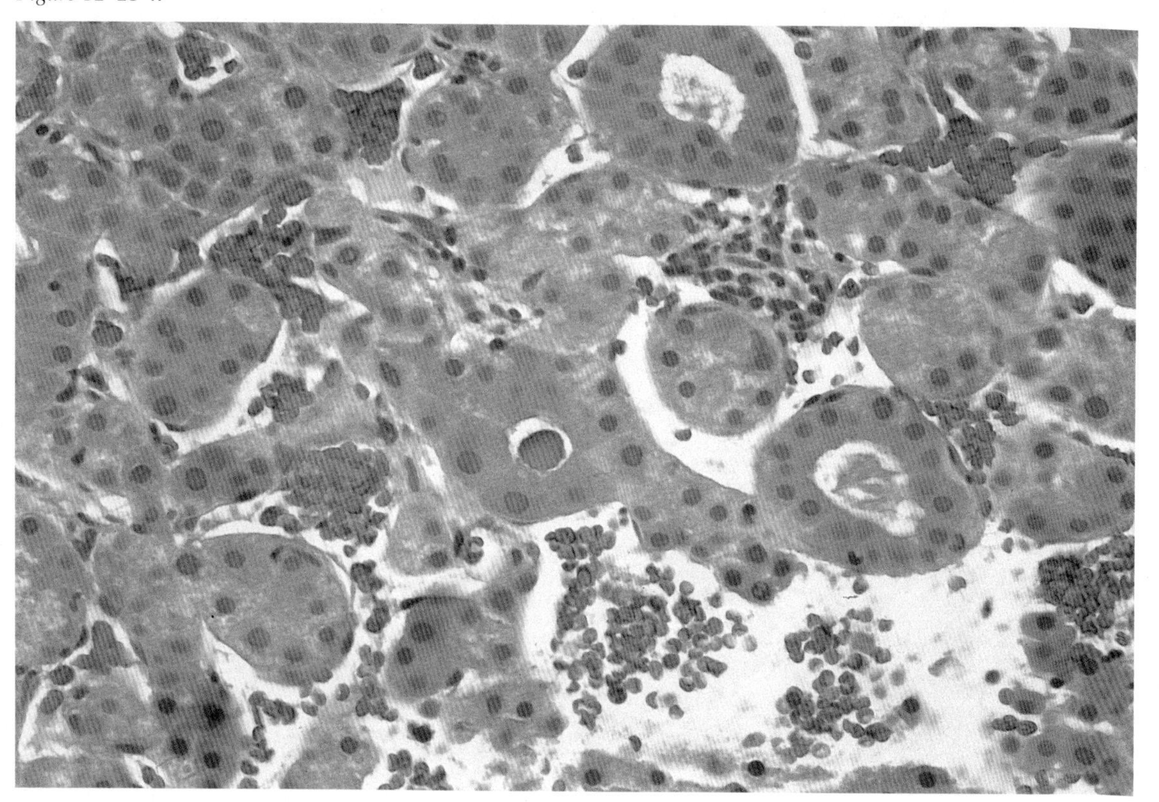

Figure 12–21 B

FIGURE 12–21. HEPATOCELLULAR CARCINOMA: A: In the glandular pattern of well-differentiated hepatocellular carcinoma, the cells are normal or slightly enlarged, with an increase in the nuclear-to-cytoplasmic ratio. In **A**, pseudorosettes are prominent. **B:** Bile production, which proves the hepatocellular origin of the tumor, is noted in this photomicrograph.

■ LIVER

MALIGNANT TUMORS

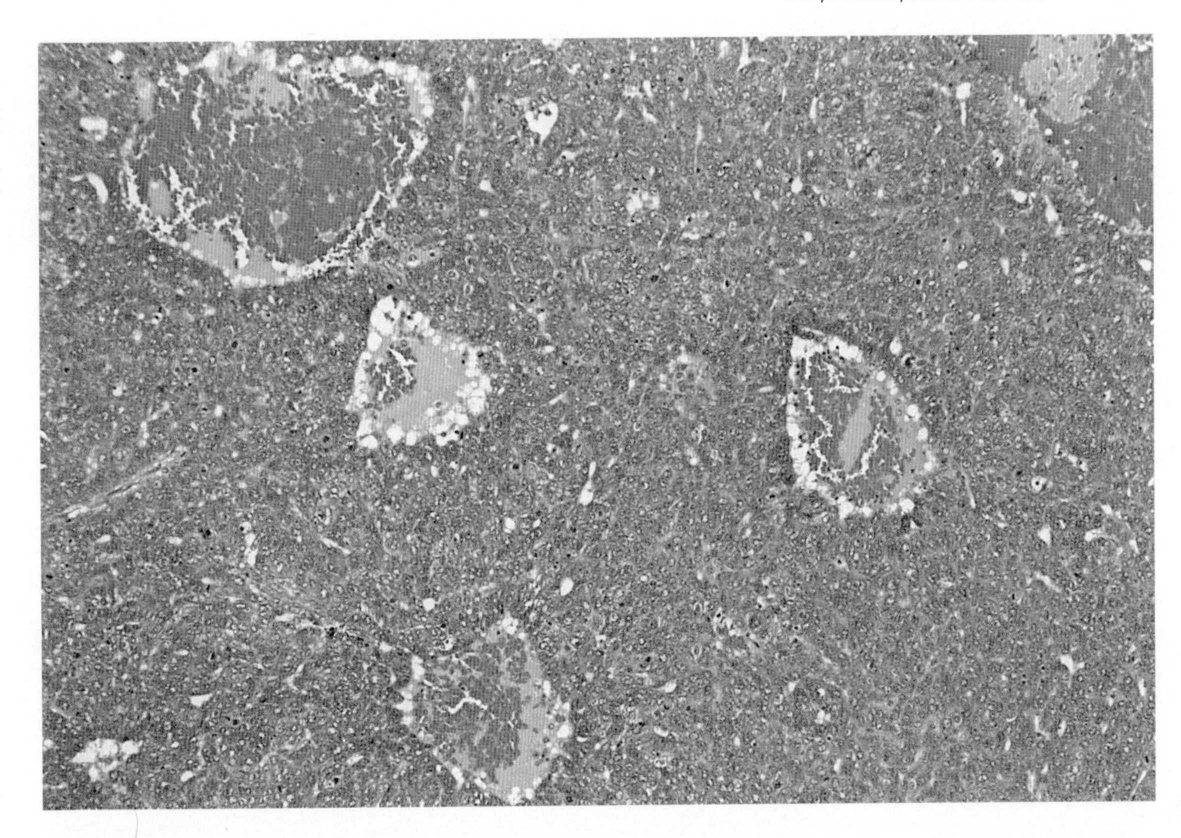

Figure 12–22 A

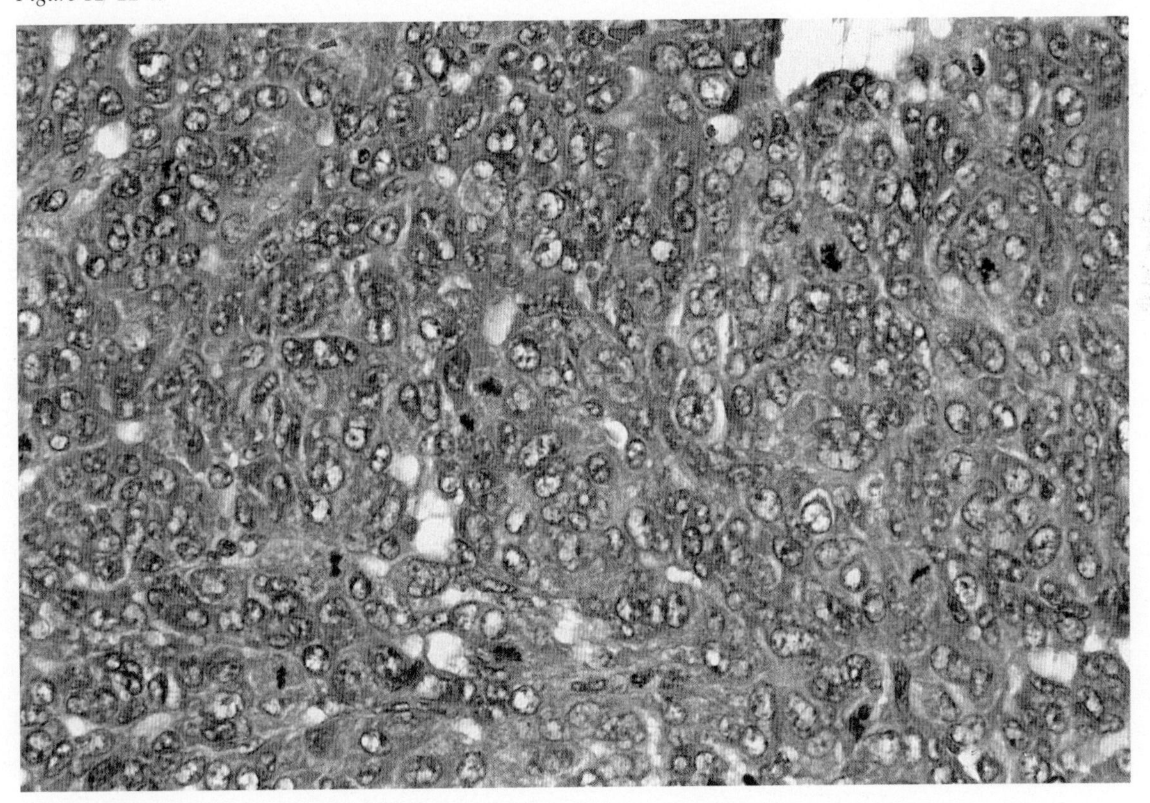

Figure 12–22 B

FIGURE 12–22. HEPATOBLASTOMA: A: This is a low-power view of the embryonal variant of hepatoblastoma, which is the most common primary hepatic tumor in children. Cells form sheets, and acini and pelioid foci are commonly found. **B:** The cells have poorly defined cell borders, basophilic cytoplasm, a high nuclear-to-cytoplasmic ratio, and prominent nucleoli. Mitotic figures are prominent.

■ LIVER

CIRCULATORY DISORDERS

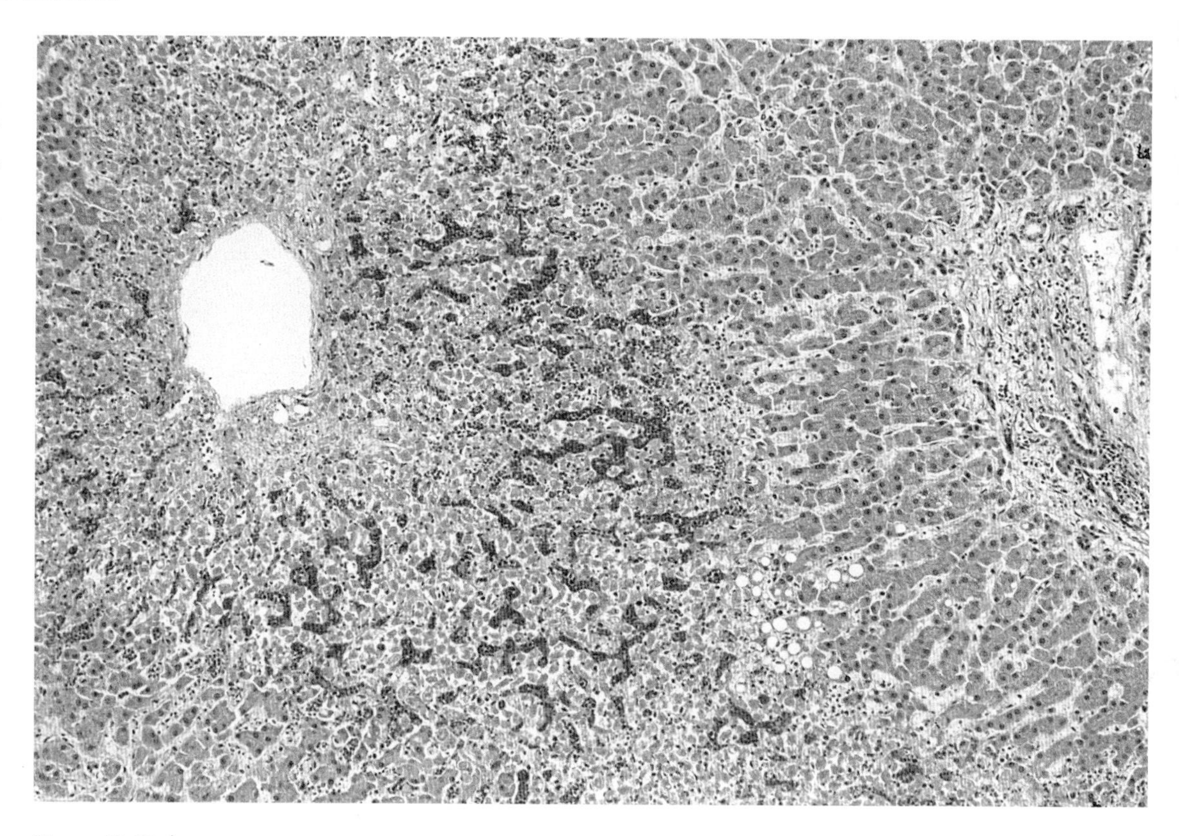

Figure 12–23 **A**

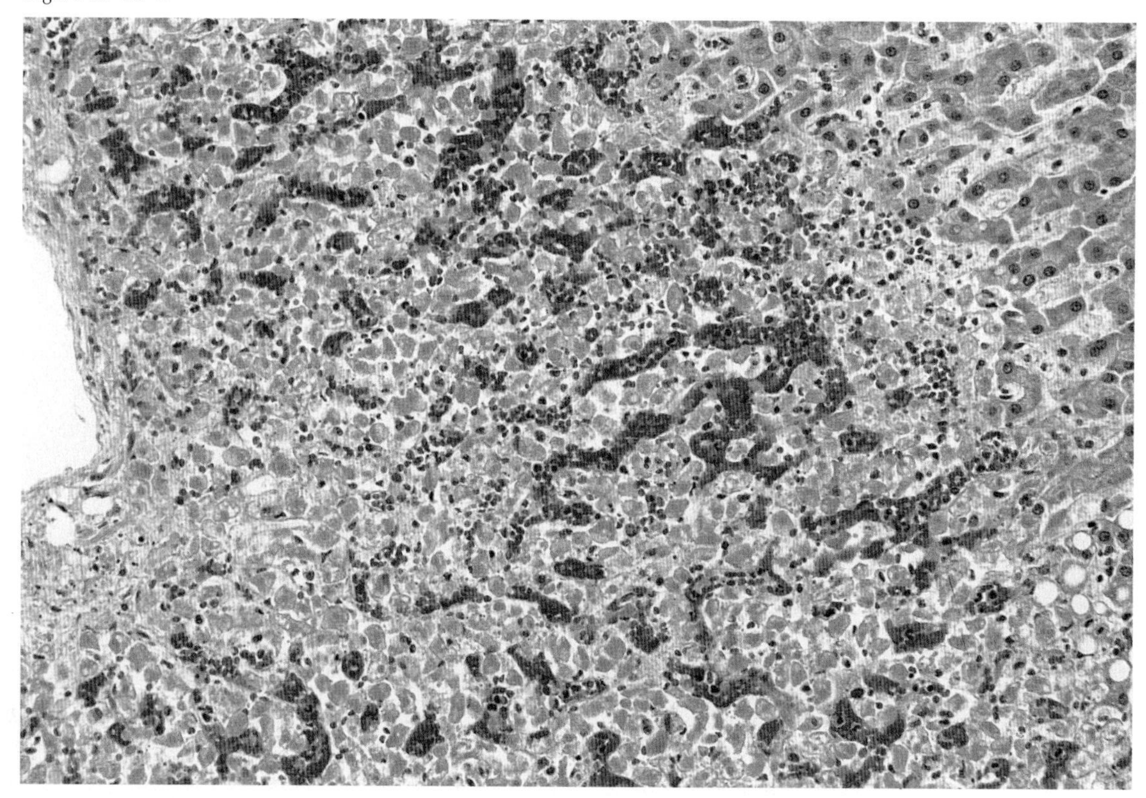

Figure 12–23 **B**

FIGURE 12–23. CHRONIC PASSIVE CONGESTION: A: Chronic passive congestion is characterized by centrilobular congestion. This is grossly seen as a "nutmeg" pattern. **B:** The hepatocytes adjacent to the central vein are atrophic and focally necrotic. Fatty change is present at the periphery of the lobules (lower right).

■ PANCREAS

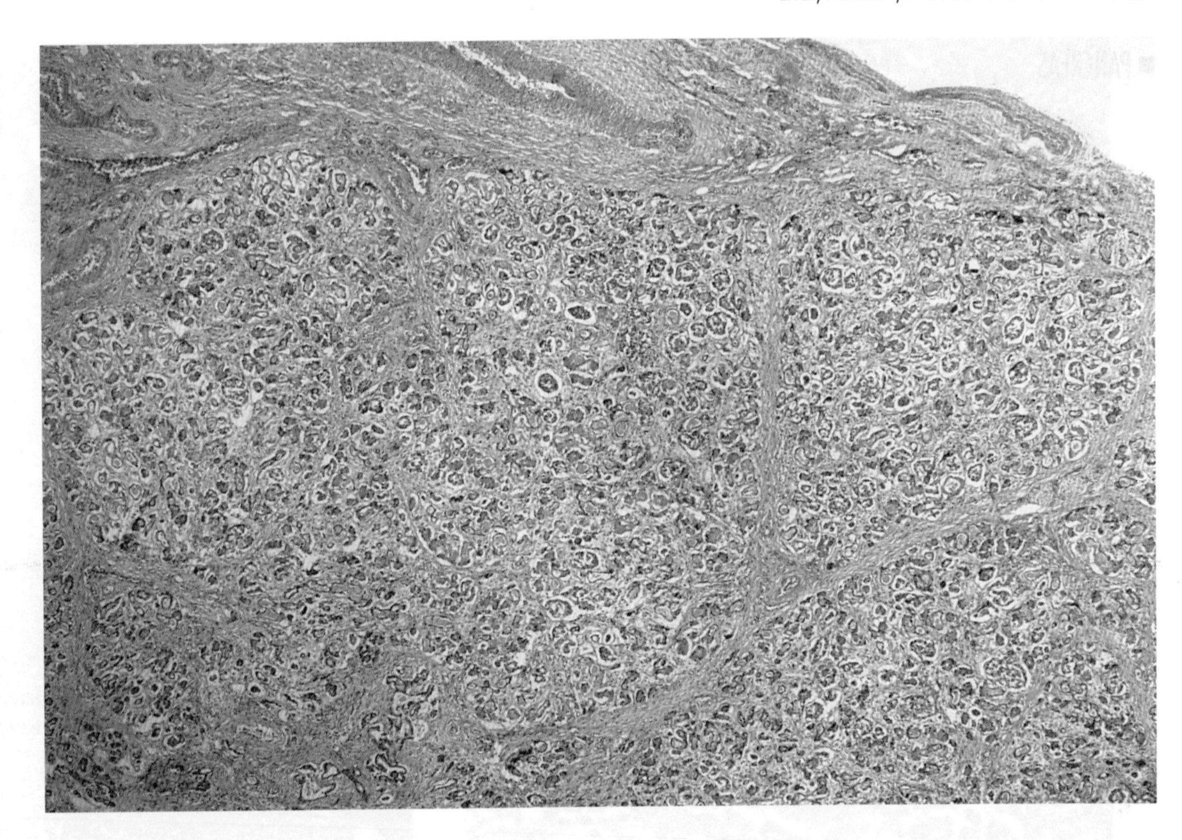

Figure 12–24 A

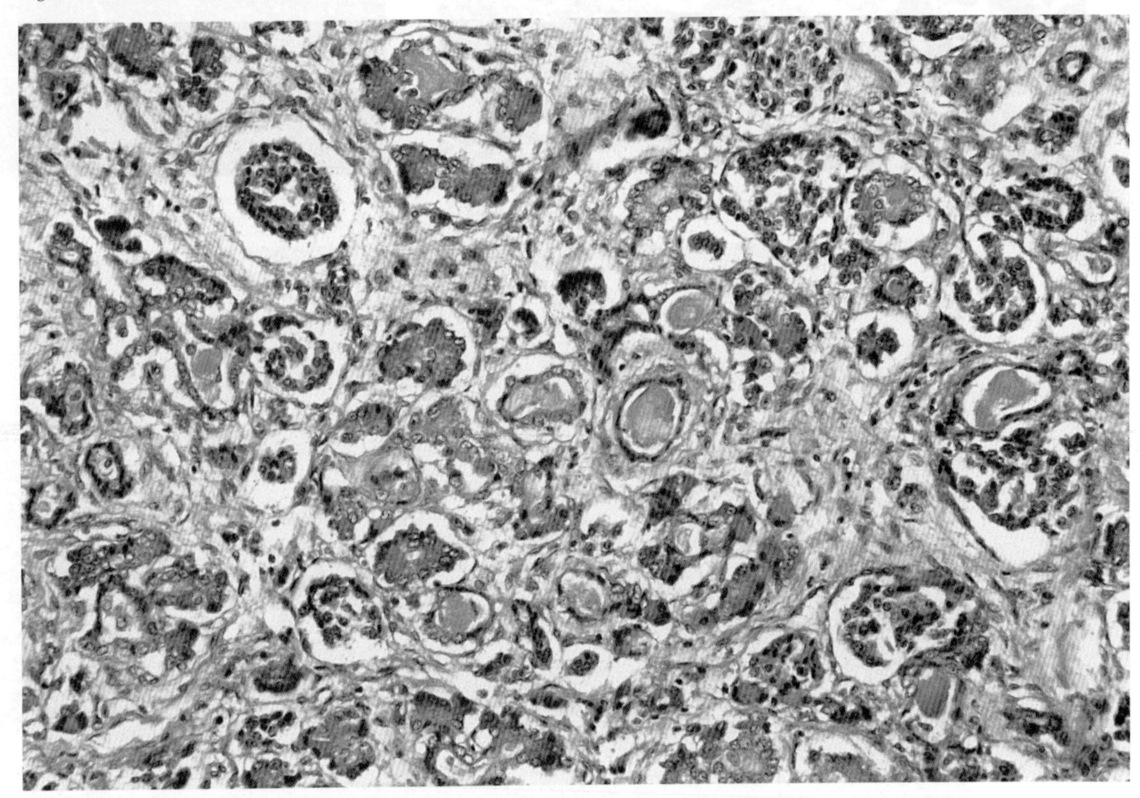

Figure 12–24 B

FIGURE 12–24. CYSTIC FIBROSIS (MUCOVISCIDOSIS): A: In cystic fibrosis, a stellate fibrous scar divides the pancreas into lobules. **B:** Ducts are dilated and filled with inspissated eosinophilic mucus plugs. The exocrine portion of the pancreas undergoes progressive atrophy.

■ GLOMERULAR DISEASES

NEPHROTIC SYNDROME

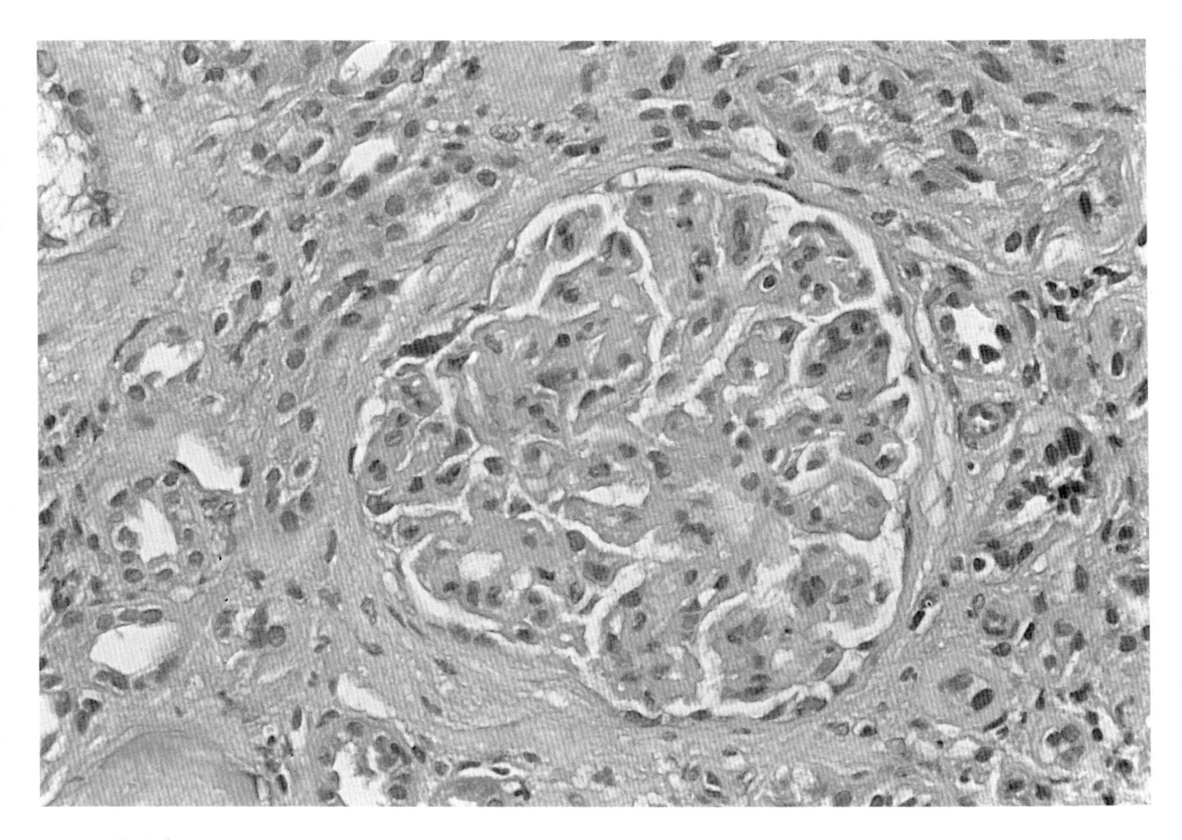

Figure 13–5 A

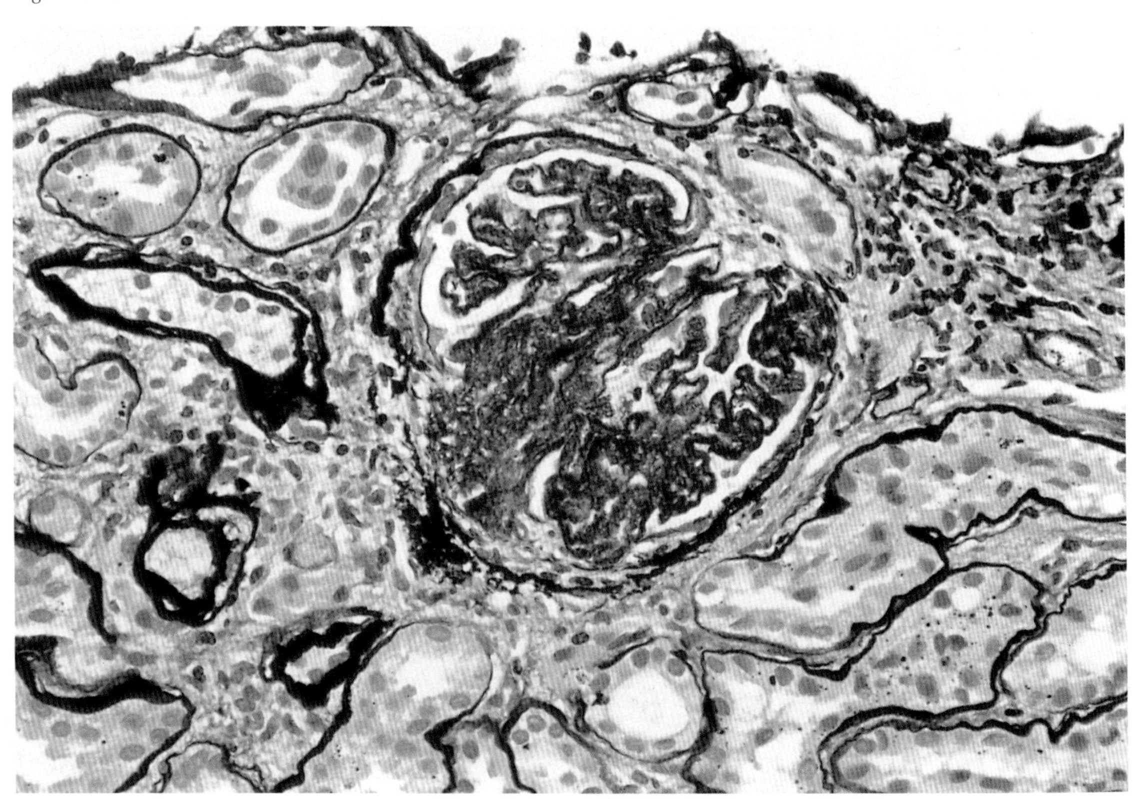

Figure 13–5 B

FIGURE 13–5. MEMBRANOUS GLOMERULONEPHRITIS: Membranous glomerulonephritis is the primary cause of the nephrotic syndrome. **A:** In membranous glomerulonephritis, there is diffuse thickening of the basement membrane of capillary walls in the glomeruli. **B:** Thickening of the basement membrane is confirmed by silver staining.

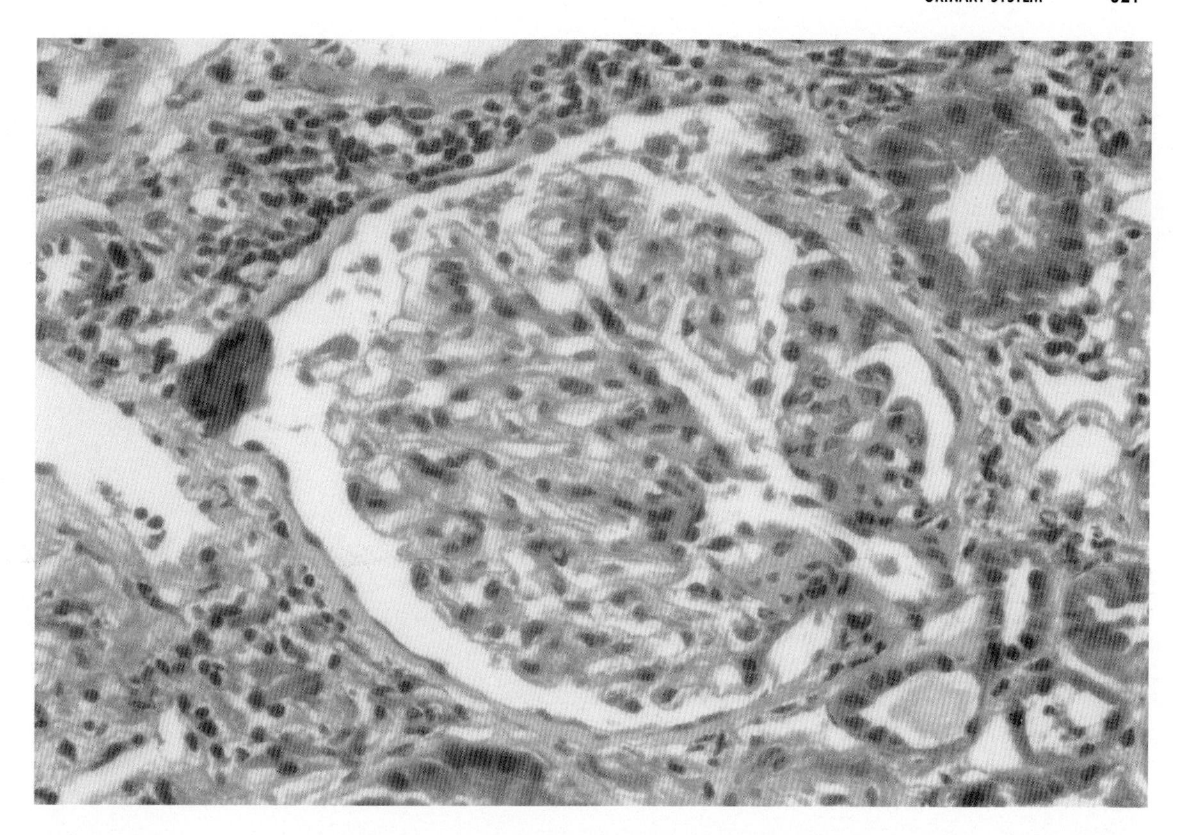

Figure 13–6 A

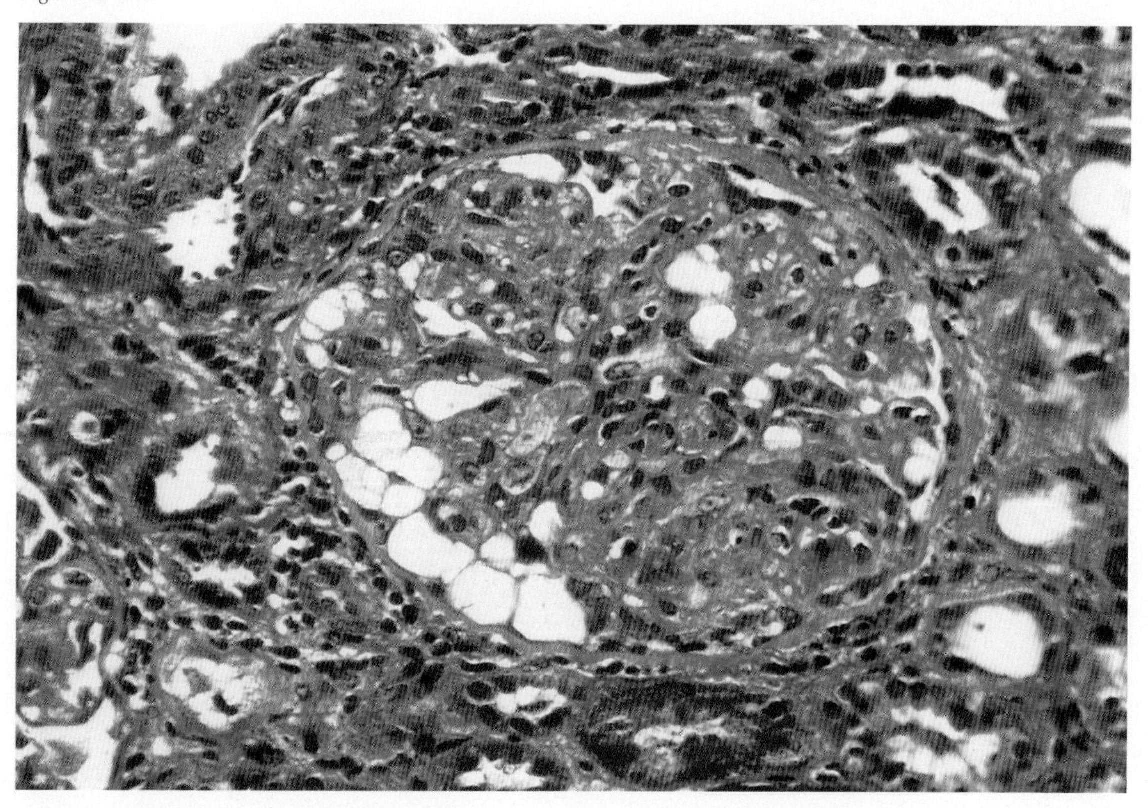

Figure 13–6 B

FIGURE 13–6. FOCAL SEGMENTAL GLOMERULOSCLEROSIS: A: Focal segmental glomerulosclerosis is characterized by sclerosis of parts of some glomeruli. **B:** This is confirmed by trichrome staining.

■ GLOMERULAR DISEASES

NEPHROTIC SYNDROME

FIGURE 13–7. MEMBRANOPROLIFERATIVE GLOMERULONEPHRITIS: The diagnostic feature in membranoproliferative glomerulonephritis is interposition of mesangial cells (*arrow*) and their processes, causing splitting—**"tramtracking"**—of capillary walls. This is demonstrated with a silver stain.

NEPHRITIC SYNDROME

FIGURE 13–8. DIFFUSE PROLIFERATIVE GLOMERULONEPHRITIS (ACUTE POSTSTREPTOCOCCAL GLOMERULONEPHRITIS): A: In diffuse proliferative glomerulonephritis, the glomeruli are enlarged and hypercellular. **B:** An electron micrograph shows subepithelial **humps** (*arrow*) composed of immune complexes.

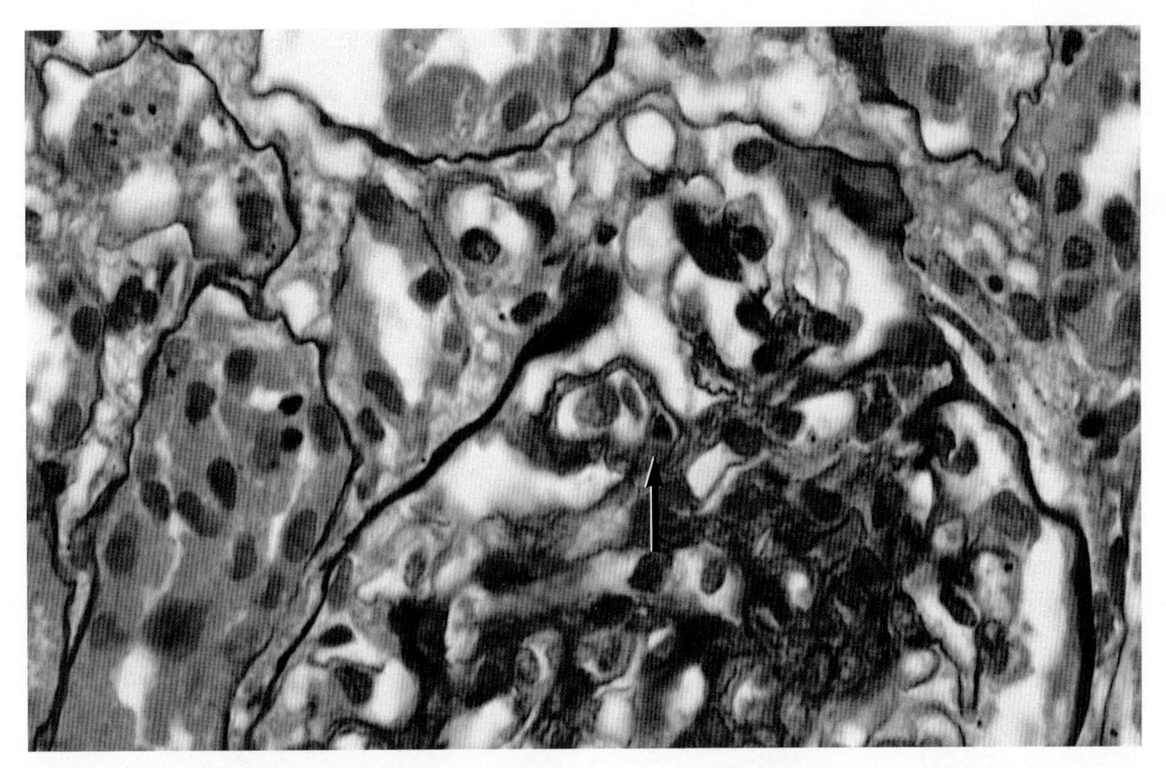

Figure 13–7

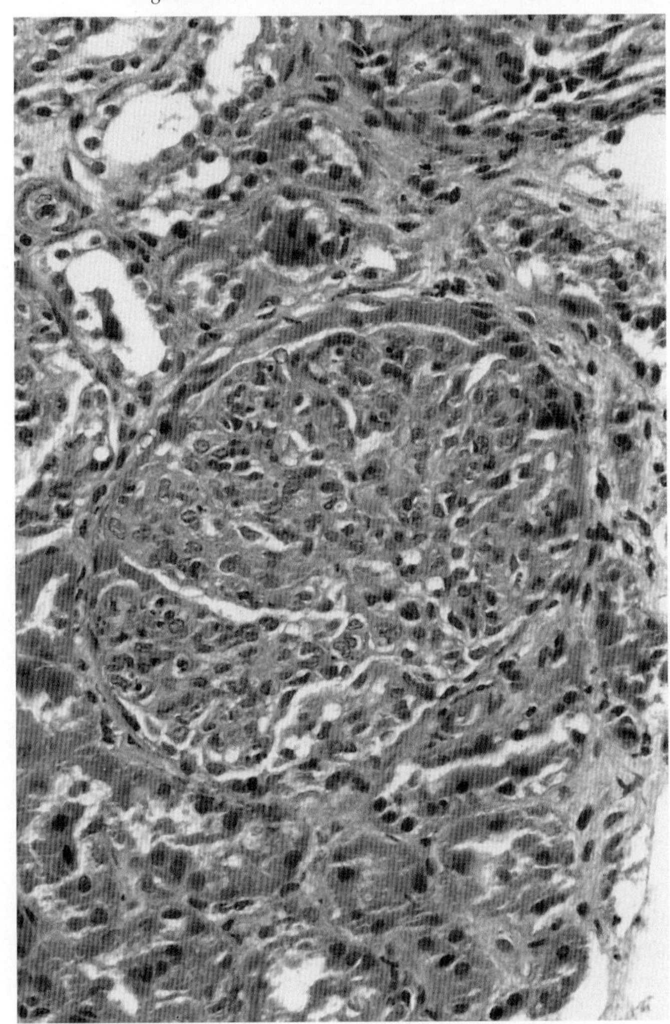

Figure 13–8 **A**

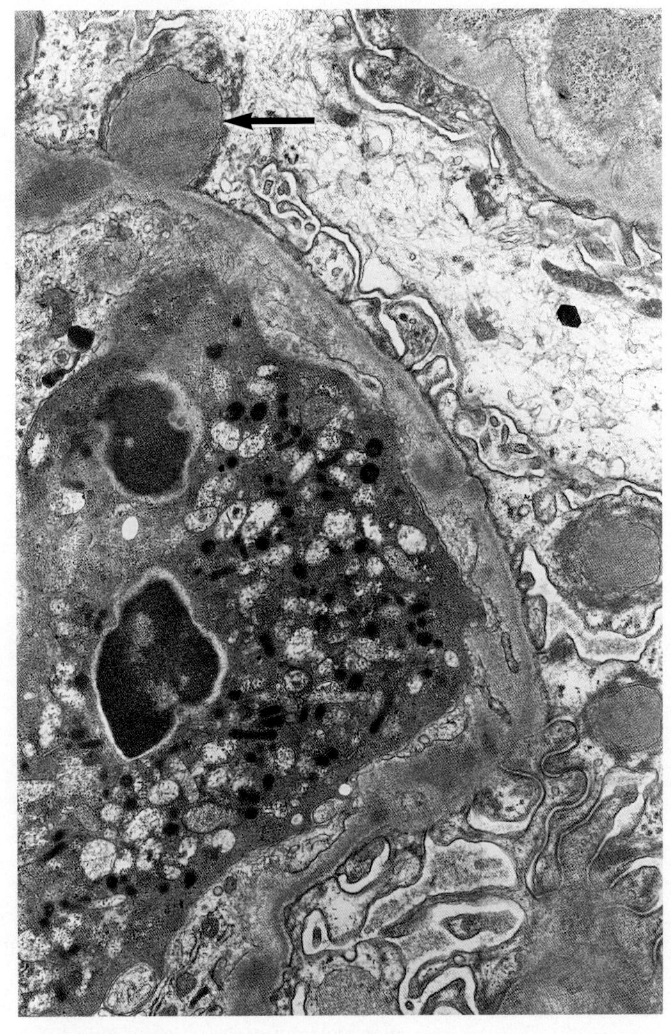

Figure 13–8 **B**

■ GLOMERULAR DISEASES

NEPHRITIC SYNDROME

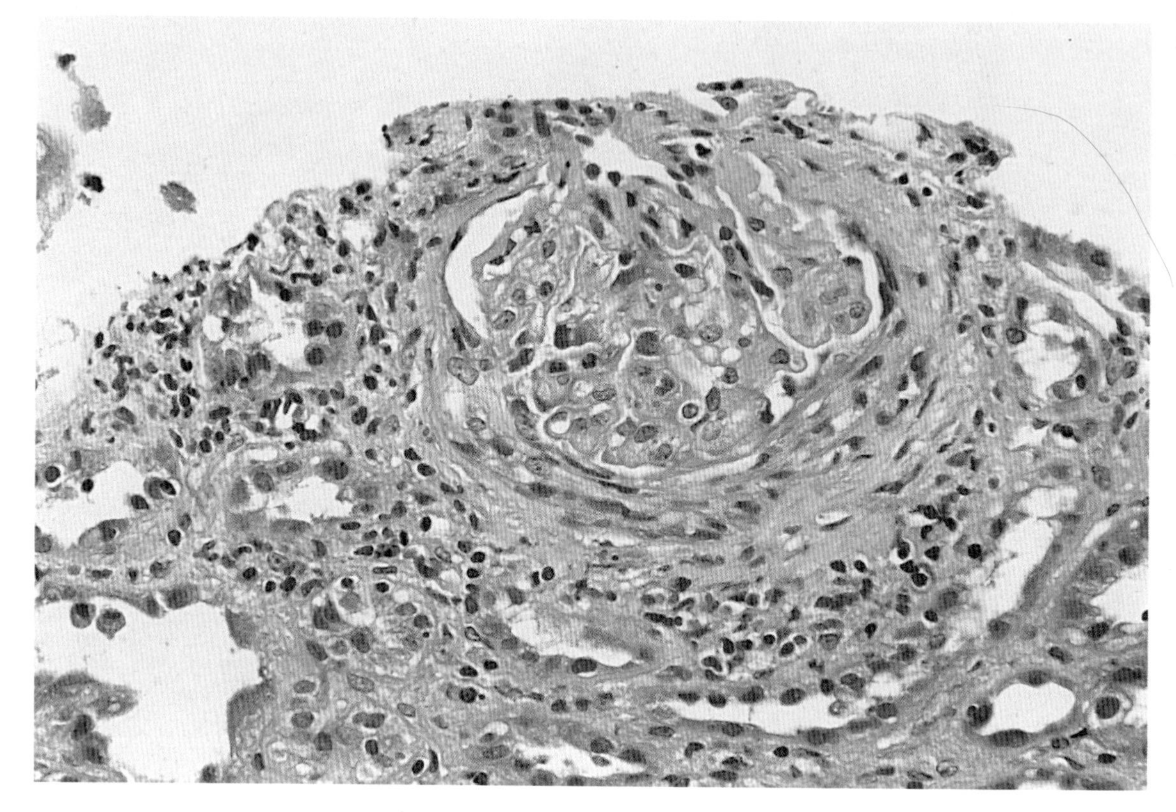

Figure 13–9 A

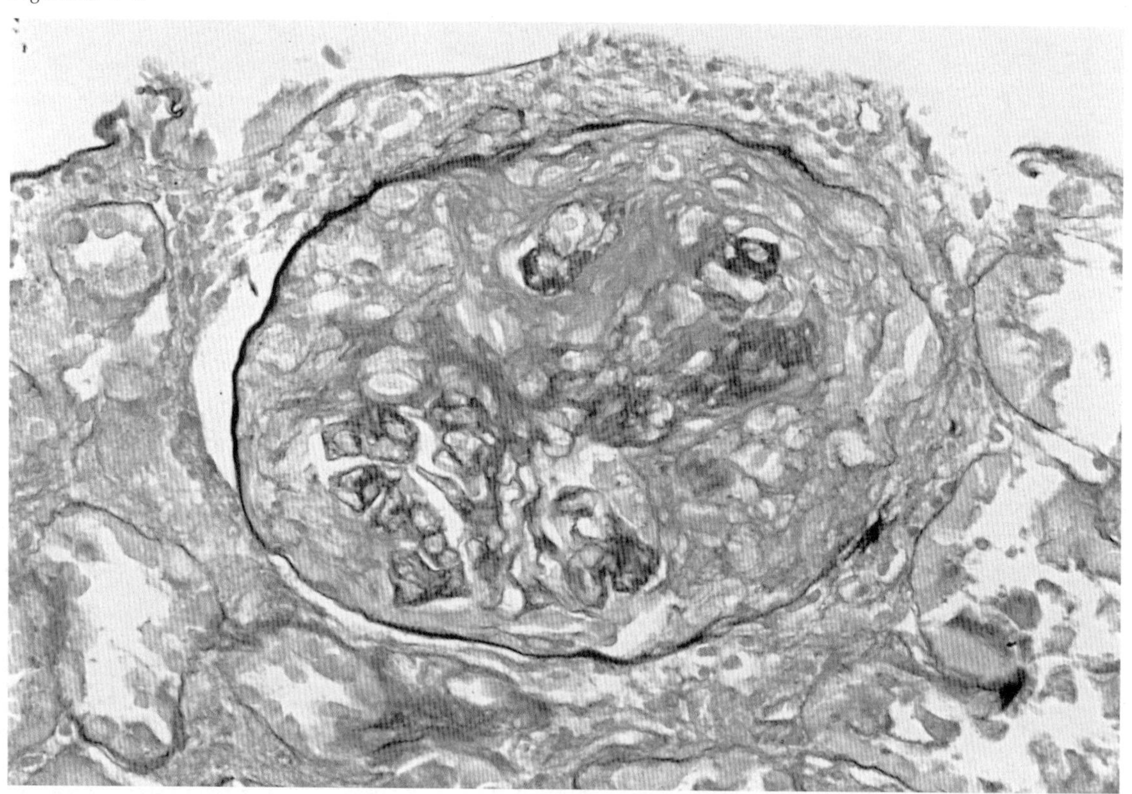

Figure 13–9 B

FIGURE 13–9. RAPIDLY PROGRESSIVE (CRESCENTIC) GLOMERULONEPHRITIS (RPGN): A: In rapidly progressive glomerulonephritis, there is proliferation of parietal epithelial cells that, when observed in the early stages of the disease process, partially obliterates Bowman's space. **B:** As the disease progresses, there is eventual obliteration of Bowman's space and collapse of the glomerular tufts (periodic acid–Schiff stain).

■ TUBULOINTERSTITIAL DISEASES

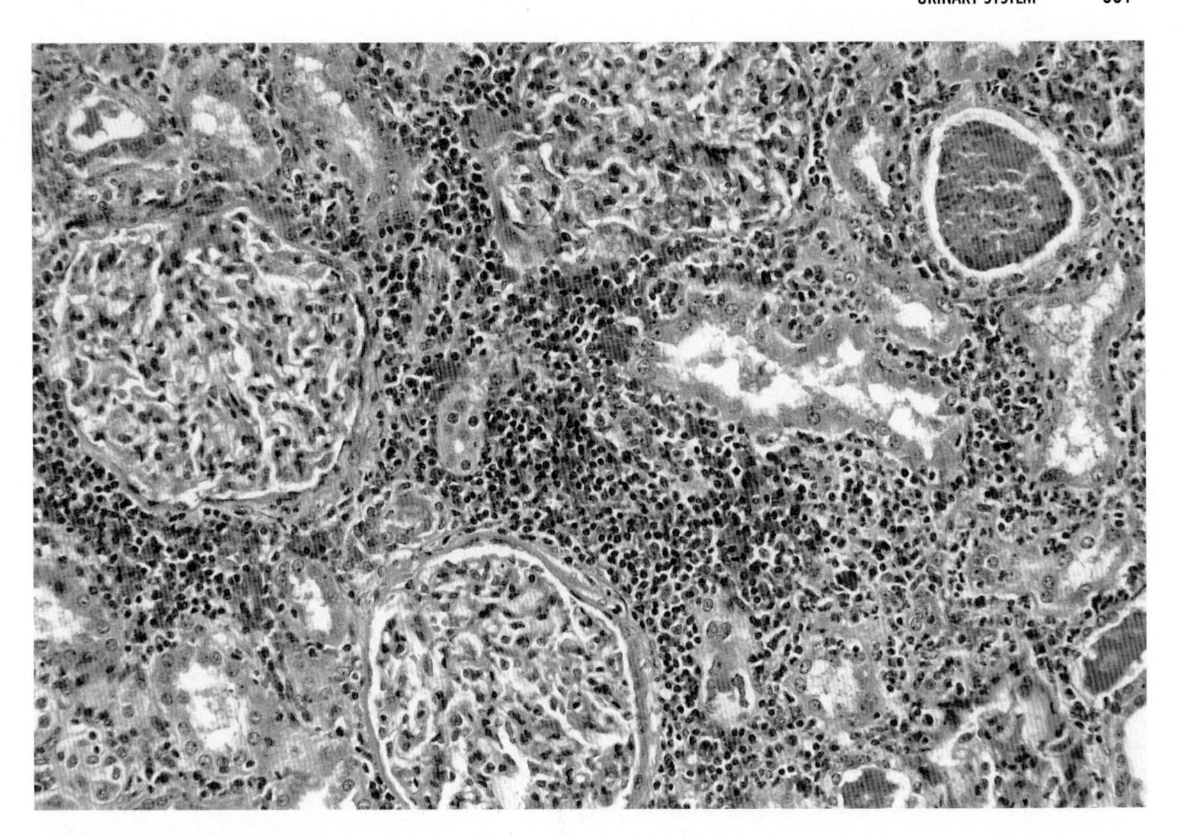

FIGURE 13–16. CHRONIC PYELONEPHRITIS: In chronic pyelonephritis, there is a lymphocytic infiltrate and associated fibrosis in the renal interstitium. In this photomicrograph, white blood cell casts are present as part of a superimposed acute process.

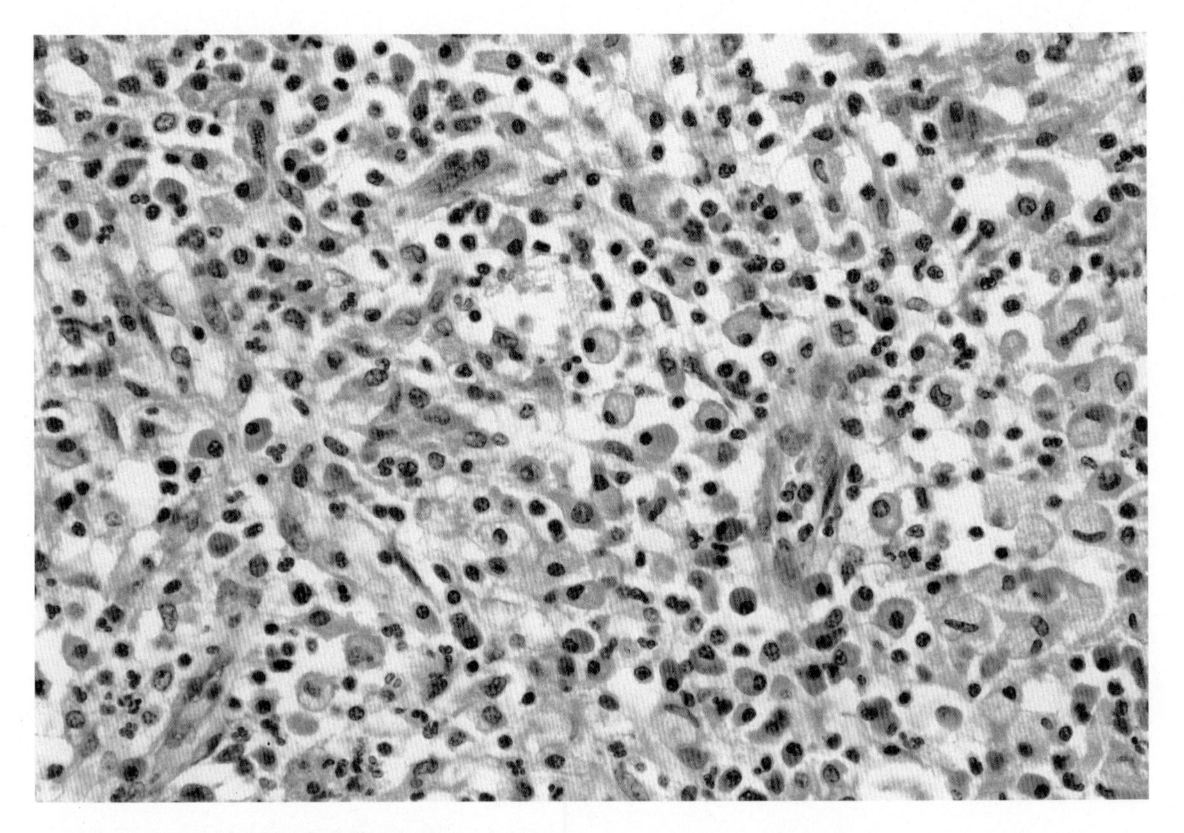

FIGURE 13–17. XANTHOGRANULOMATOUS PYELONEPHRITIS: Xanthogranulomatous pyelonephritis is characterized by lipid-laden macrophages surrounding neutrophils, lymphocytes, plasma cells, necrotic debris, and histiocytes.

■ VASCULAR DISEASES

FIGURE 13–18. BENIGN NEPHROSCLEROSIS (HYALINE ARTERIOLOSCLEROSIS): Hyaline arteriolosclerosis may be seen in elderly normotensive individuals and in persons with mild hypertension. **A:** In benign arteriolosclerosis, there is narrowing of the arteriolar lumina caused by thickening and hyalinization of the walls. The sclerosis causes ischemic atrophy of the tubules and interstitial fibrosis (**B**); in addition, there is periglomerular fibrosis and eventual total sclerosis of the glomeruli (**C**).

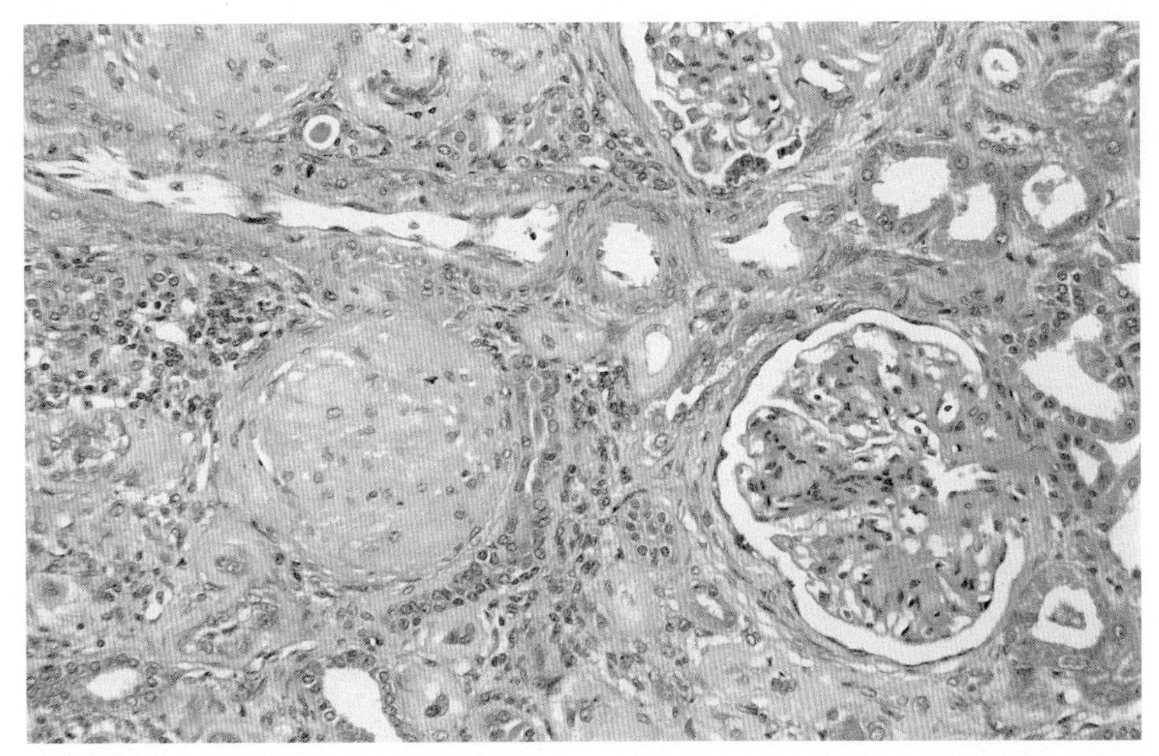

Figure 13–22 A

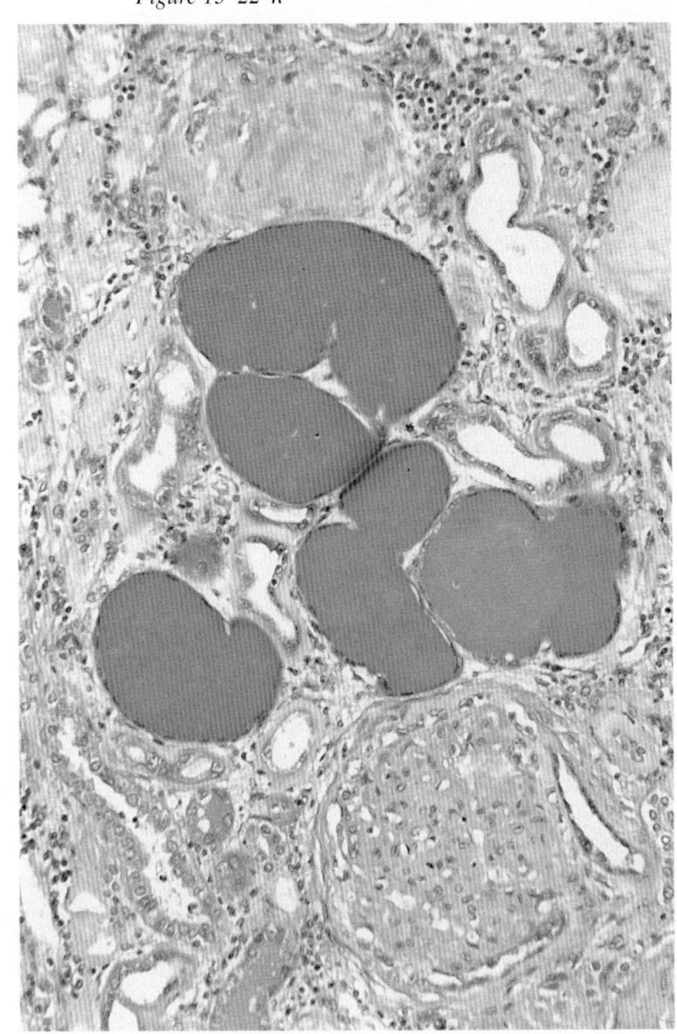

Figure 13–22 B

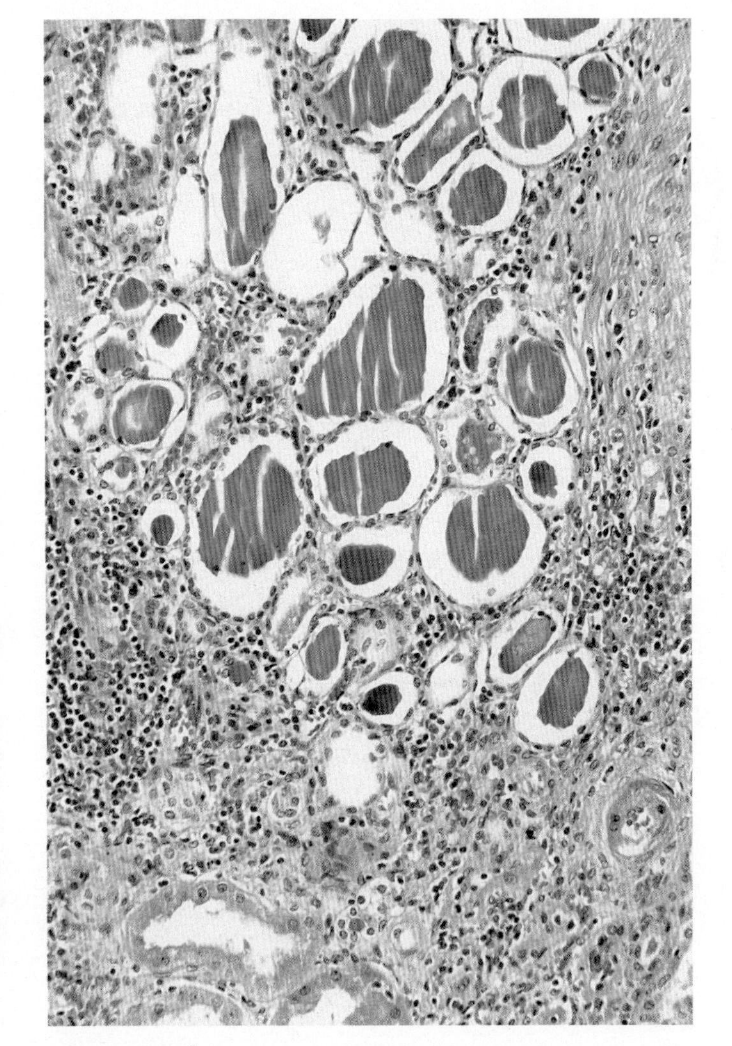

Figure 13–22 C

■ TUMORS

FIGURE 13–23. CORTICAL ADENOMA: A: Cortical adenomas are small, well-circumscribed lesions in the cortex of the kidney. B: These lesions are characterized by small irregular cuboidal cells with round uniform nuclei that lack features of anaplasia.

FIGURE 13–24. MYELOLIPOMA: Myelolipomas may be found in the retroperitoneum and are composed of hematopoietic tissue and mature adipose tissue.

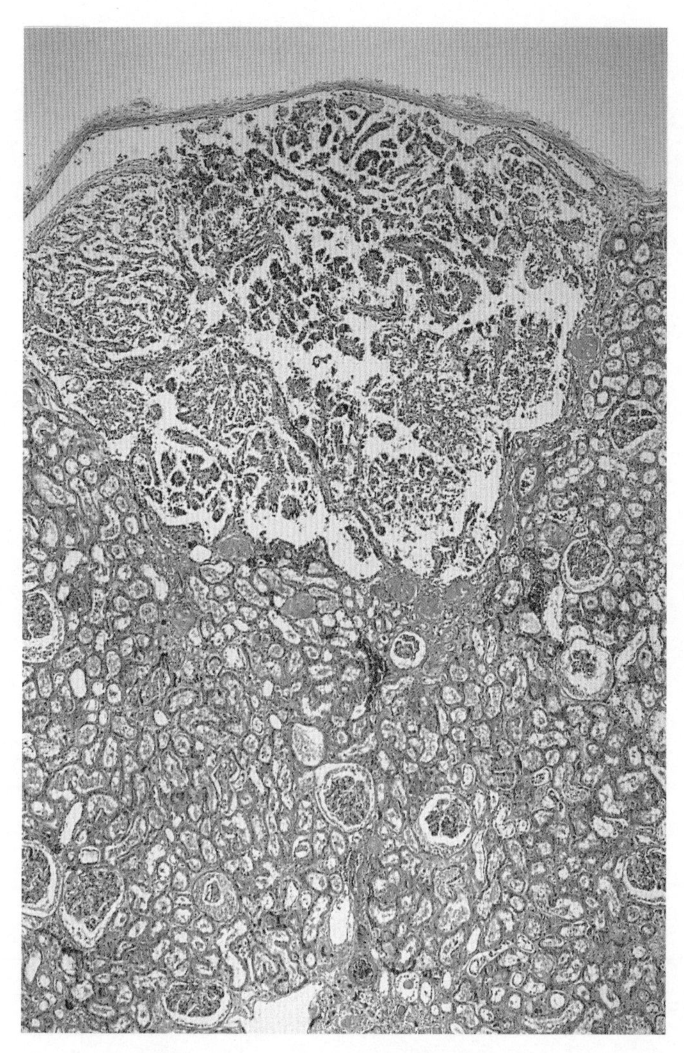

Figure 13–23 A

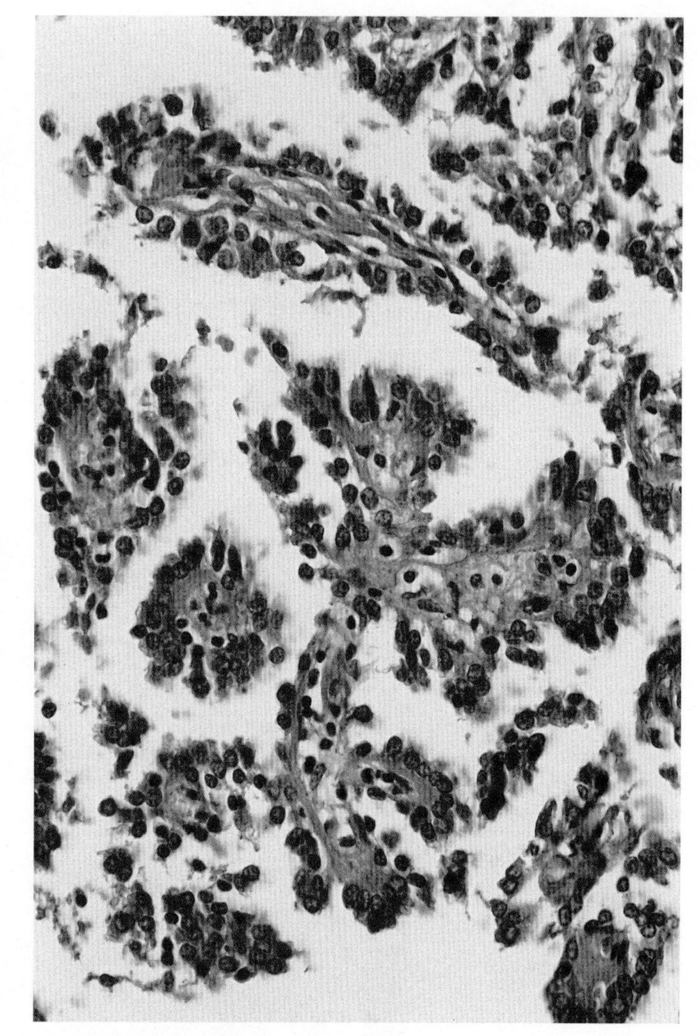

Figure 13–23 B

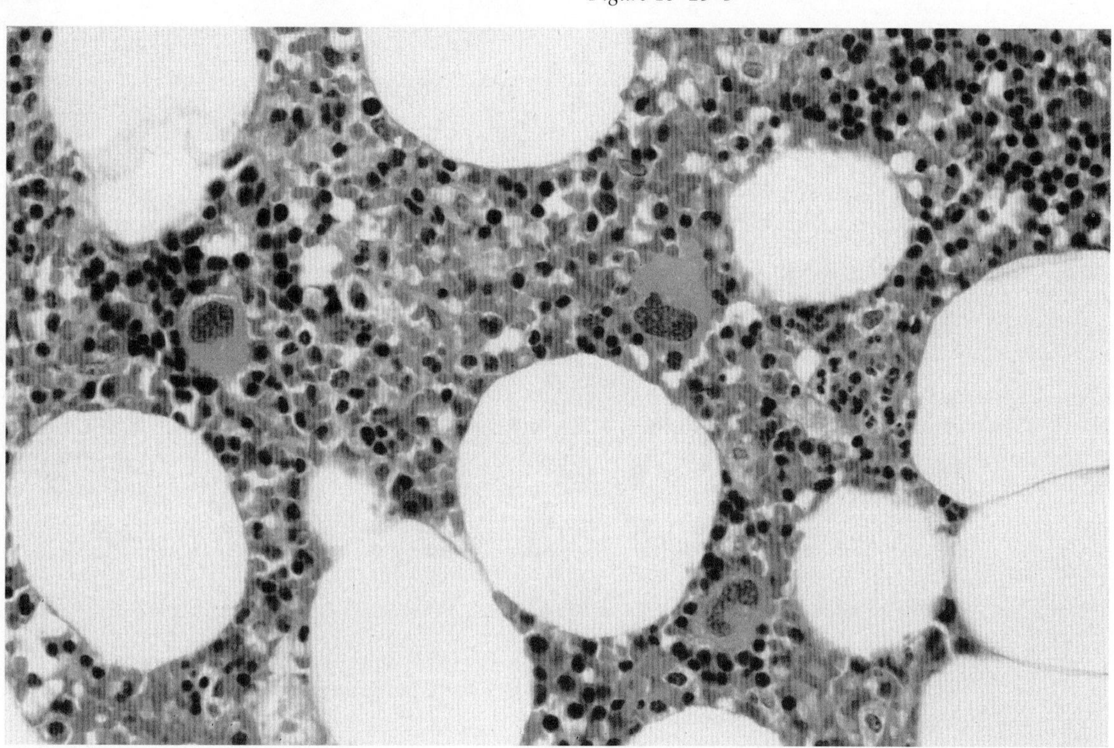

Figure 13–24

■ TUMORS

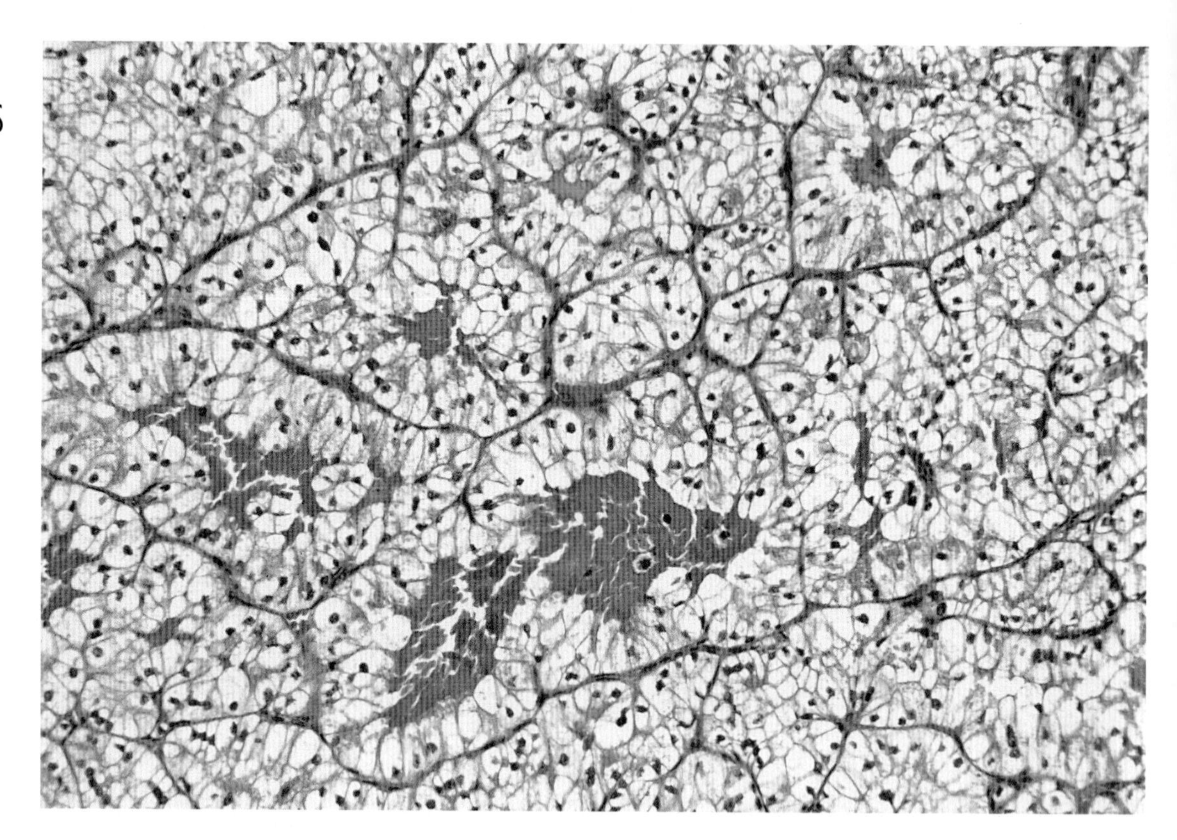

Figure 13–25 A

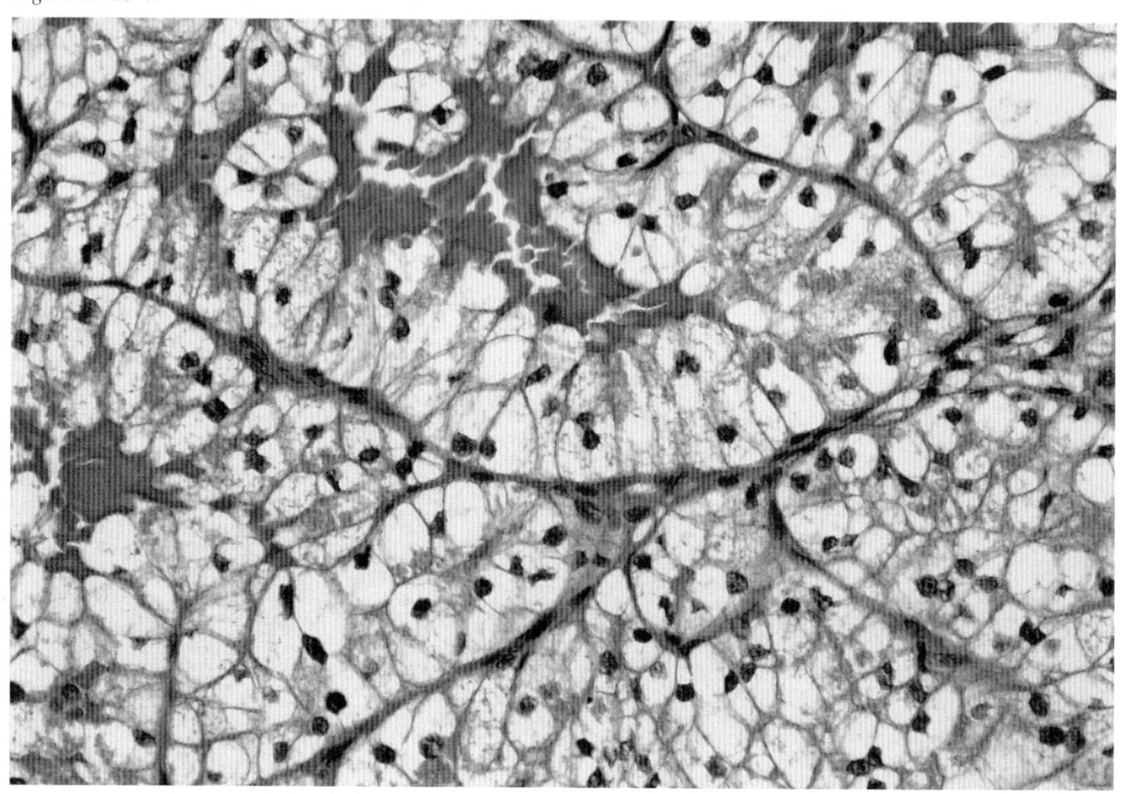

Figure 13–25 B

FIGURE 13–25. RENAL CELL CARCINOMA (HYPERNEPHROMA): A: The most common histologic pattern of renal cell carcinoma is the clear cell variant in which the cells are arranged in clusters and abortive tubules. **B:** The clear cells have distinct cell membranes and small uniform nuclei.

■ TUMORS

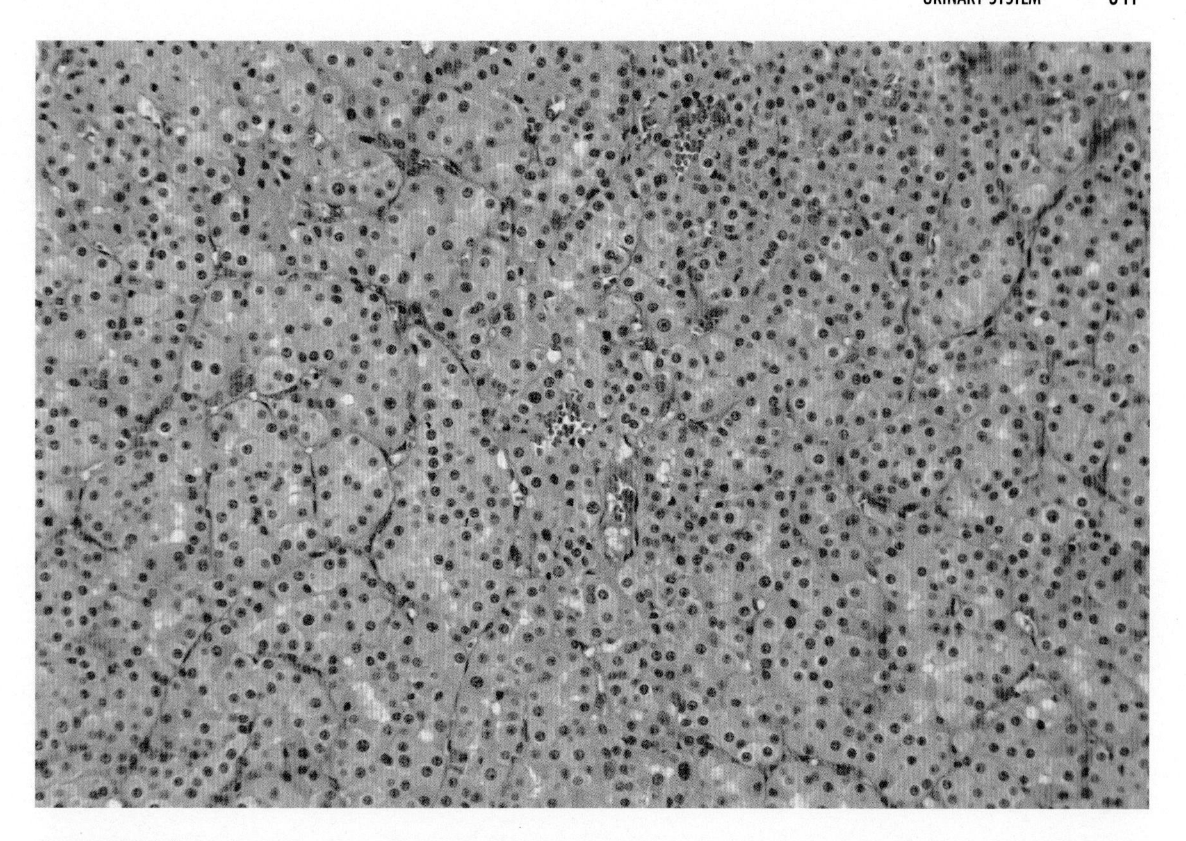

Figure 13–26 A

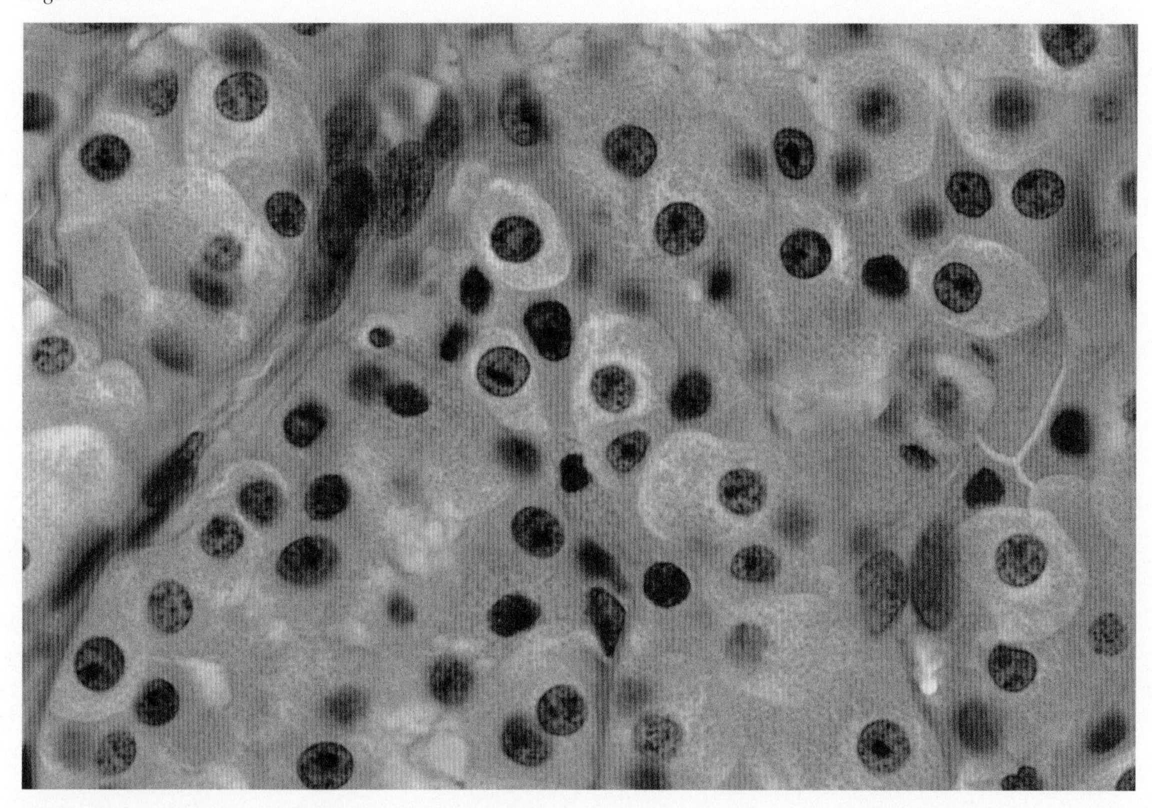

Figure 13–26 B

FIGURE 13–26. ONCOCYTOMA: A: In oncocytomas, the cells are uniform and arranged in nests. **B:** The cells have abundant granular eosinophilic cytoplasm with bland centrally located nuclei.

■ TUMORS

FIGURE 13–27. WILMS' TUMOR (NEPHROBLASTOMA): A: Wilms' tumors are triphasic tumors composed of an **epithelial** component, a **blastematous** component, and a **stromal** component. The epithelial component is composed of abortive tubules (**B**) and abortive glomeruli (**C**).

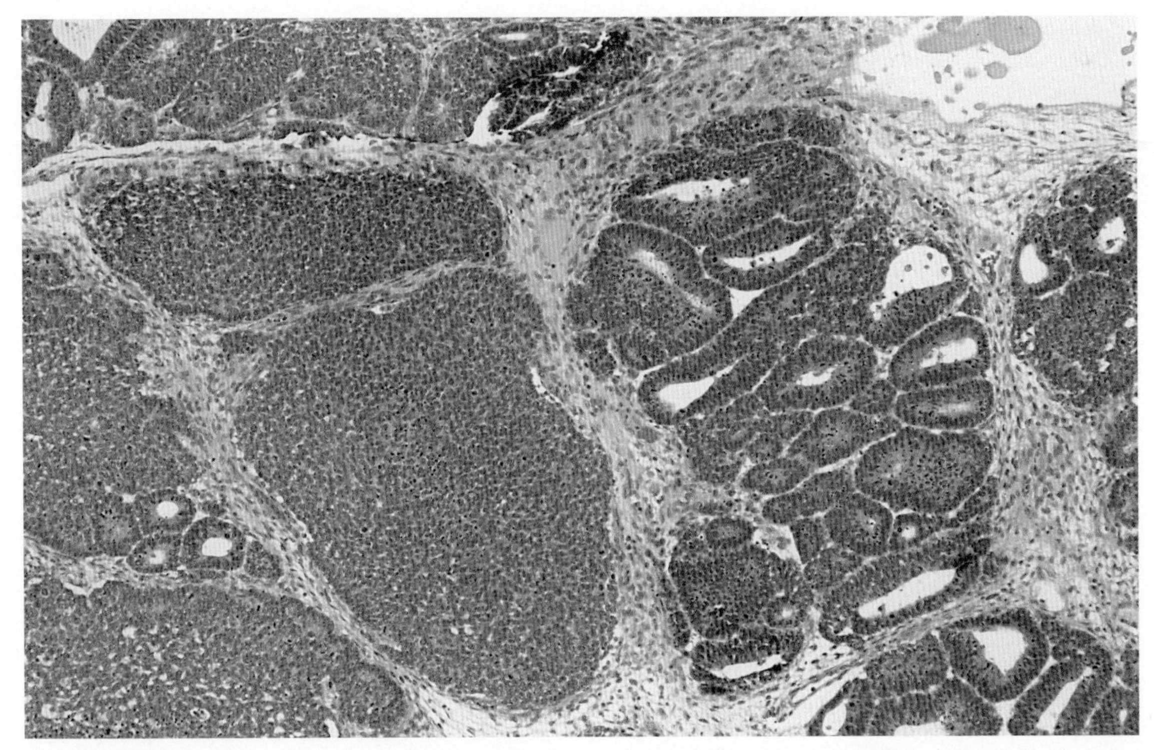

Figure 13–27 A

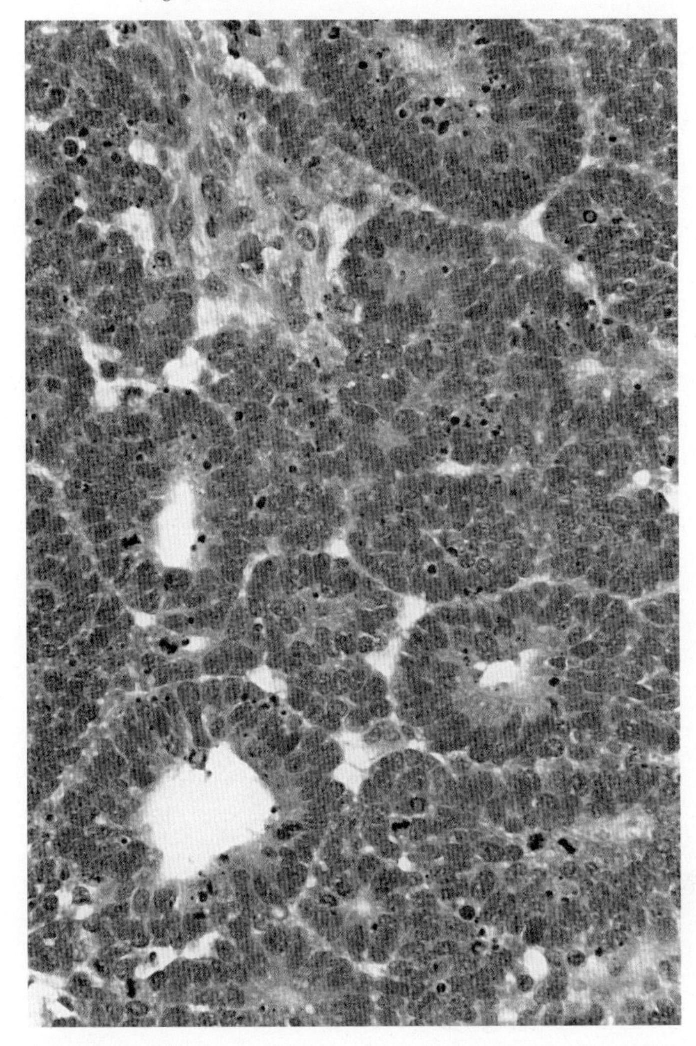

Figure 13–27 B

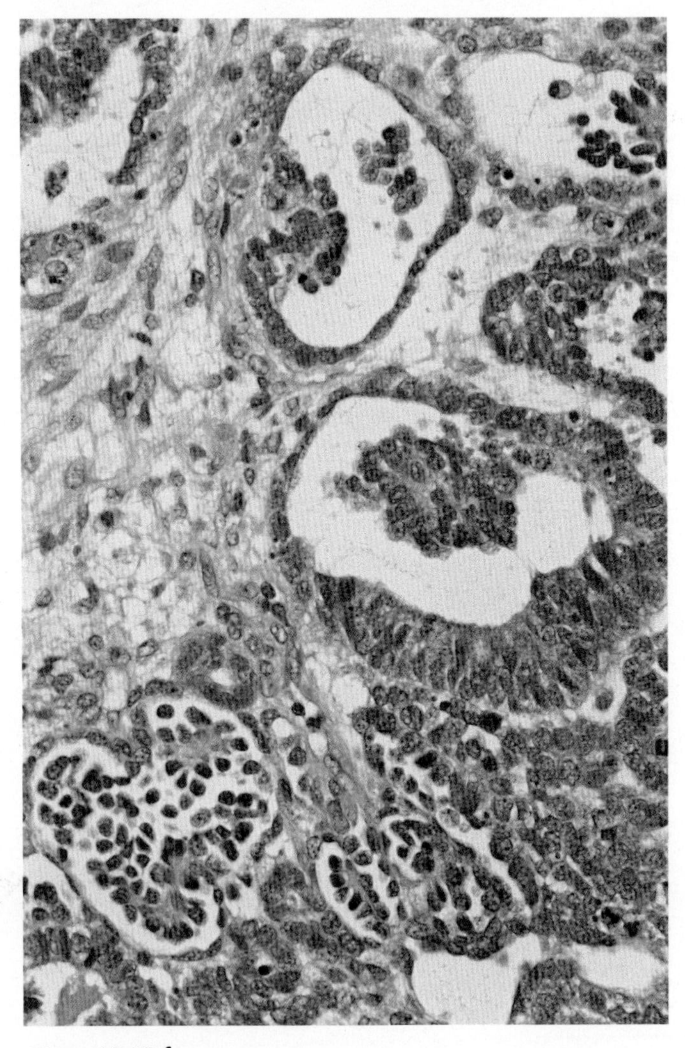

Figure 13–27 C

■ TUMORS

FIGURE 13–27. WILMS' TUMOR (NEPHROBLASTOMA): D: The blastematous component in Wilms' tumor is composed of sheets of small undifferentiated cells that have little cytoplasm. Mitotic figures are present. The stromal component separates the blastema into nodules and may be myxoid or may contain heterologous elements, such as cartilage (**E**), adipose tissue (**F**), skeletal or smooth muscle, or bone or neural tissue.

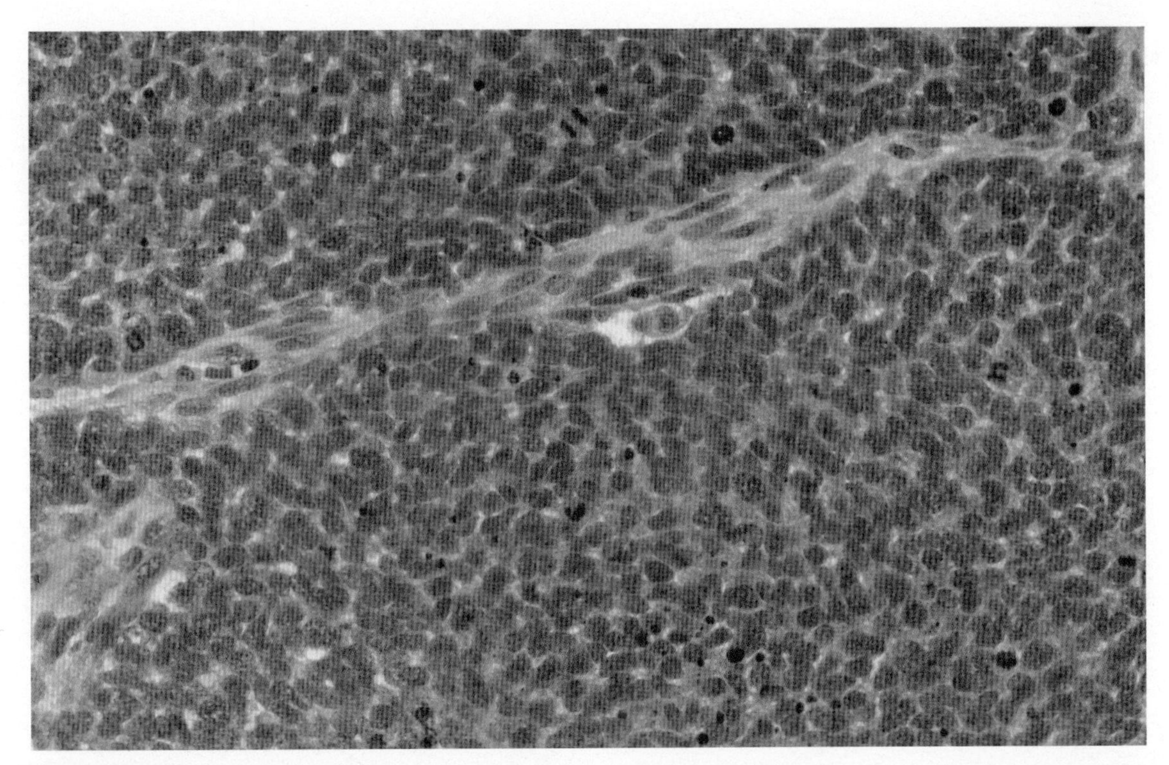

Figure 13–27 D

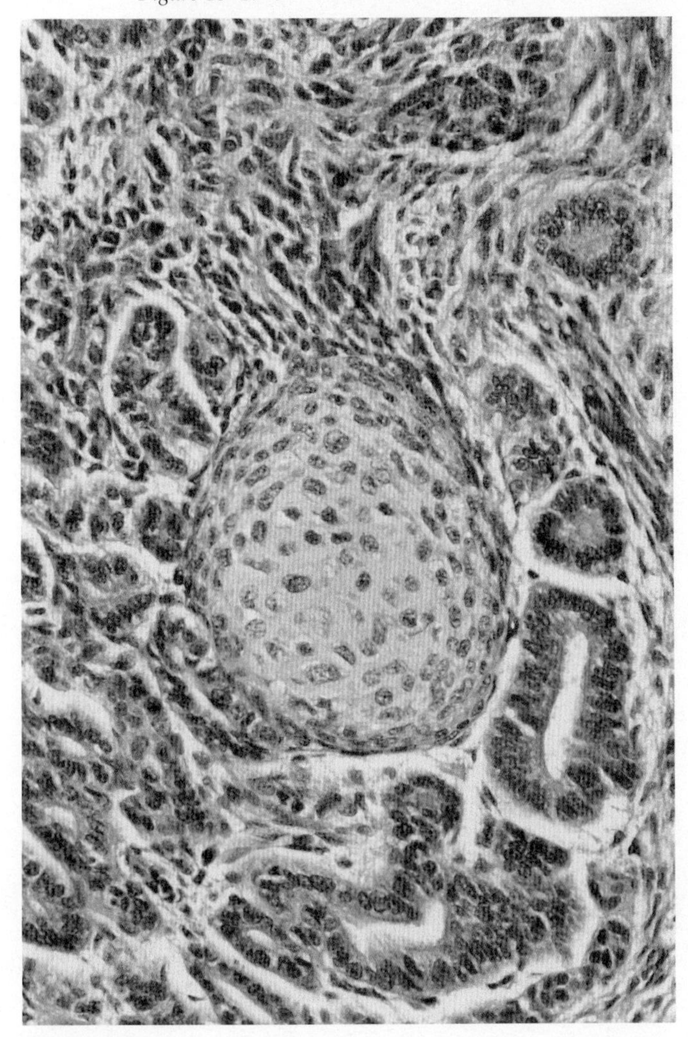

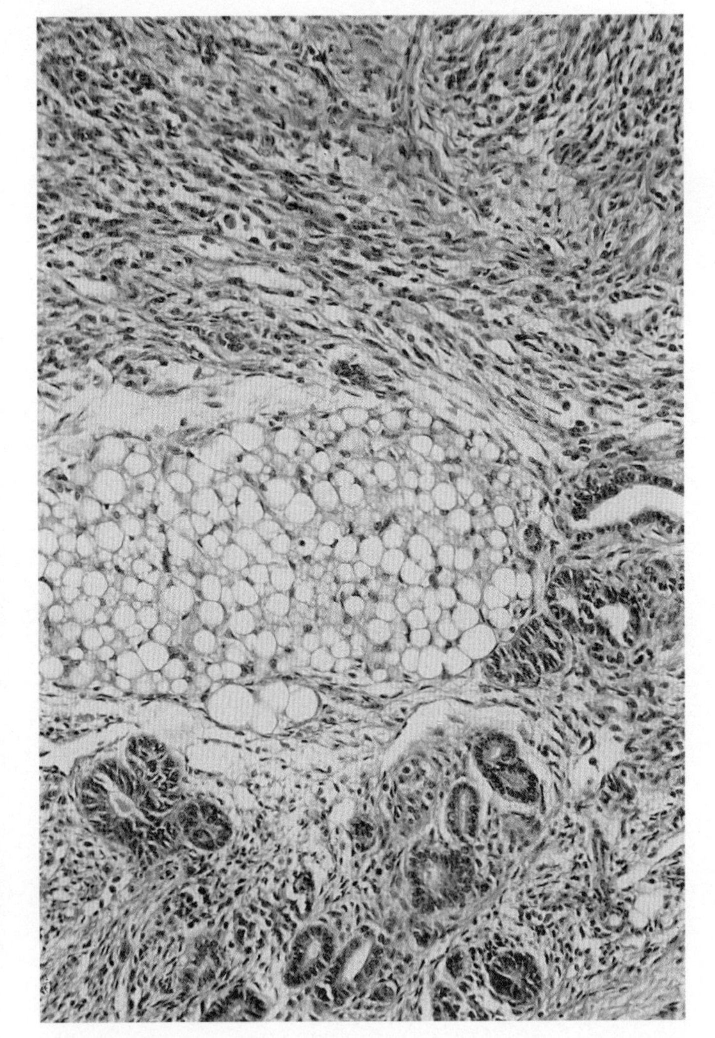

Figure 13–27 E

Figure 13–27 F

■ BLADDER AND PELVICALICEAL SYSTEM

FIGURE 13–28. INTERSTITIAL CYSTITIS (HUNNER'S ULCER): Interstitial cystitis is characterized by an ulcerated epithelium (**A**) and an inflammatory response in the lamina propria (**B**). **C**: Mast cells, best seen with a Wright-Giemsa stain, are a prominent component of the mixed inflammatory infiltrate.

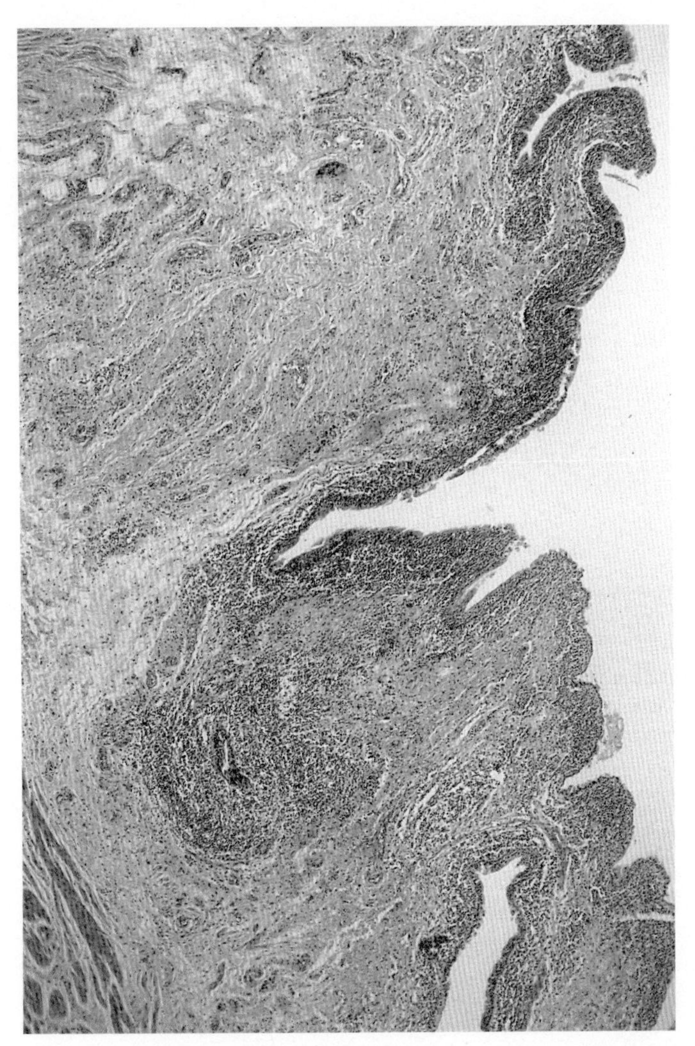

Figure 13–28 A

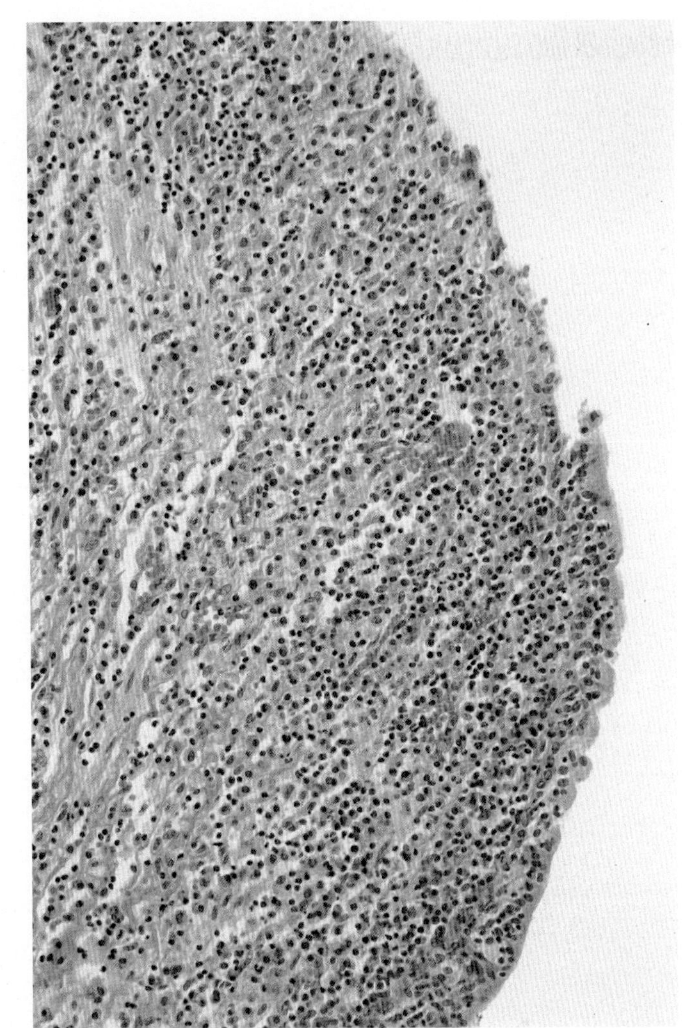

Figure 13–28 B

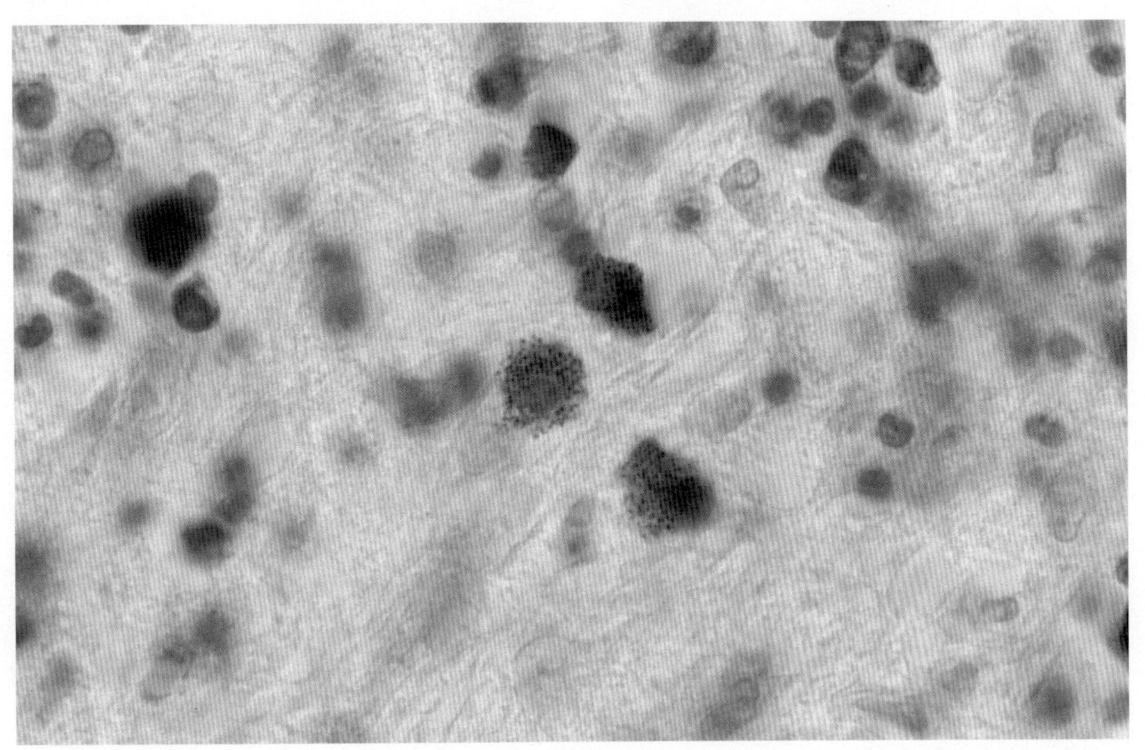

Figure 13–28 C

■ BLADDER AND PELVICALICEAL SYSTEM

FIGURE 13–29. MALAKOPLAKIA: Malakoplakia is seen in association with bacterial infections. A: In malakoplakia, the surface epithelium is denuded. B: Collections of epithelioid histiocytes are present in the lamina propria. C: Within the histiocytes are minute blue inclusions (*arrow*), which represent calcospherites—**Michaelis-Gutmann bodies**.

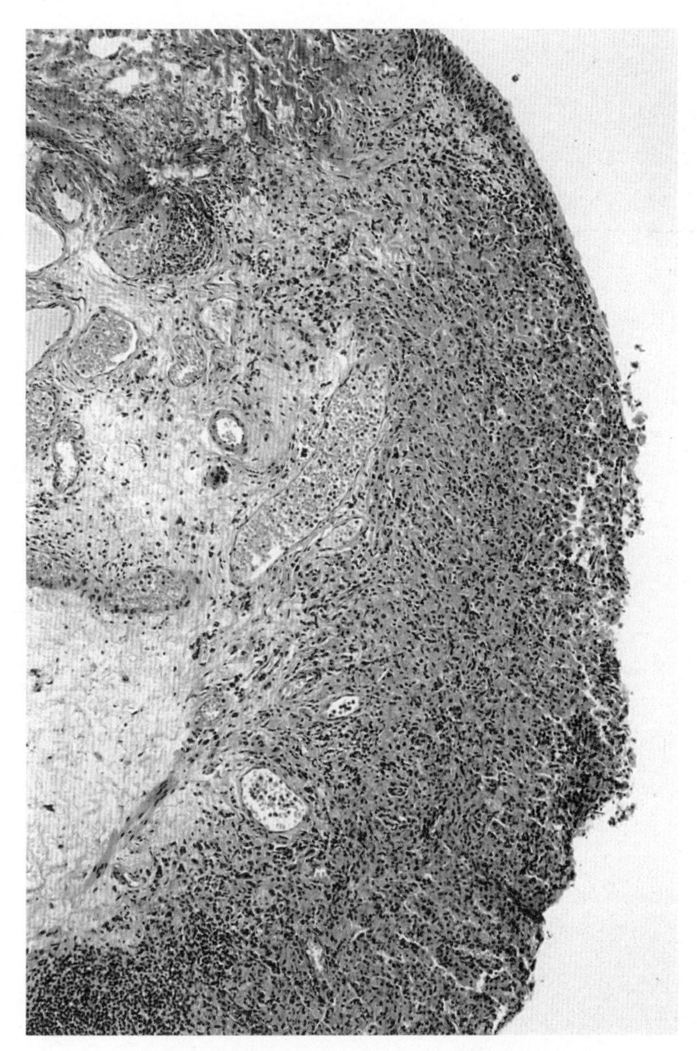

Figure 13–29 A

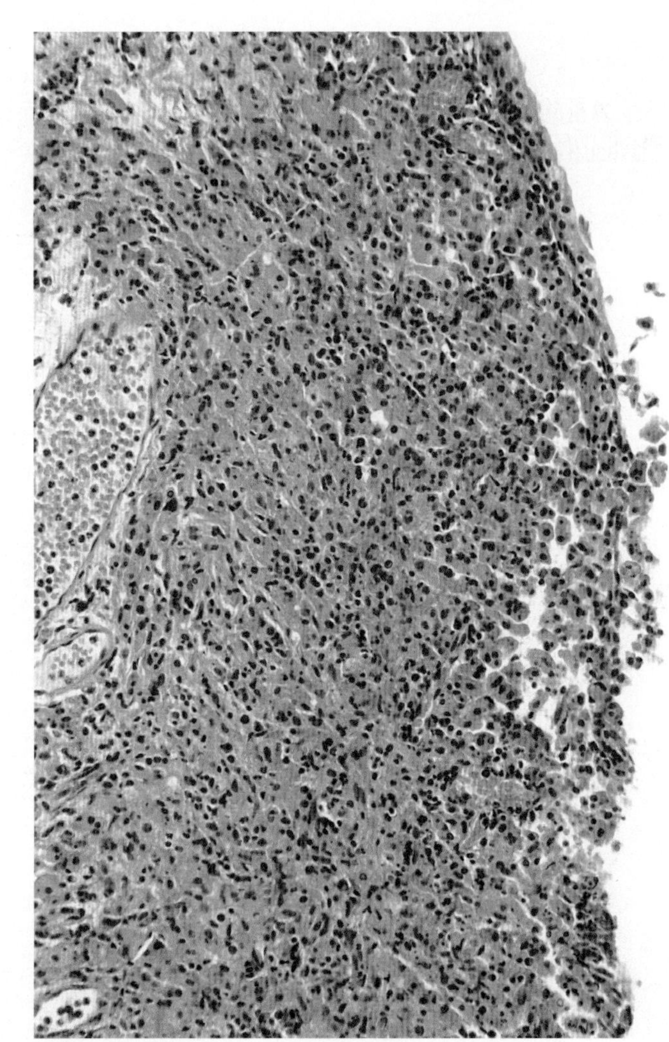

Figure 13–29 B

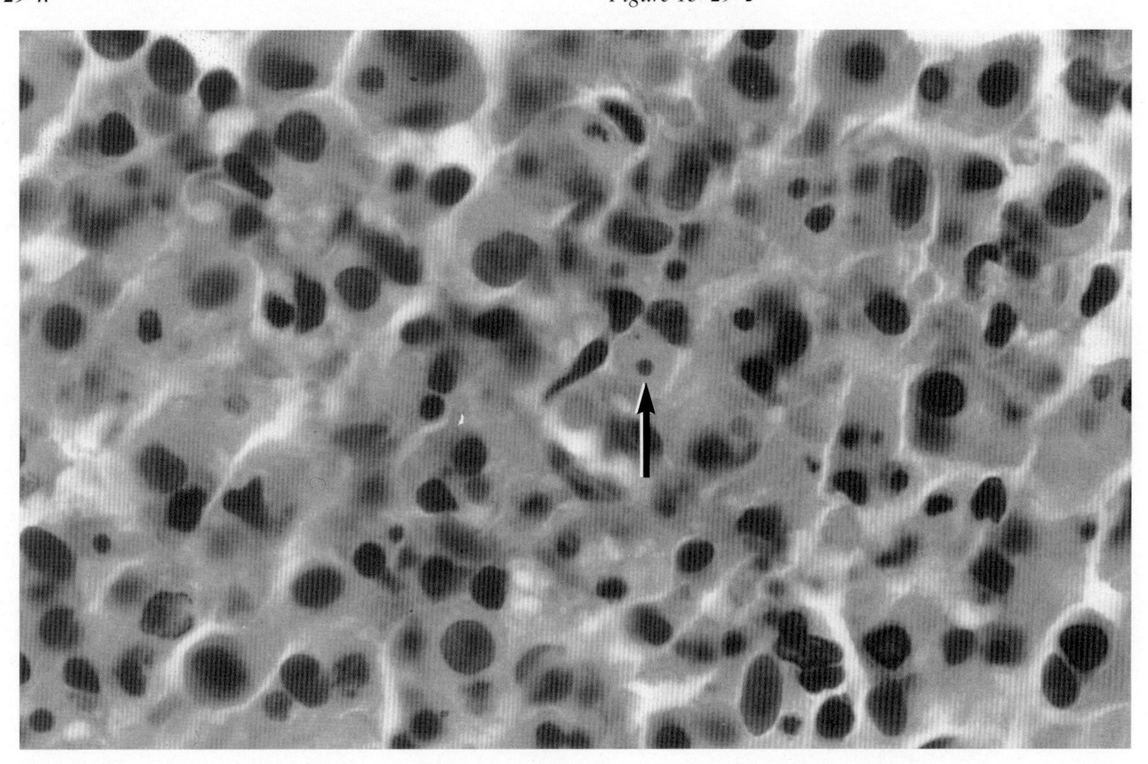

Figure 13–29 C

■ BLADDER AND PELVICALICEAL SYSTEM

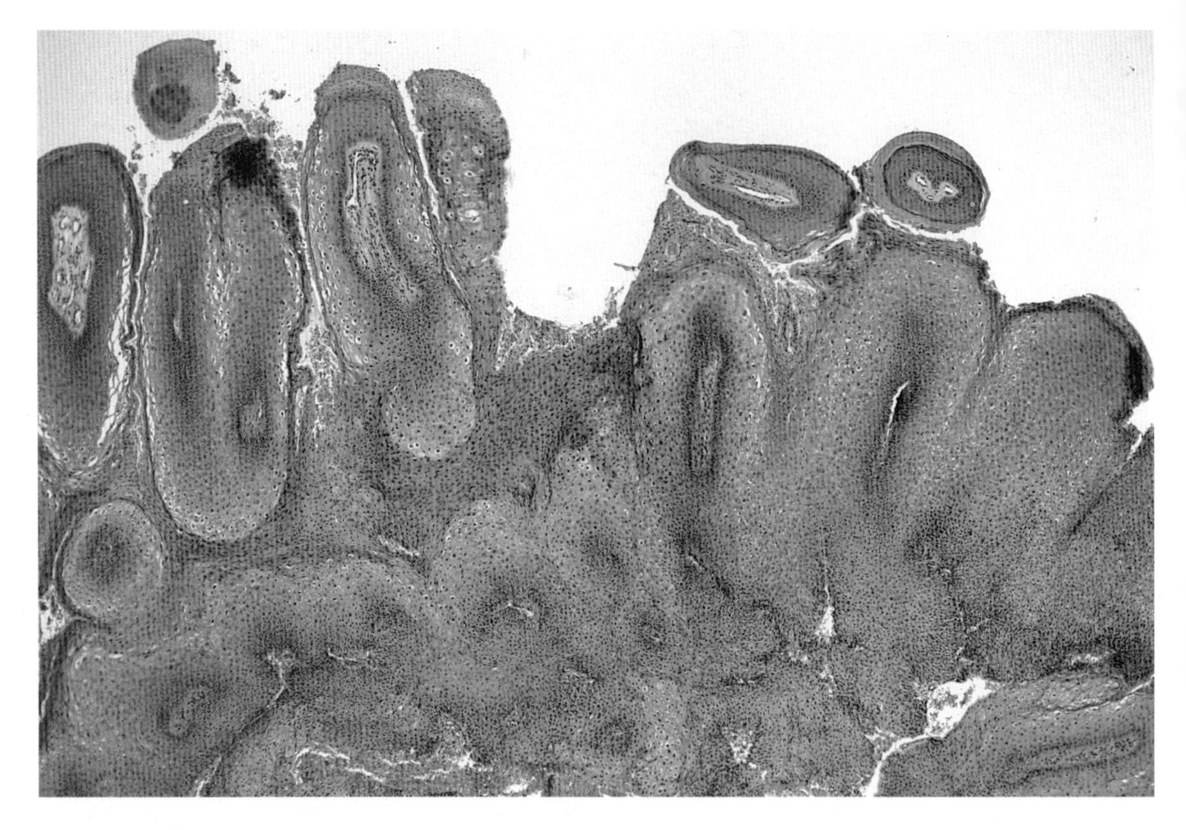

Figure 13–30 A

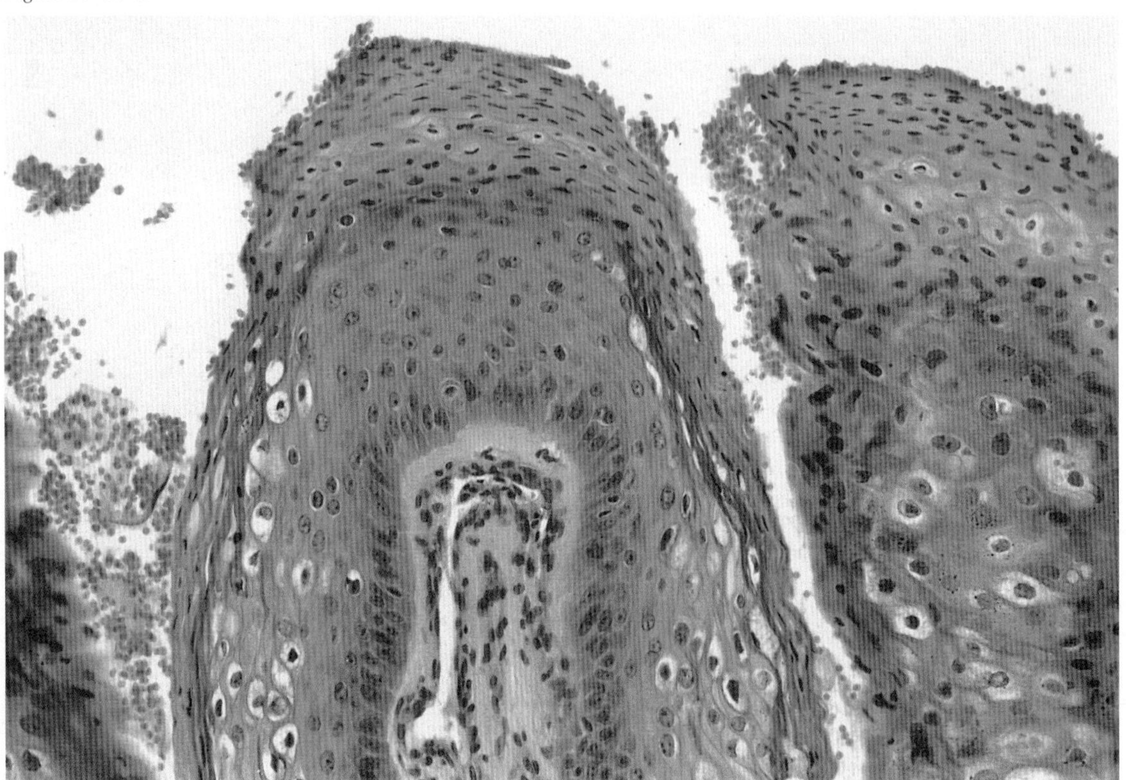

Figure 13–30 B

FIGURE 13–30. CONDYLOMA ACUMINATUM: Condyloma acuminatum is caused by human papilloma virus. A: In this photomicrograph of a condyloma taken from the urethra, papillary fronds are covered by hyperplastic and parakeratotic squamous epithelium. B: **Koilocytes**—epithelial cells with shrunken pyknotic nuclei surrounded by a clear perinuclear halo—are pathognomonic.

■ BLADDER AND PELVICALICEAL SYSTEM

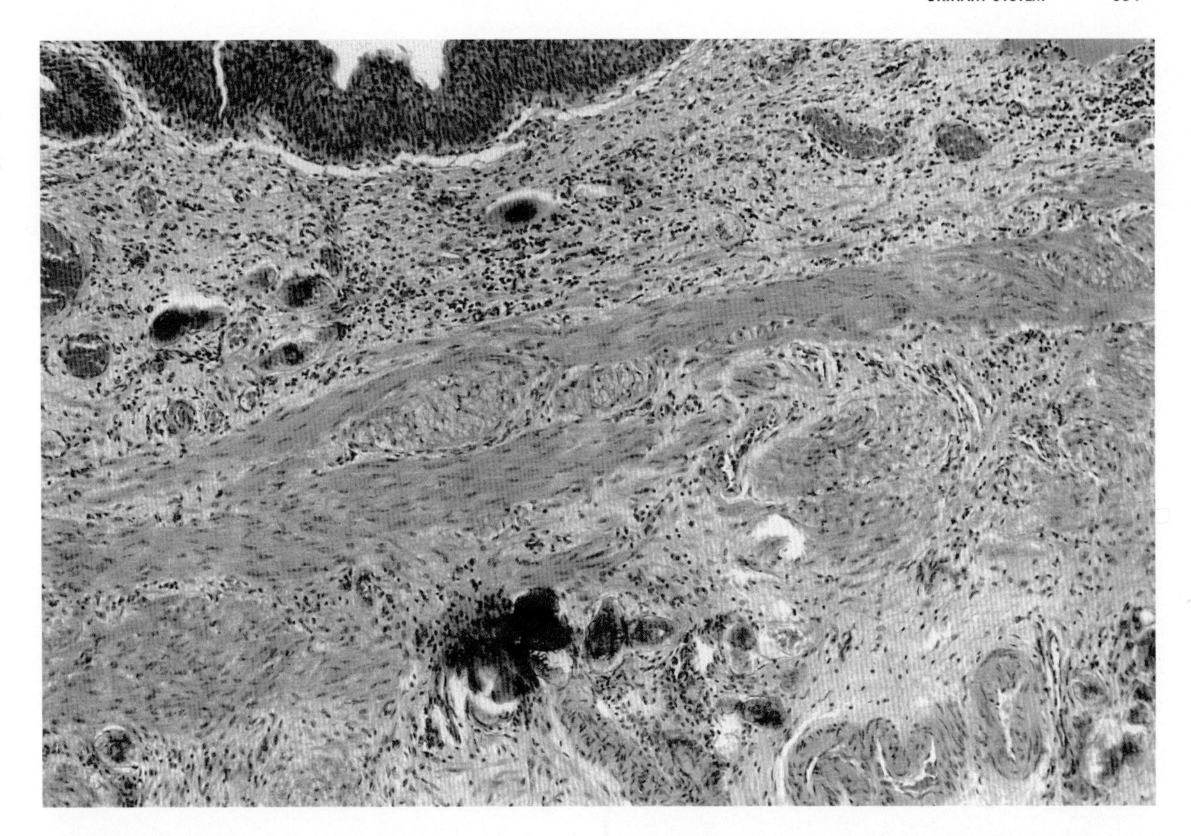

Figure 13–31 **A**

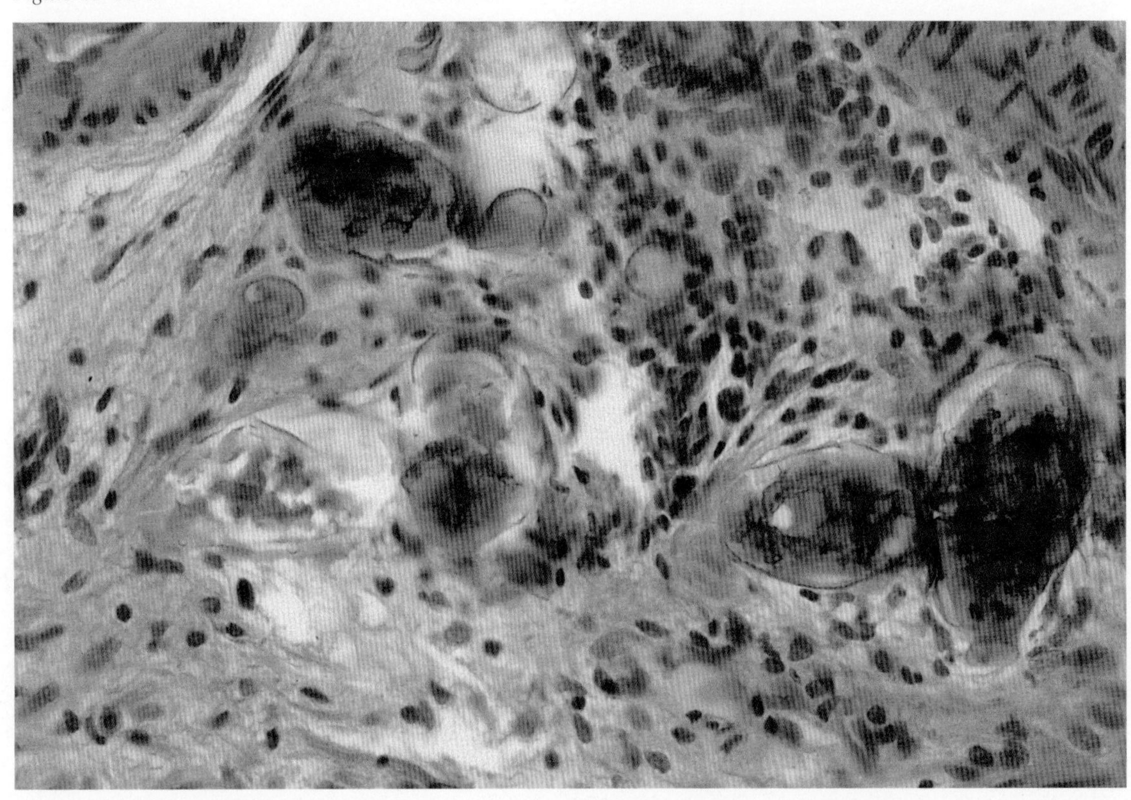

Figure 13–31 **B**

FIGURE 13–31. SCHISTOSOMIASIS: In schistosomiasis, eggs from *Schistosoma haematobium* are deposited in the bladder mucosa (**A**), where they die, calcify, and elicit a foreign body giant cell reaction (**B**).

■ BLADDER AND PELVICALICEAL SYSTEM

Reactive and proliferative changes of the urothelium can be manifested as a number of histomorphologic changes that appear to be a continuum of one condition to the next. These changes are represented by Von Brunn's nests, cystitis cystica, and cystitis glandularis.

FIGURE 13–32. VON BRUNN'S NESTS: Von Brunn's nests—invaginations of the surface urothelium—lead to the formation of urothelial nests in the lamina propria.

FIGURE 13–33. CYSTITIS CYSTICA: When Von Brunn's nests have lost continuity with the surface and have become cystically dilated, the condition is called cystitis cystica.

FIGURE 13–34. CYSTITIS GLANDULARIS: When the epithelial lining of the nests in cystitis cystica undergo glandular metaplasia to mucus-producing cells, cystitis glandularis results.

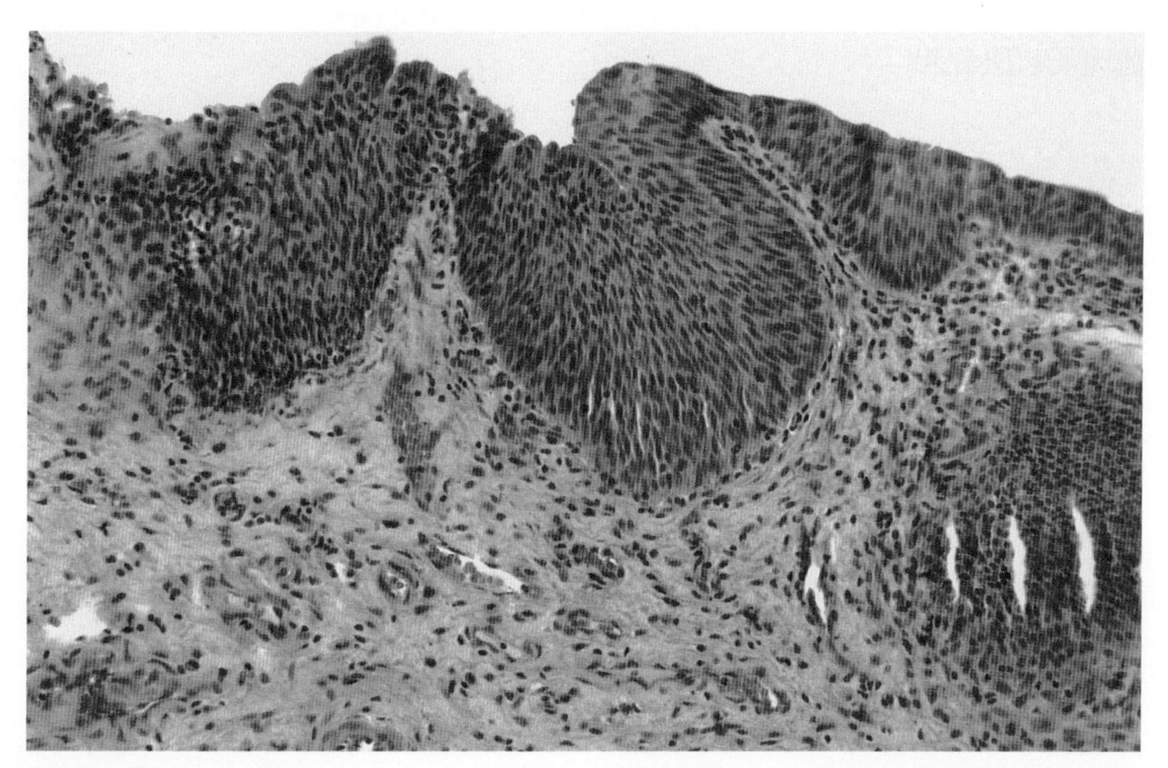

Figure 13–32

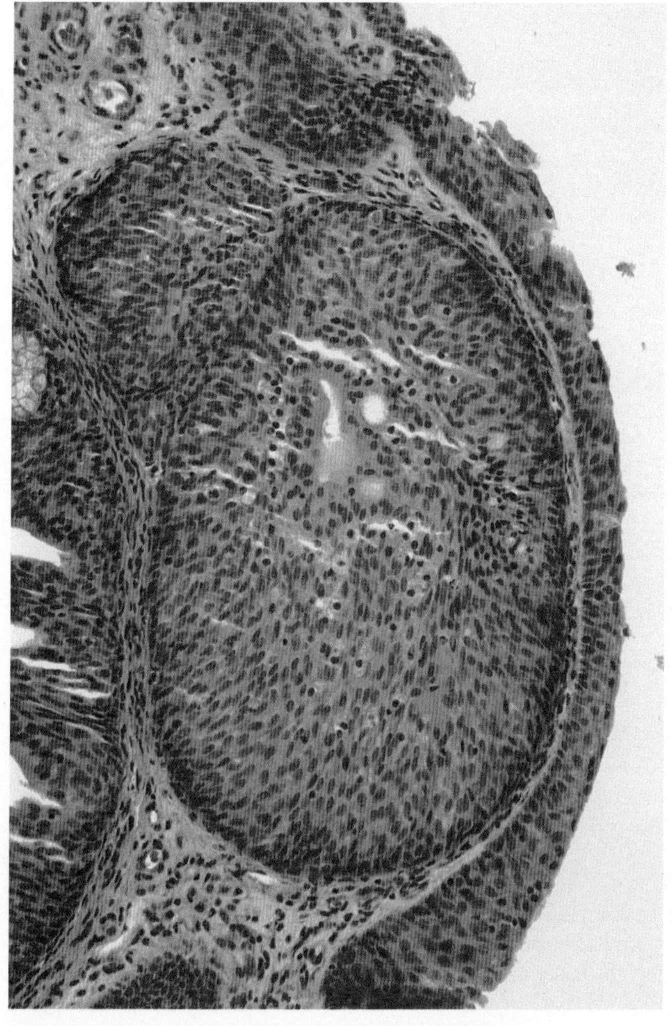

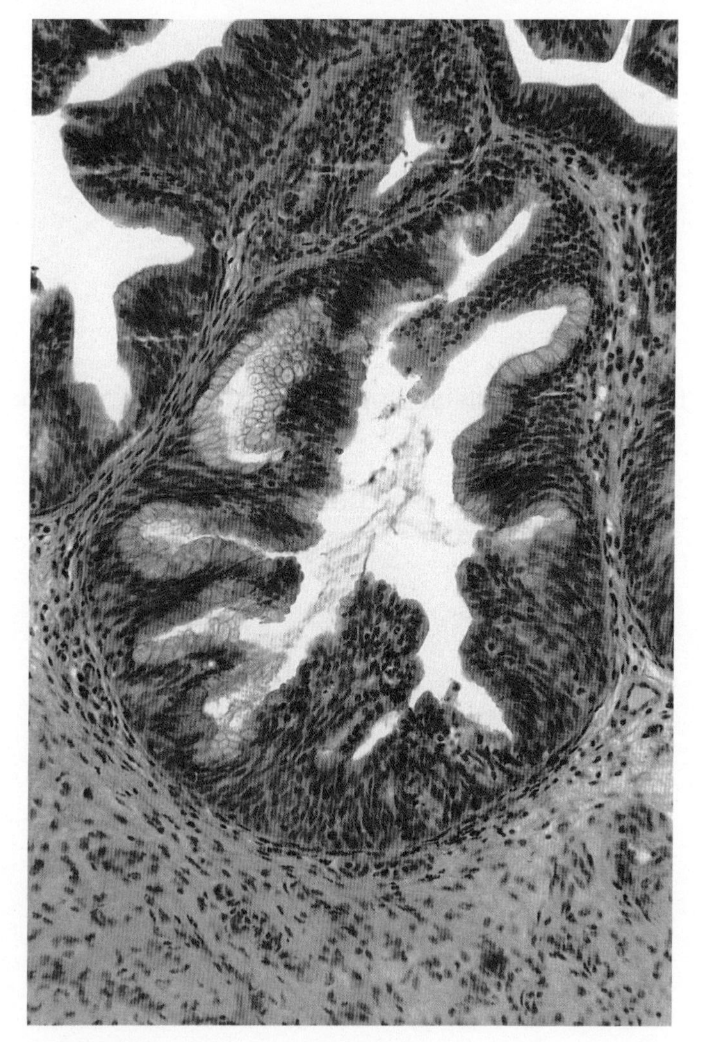

Figure 13–33

Figure 13–34

■ BLADDER AND PELVICALICEAL SYSTEM

TUMORS

TRANSITIONAL CELL CARCINOMA (TCC): In transitional cell carcinoma of the bladder and pelvicaliceal system, the histologic grade correlates with invasive potential and prognosis. Grade 1 tumors are rarely invasive but do recur. Few grade 2 tumors are invasive, and 50% of grade 3 tumors are invasive.

FIGURE 13–35. GRADE 1 PAPILLARY TRANSITIONAL CELL CARCINOMA: A: In grade 1 papillary transitional cell carcinoma, a fibrovascular connective tissue core is covered by a thickened urothelium, which is a few cell layers thicker than normal. **B:** The umbrella cell layer is preserved, and cell polarization is maintained. Few mitotic figures may be present.

Grade 2 transitional cell carcinoma has architectural and cytologic features intermediate between grades 1 and 3.

FIGURE 13–35. GRADE 3 TRANSITIONAL CELL CARCINOMA: C: Grade 3 transitional cell carcinoma usually does not maintain a papillary pattern. There is loss of cell polarization and nuclei are hyperchromatic, pleomorphic, and enlarged. Blood vessels are surrounded by a thick eosinophilic collar. Necrosis of individual cells and groups of cells and mitotic figures are seen.

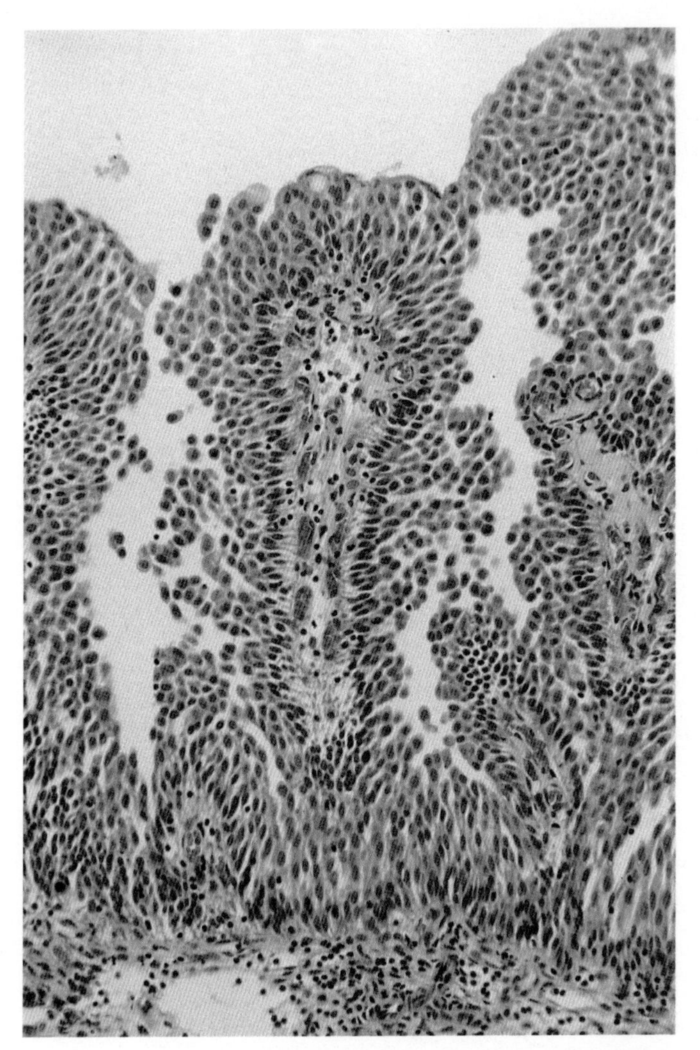

Figure 13–35 A

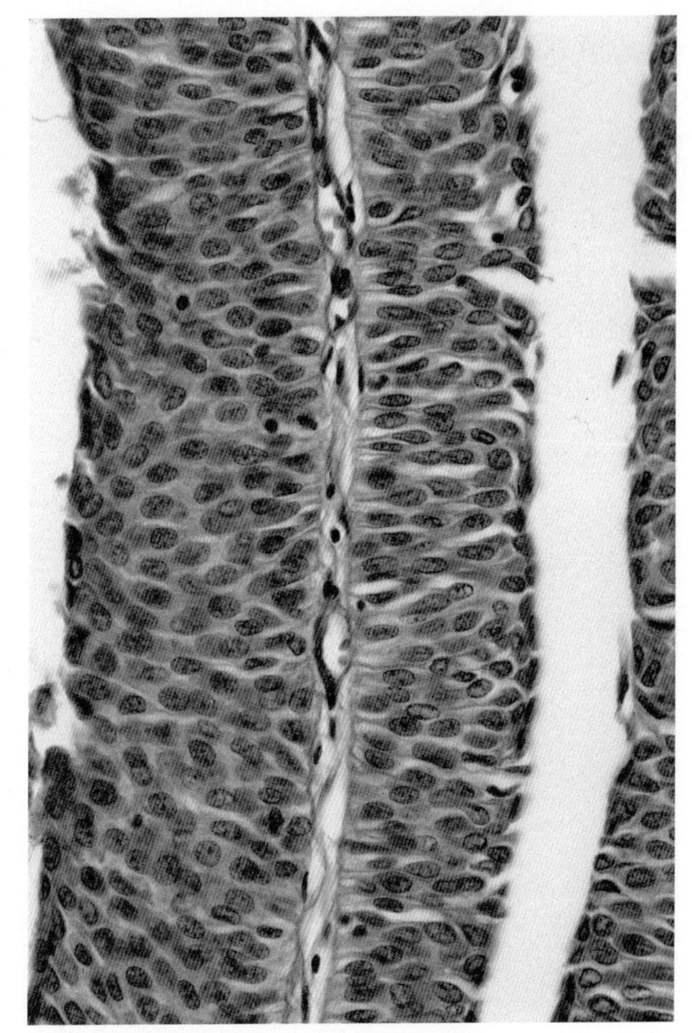

Figure 13–35 B

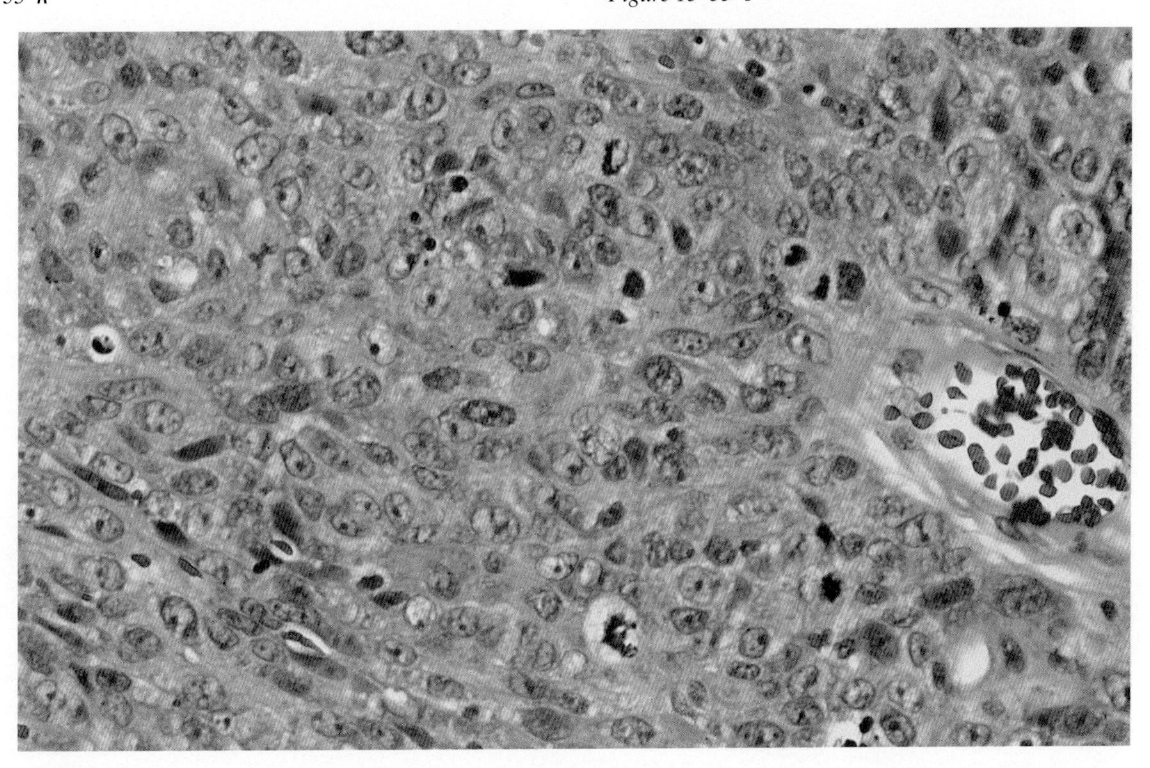

Figure 13–35 C

■ BLADDER AND PELVICALICEAL SYSTEM

TUMORS

FIGURE 13–36. TRANSITIONAL CELL CARCINOMA (TCC) IN SITU: A: Transitional cell carcinoma in situ of the urothelium, which cytologically resembles grade 3 transitional cell carcinoma, is a flat tumor. **B:** Cells do not maintain their polarity, are anaplastic and pleomorphic, and have hyperchromatic nuclei.

FIGURE 13–37. INFILTRATING TRANSITIONAL CELL CARCINOMA: In infiltrating transitional cell carcinoma, tumor cells infiltrate the muscularis propria of the bladder wall.

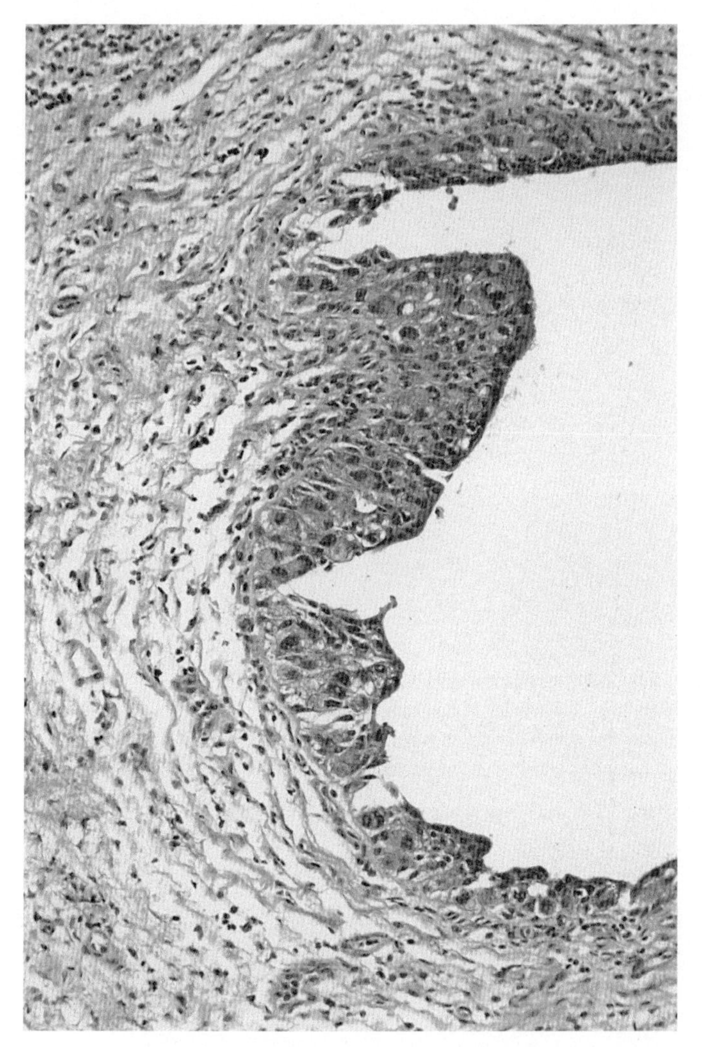

Figure 13–36 A

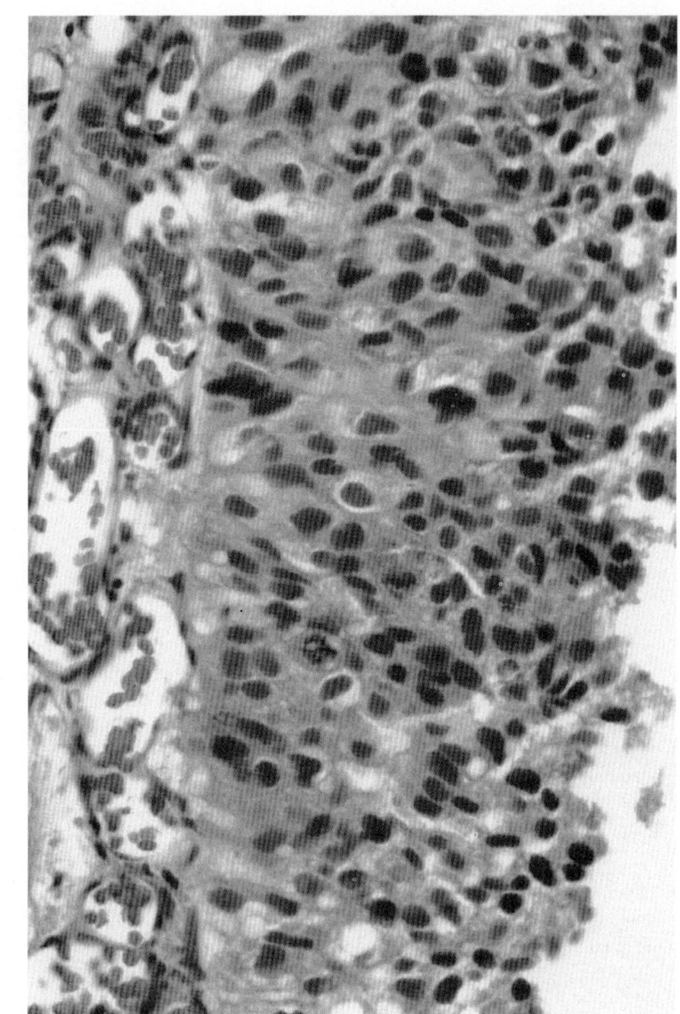

Figure 13–36 B

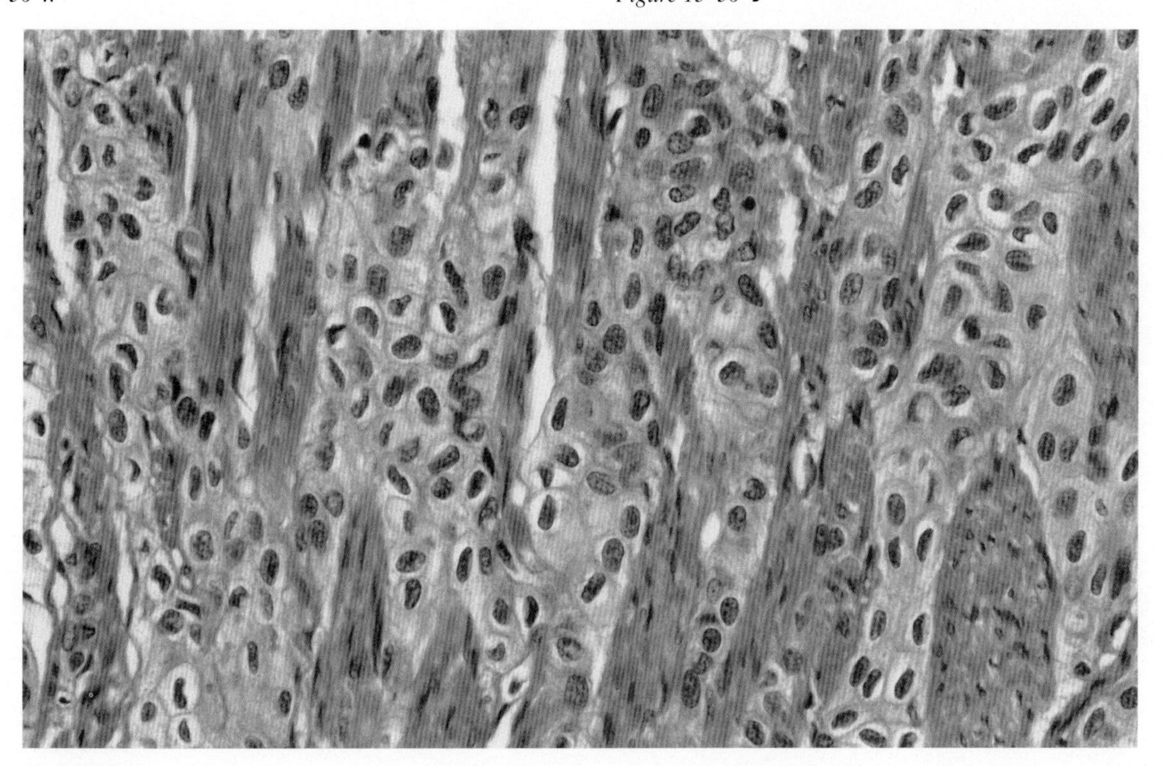

Figure 13–37

14. FEMALE REPRODUCTIVE SYSTEM

■ VULVA

INFLAMMATORY LESIONS

FIGURE 14–1. CONDYLOMA (HUMAN PAPILLOMAVIRUS INFECTION): In this photomicrograph of a verrucous condyloma, the characteristic features of infection by human papillomavirus are seen. **A:** As a consequence of hyperplasia, the epithelium is thrown into papillary projections, which contain a fibrovascular core. **B:** There is **koilocytosis,** in which a perinuclear halo surrounds an irregular and hyperchromatic nucleus. Mitoses are increased in number.

PIGMENTED LESIONS

FIGURE 14–2. LENTIGO SIMPLEX: In lentigo simplex, there is clubbing of the rete ridges, with an increase in the concentration of melanin in the basal cells of the dermal-epidermal junction. Melanophages are present in the dermis.

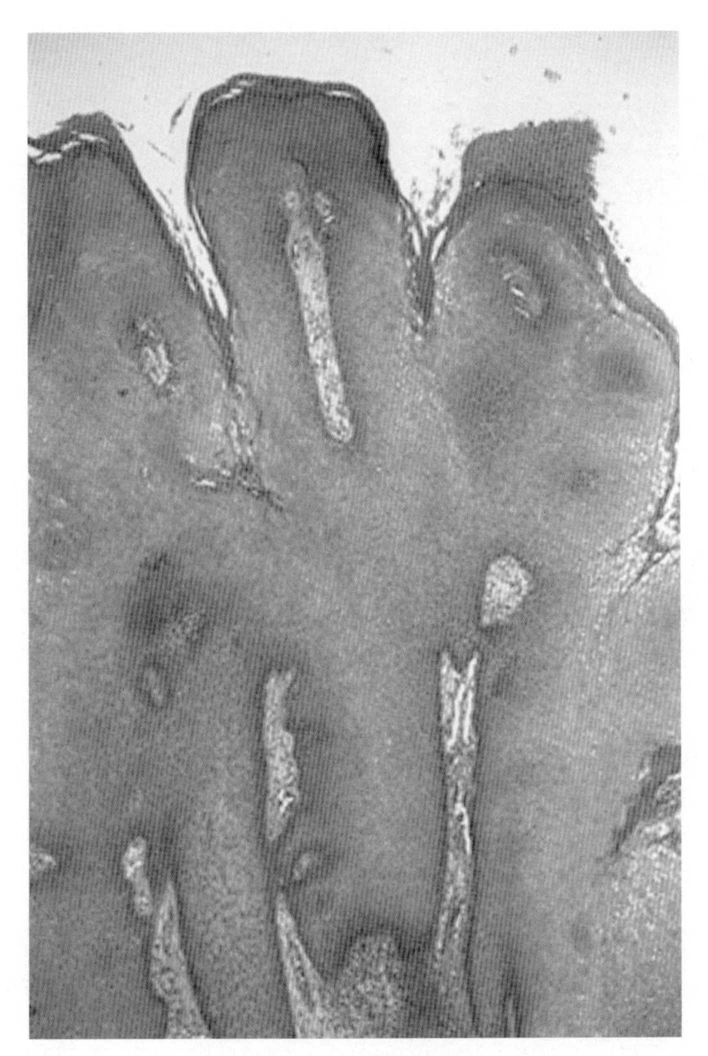

Figure 14–1 A

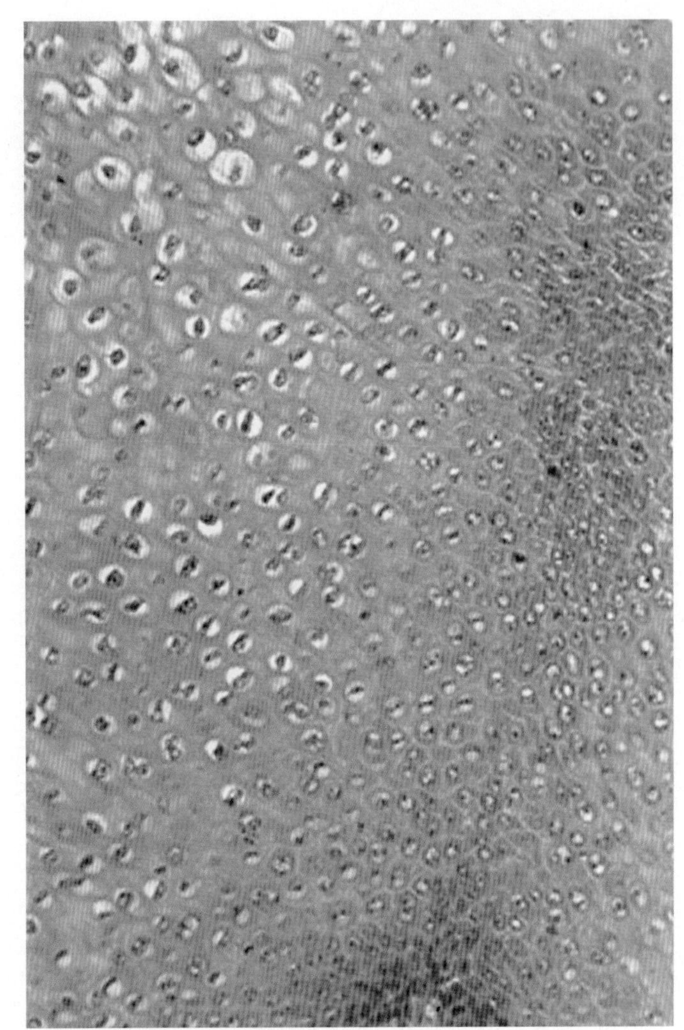

Figure 14–1 B

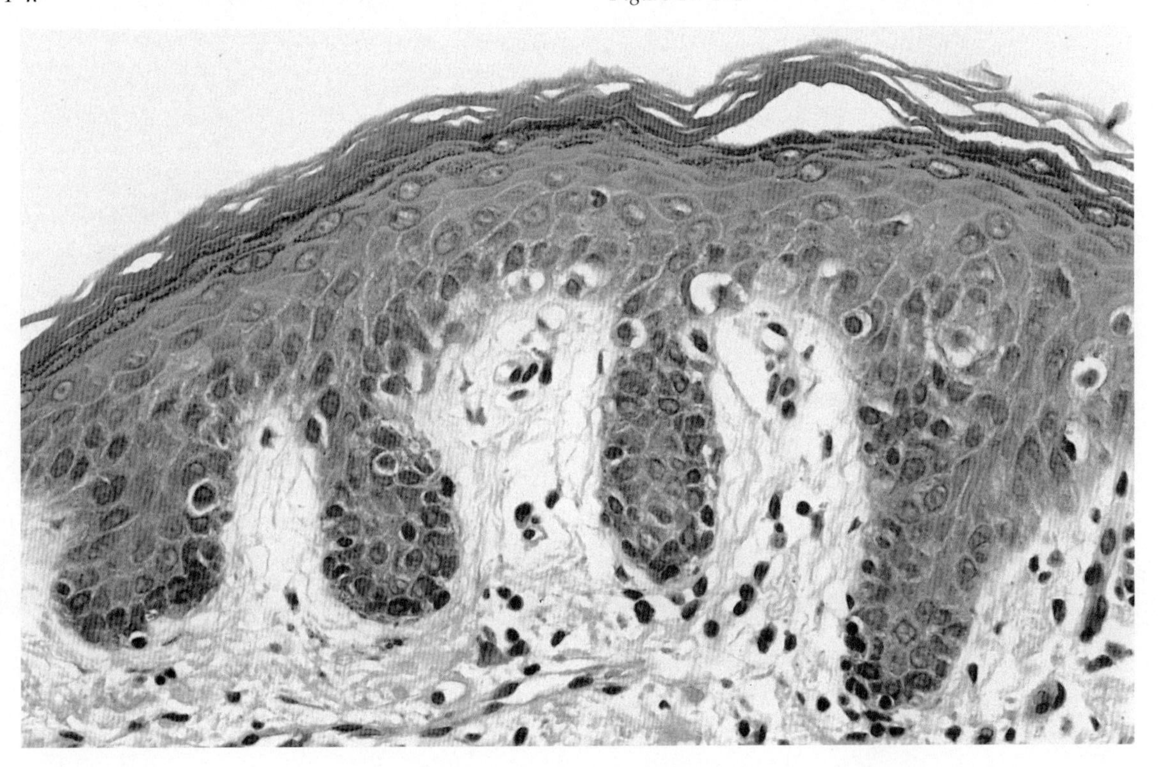

Figure 14–2

■ VULVA

BENIGN TUMORS

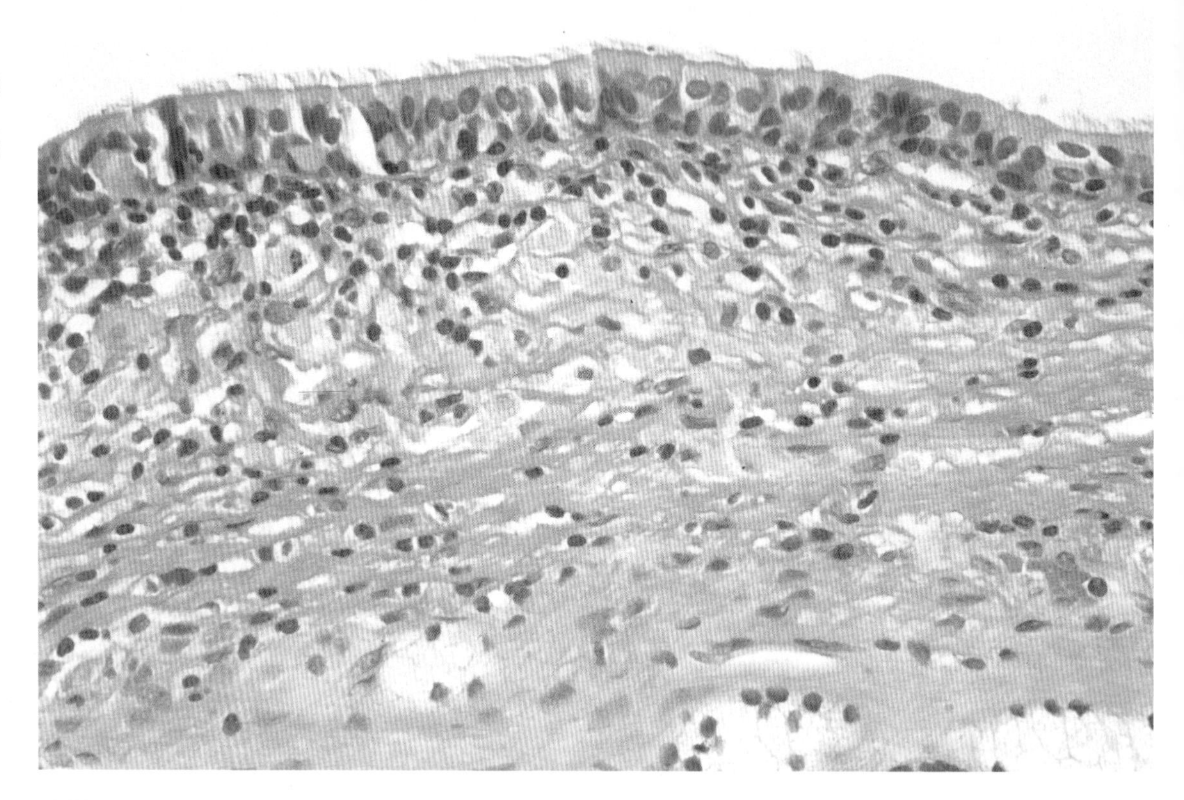

FIGURE 14–3. MUCINOUS CYST OF THE VULVA: The cysts in mucinous cysts of the vulva are lined by columnar cells. In this photomicrograph, the cells are mucus-secreting and ciliated and there is associated squamous metaplasia.

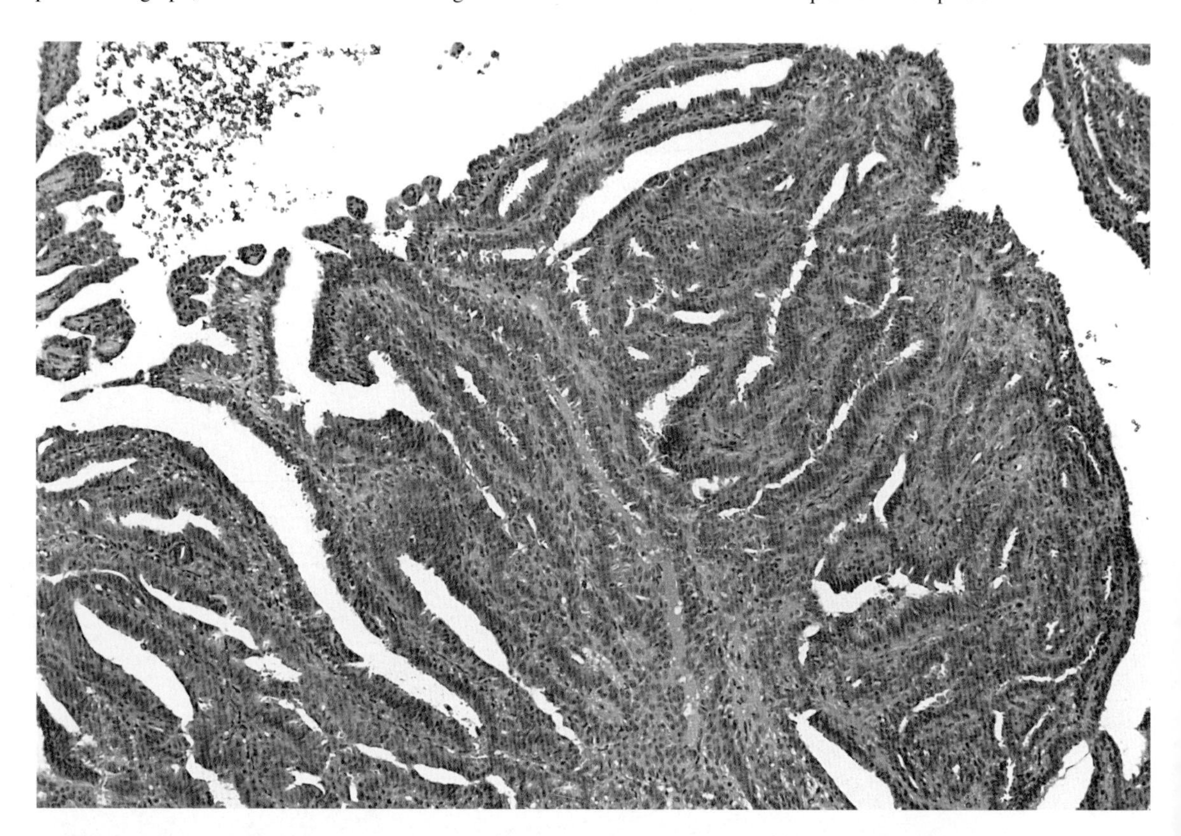

FIGURE 14–4. PAPILLARY HIDRADENOMA: Papillary hidradenoma has a complex glandular pattern. The tubules and acini are lined by a single or double layer of cuboidal cells.

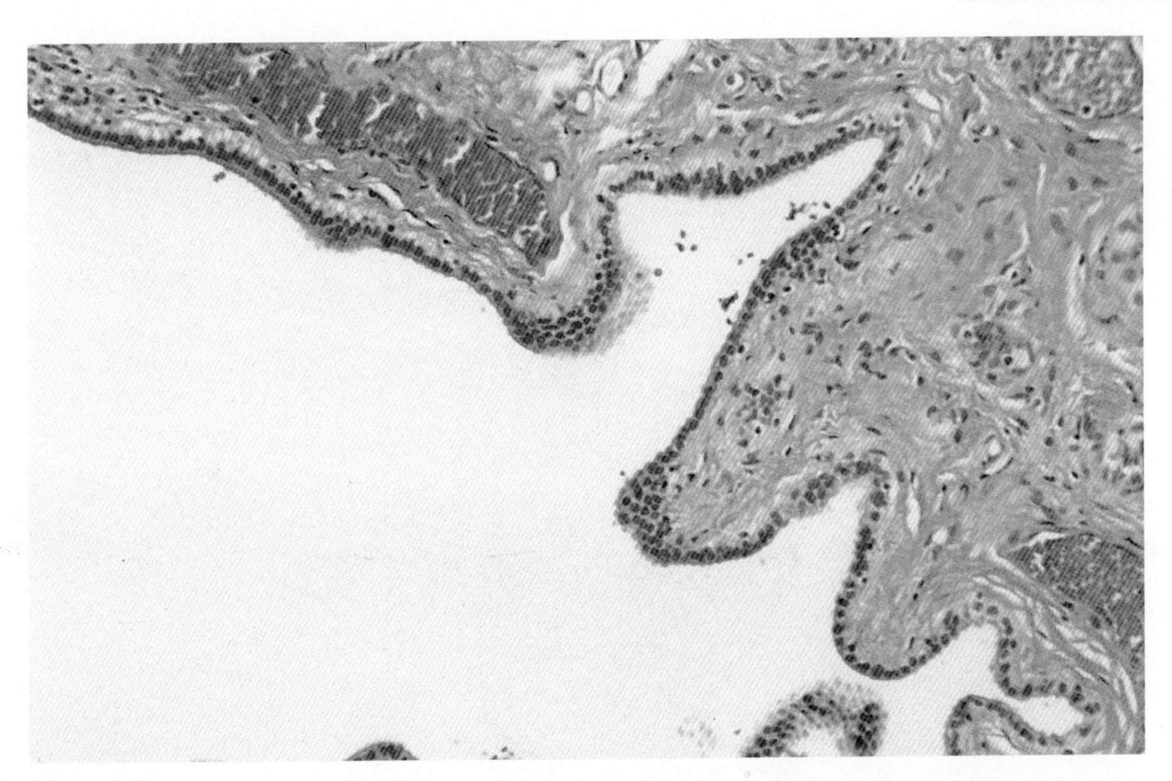

Figure 14–12

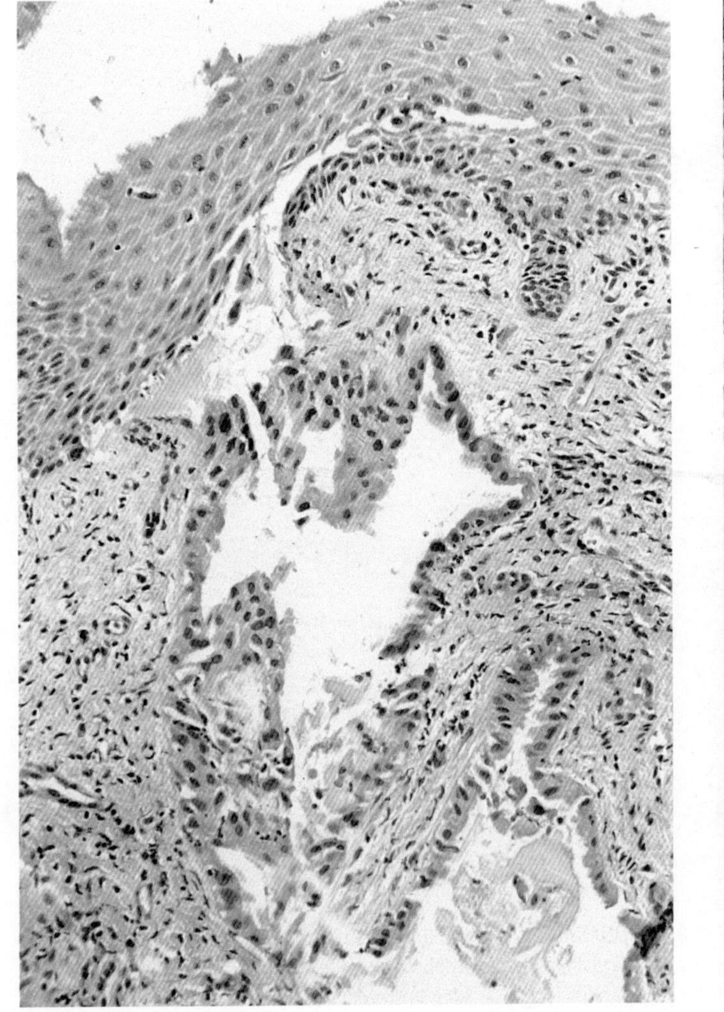

Figure 14–13

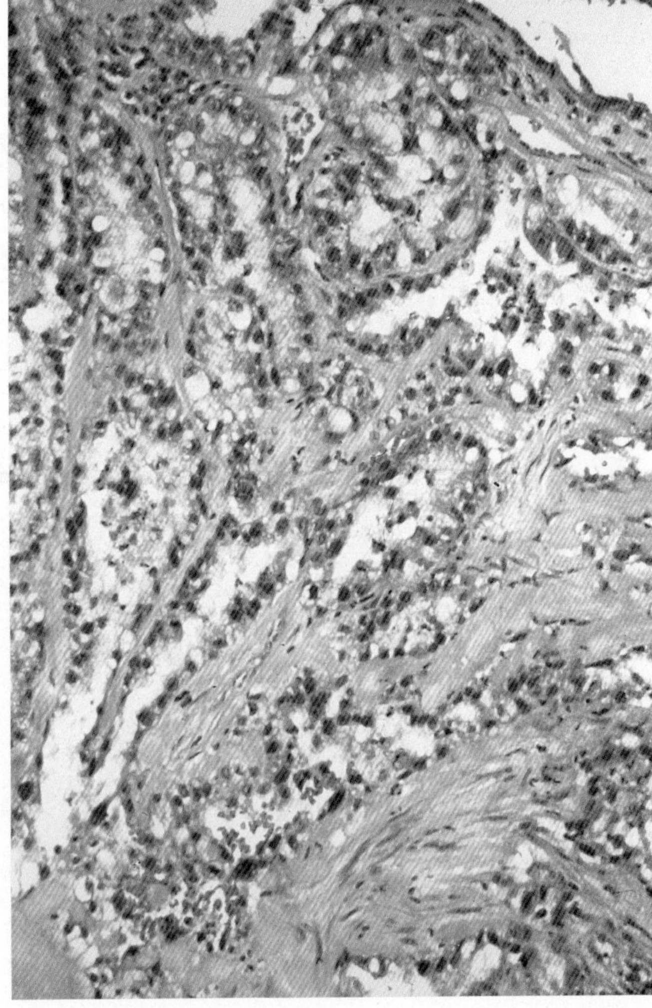

Figure 14–14

■ CERVIX

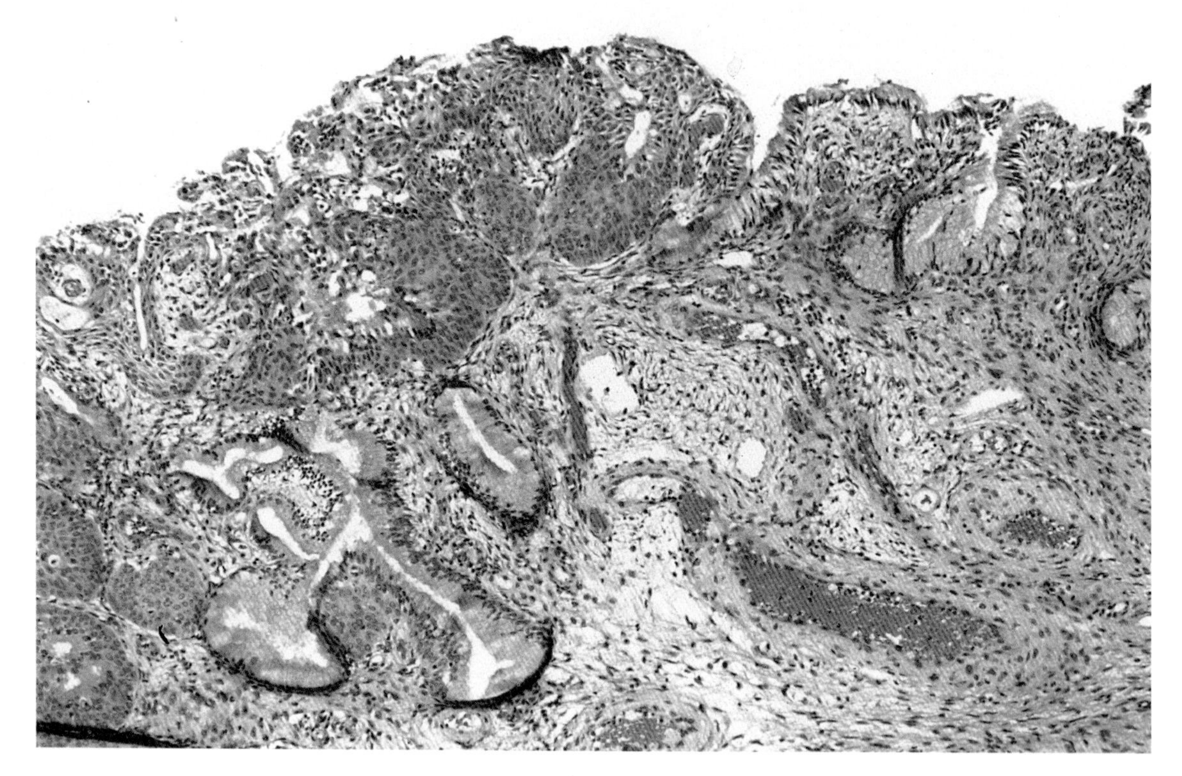

Figure 14–15 A

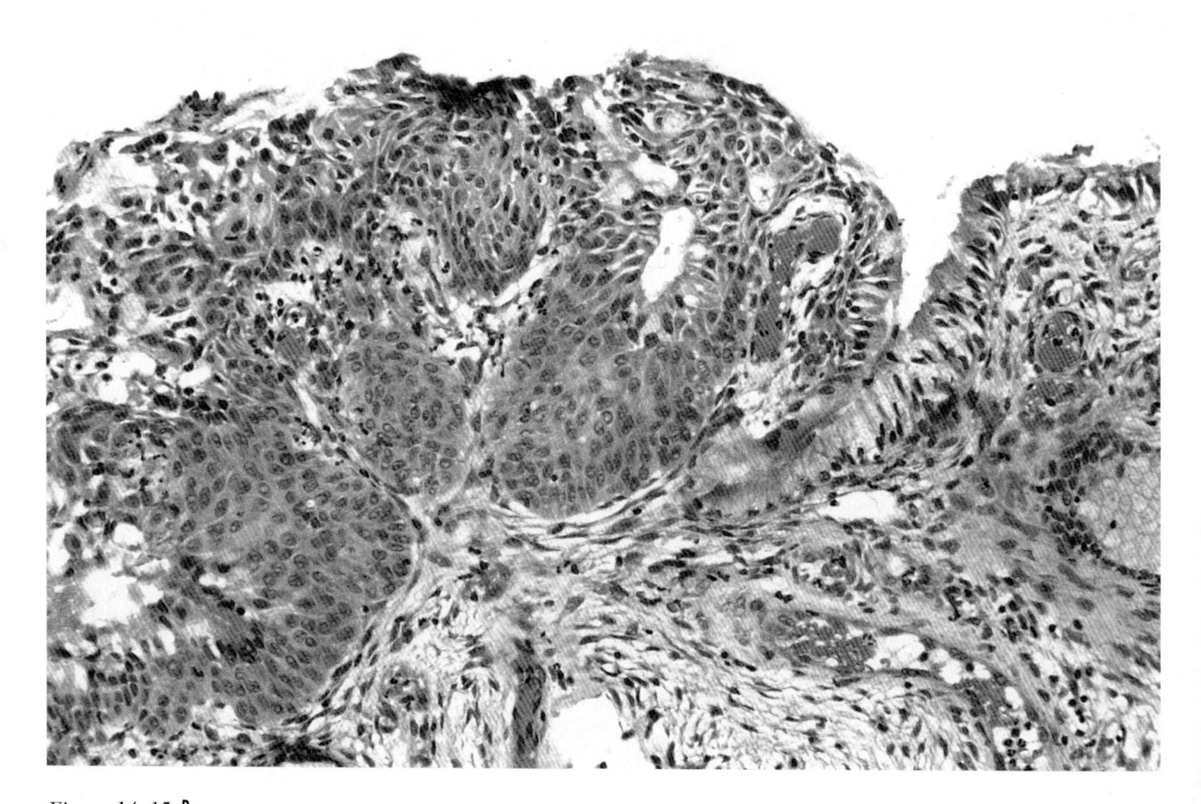

Figure 14–15 B

FIGURE 14–15. CHRONIC CERVICITIS WITH ASSOCIATED SQUAMOUS METAPLASIA: A: In chronic cervicitis, there is a nonspecific chronic inflammatory infiltrate in the mucosa. This infiltrate is composed of lymphocytes, plasma cells, and macrophages. **B:** When the endocervix is involved, as part of the reparative process, squamous metaplasia ensues.

■ CERVIX

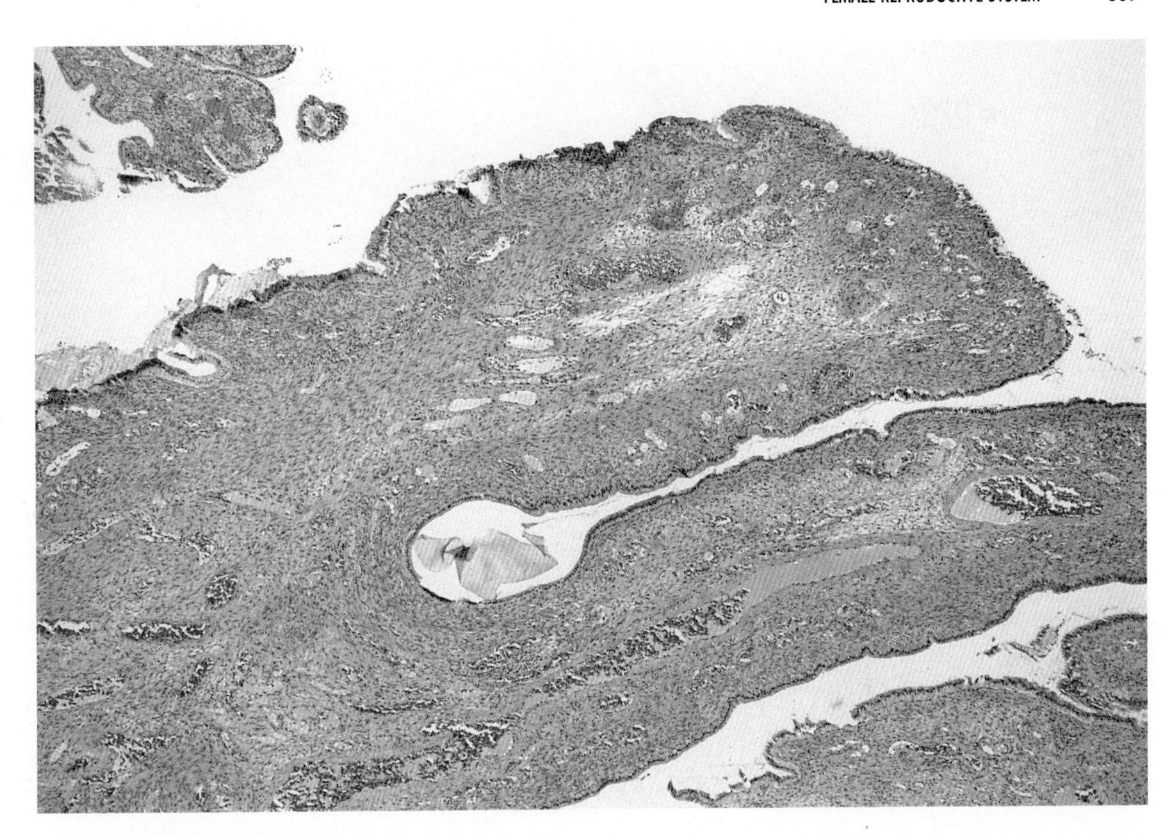

Figure 14–16 **A**

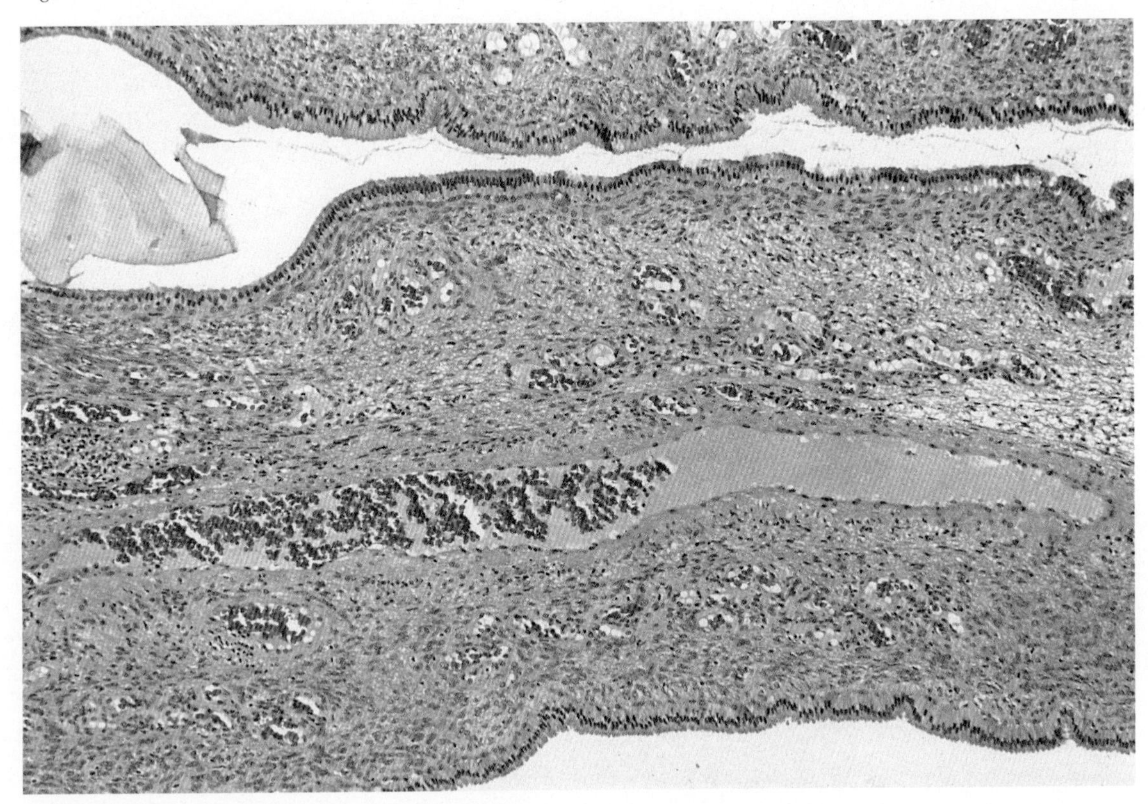

Figure 14–16 **B**

FIGURE 14–16. ENDOCERVICAL POLYP: A: In this photomicrograph of the most common type of endocervical polyp, the surface and crypts are lined by mucinous epithelium. **B:** The stroma is loose and has large ectatic feeding vessels.

■ CERVIX

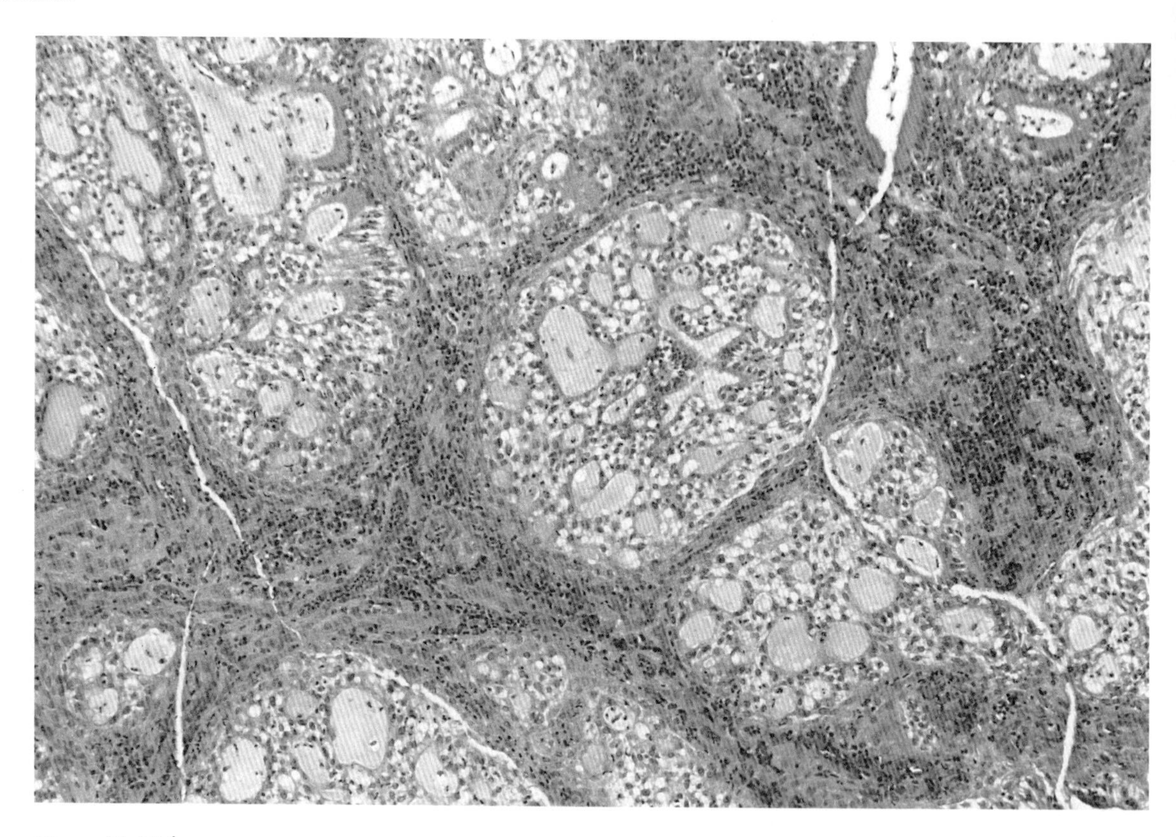

Figure 14–17 A

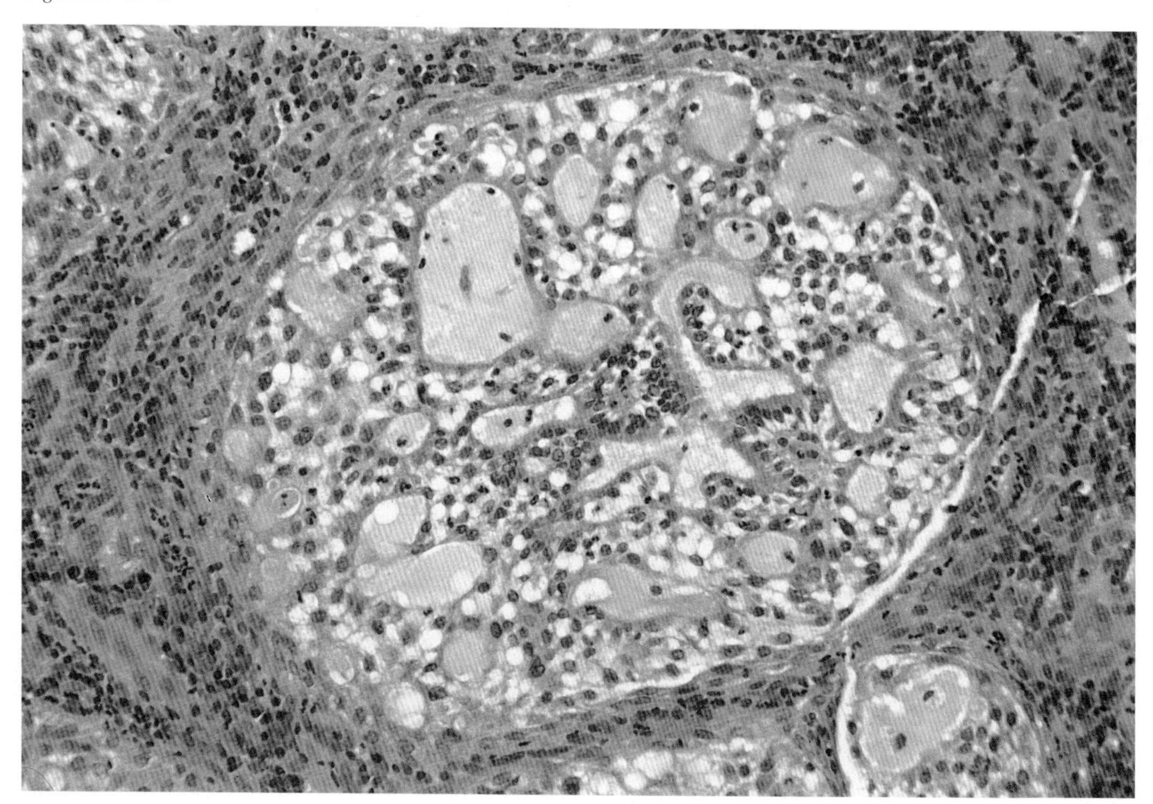

Figure 14–17 B

FIGURE 14–17. MICROGLANDULAR ENDOCERVICAL HYPERPLASIA: Microglandular endocervical hyperplasia is frequently associated with use of oral contraceptives. A: The lesion is characterized by numerous small glands arranged in a back-to-back fashion. B: The cells lining the glands are cuboidal, contain small amounts of mucin, and have uniform nuclei.

■ UTERUS

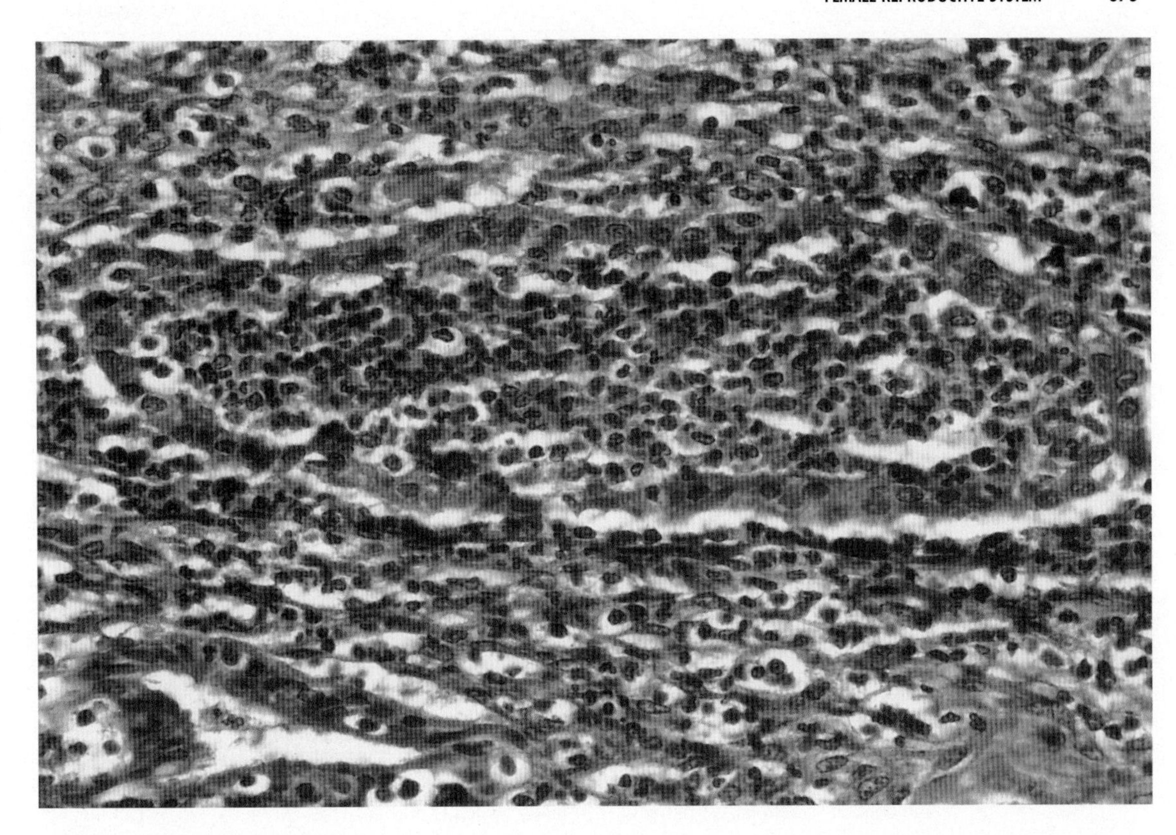

FIGURE 14–26. ACUTE ENDOMETRITIS: In acute endometritis, neutrophils are present in the stroma and fill the gland lumina.

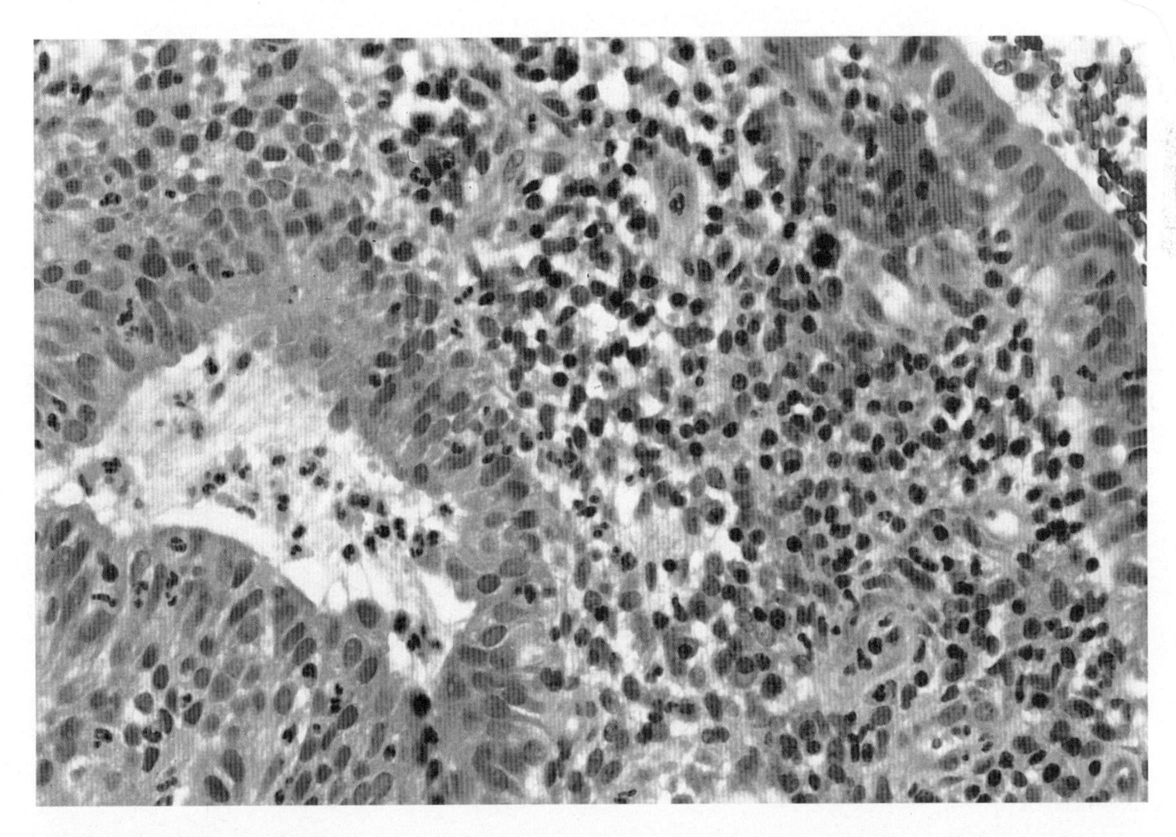

FIGURE 14–27. CHRONIC ENDOMETRITIS: In chronic endometritis, there are variable numbers of lymphocytes, plasma cells, and macrophages. The key to diagnosis is the presence of plasma cells in the stroma, because lymphocytes and macrophages are normal constituents during certain phases of the menstrual cycle.

■ UTERUS

ENDOMETRIAL DISORDERS

FIGURE 14–28. LUTEAL PHASE DEFECT: This example of a luteal phase defect shows glandular and stromal asynchrony. The stroma is consistent with day 25 of a 28-day menstrual cycle in which there is generalized stromal decidualization and infiltration by lymphocytes and few neutrophils. The glands, however, are consistent with day 21 or 22 of a 28-day menstrual cycle. The glands are angulated and somewhat distended and contain secretions. Serration of the glands, which is normally present on day 25, is lacking.

FIGURE 14–29. EXOGENOUS HORMONES: Secretory activity in the endometrial glands is lacking after use of estrogen-progestational agents. The glands are sparse, straight, and narrow. When progesterone is given late in the proliferative phase, as shown here, the stroma undergoes pseudodecidualization.

FIGURE 14–30. ENDOMETRIAL POLYP: An endometrial polyp is characterized by irregularly shaped glands that lack secretory activity. The stroma is dense. Thick-walled tortuous feeding vessels are present at the base of the polyp.

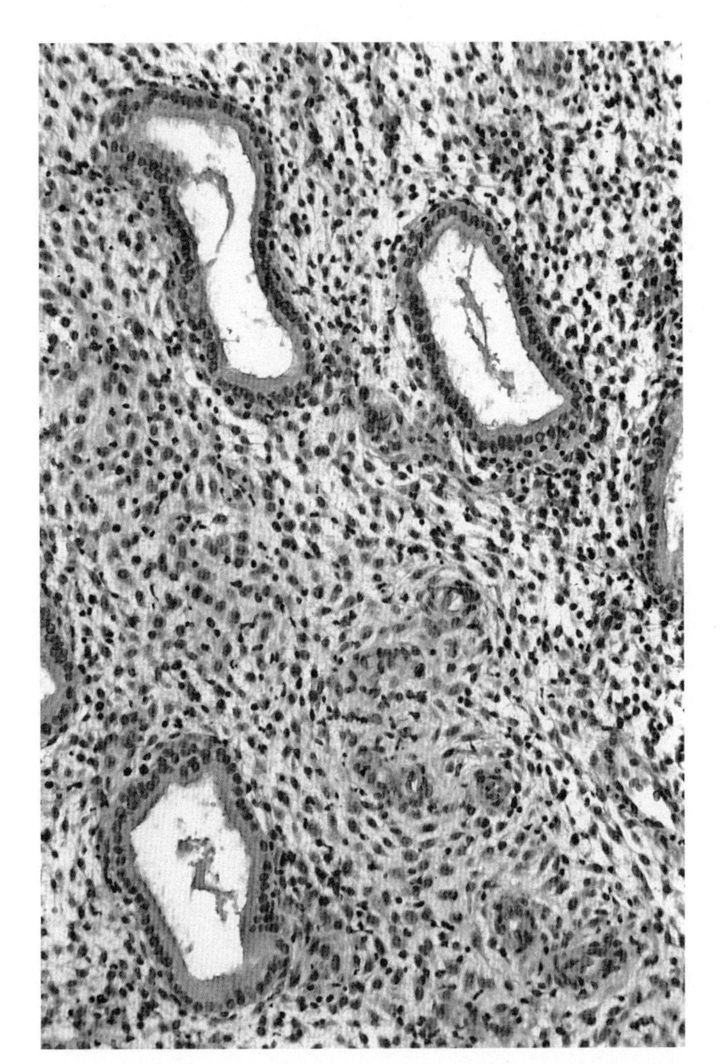

Figure 14–28

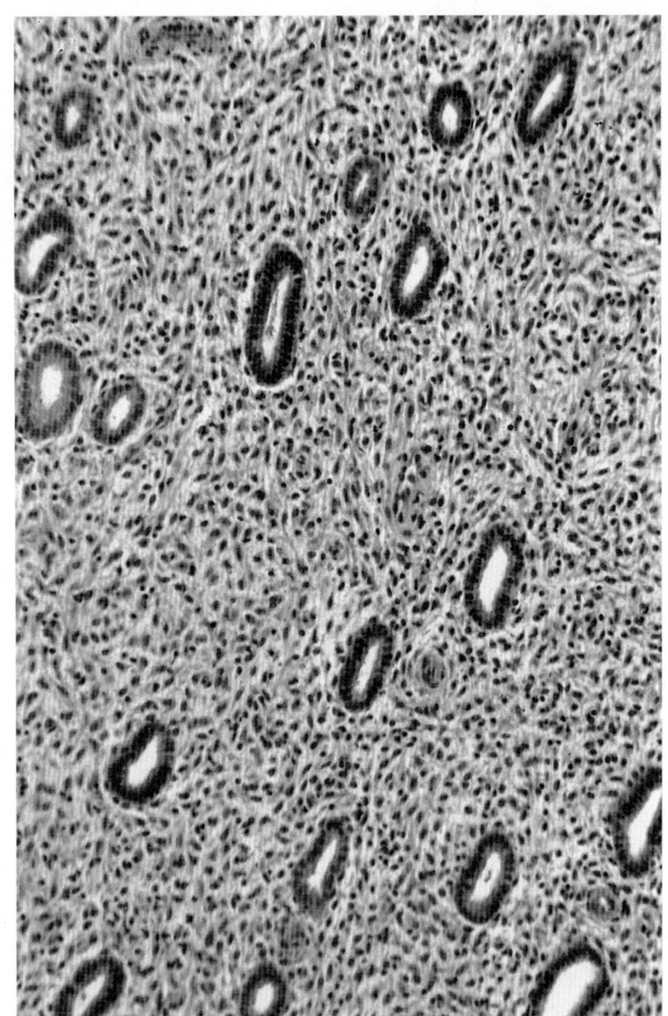

Figure 14–29

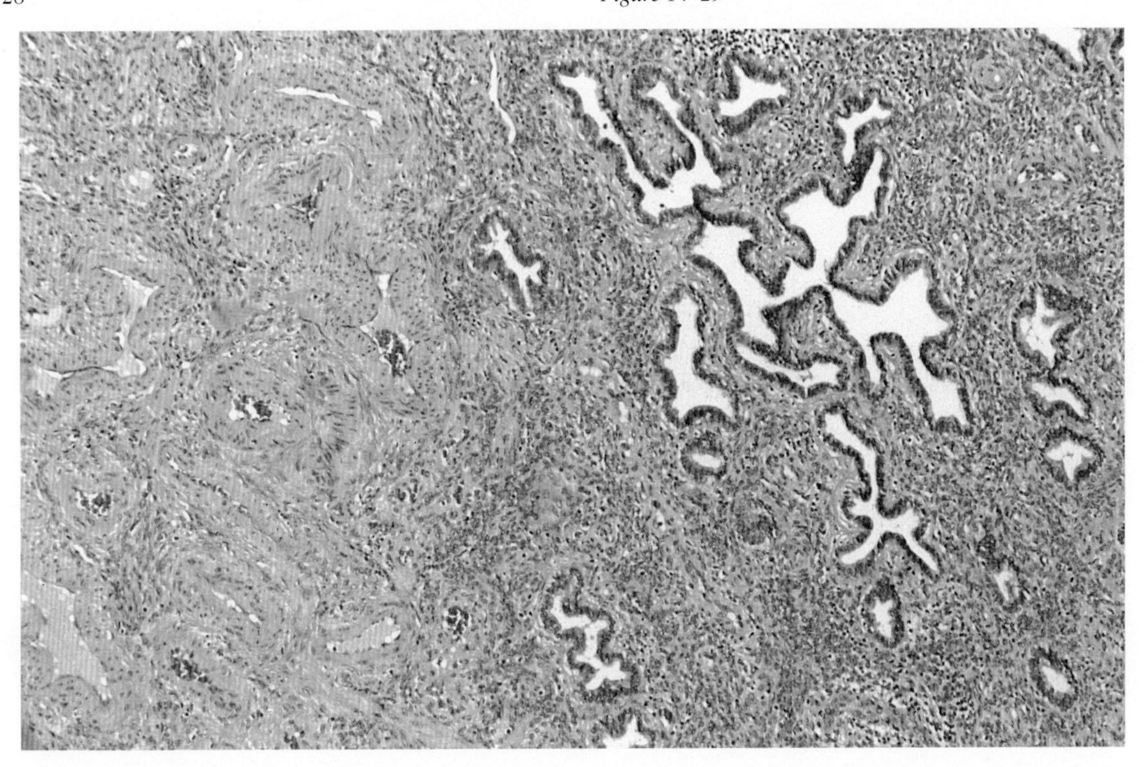

Figure 14–30

■ UTERUS

FIGURE 14–31. CYSTIC ATROPHY: In cystic atrophy, the endometrial glands are cystically dilated and lined by low cuboidal or flattened cells. Mitoses are absent.

FIGURE 14–32. ADENOMYOSIS: Adenomyosis is characterized by the presence of endometrial glands and stroma within the myometrium. There is no connection to the endometrial cavity.

FIGURE 14–33. ENDOMETRIOSIS: Endometriosis is characterized by the presence of endometrial glands and stroma anywhere in the body outside the uterus. In this photomicrograph, endometriosis is seen in the muscularis propria of the large intestine.

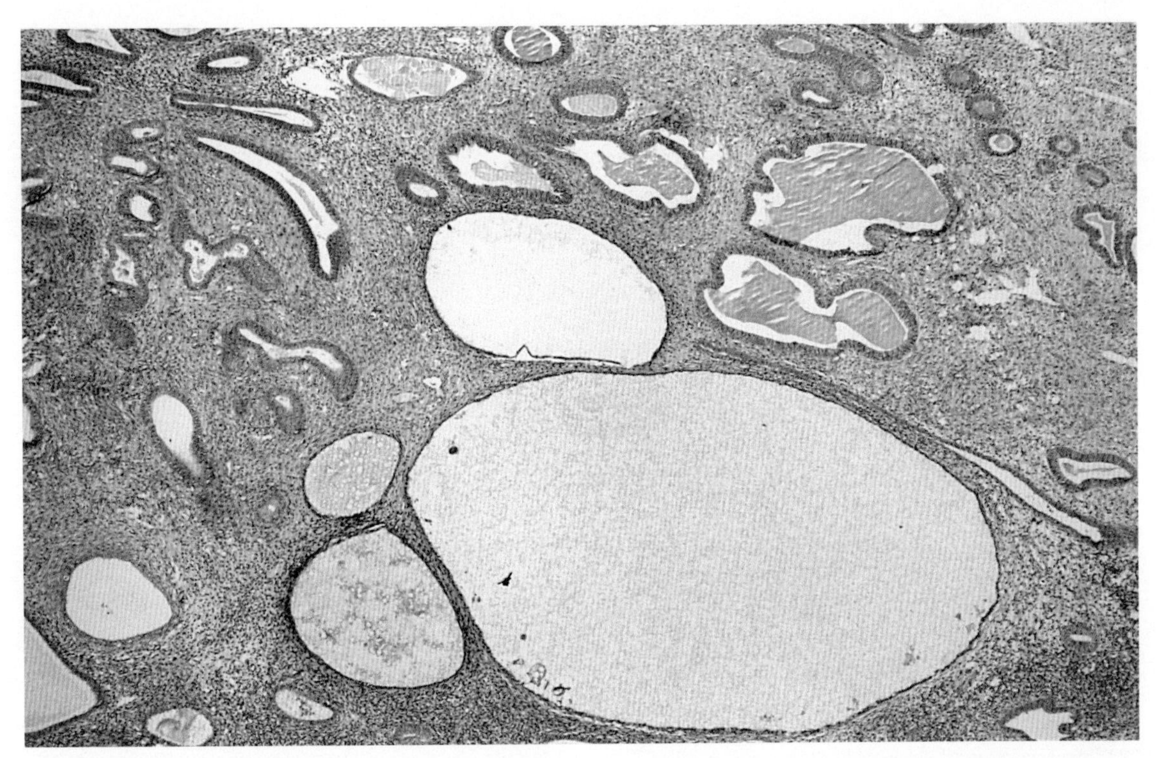

Figure 14–31

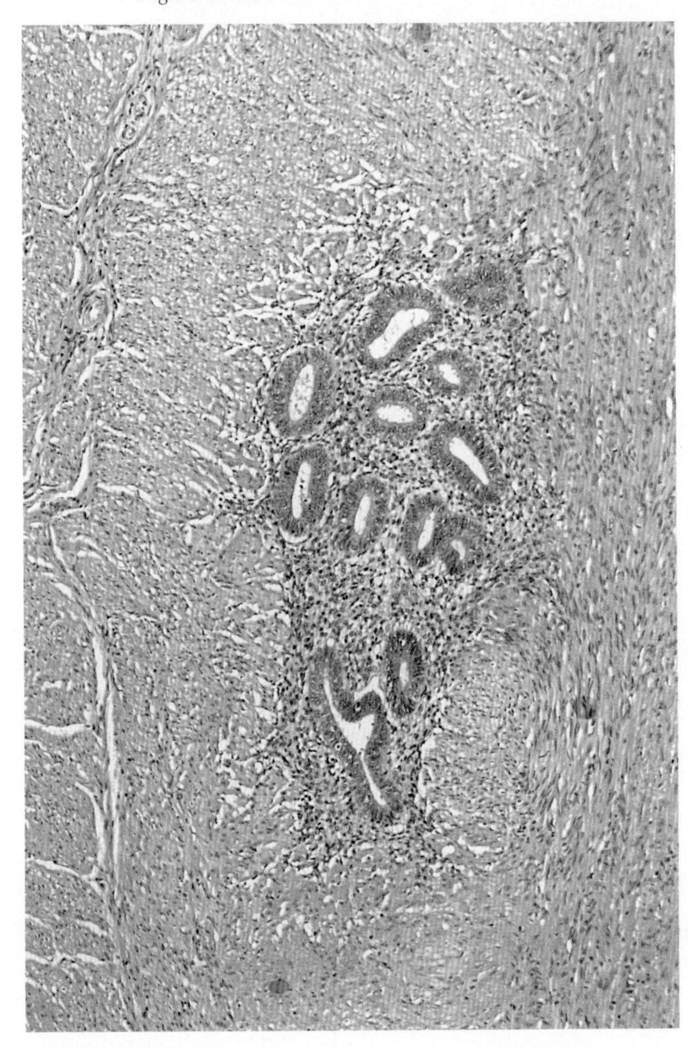

Figure 14–32

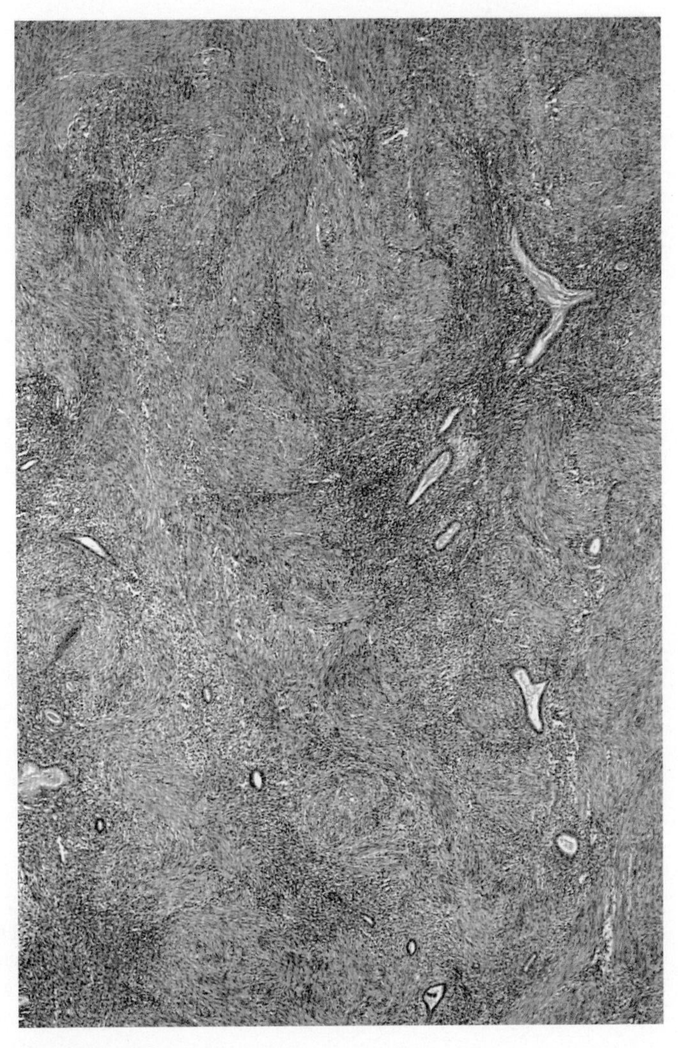

Figure 14–33

■ UTERUS

ENDOMETRIAL HYPERPLASIAS

FIGURE 14–34. SIMPLE HYPERPLASIA WITHOUT ATYPIA (CYSTIC HYPERPLASIA): In cystic hyperplasia, there is considerable variability in the caliber of the glandular lumina. Some glands are cystically dilated, and budding of the glands is minimal. The epithelium lining the glands is multilayered without loss of polarity. The stroma is hypercellular as well; thus, the normal gland-to-stroma ratio is maintained.

FIGURE 14–35. COMPLEX HYPERPLASIA WITHOUT ATYPIA (ADENOMATOUS HYPERPLASIA): In complex hyperplasia, there is a focal increase in the number of glands, resulting in glandular crowding. The glandular lumina are irregular as a result of budding. The epithelium lining the glands is multilayered without loss of polarity. The gland-to-stroma ratio is greatly increased.

FIGURE 14–36. COMPLEX HYPERPLASIA WITH ATYPIA (ATYPICAL HYPERPLASIA)/CARCINOMA IN SITU: Atypical hyperplasia has not only the architectural features of complex hyperplasia without atypia (Figure 14–35) but also cellular and nuclear enlargement. The cells are stratified with loss of polarity. Mitotic figures are increased in number. The nuclear chromatin is coarse, and nucleoli are prominent.

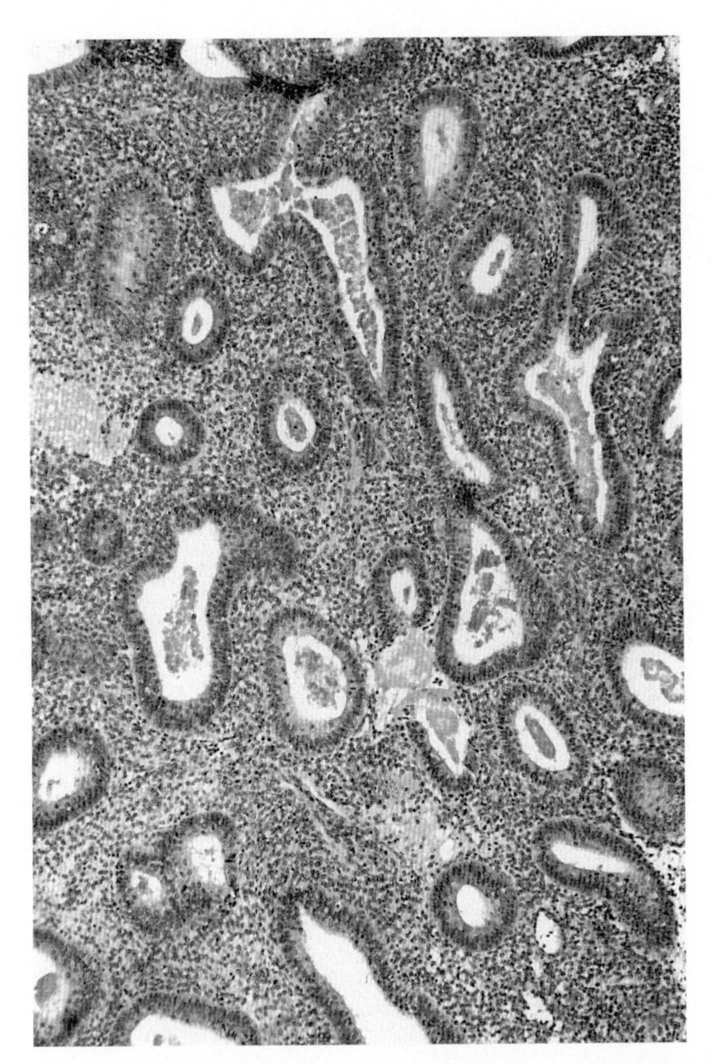

Figure 14–34

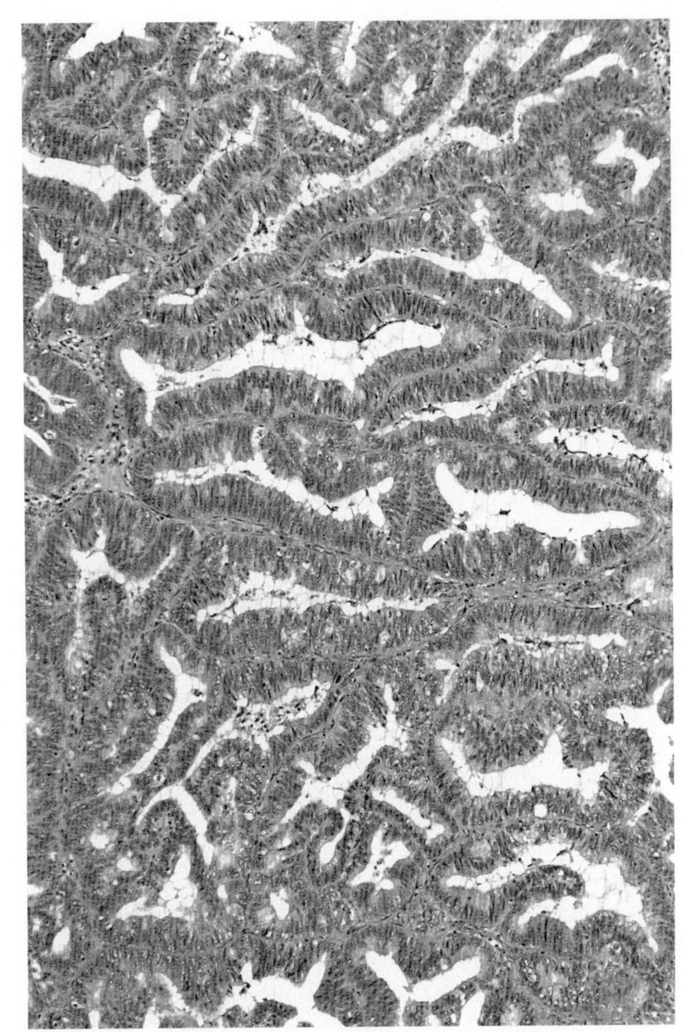

Figure 14–35

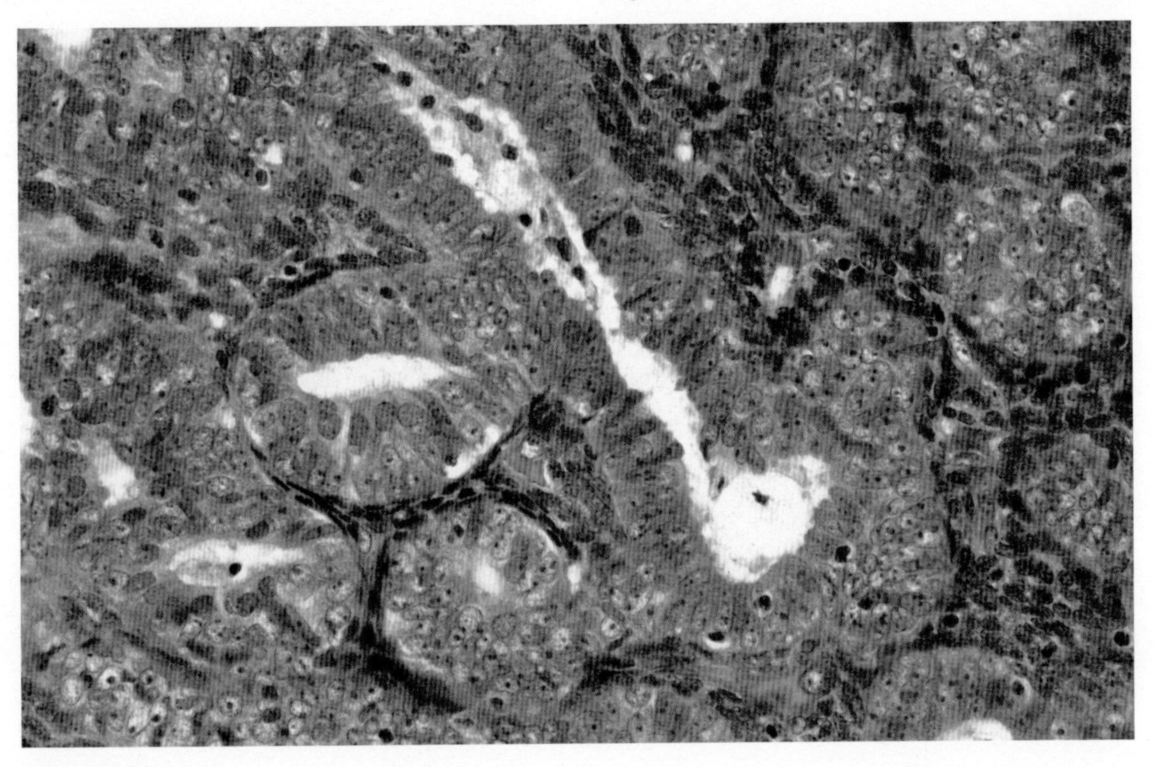

Figure 14–36

■ UTERUS

FIGURE 14–37. EXOGENOUS PROGESTERONE EFFECT: Progestational agents are one form of therapy for adenomatous hyperplasia. Progesterone causes the glands to cease proliferation. The glands remain small, inactive, and few in number. The stroma shows a pseudodecidual response and lacks mitotic figures. In this photomicrograph of a treated hyperplasia, the glands are decreased in complexity and there is an associated pseudodecidual reaction.

MALIGNANT EPITHELIAL TUMORS

FIGURE 14–38. ADENOCARCINOMA OF THE ENDOMETRIUM: This is a photomicrograph of a well-differentiated adenocarcinoma of the endometrium. The glands are back-to-back without intervening stroma, and some glands share a common epithelial wall. The epithelial lining is stratified with loss of polarity. The cells are enlarged. Their nuclei are enlarged, with chromatin clumping, prominent nucleoli, and an increase in the number of mitotic figures.

FIGURE 14–39. PAPILLARY SEROUS ADENOCARCINOMA: Papillary serous adenocarcinoma is a variant of adenocarcinoma. The papillary projections contain fibrotic stroma covered by **hobnail** epithelial cells. A desmoplastic reaction secondary to invasion is also noted and the pattern is identical to that in papillary serous adenocarcinoma of the ovary (Figure 14–44 A and B).

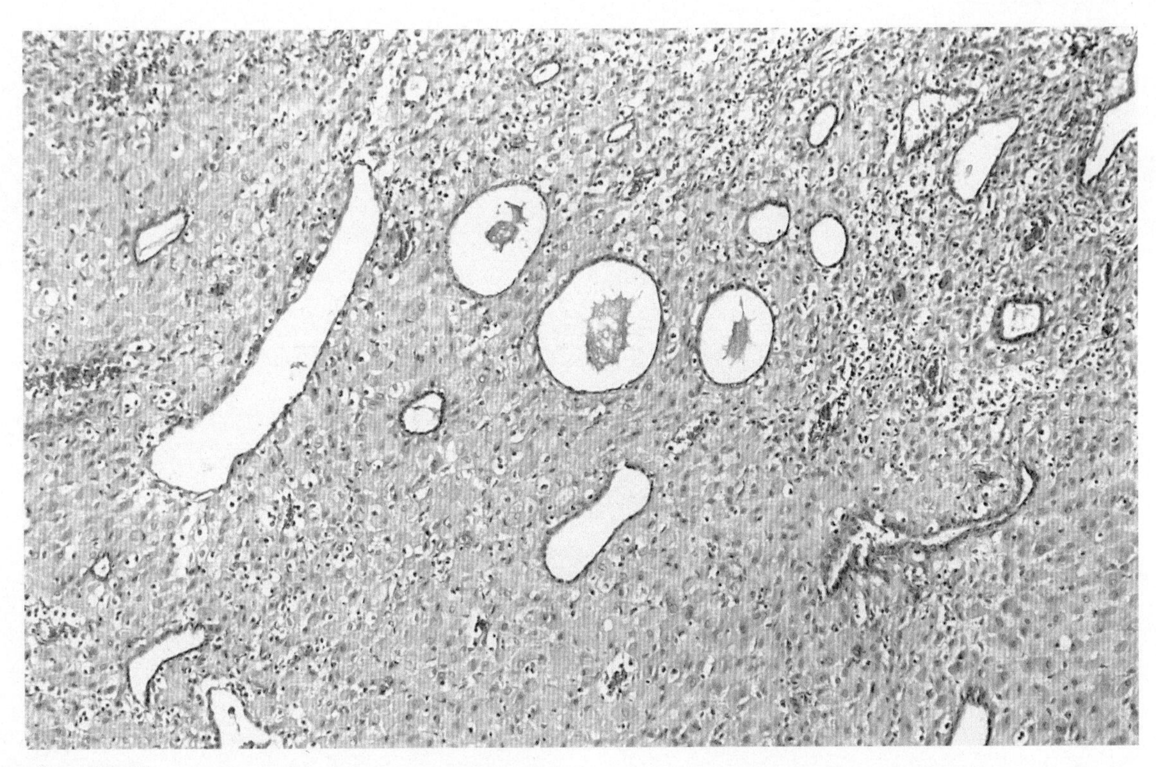

Figure 14–37

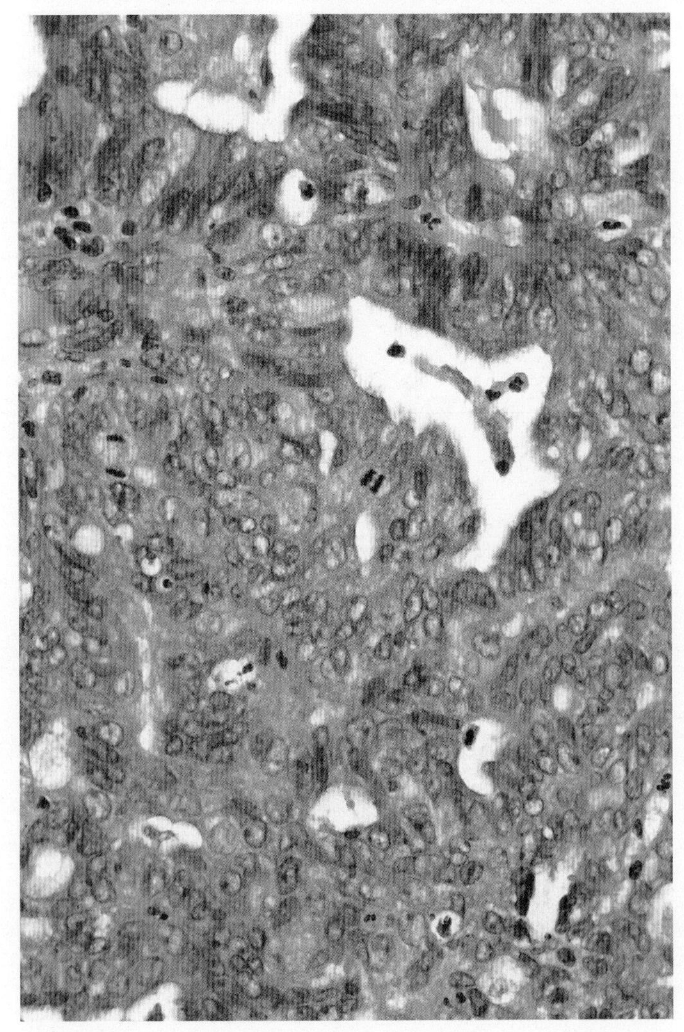

Figure 14–38

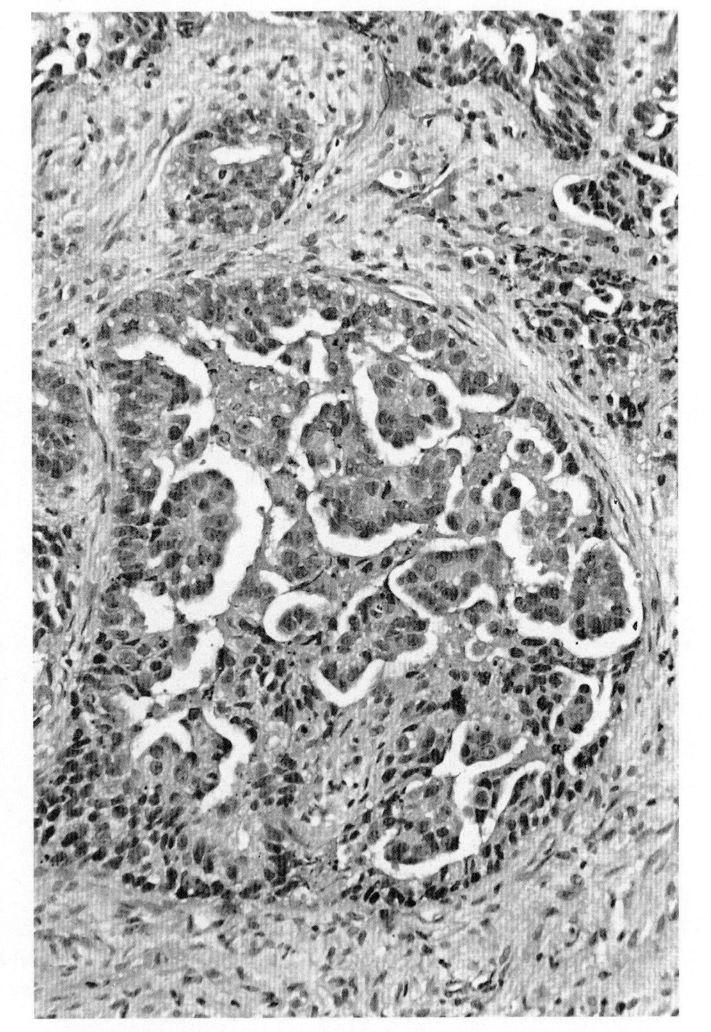

Figure 14–39

■ UTERUS

BENIGN MESENCHYMAL TUMORS

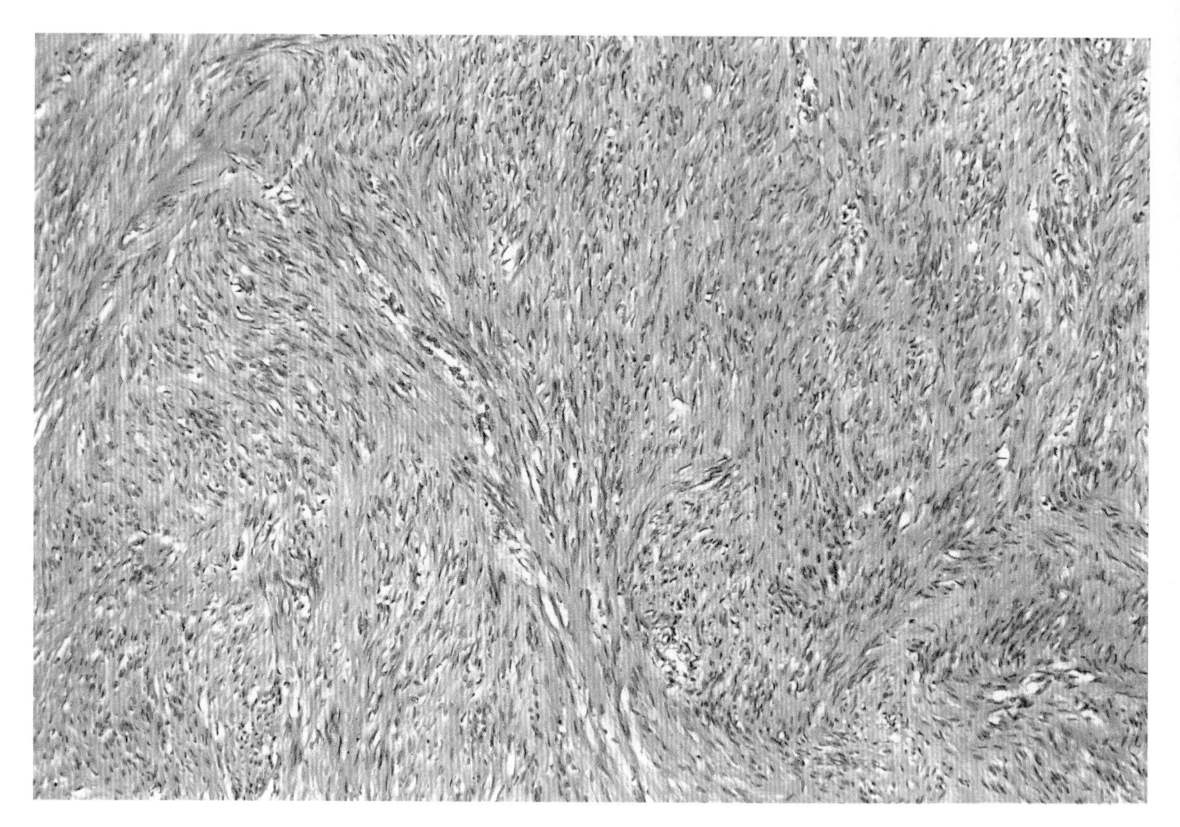

Figure 14–40 A

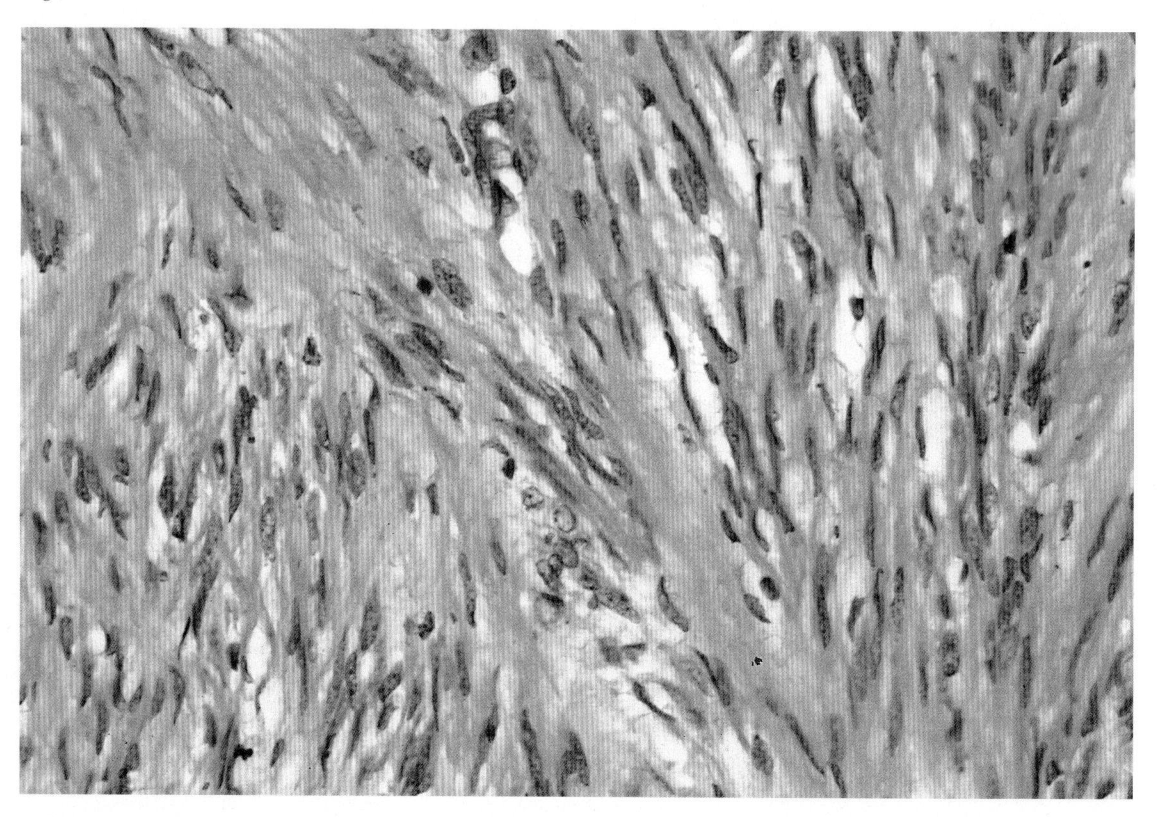

Figure 14–40 B

FIGURE 14–40. LEIOMYOMA: A: A leiomyoma is characterized by swirling bundles of spindle-shaped smooth muscle cells. **B:** The cells have eosinophilic cytoplasm and cigar-shaped nuclei. Mitotic figures are rare.

■ UTERUS

MALIGNANT MESENCHYMAL TUMORS

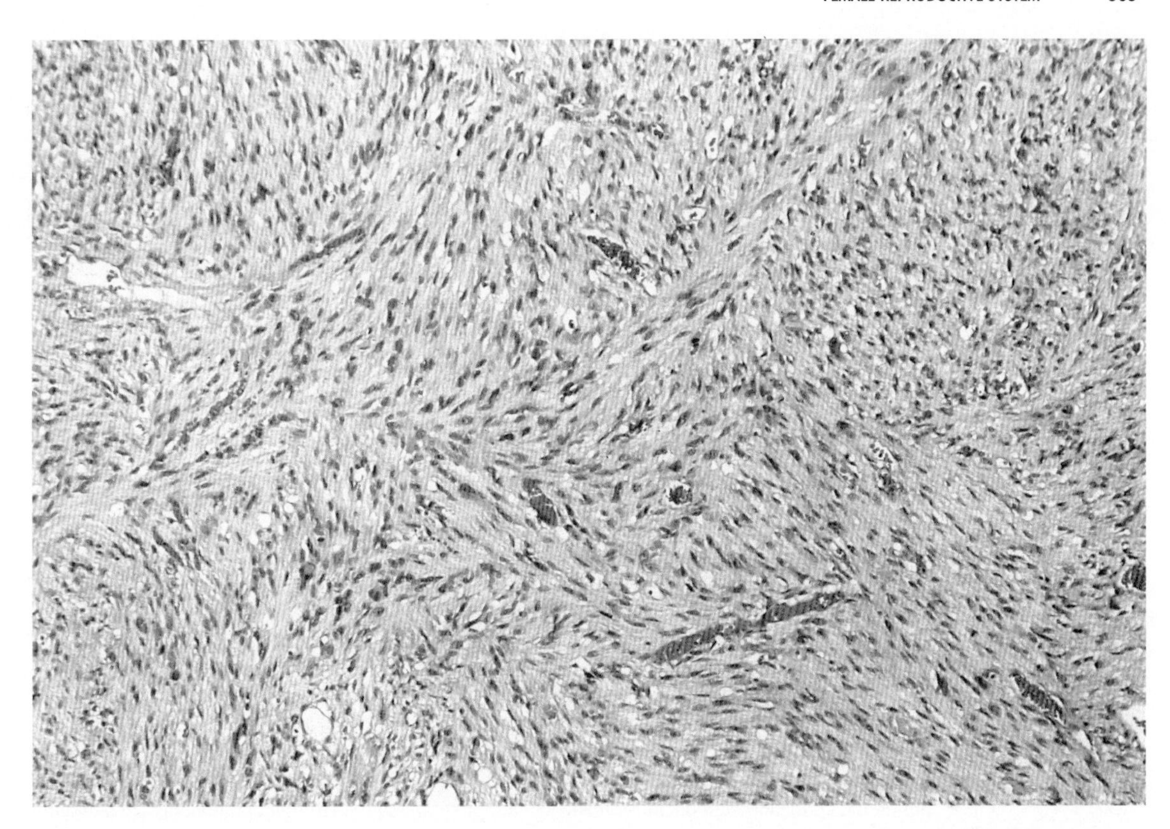

Figure 14–41 A

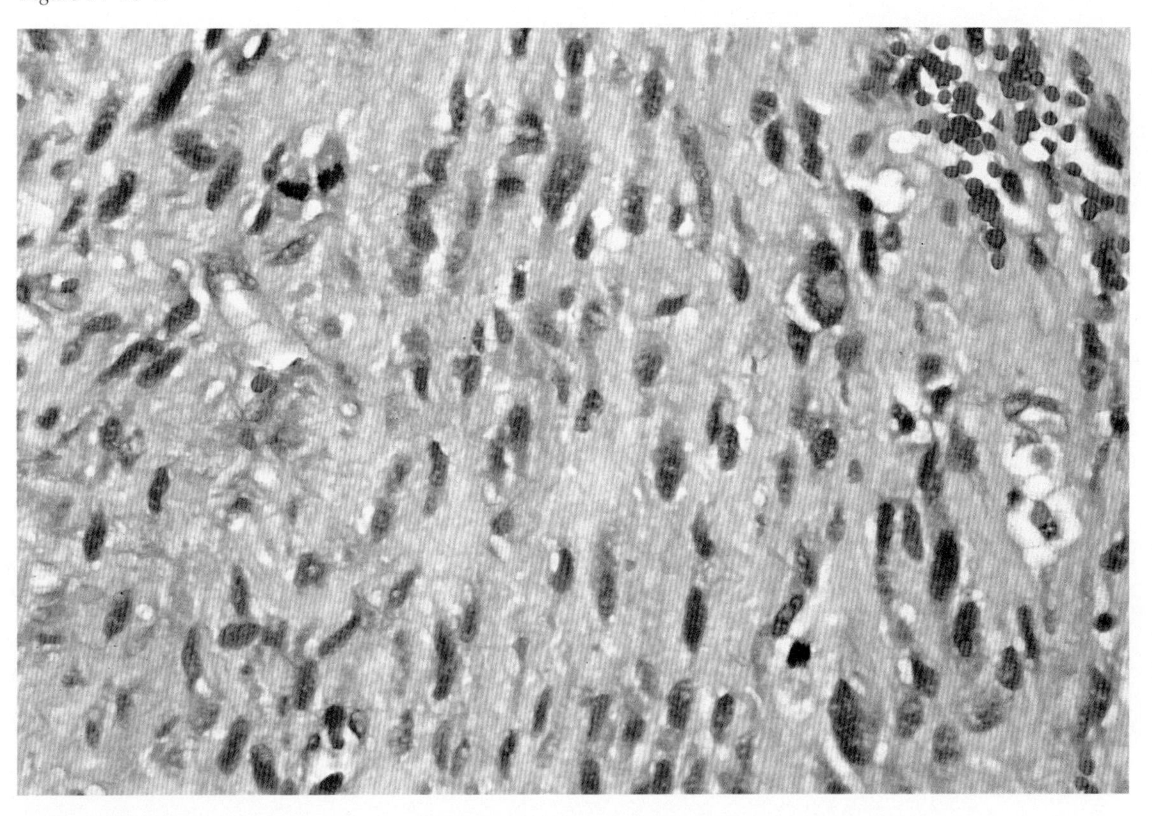

Figure 14–41 B

FIGURE 14–41. LEIOMYOSARCOMA: A: A leiomyosarcoma is characterized by fascicles of eosinophilic spindle-shaped cells. **B:** The nuclei are fusiform, with variation in size and shape. Mitotic activity is the main criterion for diagnosis. There must be at least five mitotic figures per 10 high-power fields.

■ UTERUS

FIGURE 14–42. MIXED MESODERMAL TUMOR (MALIGNANT MIXED MÜLLERIAN TUMOR): A mixed mesodermal tumor is characterized by an admixture of malignant epithelial and stromal components. When the stromal component is native to the uterus, for example, a leiomyosarcoma, the tumor is of the homologous type. When the stromal component is foreign to the uterus, the tumor is of heterologous type. In these photomicrographs of the tumor, the epithelial component is an adenocarcinoma (**A**) and the stromal components are a chondrosarcoma (**B**) and a rhabdomyosarcoma (**C**), making this a malignant mixed mesodermal tumor, heterologous type.

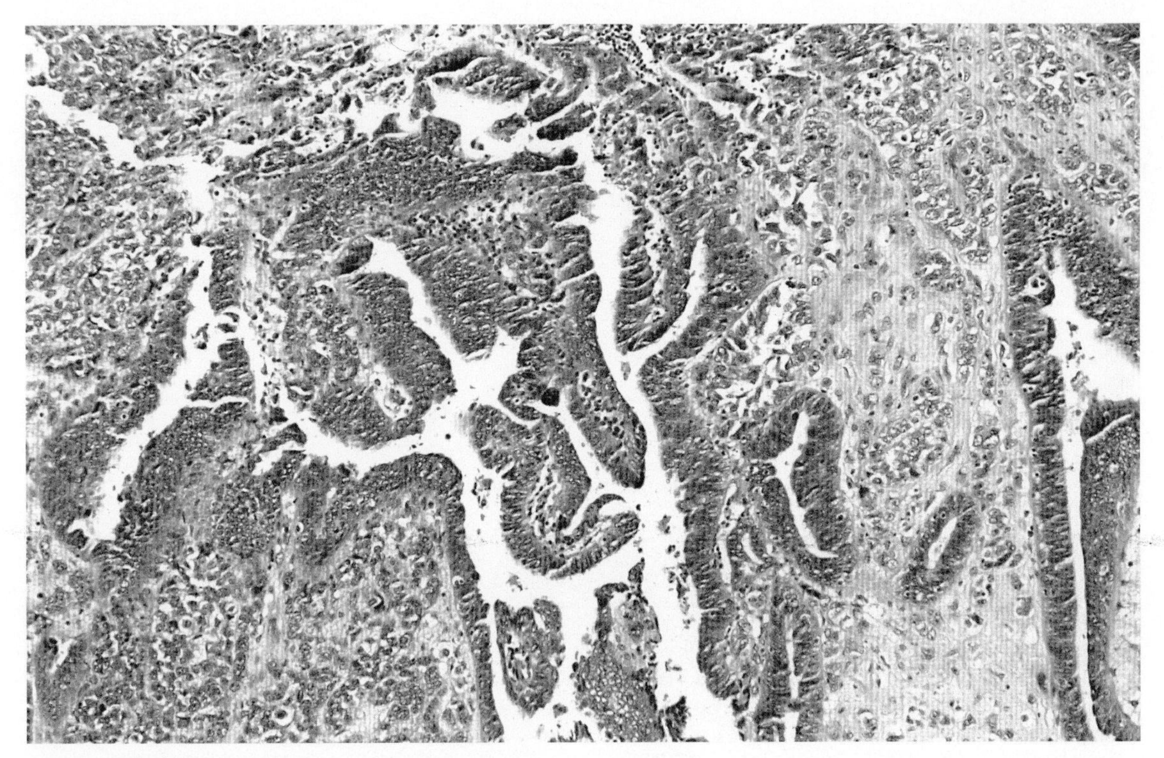

Figure 14–42 A

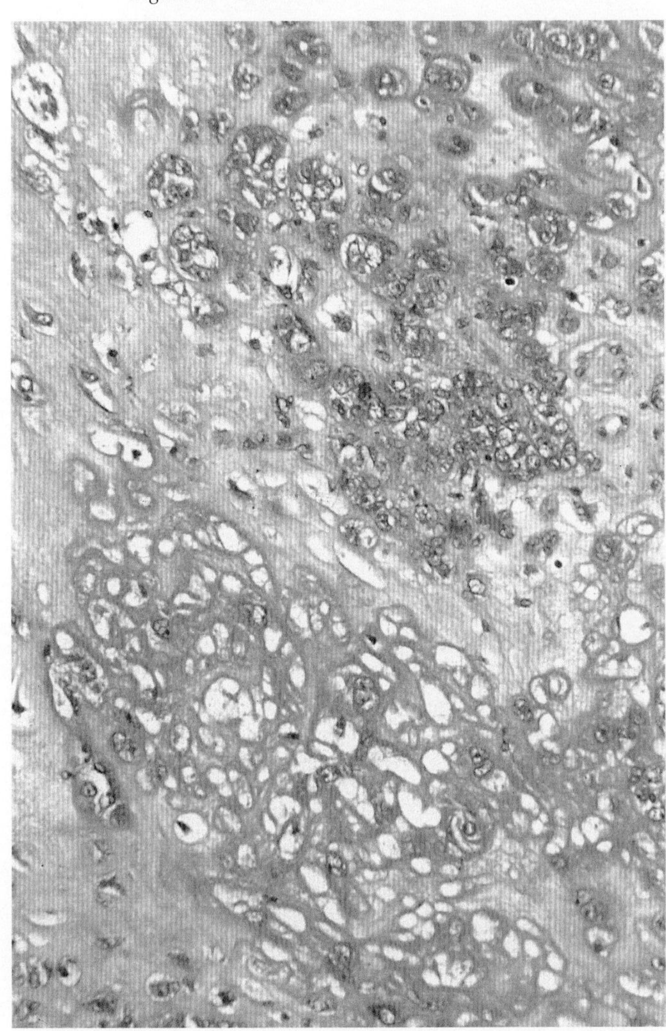

Figure 14–42 B

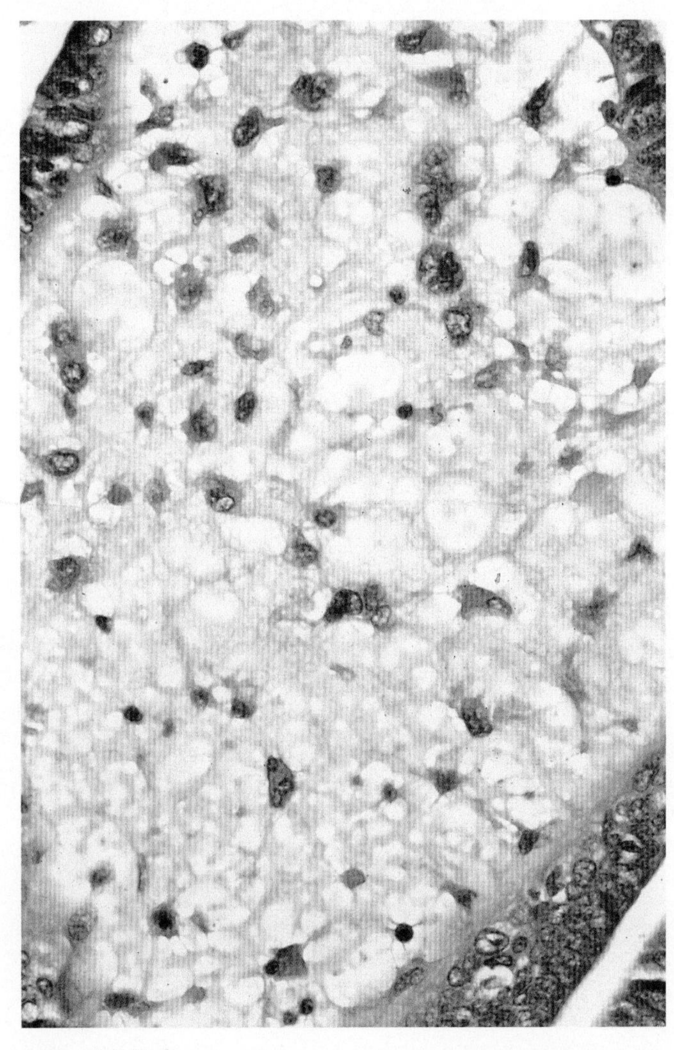

Figure 14–42 C

■ OVARIES

TUMORS OF SURFACE EPITHELIUM

FIGURE 14–43. SEROUS CYSTADENOMA: Serous cystadenomas are usually unilocular. The lining epithelium is smooth and minimally papillary. The epithelium is composed of a single layer of cuboidal cells.

FIGURE 14–44. SEROUS CYSTADENOCARCINOMA: A: Serous cystadenocarcinoma of borderline malignancy and infiltrating serous cystadenocarcinoma are multilocular tumors with papillary projections and a more complex pattern than that of their benign counterpart, serous cystadenoma (Figure 14–43). **B:** Cytologically, the epithelium lining the cystic spaces in both entities is stratified with micropapillary projections and the cells have nuclear atypia. Mitotic figures are present. In serous cystadenocarcinoma of borderline malignancy, the stroma is not infiltrated.

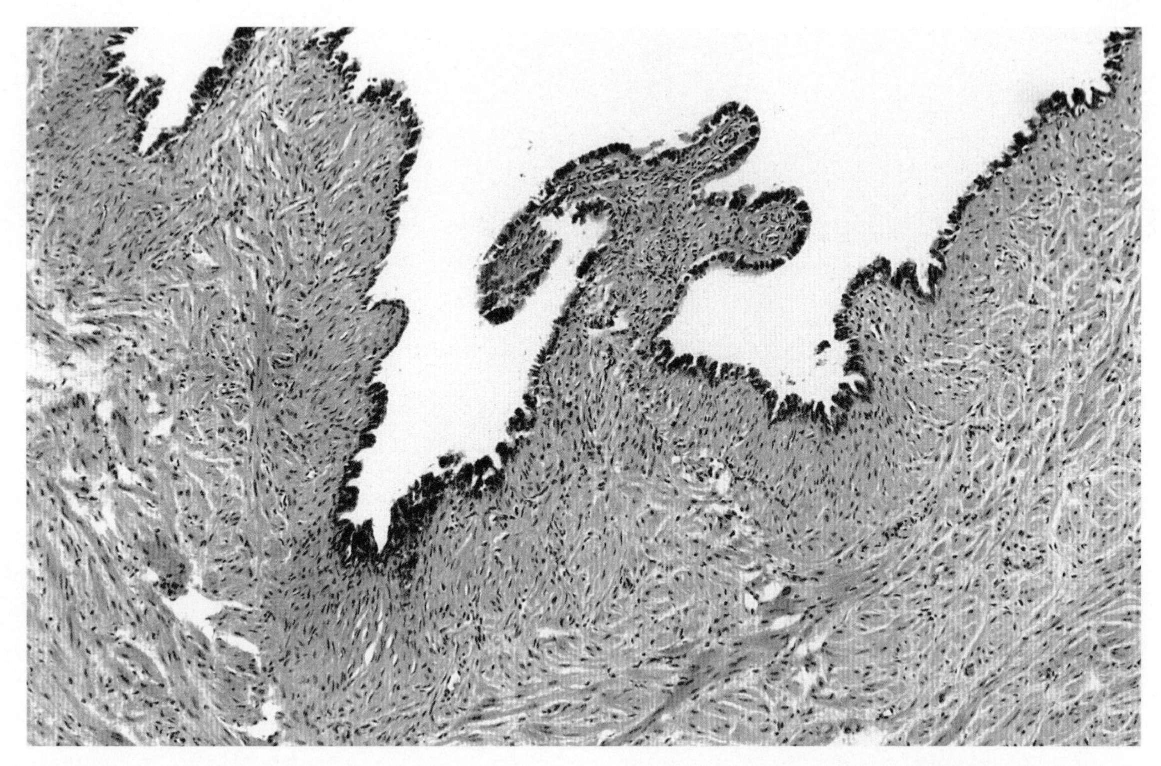

Figure 14–43

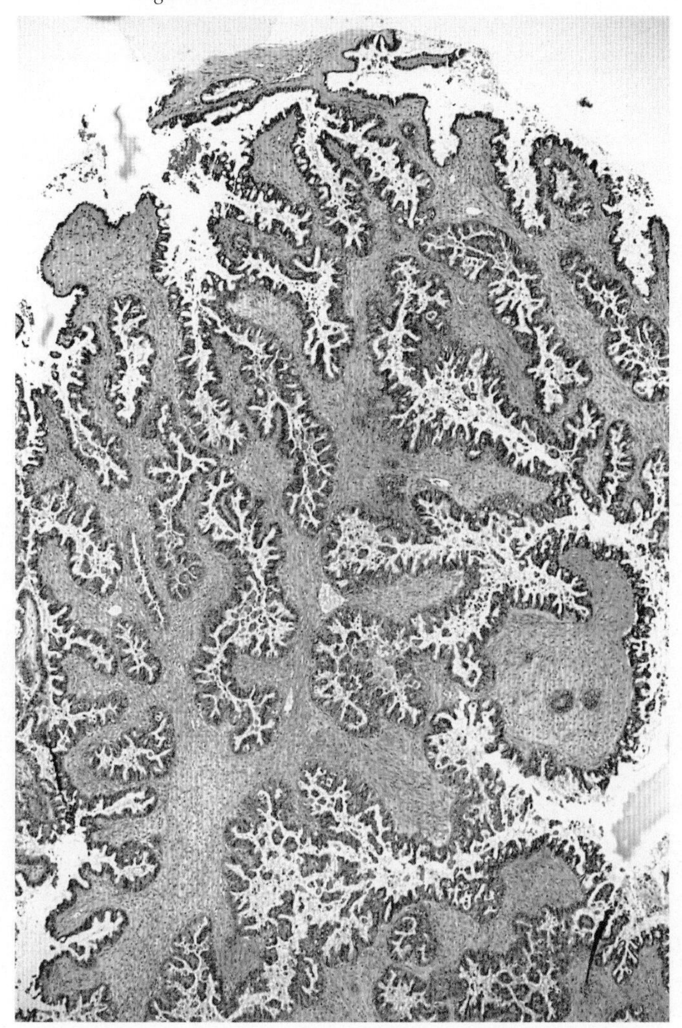

Figure 14–44 A

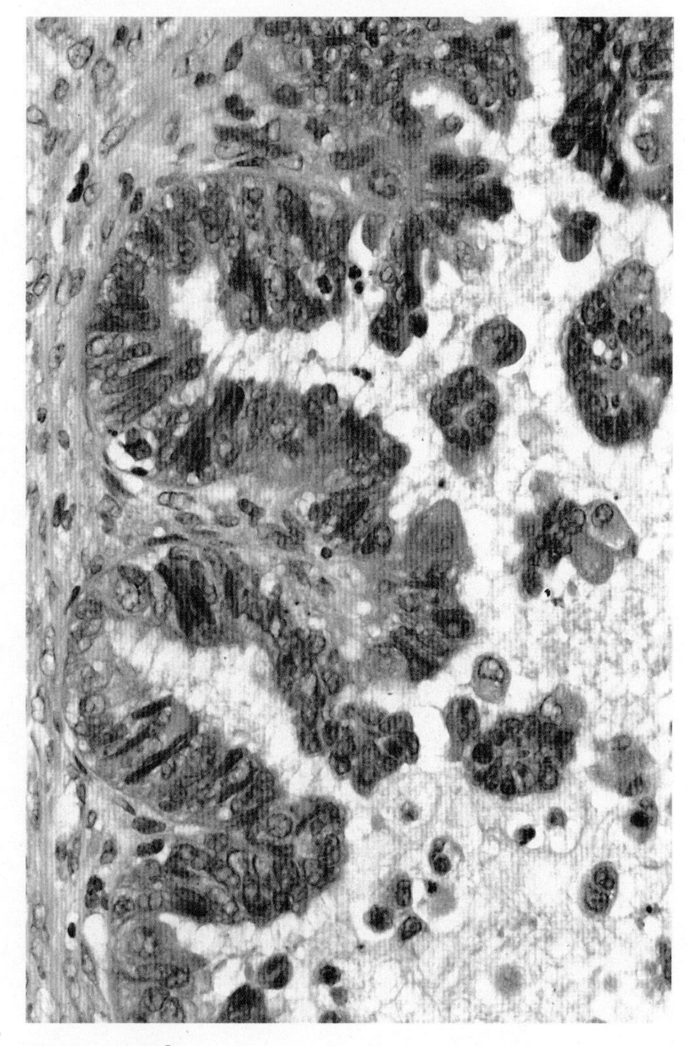

Figure 14–44 B

■ OVARIES

TUMORS OF SURFACE EPITHELIUM

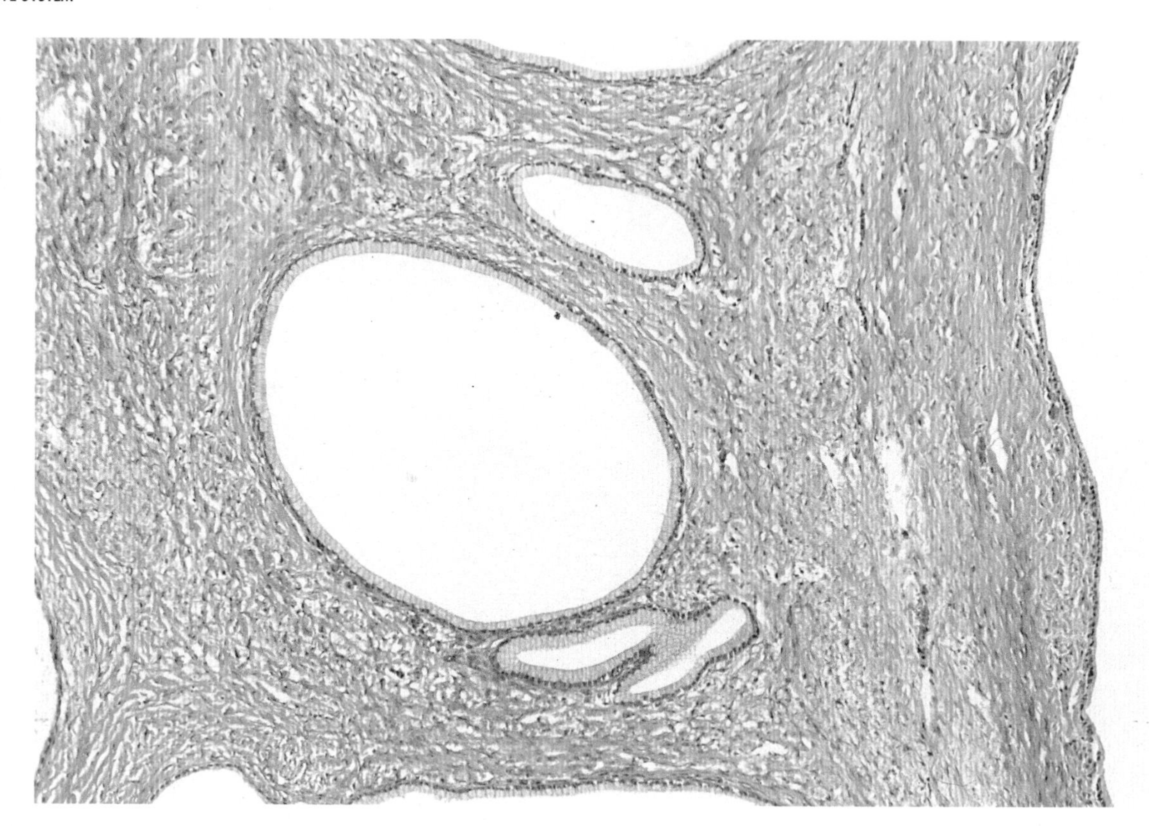

FIGURE 14–45. MUCINOUS CYSTADENOMA: Mucinous cystadenomas are multilocular. The lining epithelium is composed of a single layer of tall columnar mucin-secreting cells.

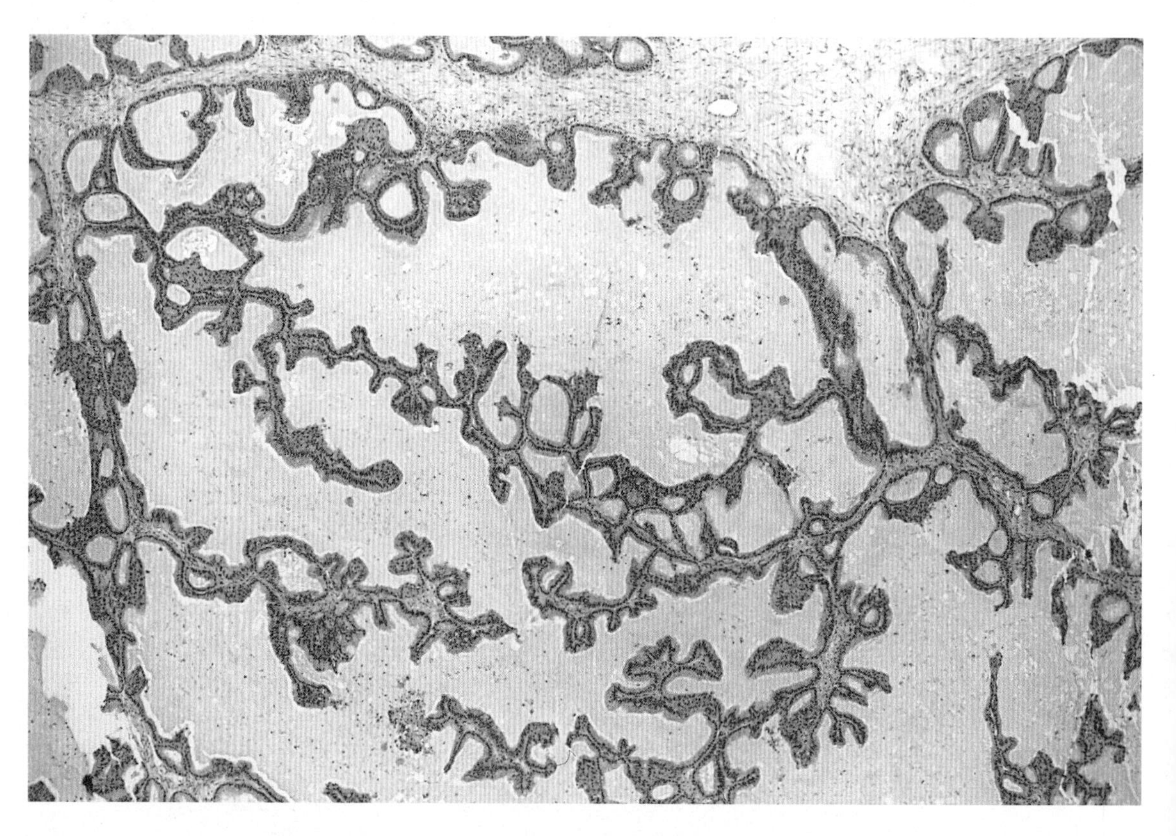

FIGURE 14–46. MUCINOUS CYSTADENOCARCINOMA: A: Mucinous cystadenocarcinoma of borderline malignancy and infiltrating mucinous cystadenocarcinoma are multilocular tumors with papillary projections and a more complex pattern than that of their benign counterpart, mucinous cystadenoma (Figure 14–45).

■ OVARIES

TUMORS OF SURFACE EPITHELIUM

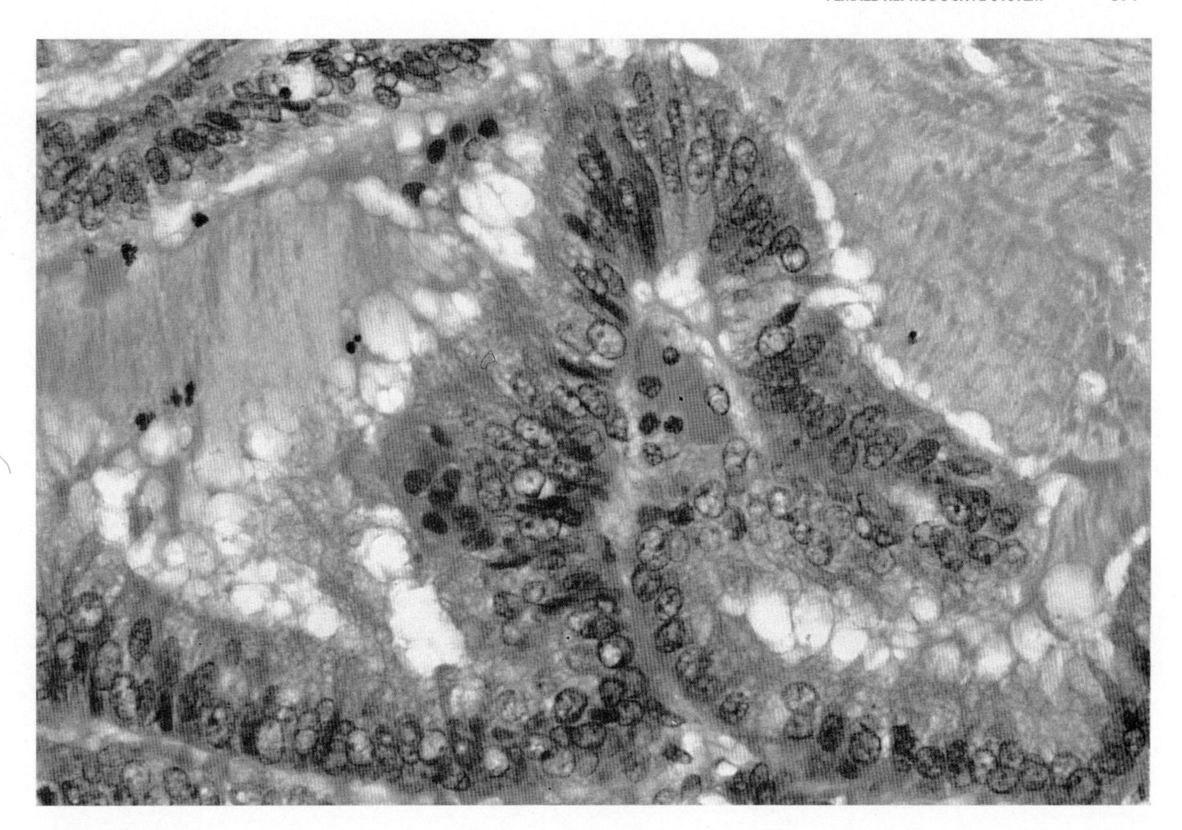

Figure 14–46 **B**

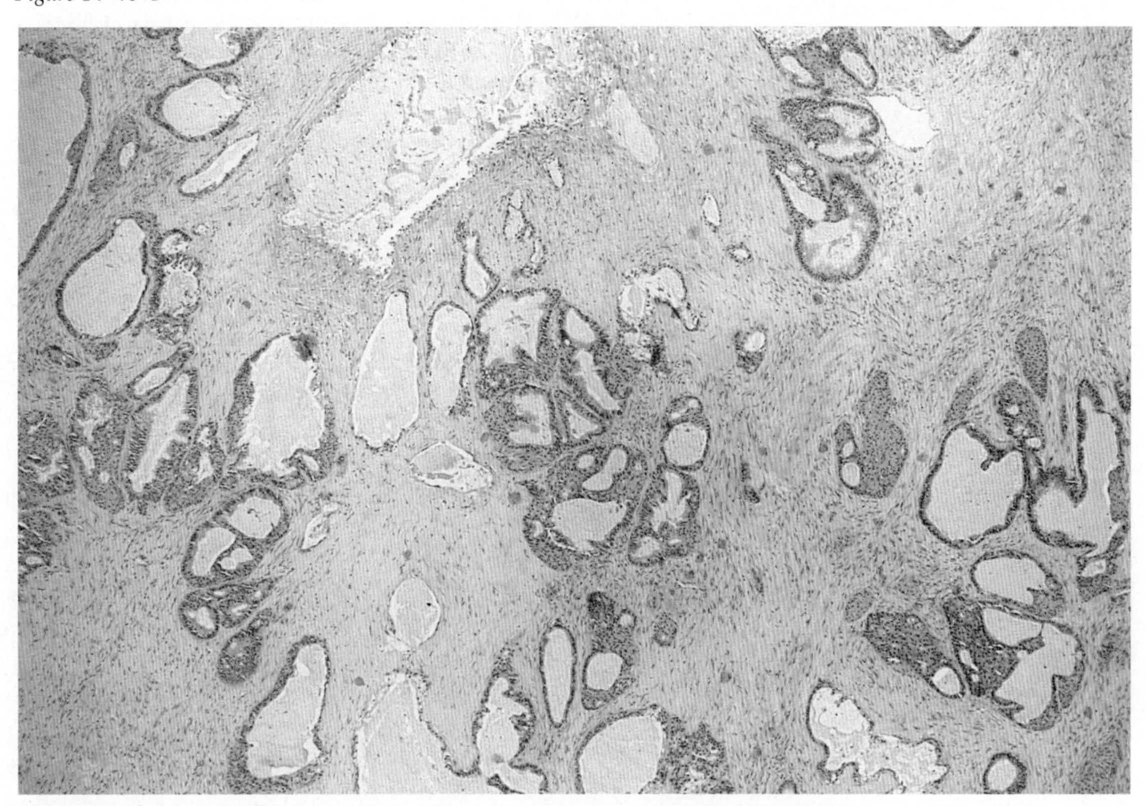

Figure 14–46 **C**

FIGURE 14–46. MUCINOUS CYSTADENOCARCINOMA: B: Cytologically, in both mucinous cystadenocarcinoma of borderline malignant potential and infiltrating mucinous cystadenocarcinoma, the epithelium lining the cystic spaces is stratified with micropapillary projections and the cells have nuclear atypia. Mitotic figures are present. **C:** In infiltrating mucinous cystadenocarcinoma, there is stromal invasion.

■ OVARIES

TUMORS OF SURFACE EPITHELIUM

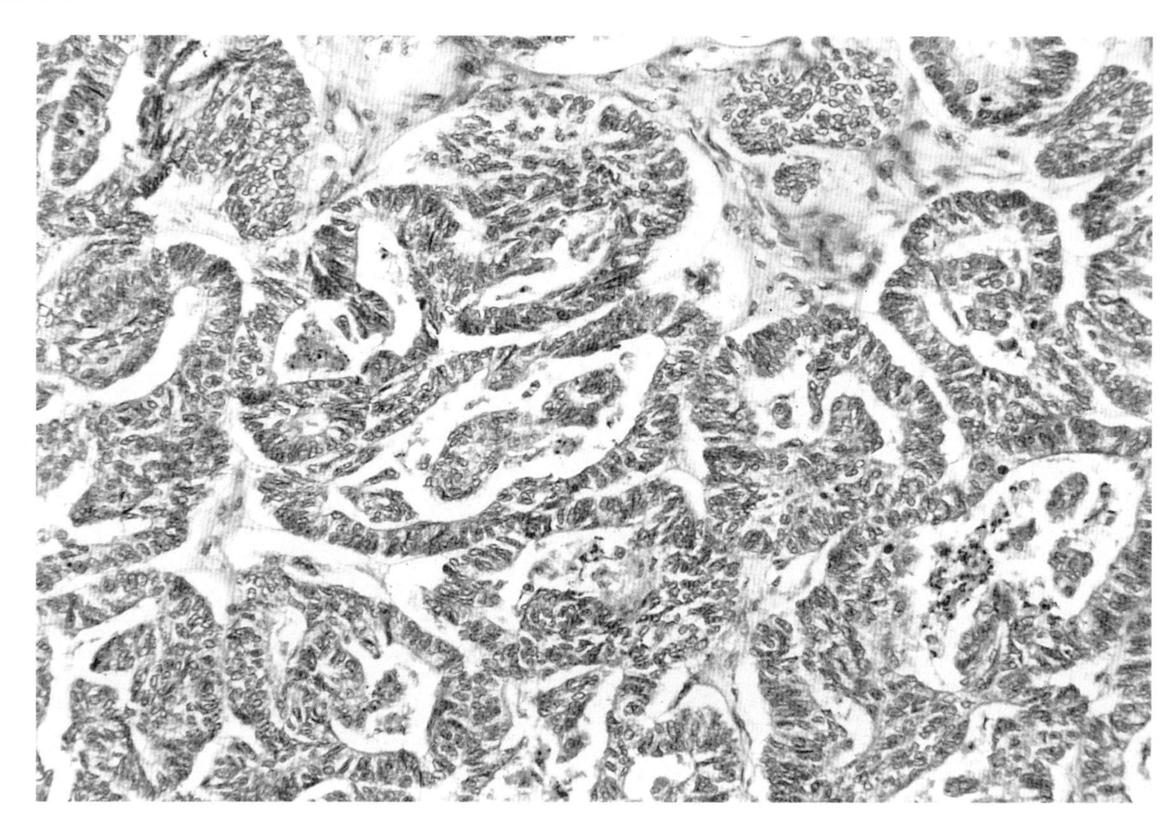

FIGURE 14–47. ENDOMETRIOID CARCINOMA: Endometrioid carcinomas can have both a borderline and infiltrative histology. The histology of these tumors is identical to that of the adenocarcinomas seen in the uterus (Figure 14–38). In this photomicrograph of an endometrioid carcinoma of the ovary, the tumor is composed of back-to-back glands. The cells lining the glands are enlarged with large hyperchromatic and pleomorphic nuclei.

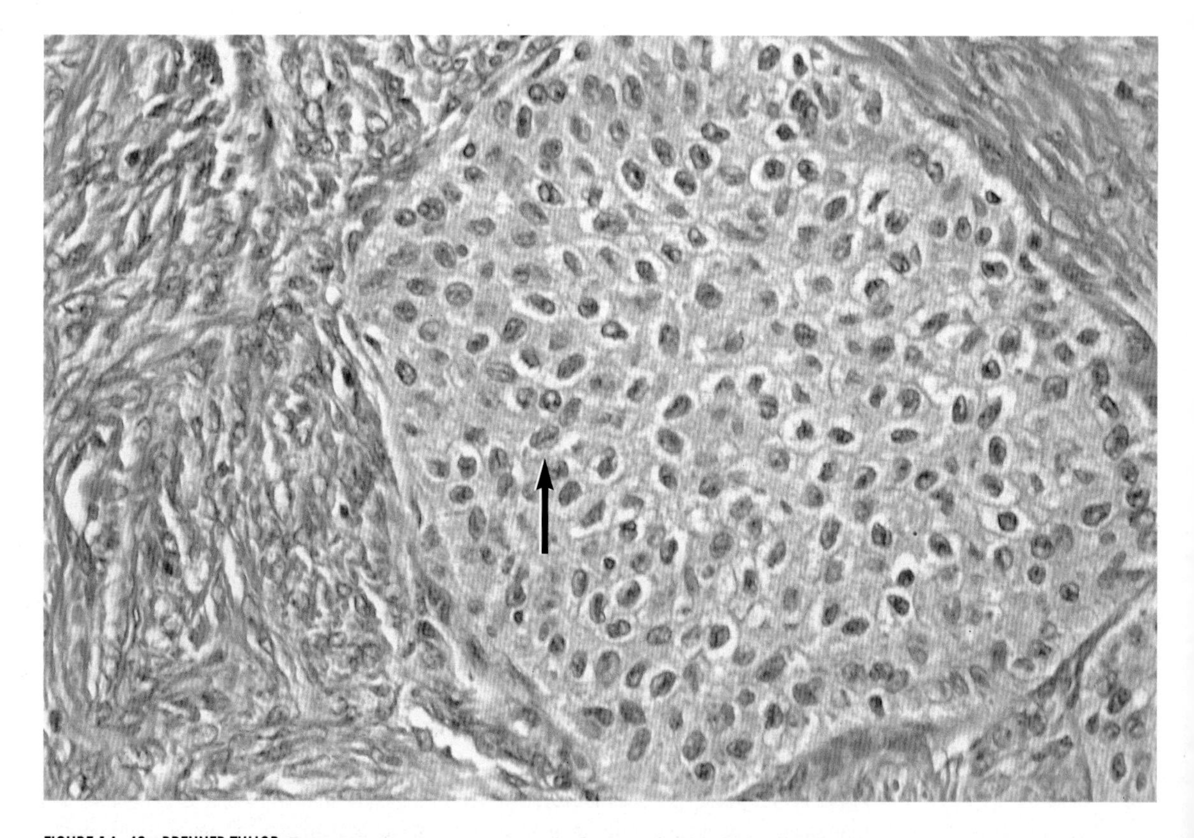

FIGURE 14–48. BRENNER TUMOR: Brenner tumors are composed of nests of epithelial cells surrounded by a spindle cell stroma. The cells are squamoid with abundant eosinophilic cytoplasm. Nuclei are oval with a longitudinal groove reminiscent of a coffee bean (*arrow*). Nucleoli are distinct.

The schematic for ovarian germ cell tumors is the same as that for testicular germ cell tumors (see page 428).

■ OVARIES

GERM CELL TUMORS

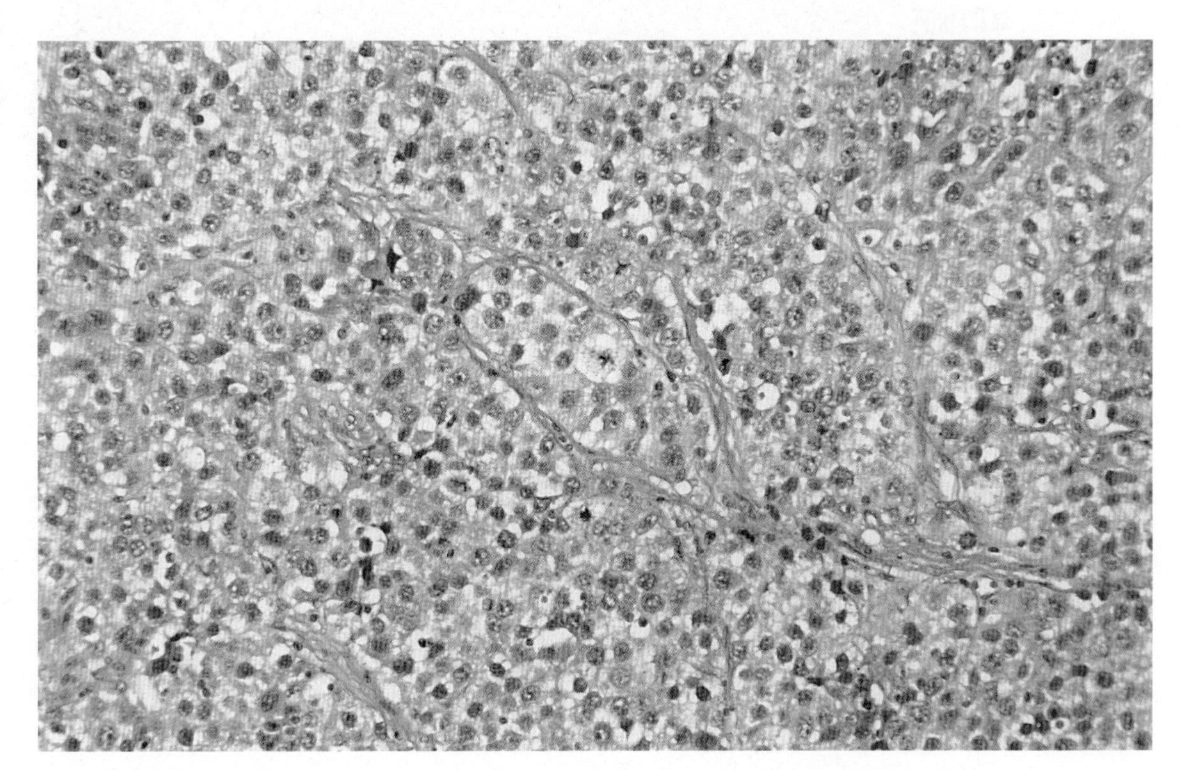

Figure 14–49 **A**

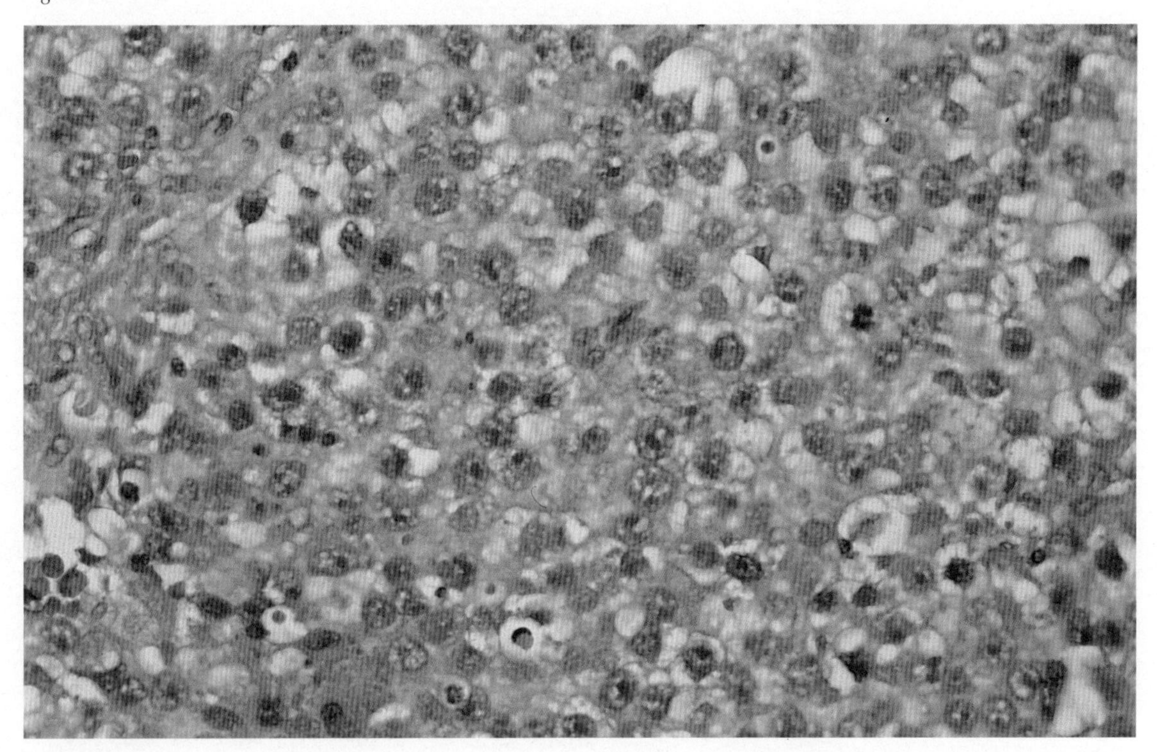

Figure 14–49 **B**

FIGURE 14–49. DYSGERMINOMA (OVARIAN SEMINOMA): Dysgerminoma is the ovarian counterpart of the classic seminoma of the testis. **A:** It is composed of groups of cells separated by variable amounts of connective tissue containing lymphocytes. In this photomicrograph, the lymphocytic infiltrate is lacking. **B:** The cells are large. When the tissue is well fixed, distinct cell borders are prominent. The cytoplasm is abundant and clear or finely granular and eosinophilic. Nuclei are large and centrally located. Nucleoli are prominent, and mitotic activity is variable.

■ OVARIES

GERM CELL TUMORS

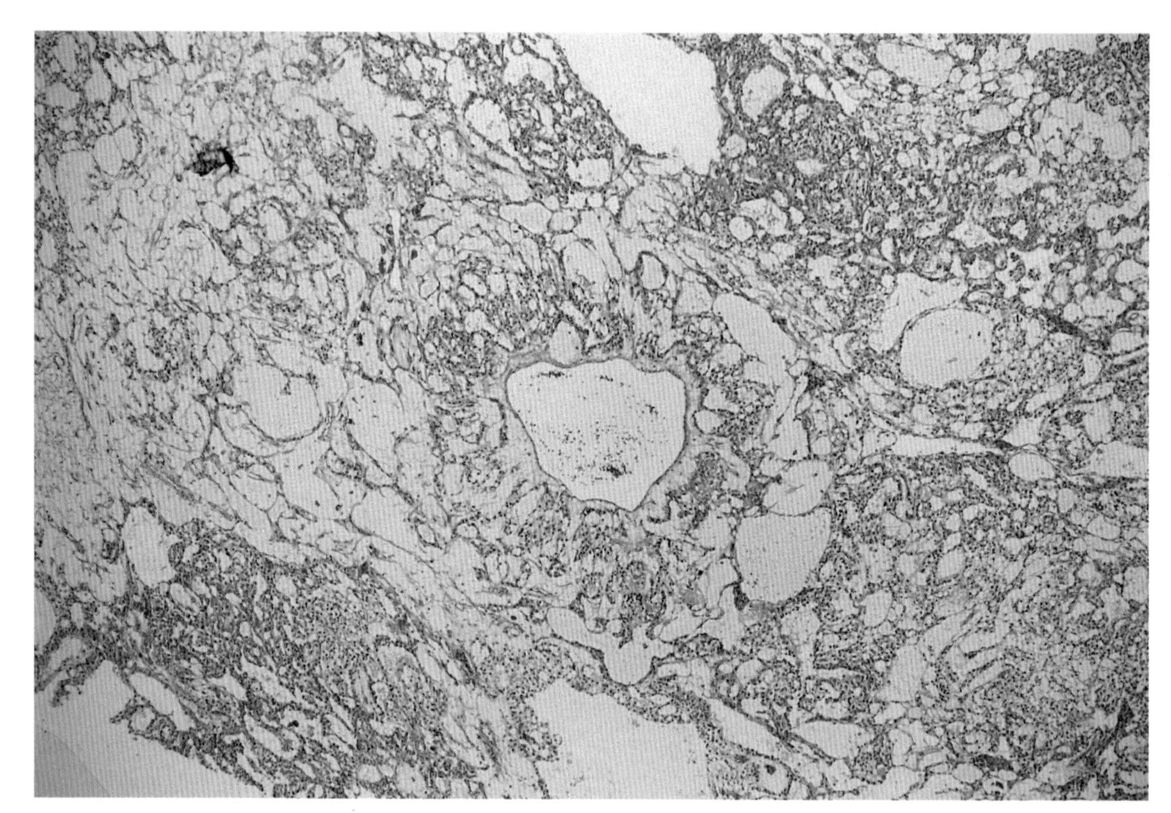

Figure 14–50 A

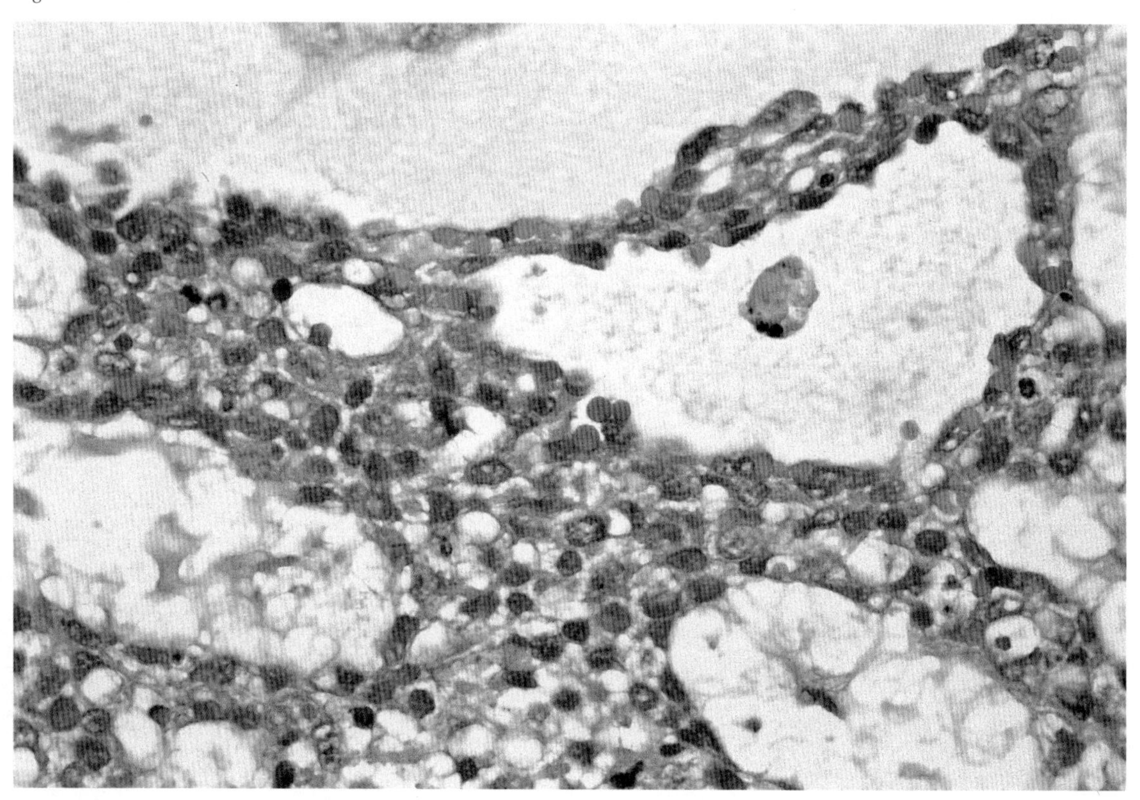

Figure 14–50 B

FIGURE 14–50. ENDODERMAL SINUS TUMOR (YOLK SAC TUMOR): A: This is an example of the microcystic pattern of an endodermal sinus tumor. In the center is a blood vessel from which microcystic structures radiate. **B:** The cysts are lined by cuboidal cells. Mitotic figures are present in variable numbers. Within the cells and the cystic spaces are eosinophilic globules.

■ OVARIES

GERM CELL TUMORS

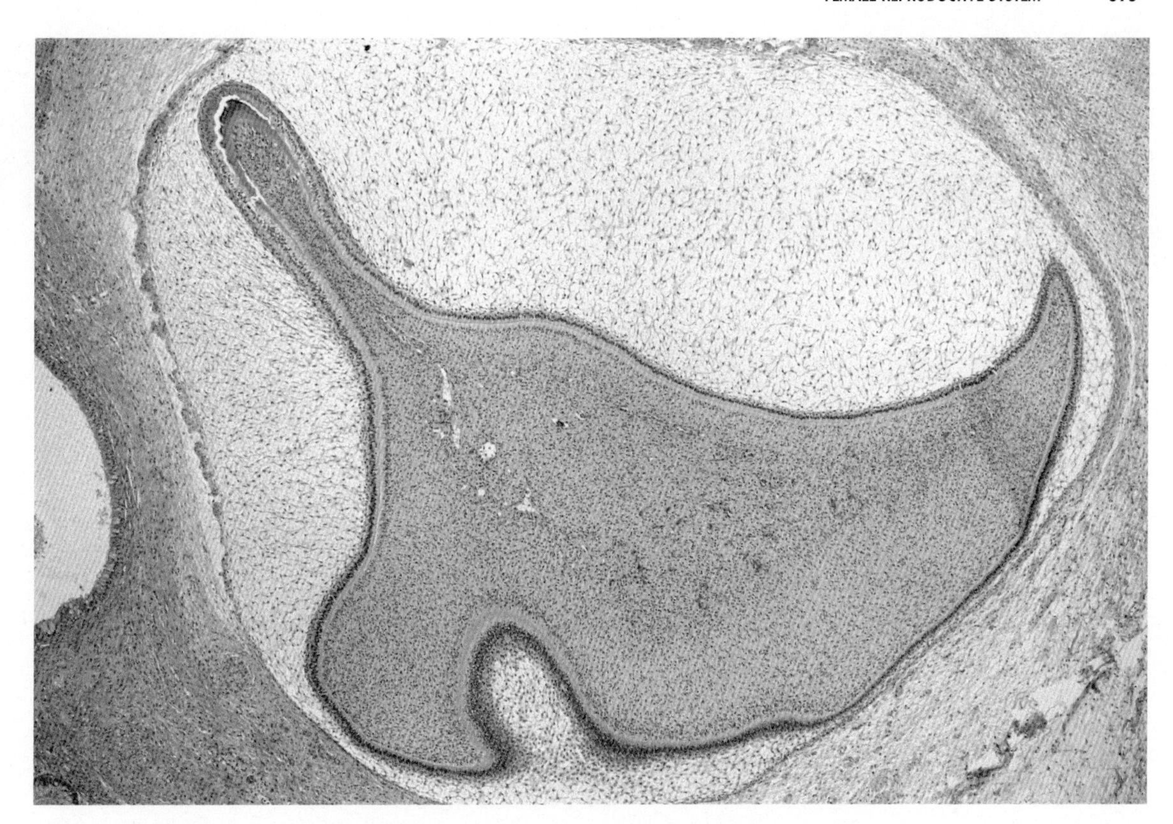

Figure 14–51 **A**

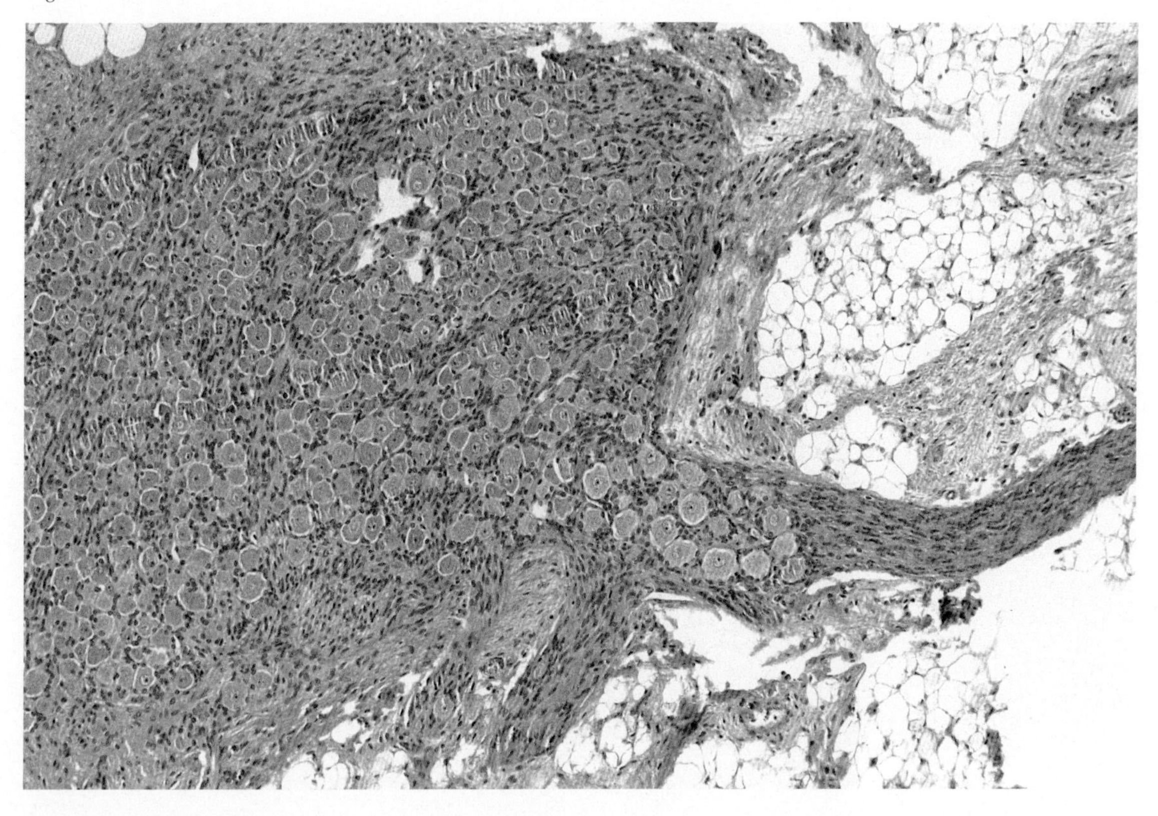

Figure 14–51 **B**

FIGURE 14–51. TERATOMA: Teratomas are composed of tissue elements derived from the three germ cell layers. In these photomicrographs of a mature teratoma, ectodermal elements are represented by the enamel of a developing tooth (**A**) and nerve and ganglia (**B**).

■ OVARIES

GERM CELL TUMORS

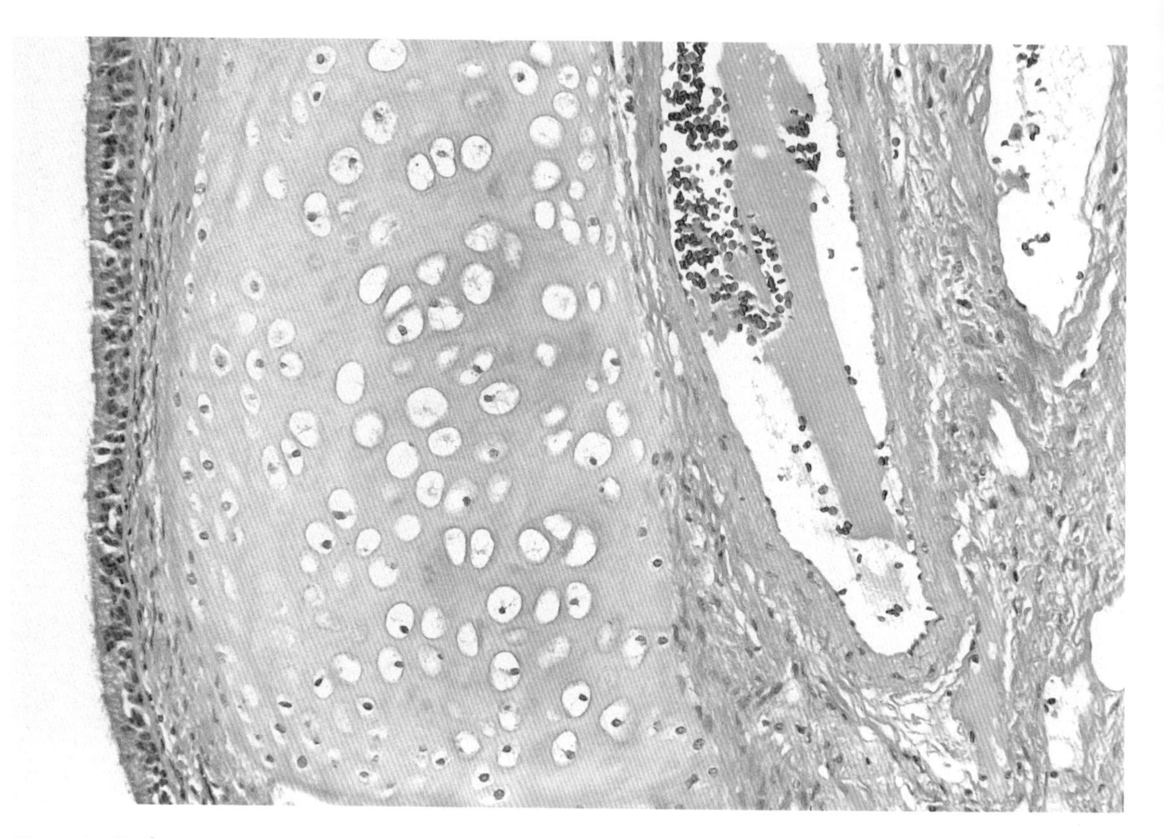

Figure 14–51 **C**

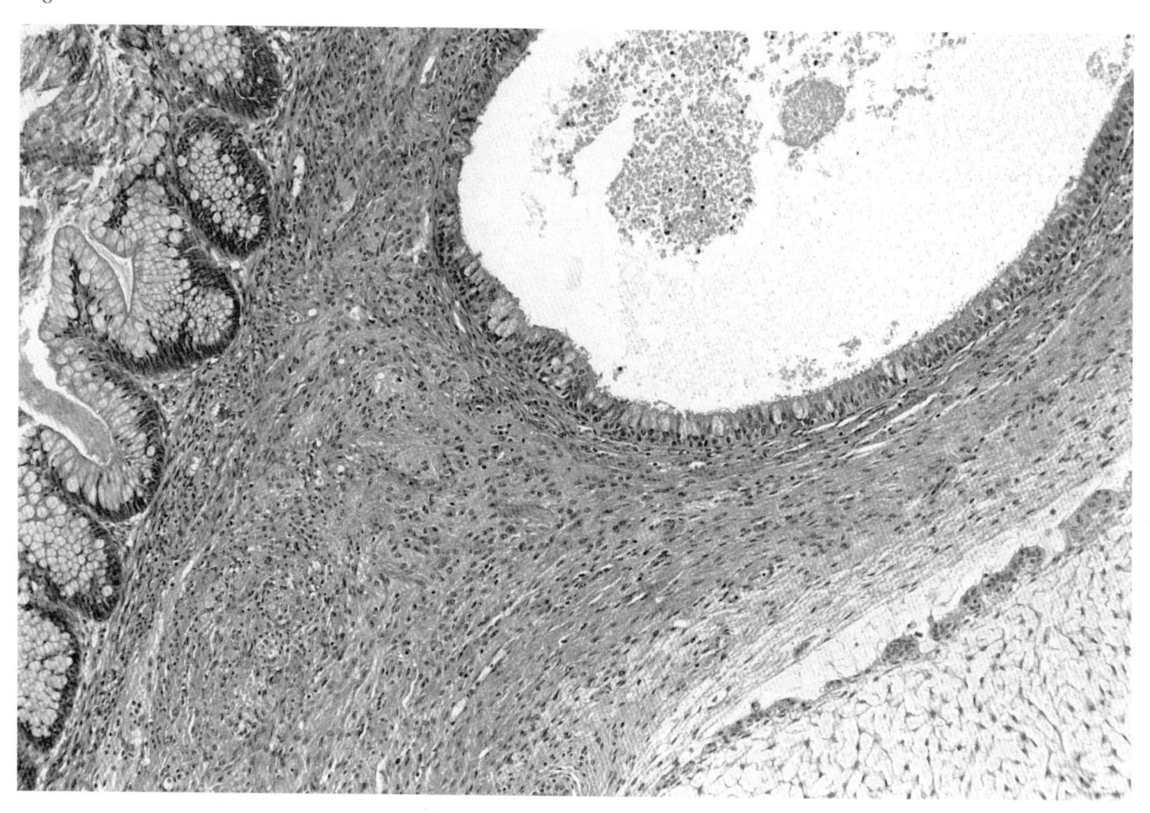

Figure 14–51 **D**

FIGURE 14–51. TERATOMA: Mesodermal elements that can be seen in teratomas are adipose tissue (**B**) and cartilage (**C**). Endodermal elements that can be observed in teratomas are respiratory epithelium (**C**) and gastrointestinal epithelium (**D**).

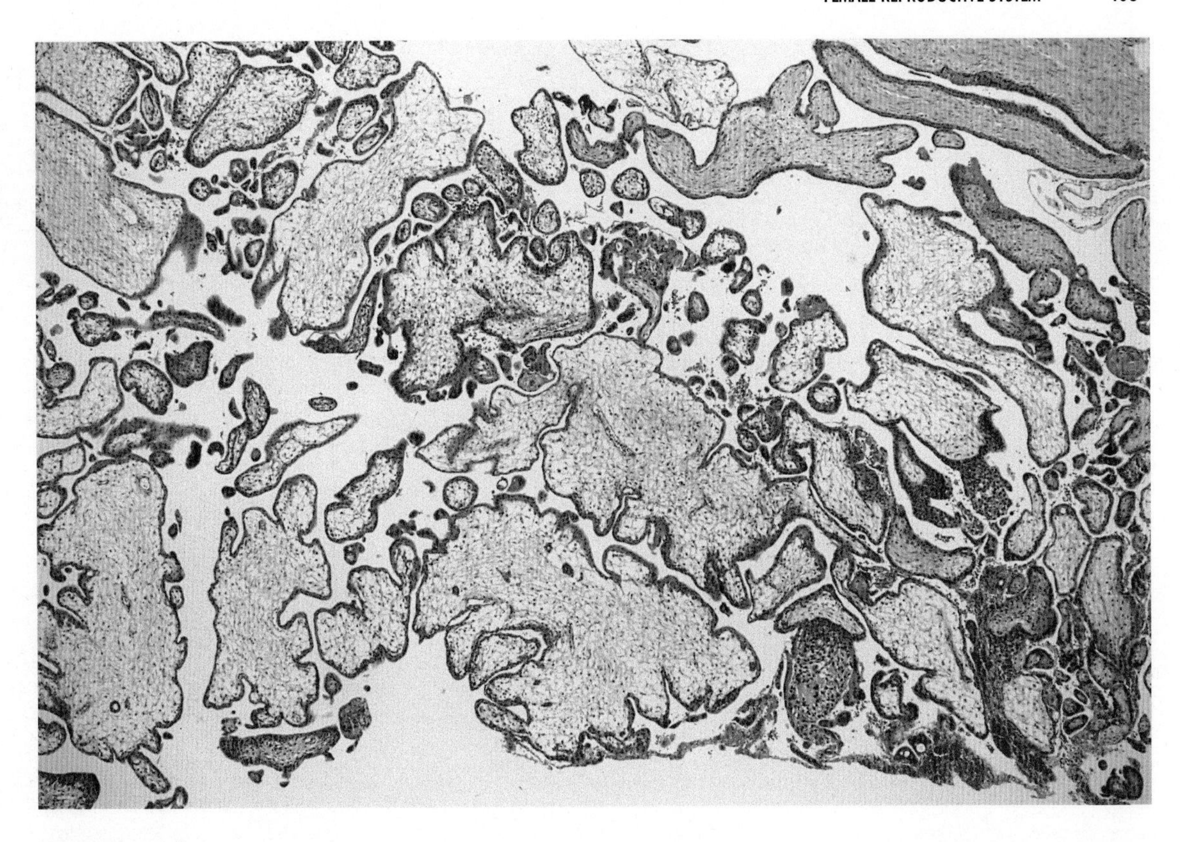

Figure 14–61 A

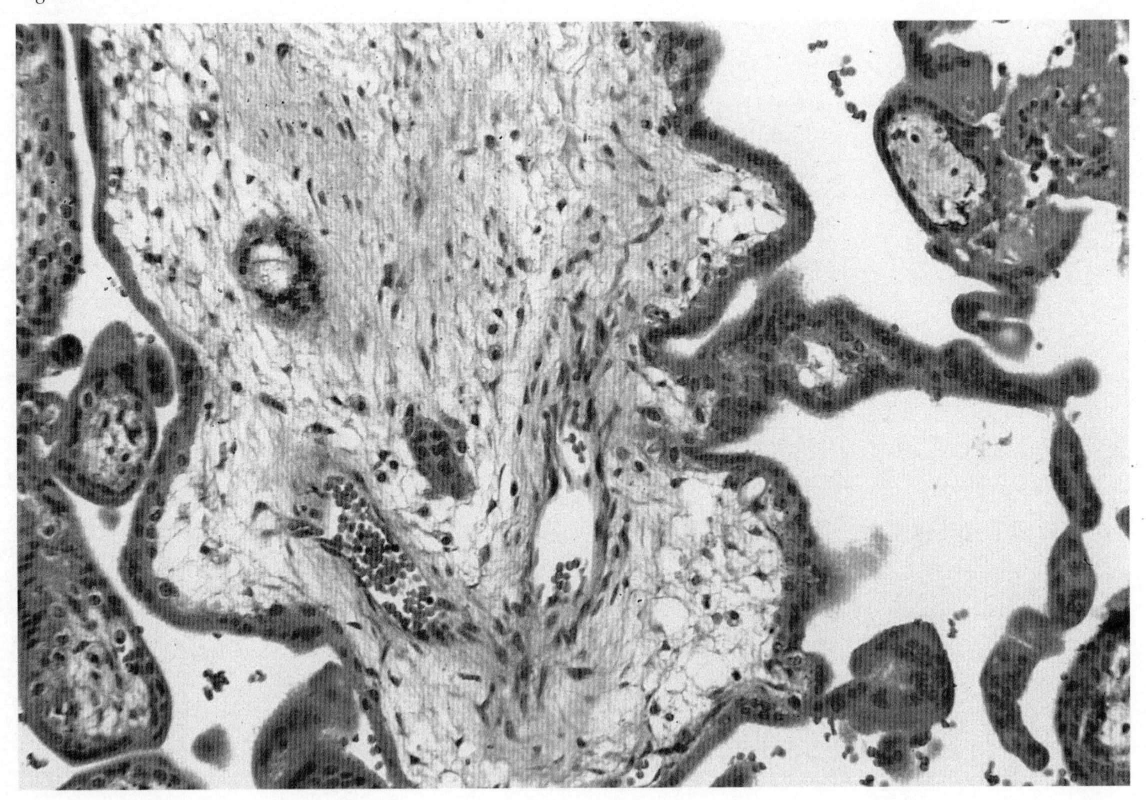

Figure 14–61 B

FIGURE 14–61. PARTIAL HYDATIDIFORM MOLE: A: In a partial mole, there is a mixture of normal and hydropic chorionic villi. The hydropic villi have a scalloped border. **B:** Because of the extensive scalloping, trophoblastic inclusions are present within the chorionic villi and trophoblastic proliferation is present focally on their surfaces. Since a fetus is present, the chorionic villi have blood vessels containing nucleated red blood cells.

PLACENTA

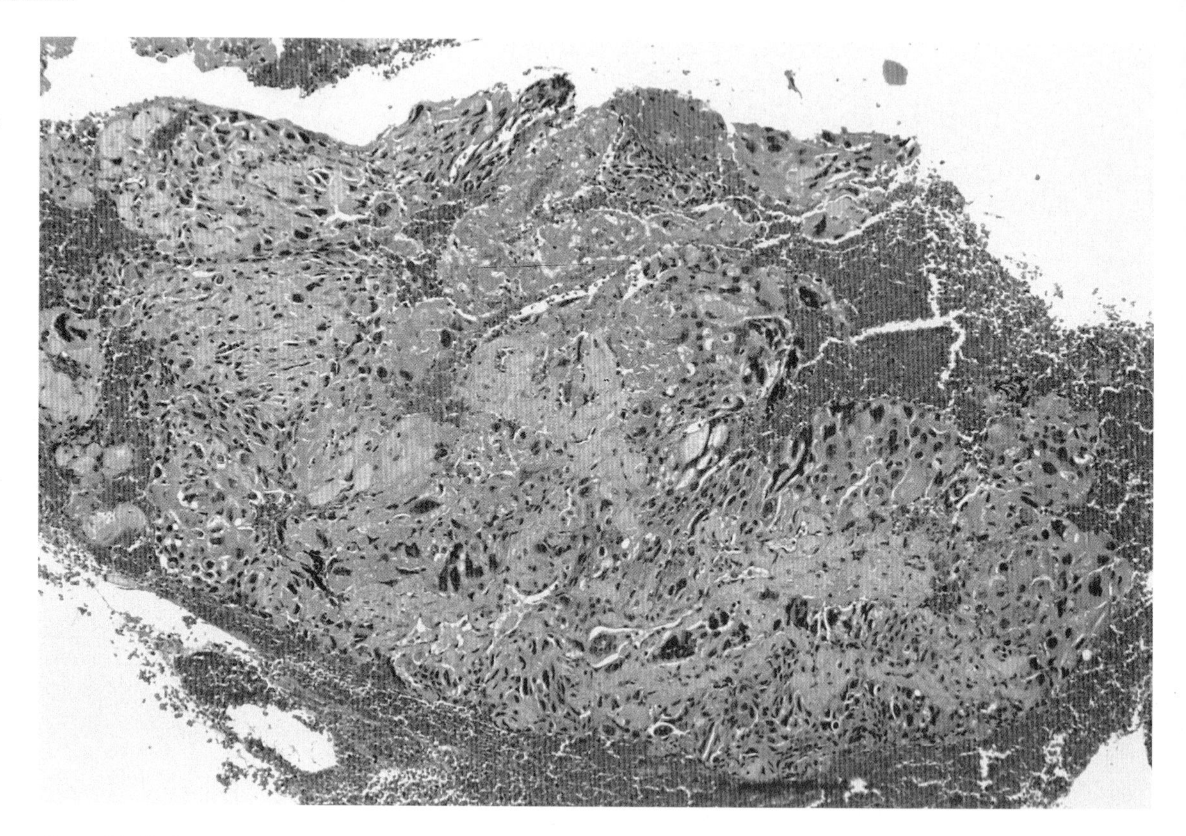

Figure 14–62 **A**

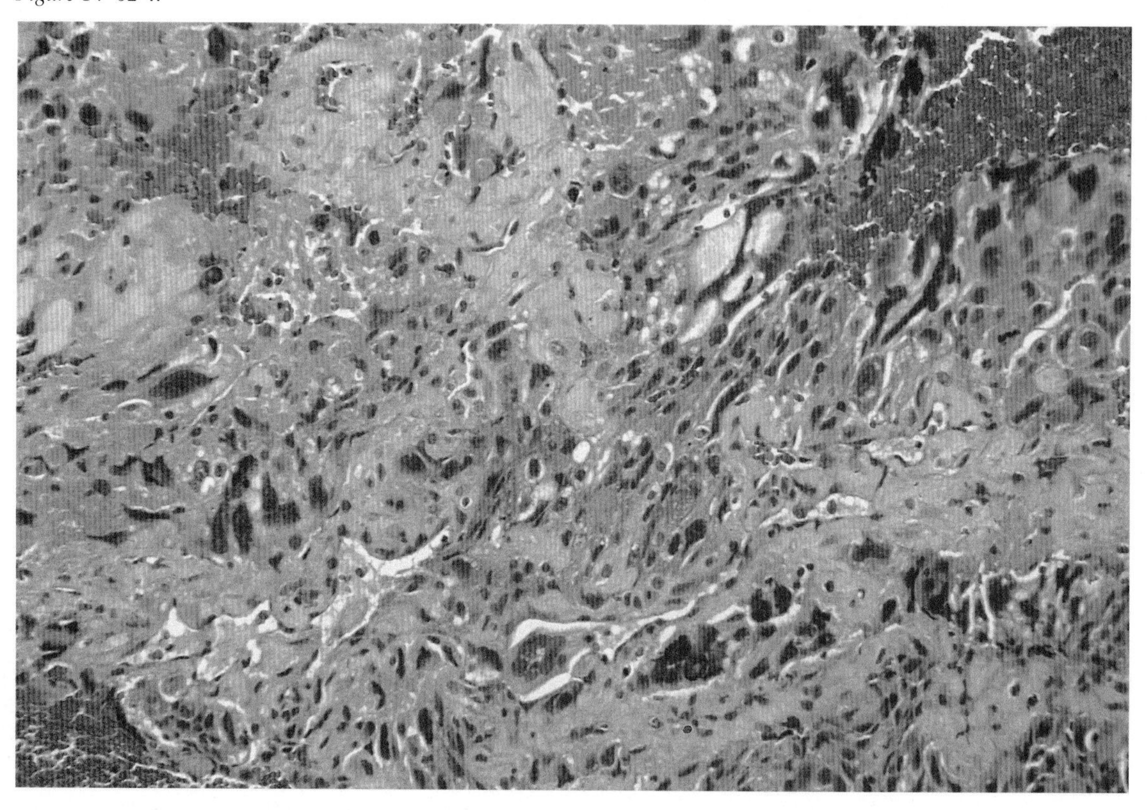

Figure 14–62 **B**

FIGURE 14–62. CHORIOCARCINOMA: A and **B**: Choriocarcinoma is characterized by an admixture of anaplastic cytotrophoblasts and syncytiotrophoblasts. Both cells have eosinophilic cytoplasm. Necrosis is characteristic. Cytotrophoblasts are mononuclear, and syncytiotrophoblasts are multinuclear.

■ INFLAMMATORY LESIONS

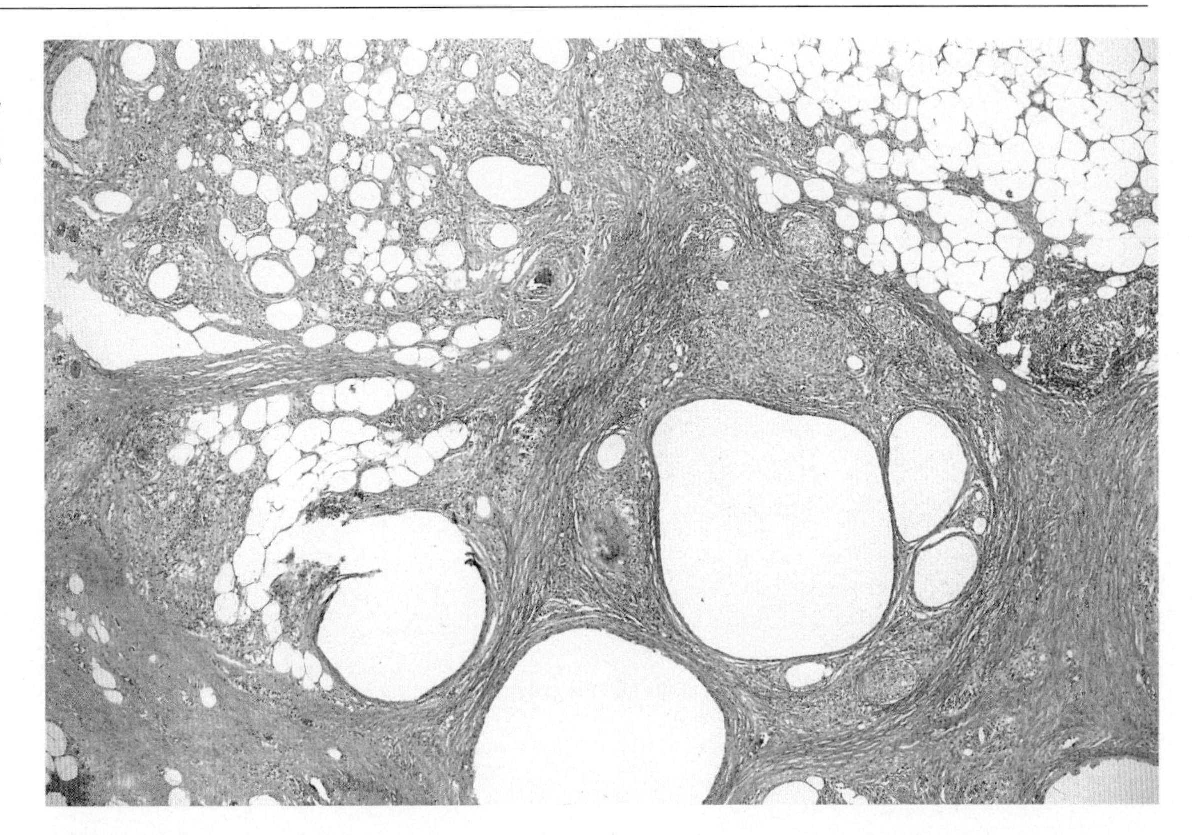

Figure 15–1 A

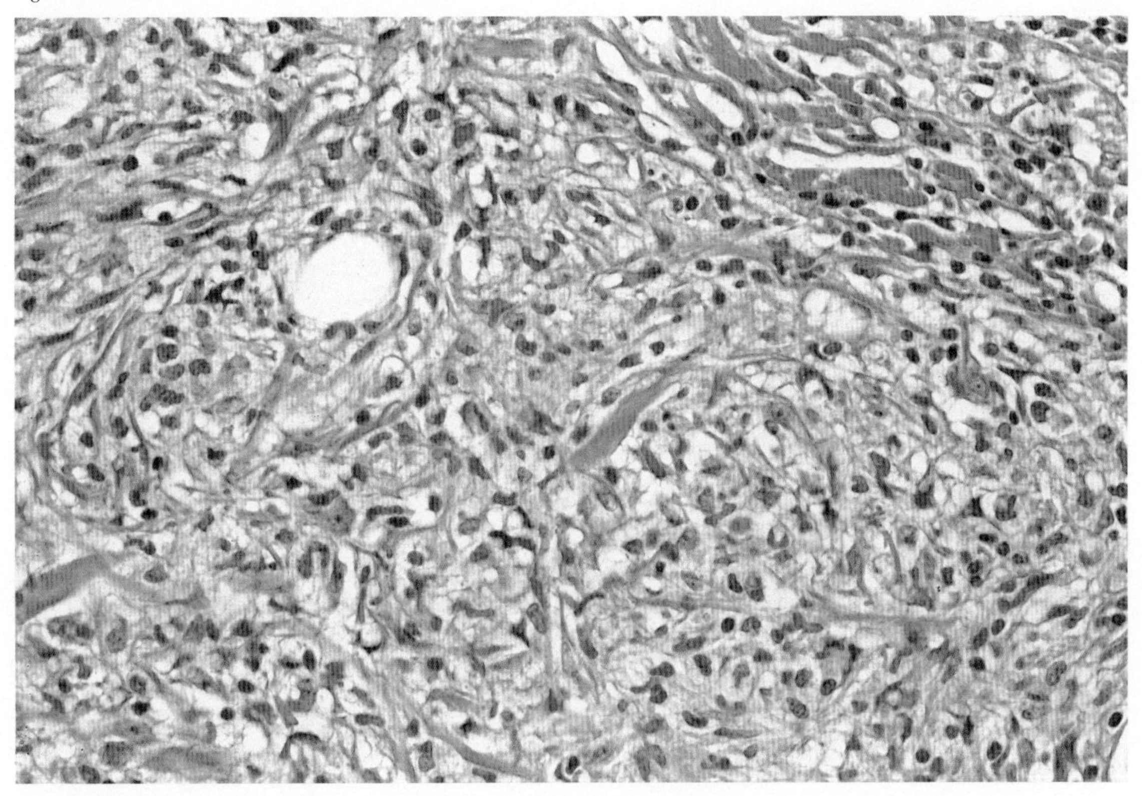

Figure 15–1 B

FIGURE 15–1. TRAUMATIC FAT NECROSIS: A: Traumatic fat necrosis is characterized by hemorrhage and necrosis in adipose tissue in the mammary glands. **B:** Higher magnification shows the inflammatory infiltrate to be composed of lipid-laden macrophages and few polymorphonuclear leukocytes.

■ INFLAMMATORY LESIONS

FIGURE 15–2. ACUTE MASTITIS: In acute mastitis, polymorphonuclear leukocytes are seen in duct walls and periductal stroma.

FIGURE 15–3. PLASMA CELL MASTITIS (MAMMARY DUCT ECTASIA): A: In plasma cell mastitis, ducts are dilated (ectatic) and filled with granular necrotic debris and the periductal stroma is infiltrated by inflammatory cells. The duct walls are thickened, and the cells lining the ducts are necrotic and/or atrophic. **B:** At higher magnification, it can be seen that the inflammatory infiltrate within the periductal stroma is composed predominantly of plasma cells, neutrophils, foamy-laden macrophages, and on occasion foreign body giant cells.

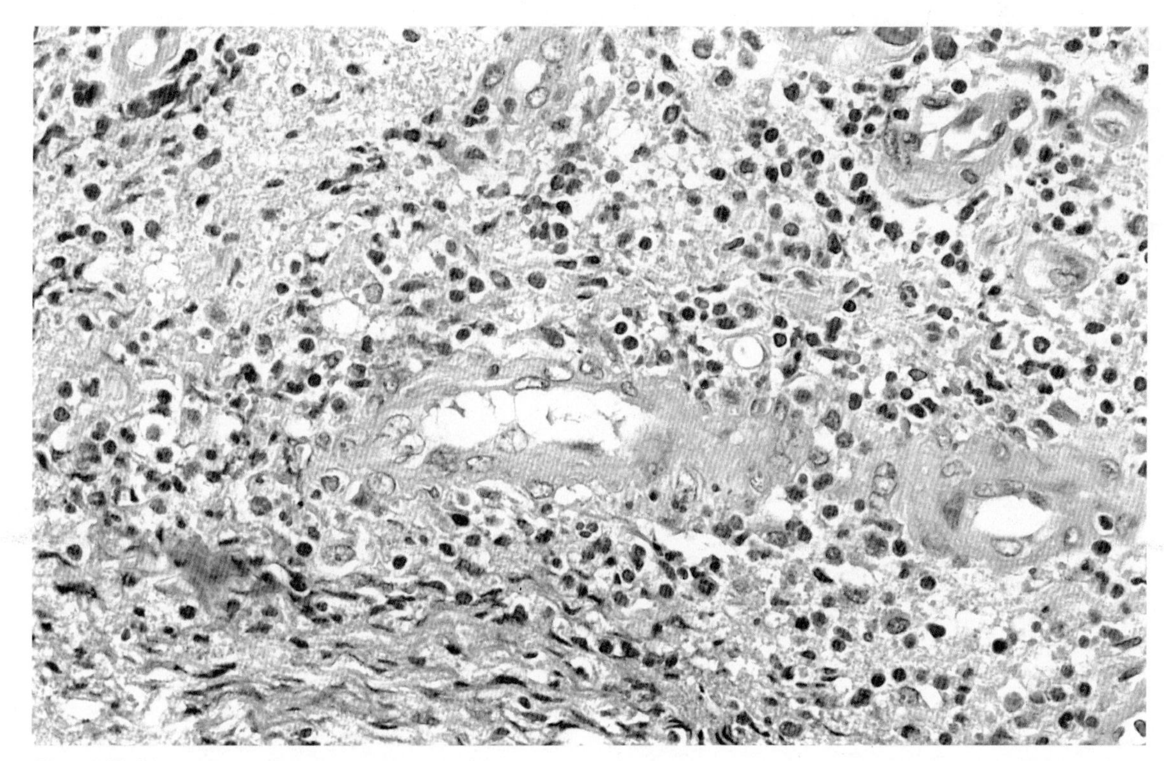

Figure 15–2

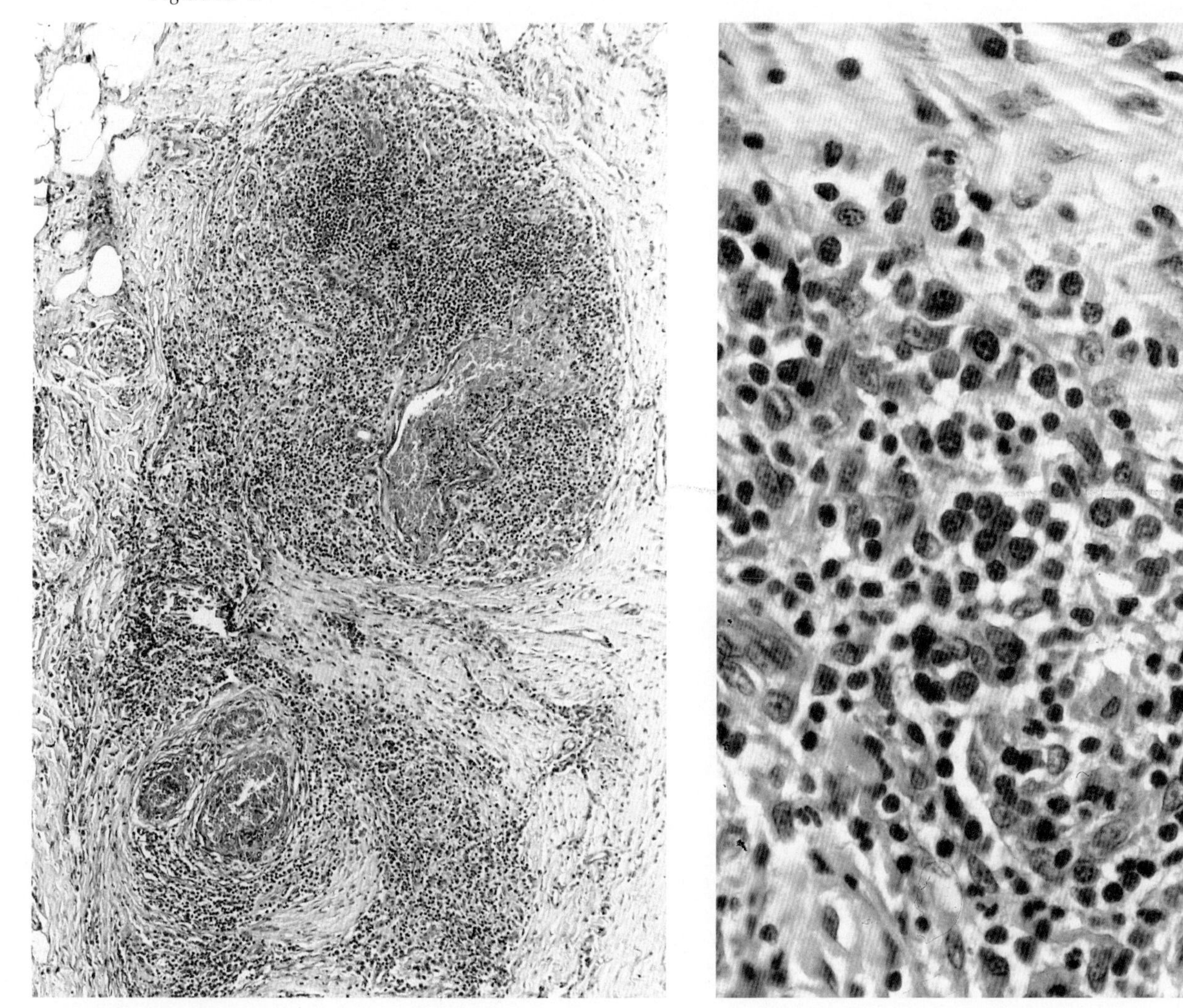

Figure 15–3 A

Figure 15–3 B

■ FIBROCYSTIC CHANGES

Fibrocystic changes of the breast encompass several histopathologic entities. The changes may be present individually or in any combination and consist of

1. cystic dilation of glands
2. apocrine metaplasia
3. fibrous mastopathy
4. intraductal hyperplasia—**papillomatosis**
5. adenosis, with or without sclerosis.

FIGURE 15–4. CYSTIC CHANGES: A: In cystic changes, ducts and ductules become dilated. The epithelium lining the ducts is cuboidal and becomes attenuated when there is considerable dilatation. Intraluminal secretions are noted.

FIGURE 15–4. APOCRINE METAPLASIA: B: In apocrine metaplasia, the epithelial cells are polygonal with abundant eosinophilic cytoplasm. At the apical aspect of the cells, buds of cytoplasm—**"apocrine snouts"**—are present. Papillary projections of the epithelium may also be present.

FIGURE 15–4. FIBROUS MASTOPATHY: C: In fibrous mastopathy, the stroma becomes fibrotic; ducts and lobules become compressed and eventually become atrophic.

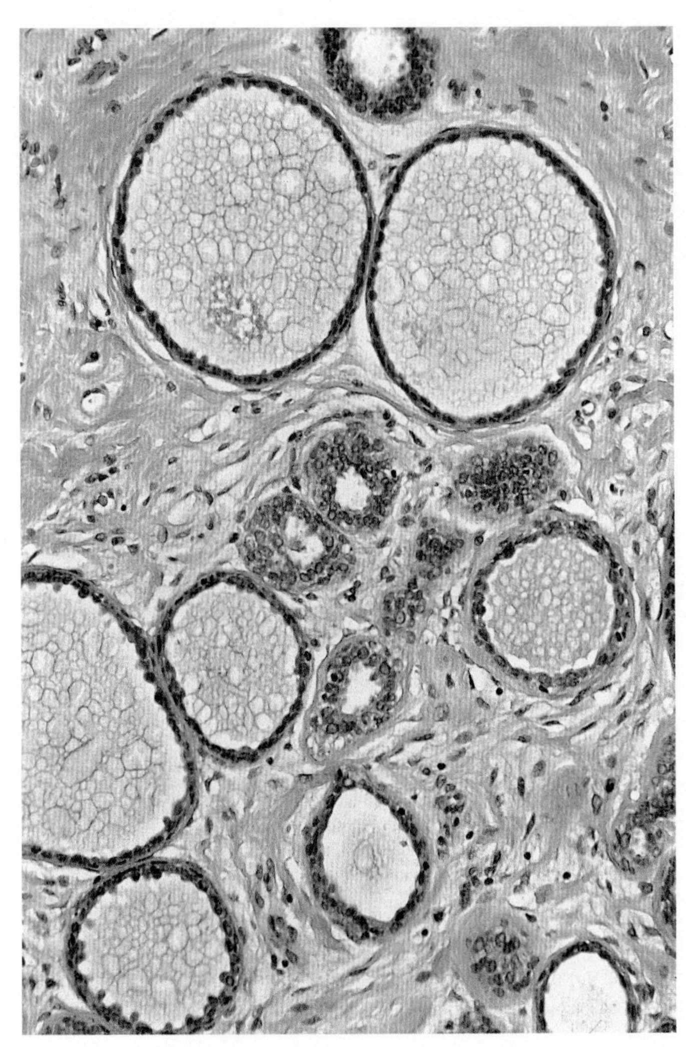

Figure 15–4 A

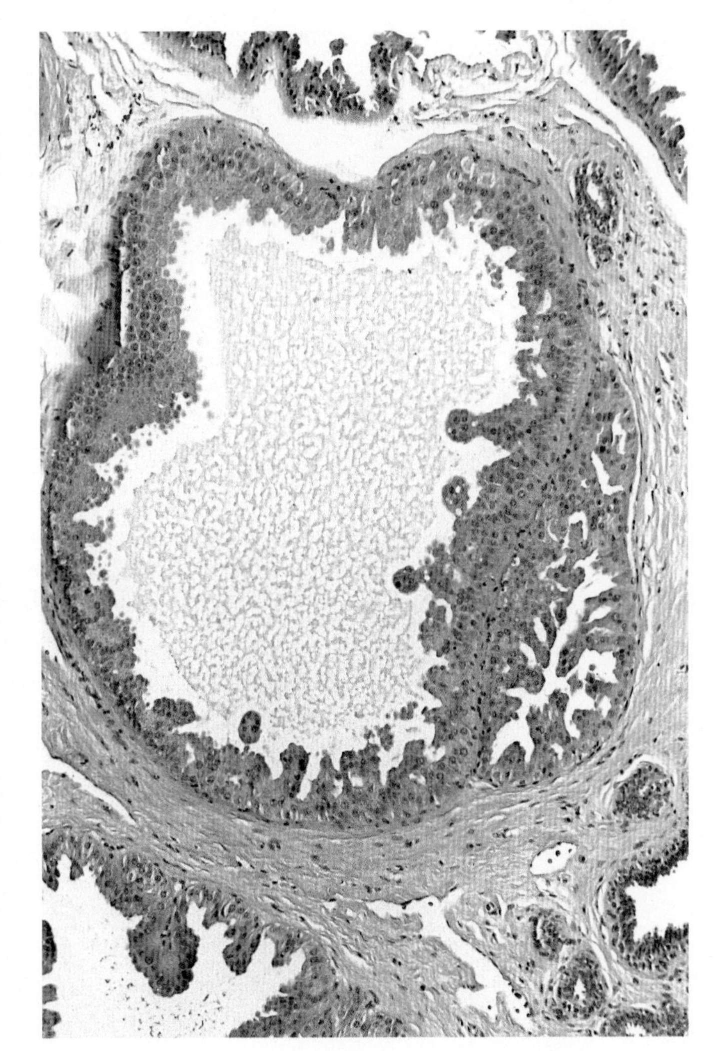

Figure 15–4 B

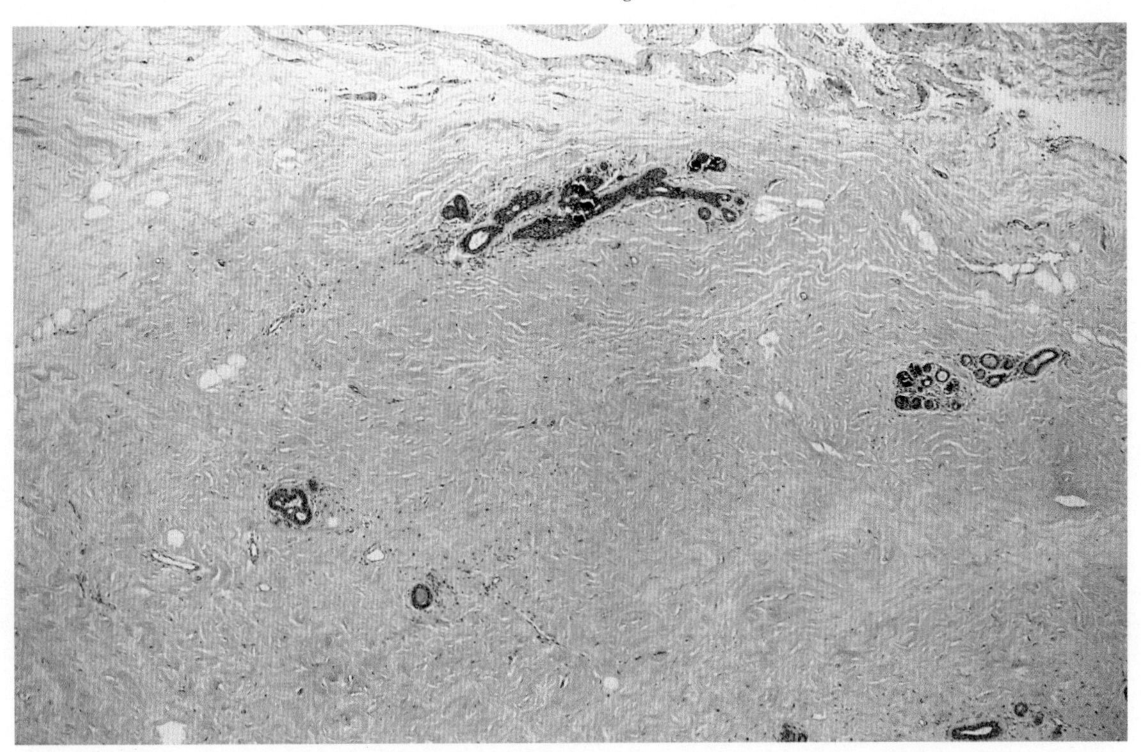

Figure 15–4 C

■ FIBROCYSTIC CHANGES

FIGURE 15–4. INTRADUCTAL HYPERPLASIA: D: In intraductal hyperplasia, there is proliferation of the epithelial cells lining ducts and ductules and intraductal lumina always remain patent. The ductal cells are oval with acidophilic cytoplasm, and cell borders are indistinct. Nuclei are normochromic.

FIGURE 15–4. ADENOSIS: E: In adenosis, ductules and acini within lobules proliferate. The intralobular stroma proliferates as well and causes distortion and compression of the glandular elements. The key features to the benign nature of this lesion are retention of the lobular configuration (**E**) and presence of a myoepithelial layer (**F**). These features should be contrasted with those of tubular carcinoma (Figure 15–11 **A** and **B**).

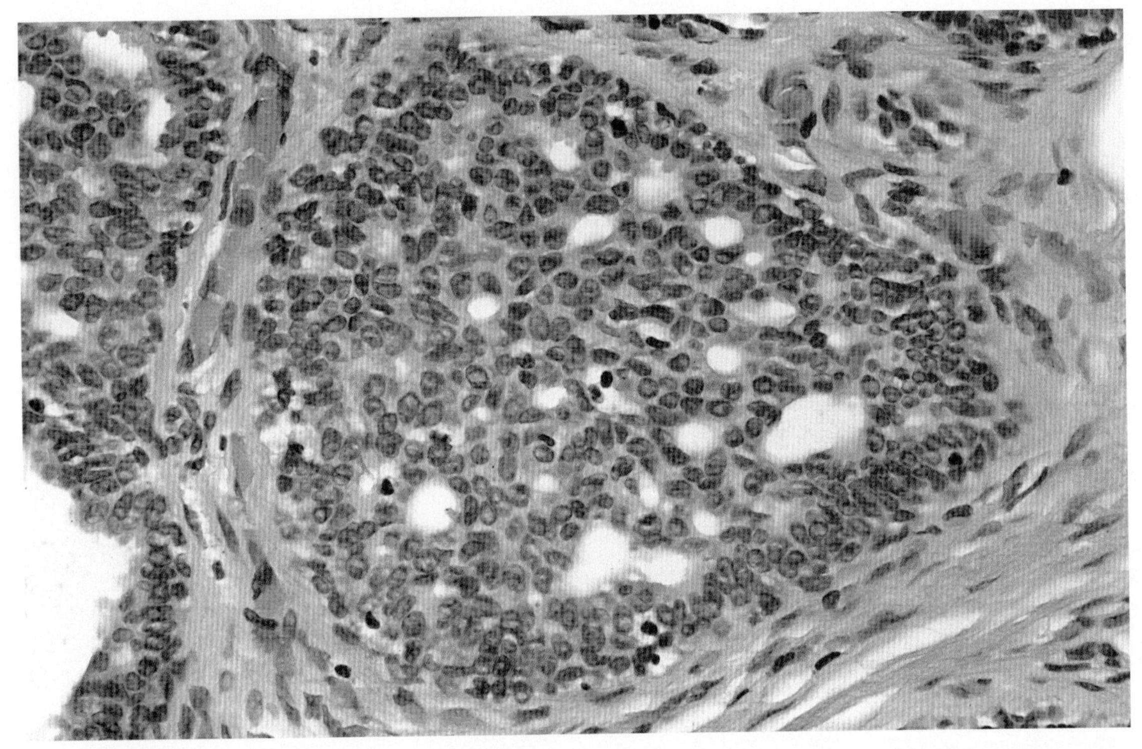

Figure 15–4 D

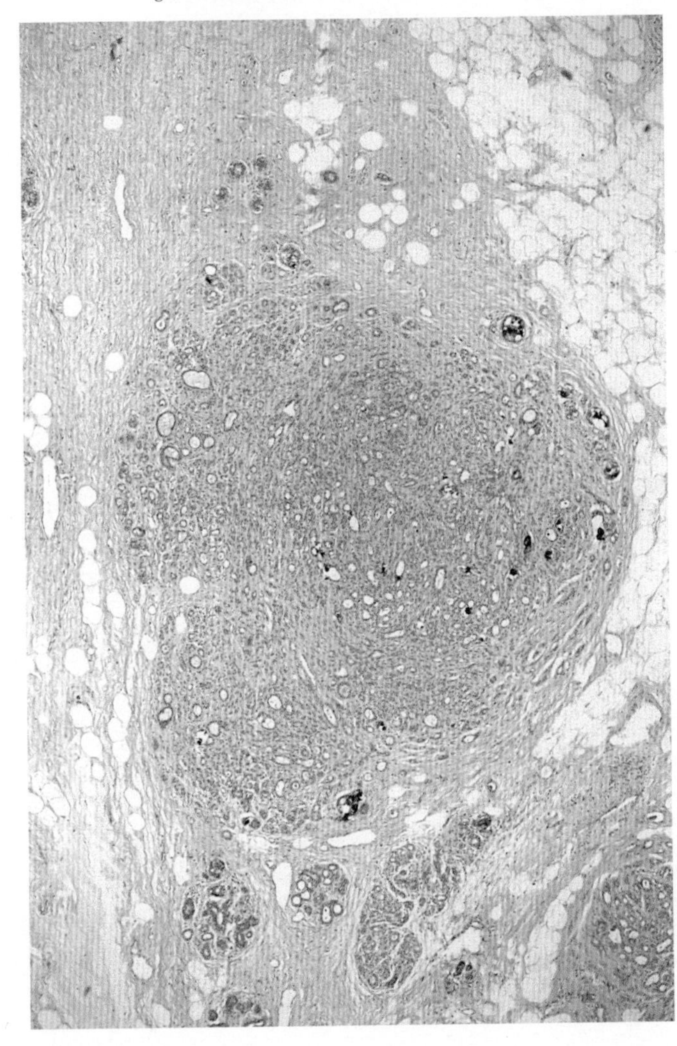

Figure 15–4 E

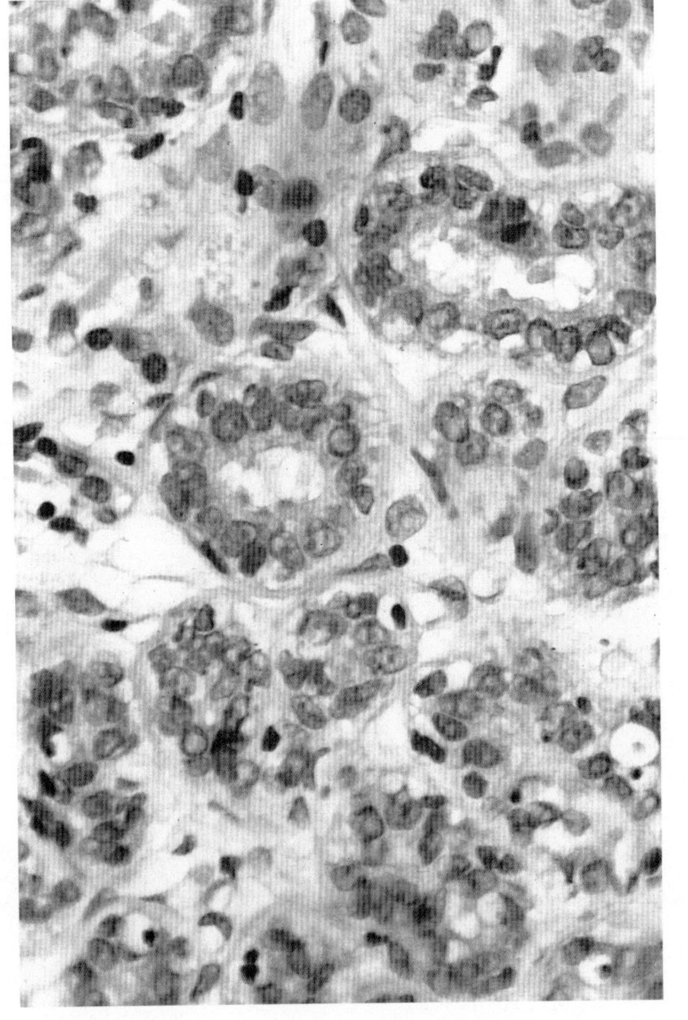

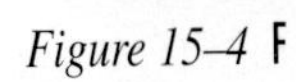

Figure 15–4 F

■ PROLIFERATIVE LESIONS

FIGURE 15–5. INTRADUCTAL PAPILLOMA: A: In intraductal papillomas, papillary lesions are seen within dilated large ducts. **B:** The papillary fronds are composed of fibrovascular cores covered by two types of cells—cuboidal epithelial and myoepithelial. Within the fibrovascular stalks, the stroma may show fibrosis and entrapped glandular elements.

■ STROMAL TUMORS OR BIPHASIC TUMORS

Stromal tumors of the breast represent a variety of lesions, ranging from the completely benign fibroadenoma to the malignant phyllodes tumor. Between the two is the benign phyllodes tumor, which although histologically benign, has a tendency to recur.

FIGURE 15–6. FIBROADENOMA: Fibroadenomas are sharply circumscribed and their predominant histologic feature is stromal proliferation in which the amount of proliferation varies from lesion to lesion. The stroma is characteristically loose and may show myxoid change. Glandular lumina are present in two patterns. The pattern in this figure is pericanalicular, in which the glandular lumina remain open. The intracanalicular pattern is shown in the benign phyllodes tumor (Figure 15–7 **B**), in which the glandular elements are compressed within the stroma.

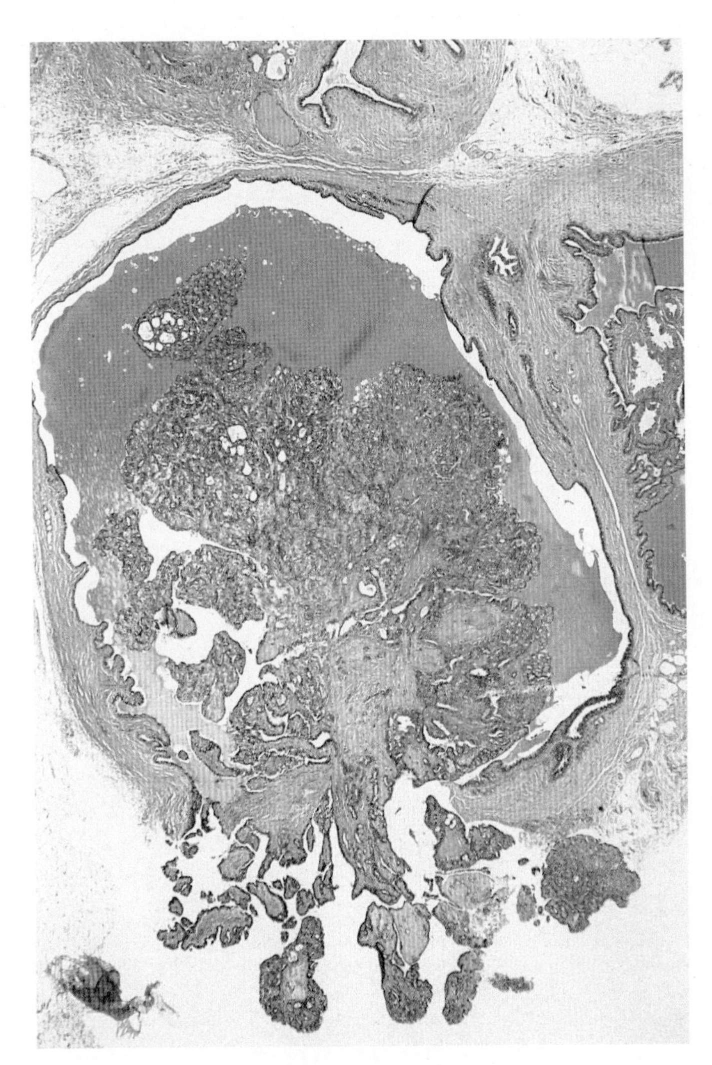

Figure 15–5 A

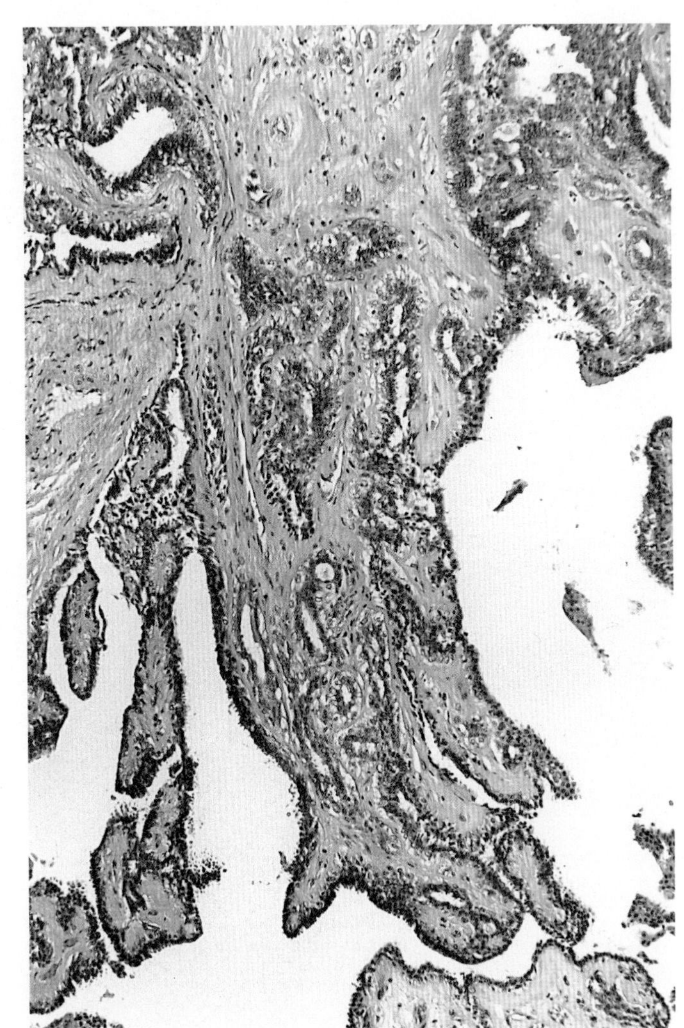

Figure 15–5 B

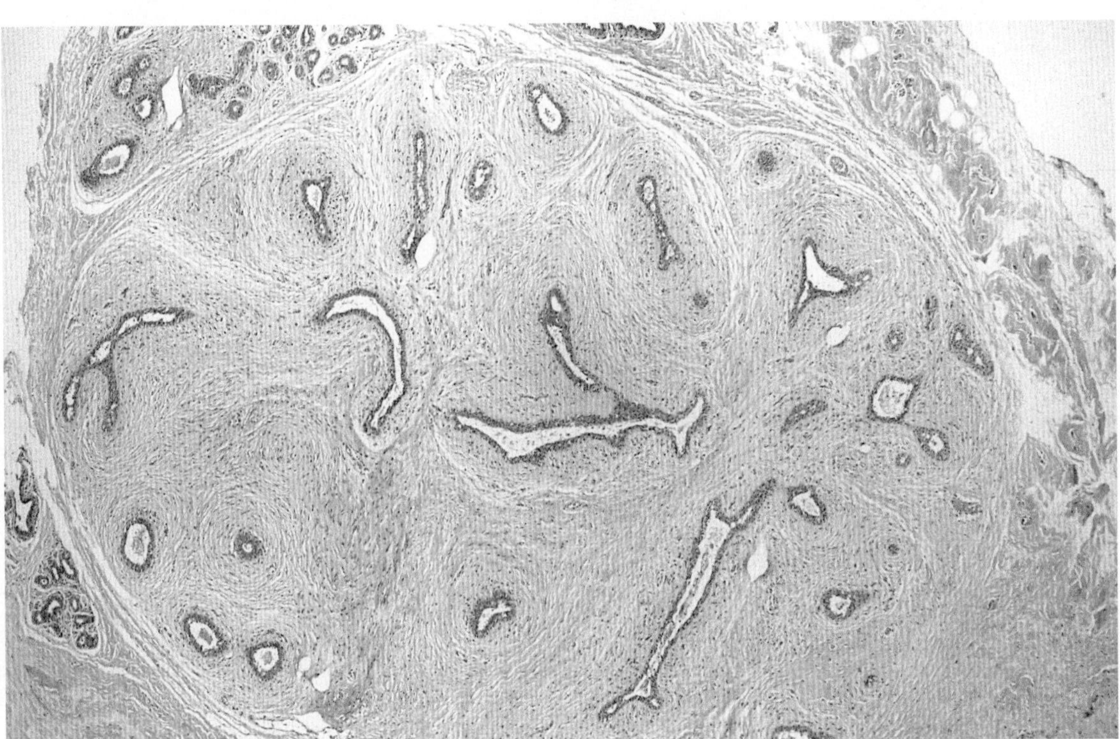

Figure 15–6

■ STROMAL TUMORS OR BIPHASIC TUMORS

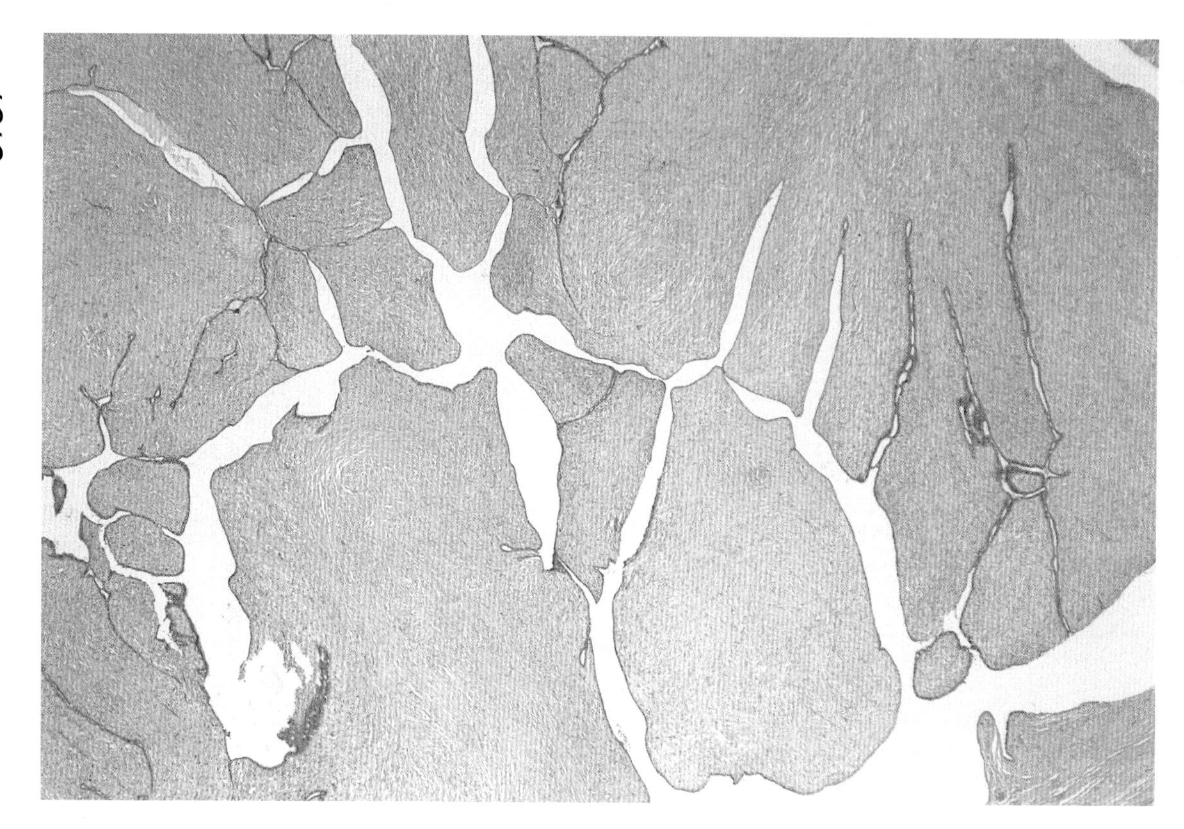

Figure 15–7 **A**

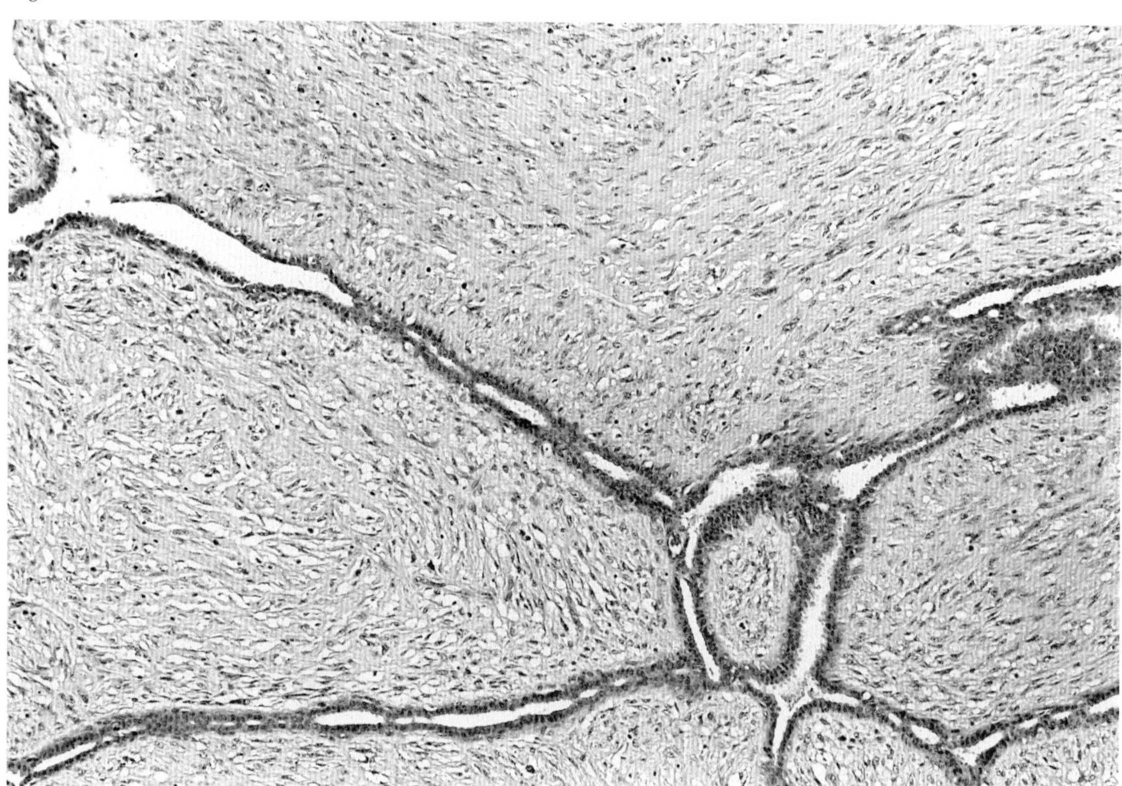

Figure 15–7 **B**

FIGURE 15–7. BENIGN PHYLLODES TUMOR: A: In benign phyllodes tumors, the solid part of the tumor has distinctive broad flower petallike projections that protrude into a cystic space. **B:** The stroma is cellular, and the cells are spindle-shaped. Note the epithelial component, which recapitulates the intracanalicular pattern of a fibroadenoma.

■ STROMAL TUMORS OR BIPHASIC TUMORS

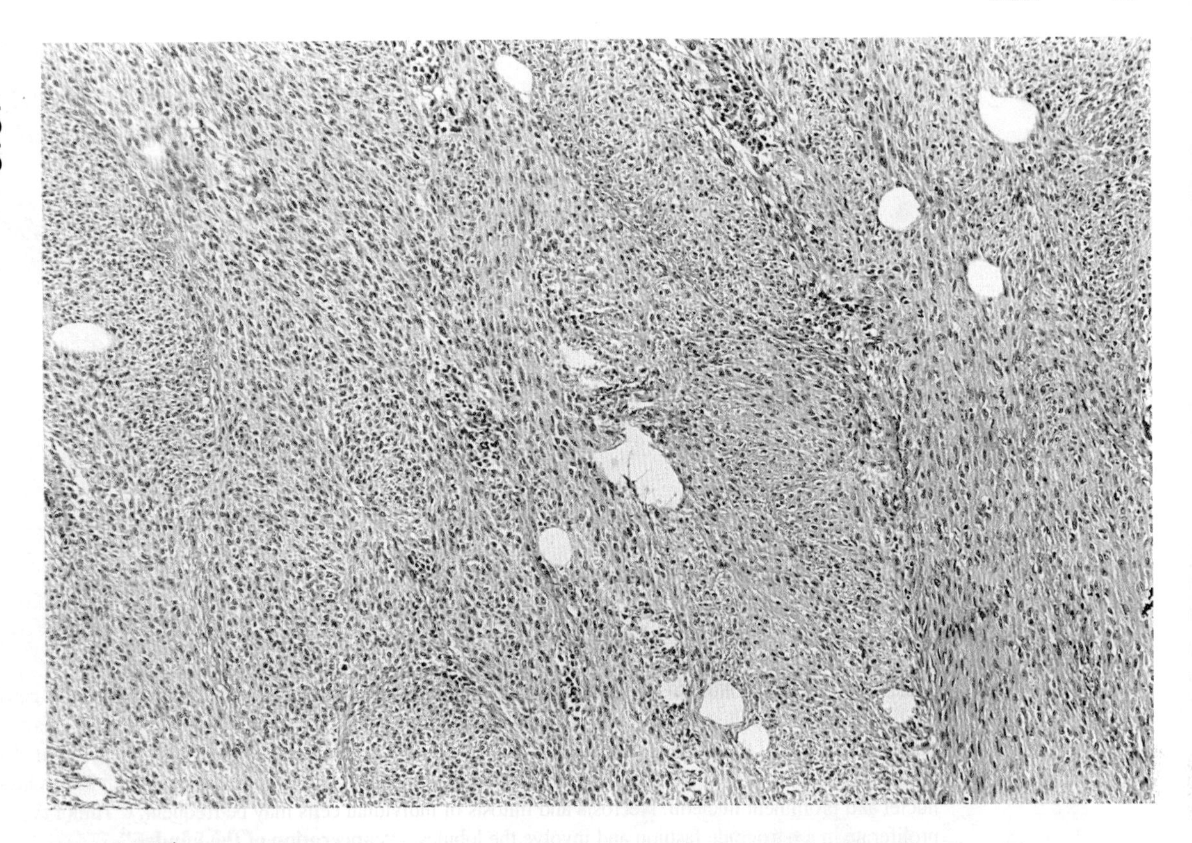

Figure 15–8 **A**

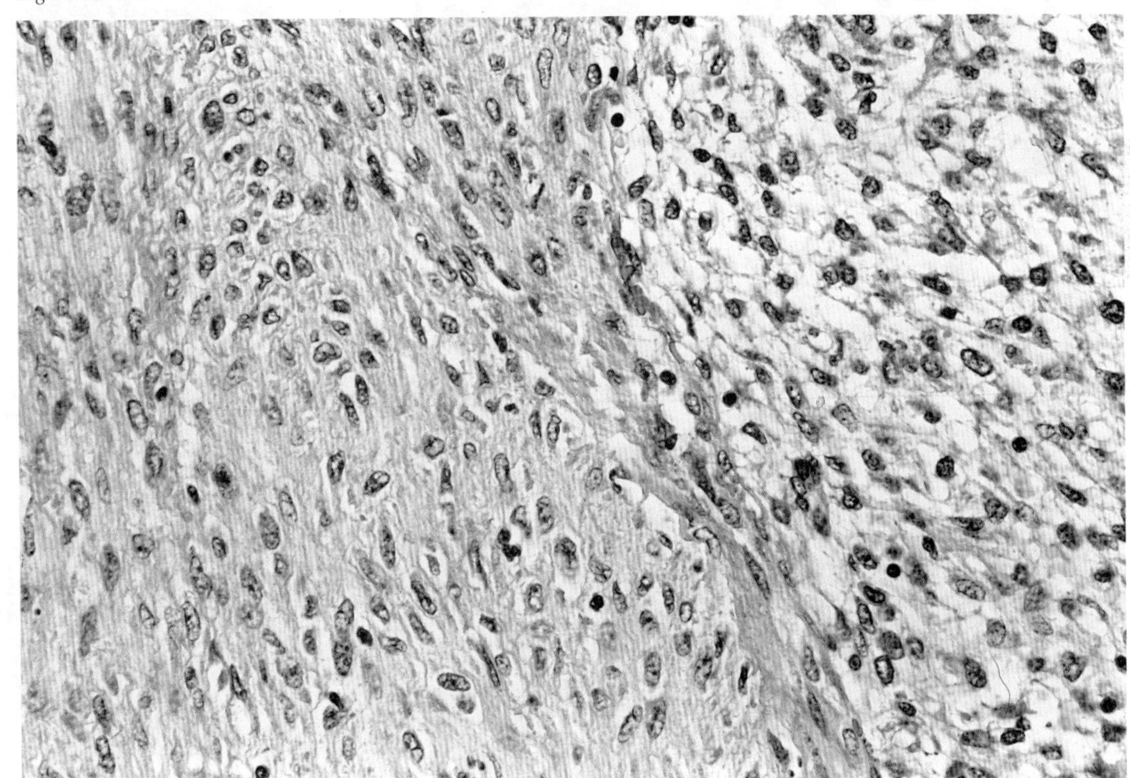

Figure 15–8 **B**

FIGURE 15–8. MALIGNANT PHYLLODES TUMORS (CYSTOSARCOMA PHYLLODES): A: Malignant phyllodes tumors, in contrast to benign phyllodes tumors (Figure 15–7 **B**), have increased stromal cellularity, which obliterates the epithelial components. **B:** The key elements of malignancy are increased cellularity, increased mitotic activity, cellular anaplasia, and infiltrative borders.

■ CARCINOMA

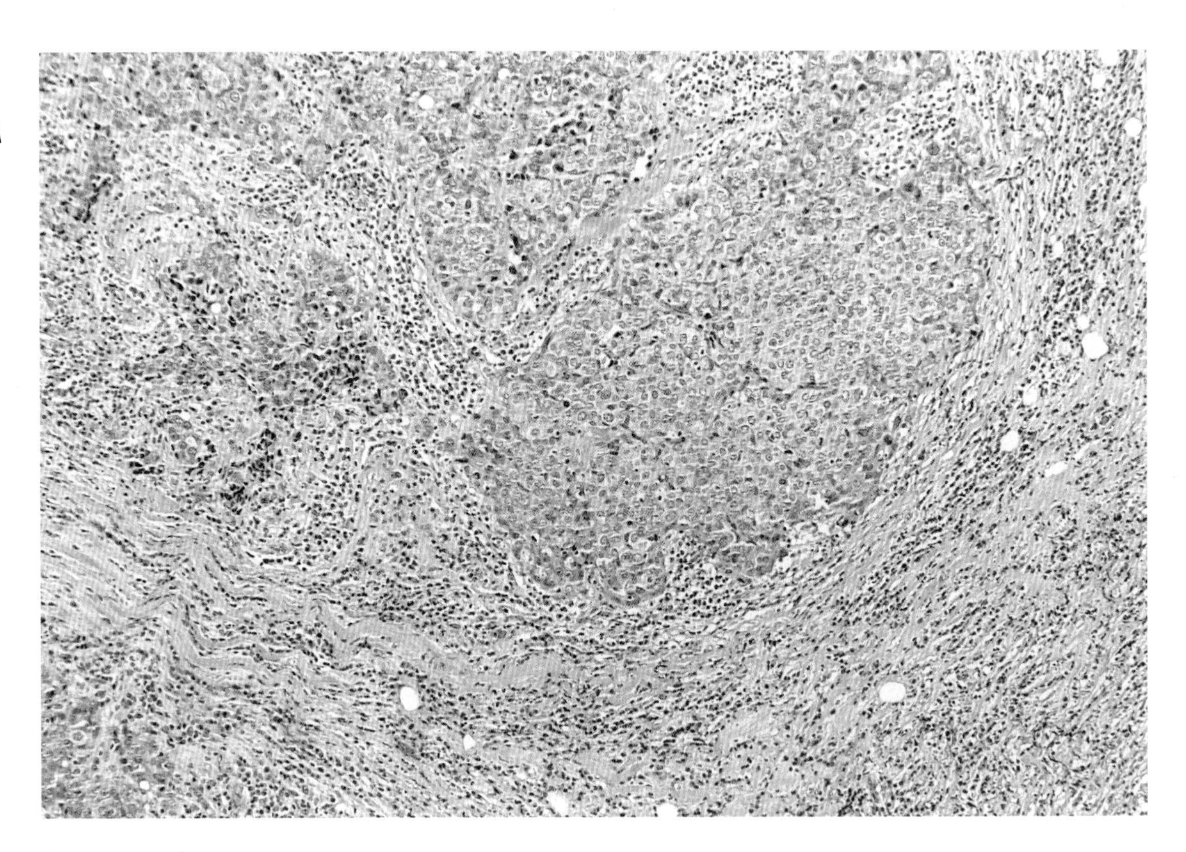

Figure 15–13 A

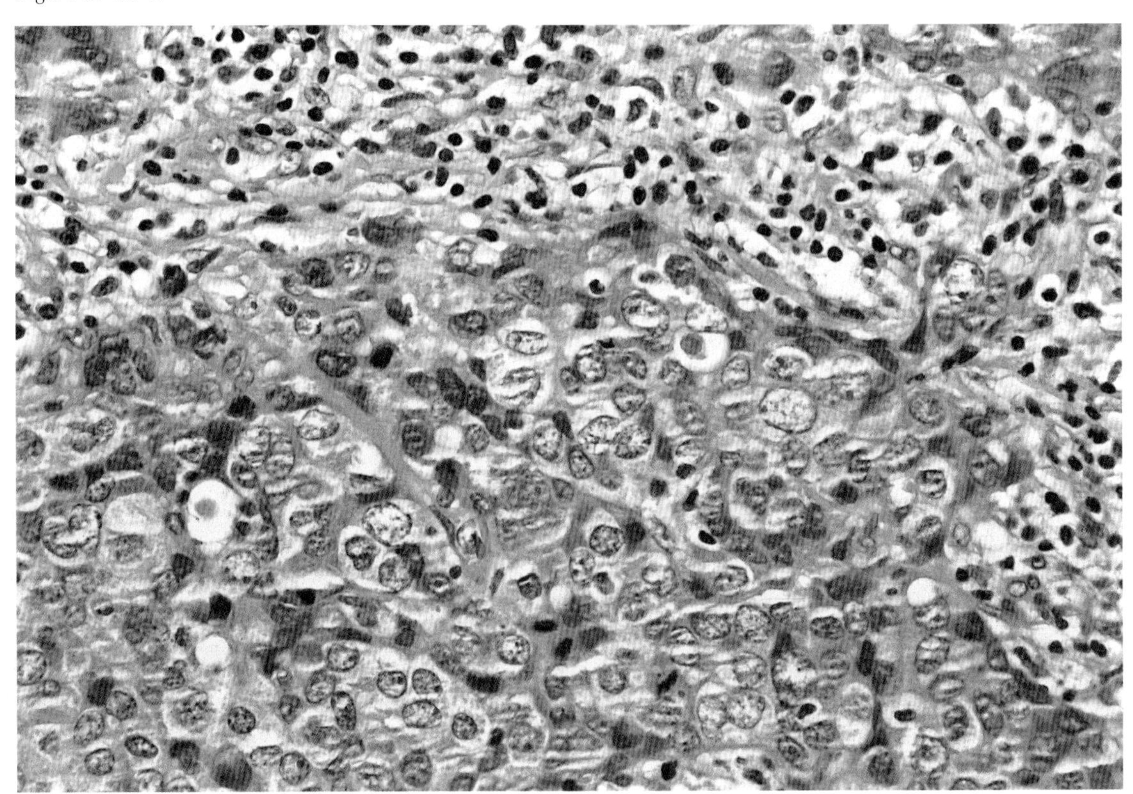

Figure 15–13 B

FIGURE 15–13. MEDULLARY CARCINOMA: A: In medullary carcinoma, sheets of neoplastic cells are divided into islands by fibrous tissue and inflammatory cells. The tumor grows by pushing borders rather than by infiltrating. **B:** The tumor cells are anaplastic and have abundant basophilic to amphophilic cytoplasm, often with indistinct cell borders. The nuclei are vesicular and pleomorphic with prominent nucleoli. Mitosis may be present. The inflammatory infiltrate is composed of lymphocytes and plasma cells.

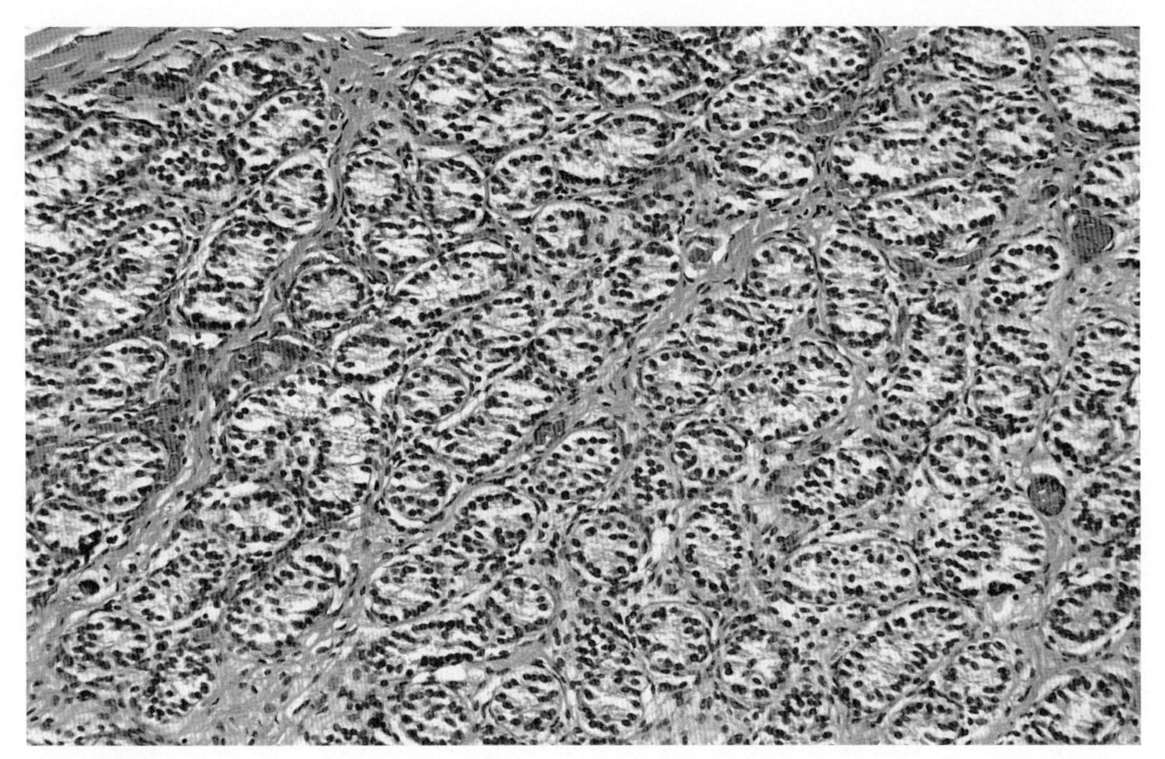

Figure 16–1

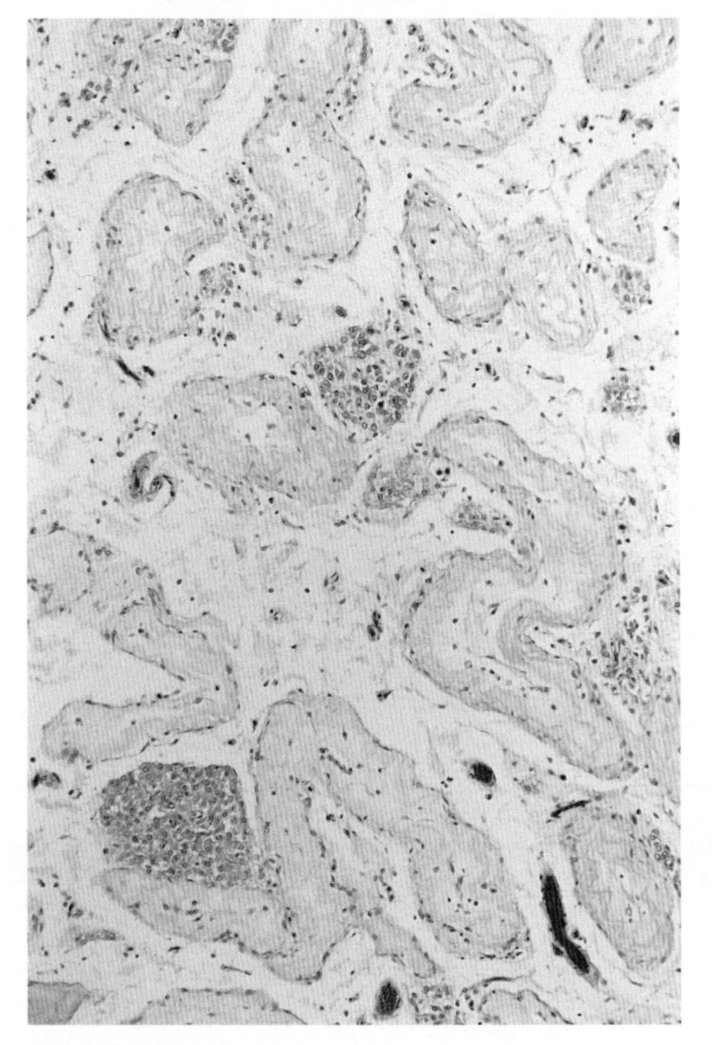

Figure 16–2 A

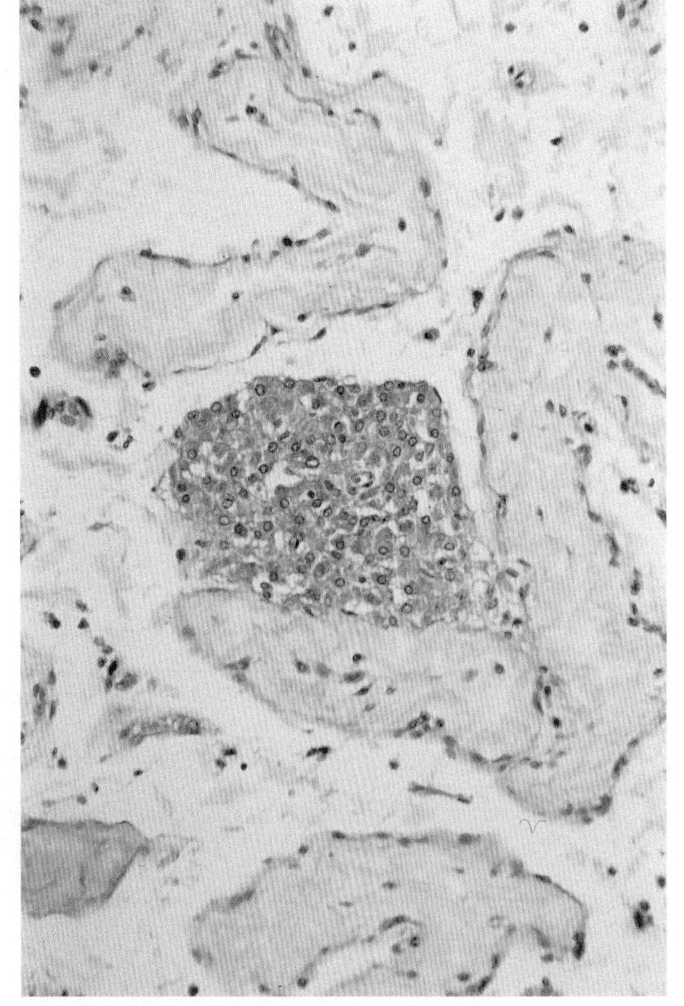

Figure 16–2 B

■ TESTIS

GERM CELL TUMORS

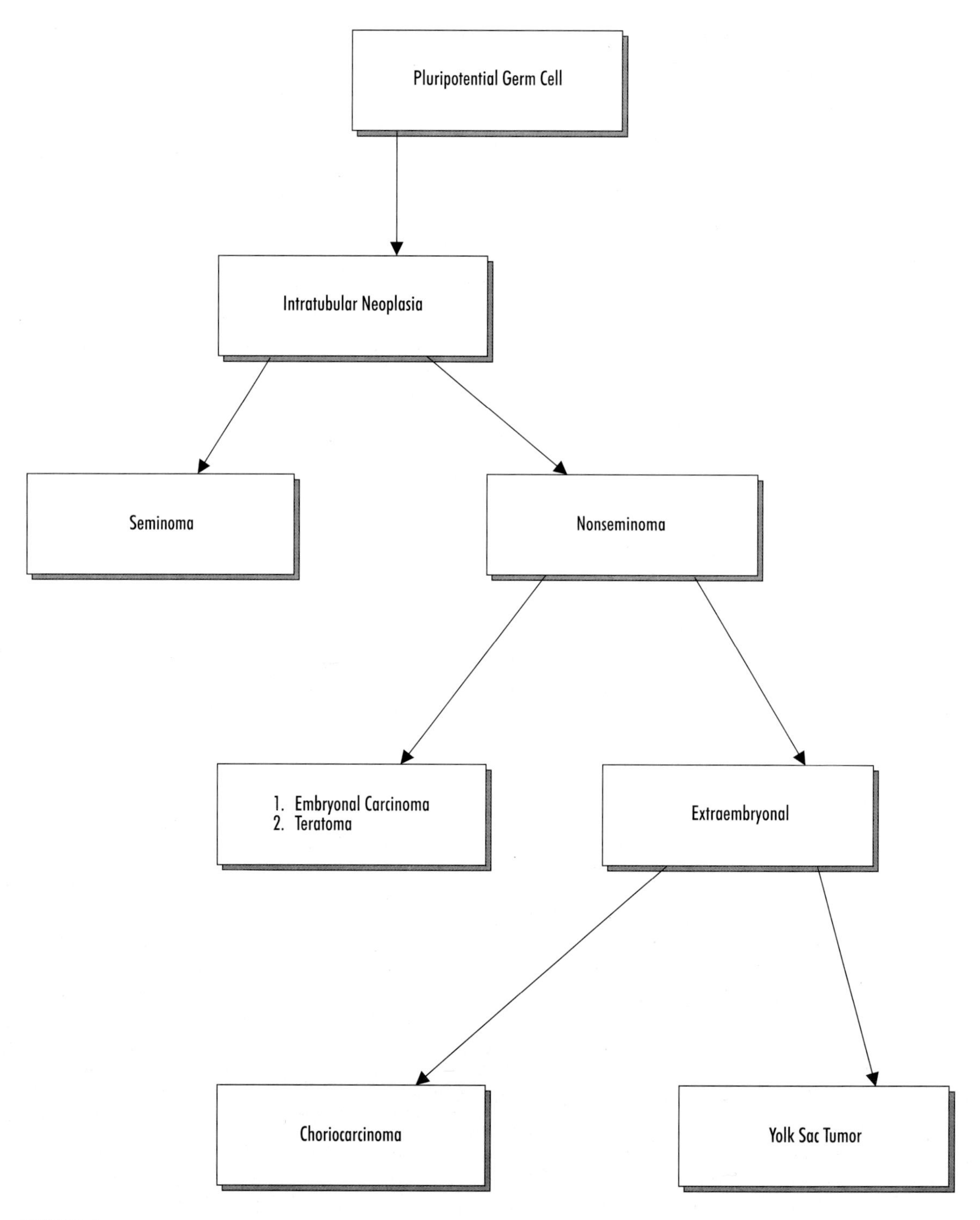

FIGURE 16–3. This is a schematic of germ cell tumors. Germ cell tumors begin as intratubular malignant germ cells. These cells are totipotent and can differentiate along several different lines. Approximately 40% of germ cell tumors are of one histologic type. The remainder are of more than one histologic type.

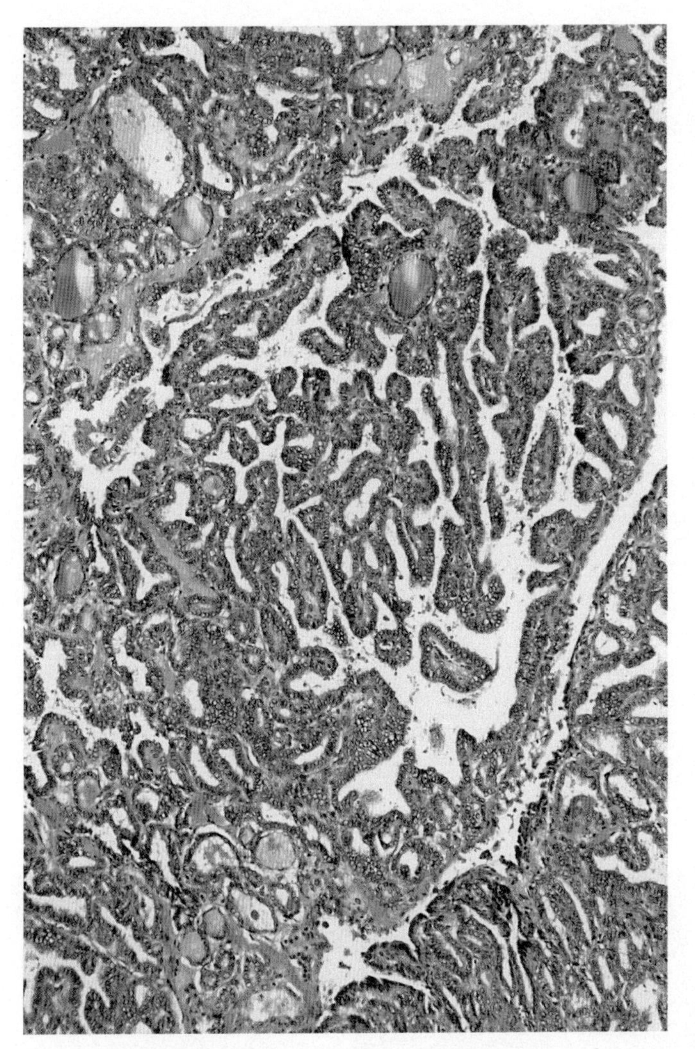

Figure 17–8 A

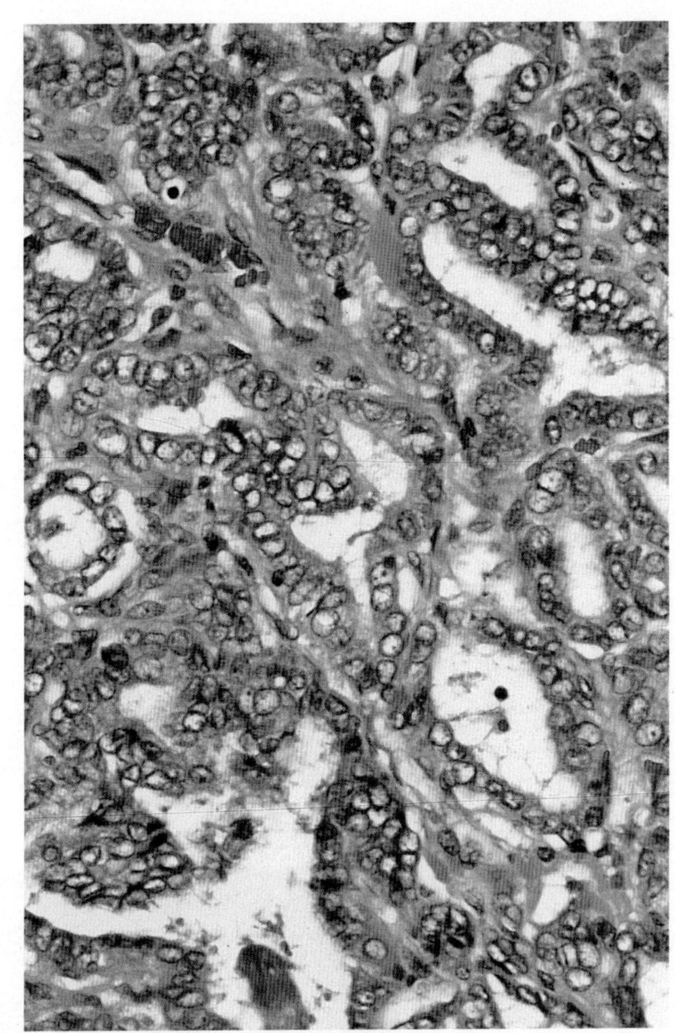

Figure 17–8 B

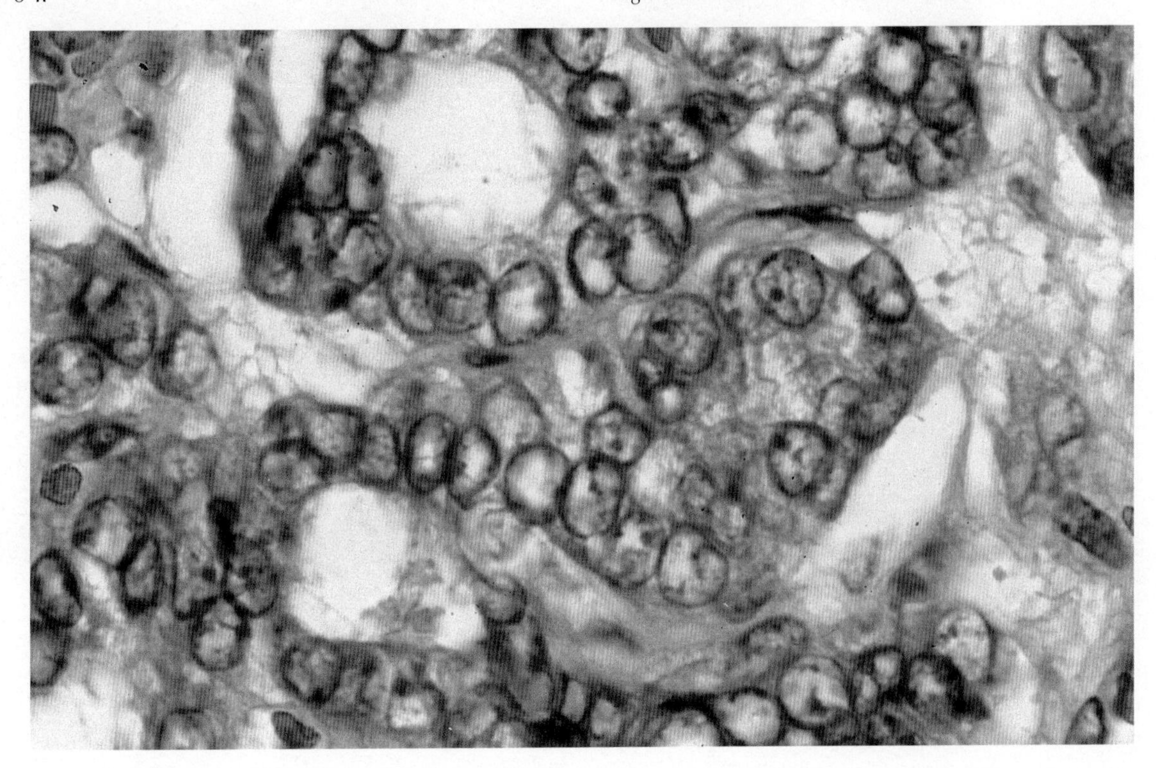

Figure 17–8 C

■ THYROID GLAND

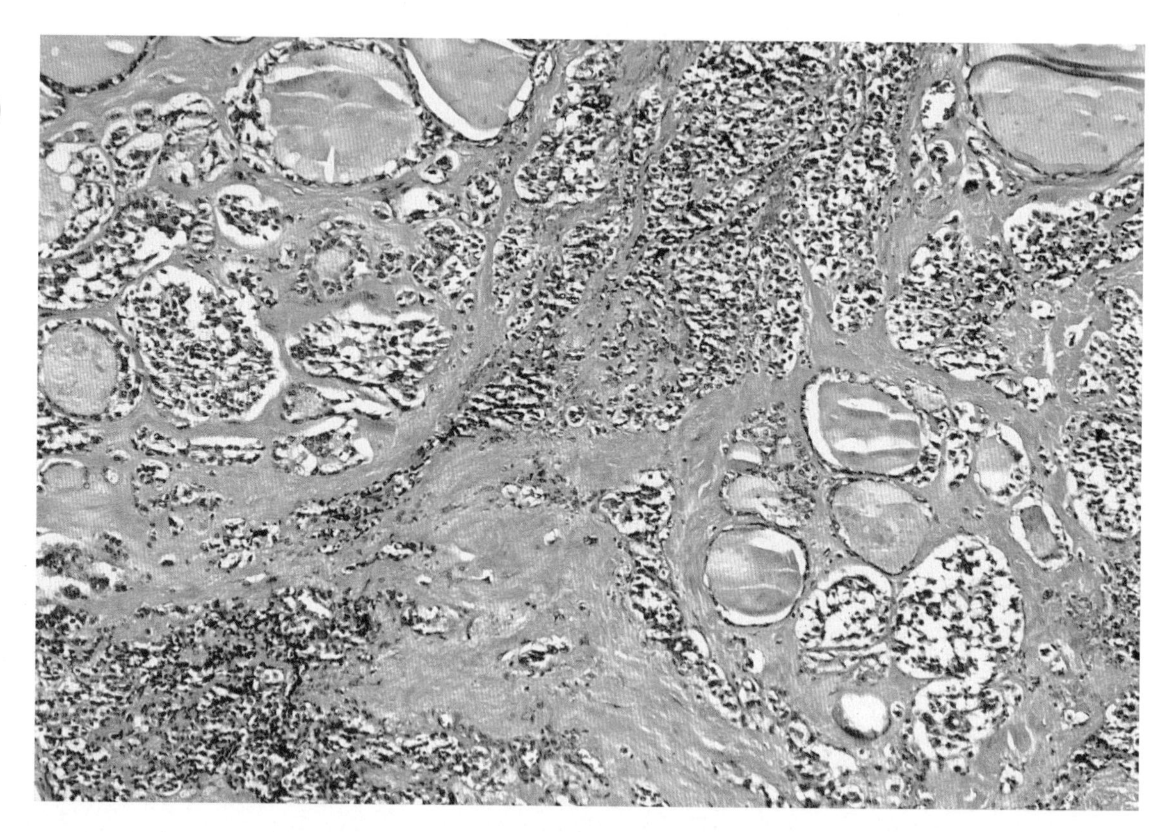

Figure 17–9 A

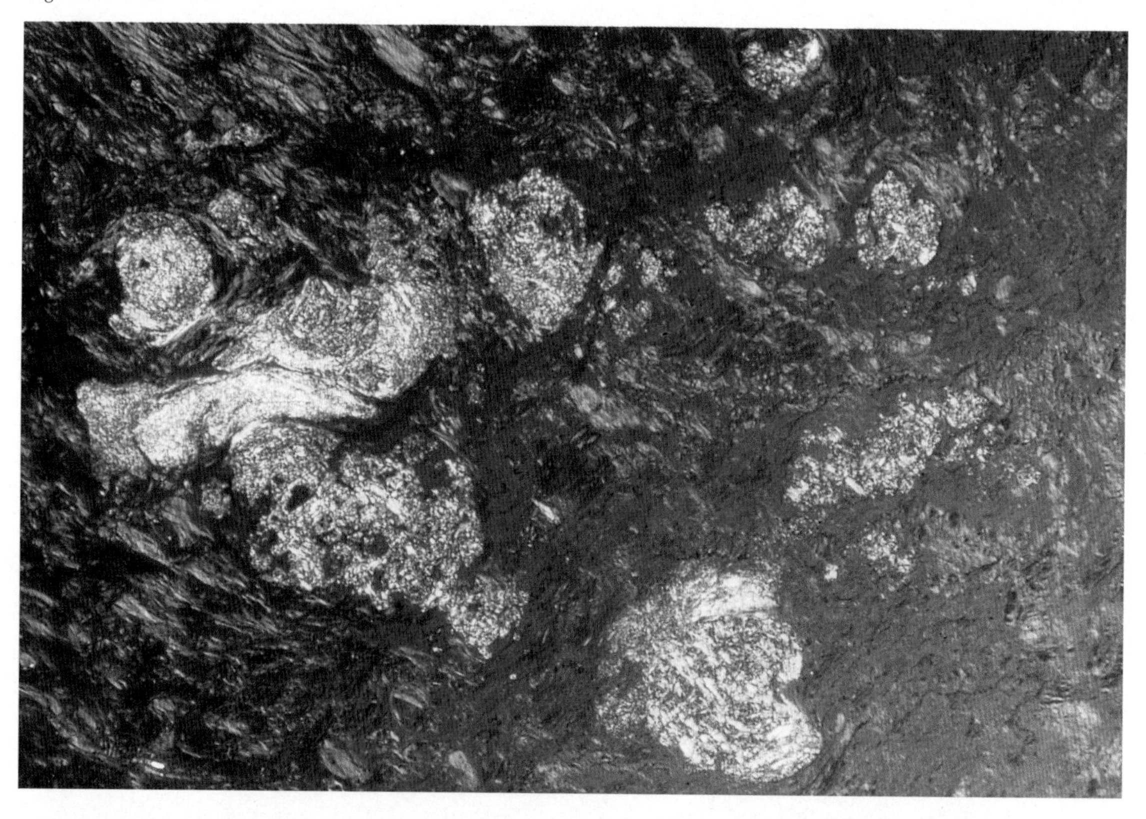

Figure 17–9 B

FIGURE 17–9. MEDULLARY CARCINOMA: A: In medullary carcinoma, nests of cells are present in a dense homogeneous eosinophilic matrix. **B:** When stained with Congo red and exposed to polarized light, the eosinophilic material shows the characteristic apple green birefringence of amyloid.

■ THYROID GLAND

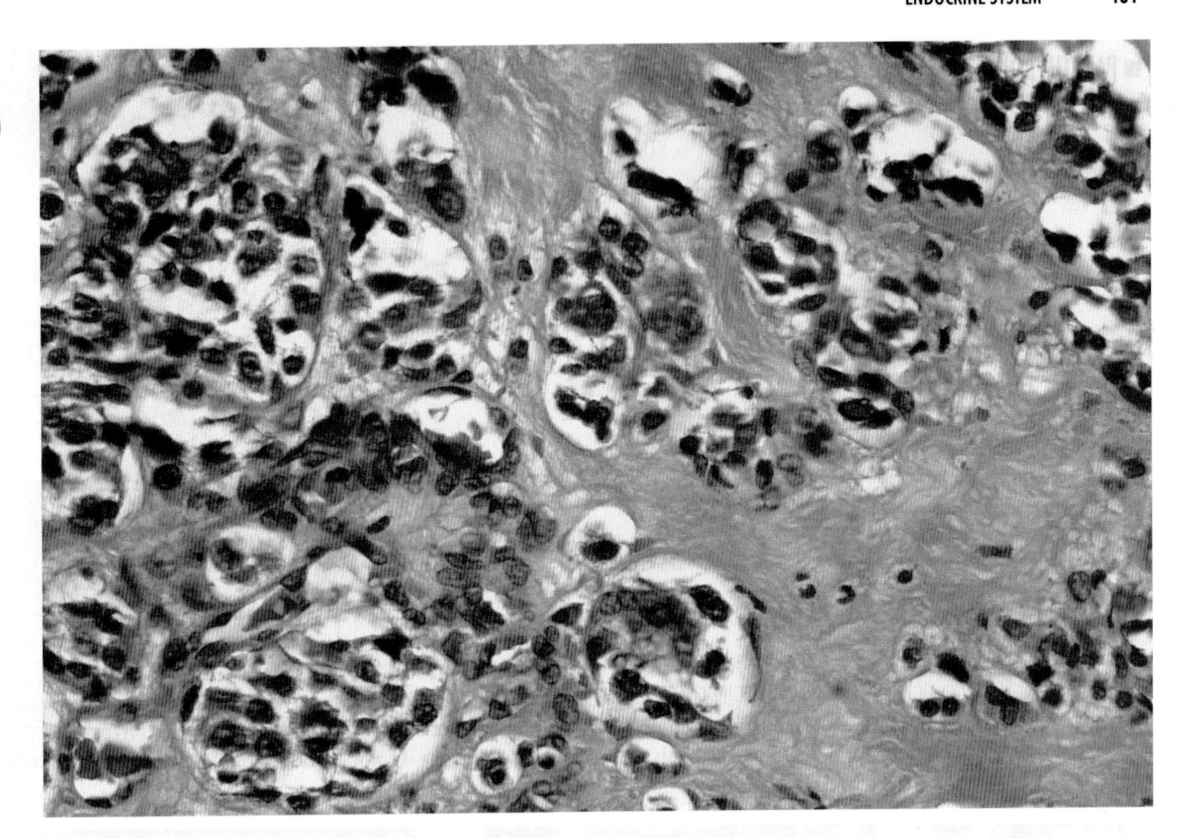

Figure 17–9 C

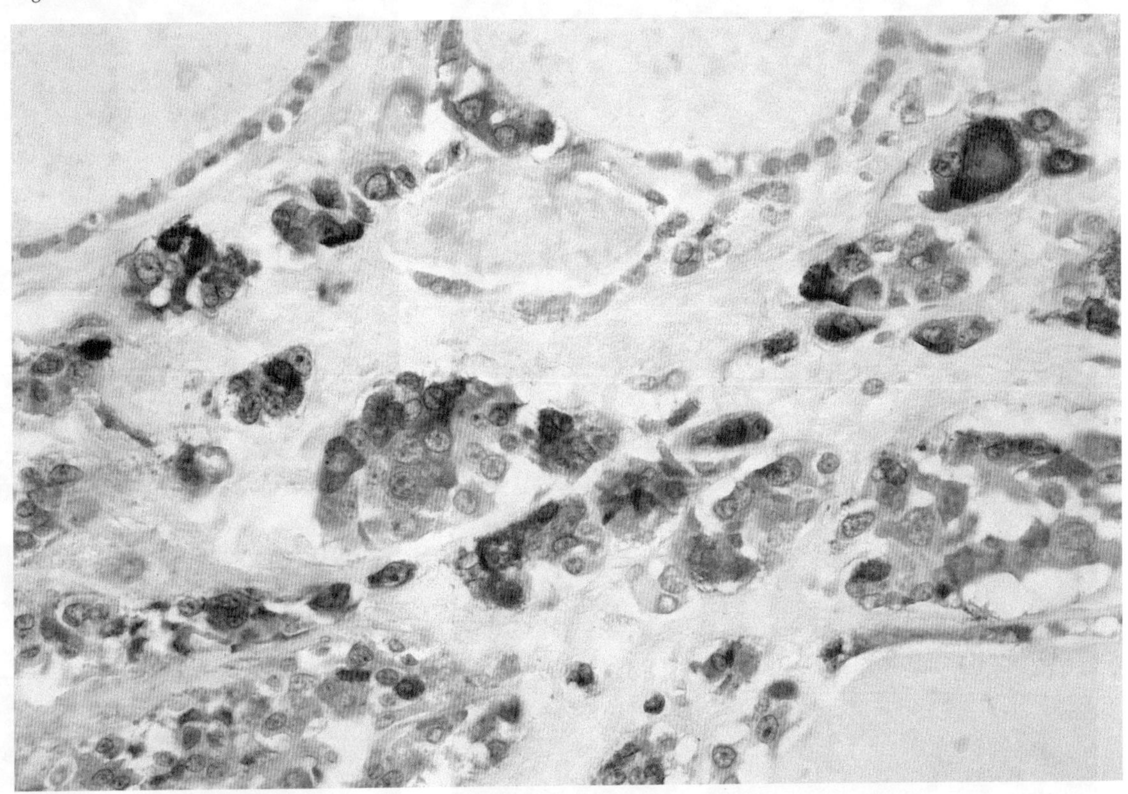

Figure 17–9 D

FIGURE 17–9. MEDULLARY CARCINOMA: C: In medullary carcinoma, the neoplastic cells are undifferentiated and anaplastic. **D:** Immunoperoxidase staining with calcitonin confirms the interfollicular C-cell origin of the neoplastic cells.

■ SKIN

BENIGN EPIDERMAL TUMORS

FIGURE 18–2. EPIDERMAL INCLUSION CYST: A: Epidermal inclusion cysts are located in the dermis. **B:** The wall of the cyst is composed of an epidermis with several layers of superficial squamous cells and a granular cell layer. The cyst contains laminated keratin.

FIGURE 18–3. TRICHILEMMAL CYST: Trichilemmal cysts are located in the dermis. The wall of the cyst is composed of an epidermis without intercellular bridges or a granular cell layer. The inner cells of the cyst have a pale eosinophilic cytoplasm. The cyst contains a homogeneous eosinophilic material.

■ SKIN

BENIGN ADNEXAL TUMC

■ SKIN

INFECTIOUS DISORDERS

BACTERIA

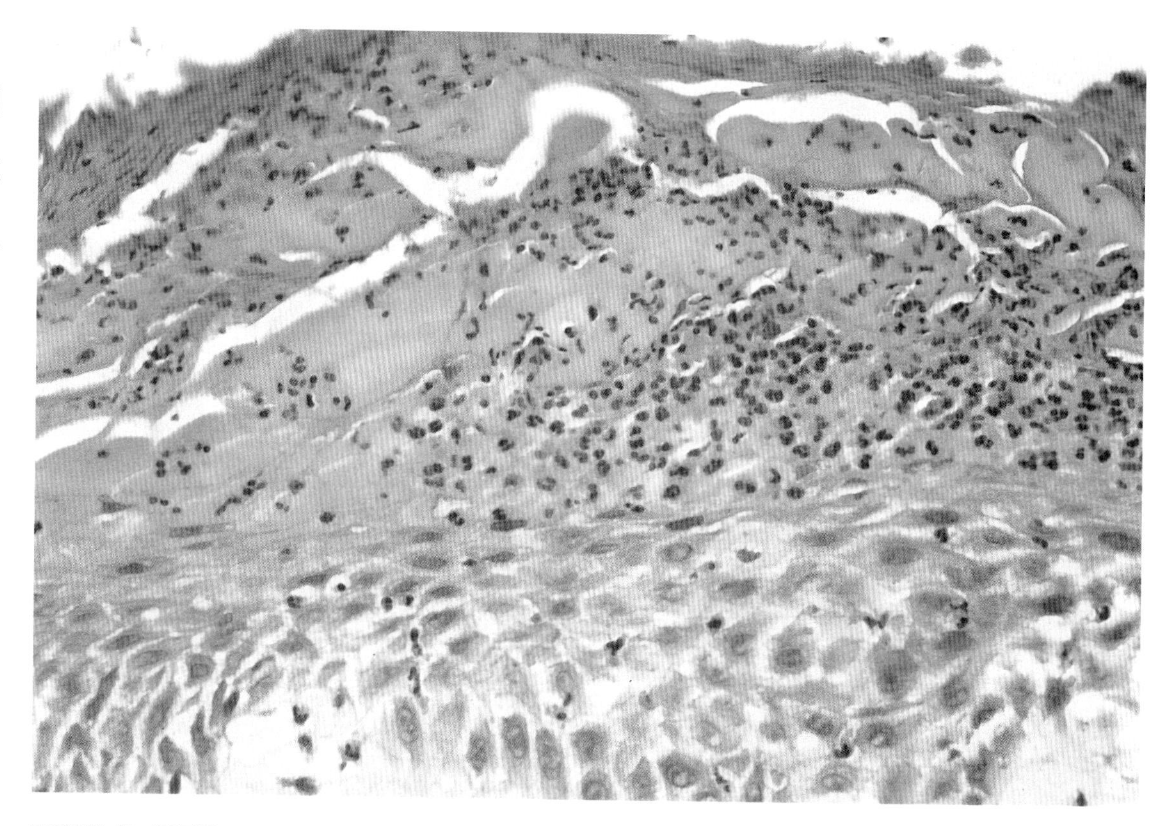

FIGURE 18–11. IMPETIGO: Impetigo is caused by streptococci or staphylococci. In the fully developed lesion, there is a crust composed of serous exudate admixed with neutrophils overlying the epidermis. Neutrophils are seen in the epidermis as well.

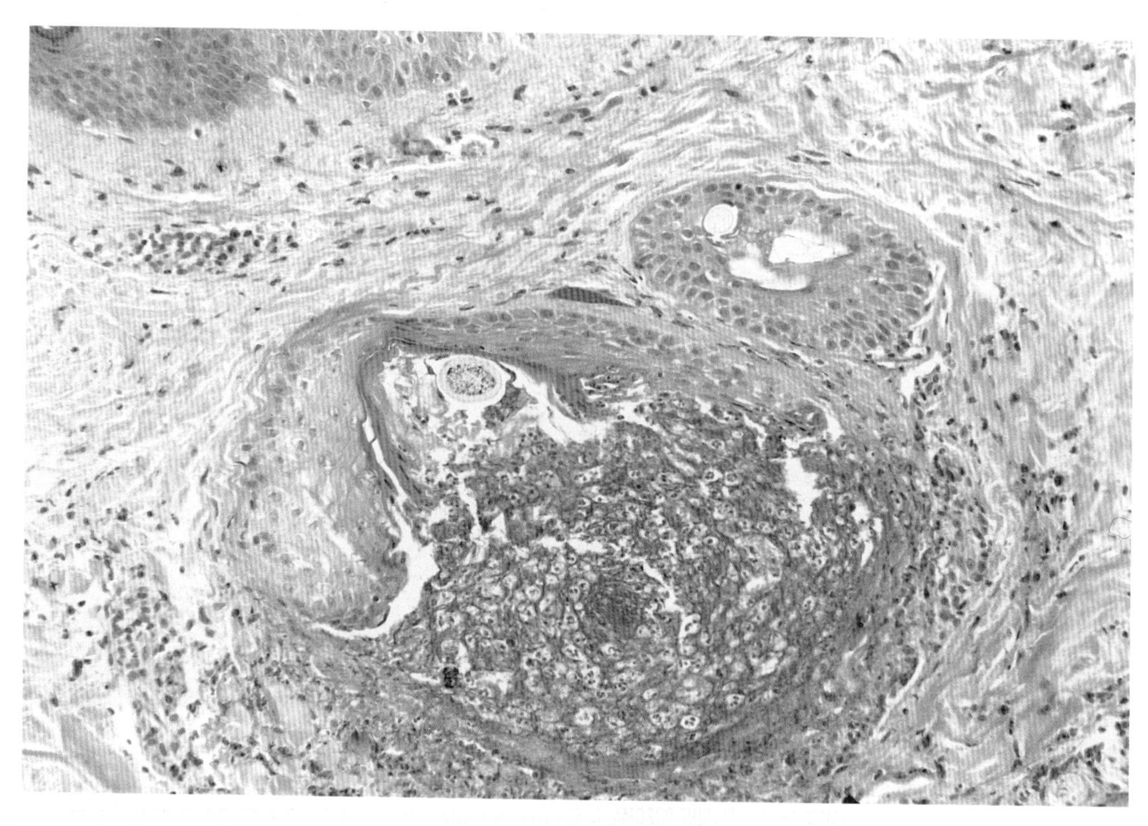

FIGURE 18–12. FOLLICULITIS: Acute folliculitis is caused by staphylococci. There is follicular and perifollicular necrosis, with deposition of fibrinoid material. Neutrophils are seen in and around the follicle.

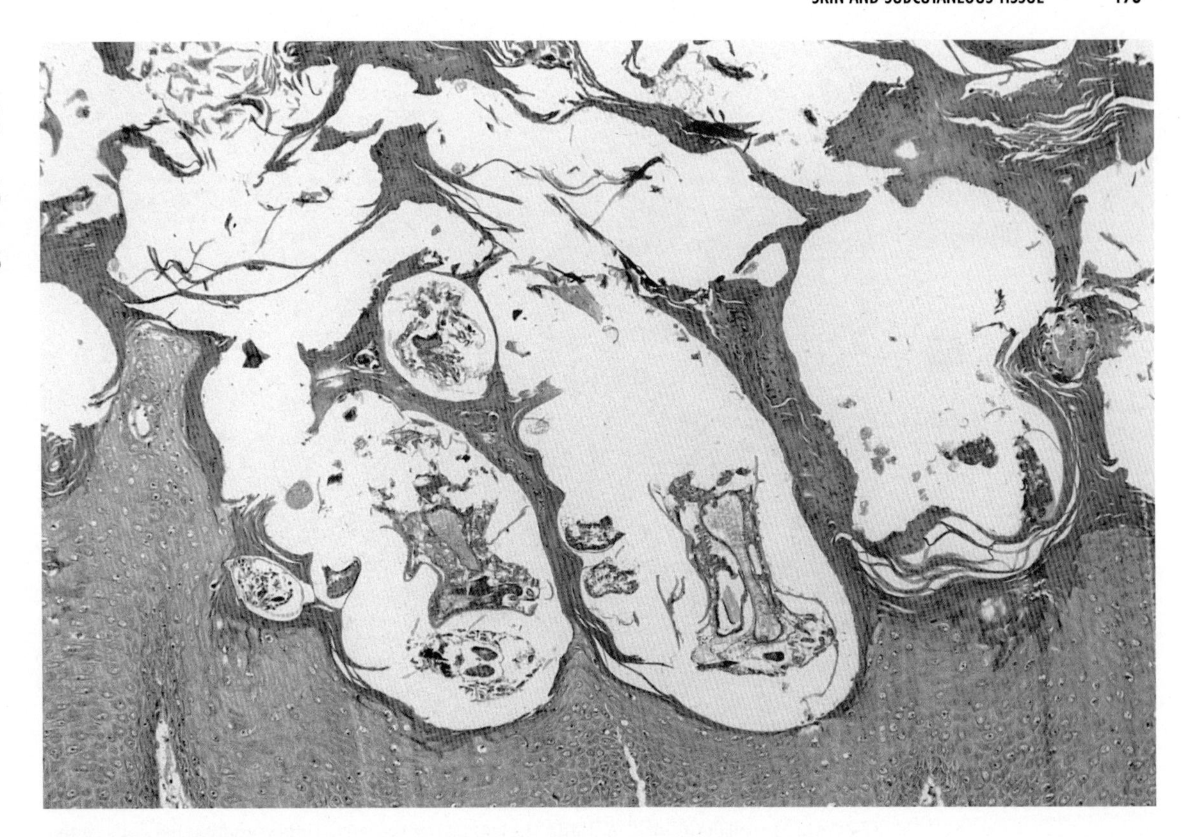

FIGURE 18–21. SCABIES: Scabies is caused by the itch mite *Sarcoptes scabiei.* The characteristic lesion is a burrow through the stratum corneum. The female mite resides in the burrow.

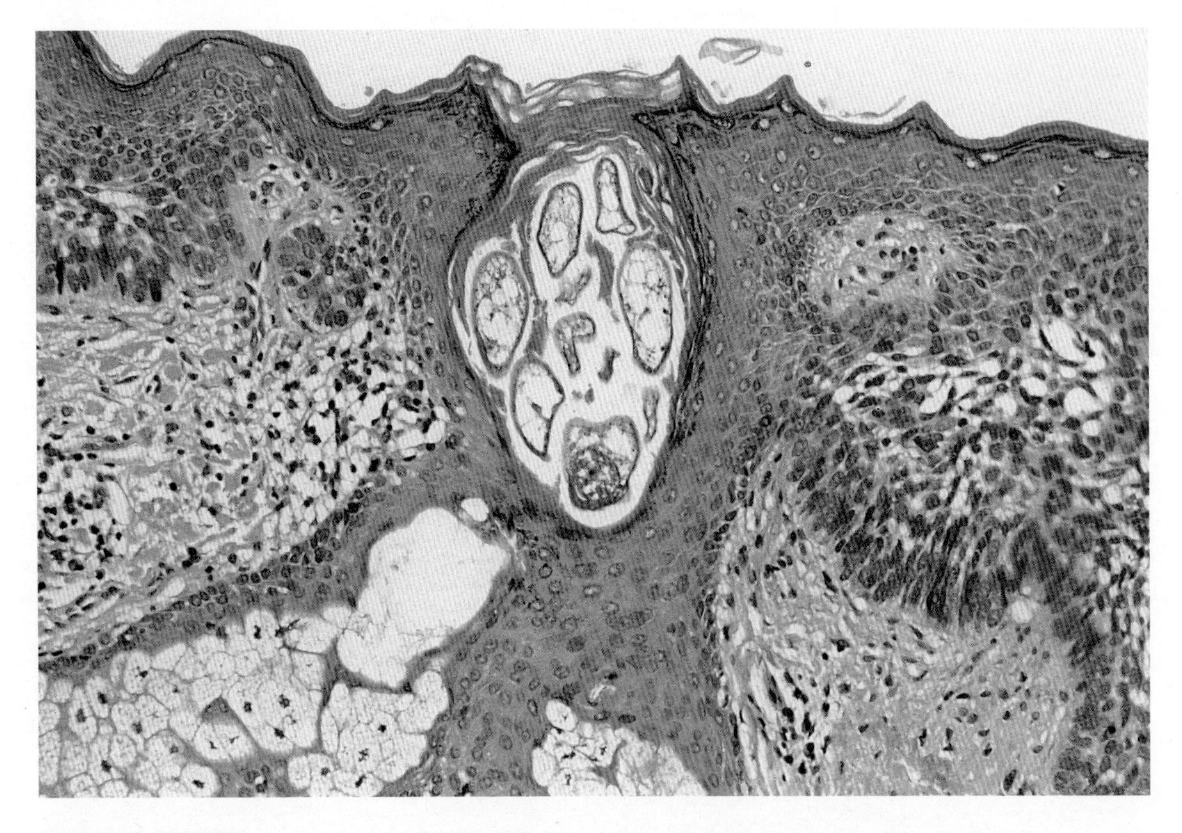

FIGURE 18–22. ***DEMODEX:*** *Demodex* species infest hair follicles or sebaceous glands. Note the small basal cell carcinoma to the right of the pilosebaceous unit.

■ SKIN

INFLAMMATORY DISORDERS

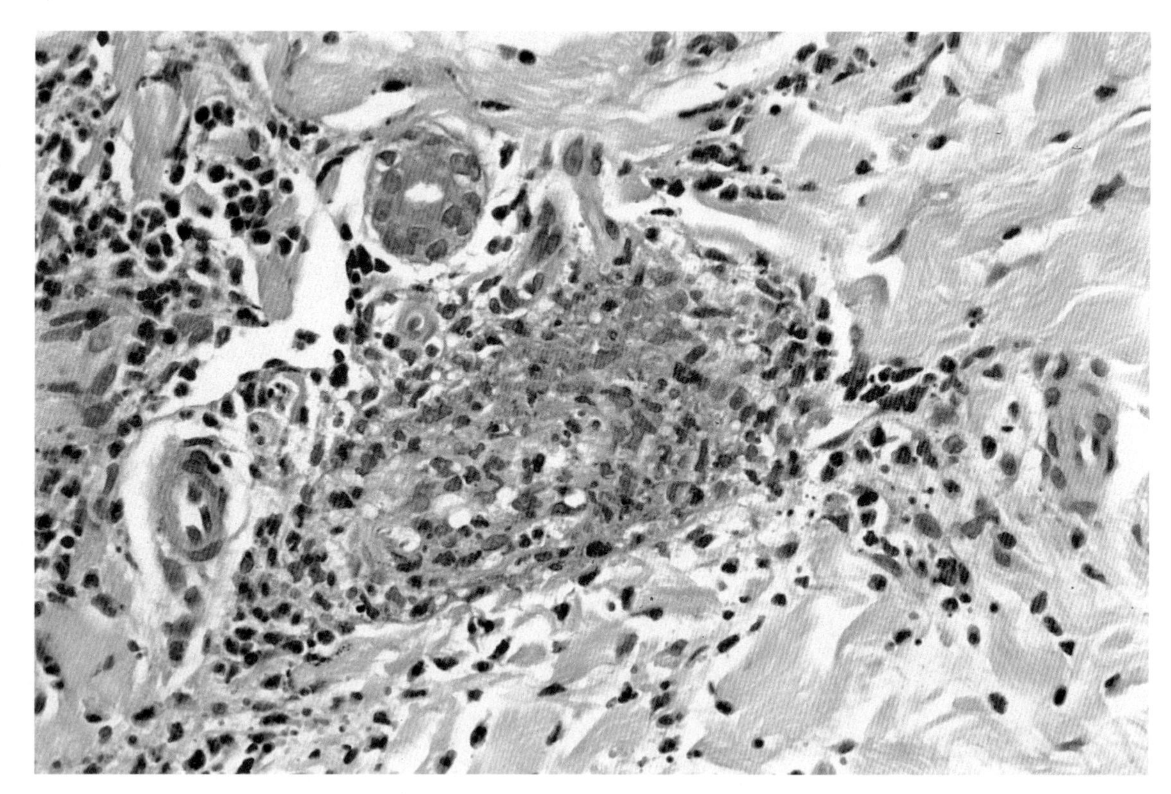

FIGURE 18–23. LEUKOCYTOCLASTIC VASCULITIS: In leukocytoclastic vasculitis, endothelial cells of the small blood vessels in the dermis are swollen. An inflammatory cell infiltrate composed mainly of neutrophils and varied amounts of eosinophils and mononuclear cells is present around and within the blood vessel walls. Characteristic of leukocytoclastic vasculitis is neutrophilic karyorrhexis as manifested by "nuclear dust."

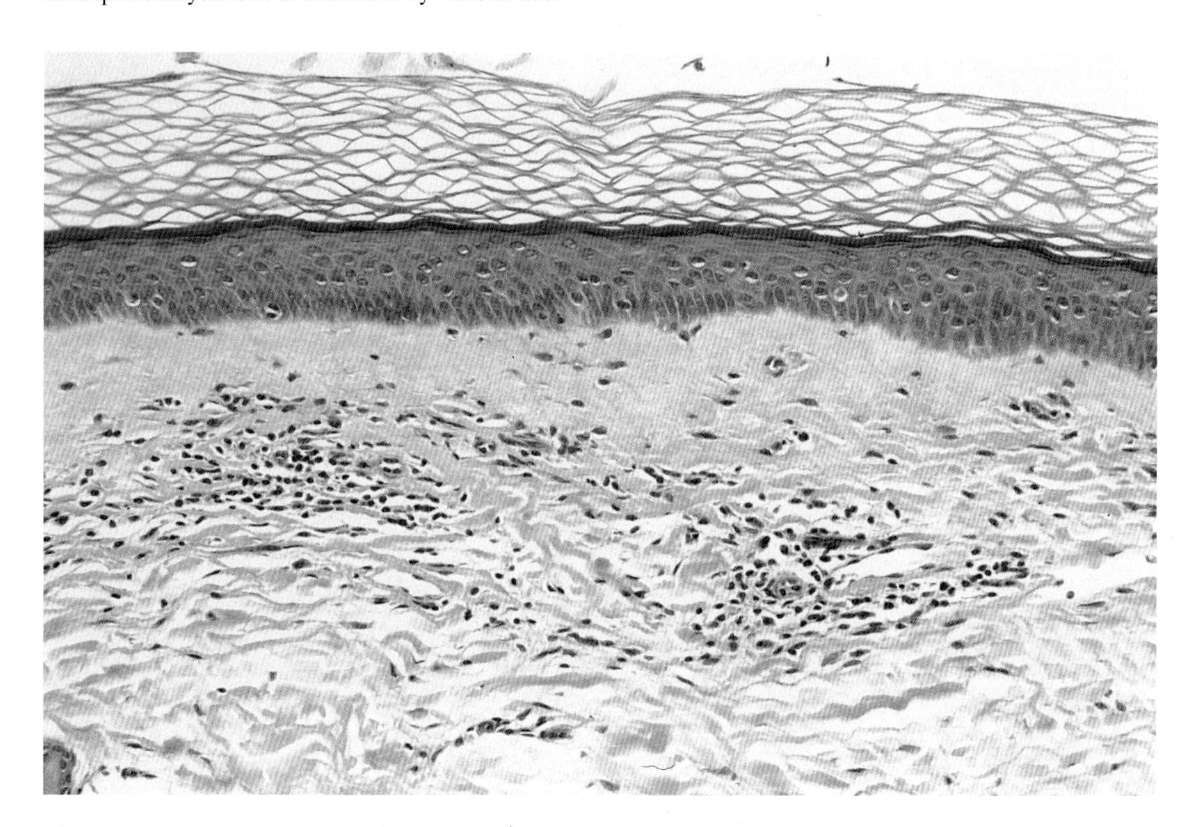

FIGURE 18–24. LICHEN SCLEROSUS ET ATROPHICUS: In lichen sclerosus et atrophicus, there is atrophy of the epidermis caused by loss of the rete ridges and atrophy of the malpighian layer. The corneal layer is thickened. There is pronounced edema and homogenization of the collagen in the upper dermis. A lymphohistiocytic inflammatory infiltrate is present in the upper dermis.

■ SKIN

INFLAMMATORY DISORDERS

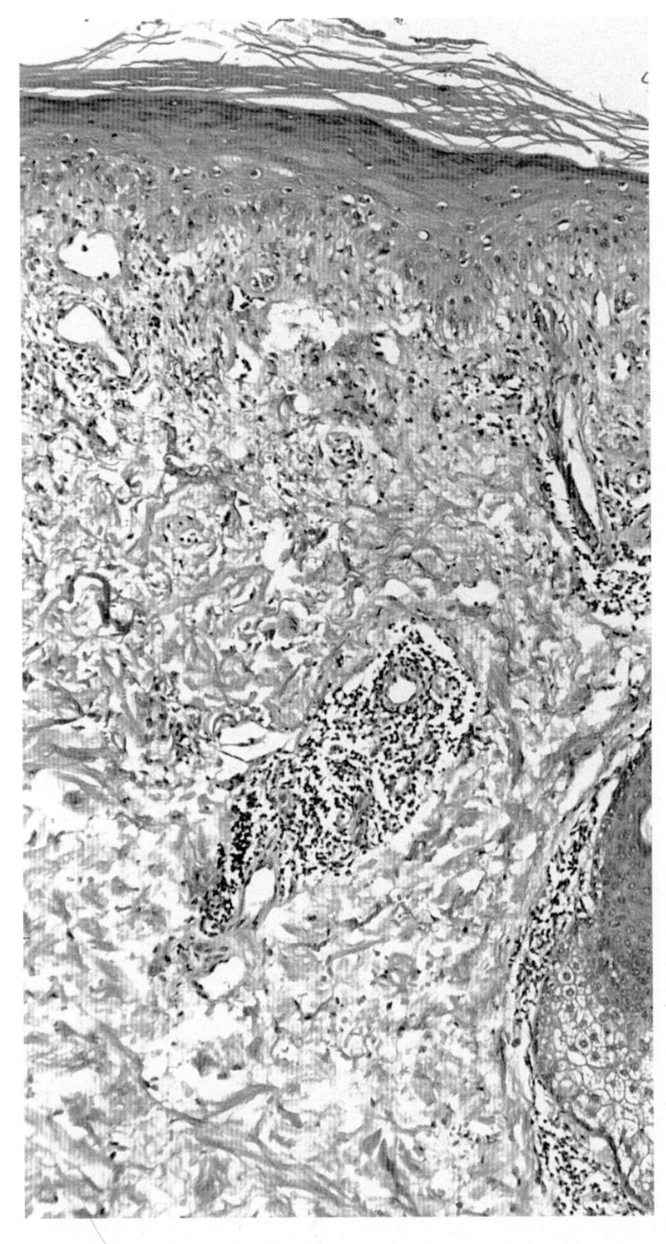

Figure 18–25 A

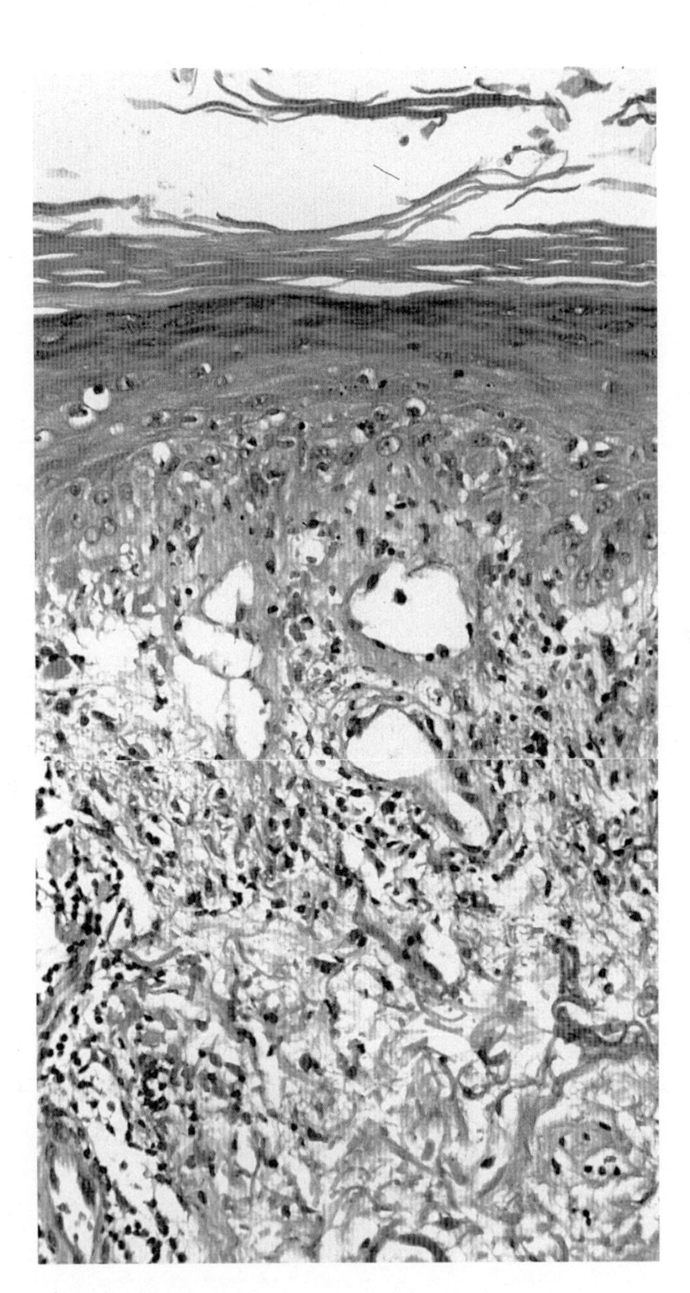

Figure 18–25 B

FIGURE 18–25. LUPUS ERYTHEMATOSUS: A and B: In lupus erythematosus, there is hydropic degeneration of the basal cell layer. In the upper dermis, there is edema, extravasation of erythrocytes, and fibrinoid deposition. There is a patchy periappendiceal lymphoid infiltrate.

■ SKIN

INFLAMMATORY DISORDERS

FIGURE 18–26. LICHEN PLANUS: A: In lichen planus, there is a bandlike inflammatory infiltrate in the upper dermis with a sharply demarcated inferior border, which extends to the dermal-epidermal junction. **B:** The upper epidermis shows focal hypergranulosis and hyperkeratosis. There is irregular acanthosis of the rete ridges. **C:** The basal cell layer shows liquefaction and appears hazy because of a lymphohistiocytic inflammatory infiltrate. Some of the epidermal cells are eosinophilic because of early necrosis.

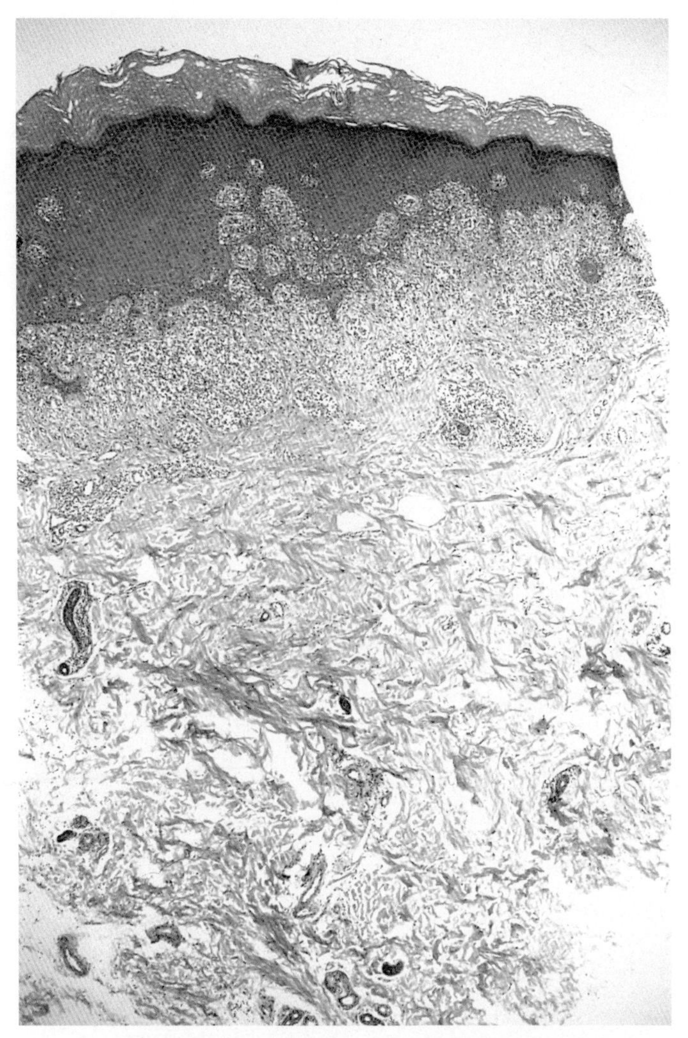

Figure 18–26 A

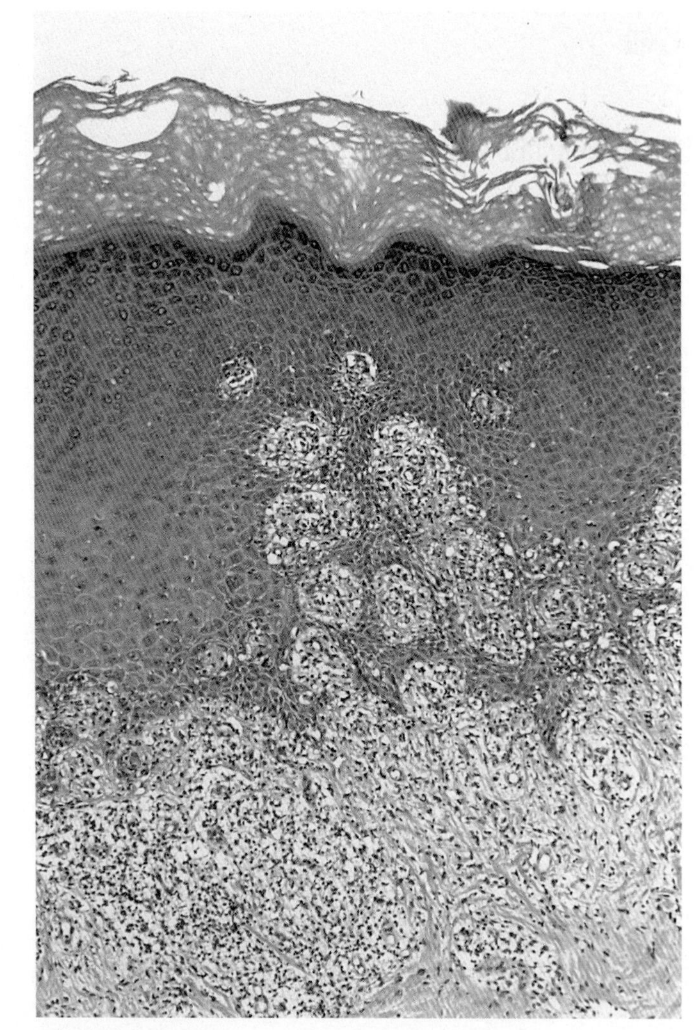

Figure 18–26 B

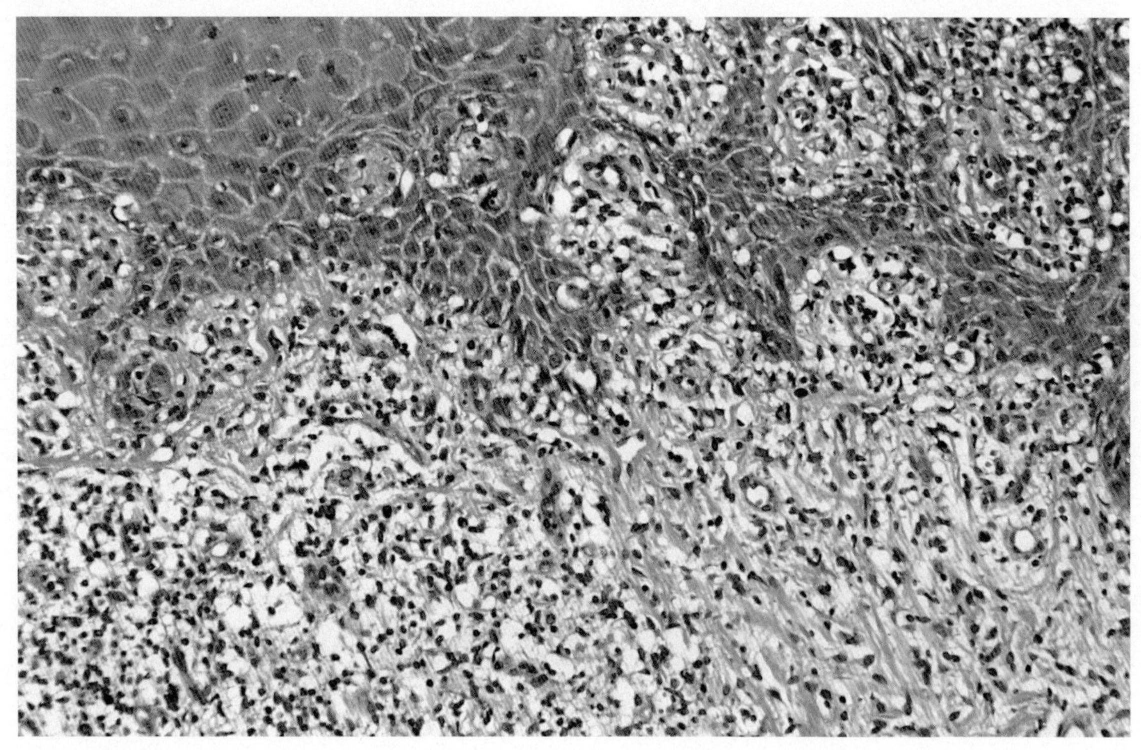

Figure 18–26 C

■ SKIN

INFLAMMATORY DISORDERS

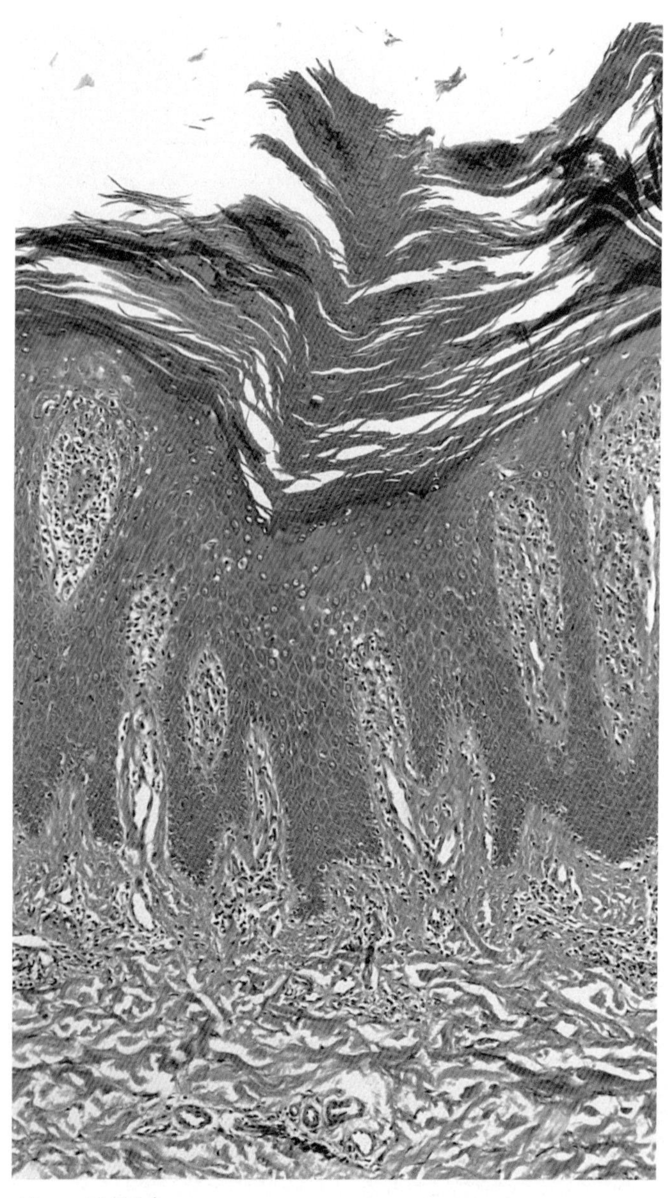

Figure 18–27 A

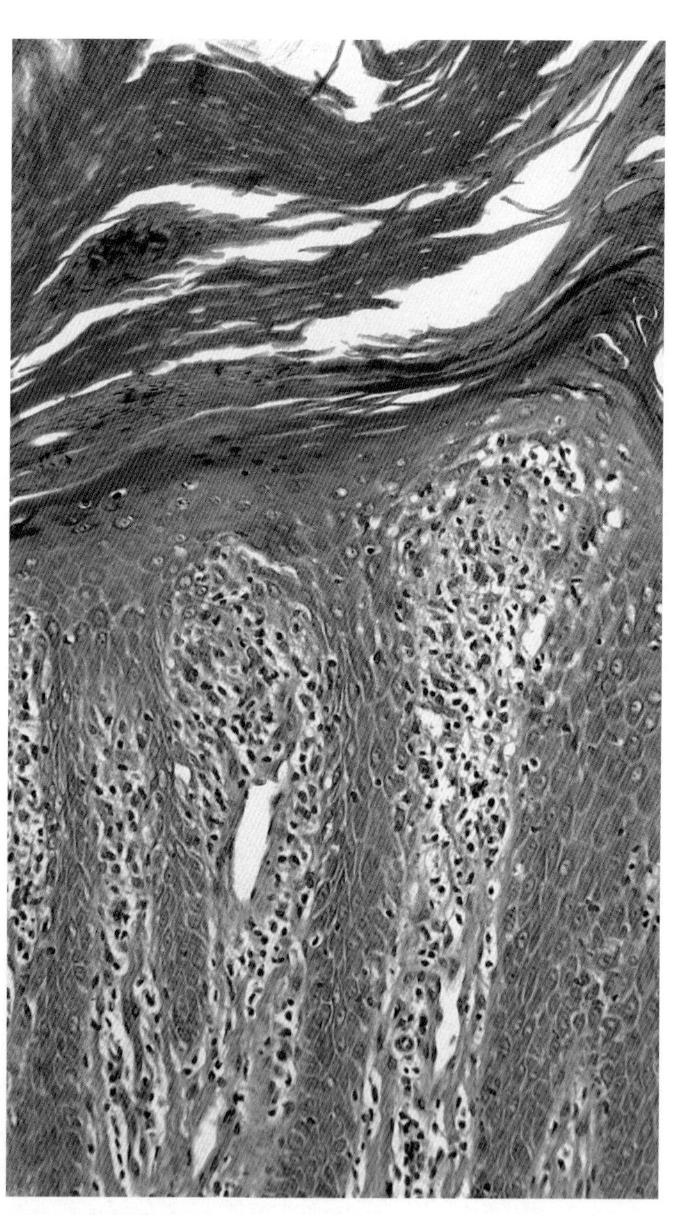

Figure 18–27 B

FIGURE 18–27. PSORIASIS: A: In psoriasis, there is elongation of the rete ridges, with thickening at their lower ends. The papillary dermis is elongated and edematous. B: The suprapapillary epidermis is thinned, and the granular cell layer is diminished or lacking. Parakeratosis is prominent. Focally, there are collections of neutrophils in the parakeratotic areas—**Munro microabscesses.**

■ SKIN

BLISTERING DISEASES

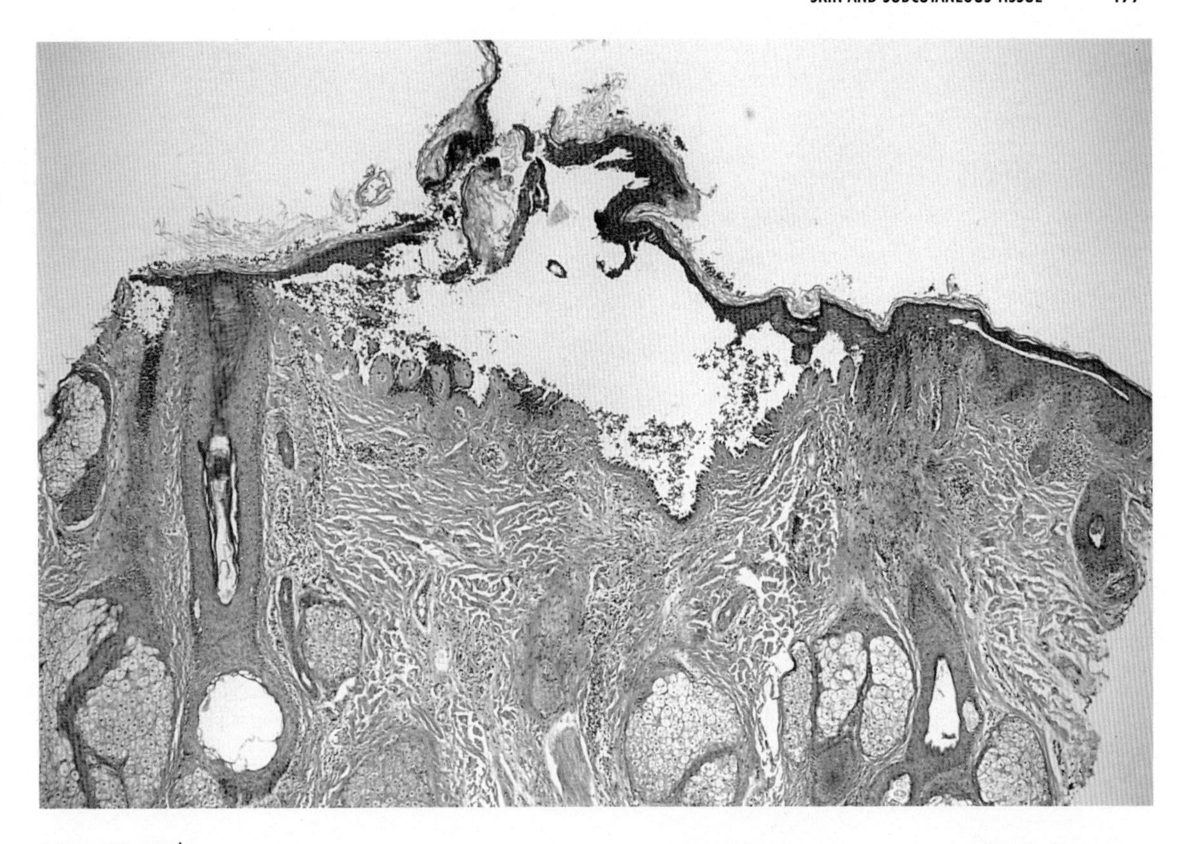

Figure 18–28 **A**

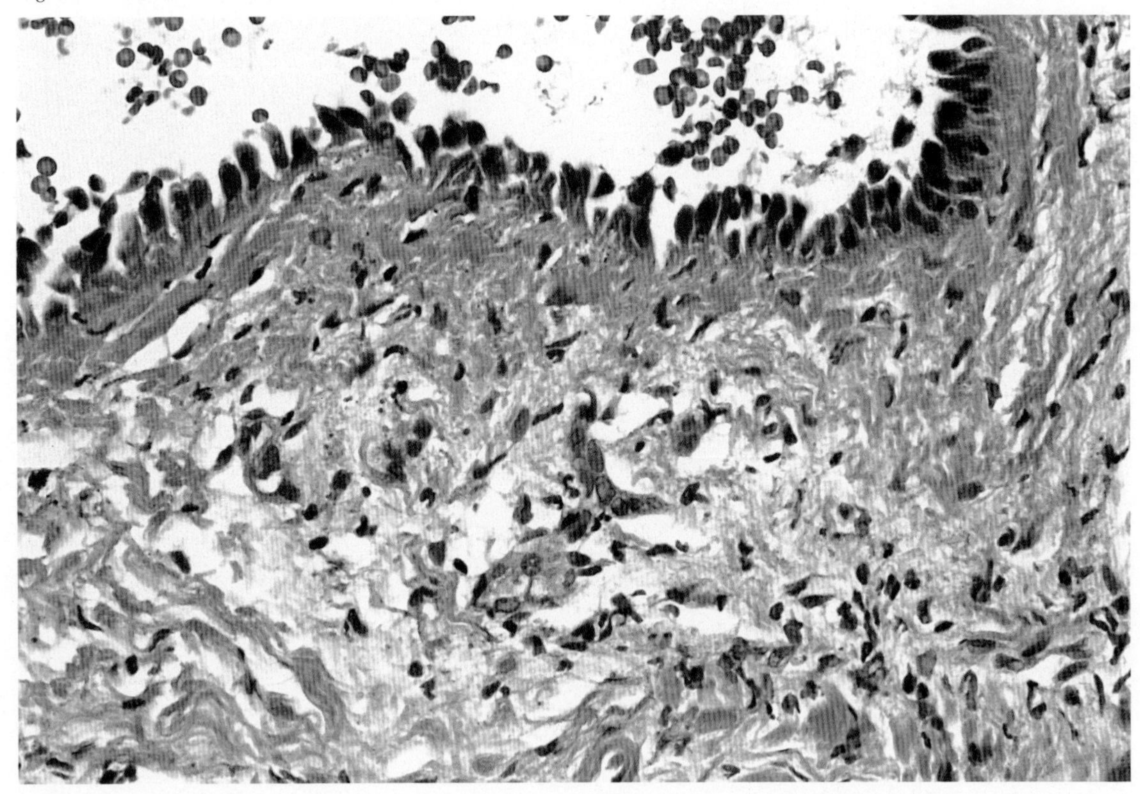

Figure 18–28 **B**

FIGURE 18–28. PEMPHIGUS VULGARIS: A: In pemphigus vulgaris, the bulla is located suprabasally. **B:** The basal cells remain attached to the dermis as a row of "tombstone" cells.

■ SKIN

BLISTERING DISEASES

FIGURE 18–29. BULLOUS PEMPHIGOID: A: In bullous pemphigoid, a blister is situated subcorneally. **B:** Inflamed lesions have a dermal infiltrate composed of neutrophils, mononuclear cells, and eosinophils.

FIGURE 18–30. DERMATITIS HERPETIFORMIS: In dermatitis herpetiformis, neutrophils and eosinophils accumulate and form microabscesses at the tips of the papillae. The microabscesses coalesce, and a blister forms in a subepidermal location. Fibrin deposition at the papillae gives the microabscesses a necrotic appearance.

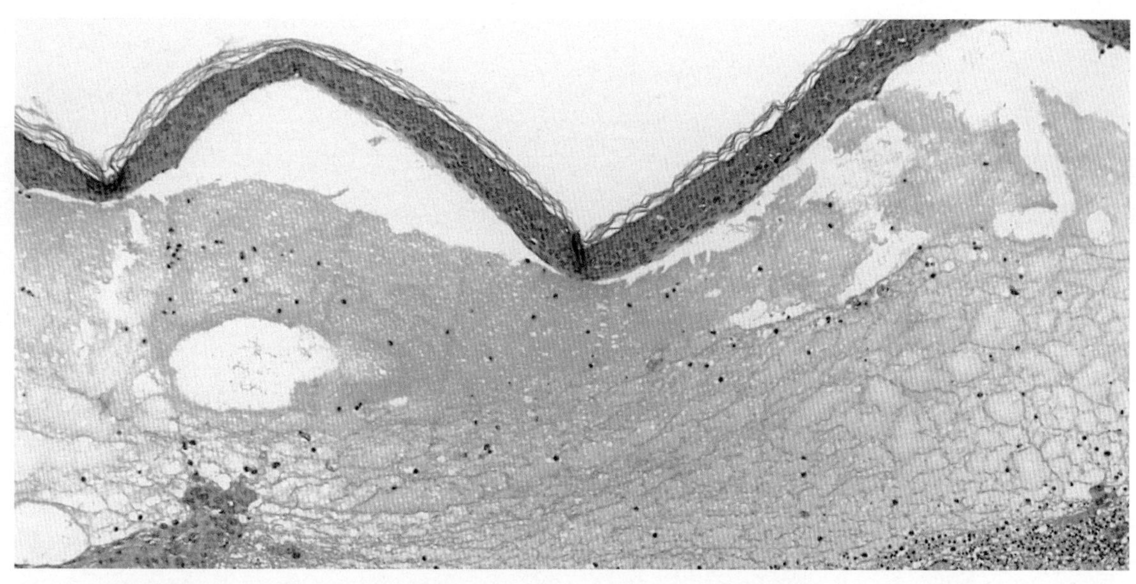

Figure 18–29 A

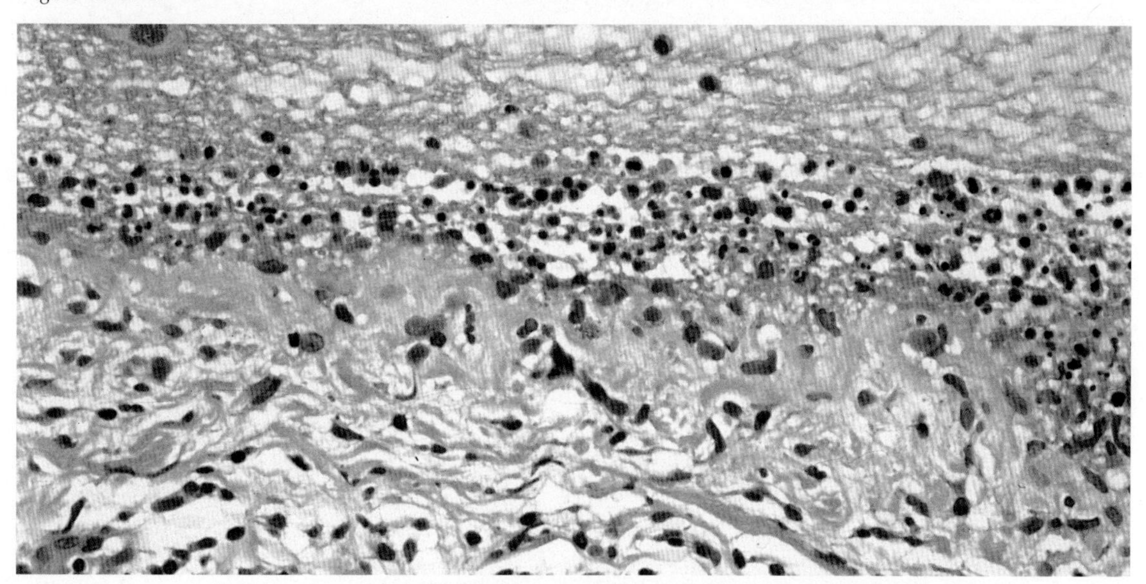

Figure 18–29 B

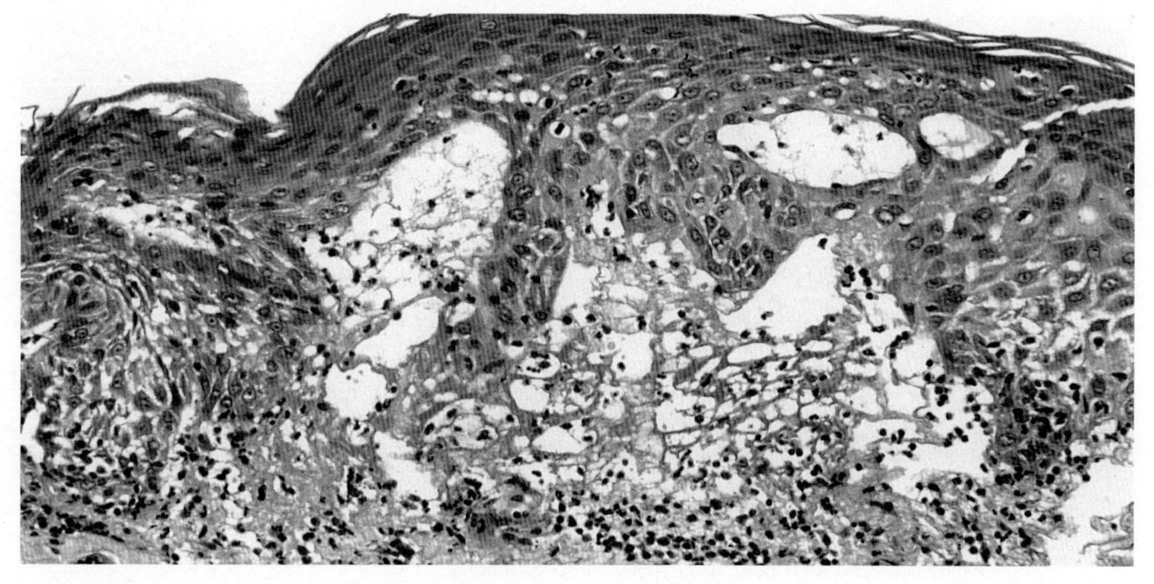

Figure 18–30

■ SKIN

PREMALIGNANT EPITHELIAL TUMORS

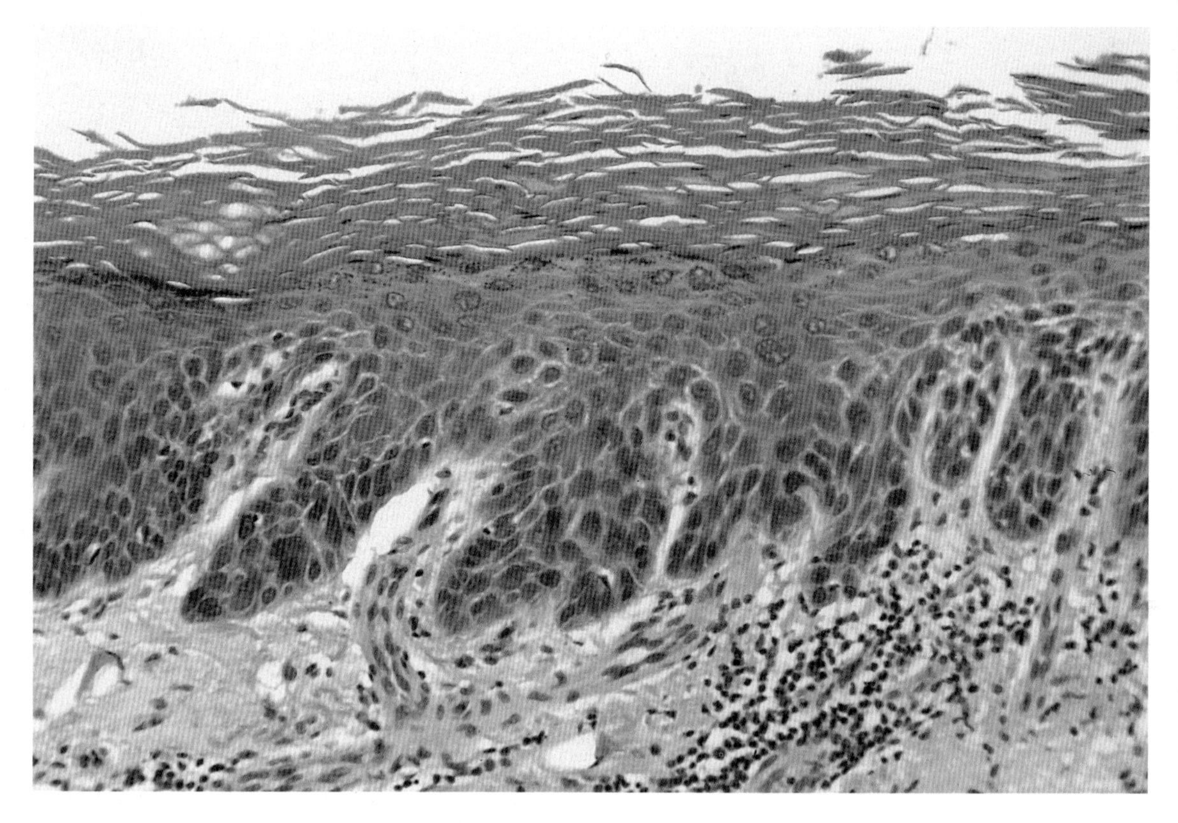

FIGURE 18–31. ACTINIC KERATOSIS, HYPERTROPHIC TYPE: In hypertrophic actinic keratosis, the epidermis is thickened secondary to irregular papillomatosis. The cells in the lower epidermis show a loss of polarity and are atypical and pleomorphic, with hyperchromatic nuclei. Hyperkeratosis is pronounced, and parakeratosis is minimal.

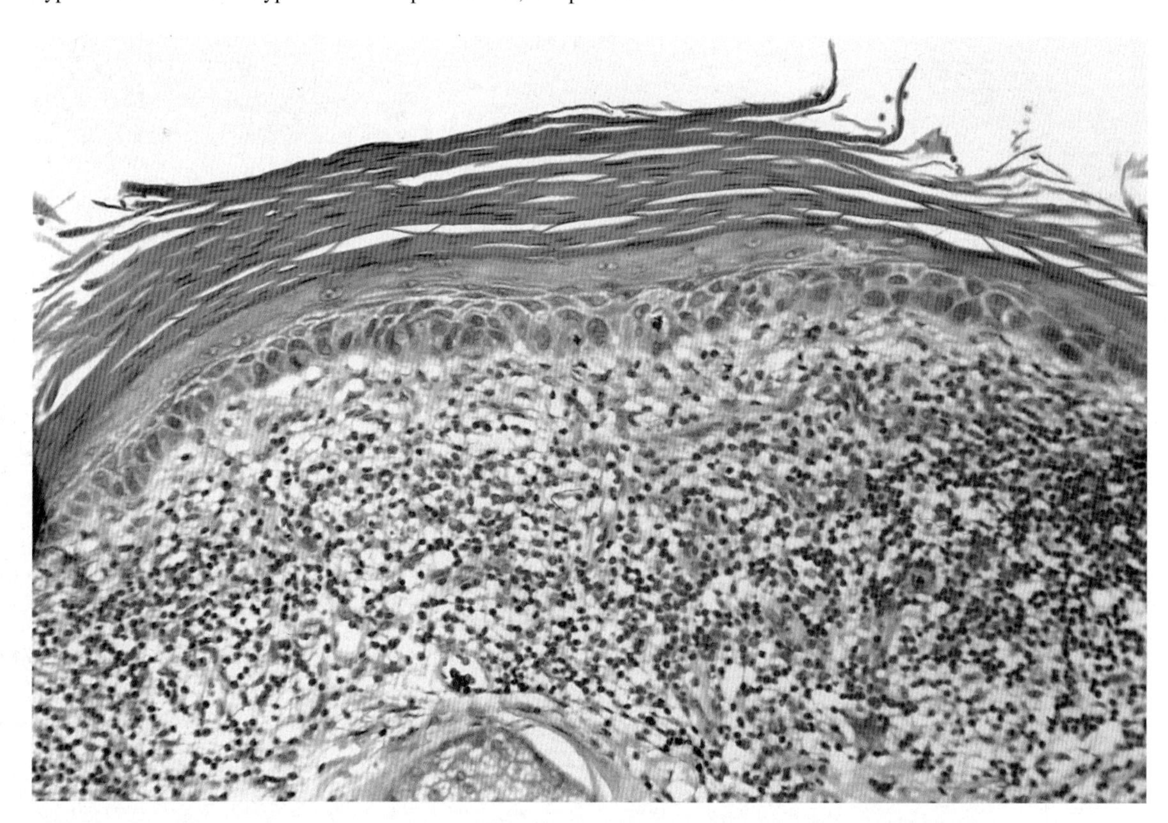

FIGURE 18–32. ACTINIC KERATOSIS, ATROPHIC TYPE: In atrophic actinic keratosis, the epidermis is thin because of the loss of the rete ridges. The cells in the lowermost portion of the epidermis are atypical and pleomorphic, with hyperchromatic nuclei. Parakeratosis is pronounced, and hyperkeratosis is minimal.

■ SKIN

MALIGNANT EPITHELIAL TUMORS

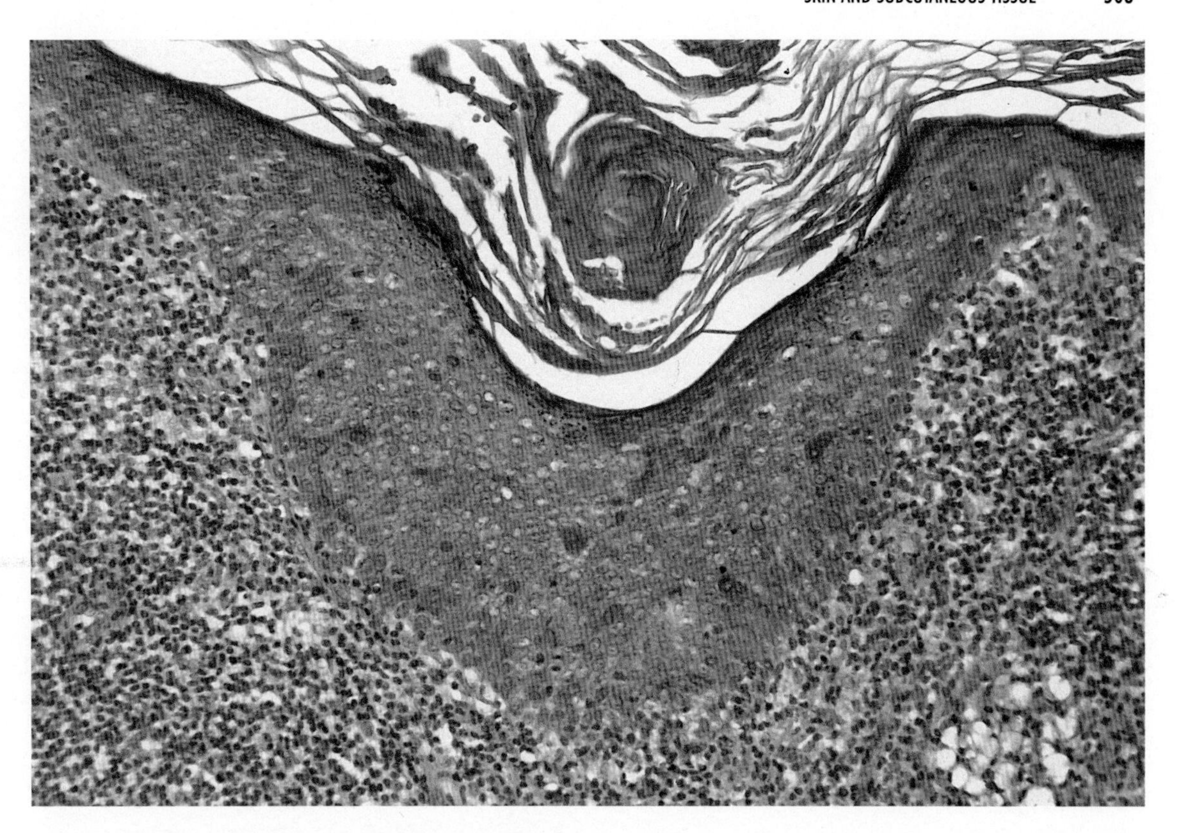

FIGURE 18–33. BOWEN'S DISEASE (IN SITU SQUAMOUS CELL CARCINOMA): In Bowen's disease, the entire thickness of the epidermis is replaced by neoplastic cells. The cells lose their polarity and are pleomorphic with hyperchromatic nuclei. Mitotic figures and dyskeratotic cells may be found at any level in the epidermis.

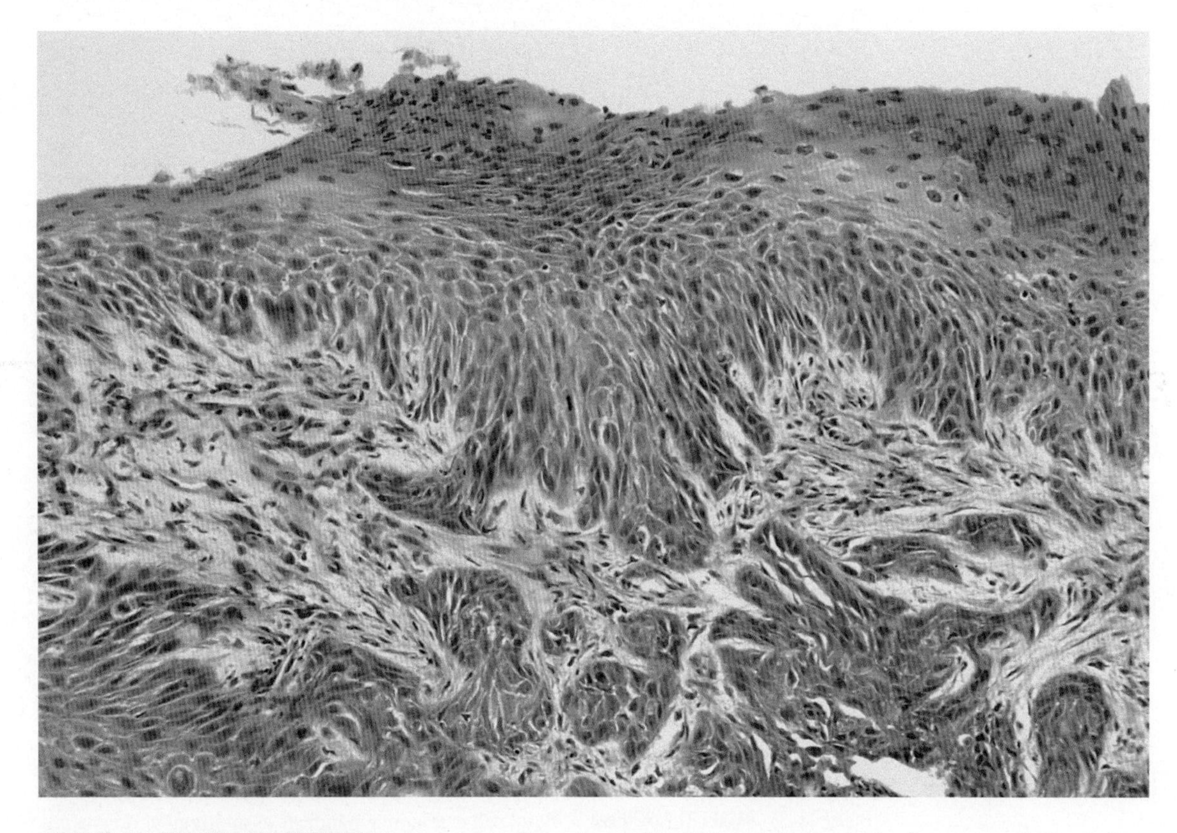

FIGURE 18–34. SQUAMOUS CELL CARCINOMA: In squamous cell carcinoma, in addition to the histologic changes seen in Bowen's disease (Figure 18–33), irregular tongues of epidermal cells invade the dermis.

■ SKIN

MALIGNANT EPITHELIAL TUMORS

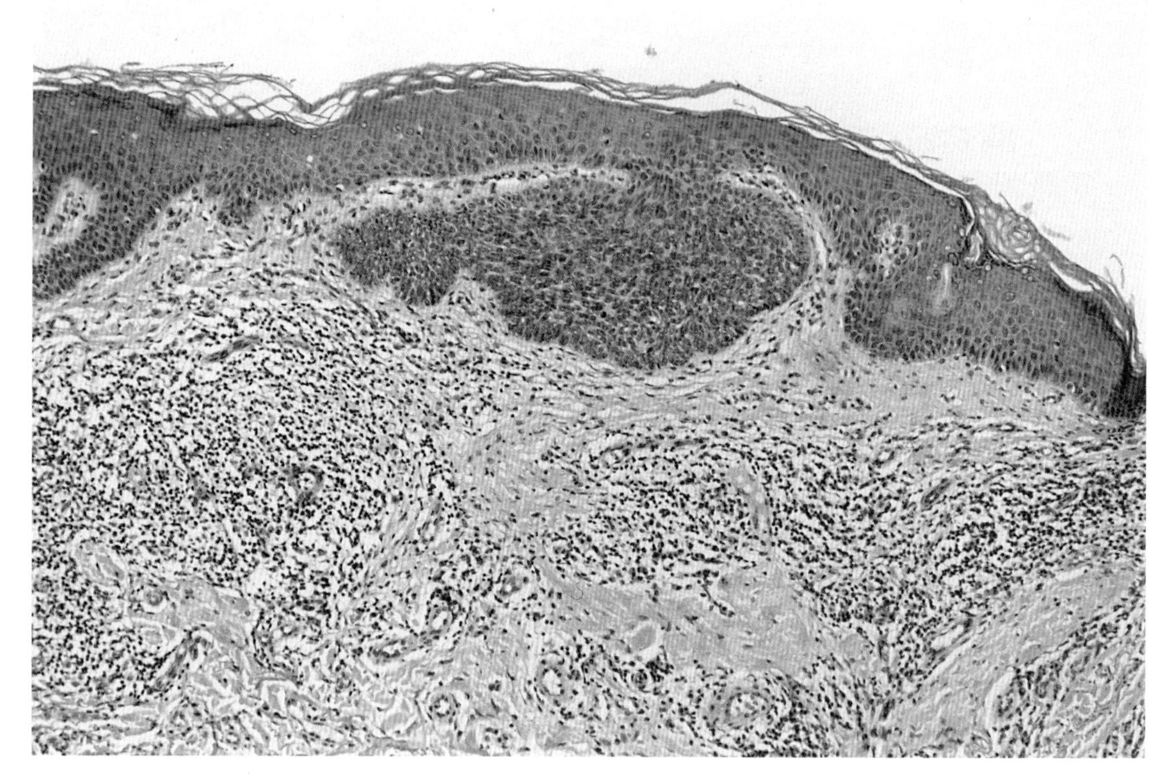

Figure 18–35 **A**

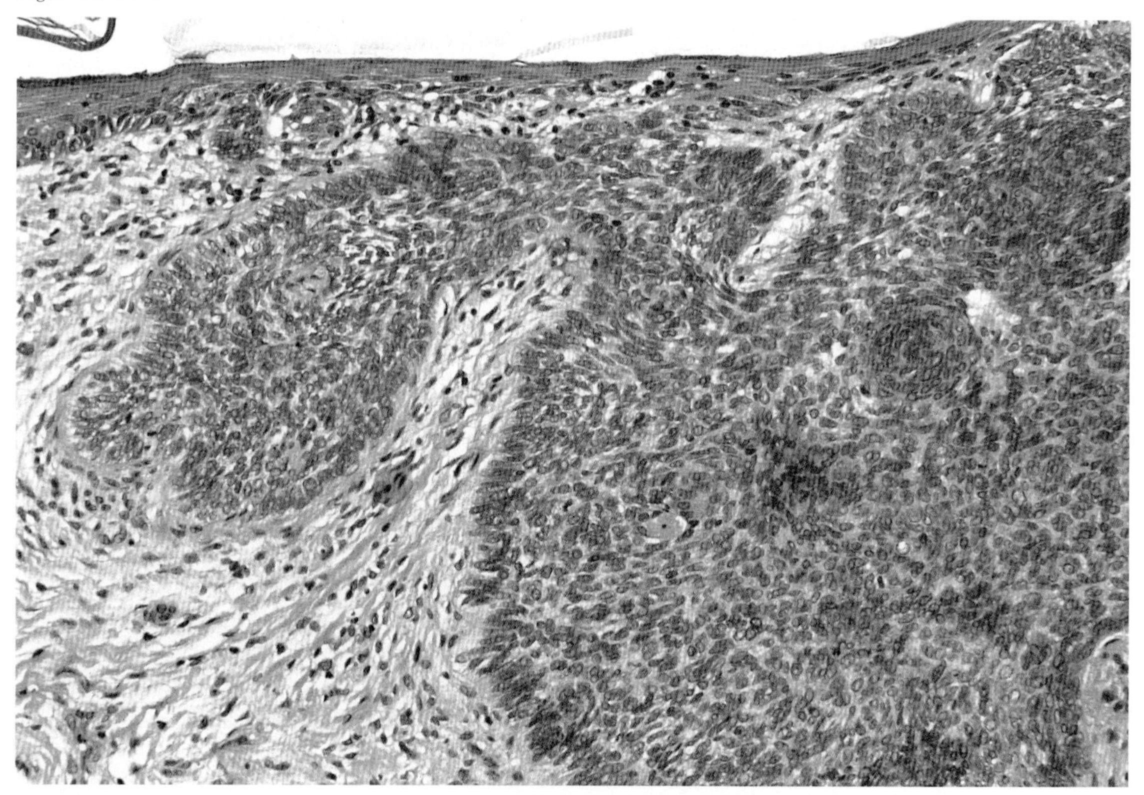

Figure 18–35 **B**

FIGURE 18–35. BASAL CELL CARCINOMA: A: Basal cell carcinoma manifests as an irregular downward proliferation of cells from the basal cell layer of the epidermis. **B:** The cells are uniform, with scant basophilic cytoplasm and large ovoid nuclei. Peripheral cells are arranged in a palisade fashion.

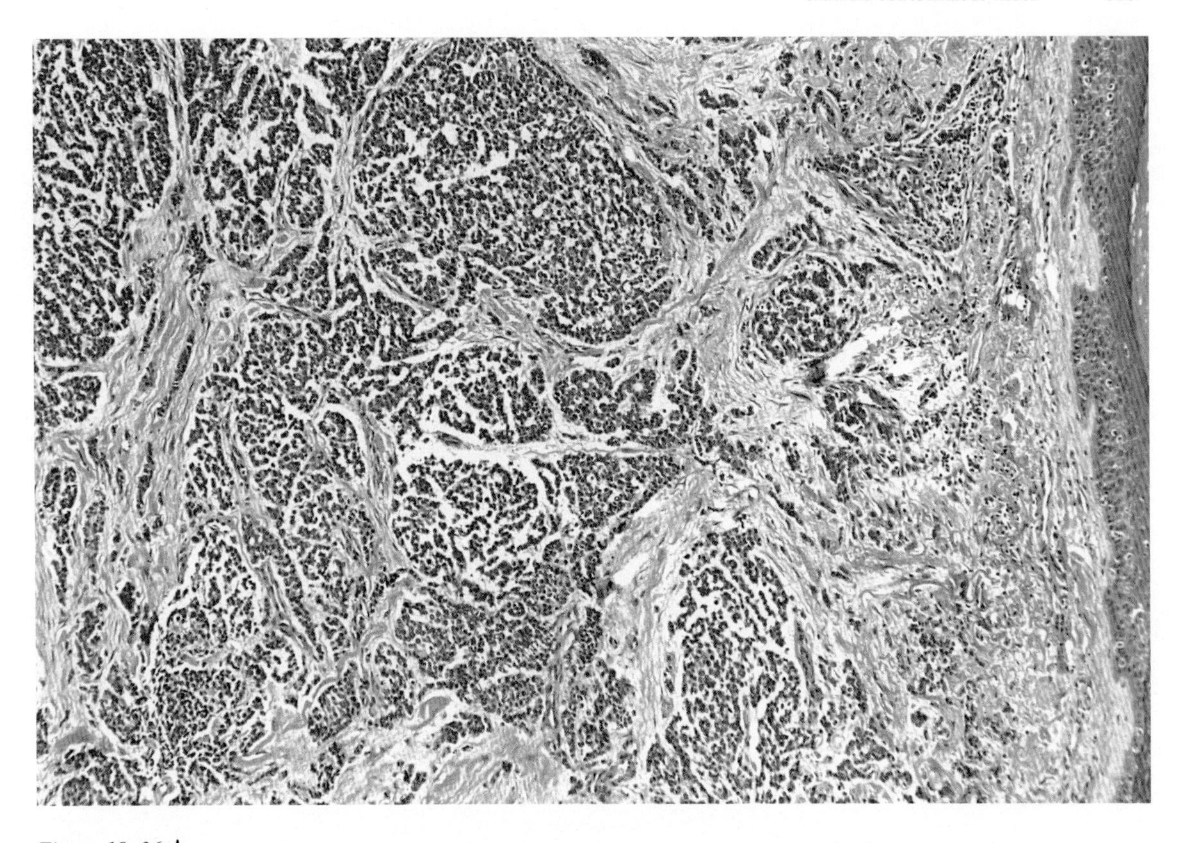

Figure 18–36 **A**

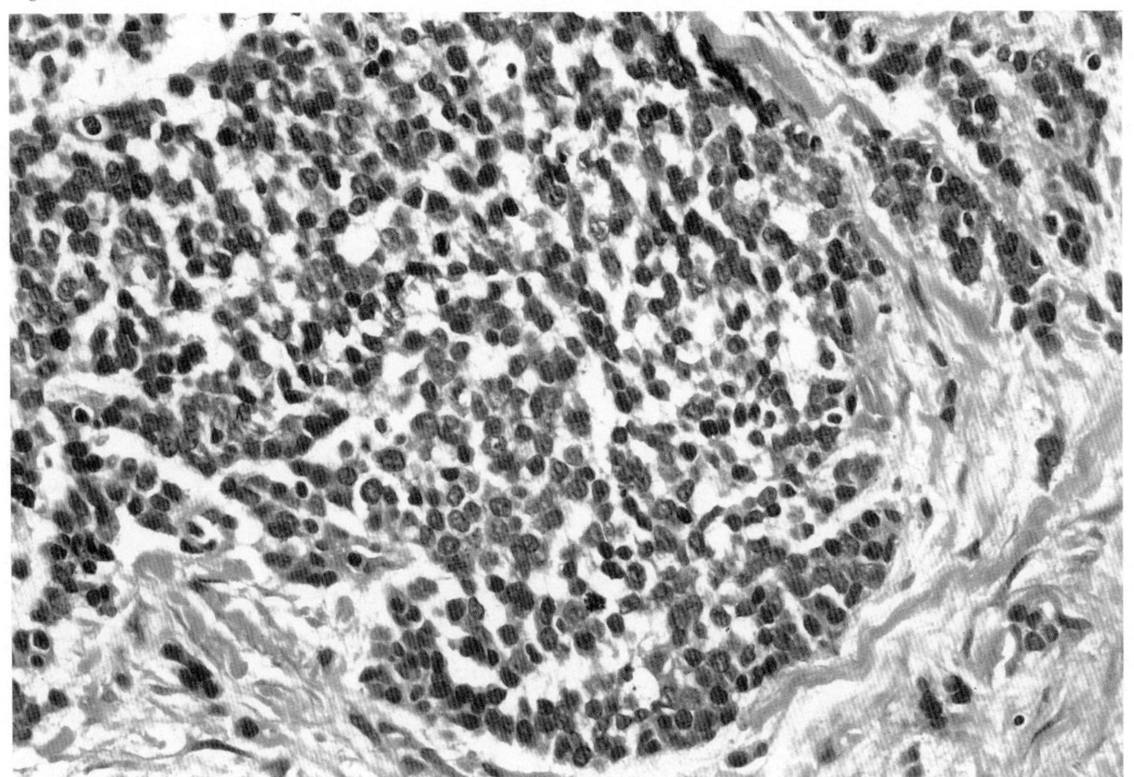

Figure 18–36 **B**

FIGURE 18–36. MERKEL CELL CARCINOMA (TRABECULAR CELL CARCINOMA): A: In Merkel cell carcinoma, cords, bands, and clusters of tumor cells are seen in the dermis and in the subcutaneous tissue. **B:** Tumor cells are uniform, with indistinct scanty cytoplasm and round, vesicular nuclei. Mitoses are present.

■ SKIN

BENIGN MELANOCYTIC LESIONS

The three major types of benign melanocytic nevi are intradermal, junctional, and compound.

FIGURE 18–37. INTRADERMAL NEVUS: An intradermal nevus is composed of nests of nevus cells in the upper dermis.

FIGURE 18–38. JUNCTIONAL NEVUS: A junctional nevus is composed of nests of nevus cells at the dermal-epidermal junction.

FIGURE 18–39. COMPOUND NEVUS: Compound nevus has features of both intradermal and junctional nevi.

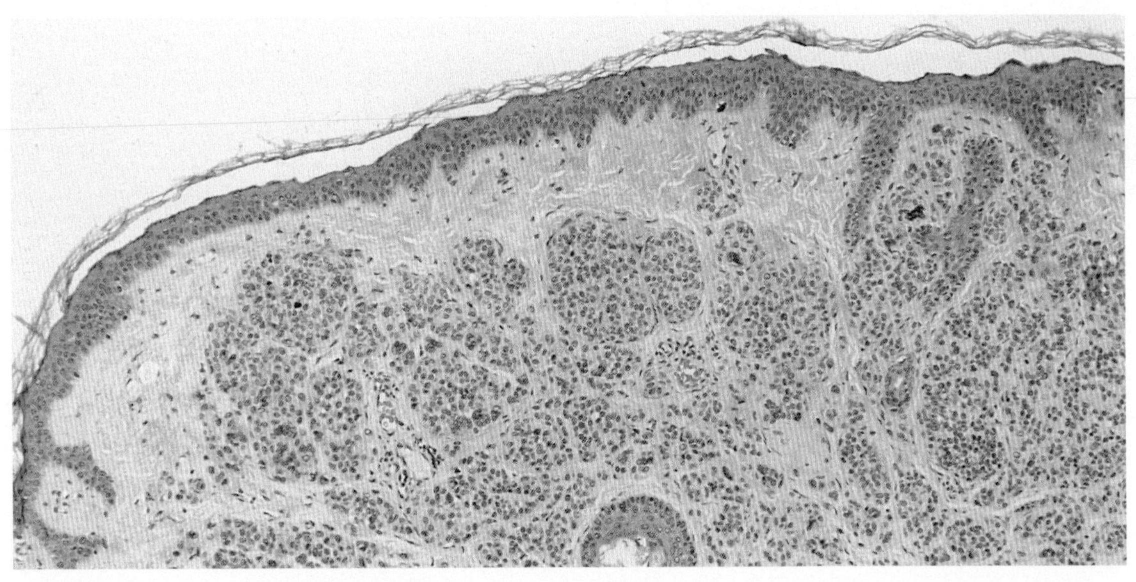

Figure 18–37

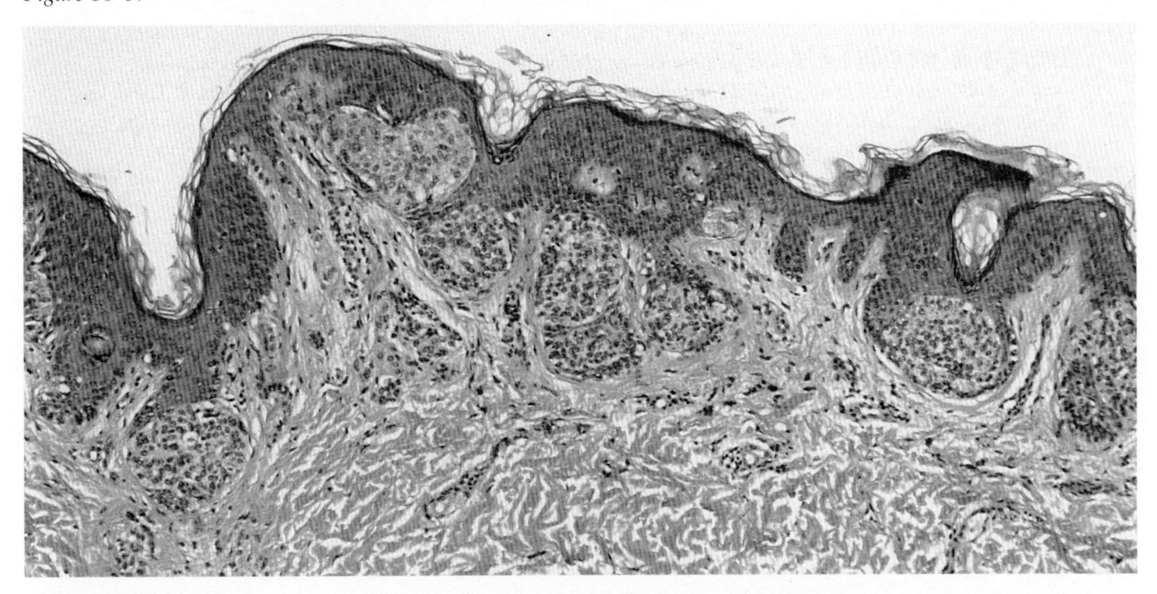

Figure 18–38

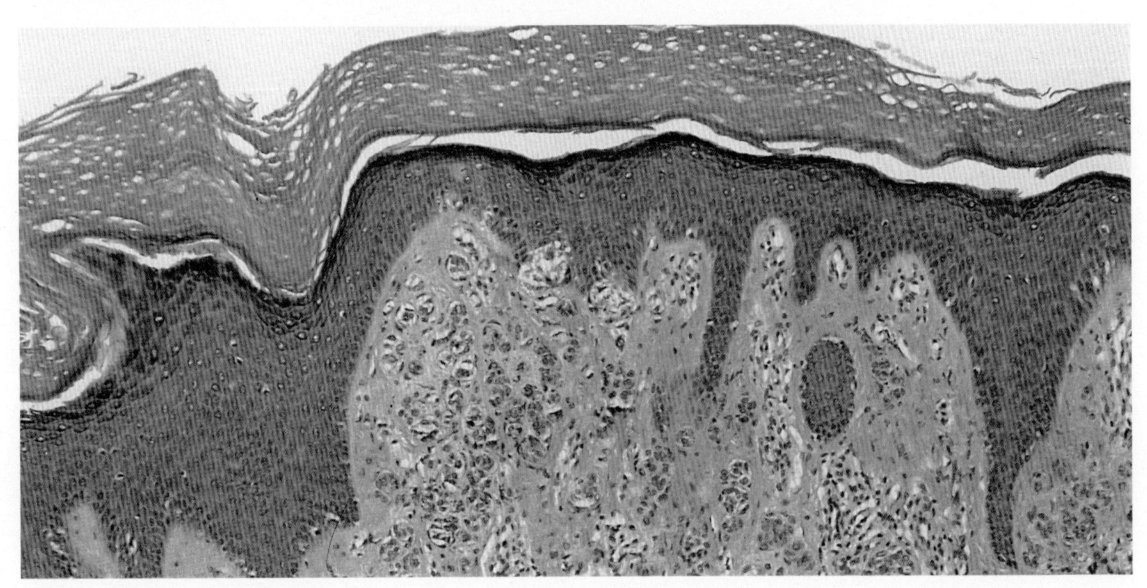

Figure 18–39

■ SKIN

BENIGN MELANOCYTIC LESIONS

FIGURE 18–40. INTRADERMAL NEVUS: A: This high-power photomicrograph of an intradermal nevus shows that some of the nevus cells have multiple nuclei arranged in a rosettelike manner—**mulberry cells**. **B:** The nevus cells in the upper dermis are cuboidal and diminish in size deeper into the dermis. This is referred to as **maturation**. Multinucleation and maturation are both signs of benignity.

FIGURE 18–41. JUNCTIONAL NEVUS: This is a high-power photomicrograph of a junctional nevus and shows uniform cuboidal cells, some of which contain melanin.

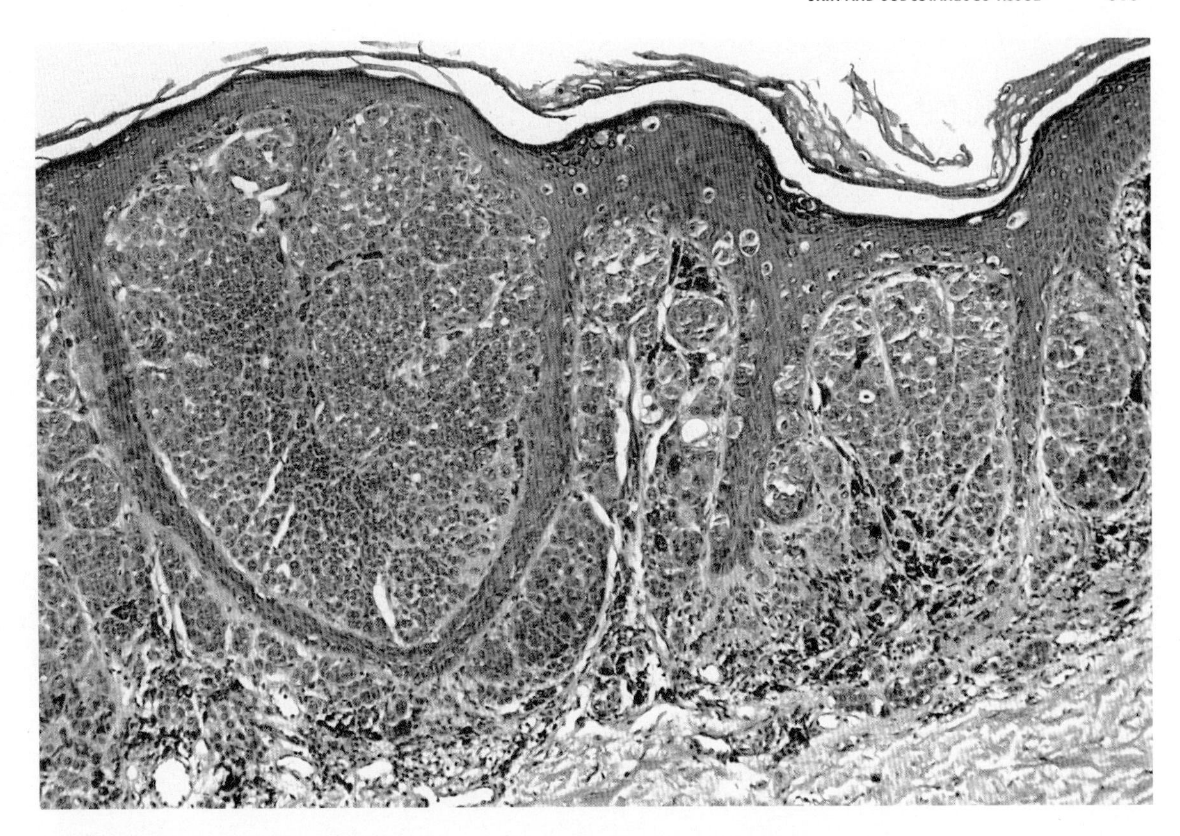

Figure 18–45 **A**

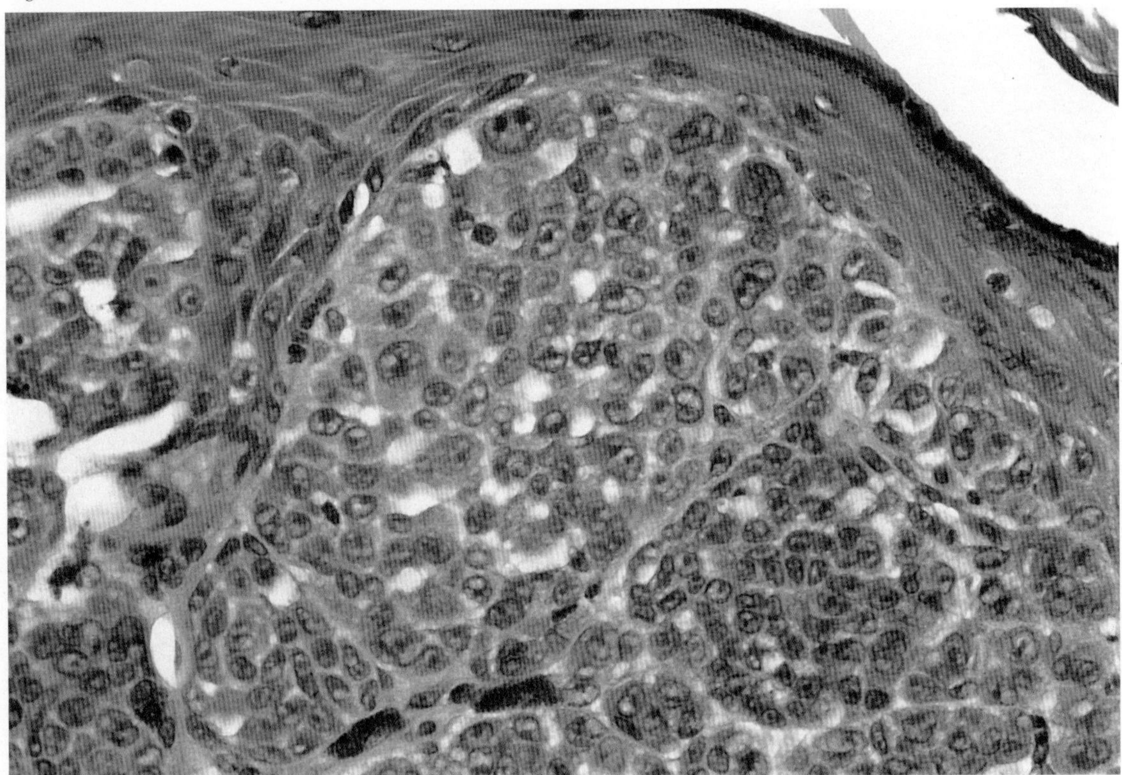

Figure 18–45 **B**

FIGURE 18–45. MALIGNANT MELANOMA: A: In malignant melanoma are seen not only the features of in situ melanoma (Figure 18–44 **A** and **B**) but also elongation of the rete ridges and tumor cells streaming downward into the dermis. **B:** In malignant melanoma, malignant nevus cells do not show maturation, in contrast to the benign nevus cells of intradermal and compound nevi (Figure 18–40 **B**).

■ SUBCUTANEOUS TISSUE

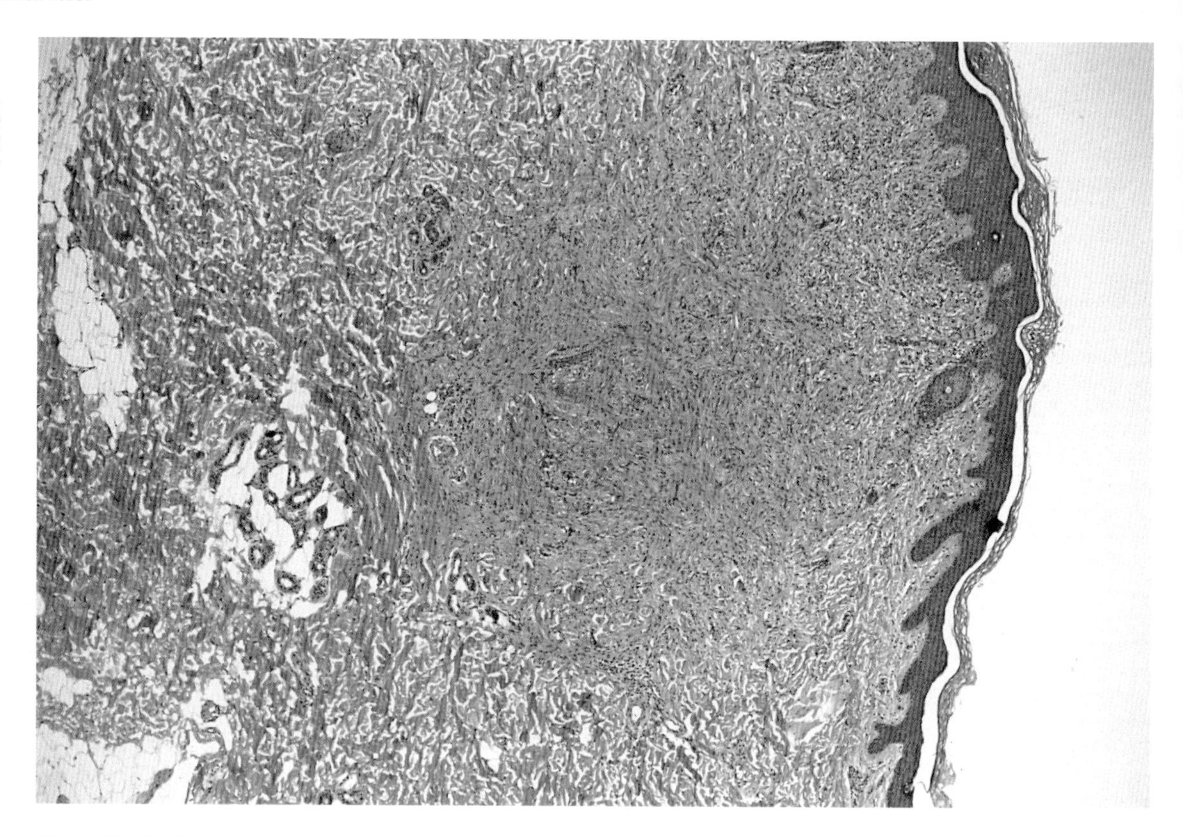

Figure 18–46 A

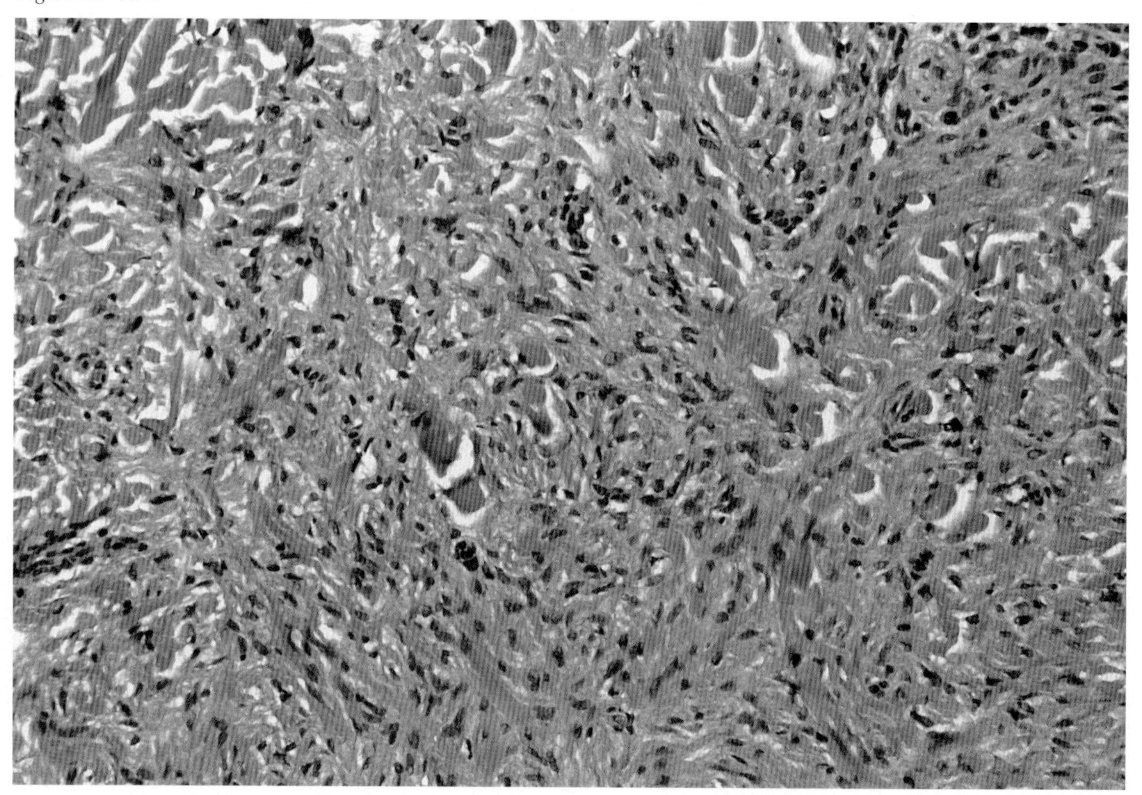

Figure 18–46 B

FIGURE 18–46. DERMATOFIBROMA (BENIGN FIBROUS HISTIOCYTOMA): A dermatofibroma is a benign fibrohistiocytic tumor. **A:** Dermatofibromas are confined to the dermis. They are composed of varied amounts of fibroblasts, histiocytes, inflammatory cells, and capillaries. **B:** The fibroblasts are arranged in fascicles and whorl from a central hub in a storiform pattern.

■ SUBCUTANEOUS TISSUE

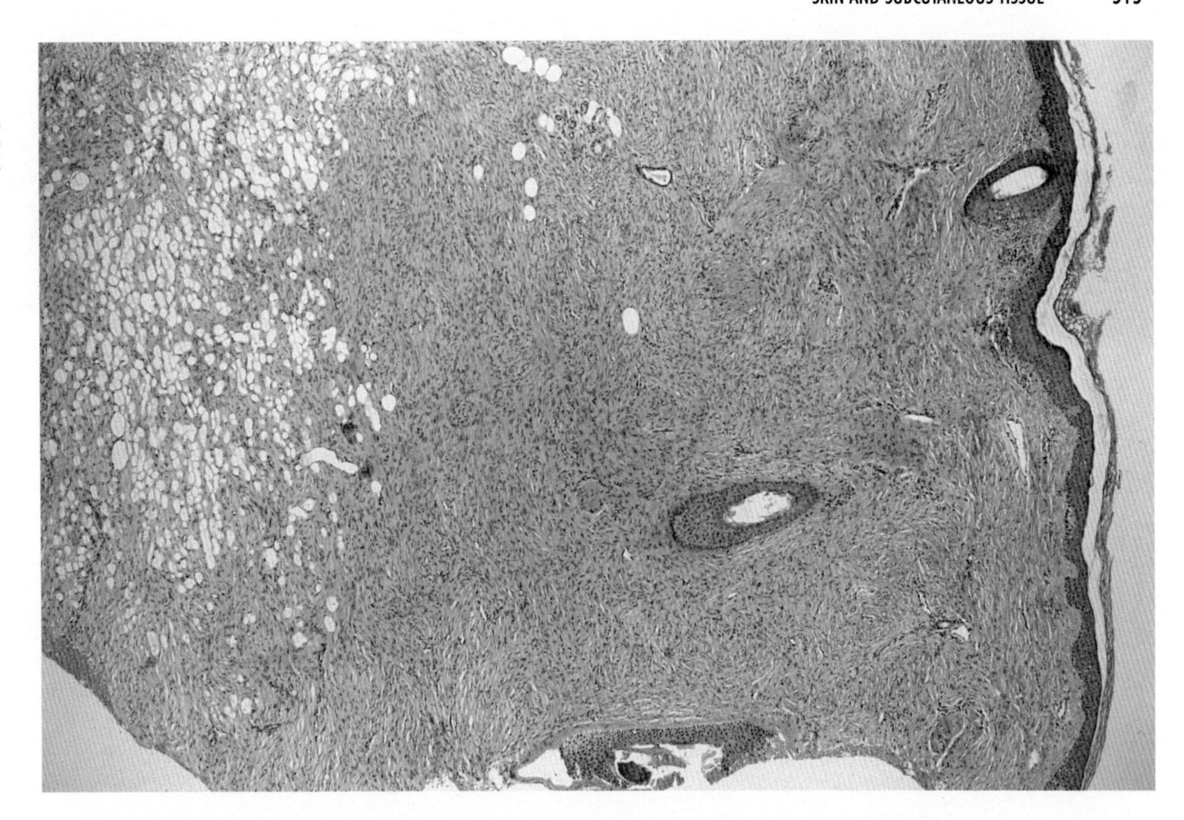

Figure 18–47 **A**

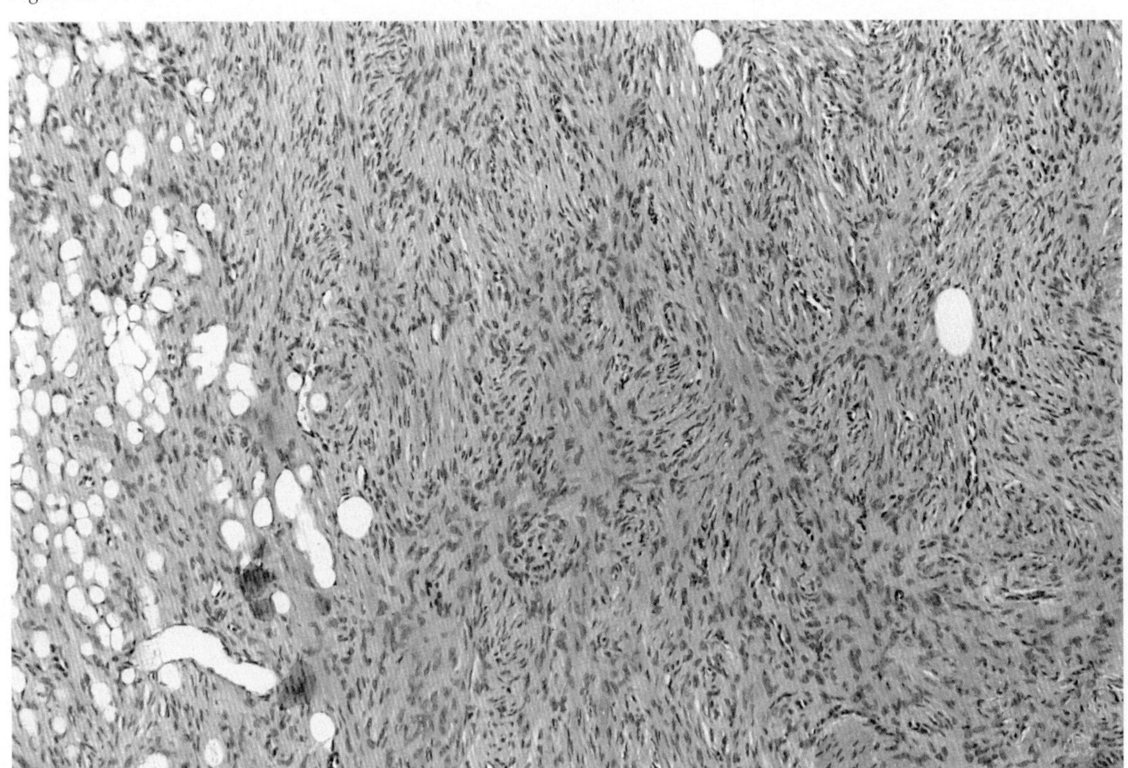

Figure 18–47 **B**

FIGURE 18–47. DERMATOFIBROSARCOMA PROTUBERANS (DFSP): A dermatofibrosarcoma protuberans is a benign fibrohistiocytic tumor; however, it has a tendency to recur if incompletely excised. **A:** The tumor extends into the subcutaneous fat. Such tumors are composed of varied amounts of fibroblasts, histiocytes, inflammatory cells, and capillaries. **B:** The fibroblasts are arranged in fascicles and whorl from a central hub in a storiform pattern. Dermatofibrosarcoma protuberans is more cellular than dermatofibroma (Figure 18–46 **A** and **B**).

■ SUBCUTANEOUS TISSUE

FIGURE 18–48. CAPILLARY HEMANGIOMA: Capillary hemangiomas are composed of small-caliber thin-walled blood vessels and pericytes in the dermis.

FIGURE 18–49. CAVERNOUS HEMANGIOMAS: Cavernous hemangiomas are composed of ectatic blood vessels.

FIGURE 18–50. XANTHOMA: Xanthomas are composed of groups of foamy histiocytes within the dermis.

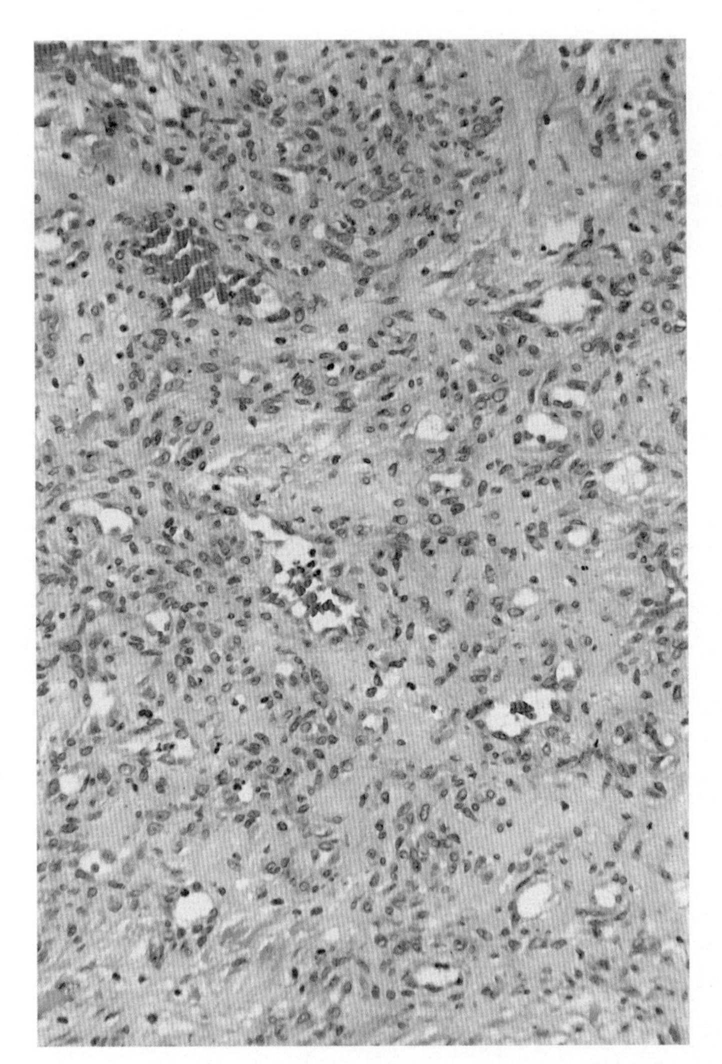

Figure 18–48

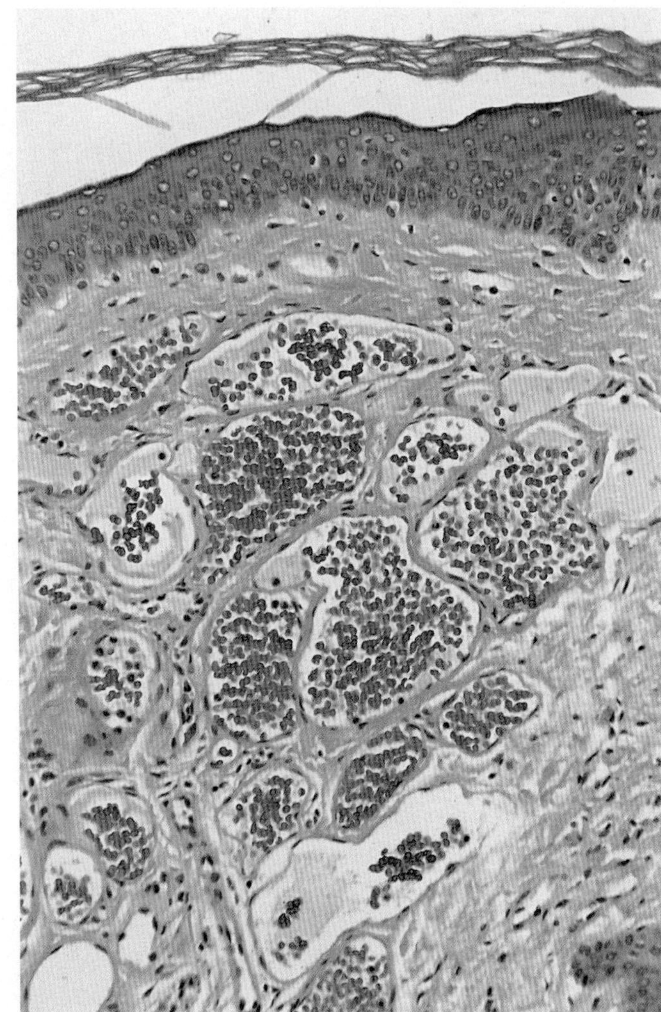

Figure 18–49

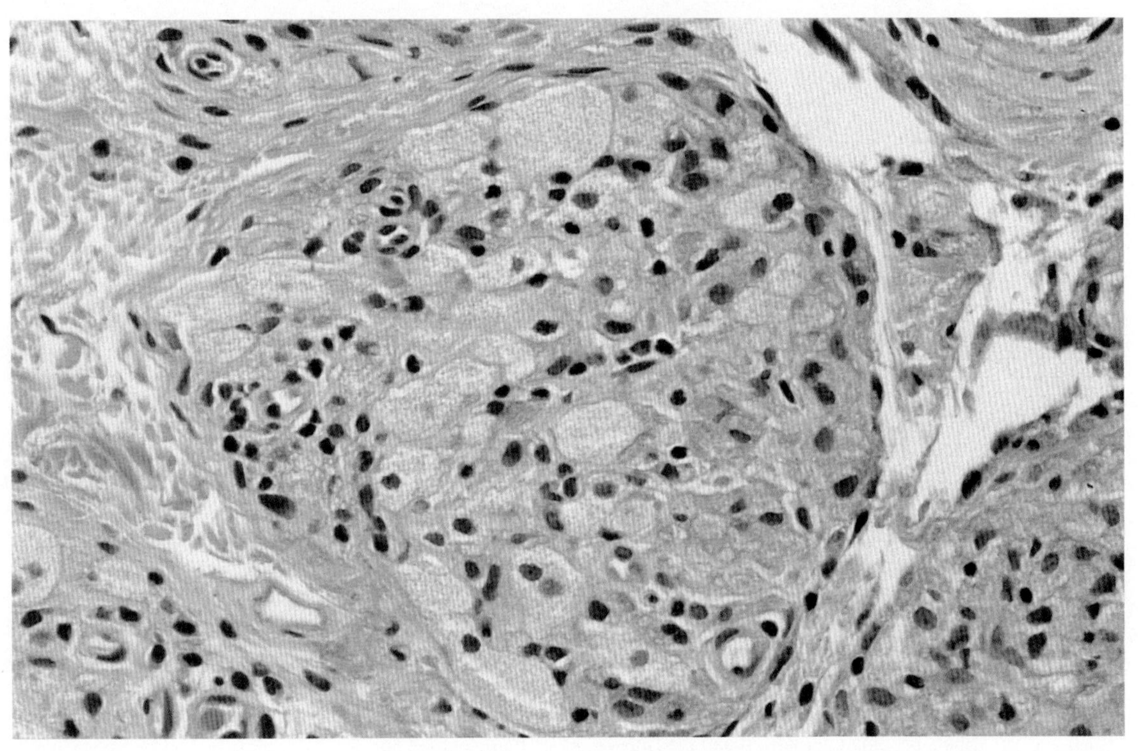

Figure 18–50

■ SKELETAL MUSCLE

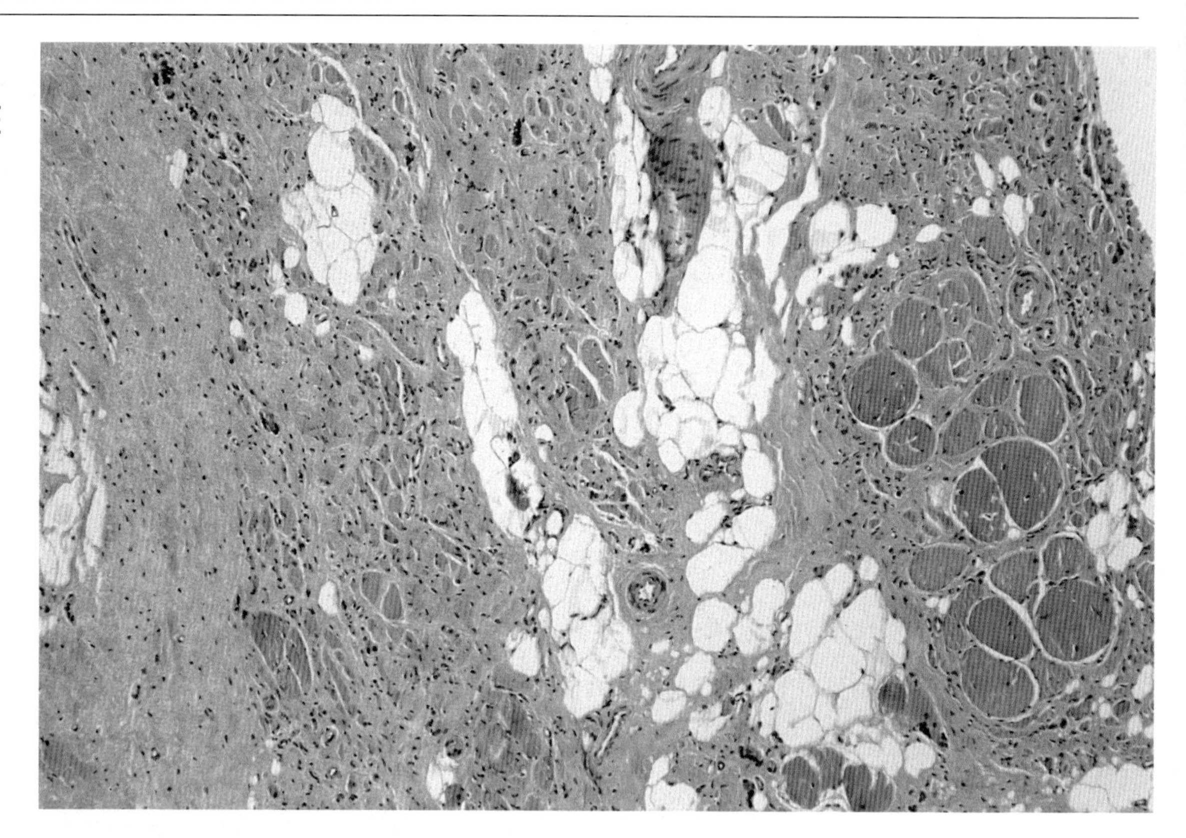

Figure 19–1 **A**

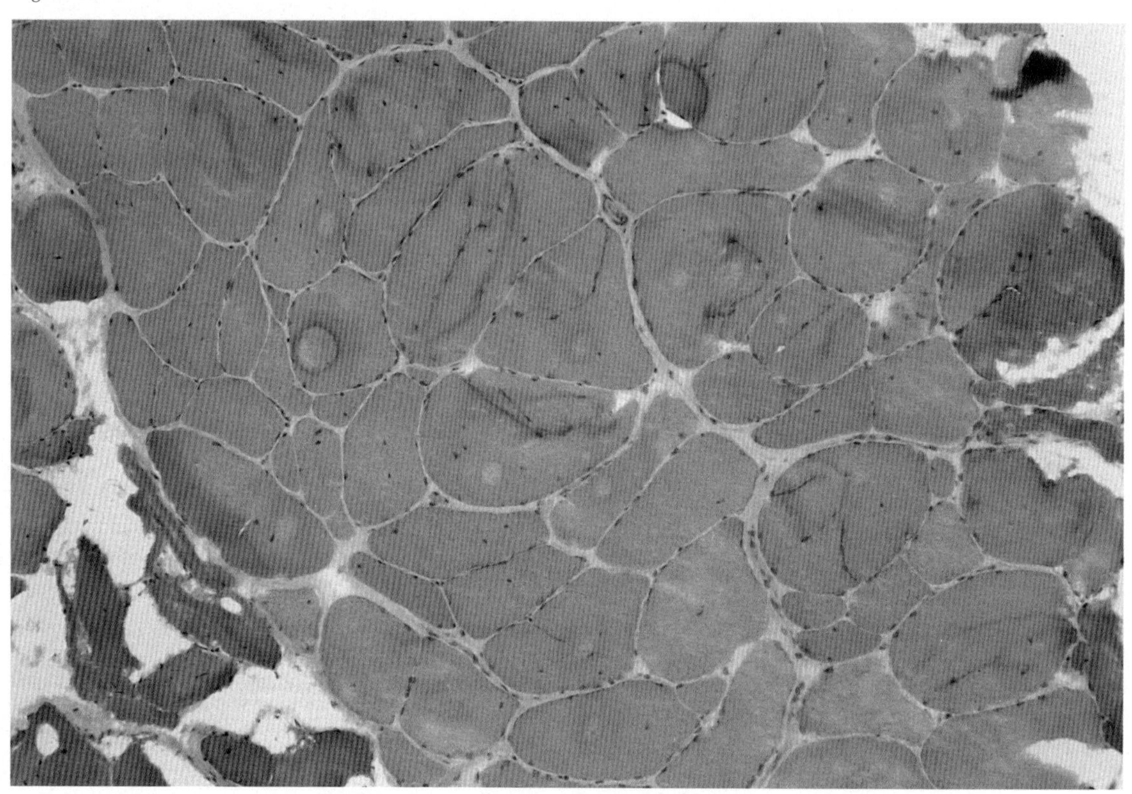

Figure 19–1 **B**

FIGURE 19–1. DUCHENNE TYPE MUSCULAR DYSTROPHY: Duchenne type muscular dystrophy is the most common of the muscular dystrophies. **A:** Atrophic and hypertrophic muscle fibers with round contours are present. Muscle fibers become necrotic and are replaced by fibrofatty tissue. **B:** In other areas, previously peripherally located nuclei of hypertrophic muscle fibers become internalized and the fiber becomes multisegmented.

■ SKELETAL MUSCLE

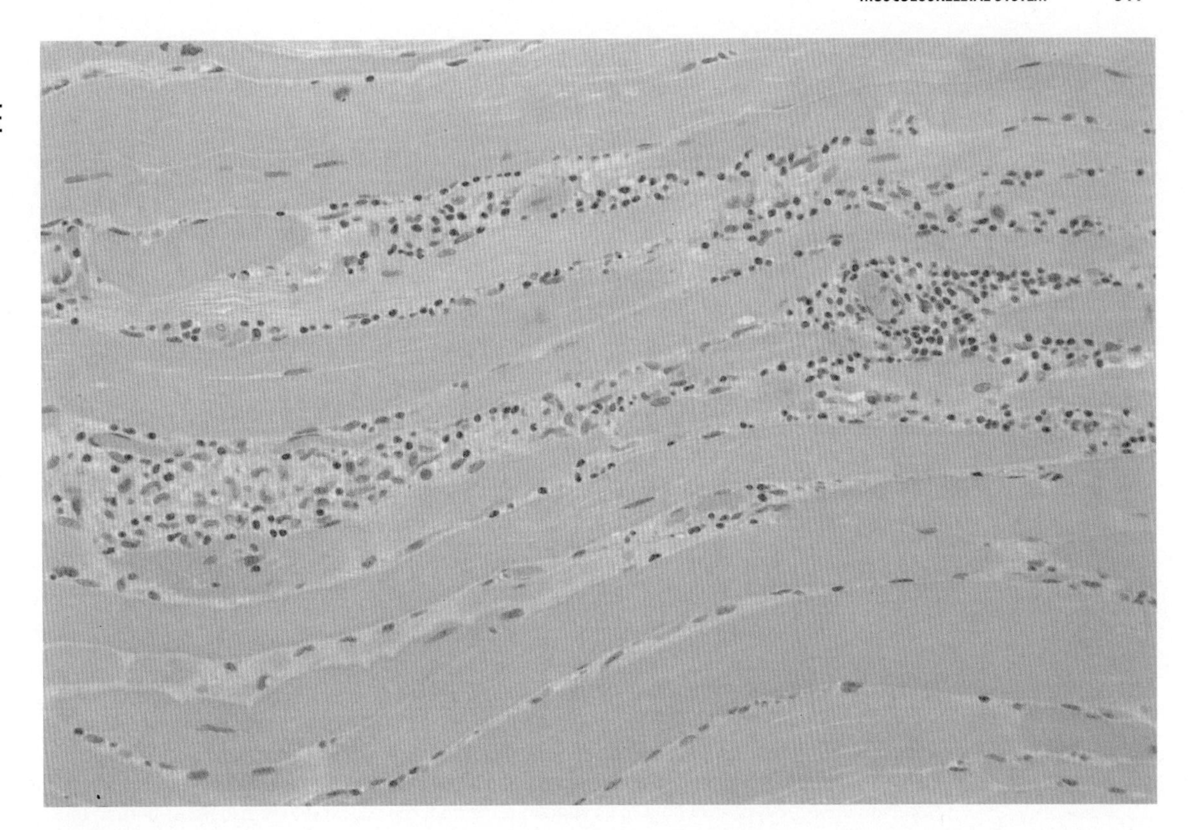

Figure 19–2 **A**

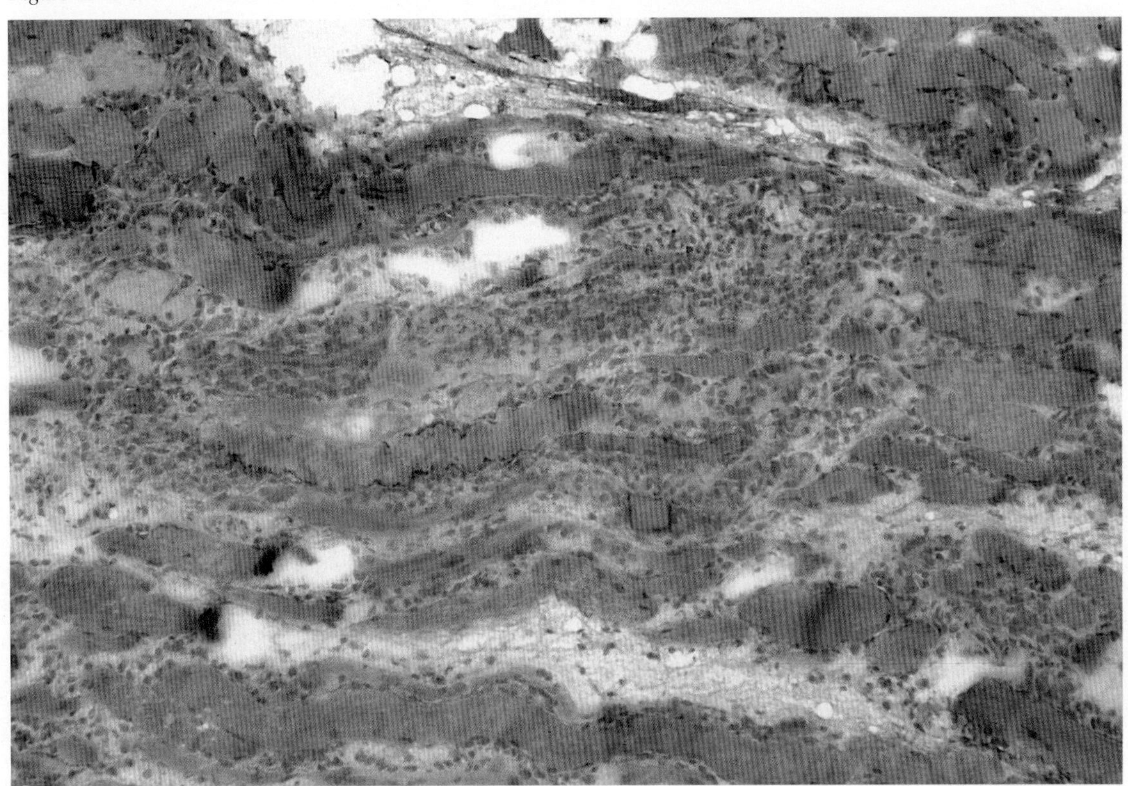

Figure 19–2 **B**

FIGURE 19–2. POLYMYOSITIS: Polymyositis is the most common myopathy in adults. **A:** This is a photomicrograph of a mild case of polymyositis in which a lymphocytic infiltrate is present in the interstitium. **B:** This photomicrograph is of a severe case of polymyositis. Muscle fibers are necrotic and are replaced by phagocytic histiocytes. Regenerating muscle fibers are basophilic, with large vesicular nuclei.

■ SKELETAL MUSCLE

FIGURE 19–3. NEUROGENIC (DENERVATION) ATROPHY: Injury to anterior horn cells of the spinal cord or axons causes denervation atrophy. **A:** Atrophic fibers are characterized by angulation, a decrease in fiber size, and an apparent increase in the number of nuclei per fiber. **B:** In chronic neurogenic atrophy with reinnervation, there is fiber type grouping. Adenosine triphosphatase stains type II fibers dark and type I fibers light. Compare the fiber-type grouping in atrophic muscle with the checkerboard pattern of normal muscle (**C**).

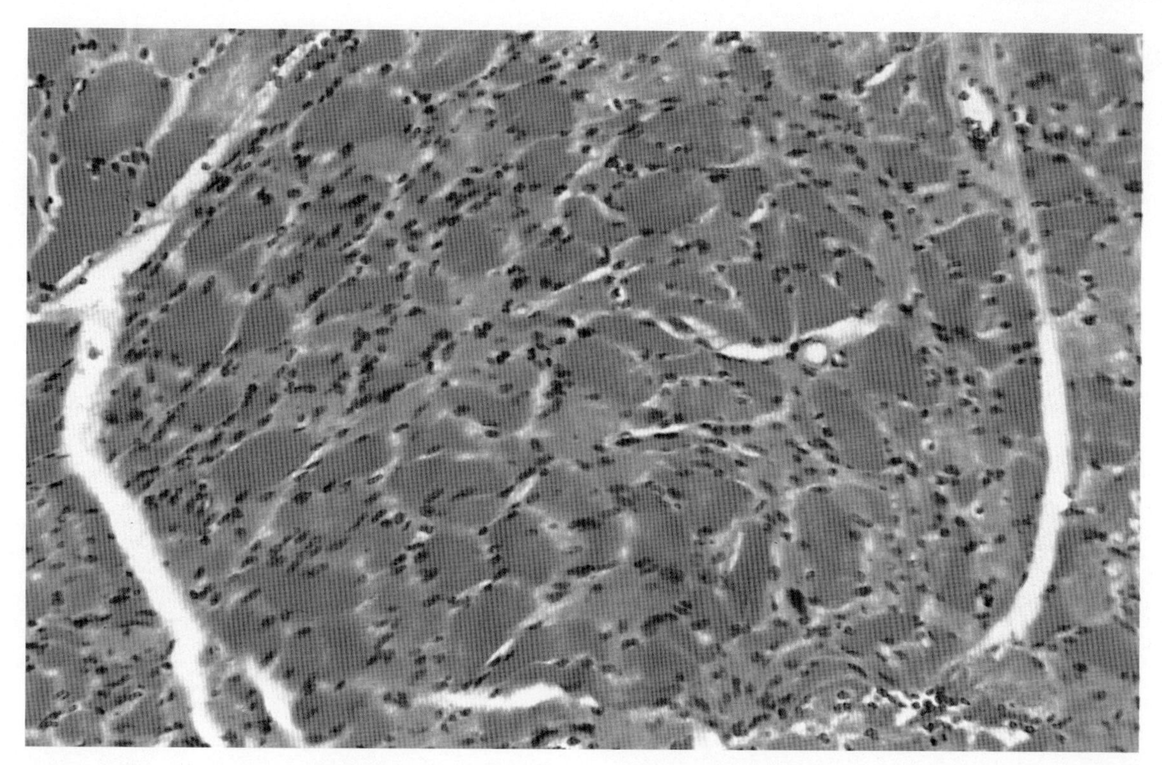

Figure 19–3 A

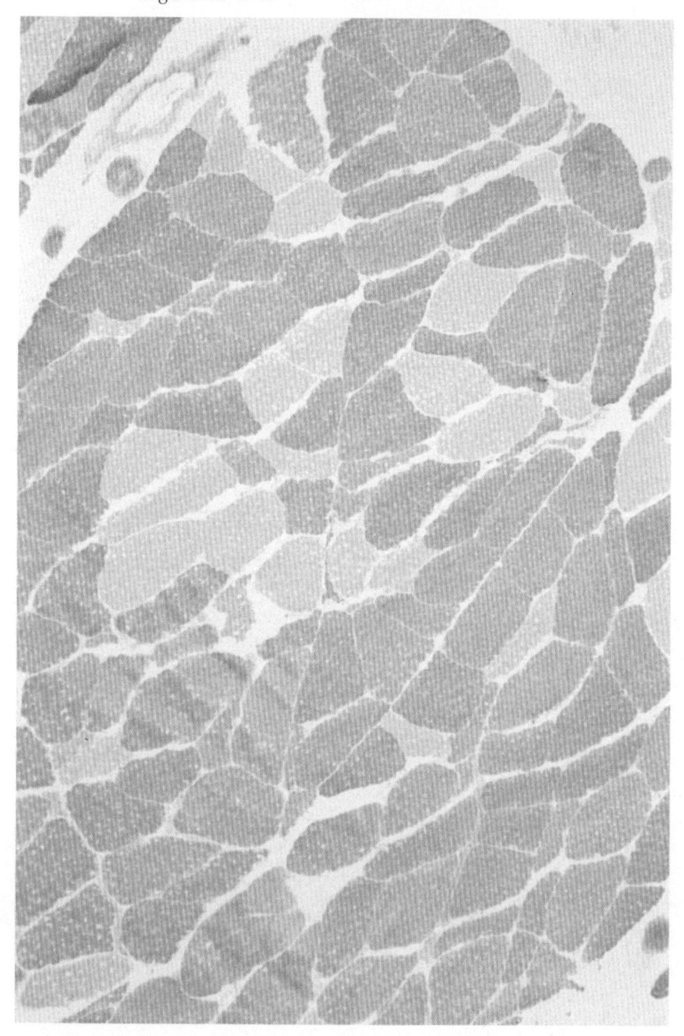

Figure 19–3 B

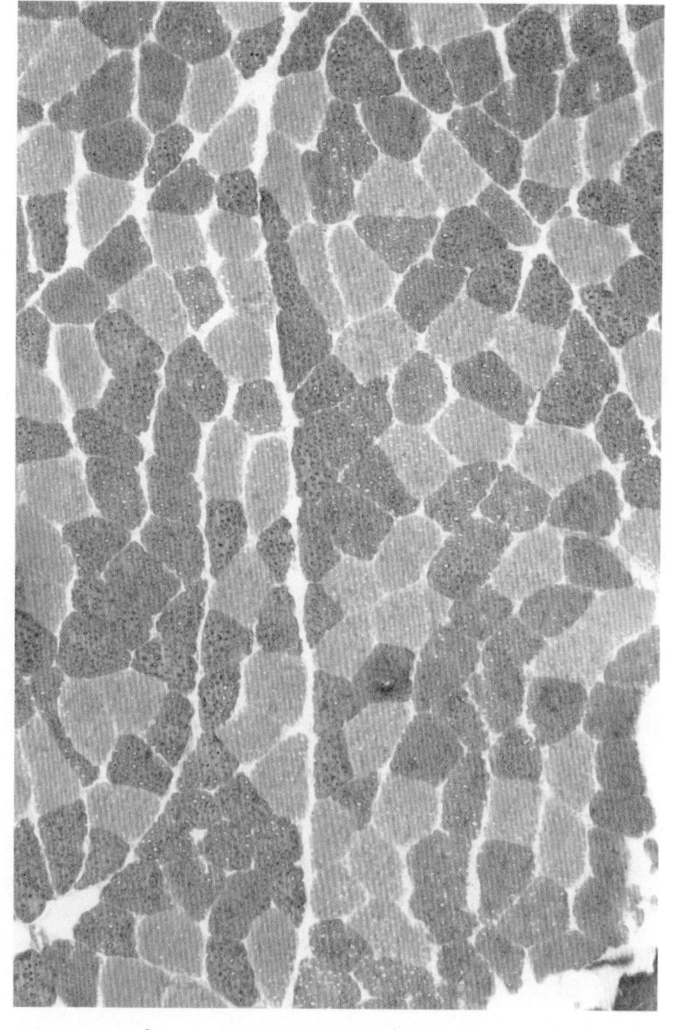

Figure 19–3 C

■ SKELETAL MUSCLE

FIGURE 19–4. WERDNIG-HOFFMANN DISEASE: Werdnig-Hoffmann disease is the most common neurogenic disease in young children. Entire fascicles of infantile muscle affected by the disease show atrophic rounded fibers. A few normal and hypertrophic fascicles may remain.

FIGURE 19–5. TRICHINOSIS: Trichinosis is caused by infestation with the nematode *Trichinella spiralis.* The most common finding is encysted larvae in skeletal muscle.

■ TENDON

FIGURE 19–6. GIANT CELL TUMOR OF TENDON SHEATH: Giant cell tumor of tendon sheath is characterized by a densely cellular stroma composed of round and spindle-shaped cells. Multinucleated cells are randomly distributed throughout the tumor.

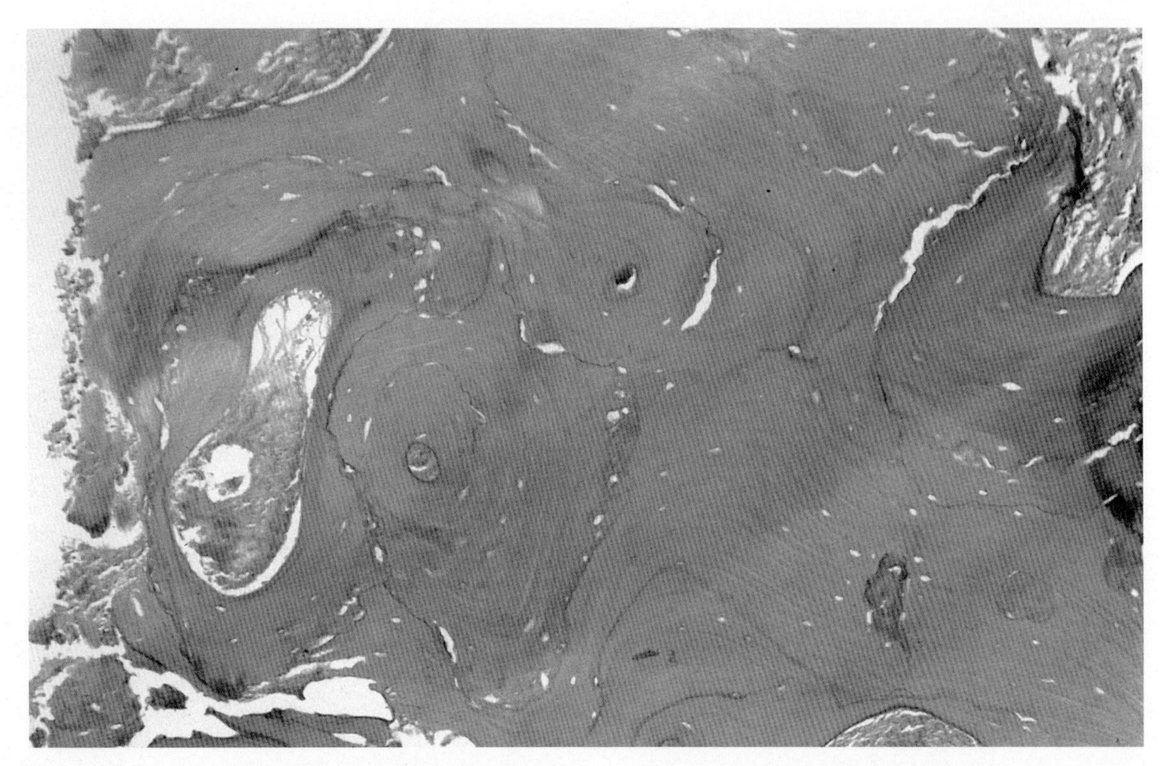

Figure 19–11

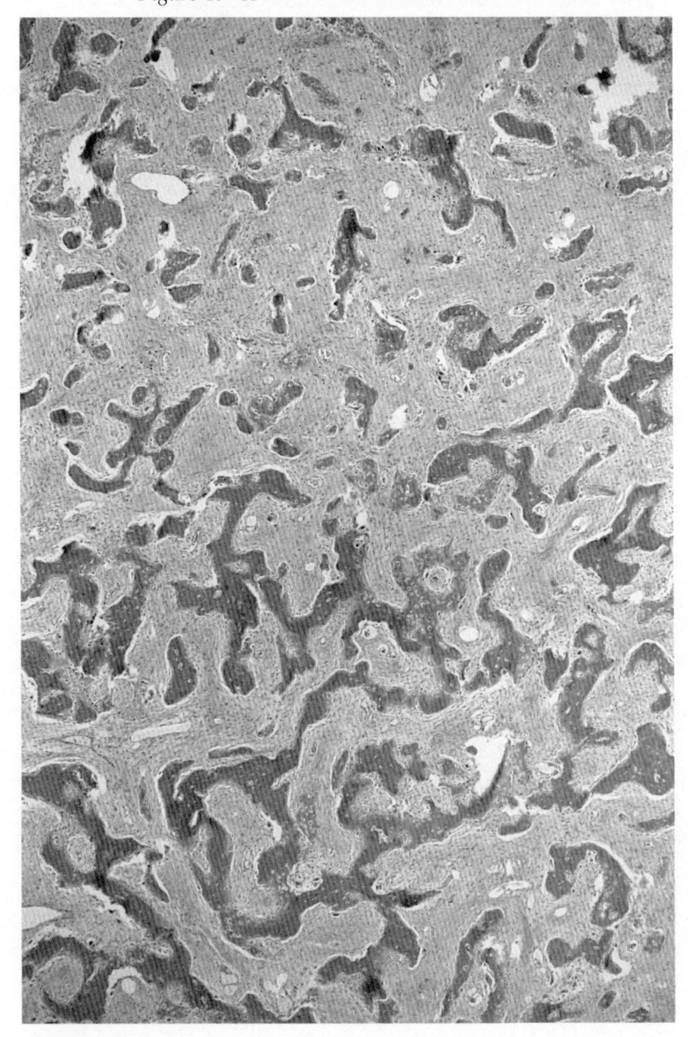

Figure 19–12 A

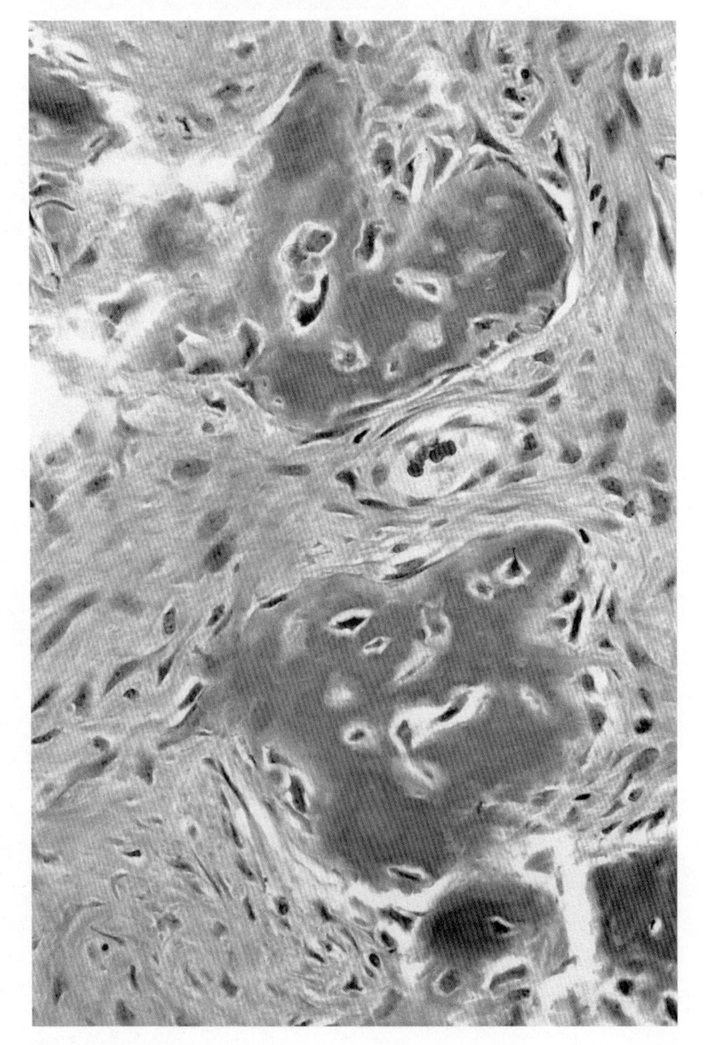

Figure 19–12 B

■ BONE

BONE-FORMING TUMORS

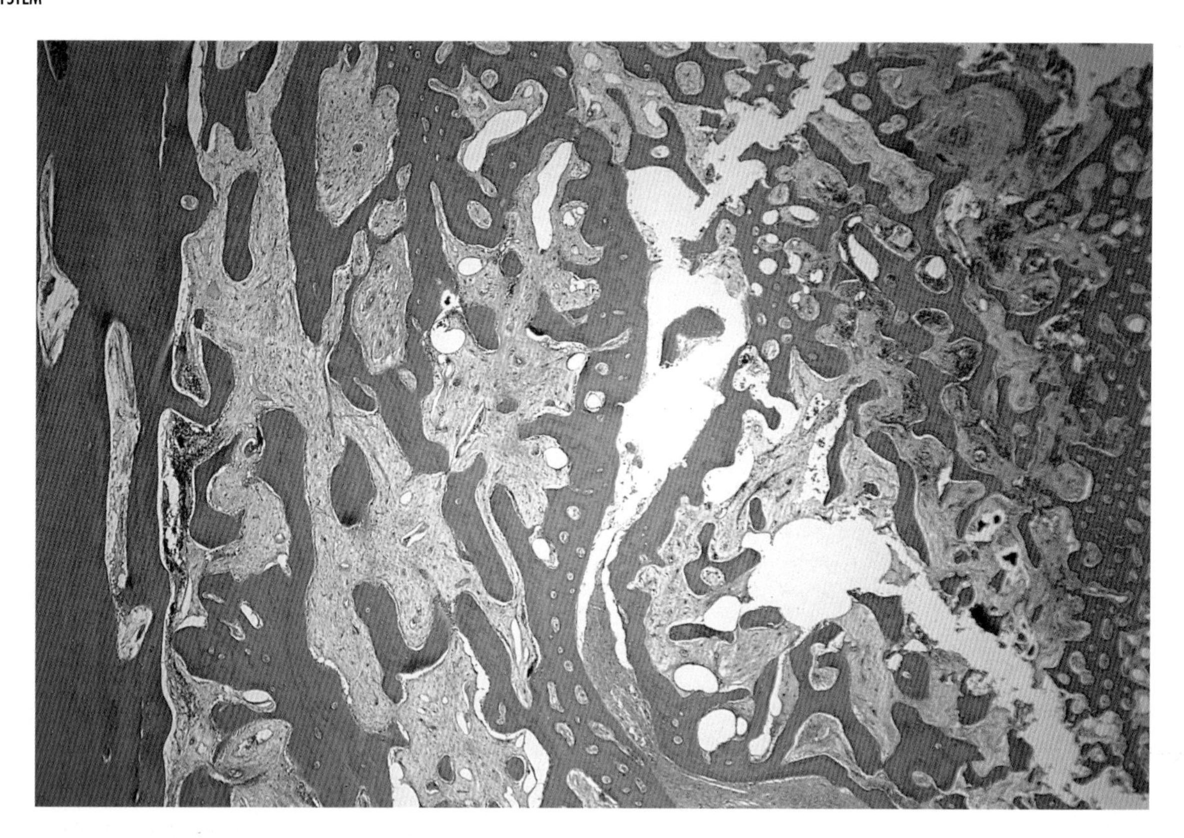

Figure 19–13 A

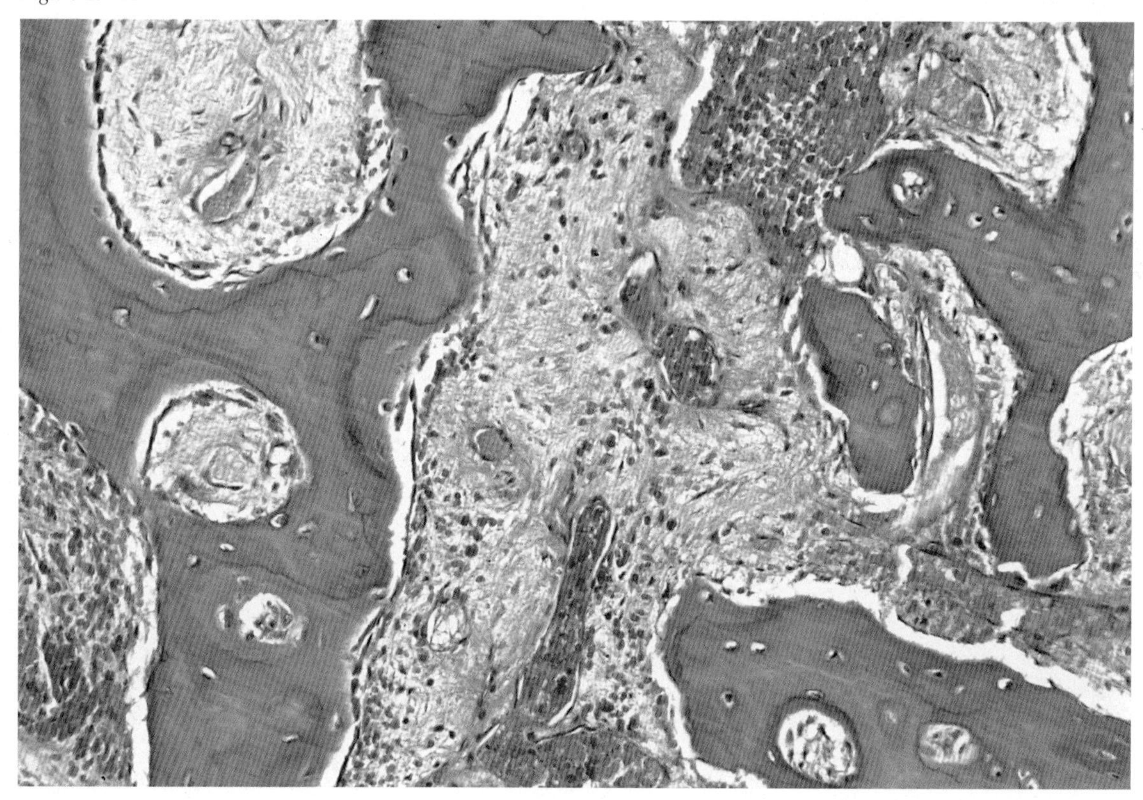

Figure 19–13 B

FIGURE 19–13. OSTEOMA: A: The shell (*far left*) of an osteoma is composed of densely sclerotic bone, and the nidus (*center*) is composed of irregular bony trabeculae. **B:** In a fibrovascular stroma, bony trabeculae of the nidus are poorly mineralized and rimmed by osteoblasts.

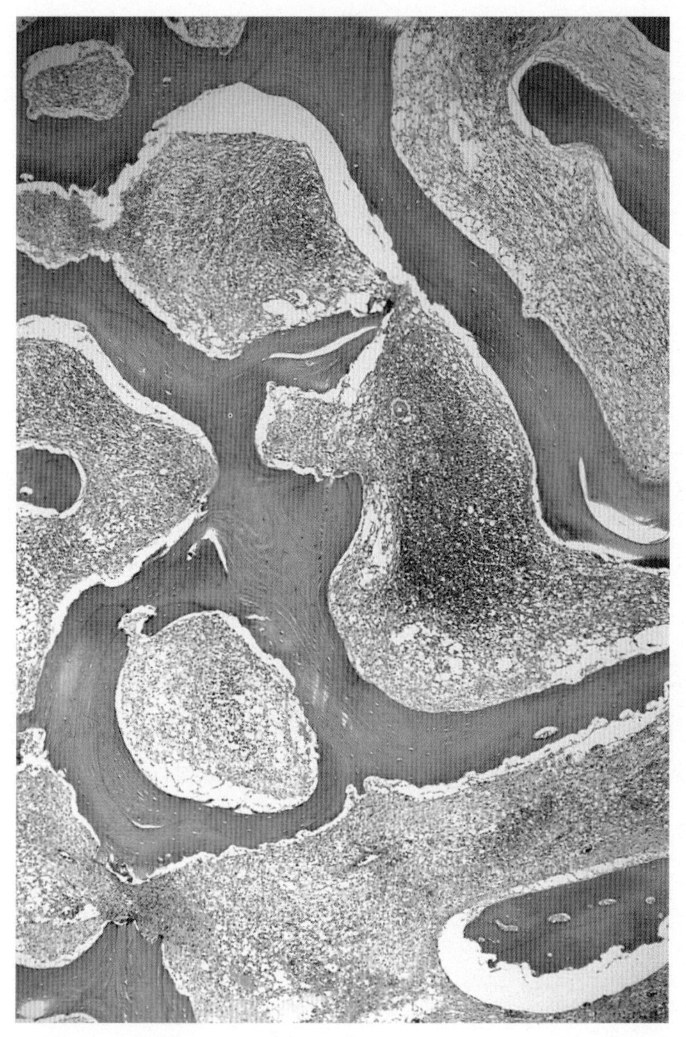

Figure 19–23 A

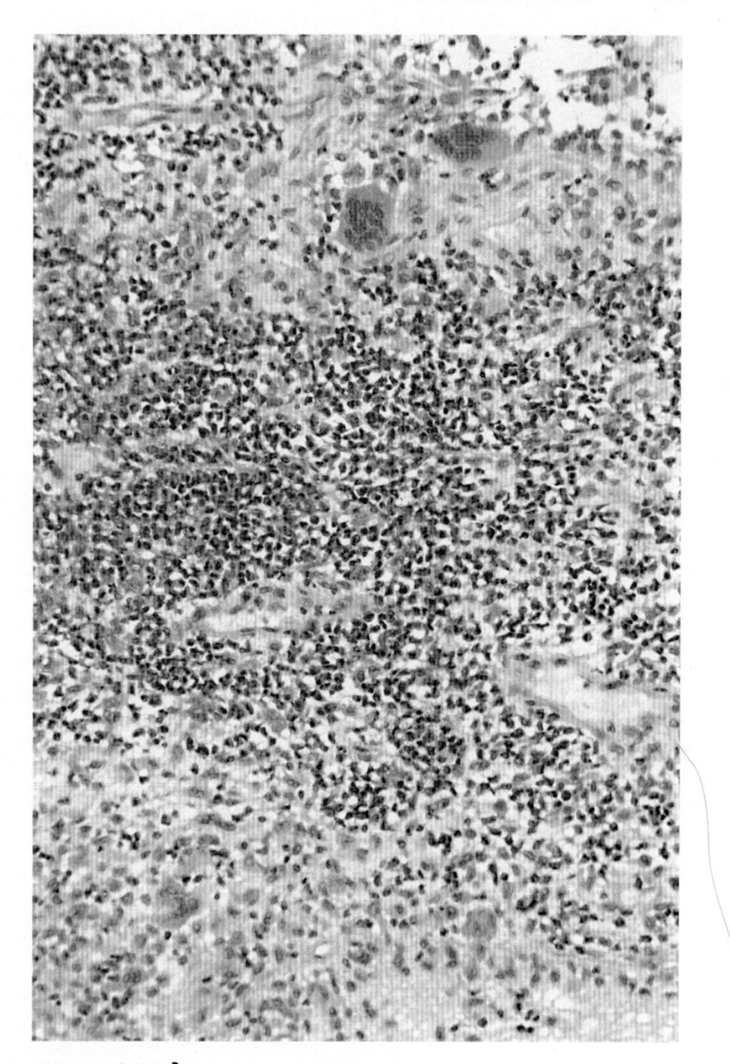

Figure 19–23 B

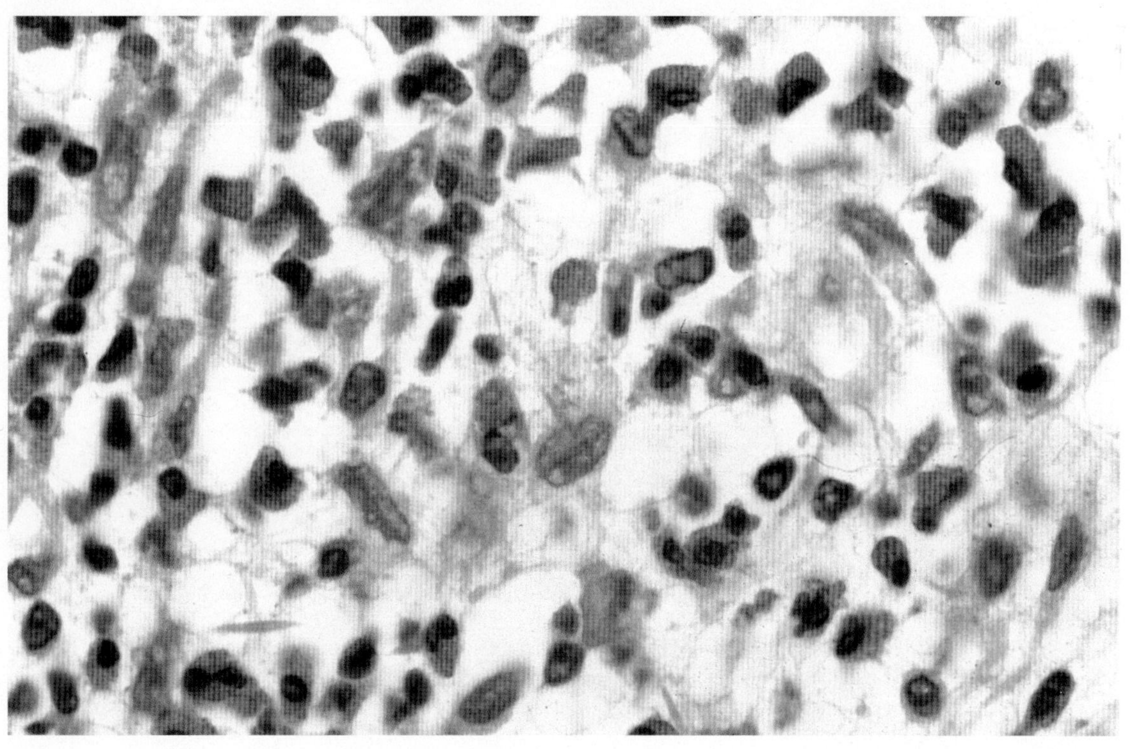

Figure 19–23 C

■ JOINTS

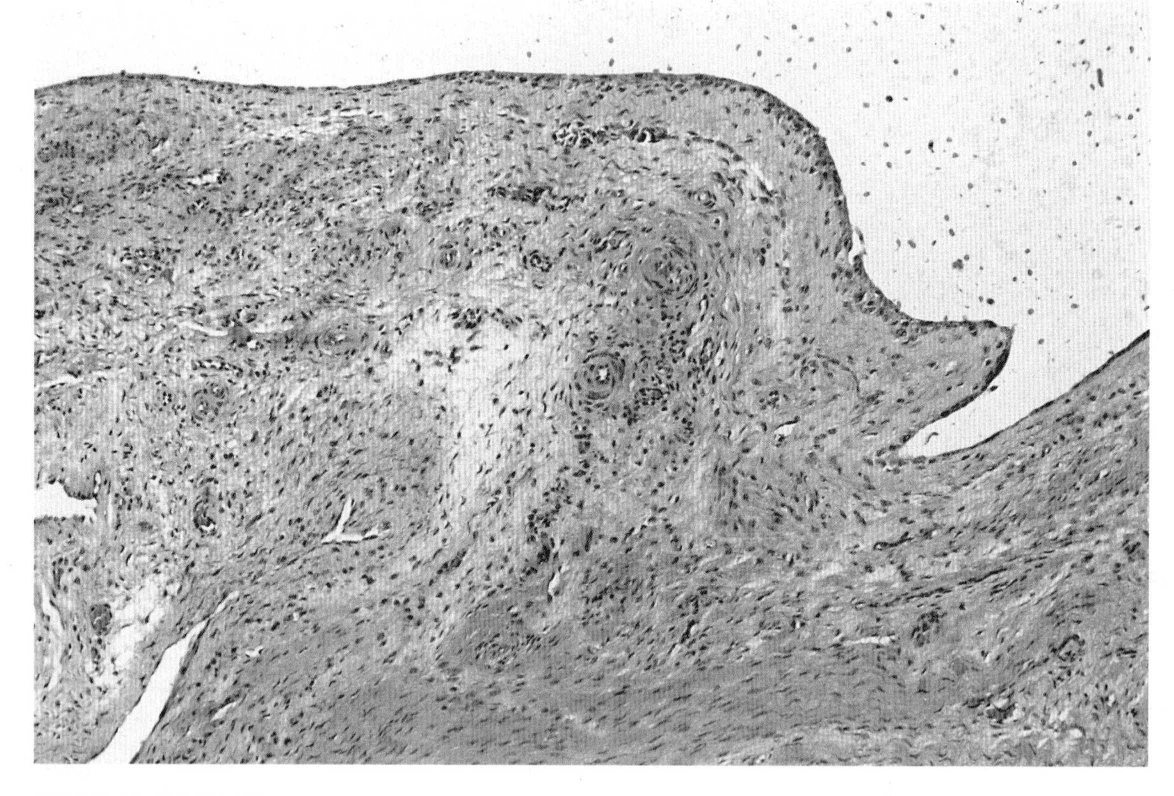

FIGURE 19–24. GANGLION CYST: A ganglion cyst is a fibrous-walled cyst filled with a clear fluid and lined by an attenuated epithelium.

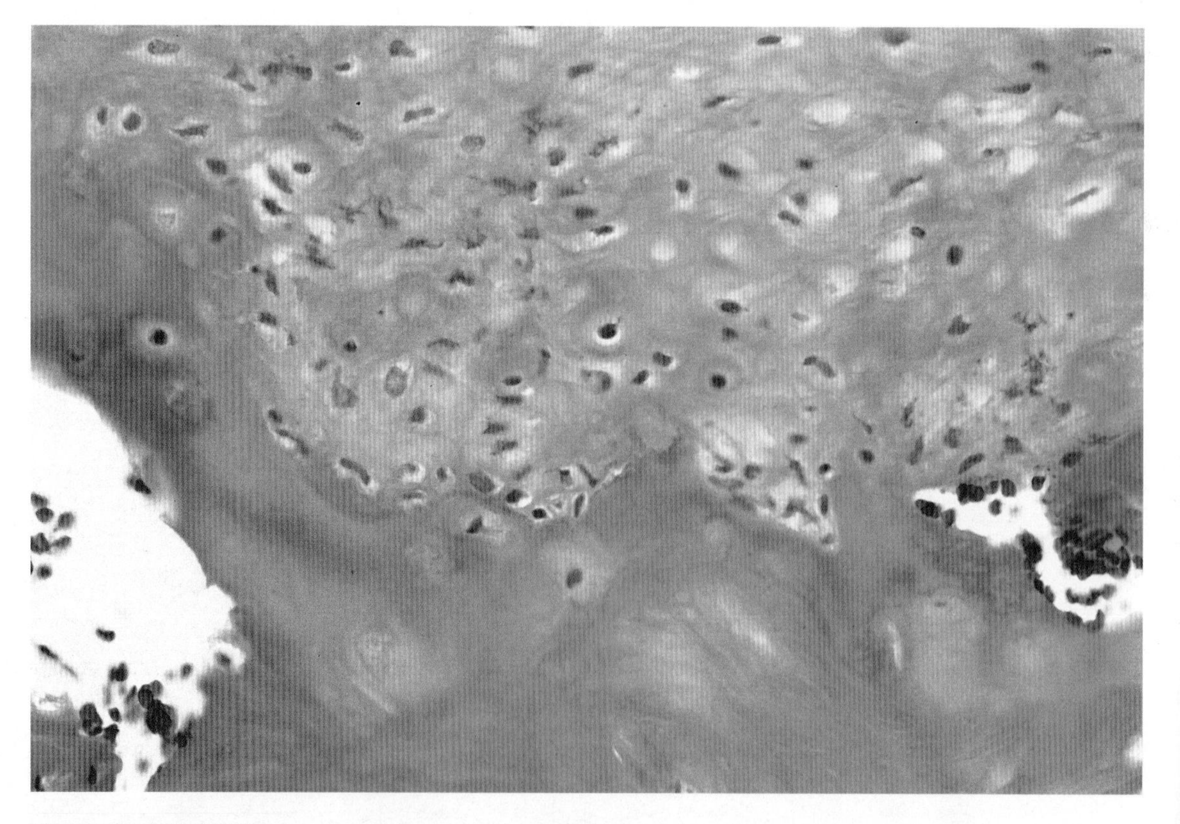

FIGURE 19–25. OSTEOARTHRITIS: This is a photomicrograph of osteoarthritis. The lower half shows degenerated cartilage, and the upper half shows regenerated cartilage. The interface between the degenerated cartilage, which contains necrotic chondrocytes, and the cellular regenerated cartilage is irregular. In the regenerating phase, the surface is covered by fibrocartilage.

■ JOINTS

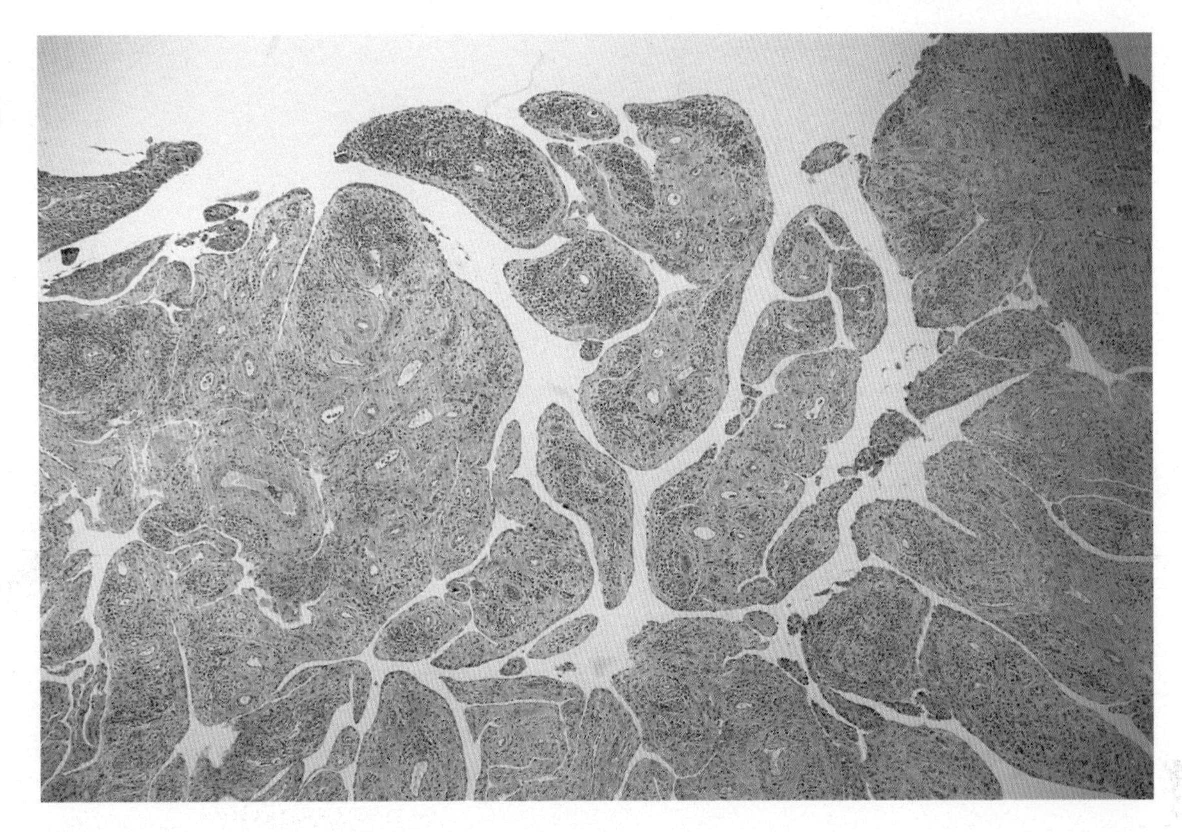

Figure 19–26 **A**

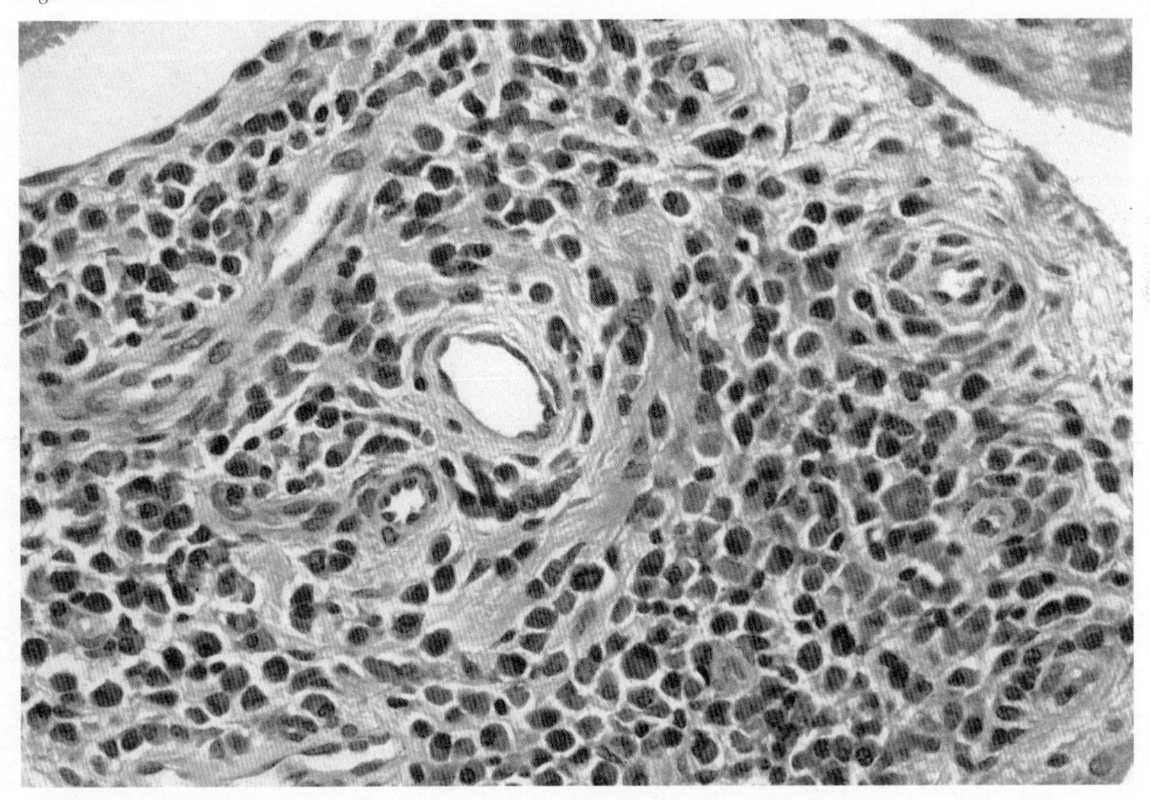

Figure 19–26 **B**

FIGURE 19–26. RHEUMATOID ARTHRITIS: A: This is a photomicrograph of the synovial changes seen in rheumatoid arthritis. Note the villous nature of the synovium, which contains inflammatory cells. **B:** Proliferation of blood vessels with an associated perivascular lymphoplasmacytic infiltrate is seen in the villous synovium. Changes in the articular surface are similar to those seen in osteoarthritis (Figure 19–25).

■ JOINTS

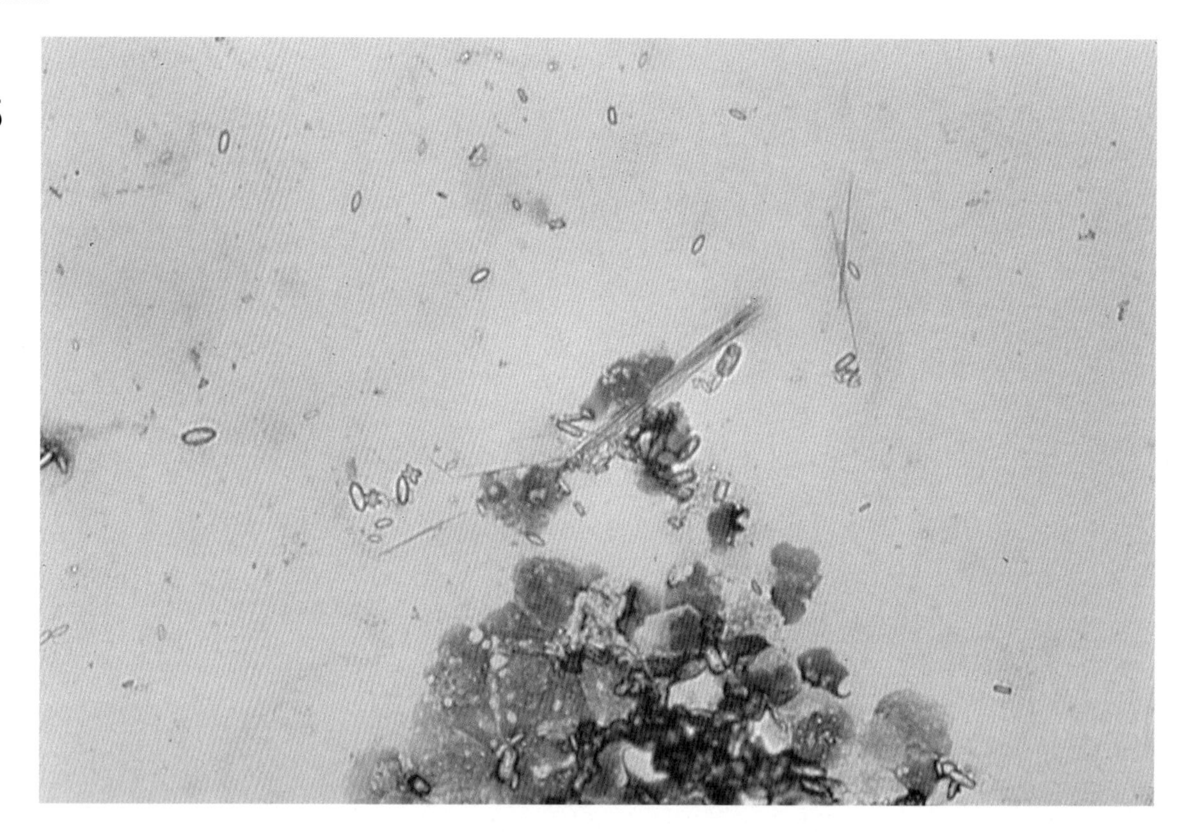

Figure 19–27 A

Figure 19–27 B

FIGURE 19–27. GOUT: A: In gout, Wright-Giemsa–stained aspirated joint fluid reveals faint needlelike monosodium urate crystals singly and in sheaths associated with an inflammatory exudate. **B:** When exposed to polarizing light, the same slide demonstrates negatively birefringent crystals.

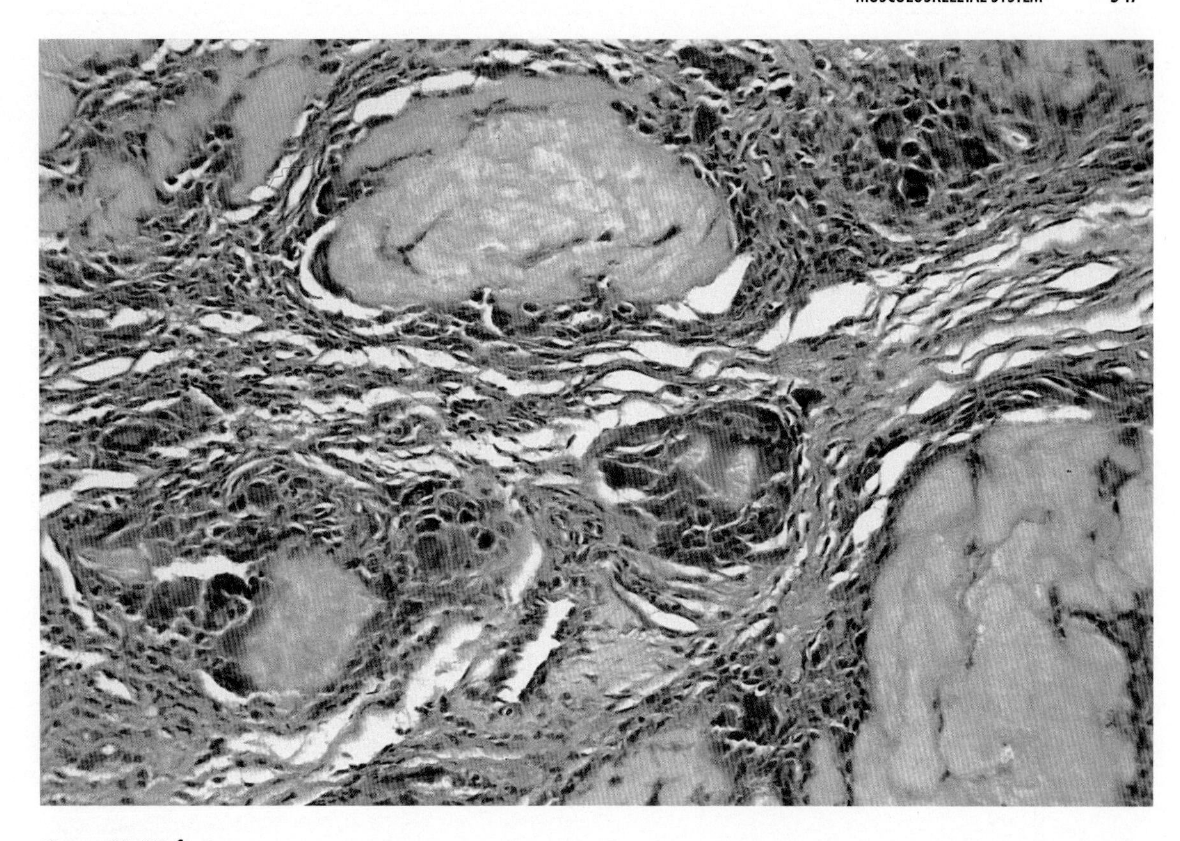

Figure 19–27 C

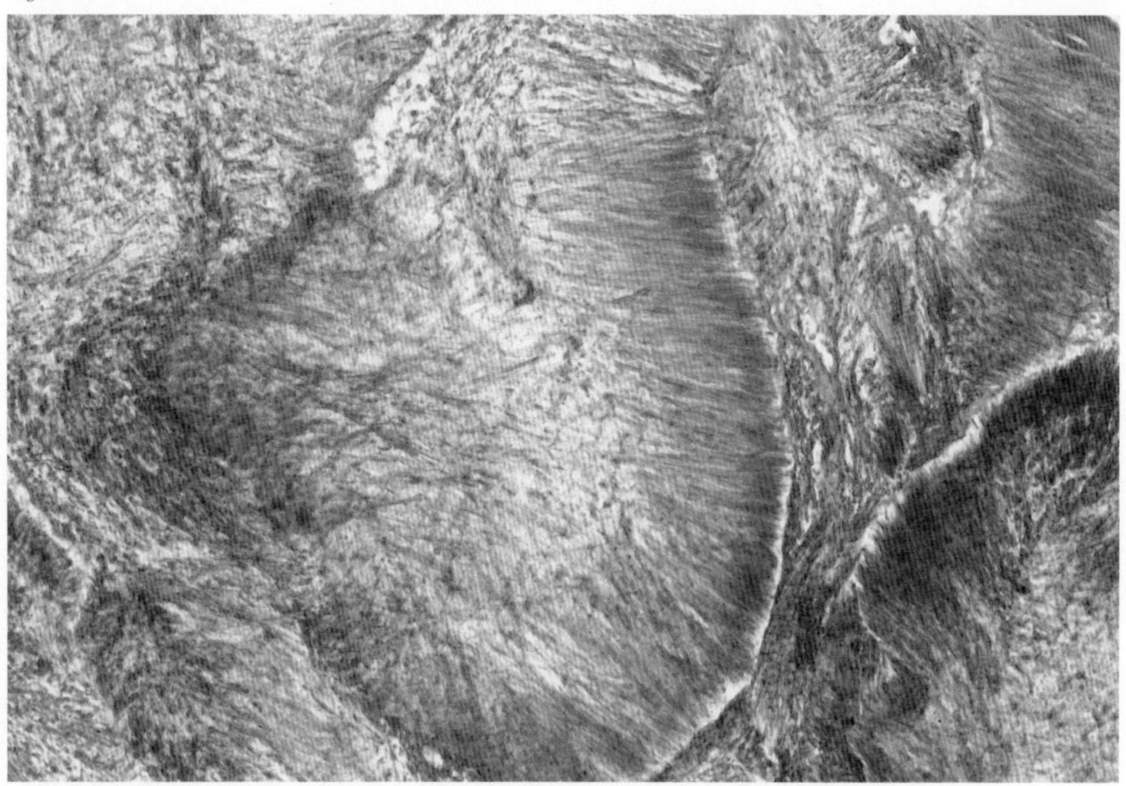

Figure 19–27 D

FIGURE 19–27. TOPHACEOUS GOUT: C: In tophaceous gout, collections of monosodium urate crystals form tophi. Tophi are amorphous and eosinophilic and surrounded by giant cells and fibrous tissue. **D:** Silver staining shows the crystalline nature of the tophi.

■ CENTRAL NERVOUS SYSTEM

BACTERIAL INFECTIONS

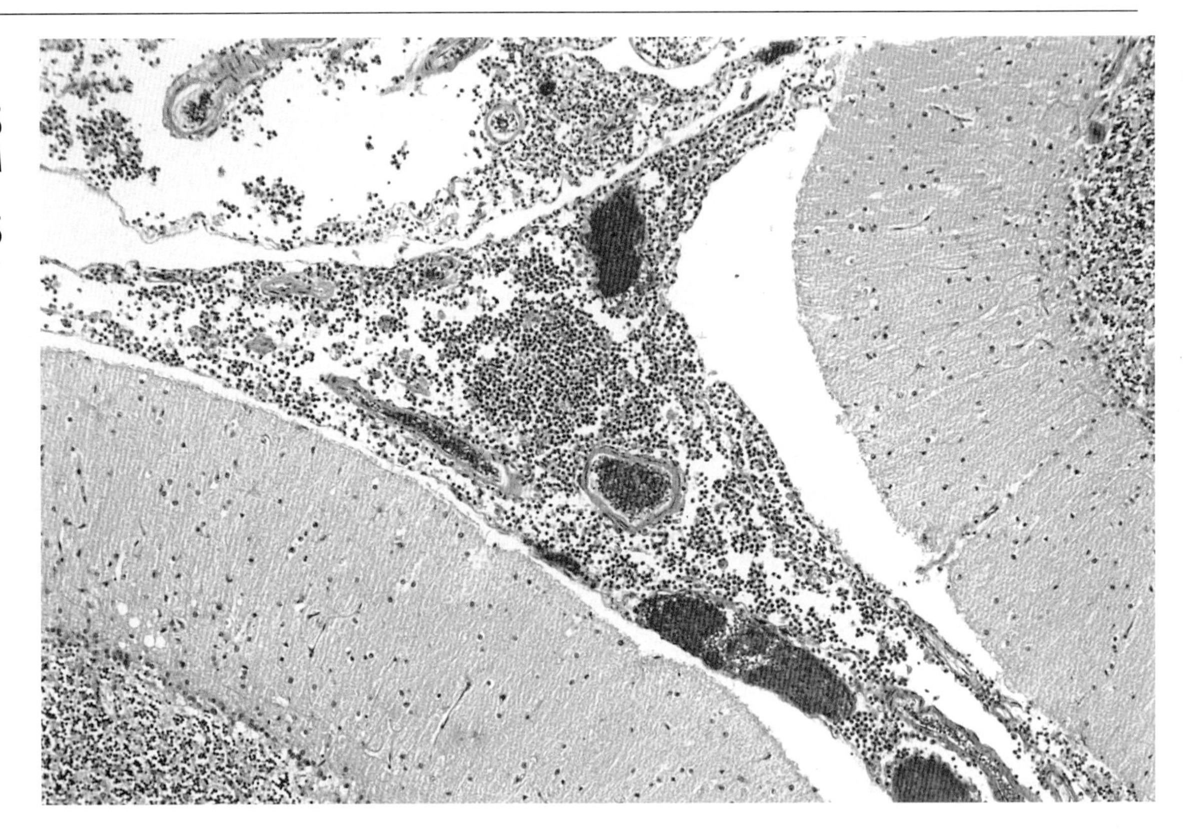

Figure 20–1 A

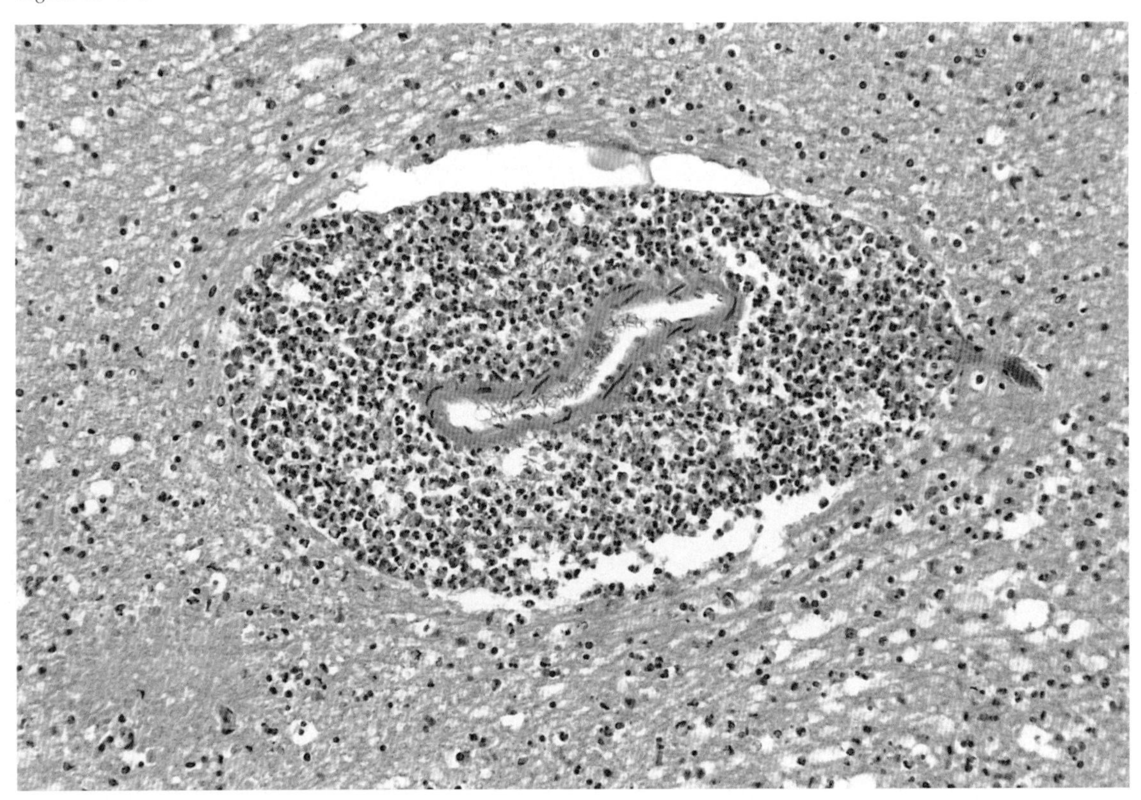

Figure 20–1 B

FIGURE 20–1. ACUTE (PURULENT) MENINGITIS: A: In acute meningitis, a purulent exudate mixed with fibrin is present in the subarachnoid space and leptomeninges. Vascular congestion is prominent, and the brain parenchyma is uninvolved. **B:** In an advanced stage of acute purulent meningitis, pus tracks down into the Virchow-Robin spaces that surround the blood vessels.

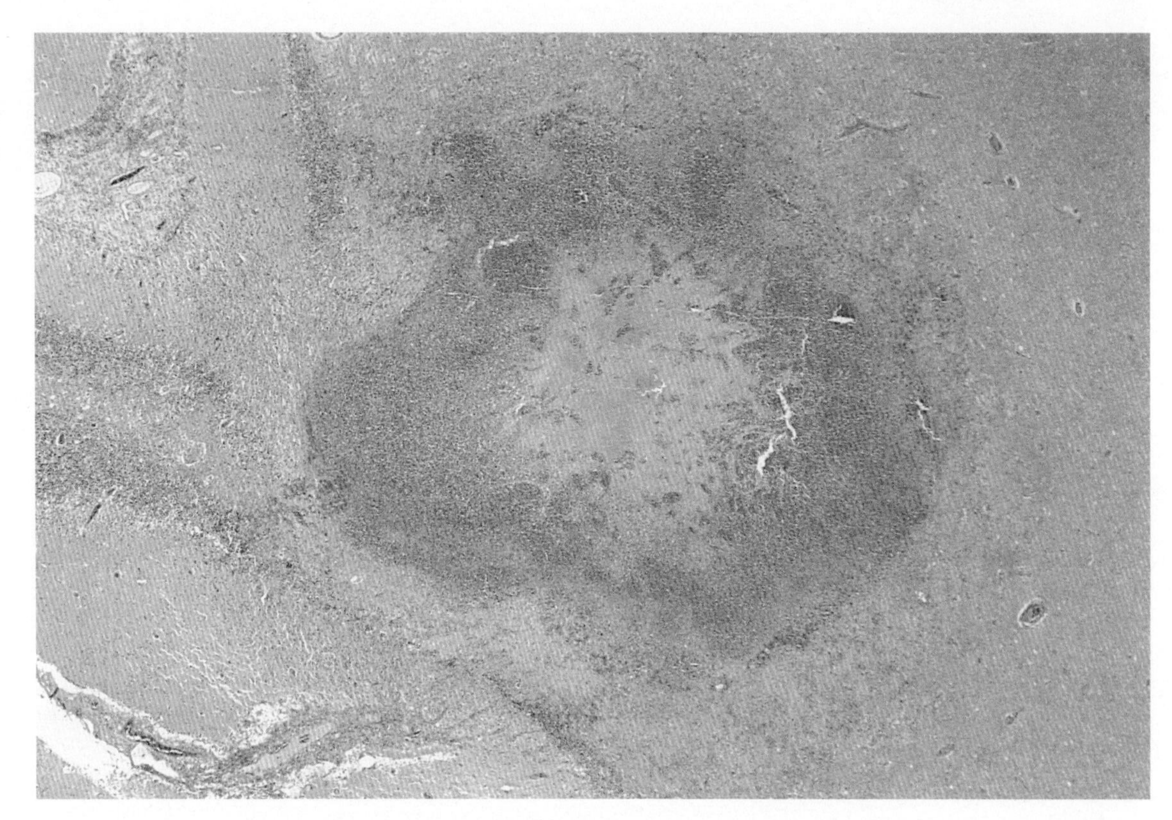

Figure 20–2 A

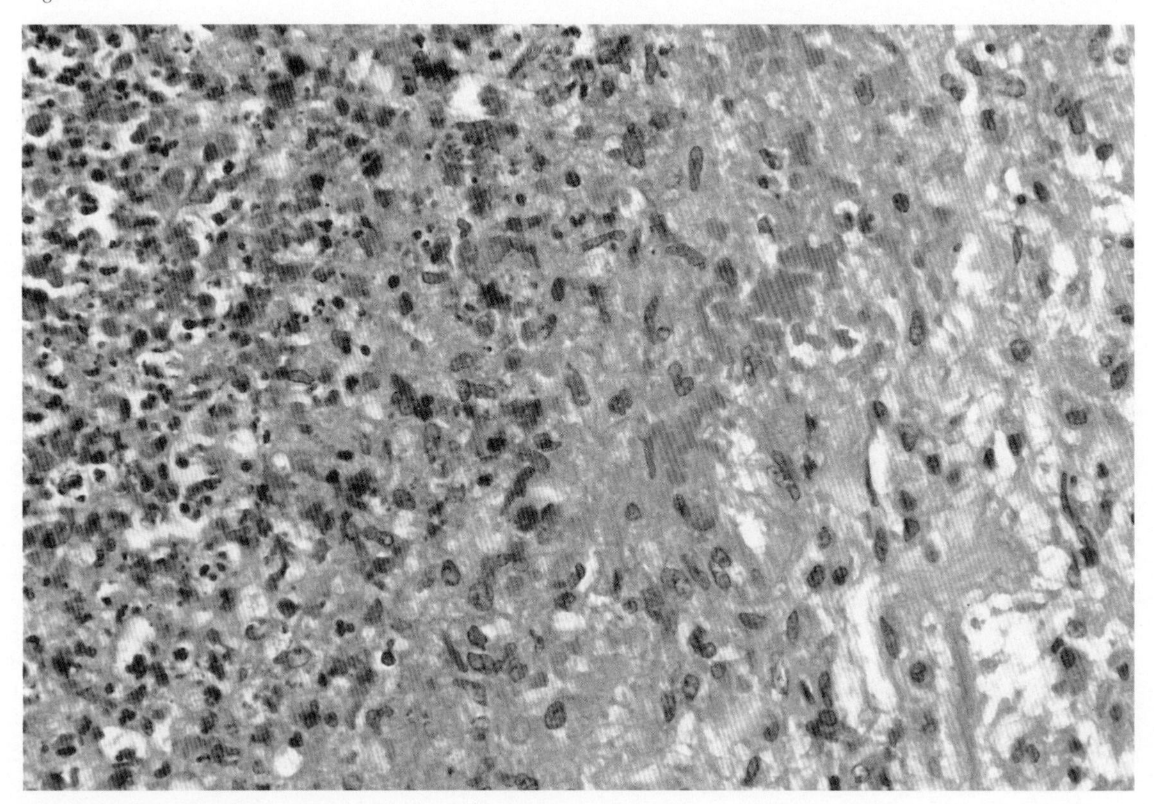

Figure 20–2 B

FIGURE 20–2. CEREBRAL ABSCESS: A: In cerebral abscess, a necrotic focus of brain parenchyma is surrounded by a collection of inflammatory cells. The wall of the abscess is composed of fibrous tissue and granulation tissue. The surrounding brain parenchyma is edematous. **B:** At high-power magnification, it is shown that the inflammatory exudate is composed predominantly of neutrophils, with varying numbers of lymphocytes and macrophages.

■ CENTRAL NERVOUS SYSTEM

VIRAL INFECTIONS

Viral encephalitis is characterized by three main histologic features: a perivascular inflammatory infiltrate, neuronophagia, and microglial nodules.

FIGURE 20–3. VIRAL ENCEPHALITIS: A: A prominent feature of viral encephalitis is **perivascular cuffing** by lymphocytes. B: In **neuronophagia**, neurons are destroyed and engulfed by macrophages. C: **Microglial nodules** are composed of clusters of rod-shaped microglial cells.

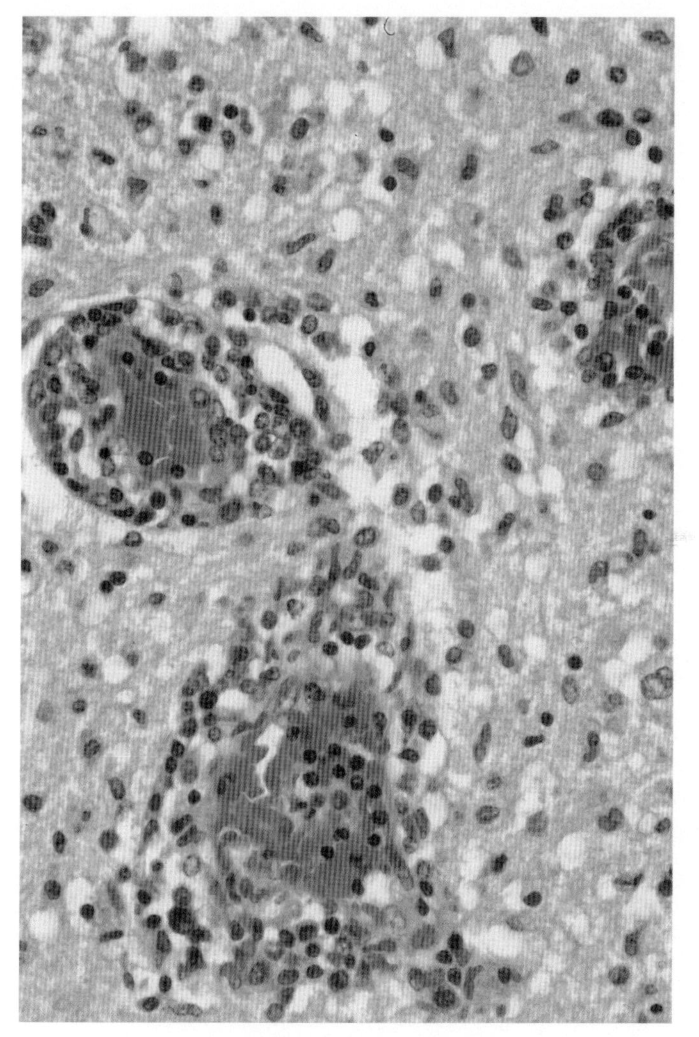

Figure 20–3 A

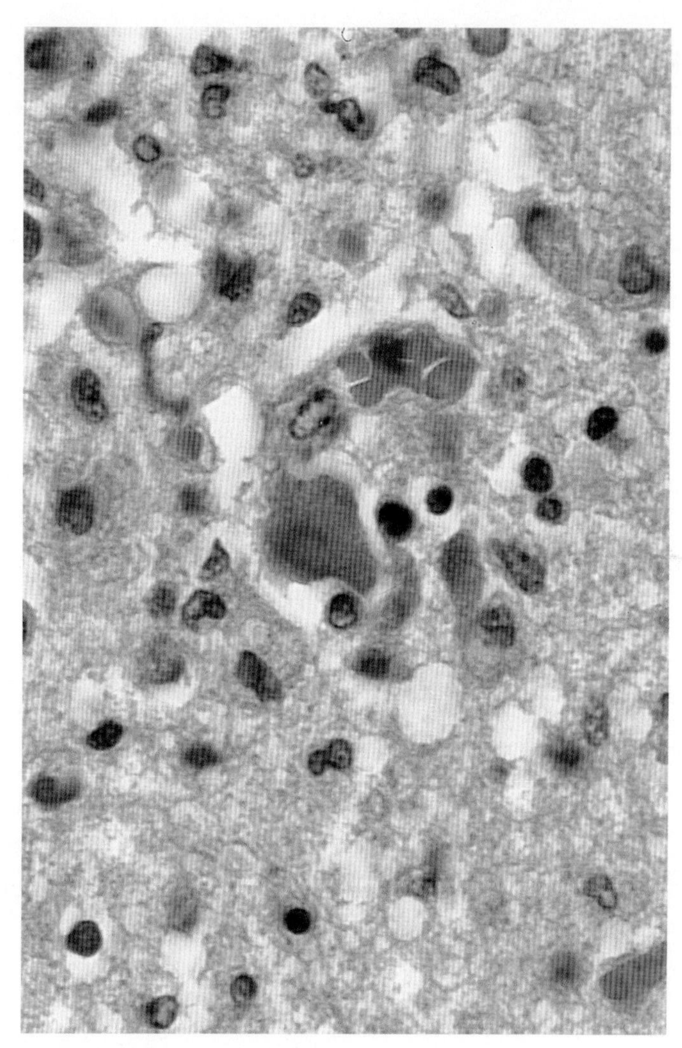

Figure 20–3 B

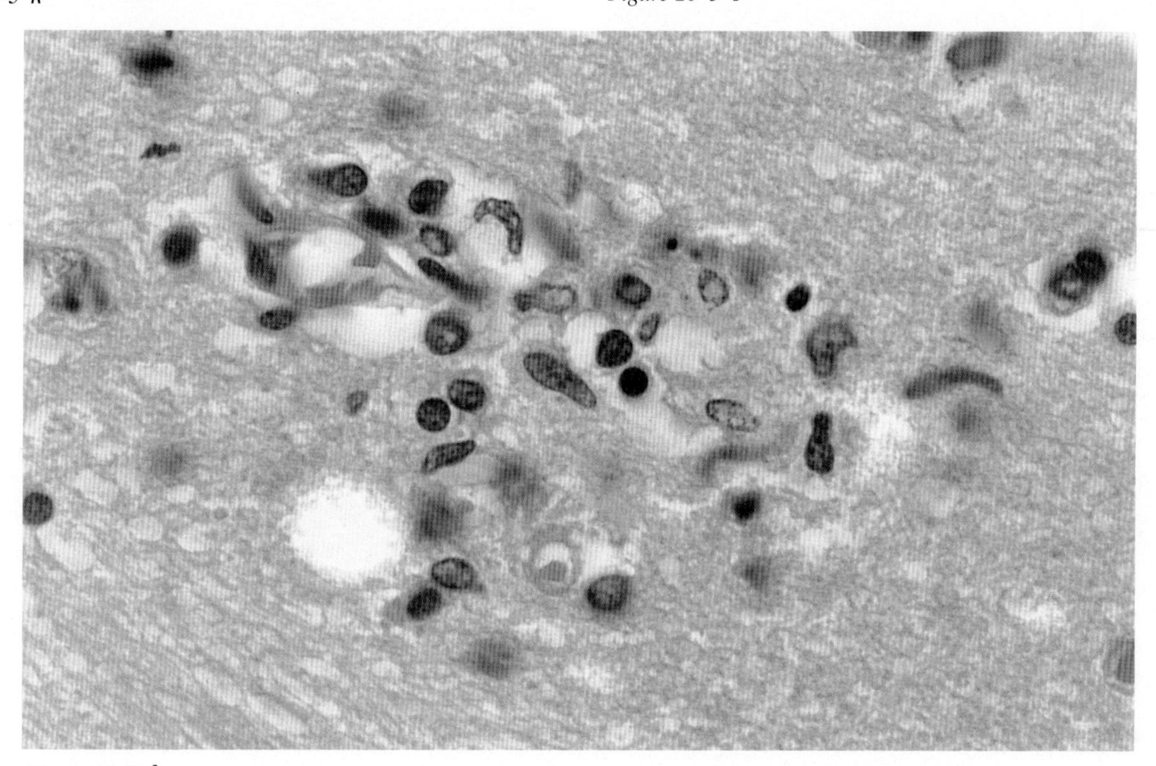

Figure 20–3 C

■ CENTRAL NERVOUS SYSTEM

VIRAL INFECTIONS

FIGURE 20–4. RABIES VIRUS ENCEPHALITIS: Infection with rabies virus has a characteristic intraneuronal eosinophilic intracytoplasmic inclusion known as a **Negri body**.

FIGURE 20–5. CREUTZFELDT-JAKOB DISEASE (SPONGIFORM DEGENERATION): This disease is caused by a slow virus. Multiple variably sized vacuoles are present in the neuropil of the gray matter, giving it the characteristic **spongiform change**. Note the lack of inflammation.

FIGURE 20–6. POLIOMYELITIS: The polio virus is neurotropic for anterior horn neurons, causing their destruction, and leads to collapse of the anterior horns.

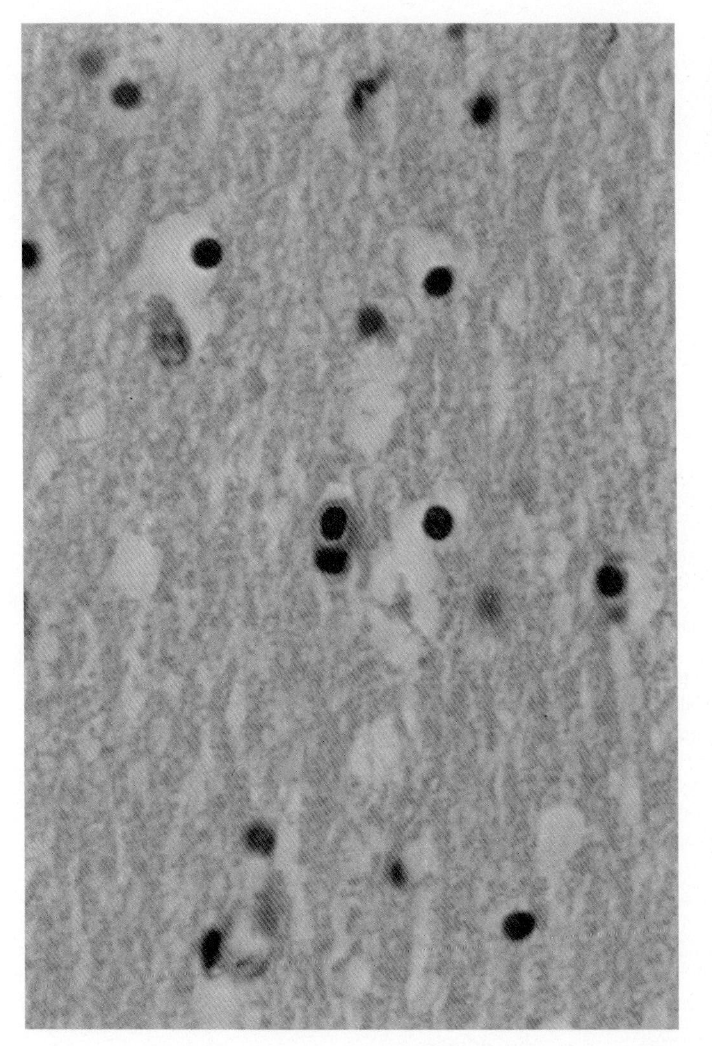

Figure 20–4

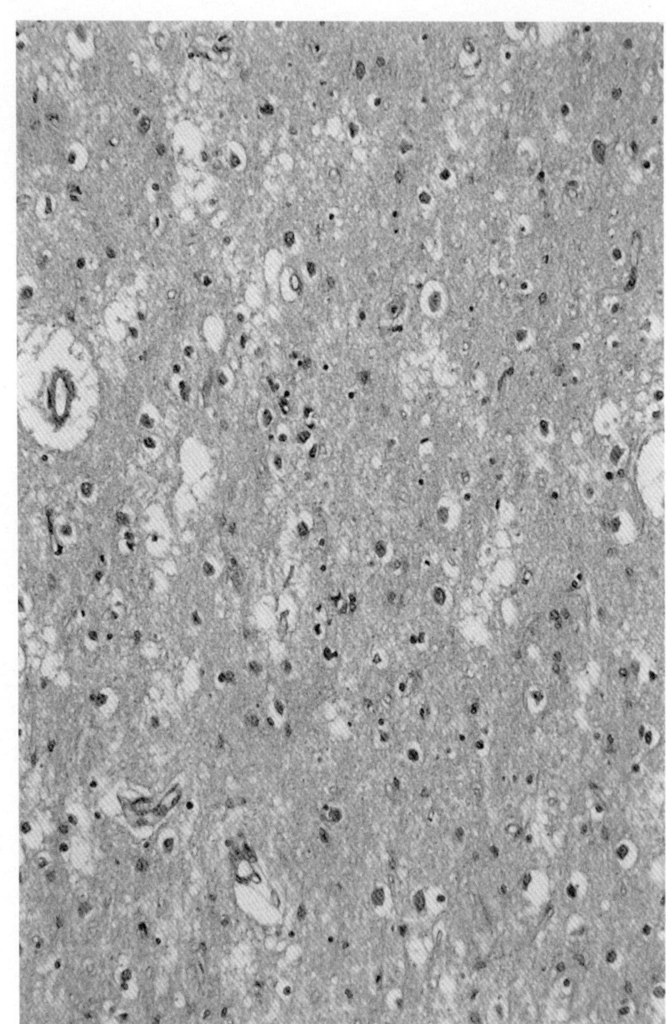

Figure 20–5

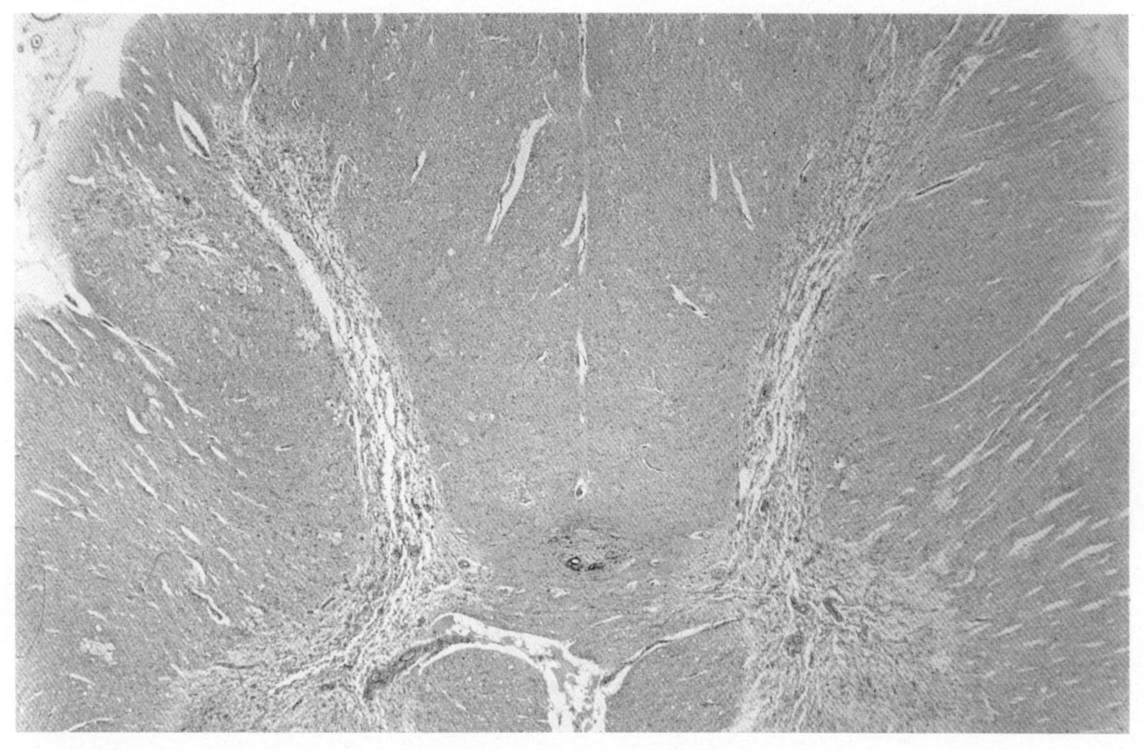

Figure 20–6

■ CENTRAL NERVOUS SYSTEM

FUNGAL INFECTIONS

FIGURE 20–7. FUNGAL ENCEPHALITIS: A: In this photomicrograph of fungal encephalitis, the brain parenchyma is necrotic and diffusely infiltrated by inflammatory cells; faint fungal elements are also noted. **B:** Staining with periodic acid–Schiff stain shows that the fungal elements in **A** are composed of pseudohyphae and budding yeast consistent with *Candida* species.

PARASITIC INFECTIONS

FIGURE 20–8. CYSTICERCOSIS: Cysticercosis is caused by the dog tapeworm *Taenia solium.* Multiple cysts containing the tapeworm are found within the brain parenchyma; when unruptured, as in this example, there is a minimal inflammatory response.

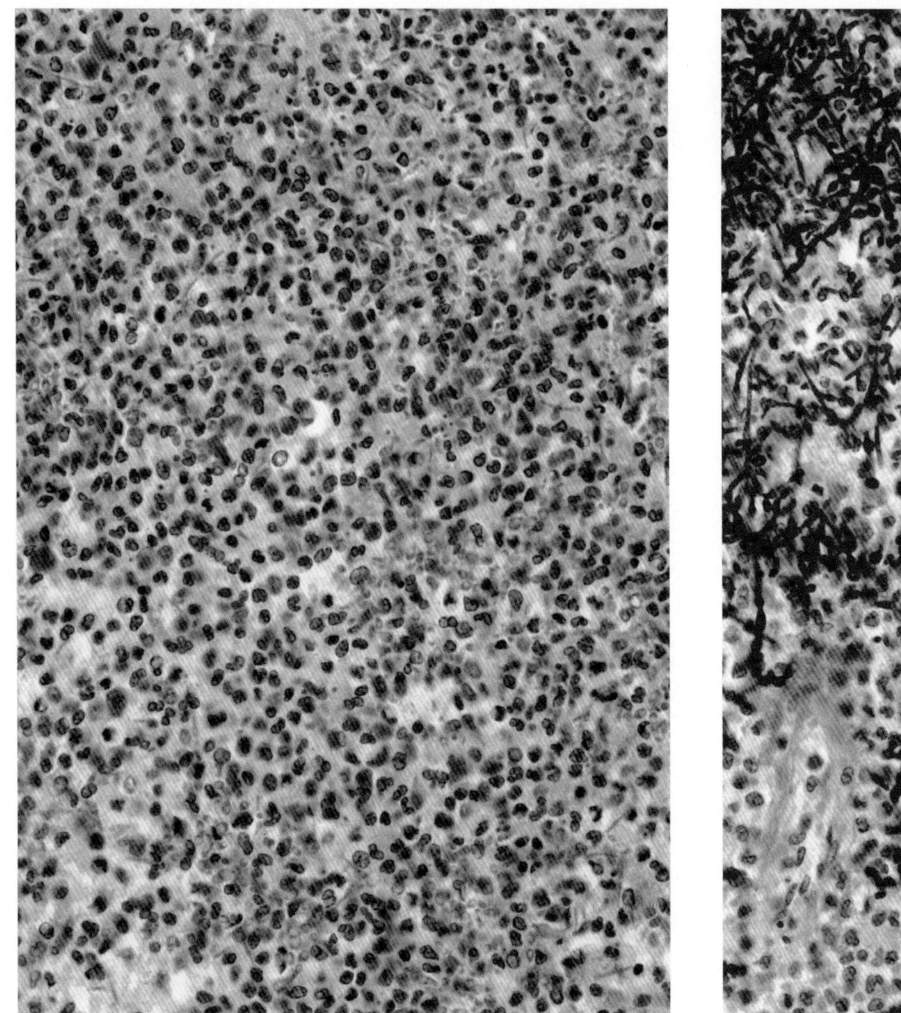

Figure 20–7 A

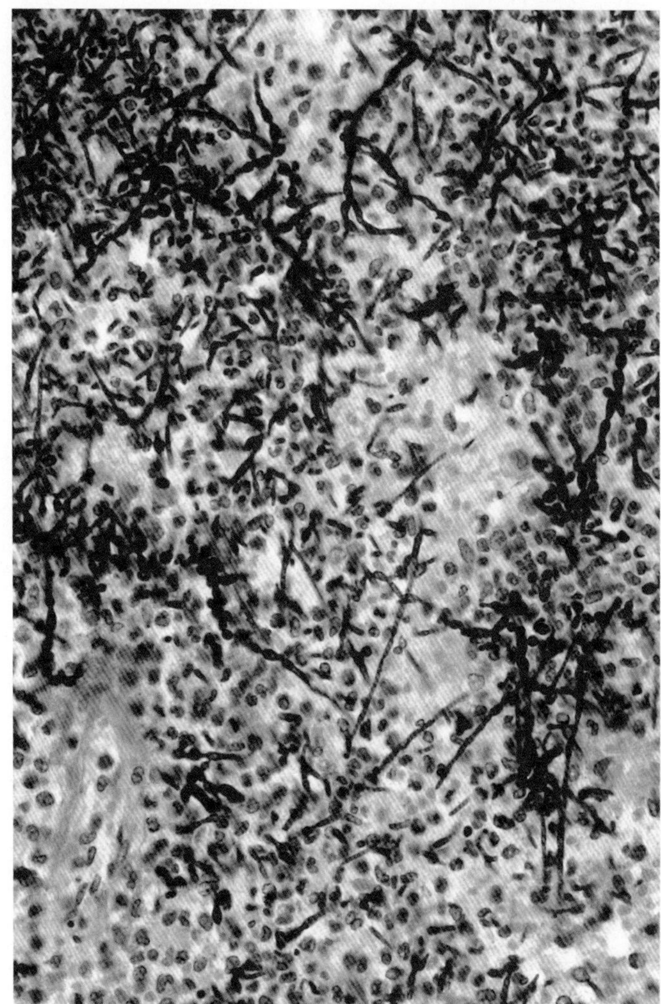

Figure 20–7 B

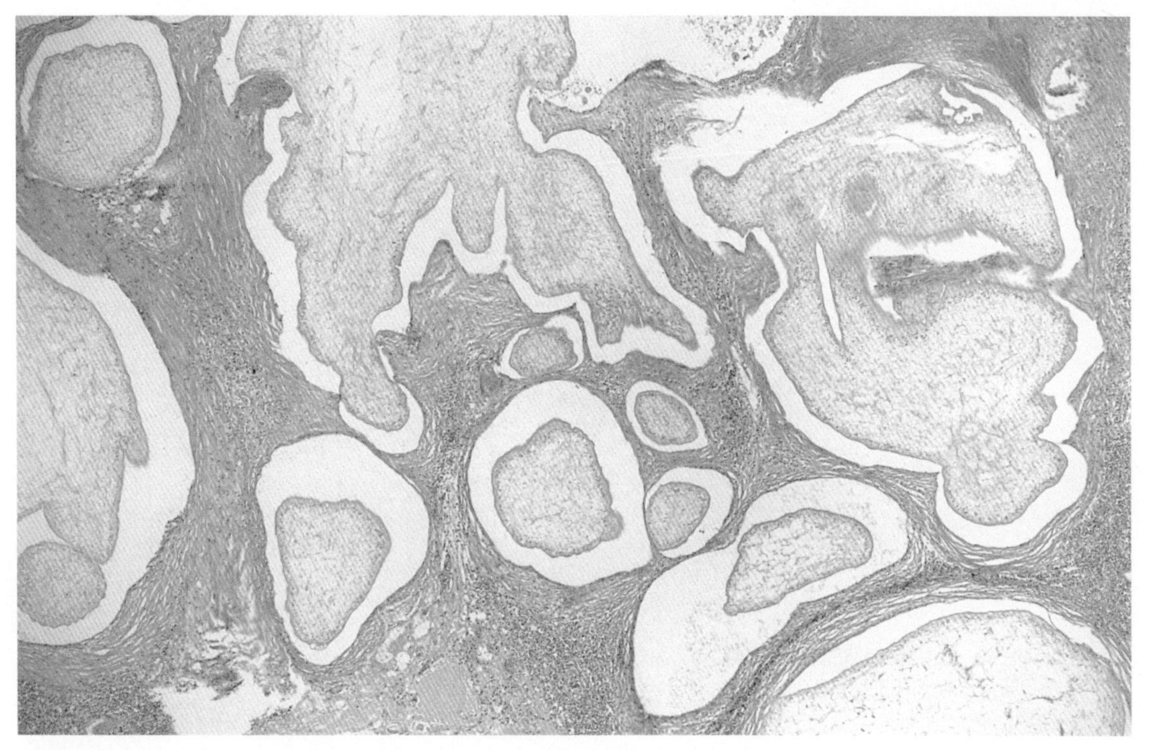

Figure 20–8

■ CENTRAL NERVOUS SYSTEM

VASCULAR DISEASES

FIGURE 20–9. ISCHEMIC (HYPOXIC) ENCEPHALOPATHY: A: In cerebral ischemia, neurons are angulated, shrunken, and eosinophilic. Nuclei are pyknotic, and the surrounding neuropil is edematous. Compare these changes with those of their normal counterparts (**B**).

FIGURE 20–10. CEREBRAL INFARCTION: A: At 24 hours, neurons show the changes of ischemia; in addition, an acute inflammatory exudate is focally present.

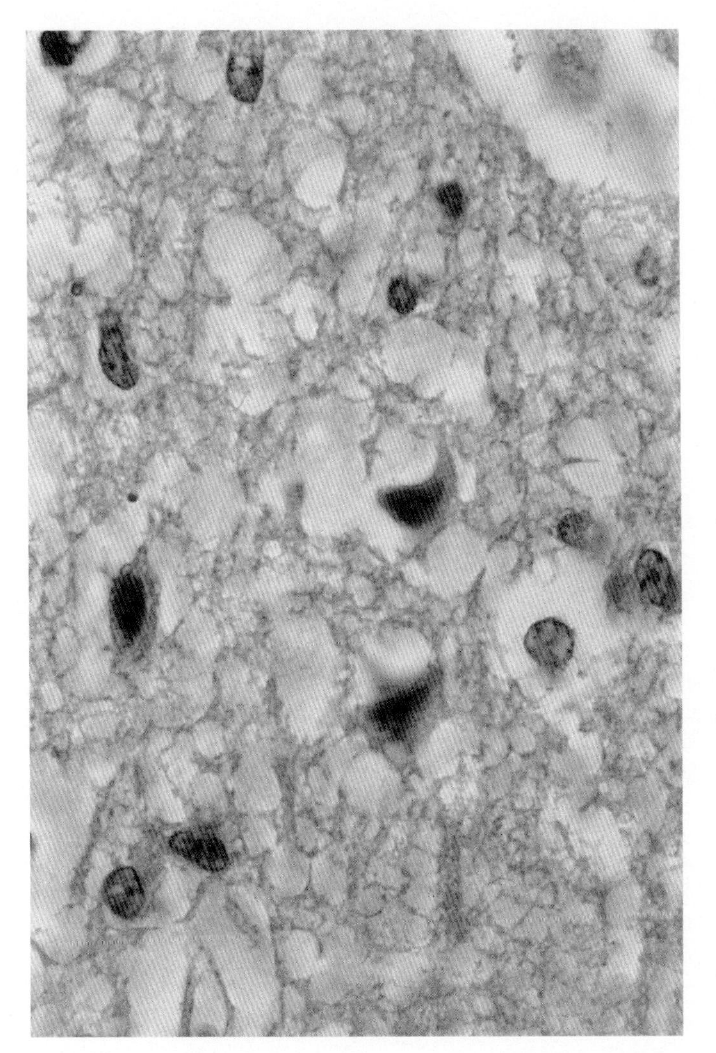

Figure 20–9 A

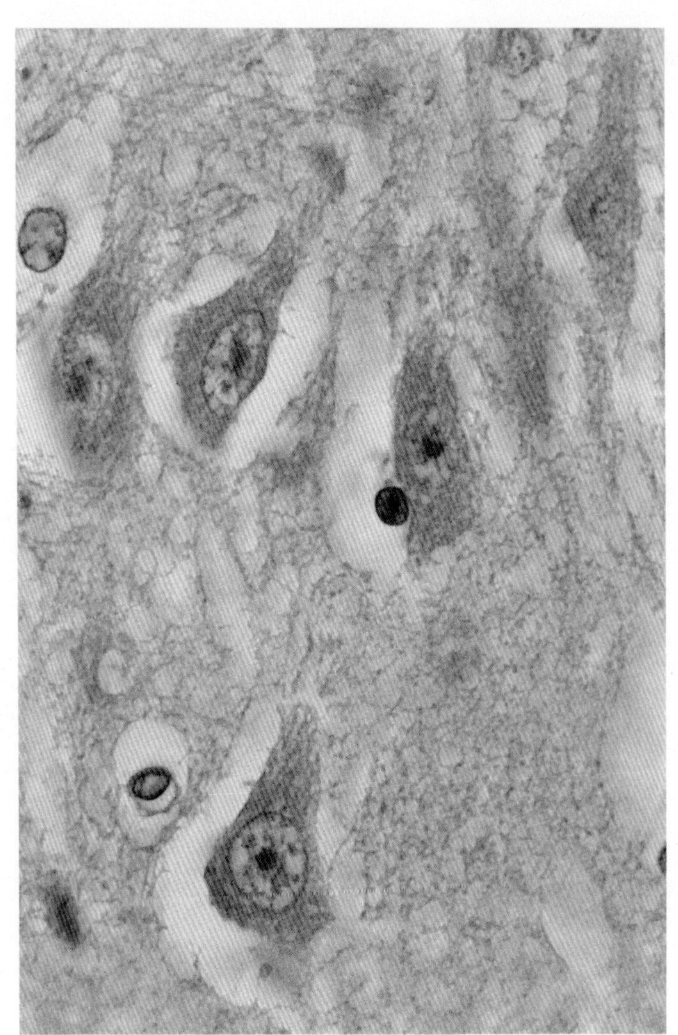

Figure 20–9 B

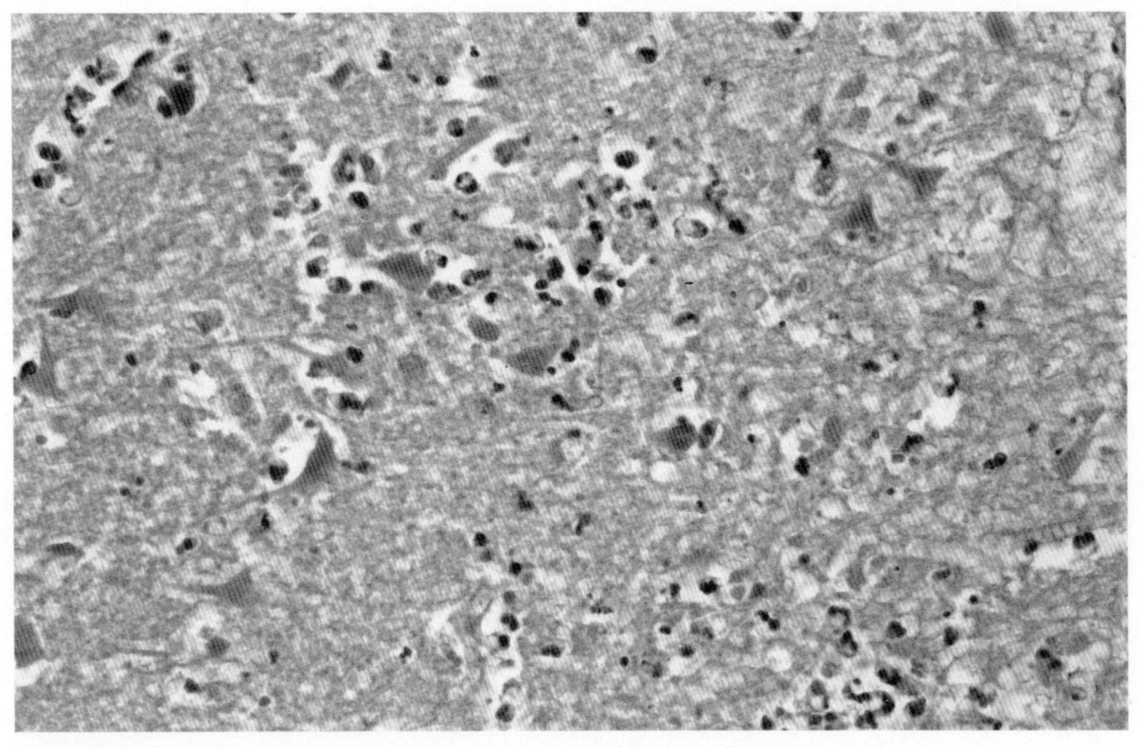

Figure 20–10 A

■ CENTRAL NERVOUS SYSTEM

VASCULAR DISEASES

FIGURE 20–10. CEREBRAL INFARCTION: B: In cerebral infarction, beginning at 72 hours and peaking by 2 weeks, macrophages, also known as **gitter cells,** replace the parenchyma and the acute inflammatory exudate. C: By 2 weeks, the viable brain at the periphery of the infarct shows reactive **gemistocytic astrocytes,** which are enlarged astrocytes with eosinophilic cytoplasm. D: In months to years, the infarct is converted to a cystic cavity in which only blood vessels remain.

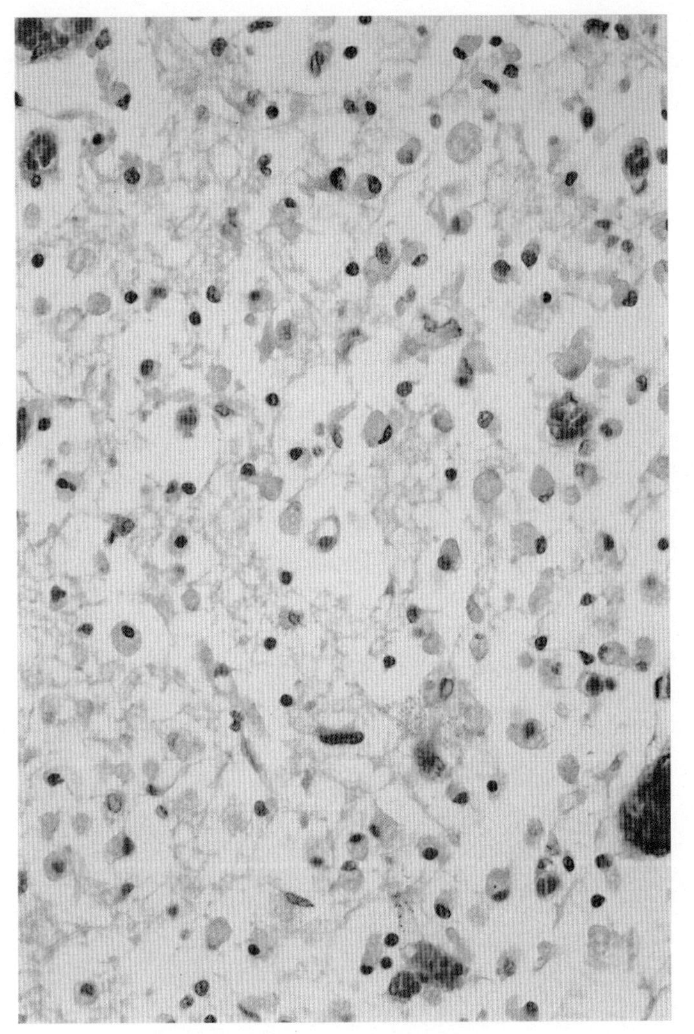

Figure 20–10 B

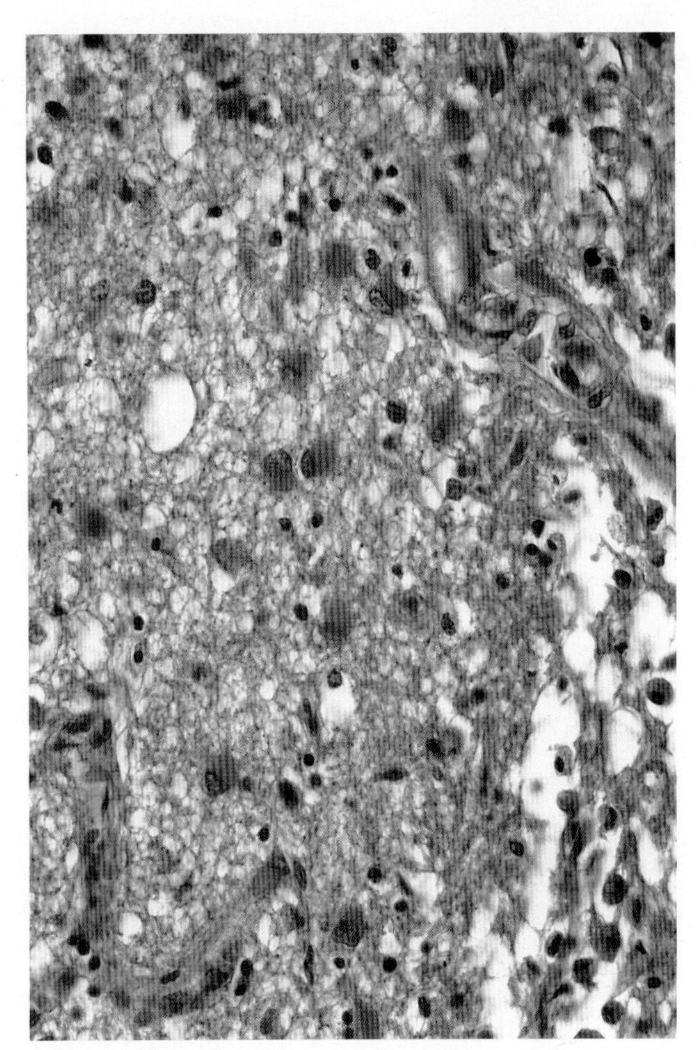

Figure 20–10 C

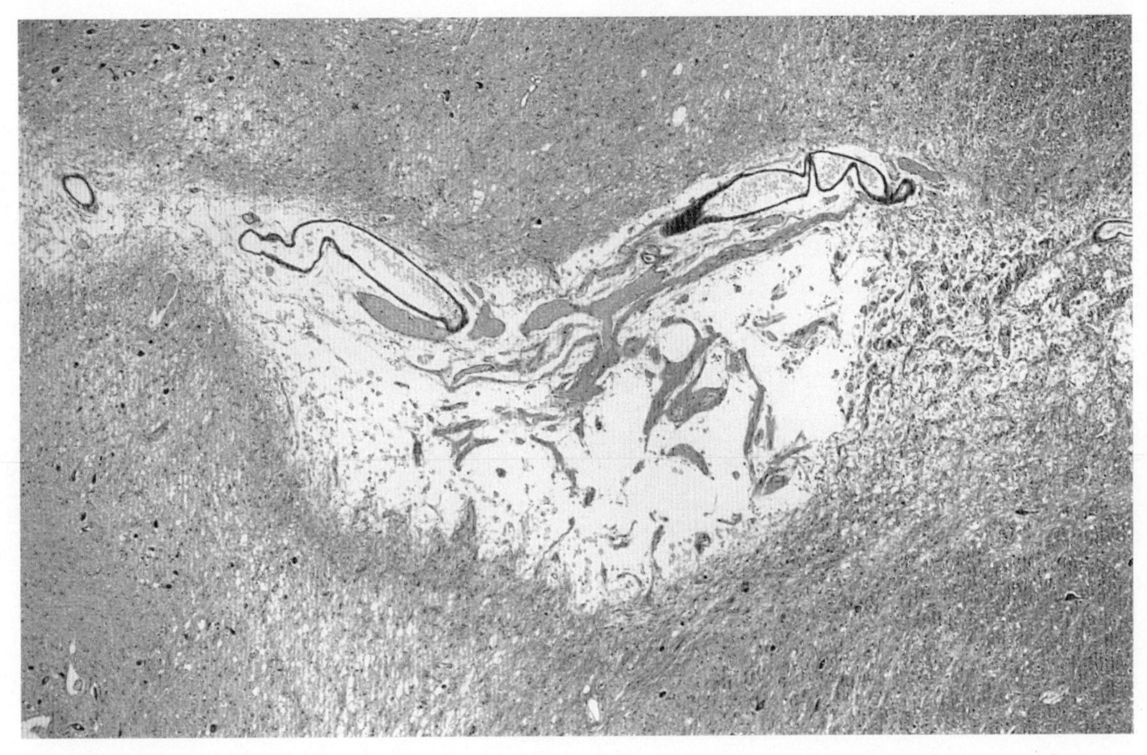

Figure 20–10 D

■ CENTRAL NERVOUS SYSTEM

VASCULAR DISEASES

FIGURE 20–11. SUBARACHNOID HEMORRHAGE: In subarachnoid hemorrhage, blood accumulates in the subarachnoid space. The most common cause is rupture of a berry aneurysm.

FIGURE 20–12. VASCULAR MALFORMATIONS: A: In this ruptured vascular malformation, tangles of abnormal vessels of variable size are present in a hemorrhagic background. **B:** Trichrome stain enhances the visualization of the vascular nature of the lesion.

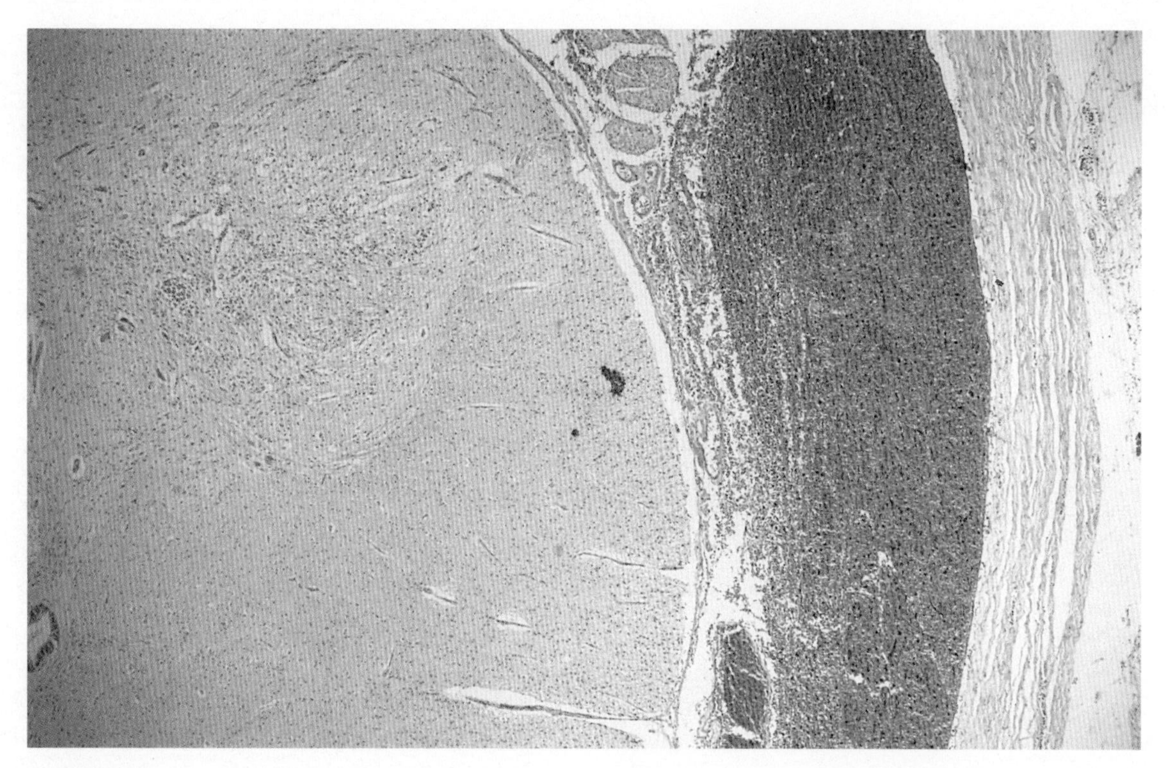

Figure 20–11

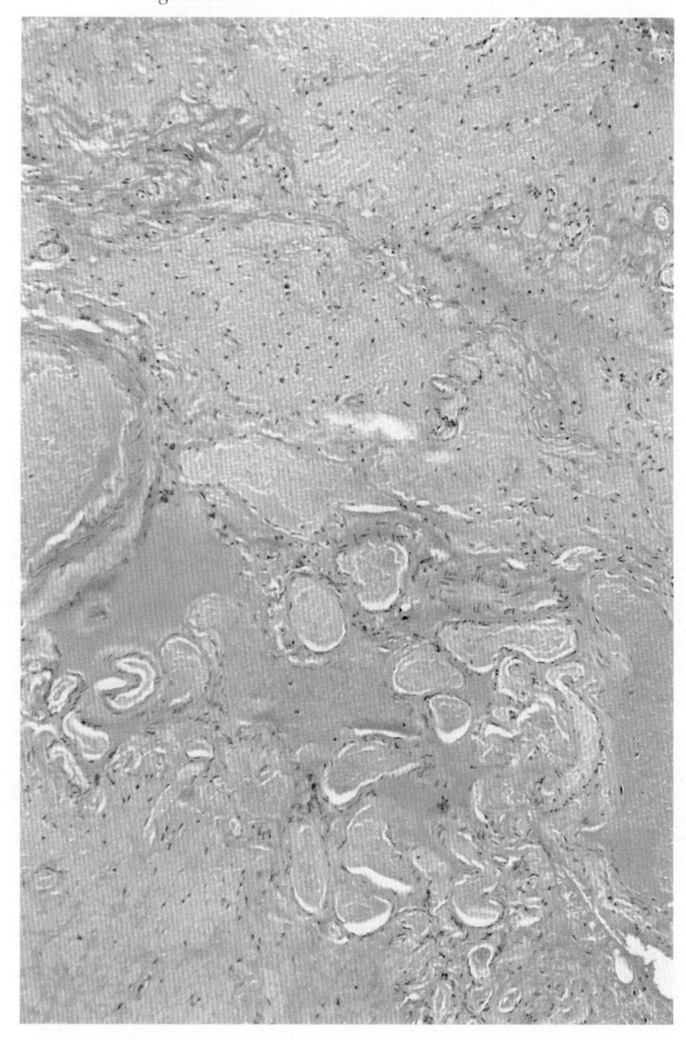

Figure 20–12 A

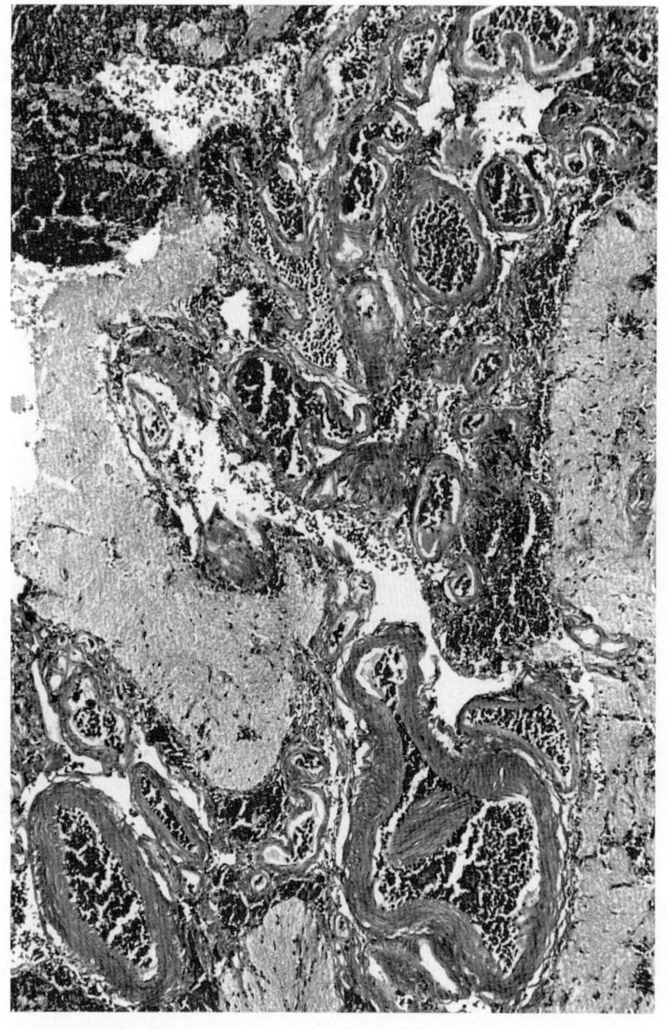

Figure 20–12 B

■ CENTRAL NERVOUS SYSTEM

DEGENERATIVE DISEASES

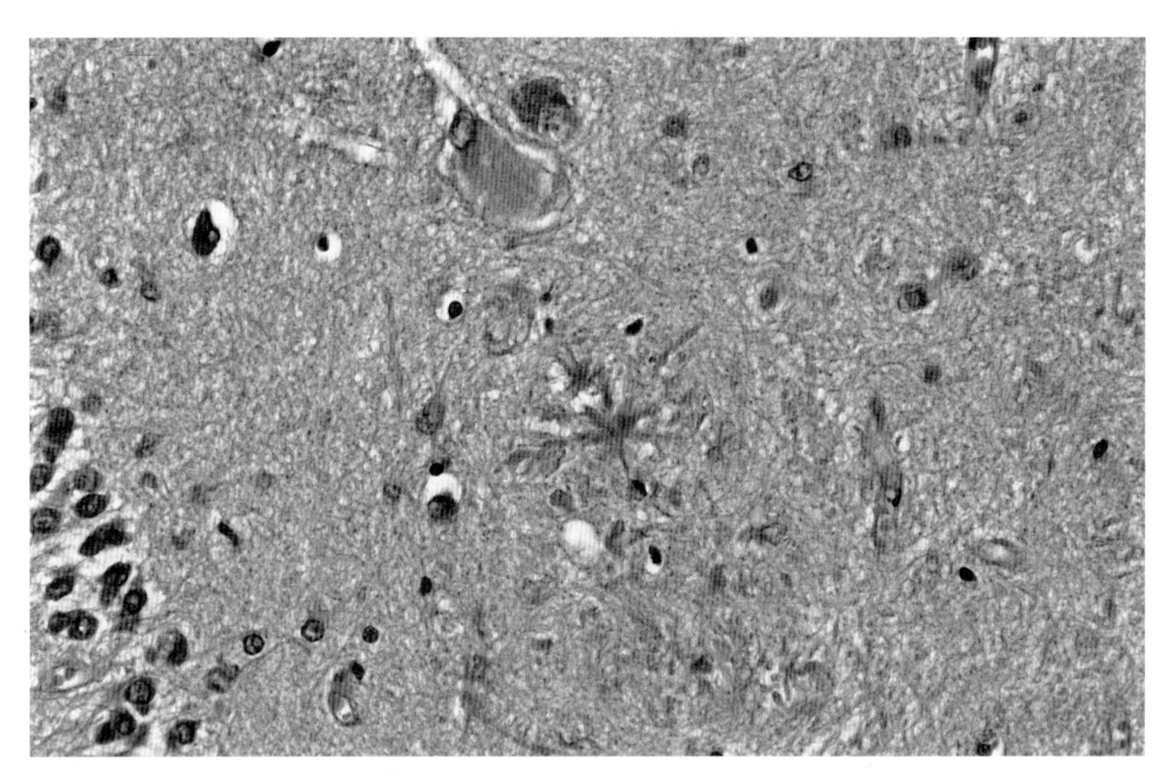

Figure 20–13 A

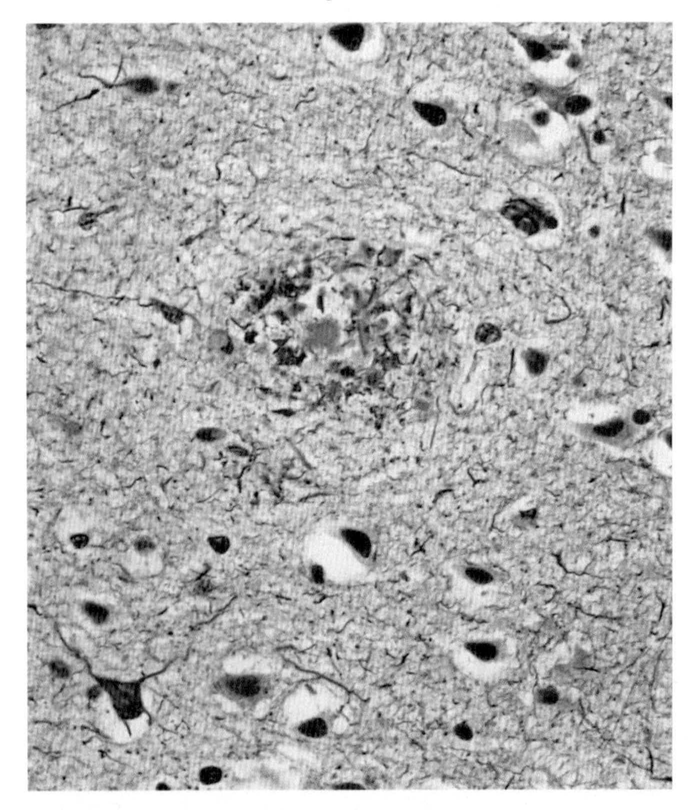

Figure 20–13 B

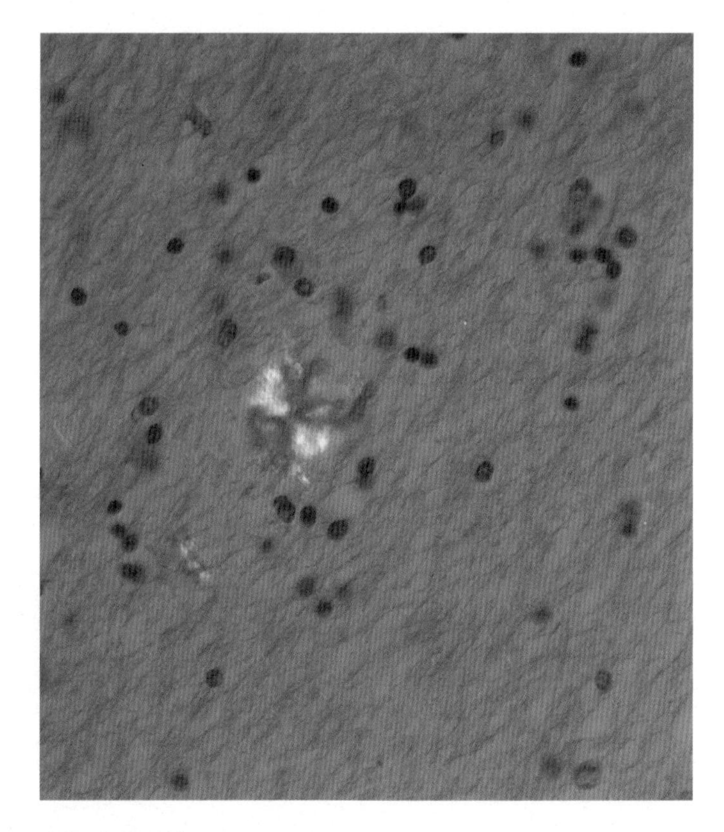

Figure 20–13 C

FIGURE 20–13. ALZHEIMER'S DISEASE: A: In Alzheimer's disease, **senile plaques**—focal collections of dilated tortuous neurites—are faintly visible in hematoxylin and eosin preparations. B: When stained with silver, plaques present as silver-positive neurites surrounding a central amyloid core. A halo separates the core from the neurites. C: The core is Congo red–positive and under polarized light exhibits apple green birefringence, demonstrating its amyloid nature.

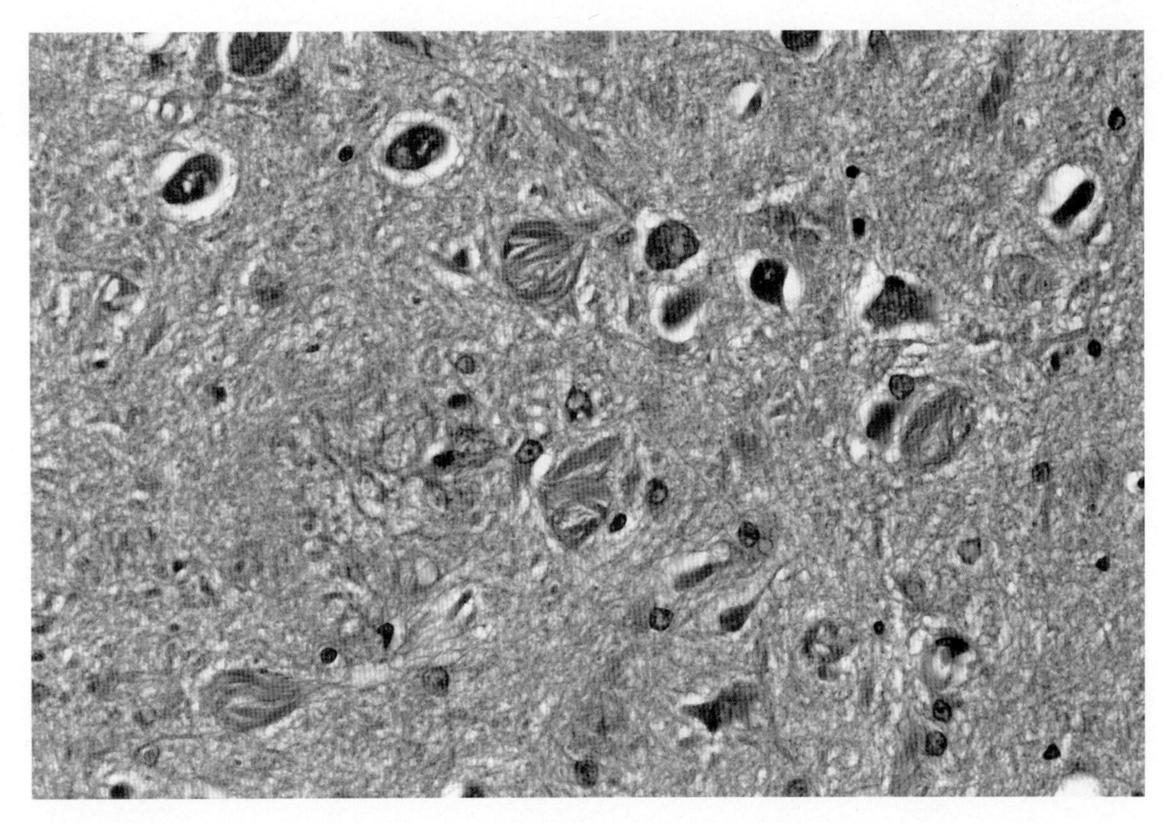

Figure 20–13 D

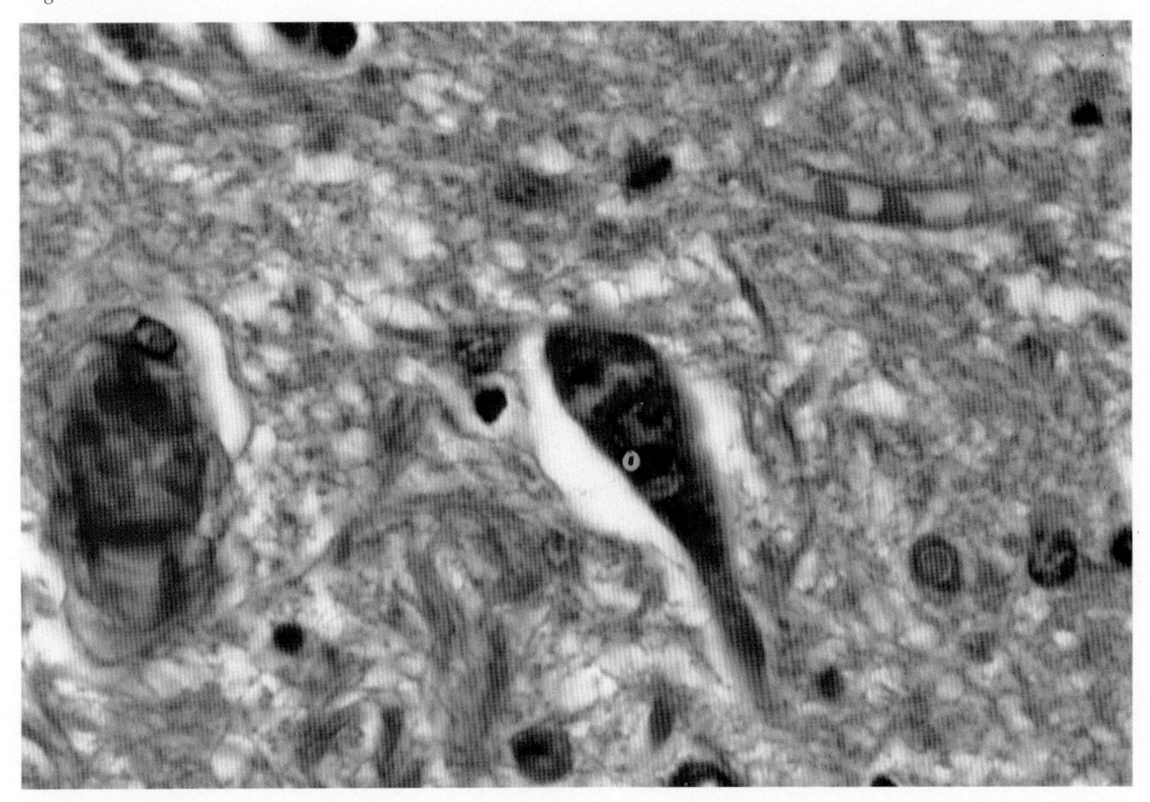

Figure 20–13 E

FIGURE 20–13. ALZHEIMER'S DISEASE: D: In this disease, pairs of helical filaments—**neurofibrillary tangles**—encircle and displace the nucleus in the cytoplasm of the neurons. After neurons degenerate, the tangles are all that remain. E: Also seen in Alzheimer's disease are intraneuronal cytoplasmic vacuoles containing argyrophilic granules—**granulovacuolar degeneration**.

■ CENTRAL NERVOUS SYSTEM

DEGENERATIVE DISEASES

FIGURE 20–14. PICK'S DISEASE: A: The most characteristic feature seen in Pick's disease is the presence of round to oval, weakly eosinophilic intracytoplasmic neuronal inclusions—**Pick bodies**. B: The Pick bodies are best seen with a silver stain.

FIGURE 20–15. IDIOPATHIC PARKINSON'S DISEASE (PARALYSIS AGITANS): In Parkinson's disease, the most characteristic feature is the presence of intracytoplasmic eosinophilic, round or elongated inclusions with a dense core and a pale rim—**Lewy bodies**. In addition, there is loss of melanin-containing neurons from the substantia nigra and locus ceruleus.

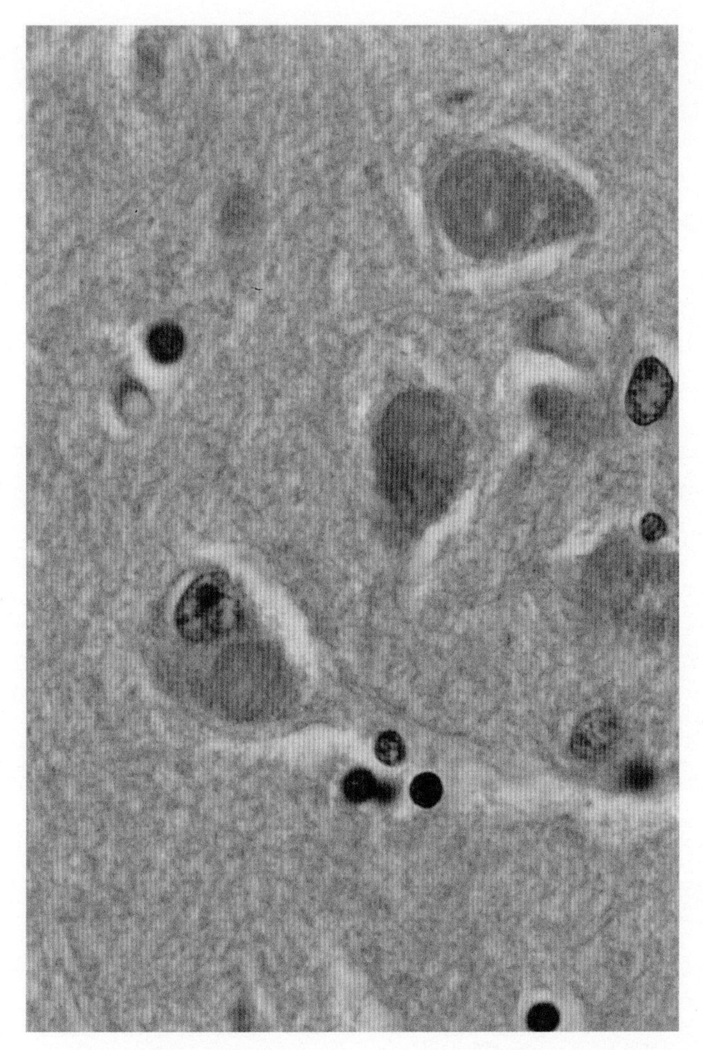

Figure 20–14 A

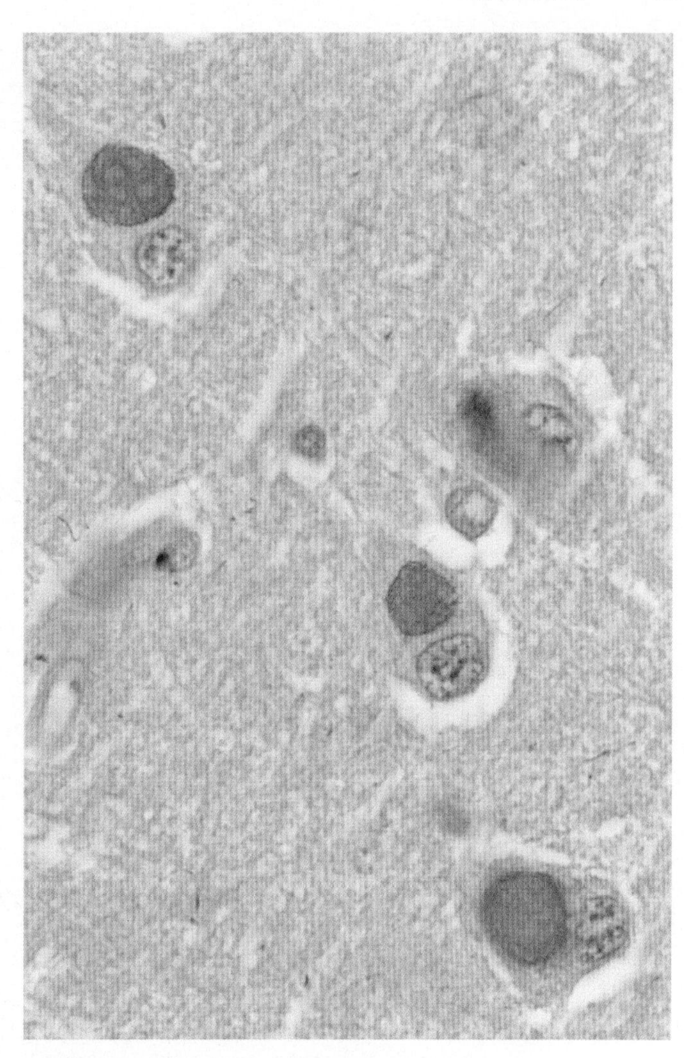

Figure 20–14 B

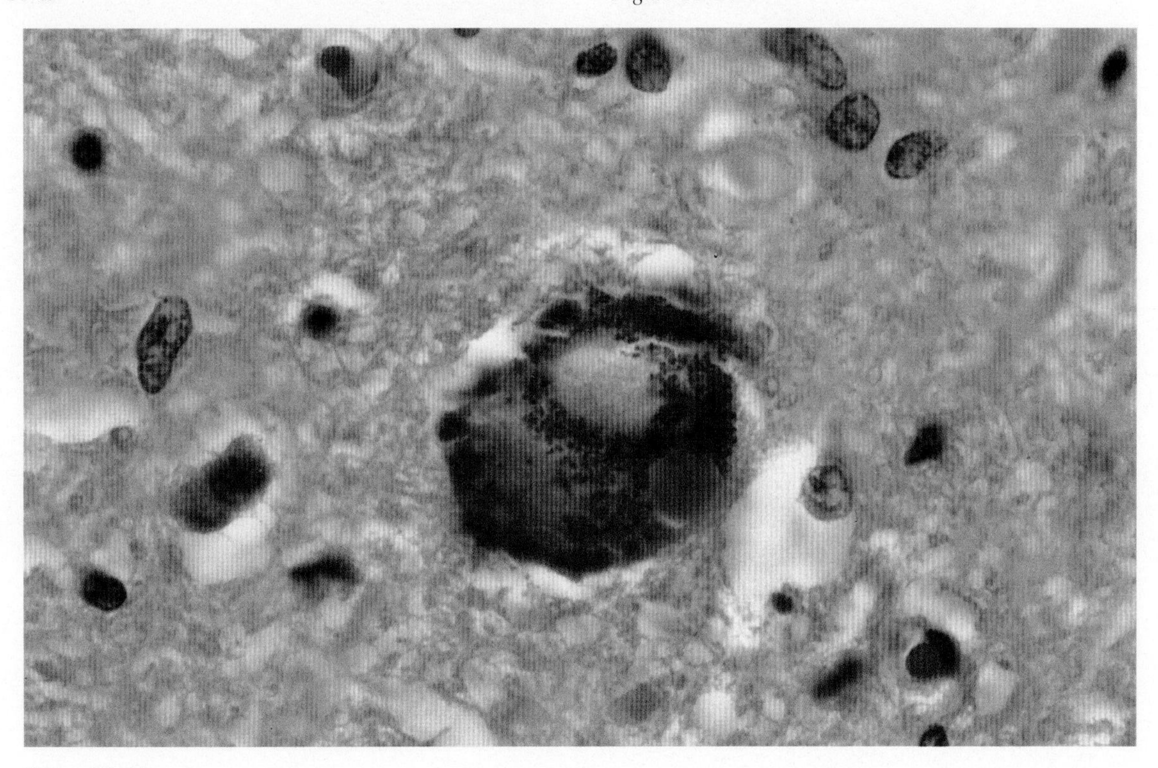

Figure 20–15

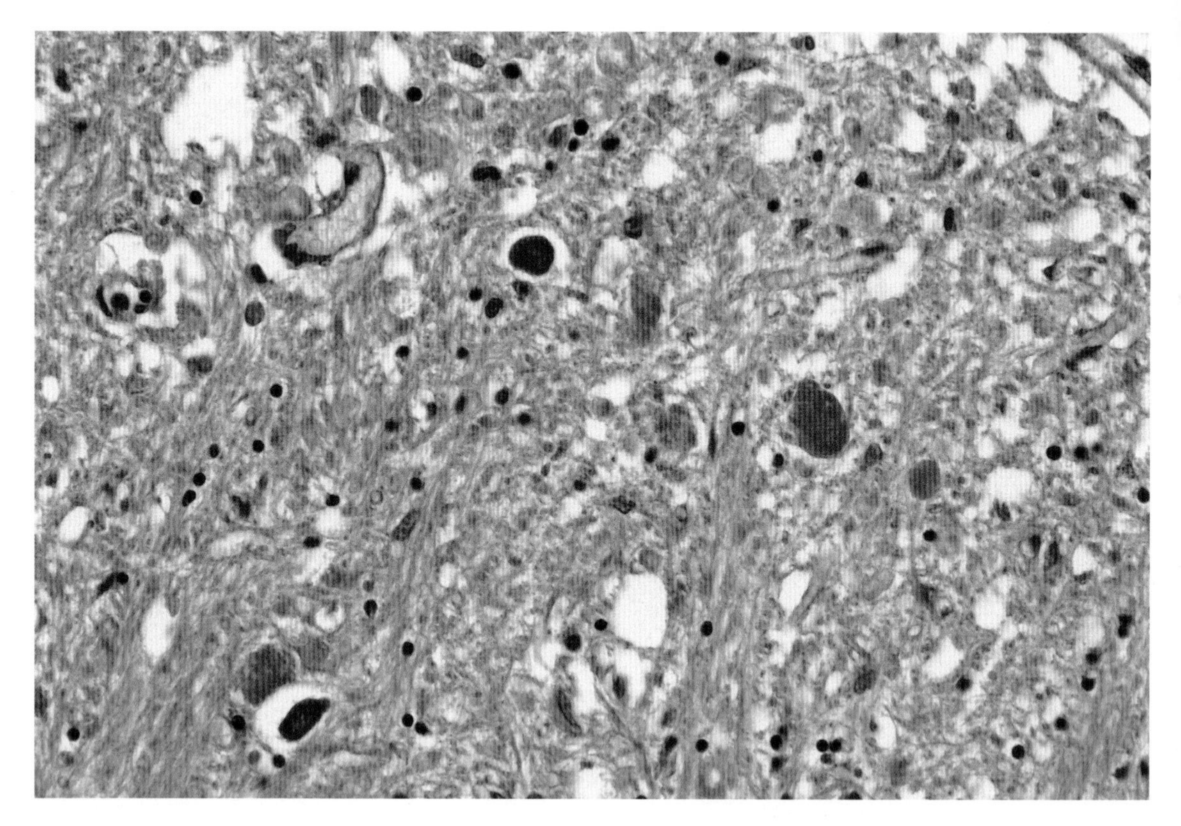

Figure 20–16 A

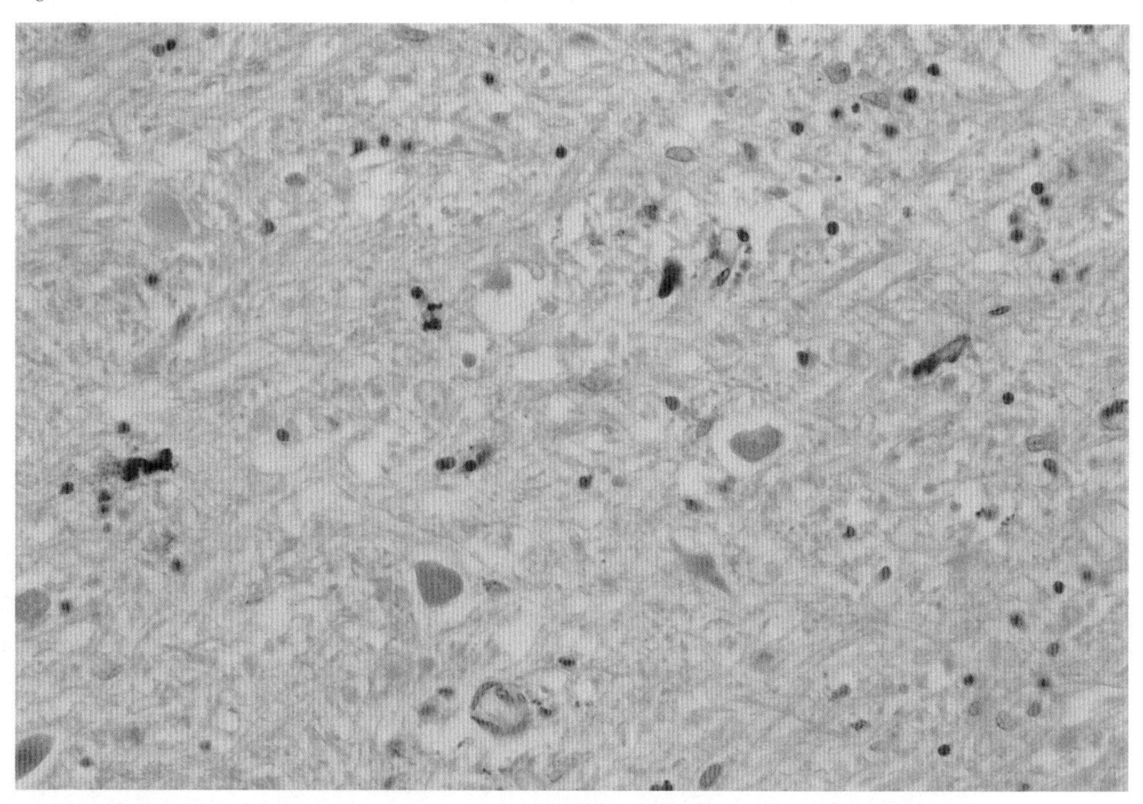

Figure 20–16 B

FIGURE 20–16. HALLERVORDEN-SPATZ DISEASE: A: In Hallervorden-Spatz disease, there are globular eosinophilic dilatations of axons—**axonal spheroids**—and deposition of a brown pigment. **B:** The brown pigment is positive for iron when stained with Prussian blue.

■ CENTRAL NERVOUS SYSTEM

DEGENERATIVE DISEASES

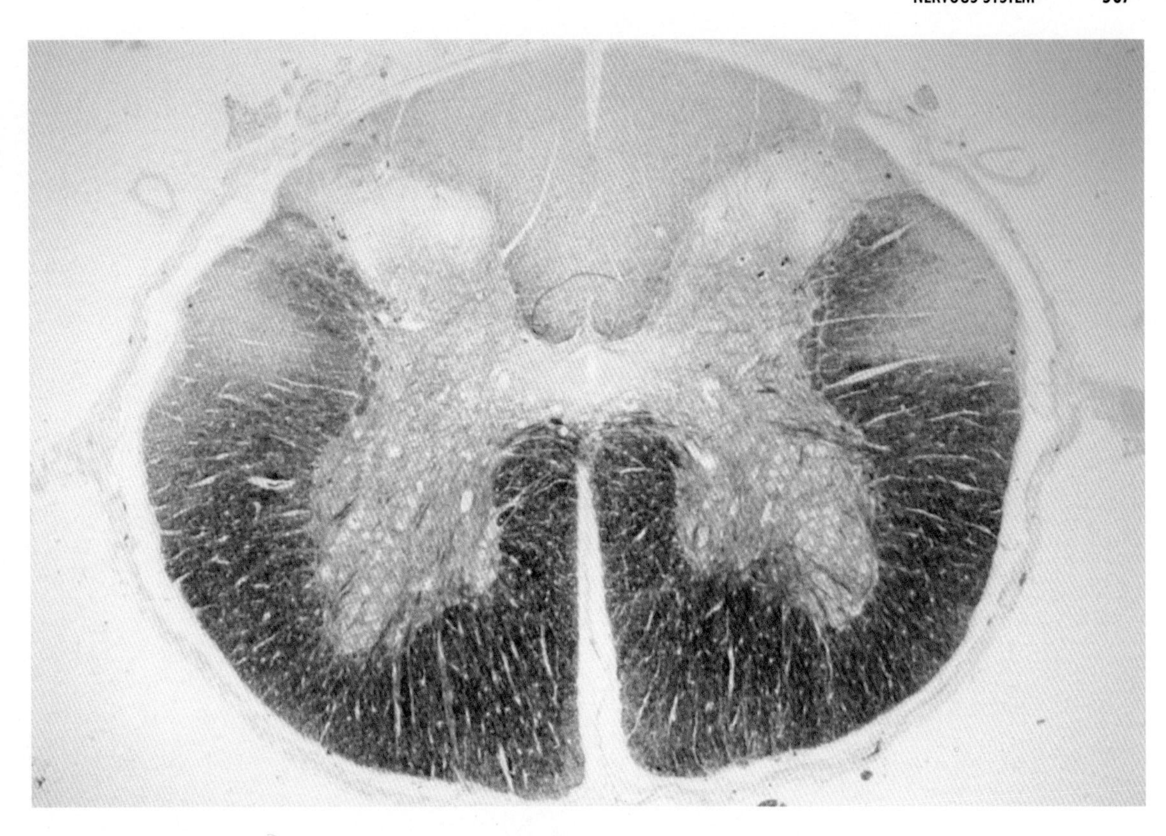

FIGURE 20–17. FRIEDRICH'S ATAXIA: In this photomicrograph of Friedrich's ataxia, note the degeneration of the posterior columns and corticospinal tracts (silver stain).

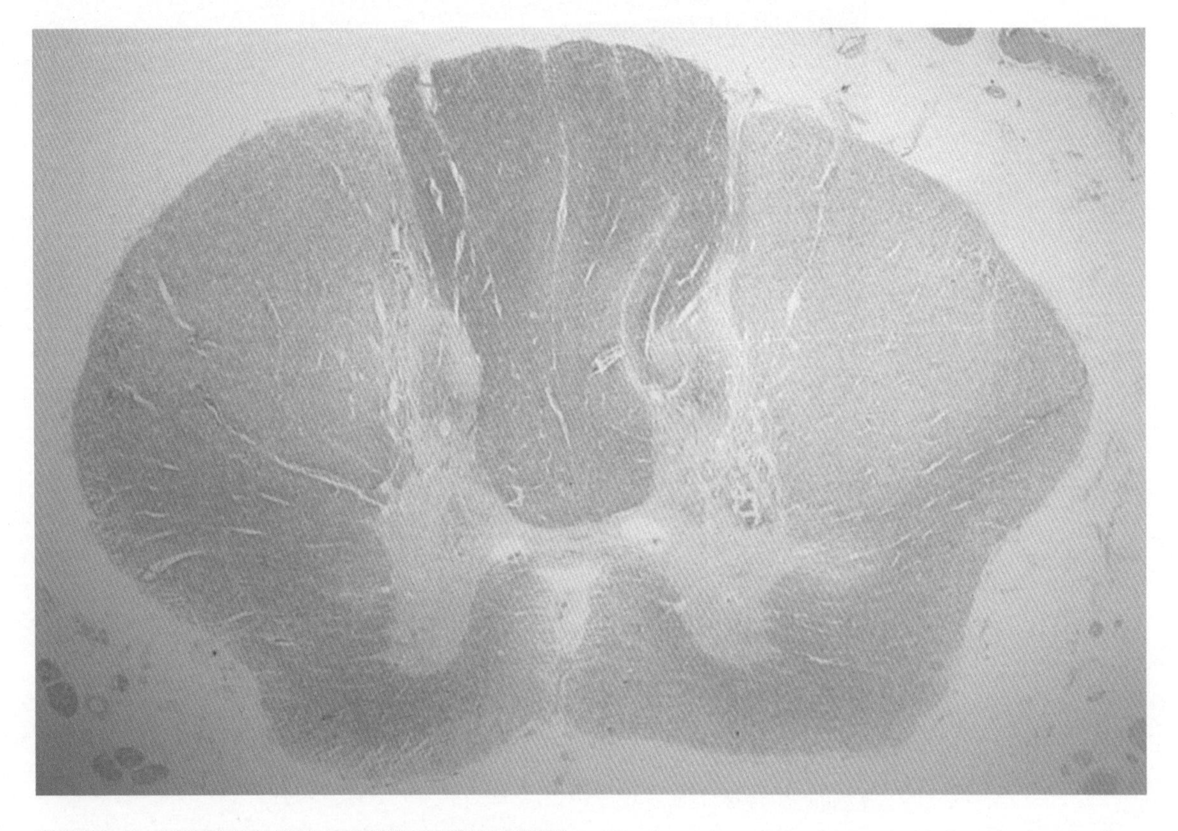

FIGURE 20–18. AMYOTROPHIC LATERAL SCLEROSIS (LOU GEHRIG DISEASE): In this photomicrograph of amyotrophic lateral sclerosis, there is degeneration of the lateral columns and corticospinal tracts (Luxol fast blue stain).

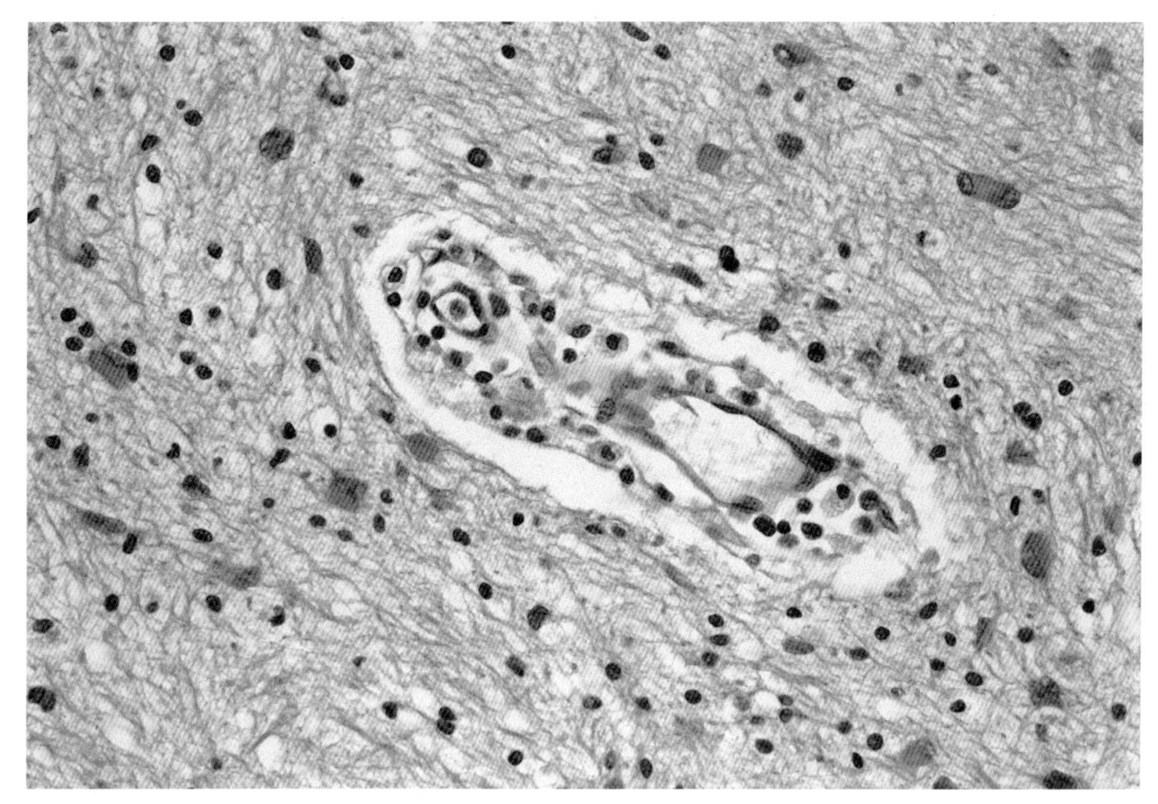

Figure 20–19 A

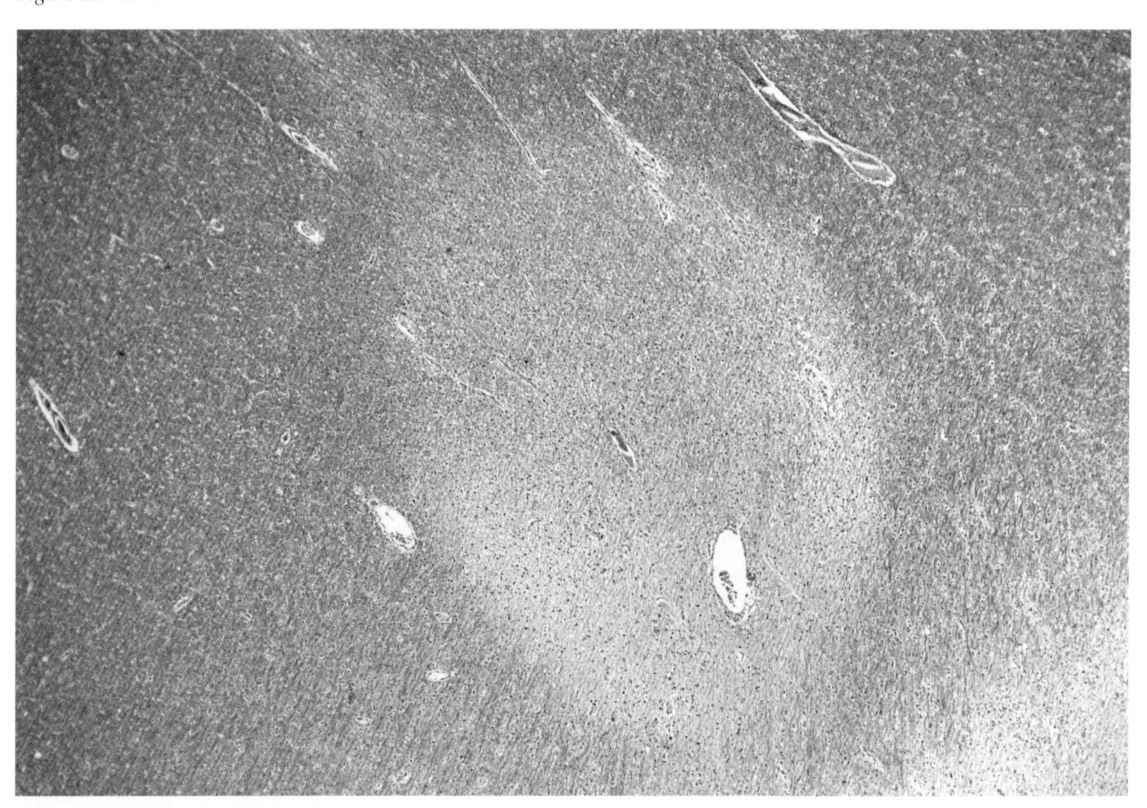

Figure 20–19 B

FIGURE 20–19. MULTIPLE SCLEROSIS: A: Early in the course of multiple sclerosis, a mononuclear perivascular lymphocytic infiltrate is present. As the disease progresses, demyelination begins around the inflamed areas, which coalesce to form **plaques.** B: The plaques are irregularly shaped pale foci of demyelination, which stand out against the dark myelinated normal parenchyma (Luxol fast blue stain).

■ CENTRAL NERVOUS SYSTEM

DEMYELINATING DISEASES

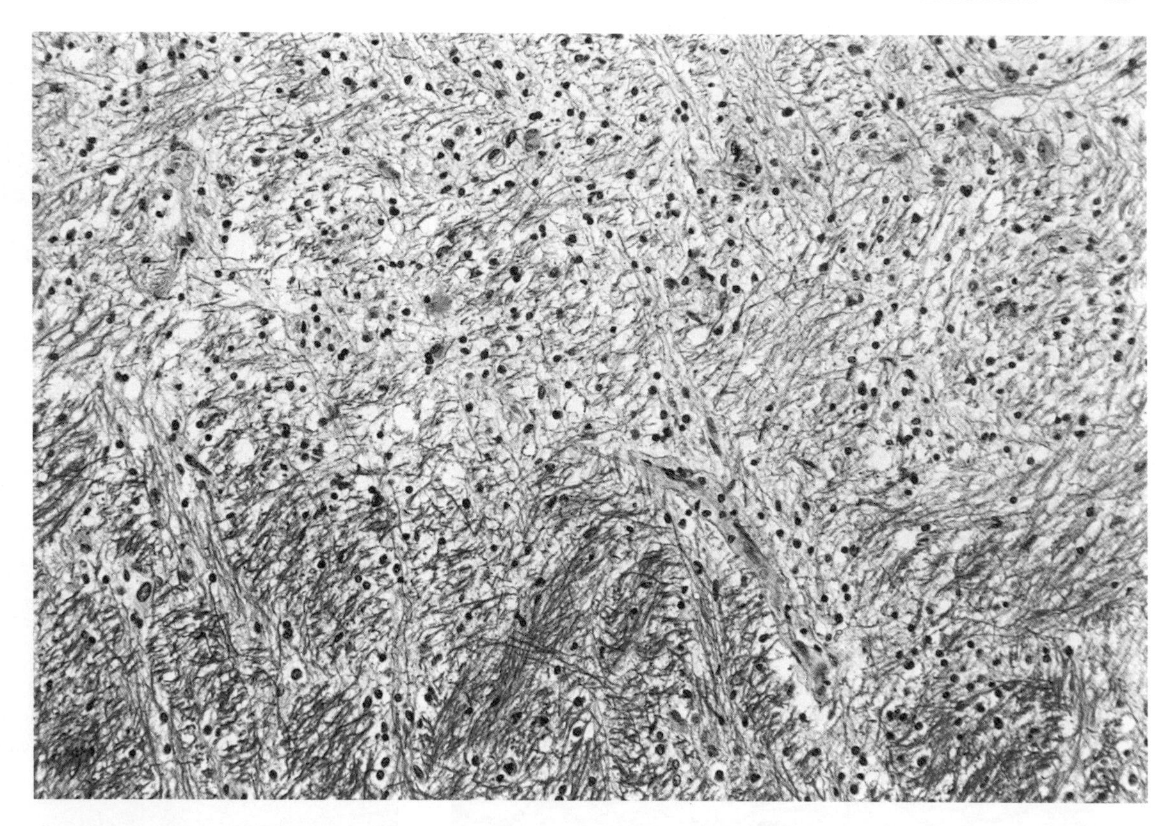

Figure 20–19 C

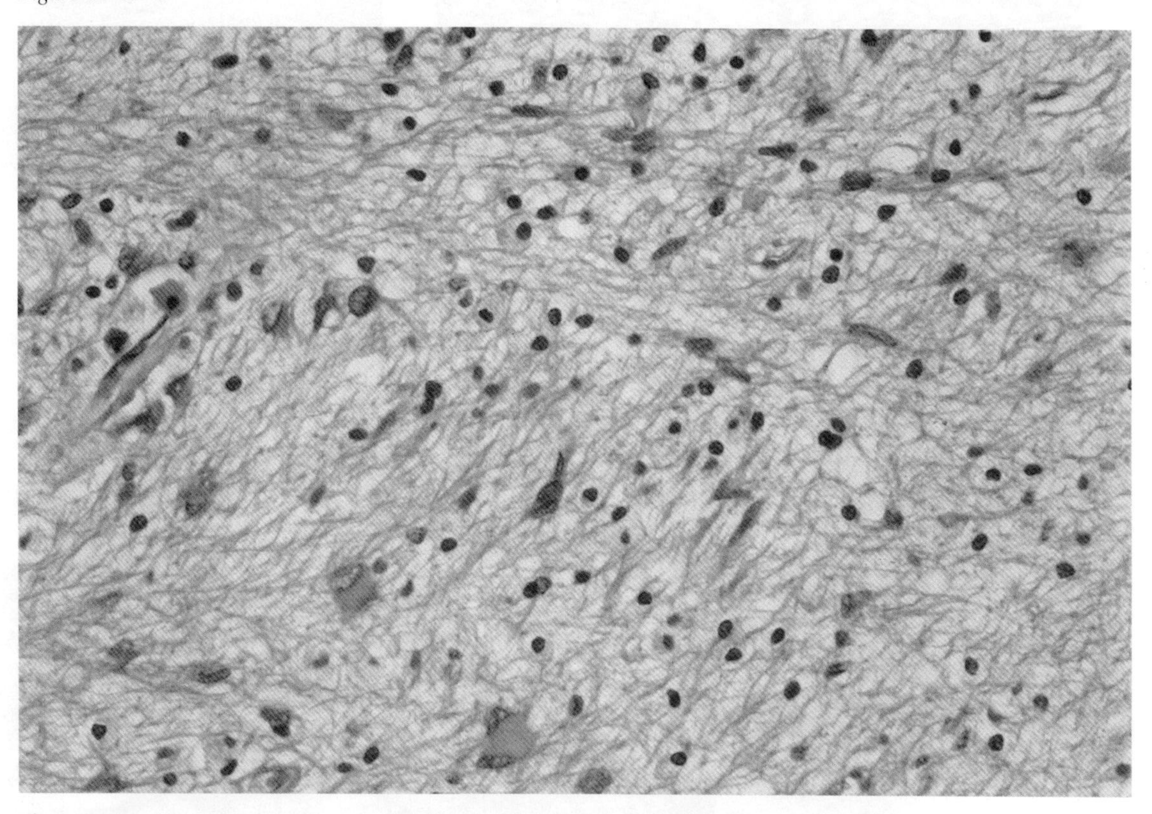

Figure 20–19 D

FIGURE 20–19. MULTIPLE SCLEROSIS: **C**: Another characteristic feature of the plaques in multiple sclerosis is myelin-axon dissociation in which there is loss of myelin and preservation of axons. This is best seen in a Luxol fast blue–Bodian stain in which Luxol fast blue stains myelin blue and Bodian stains axons black. **D**: In older plaques, there is reactive astrocytosis, lipid-laden macrophages, and loss of oligodendroglia.

■ CENTRAL NERVOUS SYSTEM

NUTRITIONAL, ENVIRONMENTAL, AND METABOLIC DISORDERS

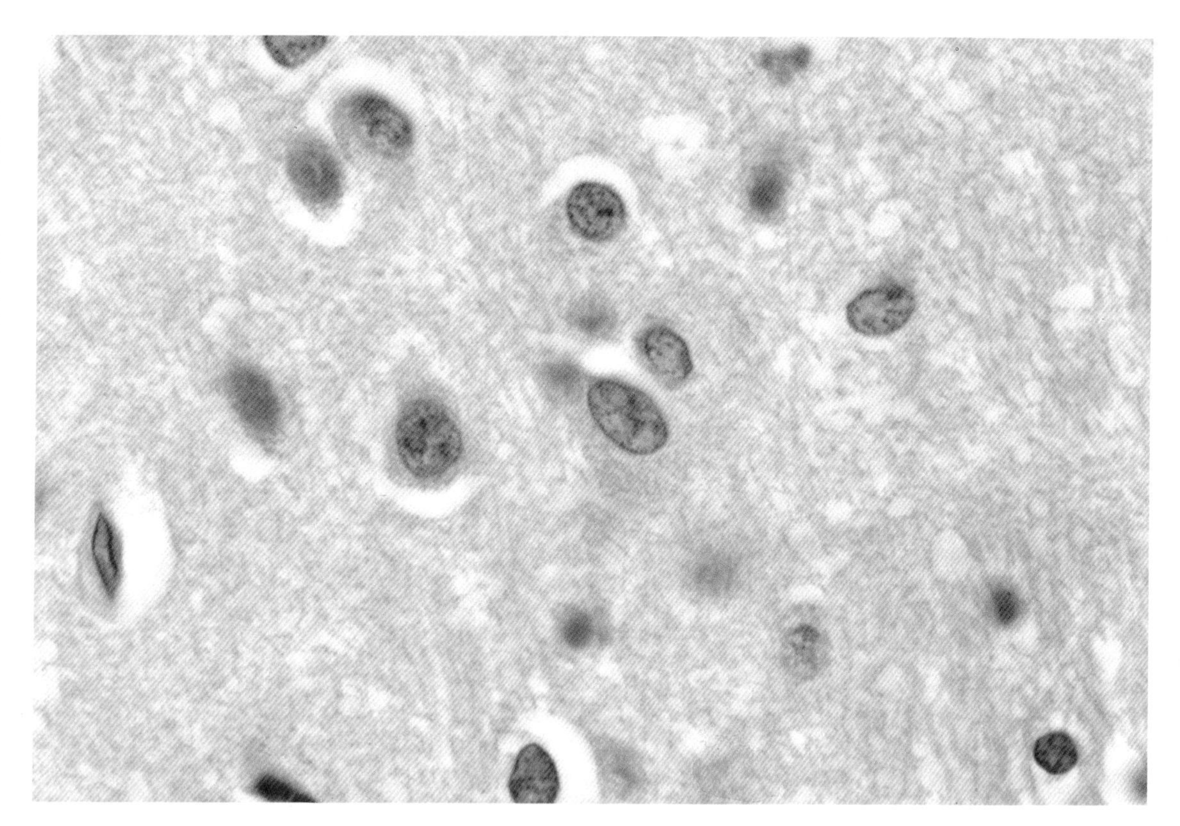

FIGURE 20–21. HEPATIC ENCEPHALOPATHY: In hepatic encephalopathy, there is proliferation of **Alzheimer's type II astrocytes,** which have enlarged hypochromatic nuclei.

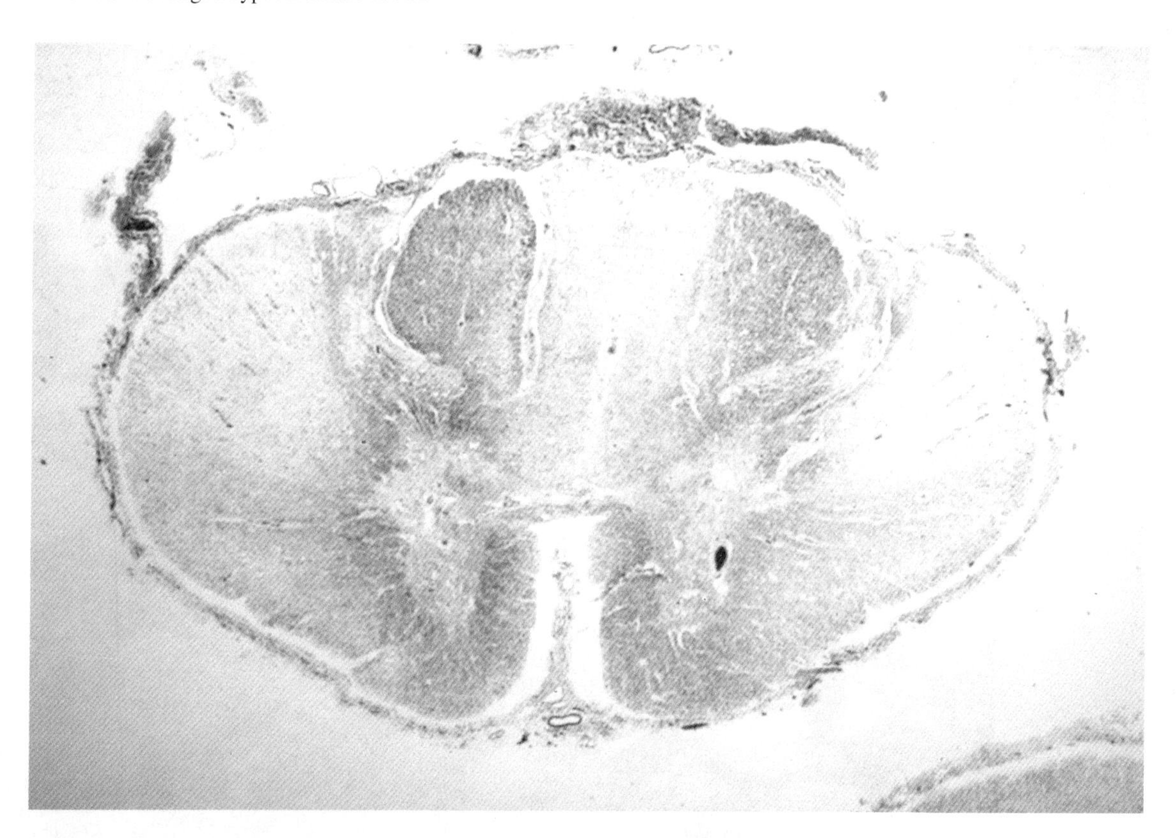

FIGURE 20–22. VITAMIN B_{12} (CYANOCOBALAMIN) DEFICIENCY, SUBACUTE COMBINED DEGENERATION (SACD): In SACD, the white matter of the dorsal and lateral columns is degenerated, as can be seen in this Luxol fast blue stain.

■ CENTRAL NERVOUS SYSTEM

DEVELOPMENTAL DISORDERS

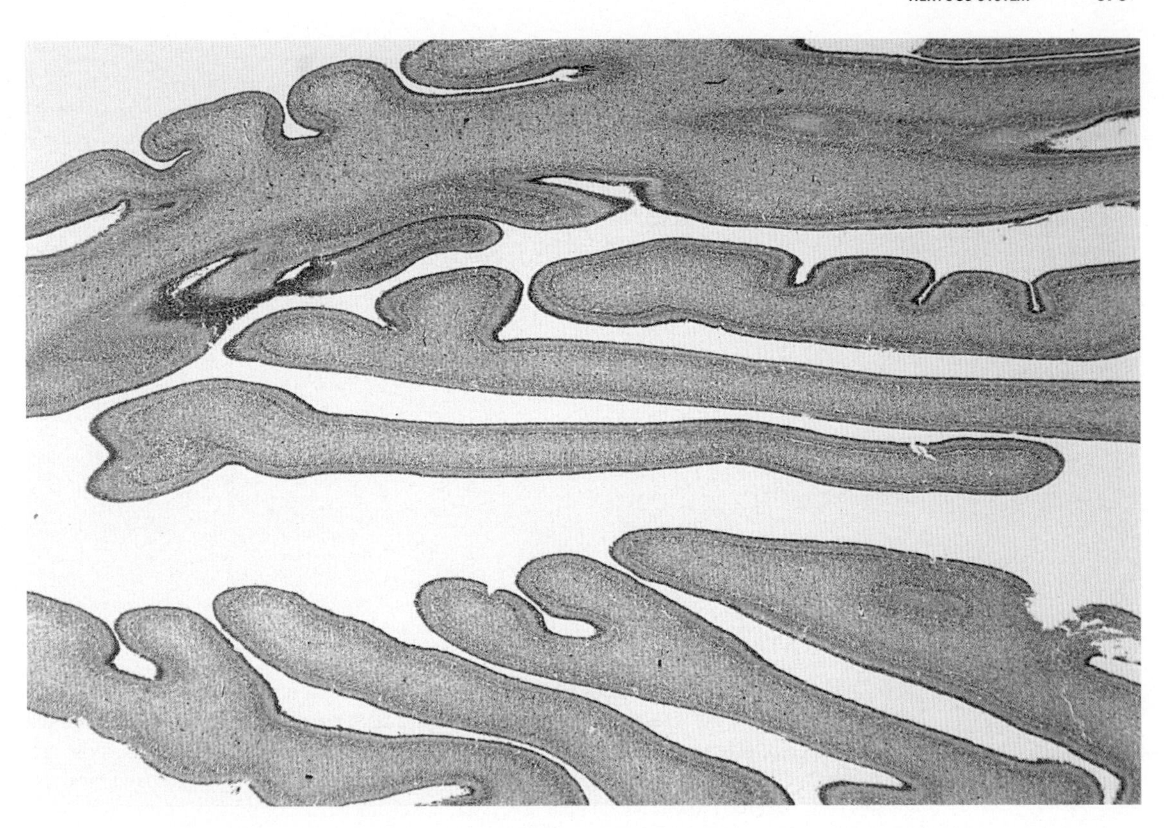

FIGURE 20–23. POLYMICROGYRIA: In polymicrogyria, there are numerous extremely small gyri.

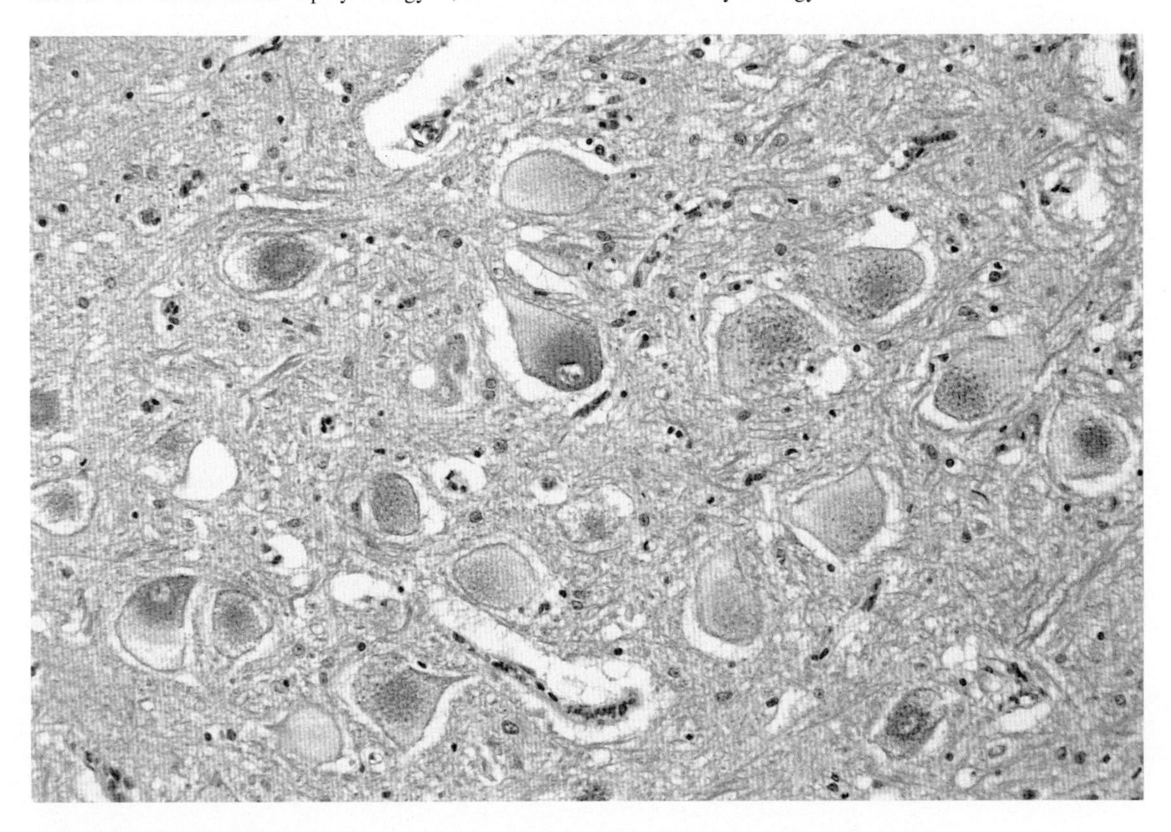

FIGURE 20–24. TAY-SACHS DISEASE: In Tay-Sachs disease, a defect in the enzyme hexosaminidase A causes an accumulation of glycoside GM_2 in neurons. Such accumulation is manifested by cytoplasmic ballooning and nuclear displacement.

■ CENTRAL NERVOUS SYSTEM

TUMORS OF ASTROCYTIC ORIGIN

FIGURE 20–25. PILOCYTIC ASTROCYTOMA, GRADE I: Most pilocytic astrocytomas occur in children and young adults, with a predilection for the posterior fossa. A: The astrocytes have parallel bundles of fibrillar cytoplasmic processes that resemble mats of hair, hence the name **"hair cell"** or **pilocyte. Microcysts,** loose areas, alternate with dense areas. B: **Rosenthal fibers,** which are hyaline, eosinophilic, and round, oval, or irregular proteinaceous material present within glial processes, are characteristic of pilocytic astrocytomas. Extracellular eosinophilic proteinaceous droplets are also present.

FIGURE 20–26. ASTROCYTOMA, GRADE I–II: In low-grade astrocytoma, there is an increase in the number of neoplastic astrocytes *(left)* that blend into the normal adjacent parenchyma.

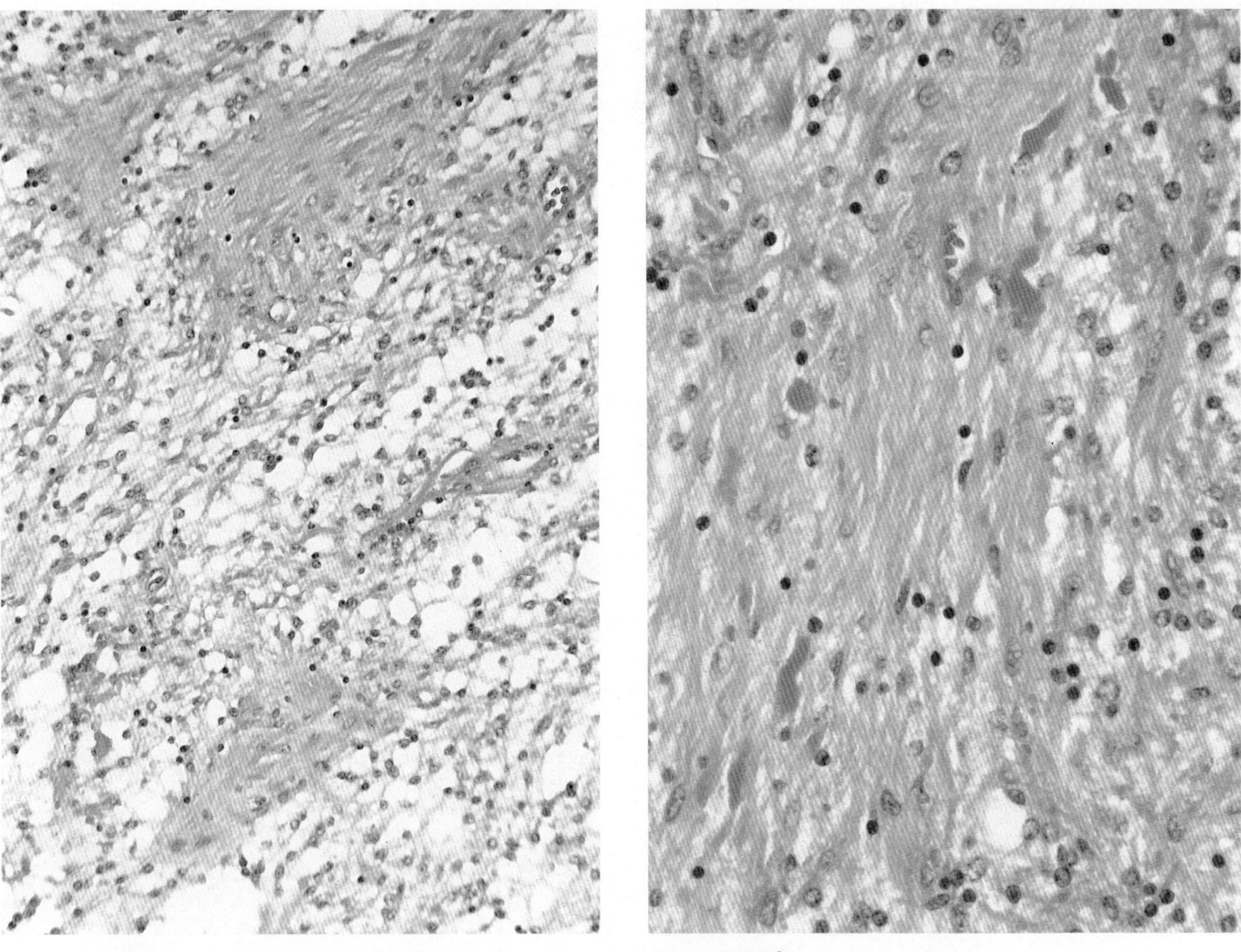

Figure 20–25 A

Figure 20–25 B

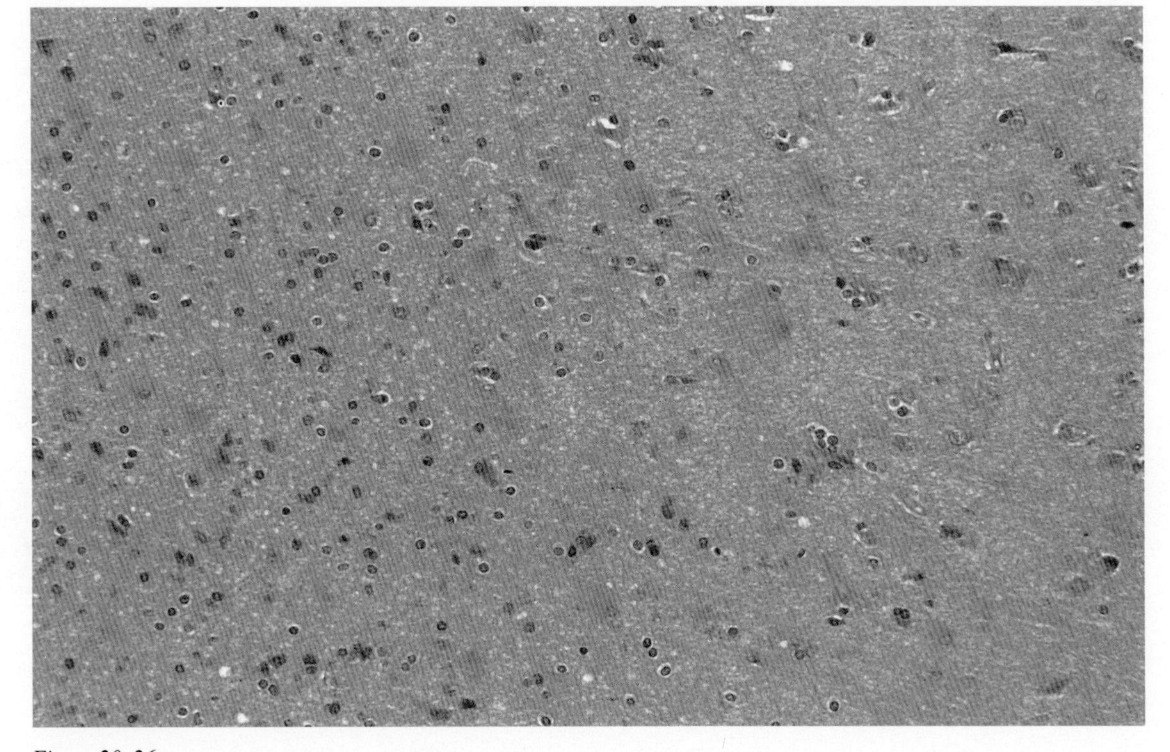

Figure 20–26

■ CENTRAL NERVOUS SYSTEM

TUMORS OF ASTROCYTIC ORIGIN

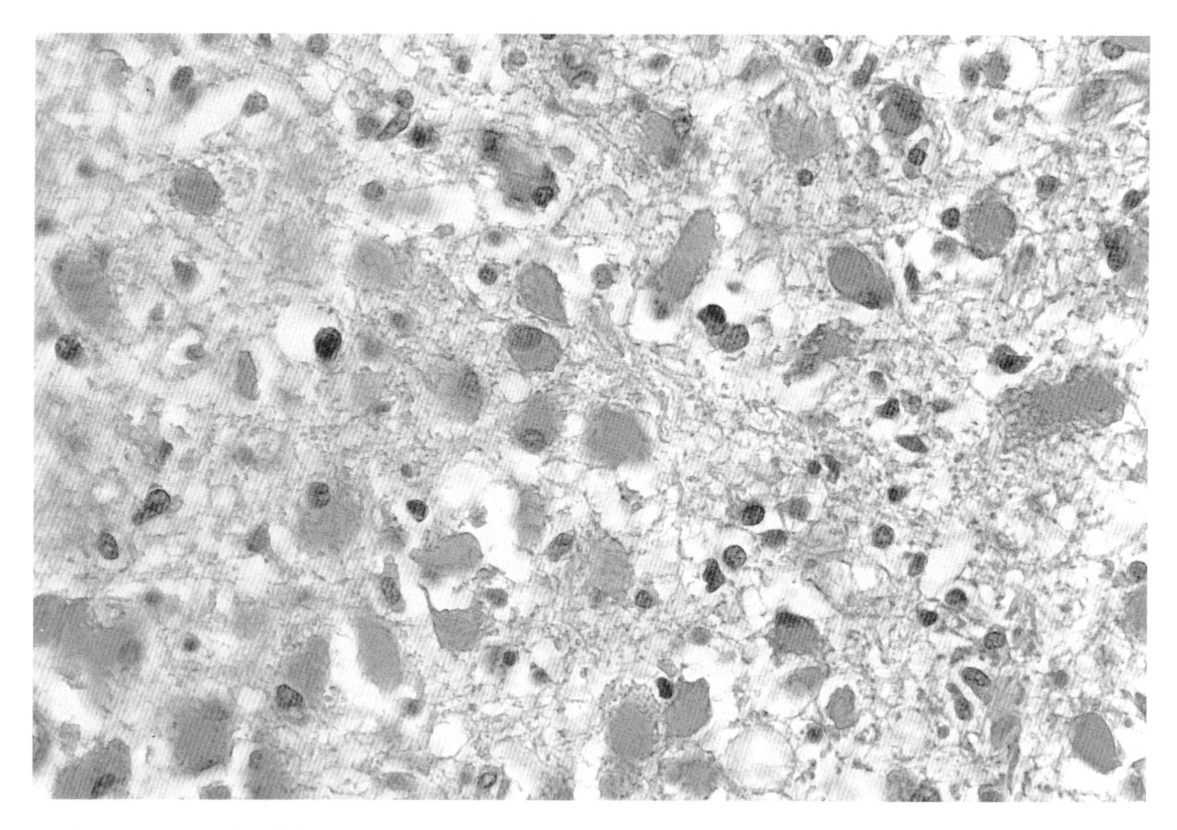

FIGURE 20–27. GEMISTOCYTIC ASTROCYTOMA, GRADE II–III: In gemistocytic astrocytoma, the neoplastic cells are swollen with abundant eosinophilic cytoplasm. Nuclei are hyperchromatic and situated at the periphery of the cell.

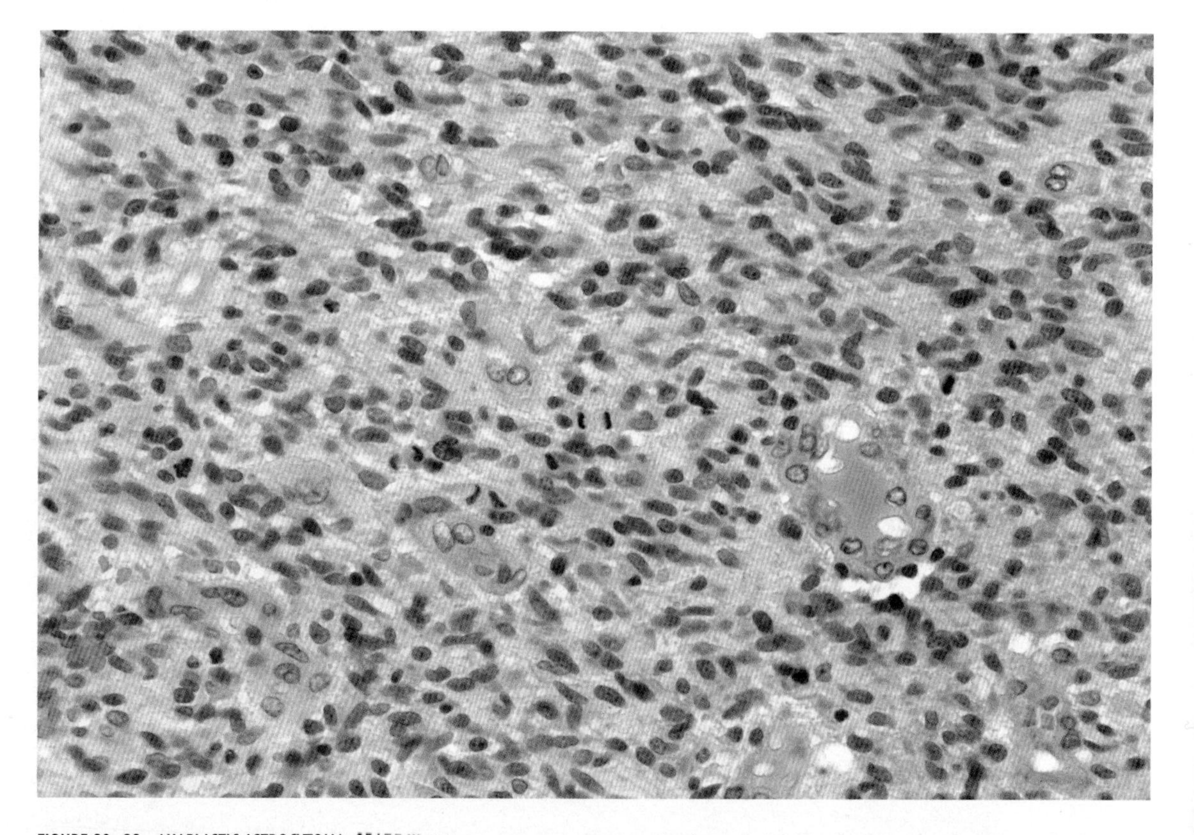

FIGURE 20–28. ANAPLASTIC ASTROCYTOMA, GRADE III: High-grade astrocytoma is characterized by increased cellularity, mitotic figures, pleomorphic and hyperchromatic nuclei, and endothelial proliferation. Compare these features with those of low-grade astrocytoma (Figure 20–26).

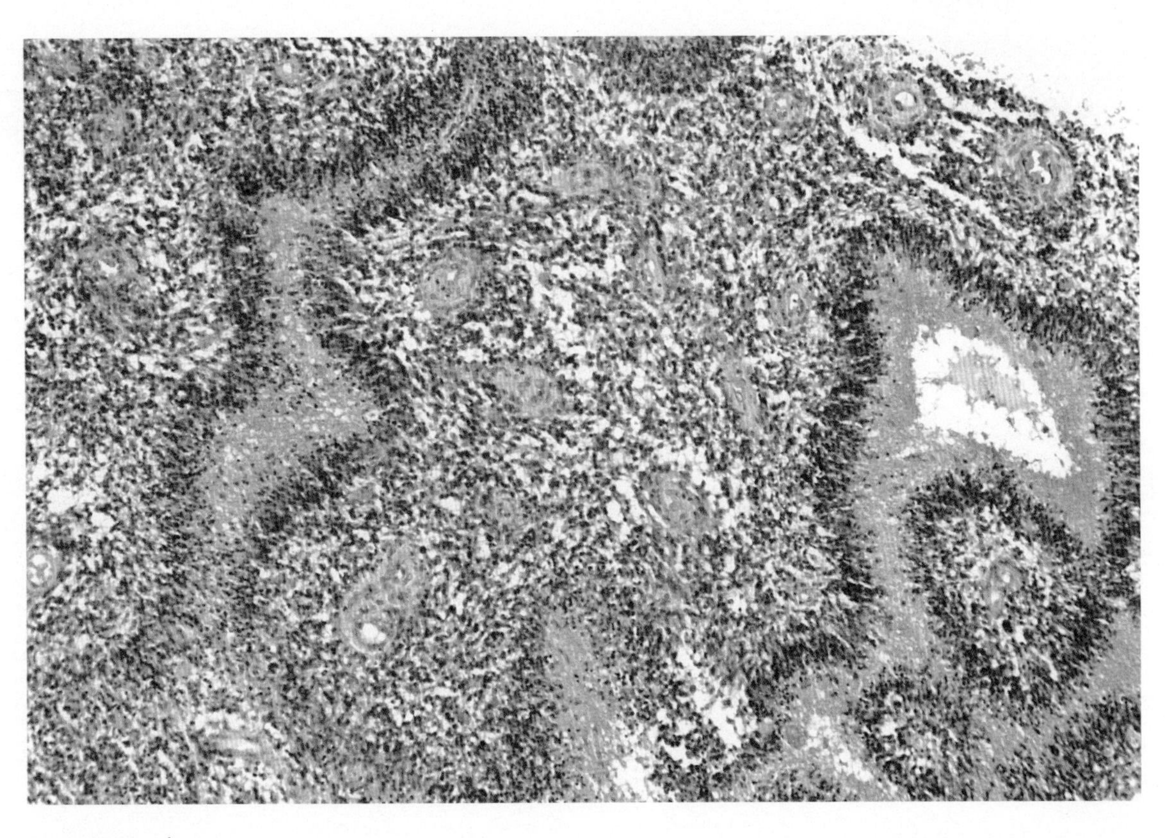

Figure 20–29 A

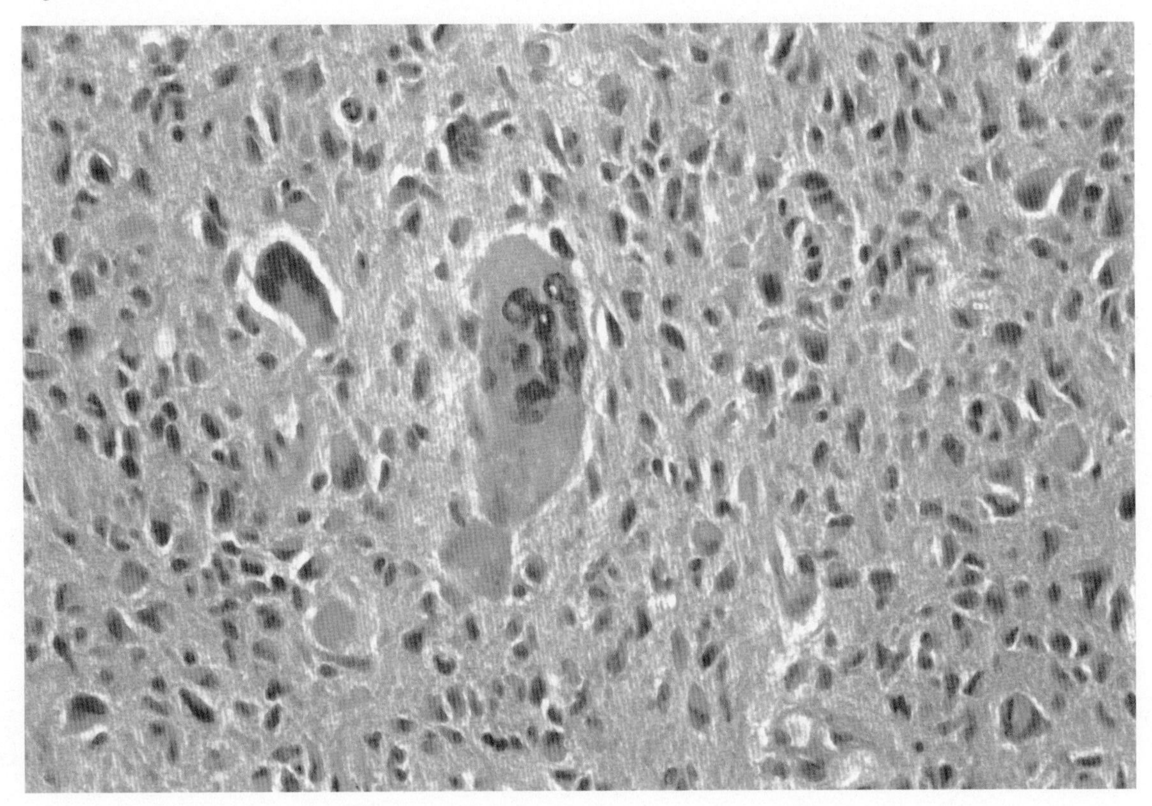

Figure 20–29 B

FIGURE 20–29. GLIOBLASTOMA MULTIFORME (ASTROCYTOMA, GRADE IV): A: In addition to the features seen in high-grade astrocytoma (Figure 20–28), **pseudopalisading** of nuclei around areas of coagulative necrosis is virtually diagnostic of glioblastoma multiforme. B: In addition, multinucleated giant cells are present.

■ CENTRAL NERVOUS SYSTEM

TUMORS OF EPENDYMAL CELLS

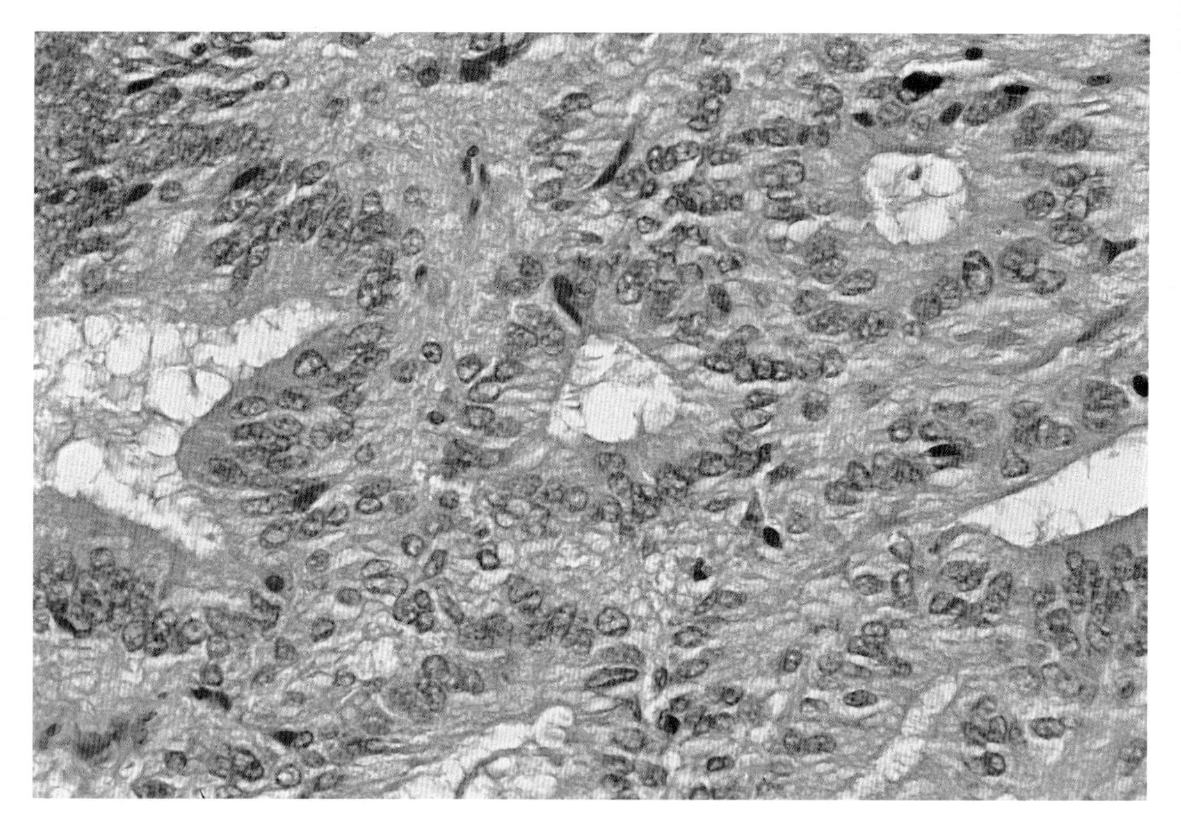

Figure 20–30 A

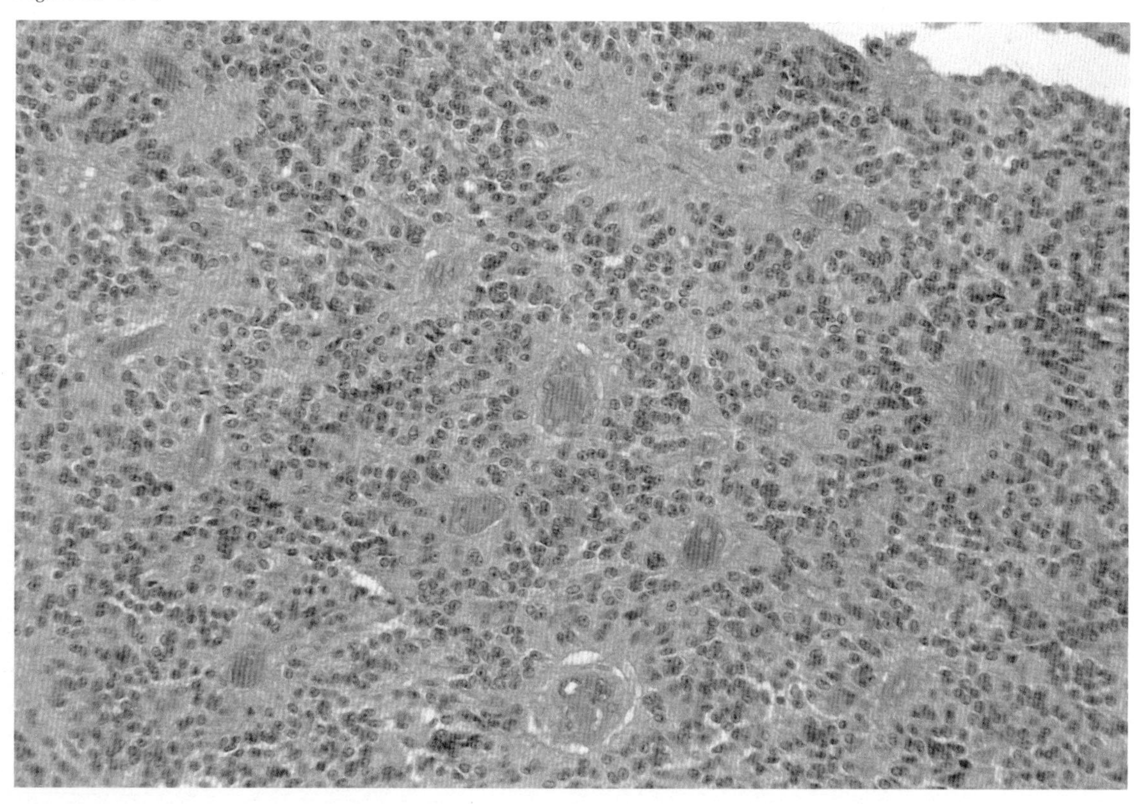

Figure 20–30 B

FIGURE 20–30. EPENDYMOMA: A: The diagnostic feature of ependymomas is the presence of **true rosettes,** which are lumina evenly lined by cells resembling ciliated columnar epithelium. True rosettes represent an attempt to recapitulate the central canal; however, they are infrequently found. B: **Pseudorosettes,** an additional and more commonly found feature, are nucleus-free halos around blood vessels.

■ CENTRAL NERVOUS SYSTEM

TUMORS OF EPENDYMAL CELLS

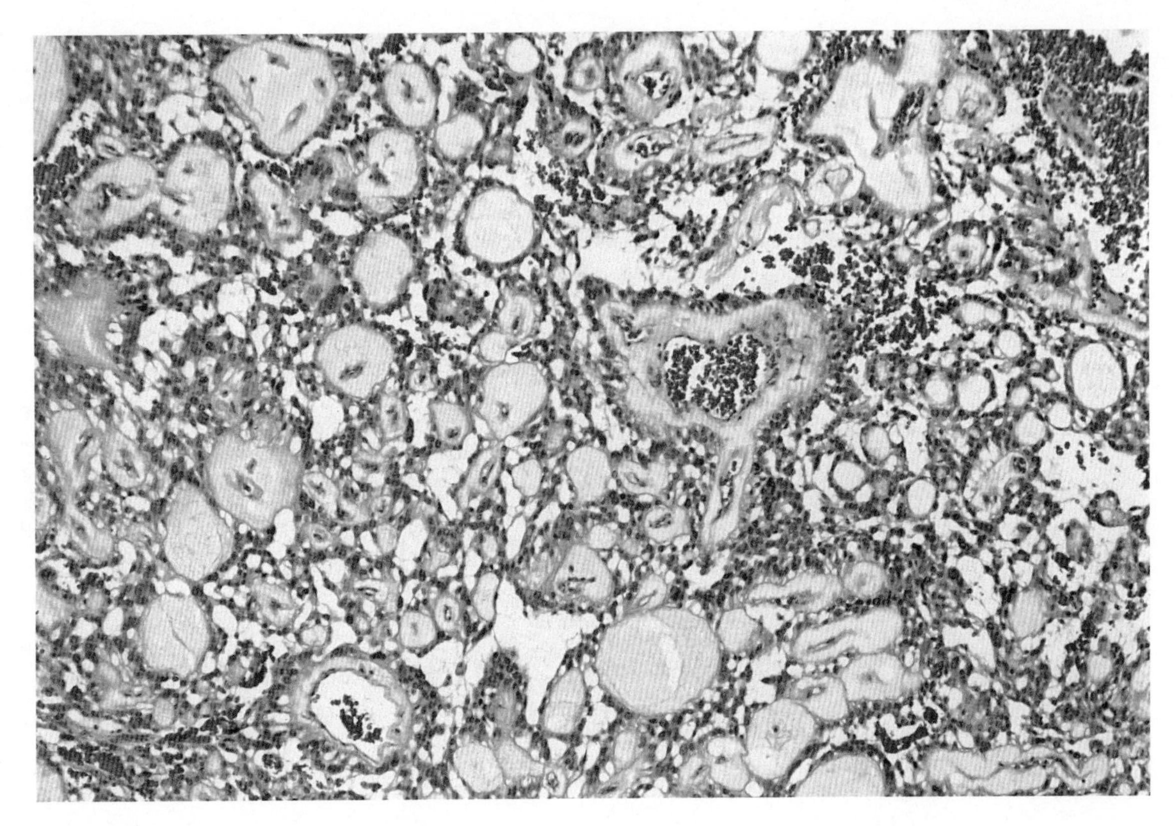

Figure 20–31 A

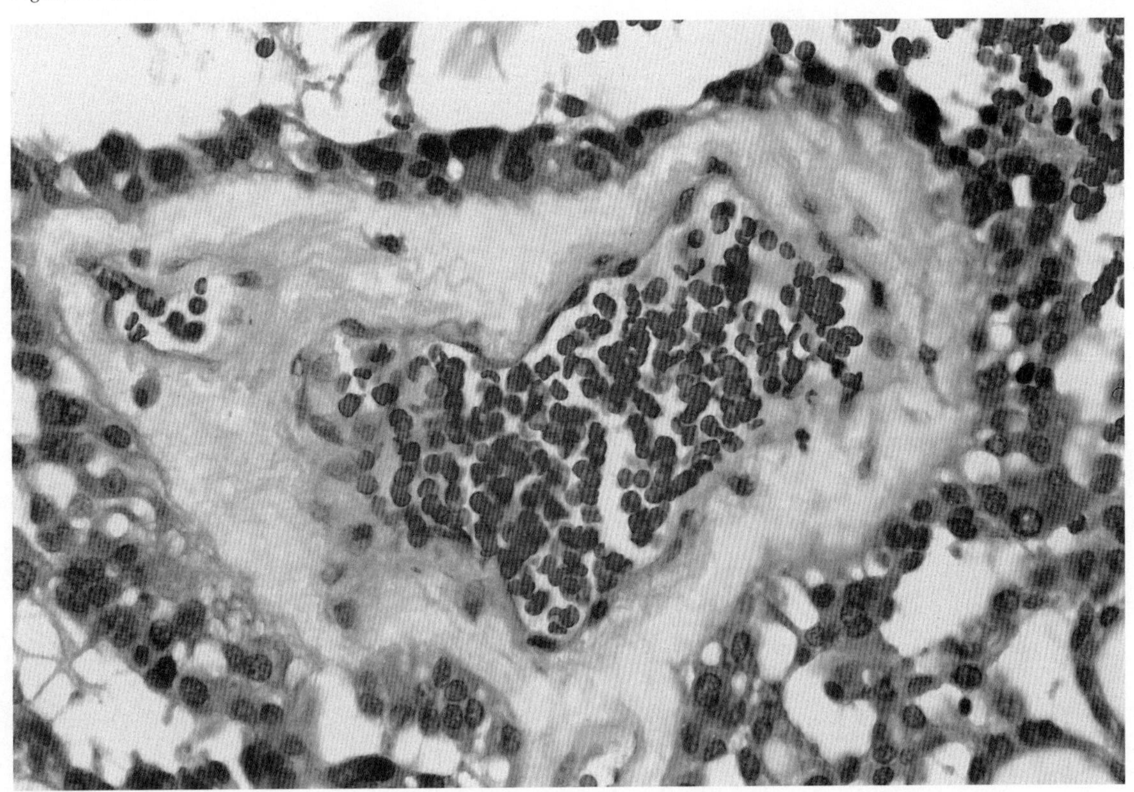

Figure 20–31 B

FIGURE 20–31. MYXOPAPILLARY EPENDYMOMA: These tumors are found in the filum terminale and conus medullaris. **A:** The cells are present in papillary formations. **B:** The cells are low columnar and surround a central core of acellular hyaline connective tissue rich in blood vessels.

■ CENTRAL NERVOUS SYSTEM

TUMORS OF OLIGODENDROCYTES

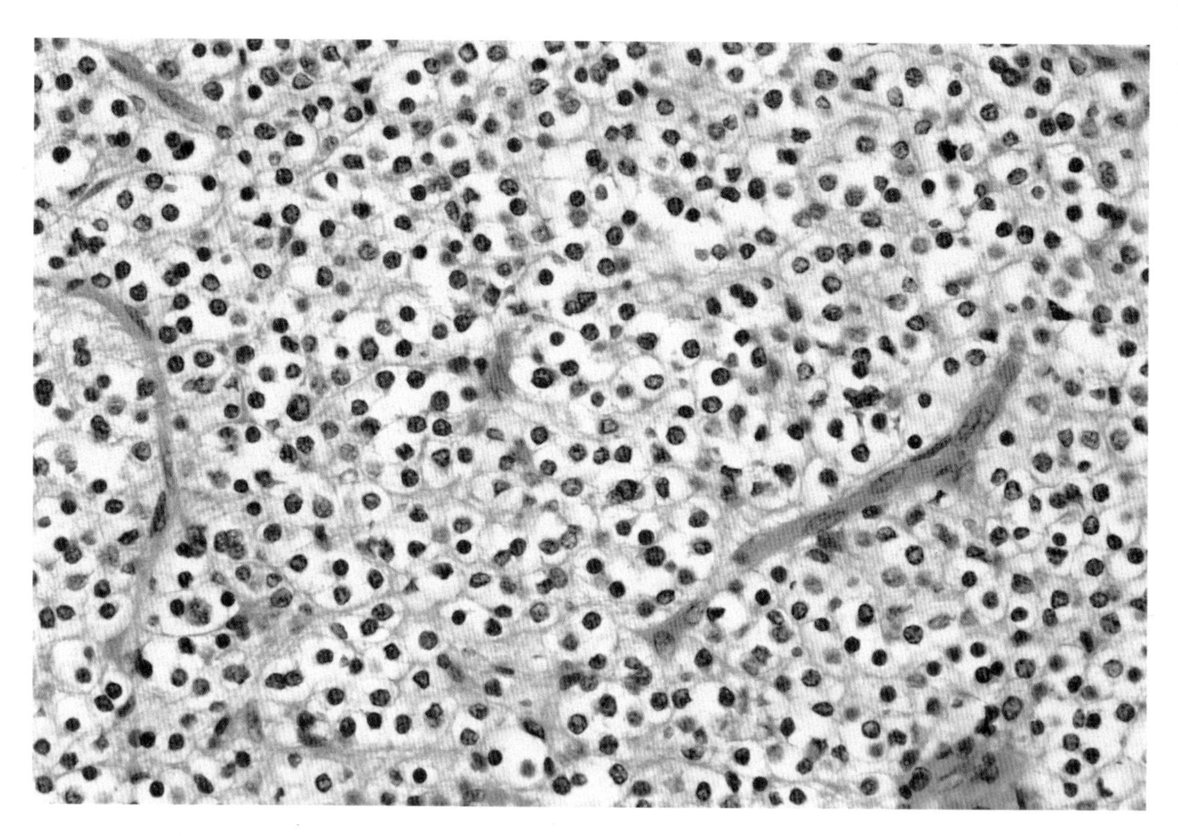

FIGURE 20–32. OLIGODENDROGLIOMA: In oligodendroglioma, sheets of uniform cells are separated into lobules by a fine fibrovascular stroma. The cells are round and uniform, with centrally located nuclei. As a result of an artifact of fixation, the cytoplasm is clear and the cells resemble fried eggs.

TUMORS OF NEURONAL ORIGIN

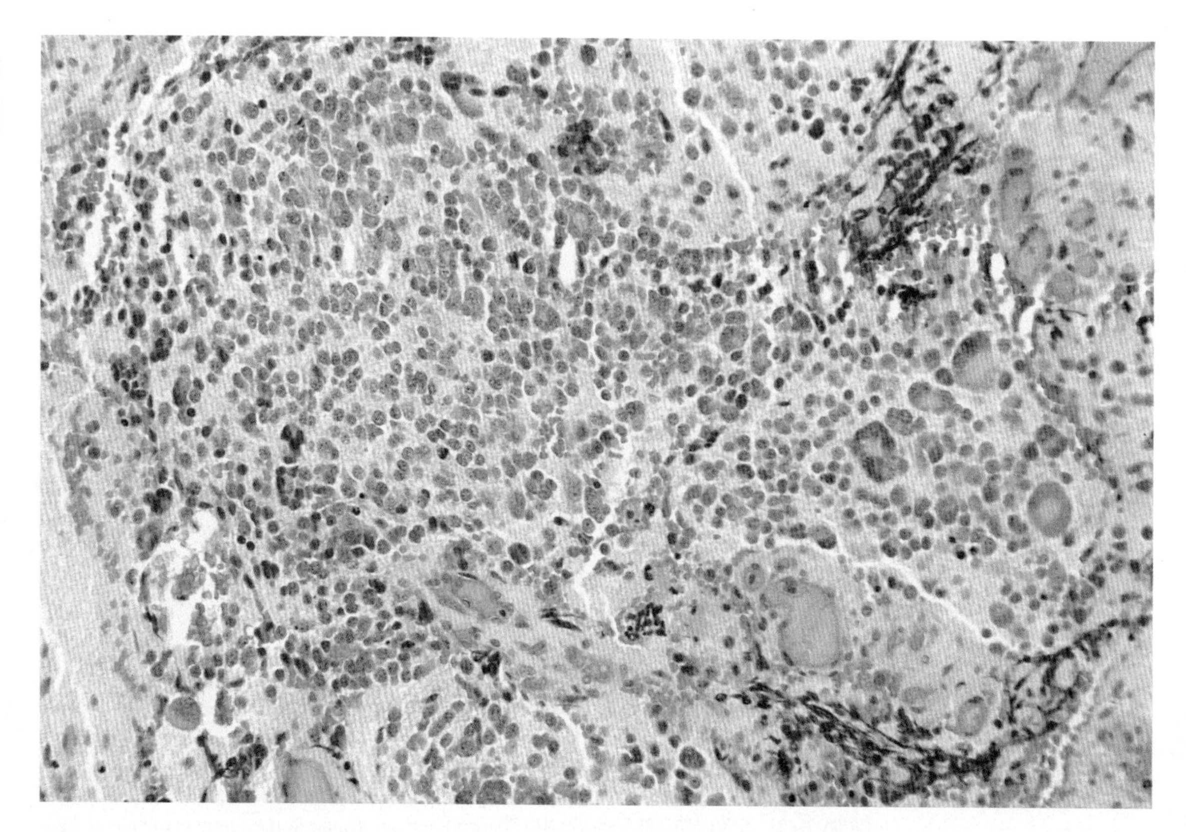

FIGURE 20–33. GANGLIONEUROBLASTOMA: A: The two components of ganglioneuroblastoma—ganglion cells and neuroblastoma—are seen at low-power magnification in this photomicrograph.

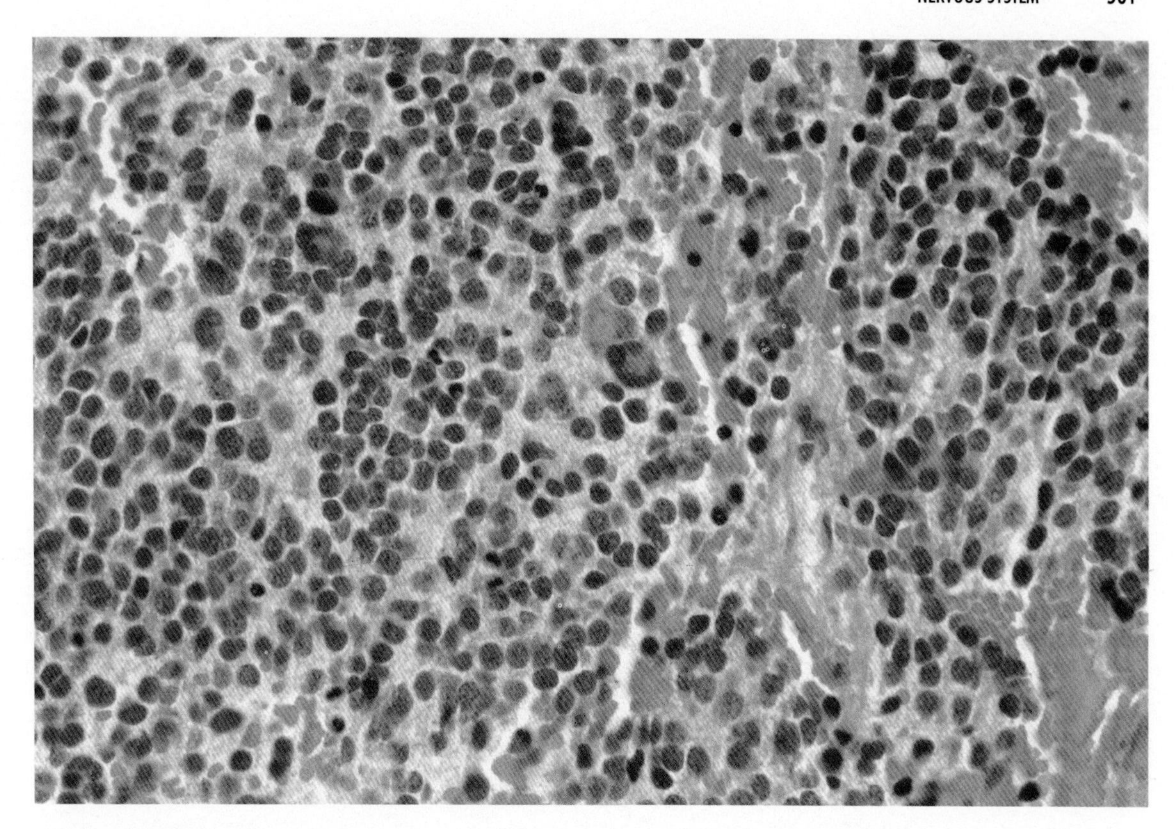

Figure 20–33 **B**

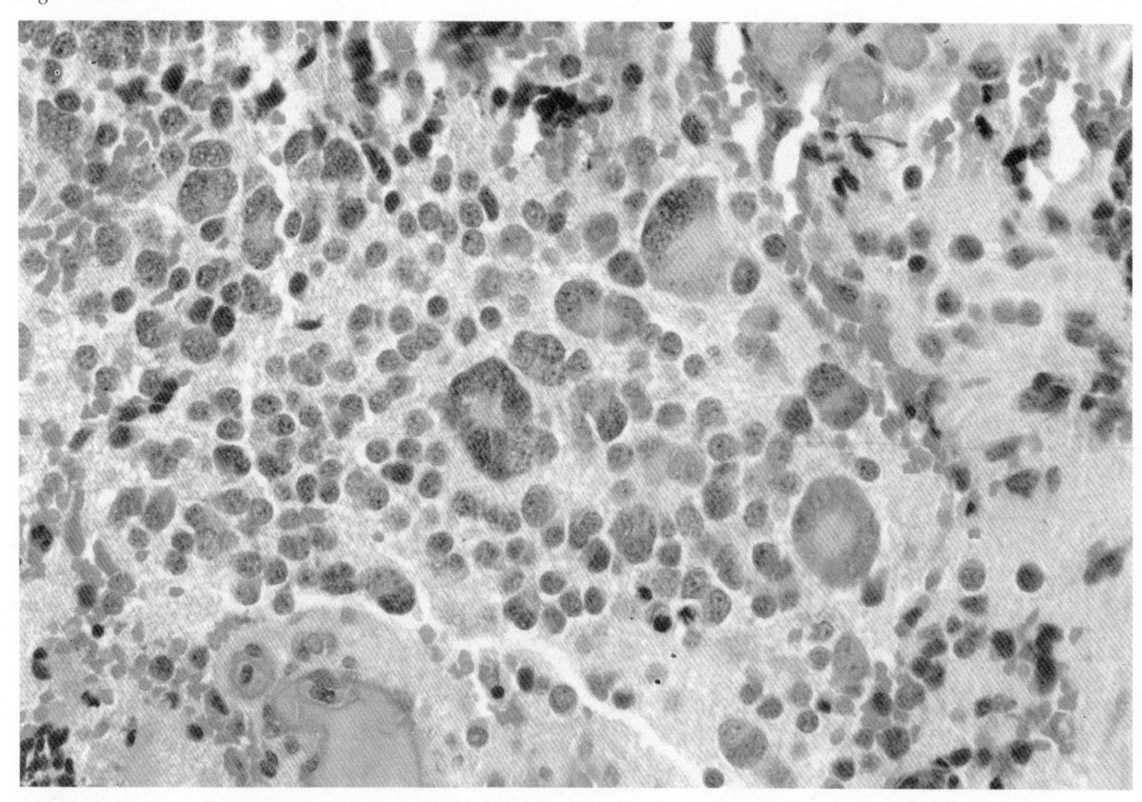

Figure 20–33 **C**

FIGURE 20–33. GANGLIONEUROBLASTOMA: B: This is a photomicrograph of the neuroblastoma component of the ganglioneuroblastoma shown in Figure 20–33 **A**. The cells, which are arranged in sheets and nests, are small and undifferentiated, with scant cytoplasm and hyperchromatic nuclei. **C:** This is a photomicrograph of the ganglion cell component. The ganglionlike cells are present in a background of neuroblastoma, from which they are derived.

■ CENTRAL NERVOUS SYSTEM

UNDIFFERENTIATED AND MULTIPOTENTIAL TUMORS

FIGURE 20–34. MEDULLOBLASTOMA: Medulloblastoma is the second most common central nervous system tumor in children and is most frequently found in the cerebellum. The tumor is composed of sheets of primitive cells with scant cytoplasm and small hyperchromatic nuclei. **Homer Wright rosettes** filled with fibrillar material, which extends centrally from the neoplastic cells, are a characteristic feature.

MENINGEAL AND OTHER TUMORS

FIGURE 20–35. MENINGIOMA: A: In the **syncytial, meningotheliomatous** variant of meningioma, whorls of meningotheliallike cells with abundant cytoplasm and oval nuclei are present. B: In the **transitional** variant of meningioma, laminated spherical calcified structures—**psammoma bodies**—are present in addition to the meningothelial whorls.

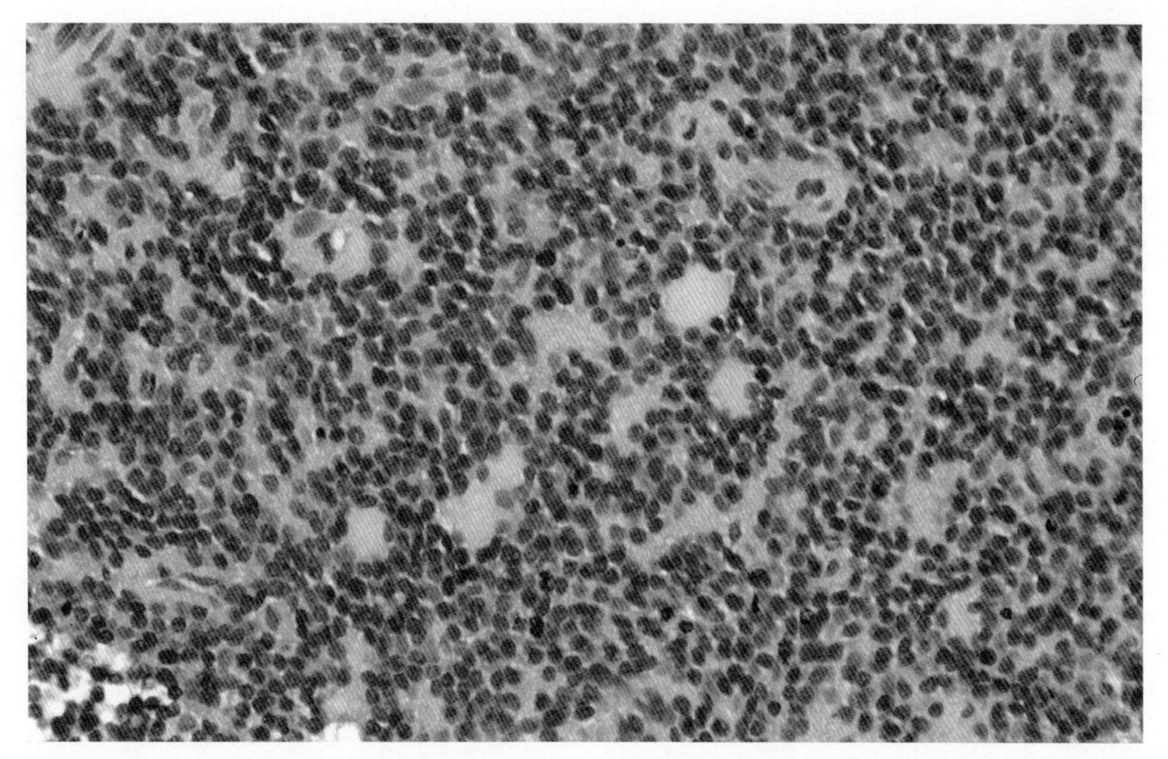

Figure 20–34

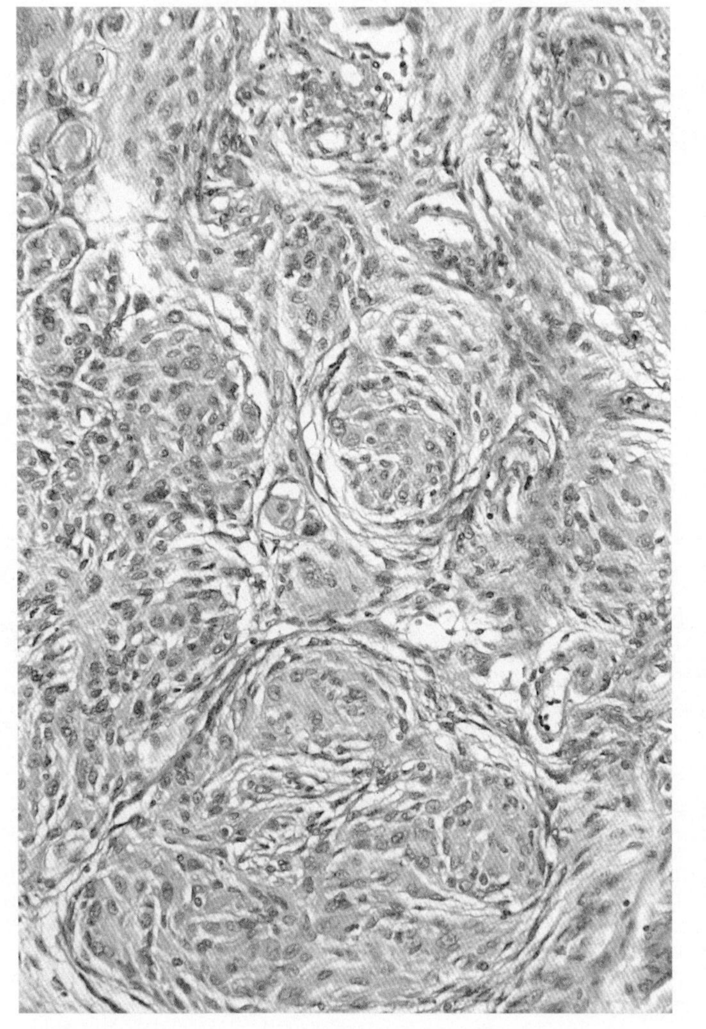

Figure 20–35 **A**

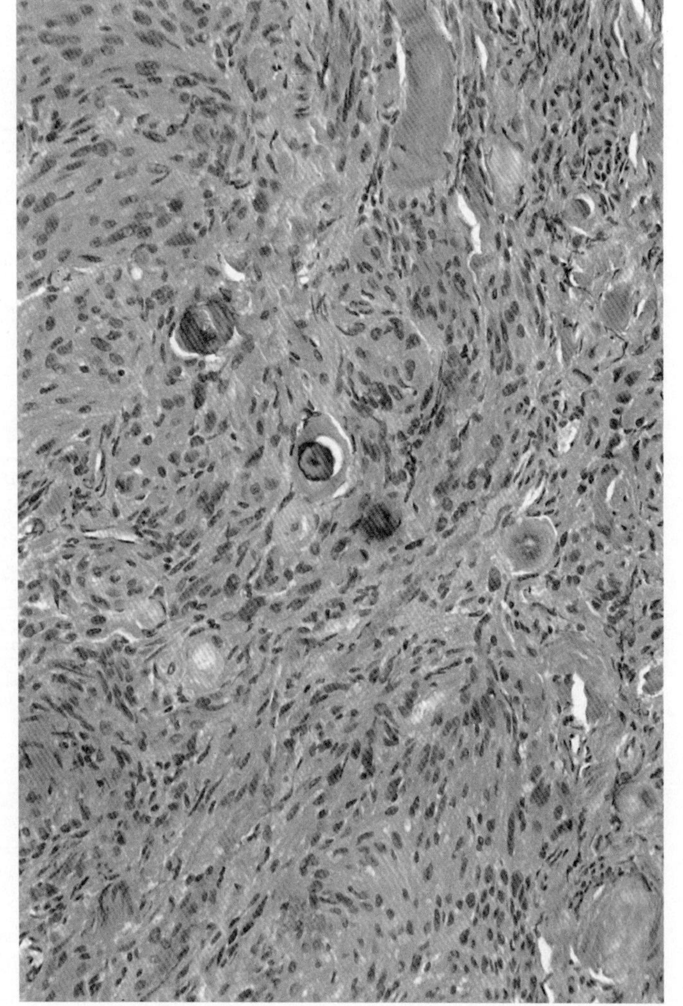

Figure 20–35 **B**

■ CENTRAL NERVOUS SYSTEM

MENINGEAL AND OTHER TUMORS

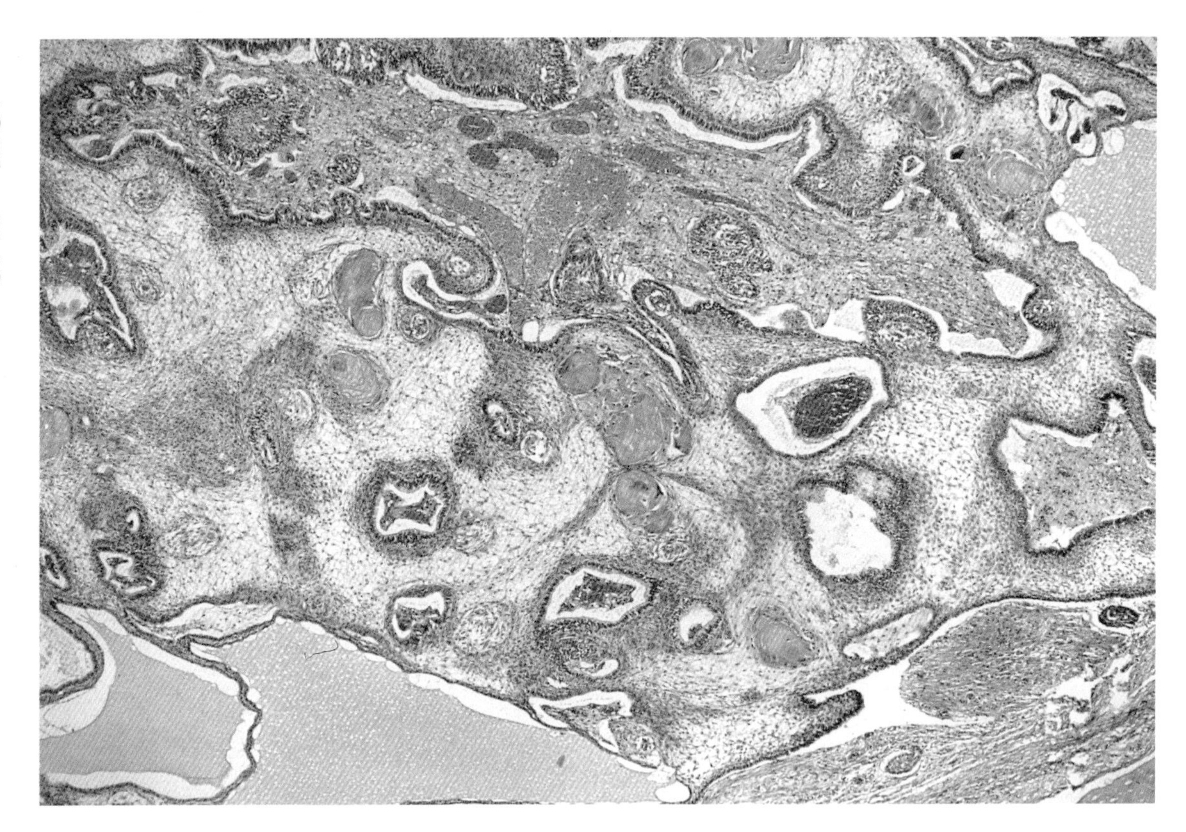

Figure 20–36 A

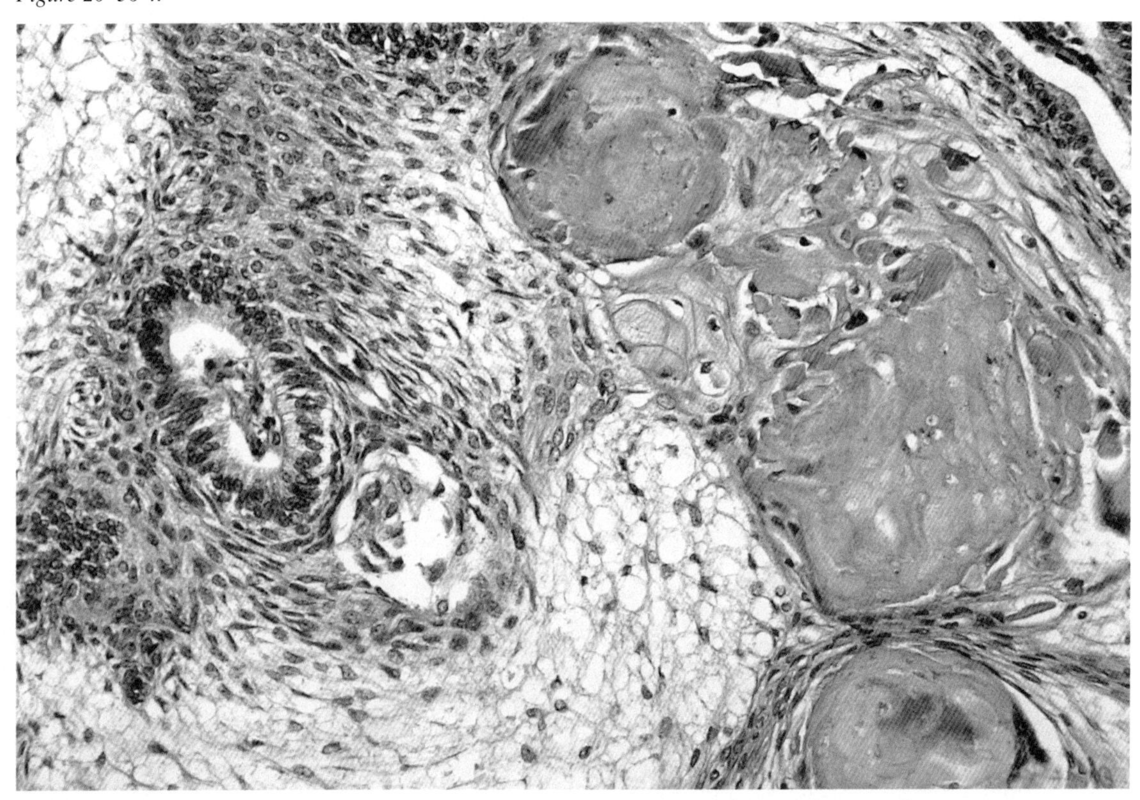

Figure 20–36 B

FIGURE 20–36. CRANIOPHARYNGIOMA: A: Craniopharyngiomas are composed of distinctive cords and sheets of epithelial cells with a layer of basaloid cells palisading at the periphery. **B:** The epithelial islands are composed of keratinizing squamous cells with foci of stellate cells in a loose fibrillar background.

■ CENTRAL NERVOUS SYSTEM

MENINGEAL AND OTHER TUMORS

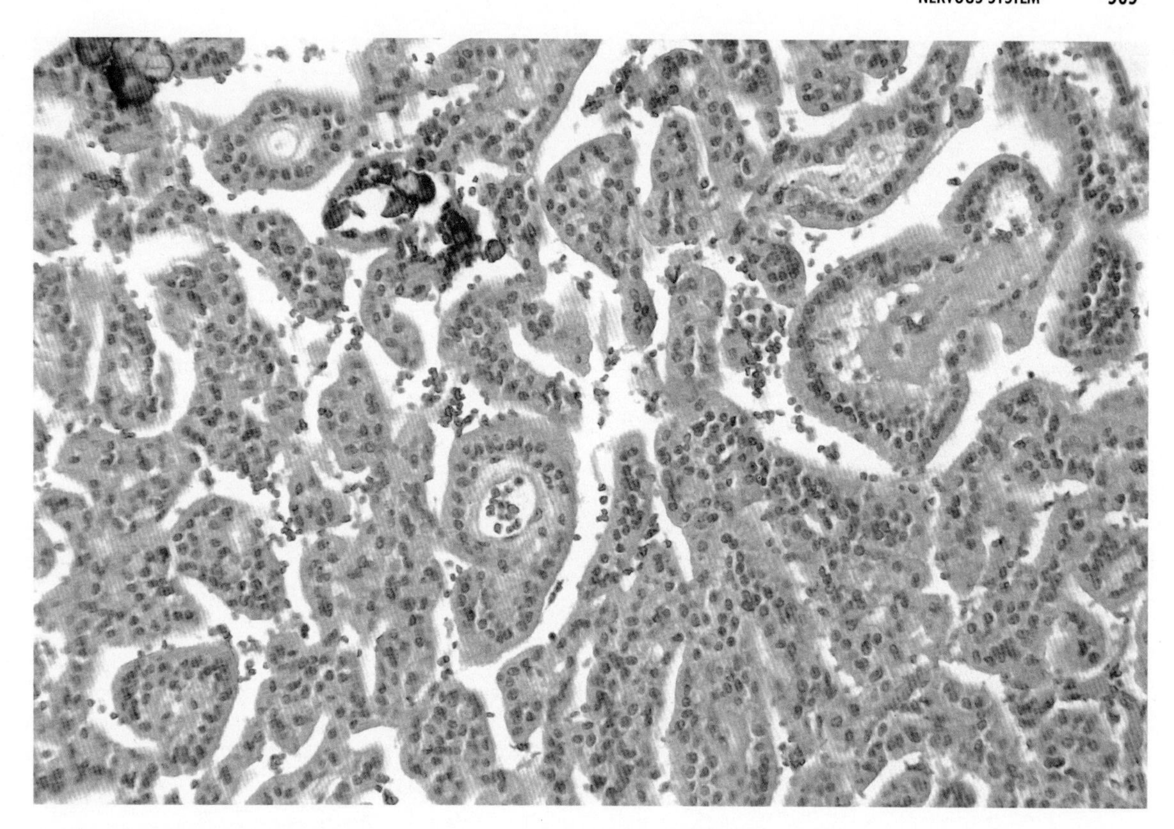

FIGURE 20–37. CHOROID PLEXUS PAPILLOMA: In choroid plexus papilloma, a fibrovascular stroma is covered by cuboidal to low columnar epithelial cells. In this photomicrograph, calcification is present as well.

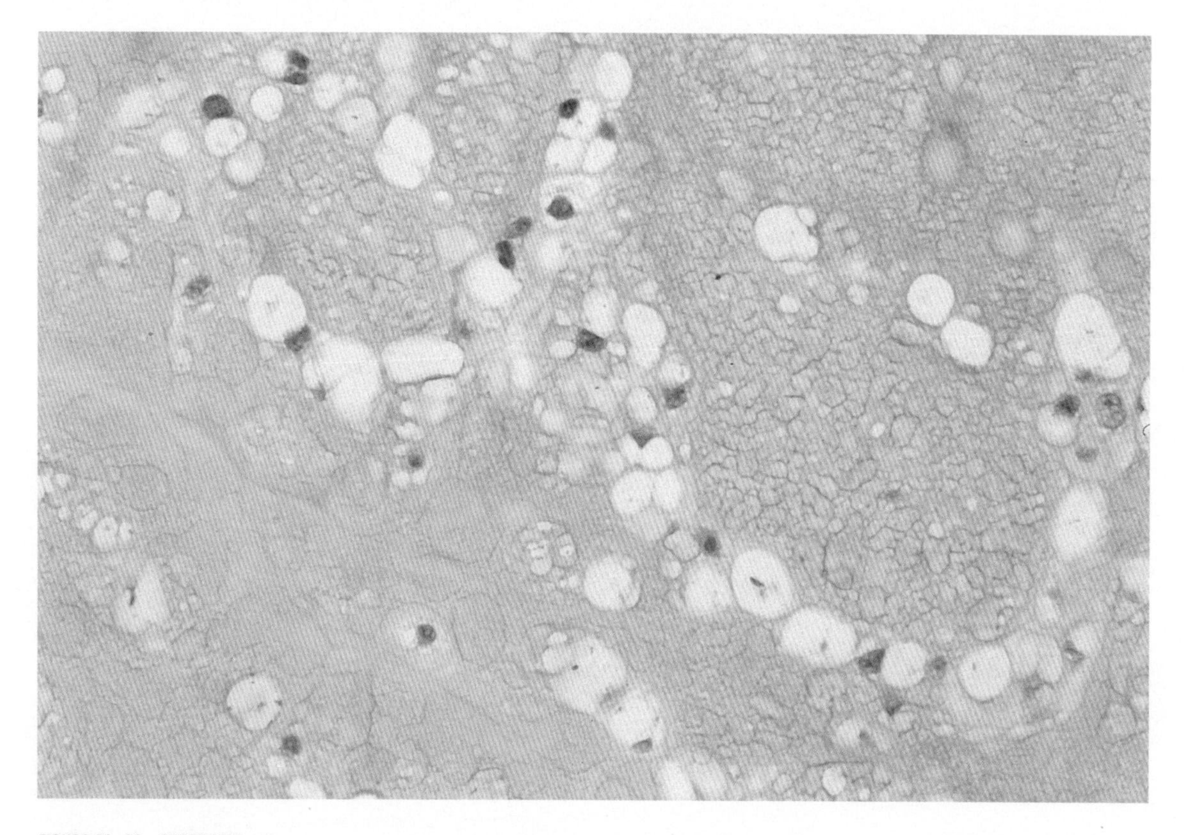

FIGURE 20–38. CHORDOMA: Chordomas arise from the embryonic notochord and therefore are always midline tumors. The cells grow in cords and are surrounded by a chondroid matrix. **Physaliphorous cells** contain large characteristic intracytoplasmic vacuoles and look like a string of beads.

■ PERIPHERAL NERVOUS SYSTEM

FIGURE 20–39. SCHWANNOMA (NEURILEMMOMA): Schwannomas are benign spindle cell tumors composed of **Antoni A** and **Antoni B** areas. A and B: In the Antoni A areas, cell nuclei are tightly packed and alternate with anuclear areas—**Verocay bodies.** C: The Antoni B areas appear spongy, and the cells are randomly arranged.

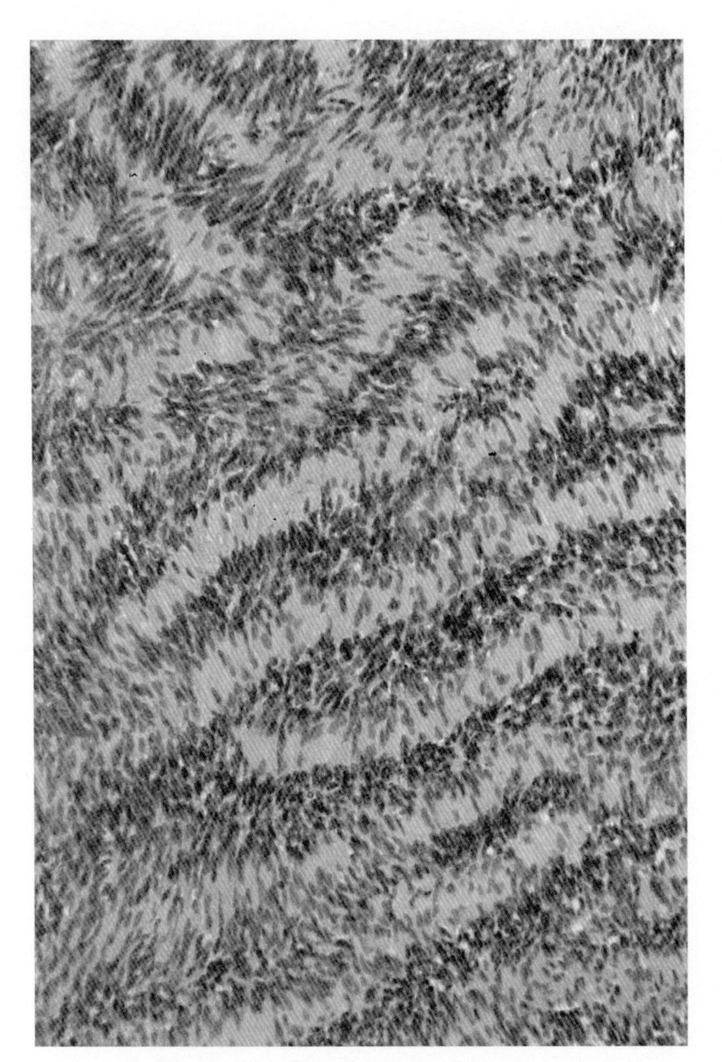

Figure 20–39 A

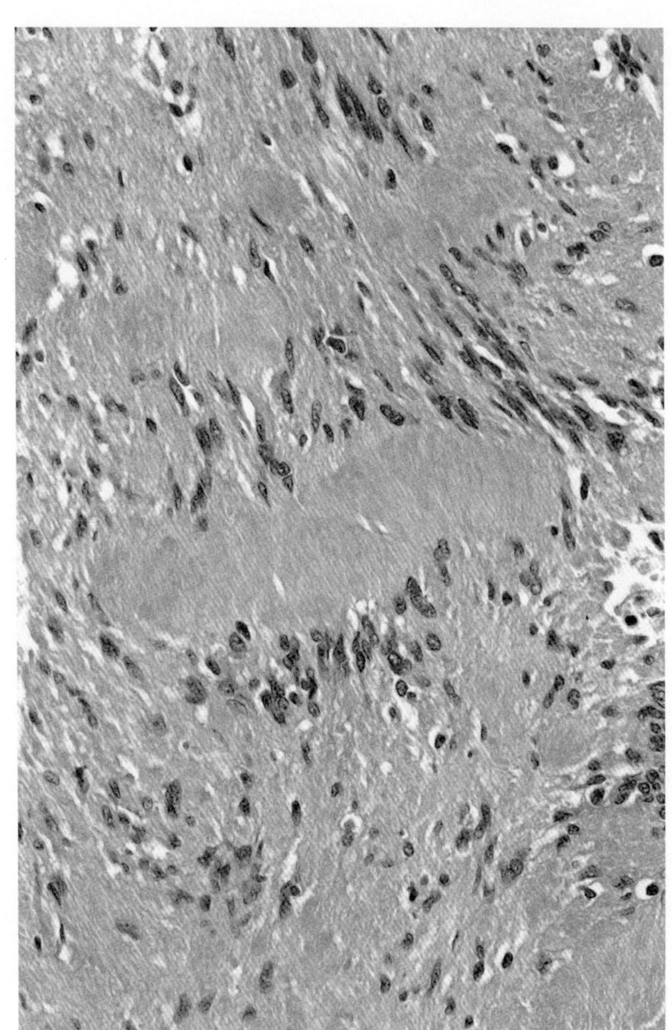

Figure 20–39 B

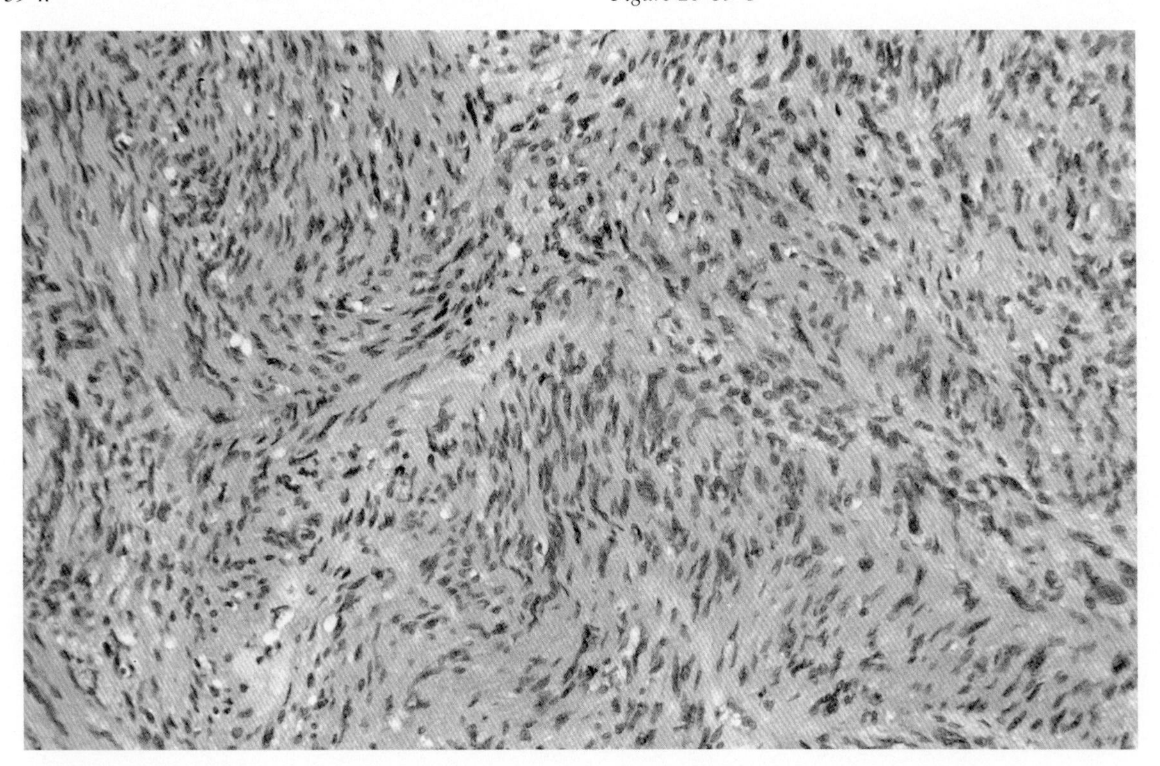

Figure 20–39 C

■ PERIPHERAL NERVOUS SYSTEM

FIGURE 20–40. NEUROFIBROMA: A: In this photomicrograph of a plexiform neurofibroma, multiple nerve fibers are entrapped by an overgrowth of neural tissue. **B:** Nuclei are elongated and wavy.

FIGURE 20–41. TRAUMATIC NEUROMA: A traumatic neuroma is characterized by a mass of disorganized nerve fibers and varying amounts of perineural tissue.

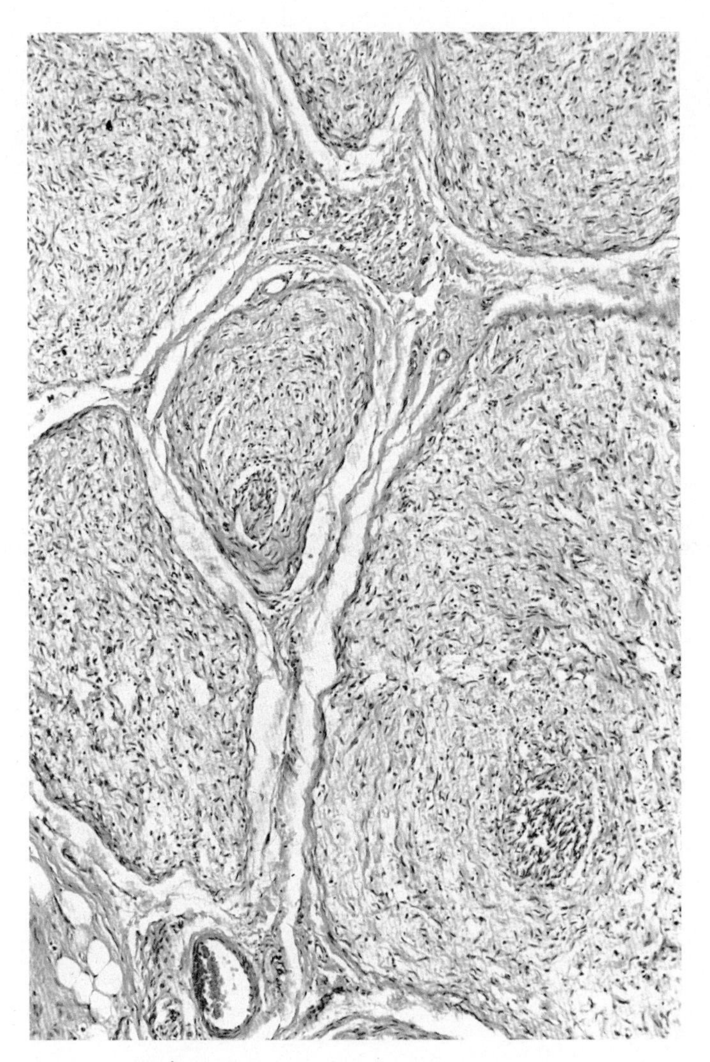

Figure 20–40 **A**

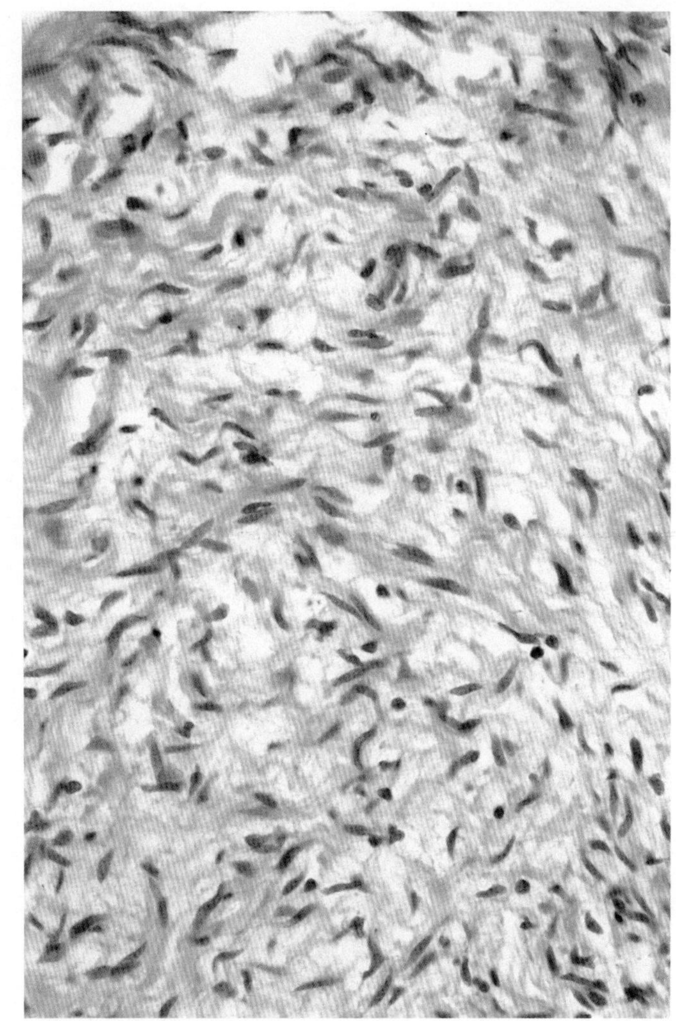

Figure 20–40 **B**

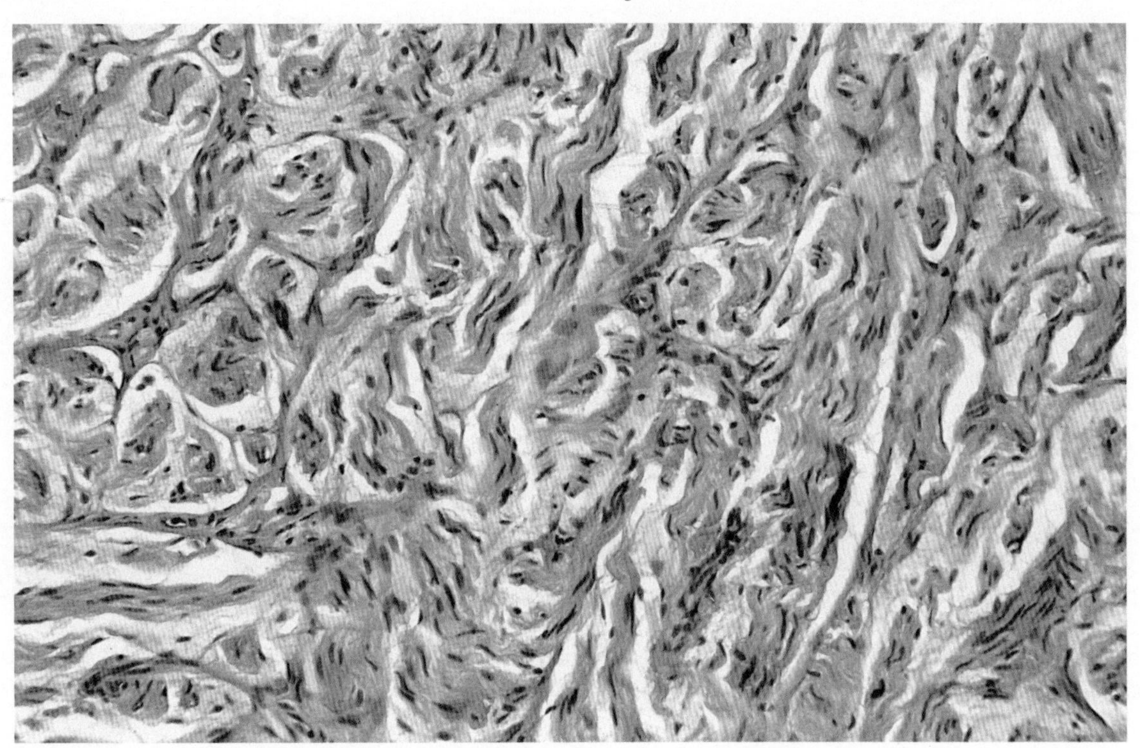

Figure 20–41

■ PERIPHERAL NERVOUS SYSTEM

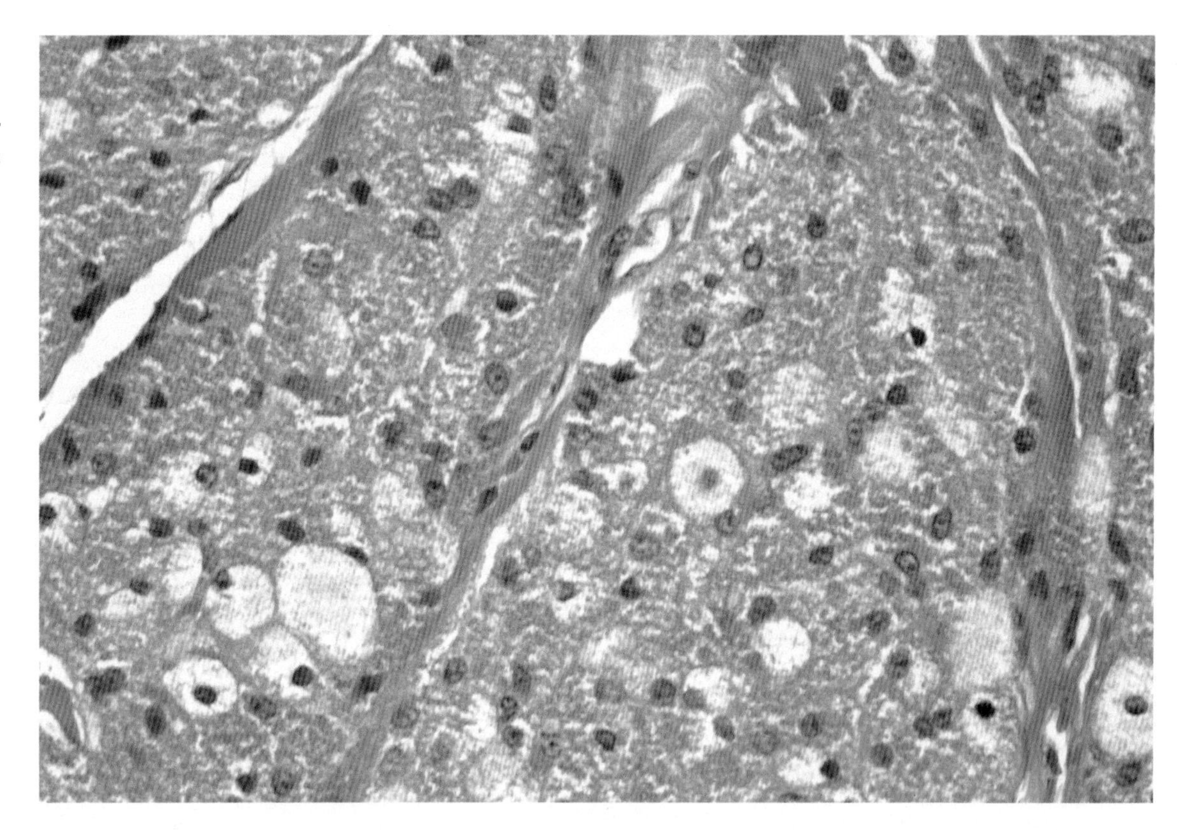

FIGURE 20–42. GRANULAR CELL TUMOR (GRANULAR CELL MYOBLASTOMA): Granular cell tumors are composed of polyhedral cells with granular eosinophilic cytoplasm and bland nuclei.

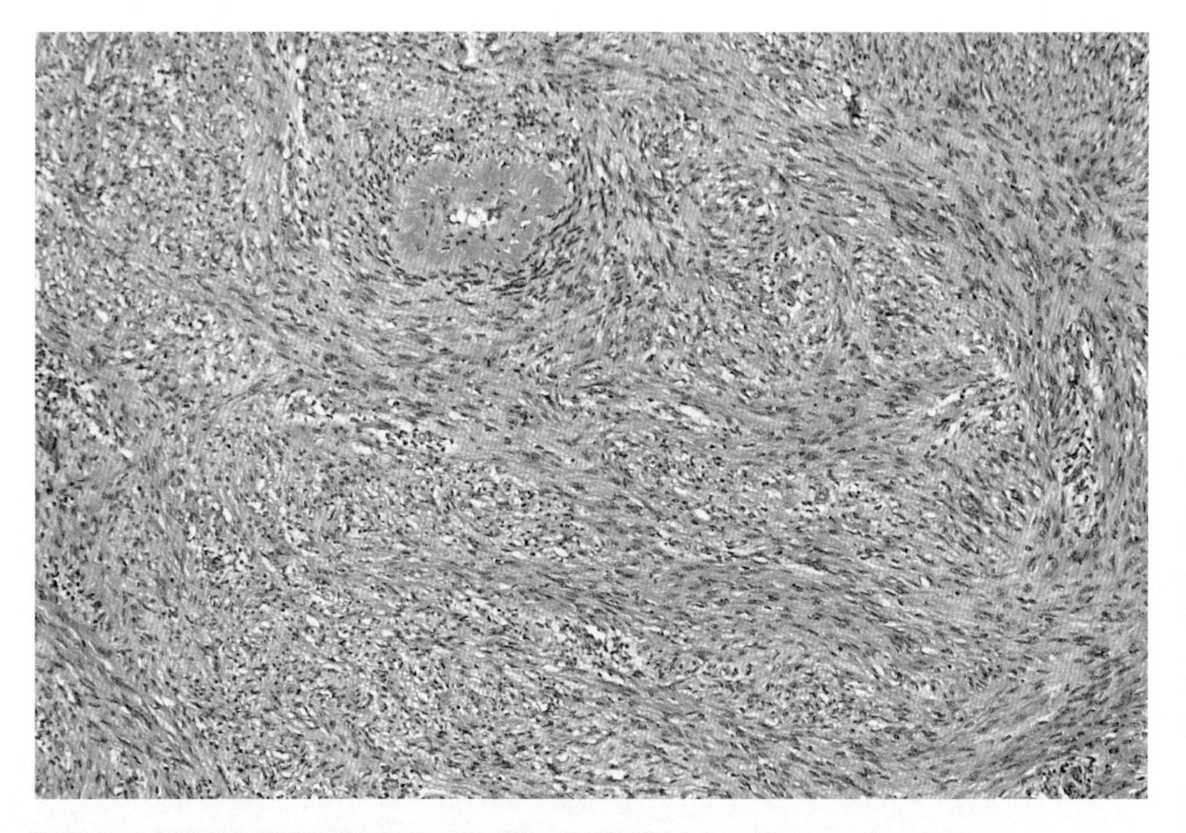

FIGURE 20–43. MALIGNANT PERIPHERAL NERVE SHEATH TUMOR (NEUROGENIC SARCOMA): A: A malignant peripheral nerve sheath tumor is composed of interlacing bundles of spindle cells. Abnormal and increased numbers of mitotic figures are noted.

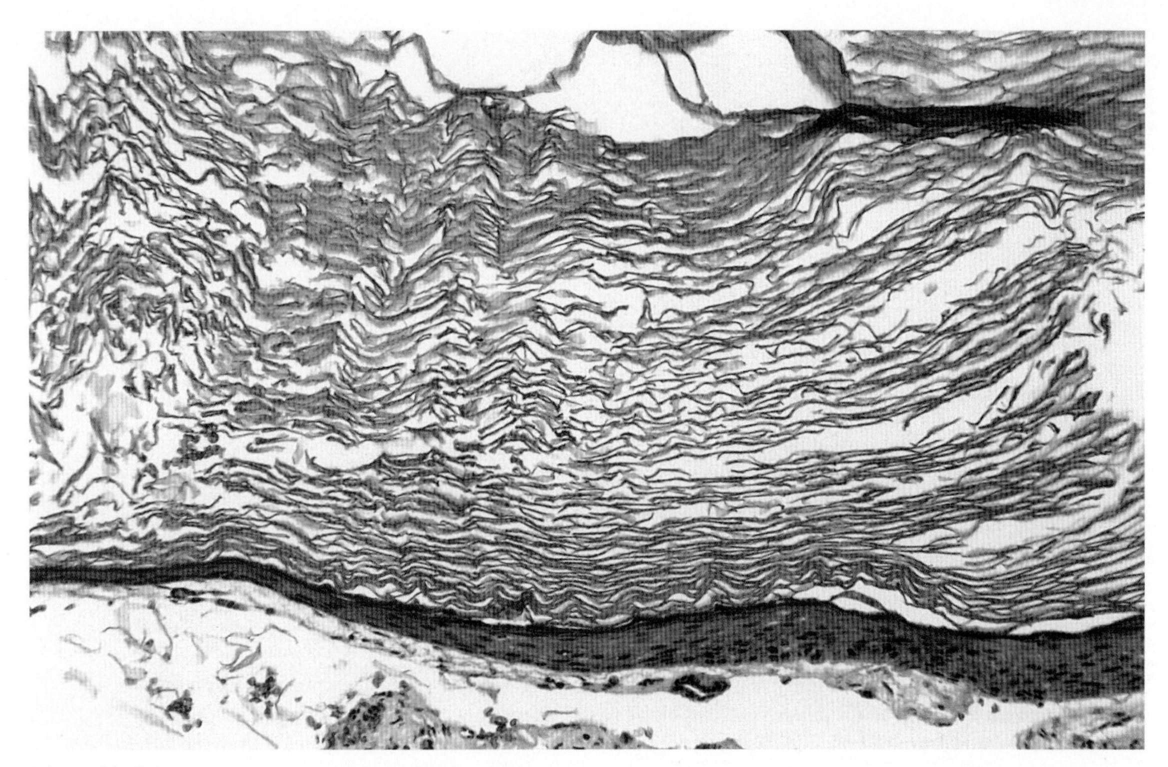

Figure 21–1 A

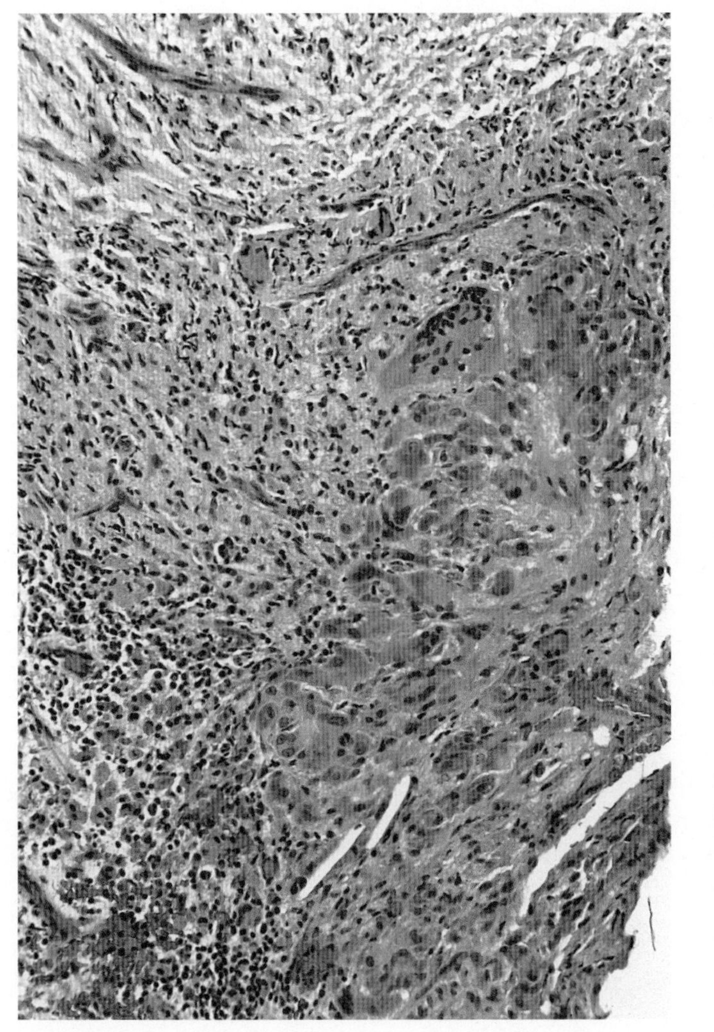

Figure 21–1 B

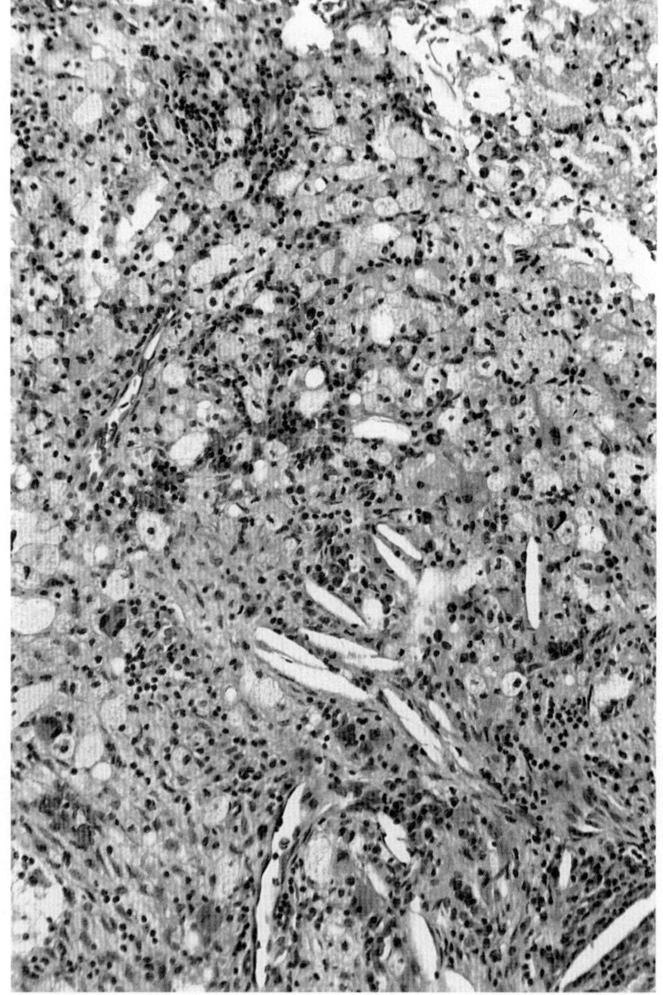

Figure 21–1 C

■ EAR

FIGURE 21–2. OTOSCLEROSIS: A: Early in the disease process of otosclerosis, bone resorption—**otospongiosis**—is prominent. B: Later in the disease process, blood vessels are seen within marrow spaces surrounded by newly formed woven bone. C: Late lesions are characterized by sclerotic bone separated from the woven bone by prominent cement lines.

Basic Gynecology and Obstetrics
Gant & Cunningham
1993, ISBN 0-8385-9633-9, A9633-7

Review of General Psychiatry, 4/e
Goldman
1995, ISBN 0-8385-8421-7, A8421-8

Principles of Clinical Electrocardiography, 13/e
Goldschlager & Goldman
1990, ISBN 0-8385-7951-5, A7951-5

Medical Epidemiology, 2/e
Greenberg, Daniels, Flanders, Eley, & Boring
1995, ISBN 0-8385-6206-X, A6206-5

Basic and Clinical Endocrinology, 5/e
Greenspan & Strewler
1997, ISBN 0-8385-0588-0, A0588-2

Occupational Medicine
LaDou
1990, ISBN 0-8385-7207-3, A7207-2

Clinical Anesthesiology, 2/e
Morgan & Mikhail
1996, ISBN 0-8385-1381-6, A1381-1

Basic Surgery
Niederhuber
1997, ISBN 0-8385-0509-0, A0509-8

Dermatology
Orkin, Maibach, & Dahl
1991, ISBN 0-8385-1288-7, A1288-8

Rudolph's Fundamentals of Pediatrics
Rudolph & Kamei
1994, ISBN 0-8385-8233-8, A8233-7

Genetics in Primary Care & Clinical Medicine
Seashore
1995, ISBN 0-8385-3128-8, A3128-4

Clinical Thinking in Surgery
Sterns
1989, ISBN 0-8385-5686-8, A5686-9

The Principles and Practice of Medicine, 23/e
Stobo, Hellman, Ladenson, Petty, & Traill
1996, ISBN 0-8385-7963-9, A7963-0

Smith's General Urology, 14/e
Tanagho & McAninch
1995, ISBN 0-8385-8612-0, A8612-2

General Ophthalmology, 14/e
Vaughan, Asbury, & Riordan-Eva
1995, ISBN 0-8385-3127-X, A3127-6

Clinical Oncology
Weiss
1993, ISBN 0-8385-1325-5, A1325-8

CURRENT Clinical References

CURRENT Critical Care Diagnosis & Treatment, 2/e
Bongard & Sue
1997, ISBN 0-8385-1454-5, A1454-6

CURRENT Diagnosis & Treatment in Cardiology
Crawford
1995, ISBN 0-8385-1444-8, A1444-7

CURRENT Diagnosis & Treatment in Vascular Surgery
Dean, Yao, & Brewster
1995, ISBN 0-8385-1351-4, A1351-4

CURRENT Obstetric & Gynecologic Diagnosis & Treatment, 9/e
DeCherney
1997, ISBN 0-8385-1401-4, A1401-7

CURRENT Diagnosis & Treatment in Gastroenterology
Grendell, McQuaid, & Friedman
1996, ISBN 0-8385-1448-0, A1448-8

CURRENT Pediatric Diagnosis & Treatment, 13/e
Hay, Groothius, Hayward, & Levin
1997, ISBN 0-8385-1400-6, A1400-9

CURRENT Emergency Diagnosis & Treatment, 4/e
Saunders & Ho
1992, ISBN 0-8385-1347-6, A1347-2

CURRENT Diagnosis & Treatment in Orthopedics
Skinner
1995, ISBN 0-8385-1009-4, A1009-8

CURRENT Medical Diagnosis & Treatment 1997, 36/e
Tierney, McPhee, & Papadakis
1997, ISBN 0-8385-1489-8, A1489-2

CURRENT Surgical Diagnosis & Treatment, 11/e
Way
1997, ISBN 0-8385-1456-1, A1456-1

LANGE Clinical Manuals

Dermatology
Diagnosis and Therapy
Bondi, Jegasothy, & Lazarus
1991, ISBN 0-8385-1274-7, A1274-8

Practical Oncology
Cameron
1994, ISBN 0-8385-1326-3, A1326-6

Office & Bedside Procedures
Chesnutt, Dewar, & Locksley
1993, ISBN 0-8385-1095-7, A1095-7

Psychiatry
Diagnosis & Treatment, 2/e
Flaherty, Davis, & Janicak
1993, ISBN 0-8385-1267-4, A126 7-2

Practical Gynecology
Jacobs & Gast
1994, ISBN 0-8385-1336-0, A1336-5

Drug Therapy, 2/e
Katzung
1991, ISBN 0-8385-1312-3, A1312-6

Geriatrics
Lonergan
1996, ISBN 0-8385-1094-9, A1094-0

Ambulatory Medicine
The Primary Care of Families, 2/e
Mengel & Schwiebert
1996, ISBN 0-8385-1466-9, A1466-0

Practical Pain Management
Minzter
1997, ISBN 0-8385-8116-1, A8116-4

Poisoning & Drug Overdose, 2/e
Olson
1994, ISBN 0-8385-1108-2, A1108-8

Internal Medicine
Diagnosis and Therapy, 3/e
Stein
1993, ISBN 0-8385-1112-0, A1112-0

Surgery
Diagnosis and Therapy
Stillman
1989, ISBN 0-8385-1283-6, A1283-9

Medical Perioperative Management
Wolfsthal
1989, ISBN 0-8385-1298-4, A1298-7

LANGE Handbooks

Handbook of Gynecology & Obstetrics
Brown & Crombleholme
1993, ISBN 0-8385-3608-5, A3608-5

HIV/AIDS Primary Care Handbook
Carmichael, Carmichael, & Fischl
1995, ISBN 0-8385-3557-7, A3557-4

Handbook of Poisoning, 12/e
Prevention, Diagnosis, & Treatment
Dreisbach & Robertson
1987, ISBN 0-8385-3643-3, A3643-2

Handbook of Clinical Endocrinology
Fitzgerald
1992, ISBN 0-8385-3615-8, A3615-0

Clinician's Pocket Reference, 8/e
Gomella
1997, ISBN 0-8385-1476-6, A1476-9

Neonatology
Management, Procedures, On-Call Problems, Diseases, and Drugs, 3/e
Gomella
1994, ISBN 0-8385-1331-X, A1331-6

Surgery on Call, 2/e
Gomella & Lefor
1995, ISBN 0-8385-8746-1, A8746-8

Internal Medicine On Call, 2/e
Haist, Robbins, & Gomella
1997, ISBN 0-8385-4056-2, A4056-6

Obstetrics & Gynecology On Call
Horowitz & Gomella
1993, ISBN 0-8385-7174-3, A7174-4

Pocket Guide to Commonly Prescribed Drugs, 2/e
Levine
1996, ISBN 0-8385-8099-8, A8099-2

Handbook of Pediatrics, 18/e
Merenstein, Kaplan, & Rosenberg
1997, ISBN 0-8385-3625-5, A3625-9

Pocket Guide to Diagnostic Tests, 2/e
Nicoll, McPhee, Chou, & Detmer
1997, ISBN 0-8385-8100-5, A8100-8

Quick Medical Spanish
Rogers
1996, ISBN 0-8385-8258-3, A8258-4

Pocket Guide to the Essentials of Diagnosis & Treatment
Tierney
1997, ISBN 0-8385-3605-0, A3605-1

Appleton & Lange • P.O. Box 120041 • Stamford, CT • 06912-0041